Transactions on Computational Science and Computational Intelligence

Series Editor
Hamid Arabnia
Department of Computer Science
The University of Georgia
Athens, GA, USA

Computational Science (CS) and Computational Intelligence (CI) both share the same objective: finding solutions to difficult problems. However, the methods to the solutions are different. The main objective of this book series, "Transactions on Computational Science and Computational Intelligence", is to facilitate increased opportunities for cross-fertilization across CS and CI. This book series publishes monographs, professional books, contributed volumes, and textbooks in Computational Science and Computational Intelligence. Book proposals are solicited for consideration in all topics in CS and CI including, but not limited to, Pattern recognition applications; Machine vision; Brain-machine interface; Embodied robotics; Biometrics; Computational biology; Bioinformatics; Image and signal processing; Information mining and forecasting; Sensor networks; Information processing; Internet and multimedia; DNA computing; Machine learning applications; Multi-agent systems applications; Telecommunications; Transportation systems; Intrusion detection and fault diagnosis; Game technologies; Material sciences; Space, weather, climate systems, and global changes; Computational ocean and earth sciences; Combustion system simulation; Computational chemistry and biochemistry; Computational physics; Medical applications; Transportation systems and simulations; Structural engineering; Computational electro-magnetic; Computer graphics and multimedia; Face recognition; Semiconductor technology, electronic circuits, and system design; Dynamic systems; Computational finance; Information mining and applications; Biometric modeling; Computational journalism; Geographical Information Systems (GIS) and remote sensing; Ubiquitous computing; Virtual reality; Agent-based modeling; Computational psychometrics; Affective computing; Computational economics; Computational statistics; and Emerging applications. For further information, please contact Mary James, Senior Editor, Springer, mary.james@springer.com.

More information about this series at http://www.springer.com/series/11769

Kevin Daimi • Hamid R. Arabnia
Leonidas Deligiannidis • Min-Shiang Hwang
Fernando G. Tinetti
Editors

Advances in Security, Networks, and Internet of Things

Proceedings from SAM'20, ICWN'20, ICOMP'20, and ESCS'20

Springer

Editors
Kevin Daimi
Electrical and Computer Engineering, and Computer Science
University of Detroit Mercy
Detroit, MI, USA

Leonidas Deligiannidis
School of Computing and Data Sciences
Wentworth Institute of Technology
Boston, MA, USA

Fernando G. Tinetti
Facultad de Informática – CIC PBA
Universidad Nacional de La Plata
La Plata, Argentina

Hamid R. Arabnia
Department of Computer Science
University of Georgia
Athens, GA, USA

Min-Shiang Hwang
Computer Science and Information Engineering
Asian University
Taichung City, Taiwan

ISSN 2569-7072 ISSN 2569-7080 (electronic)
Transactions on Computational Science and Computational Intelligence
ISBN 978-3-030-71016-3 ISBN 978-3-030-71017-0 (eBook)
https://doi.org/10.1007/978-3-030-71017-0

This Springer imprint is published by the registered company Springer Nature Switzerland AG
The registered company address is: Gewerbestrasse 11, 6330 Cham, Switzerland

Preface

It gives us great pleasure to introduce this collection of papers that were presented at the following international conferences: Security and Management (SAM 2020); Wireless Networks (ICWN 2020); Internet Computing & IoT (ICOMP 2020); and Embedded Systems, Cyber-physical Systems, and Applications (ESCS 2020). These four conferences were held simultaneously (same location and dates) at Luxor Hotel (MGM Resorts International), Las Vegas, USA, July 27–30, 2020. This international event was held using a hybrid approach, that is, "in-person" and "virtual/online" presentations and discussions.

This book is composed of seven parts. Parts 1 through 4 (composed of 33 chapters) include articles that address various challenges with security and management (SAM). Part 5 (composed of 8 chapters) presents novel methods and applications in the areas of wireless networks (ICWN). Part 6 (composed of 8 chapters) discusses advancements in Internet computing and Internet of Things (ICOMP). Lastly, Part 7 (composed of 11 chapters) presents emerging trends in the areas of embedded systems and cyber-physical systems (ESCS).

An important mission of the World Congress in Computer Science, Computer Engineering, and Applied Computing, CSCE (a federated congress to which this event is affiliated with), includes *"Providing a unique platform for a diverse community of constituents composed of scholars, researchers, developers, educators, and practitioners. The Congress makes concerted effort to reach out to participants affiliated with diverse entities (such as: universities, institutions, corporations, government agencies, and research centers/labs) from all over the world. The congress also attempts to connect participants from institutions that have* ***teaching*** *as their main mission with those who are affiliated with institutions that have* ***research*** *as their main mission. The congress uses a quota system to achieve its institution and geography diversity objectives."* By any definition of diversity, this congress is among the most diverse scientific meetings in the USA. We are proud to report that this federated congress had authors and participants from 54 different nations, representing variety of personal and scientific experiences that arise from differences in culture and values.

The program committees (refer to subsequent pages for the list of the members of committees) would like to thank all those who submitted papers for consideration. About 50% of the submissions were from outside the USA. Each submitted paper was peer reviewed by two experts in the field for originality, significance, clarity, impact, and soundness. In cases of contradictory recommendations, a member of the conference program committee was charged to make the final decision; often, this involved seeking help from additional referees. In addition, papers whose authors included a member of the conference program committee were evaluated using the double-blind review process. One exception to the above evaluation process was for papers that were submitted directly to chairs/organizers of pre-approved sessions/workshops; in these cases, the chairs/organizers were responsible for the evaluation of such submissions. The Congress (the joint conferences) received many good submissions. The overall acceptance rate for regular papers was 20%; 18% of the remaining papers were accepted as short and/or poster papers.

We are grateful to the many colleagues who offered their services in preparing this book. In particular, we would like to thank the members of the Program Committees of individual research tracks as well as the members of the Steering Committees of SAM 2020, ICWN 2020, ICOMP 2020, and ESCS 2020; their names appear in the subsequent pages. We would also like to extend our appreciation to over 500 referees.

As sponsors-at-large, partners, and/or organizers, each of the following (separated by semicolons) provided help for at least one research track: Computer Science Research, Education, and Applications (CSREA); US Chapter of World Academy of Science; American Council on Science and Education & Federated Research Council; and Colorado Engineering Inc. In addition, a number of university faculty members and their staff, several publishers of computer science and computer engineering books and journals, chapters and/or task forces of computer science associations/organizations from three regions, and developers of high-performance machines and systems provided significant help in organizing the event as well as providing some resources. We are grateful to them all.

We express our gratitude to all authors of the articles published in this book and the speakers who delivered their research results at the congress. We would also like to thank the following: UCMSS (Universal Conference Management Systems & Support, California, USA) for managing all aspects of the conference; Dr. Tim Field of APC for coordinating and managing the printing of the programs; the staff at Luxor Hotel (MGM Convention) for the professional service they provided; and Ashu M. G. Solo for his help in publicizing the congress. Last but not least, we would like to thank Ms. Mary James (Springer Senior Editor in New York) and Arun Pandian KJ (Springer Production Editor) for the excellent professional service they provided for this book project.

Book Co-editors and Chapter Co-editors: SAM 2020, ICWN 2020, ICOMP 2020, ESCS 2020

Detroit, MI, USA — Kevin Daimi

Athens, GA, USA — Hamid R. Arabnia

Boston, MA, USA — Leonidas Deligiannidis

La Plata, Argentina — Fernando G. Tinetti

Security and Management

SAM 2020 – Program Committee

- Dr. Jacques Bou Abdo, Computer Science Department, Notre Dame University – Louaize, Lebanon
- Dr. Hanaa Ahmed, Computer Science Department, University of Technology, Iraq
- Dr. Mohammed Akour, Department of Computer and Information Systems, Yarmouk University, Jordan
- Professor Emeritus Nizar Al Holou, Department of Electrical and Computer Engineering, University of Detroit Mercy, USA
- Professor Nadia Alsaidi, Department of Applied Mathematics & Computing, University of Technology, Iraq
- Allen Ashourian, ZRD Technology, USA
- Professor Emeritus Hamid Arabnia, (Vice Chair and Coordinator, SAM'20), Department of Computer Science, University of Georgia, USA
- Dr. David Arroyo, Researcher, Spanish National Research Council (CSIC), Spain
- Dr. Shadi Banitaan, (Sessions/Workshops Co-Chair, SAM'20), Computer Science and Software Engineering, University of Detroit Mercy, USA
- Dr. Clive Blackwell, Innovation Works, Airbus Group, United Kingdom
- Dr. Violeta Bulbenkiene, Department of Informatics Engineering, Klaipeda University, Lithuania
- Dr. María Calle, Department of Electrical and Electronics Engineering, Universidad del Norte, Barranquilla, Colombia
- Eralda Caushaj, School of Business Administration, Oakland University, USA
- Dr. Feng Cheng, Internet Technologies and Systems, Hasso-Plattner-Institute, University of Potsdam, Germany
- Professor Emeritus Kevin Daimi, (Conference Chair, SAM'20), Computer Science and Software Engineering, University of Detroit Mercy, USA
- Dr. Ioanna Dionysiou, Department of Computer Science, University of Nicosia, Cyprus

- Dr. Esmiralda Moradian, (Posters Co-Chair, SAM'20), Department of Computer and Systems Sciences, Stockholm University, Sweden
- Dr. Nader M Nassar, Innovation for Security and Compliance Group, IBM Corp, USA
- Dr. Ana Nieto, Computer Science Department, University of Malaga, Spain
- Dr. Mais Nijim, Department of Electrical Engineering and Computer Science, Texas A&M University-Kingsville, USA
- Dr. Eugenia Nikolouzou, Internet Application Department, Hellenic Authority for Communication Security and Privacy, Greece
- Dr. Saibal K. Pal, DRDO & University of Delhi, India
- Dr. Cathryn Peoples, (Posters Co-Chair, SAM'20), School of Computing and Communications, Faculty of Science, Technology, Engineering & Mathematics, The Open University, United Kingdom
- Dr. Junfeng Qu, Deptartment of Information Technology, Clayton State University, USA
- Dr. Peter Schartner, System Security Research Group, Alpen-Adria-Universität Klagenfurt, Austria
- Dr. Karpoor Shashidhar, Computer Science Department, Sam Houston State University, USA
- Dr. Nicolas Sklavos, Computer Engineering & Informatics Department, University of Patras, Greece
- Ashu M.G. Solo, (Publicity Chair, SAM'20) Maverick Technologies America, USA.
- Dr. Cristina Soviany, (Program Co-Chair, SAM'20), Features Analytics SA, Belgium
- Professor Hung-Min Sun, Information Security, Department of Computer Science, National Tsing Hua University, Taiwan
- Professor Woei-Jiunn Tsaur, Department of Computer Science, National Taipei University, Taiwan
- Professor Shengli Yuan, Department of Computer Science and Engineering Technology, University of Houston-Downtown, USA

Wireless Networks

ICWN 2020 – Program Committee

- Prof. Emeritus Nizar Al-Holou (Congress Steering Committee); ECE Department; Vice Chair, IEEE/SEM-Computer Chapter; University of Detroit Mercy, Detroit, Michigan, USA
- Prof. Emeritus Hamid R. Arabnia (Congress Steering Committee); The University of Georgia, USA; Editor-in-Chief, Journal of Supercomputing (Springer); Fellow, Center of Excellence in Terrorism, Resilience, Intelligence & Organized Crime Research (CENTRIC).
- Dr. Travis Atkison; Director, Digital Forensics and Control Systems Security Lab, Department of Computer Science, College of Engineering, The University of Alabama, Tuscaloosa, Alabama, USA
- Prof. Dr. Juan-Vicente Capella-Hernandez; Universitat Politecnica de Valencia (UPV), Department of Computer Engineering (DISCA), Valencia, Spain
- Prof. Emeritus Kevin Daimi (Congress Steering Committee); Department of Mathematics, Computer Science and Software Engineering, University of Detroit Mercy, Detroit, Michigan, USA
- Prof. Leonidas Deligiannidis (Congress Steering Committee); Department of Computer Information Systems, Wentworth Institute of Technology, Boston, Massachusetts, USA
- Prof. Mary Mehrnoosh Eshaghian-Wilner (Congress Steering Committee); Professor of Engineering Practice, University of Southern California, California, USA; Adjunct Professor, Electrical Engineering, University of California Los Angeles, Los Angeles (UCLA), California, USA
- Prof. Tai-hoon Kim; School of Information and Computing Science, University of Tasmania, Australia
- Prof. Louie Lolong Lacatan; Chairperson, Computer Engineerig Department, College of Engineering, Adamson University, Manila, Philippines; Senior Member, International Association of CS & IT (IACSIT), Singapore; Member, International Association of Online Engineering (IAOE), Austria

- Prof. Dr. Guoming Lai; Computer Science and Technology, Sun Yat-Sen University, Guangzhou, P. R. China
- Dr. Andrew Marsh (Congress Steering Committee); CEO, HoIP Telecom Ltd (Healthcare over Internet Protocol), UK; Secretary General of World Academy of BioMedical Sciences and Technologies (WABT) a UNESCO NGO, The United Nations
- Prof. Salahuddin Mohammad Masum; Computer Engineering Technology, Southwest Tennessee Community College, Memphis, Tennessee, USA
- Prof. Dr., Eng. Robert Ehimen Okonigene (Congress Steering Committee); Department of Electrical & Electronics Engineering, Faculty of Engineering and Technology, Ambrose Alli University, Nigeria
- Prof. James J. (Jong Hyuk) Park (Congress Steering Committee); Department of Computer Science and Engineering (DCSE), SeoulTech, Korea; President, FTRA, EiC, HCIS Springer, JoC, IJITCC; Head of DCSE, SeoulTech, Korea
- Dr. Akash Singh (Congress Steering Committee); IBM Corporation, Sacramento, California, USA; Chartered Scientist, Science Council, UK; Fellow, British Computer Society; Member, Senior IEEE, AACR, AAAS, and AAAI; IBM Corporation, USA
- Ashu M. G. Solo (Publicity), Fellow of British Computer Society, Principal/R&D Engineer, Maverick Technologies America Inc.
- Prof. Fernando G. Tinetti (Congress Steering Committee); School of CS, Universidad Nacional de La Plata, La Plata, Argentina; also at Comision Investigaciones Cientificas de la Prov. de Bs. As., Argentina
- Prof. Hahanov Vladimir (Congress Steering Committee); Vice Rector, and Dean of the Computer Engineering Faculty, Kharkov National University of Radio Electronics, Ukraine and Professor of Design Automation Department, Computer Engineering Faculty, Kharkov; IEEE Computer Society Golden Core Member; National University of Radio Electronics, Ukraine
- Prof. Shiuh-Jeng Wang (Congress Steering Committee); Director of Information Cryptology and Construction Laboratory (ICCL) and Director of Chinese Cryptology and Information Security Association (CCISA); Department of Information Management, Central Police University, Taoyuan, Taiwan; Guest Ed., IEEE Journal on Selected Areas in Communications.
- Dr. Yunlong Wang; Advanced Analytics at QuintilesIMS, Pennsylvania, USA
- Prof. Layne T. Watson (Congress Steering Committee); Fellow of IEEE; Fellow of The National Institute of Aerospace; Professor of Computer Science, Mathematics, and Aerospace and Ocean Engineering, Virginia Polytechnic Institute & State University, Blacksburg, Virginia, USA
- Prof. Hyun Yoe; Director of Agrofood IT Research Center and Vice President of Korea Association of ICT Convergence in the Agriculture and Food Business (KAICAF); Director of Agriculture IT Convergence Support Center (AITCSC); Department of Information and Communication Engineering, Sunchon National University, Suncheon, Republic of Korea (South Korea)
- Prof. Jane You (Congress Steering Committee); Associate Head, Department of Computing, The Hong Kong Polytechnic University, Kowloon, Hong Kong

Internet Computing & IoT

ICOMP 2020 – Program Committee

- Prof. Afrand Agah; Department of Computer Science, West Chester University of Pennsylvania, West Chester, PA, USA
- Prof. Emeritus Nizar Al-Holou (Congress Steering Committee); Professor and Chair, Electrical and Computer Engineering Department; Vice Chair, IEEE/SEM-Computer Chapter; University of Detroit Mercy, Detroit, Michigan, USA
- Prof. Emeritus Hamid R. Arabnia (Congress Steering Committee); The University of Georgia, USA; Editor-in-Chief, Journal of Supercomputing (Springer); Fellow, Center of Excellence in Terrorism, Resilience, Intelligence & Organized Crime Research (CENTRIC).
- Prof. Dr. Juan-Vicente Capella-Hernandez; Universitat Politecnica de Valencia (UPV), Department of Computer Engineering (DISCA), Valencia, Spain
- Prof. Emeritus Kevin Daimi (Congress Steering Committee); Department of Mathematics, Computer Science and Software Engineering, University of Detroit Mercy, Detroit, Michigan, USA
- Prof. Zhangisina Gulnur Davletzhanovna; Vice-rector of the Science, Central-Asian University, Kazakhstan, Almaty, Republic of Kazakhstan; Vice President of International Academy of Informatization, Kazskhstan, Almaty, Republic of Kazakhstan
- Prof. Leonidas Deligiannidis (Congress Steering Committee); Department of Computer Information Systems, Wentworth Institute of Technology, Boston, Massachusetts, USA
- Prof. Mary Mehrnoosh Eshaghian-Wilner (Congress Steering Committee); Professor of Engineering Practice, University of Southern California, California, USA; Adjunct Professor, Electrical Engineering, University of California Los Angeles, Los Angeles (UCLA), California, USA
- Prof. Houcine Hassan; Department of Computer Engineering (Systems Data Processing and Computers), Universitat Politecnica de Valencia, Spain

- Prof. Tai-hoon Kim; School of Information and Computing Science, University of Tasmania, Australia
- Prof. Louie Lolong Lacatan; Chairperson, Computer Engineerig Department, College of Engineering, Adamson University, Manila, Philippines; Senior Member, International Association of CS & IT (IACSIT), Singapore; Member, International Association of Online Engineering (IAOE), Austria
- Prof. Dr. Guoming Lai; Computer Science and Technology, Sun Yat-Sen University, Guangzhou, P. R. China
- Dr. Andrew Marsh (Congress Steering Committee); CEO, HoIP Telecom Ltd (Healthcare over Internet Protocol), UK; Secretary General of World Academy of BioMedical Sciences and Technologies (WABT) a UNESCO NGO, The United Nations
- Dr. Ali Mostafaeipour; Industrial Engineering Department, Yazd University, Yazd, Iran
- Prof. Dr., Eng. Robert Ehimen Okonigene (Congress Steering Committee); Department of Electrical & Electronics Engineering, Faculty of Engineering and Technology, Ambrose Alli University, Nigeria
- Prof. James J. (Jong Hyuk) Park (Congress Steering Committee); Department of Computer Science and Engineering (DCSE), SeoulTech, Korea; President, FTRA, EiC, HCIS Springer, JoC, IJITCC; Head of DCSE, SeoulTech, Korea
- Dr. Xuewei Qi; Research Faculty & PI, Center for Environmental Research and Technology, University of California, Riverside, California, USA
- Dr. Akash Singh (Congress Steering Committee); IBM Corporation, Sacramento, California, USA; Chartered Scientist, Science Council, UK; Fellow, British Computer Society; Member, Senior IEEE, AACR, AAAS, and AAAI; IBM Corporation, USA
- Ashu M. G. Solo (Publicity), Fellow of British Computer Society, Principal/R&D Engineer, Maverick Technologies America Inc.
- Prof. Fernando G. Tinetti (Congress Steering Committee); School of CS, Universidad Nacional de La Plata, La Plata, Argentina; also at Comision Investigaciones Cientificas de la Prov. de Bs. As., Argentina
- Prof. Hahanov Vladimir (Congress Steering Committee); Vice Rector, and Dean of the Computer Engineering Faculty, Kharkov National University of Radio Electronics, Ukraine and Professor of Design Automation Department, Computer Engineering Faculty, Kharkov; IEEE Computer Society Golden Core Member; National University of Radio Electronics, Ukraine
- Varun Vohra; Certified Information Security Manager (CISM); Certified Information Systems Auditor (CISA); Associate Director (IT Audit), Merck, New Jersey, USA
- Prof. Shiuh-Jeng Wang (Congress Steering Committee); Director of Information Cryptology and Construction Laboratory (ICCL) and Director of Chinese Cryptology and Information Security Association (CCISA); Department of Information Management, Central Police University, Taoyuan, Taiwan; Guest Ed., IEEE Journal on Selected Areas in Communications.
- Dr. Yunlong Wang; Advanced Analytics at QuintilesIMS, Pennsylvania, USA

- Prof. Layne T. Watson (Congress Steering Committee); Fellow of IEEE; Fellow of The National Institute of Aerospace; Professor of Computer Science, Mathematics, and Aerospace and Ocean Engineering, Virginia Polytechnic Institute & State University, Blacksburg, Virginia, USA
- Prof. Hyun Yoe; Director of Agrofood IT Research Center and Vice President of Korea Association of ICT Convergence in the Agriculture and Food Business (KAICAF); Director of Agriculture IT Convergence Support Center (AITCSC); Department of of Information and Communication Engineering, Sunchon National University, Suncheon, Republic of Korea (South Korea)
- Prof. Jane You (Congress Steering Committee); Associate Head, Department of Computing, The Hong Kong Polytechnic University, Kowloon, Hong Kong
- Dr. Farhana H. Zulkernine; Coordinator of the Cognitive Science Program, School of Computing, Queen's University, Kingston, ON, Canada

Embedded Systems, Cyber-physical Systems, & Applications

ESCS 2020 – Program Committee

- Prof. Emeritus Nizar Al-Holou (Congress Steering Committee); Professor and Chair, ECE Department; Vice Chair, IEEE/SEM-Computer Chapter; University of Detroit Mercy, Detroit, Michigan, USA
- Prof. Emeritus Hamid R. Arabnia (Congress Steering Committee); The University of Georgia, USA; Editor-in-Chief, Journal of Supercomputing (Springer); Fellow, Center of Excellence in Terrorism, Resilience, Intelligence & Organized Crime Research (CENTRIC).
- Dr. P. Balasubramanian; School of CSE, Nanyang Technological University, Singapore
- Prof. Dr. Juan-Vicente Capella-Hernandez; Universitat Politecnica de Valencia (UPV), Department of Computer Engineering (DISCA), Valencia, Spain
- Prof. Emeritus Kevin Daimi (Congress Steering Committee); Director, Computer Science and Software Engineering Programs, Department of Mathematics, Computer Science and Software Engineering, University of Detroit Mercy, Detroit, Michigan, USA
- Prof. Leonidas Deligiannidis (Congress Steering Committee); Department of Computer Information Systems, Wentworth Institute of Technology, Boston, Massachusetts, USA
- Prof. Mary Mehrnoosh Eshaghian-Wilner (Congress Steering Committee); Professor of Engineering Practice, University of Southern California, California, USA; Adjunct Professor, Electrical Engineering, University of California Los Angeles, Los Angeles (UCLA), California, USA
- Prof. Houcine Hassan; Department of Computer Engineering (Systems Data Processing and Computers), Universitat Politecnica de Valencia, Spain
- Prof. Dr. Guoming Lai; Computer Science and Technology, Sun Yat-Sen University, Guangzhou, P. R. China
- Dr. Andrew Marsh (Congress Steering Committee); CEO, HoIP Telecom Ltd (Healthcare over Internet Protocol), UK; Secretary General of World Academy of

Contents

Part V Wireless Networks, Novel Technologies and Applications

Part VI Internet Computing, Internet of Things, and Applications

Part I
Authentication, Biometrics, and Cryptographic Technologies

Statistical Analysis of Prime Number Generators Putting Encryption at Risk

Aykan Inan

1 Introduction

When it comes down to investigating the security properties of a cryptographic procedure there are different methods to do so, depending on the cryptoscheme itself. This includes inter alia, protocol, side-channel, and mathematical attacks. But the security of a strong cryptographically system is primarily based on the secure management of the secret key. If this key can be easily accessed or even worse guessed by the attacker the system is compromised. No matter how strong the used encryption method itself is. Therefore, it is of great importance that the secret key cannot be revealed in any case. Storing the key securely is one matter but the unpredictability is another [1].

Some cryptoschemes, such as RC4, rely on a random stream. Others, such as RSA need a PNG in order to generate two primes for generating the public and corresponding private key. Therefore, randomness for both "normal" random numbers and primes plays a major role.

Random number and prime number generators (RNGs and PNGs) are typically used when a cryptographic scheme needs a random number or random prime number to some extent. Random prime number generators have additional features as general random number generators: Any number generated needs to be odd and needs to be tested for primality afterwards. The following section gives an overview over the current state of PNGs. Section 3 demonstrates the approach used in this paper to verify the PNGs randomness.Sections 4 and 5 analyze specific outliers

A. Inan (✉)
Ravensburg-Weingarten University, Weingarten, Baden-Württemberg, Germany
e-mail: aykan.inan@hs-weingarten.de

K. Daimi et al. (eds.), *Advances in Security, Networks, and Internet of Things*, Transactions on Computational Science and Computational Intelligence,
https://doi.org/10.1007/978-3-030-71017-0_1

within the randomness spectrum. Section 6 looks for patterns within primes, and Sect. 7 for patterns in the last 32 to 64 bits. The last Sect. 8 concludes and gives an outlook.

2 Related Work and Basics

This section is splitted in two parts: while Sects. 2.1 and 2.2 discusses deterministic (Sect. 2.1) and non-deterministic (Sect. 2.2) random number generators, Sect. 2.3 gives an overview on prime-number generators. A key requirement for both types of random number generators is that their output cannot be reproduced or predicted [2]. There are many mathematical tests such as the chi-square test, which can verify the statistical behavior of RNGs or PNGs sequences.

2.1 Deterministic RNG

A deterministic random number generator is always producing the same sequences of random numbers under the same circumstances. That is why they are also called pseudo-random number generators (PRNG). But the produced consecutive numbers appear to be random enough for most applications. The generated sequence of a PRNG is computed recursively from an initial seed value initializing a function $f(s_0)$:

$$\begin{aligned} s_0 &= \text{seed} \\ s_{i+1} &= f(s_i),\ i = 0, 1, \ldots\ i \in \mathbb{N}_0 \end{aligned} \tag{1}$$

In general, the generated sequence can be described as:

$$s_{i+1} = f(s_i, s_{i-1}, \ldots, s_{i-t}) \tag{2}$$

In this case t is representing an integer constant.

Consequently, a PRNG does not generate true random numbers in a proper or true sense because it is computing its random numbers initialized from a starting (seed) value. Thus, it is completely deterministic [2]. According to Manuel Blum and Silvio Micali [3] a polynomial algorithm should not be capable of predicting and computing the next sequence better than 0.5 (50%) chance of success without knowing the initial seed value. For this, different mathematical tests are being used to prove the correctness.

2.2 Non-deterministic RNG

In contrast to the deterministic RNG a non-deterministic random number generator includes external source of randomness (entropy) such as hardware noise or the current time [1]. They are also known as cryptographically secure pseudo-random number generators (CSPRNG) and can be seen as a special type of PRNG which represents an unpredictable PRNG [2].

Assuming we have the following output sequence of n bits, where n is representing some integer:

$$s_i, s_{i+1}, \ldots, s_{i+n-1} \tag{3}$$

Then it must be computationally infeasible to compute the subsequent bits:

$$s_{i+n}, s_{i+n+1} \cdots \tag{4}$$

PNRGs and CSPSRNGs are described and defined as an algorithm that is producing an unpredictable sequence of random numbers in such a way that an attacker is not capable of computing or guessing them. This means that all generated random numbers must have the same likelihood of occurrence.

The characteristic of fully randomness can only be fulfilled with a One-time-pad which is, however, unsuitable for practical applications. So, the solution is to use pseudo-random numbers or pseudo-random sequences which are based on a deterministic process. Despite of this deterministic behavior and the use of an initialization seed value the produced output still must have the property of a truly random sequence [1].

In this context it is important to point out that the key generation in asymmetrical procedures usually requires some more effort than the generation process of pseudo-random numbers used in symmetrical systems. This is because an asymmetric cryptoscheme, such as RSA, requires large primes. Thus, the produced random numbers must fulfill the above-mentioned properties as well as the category of being prime at the same time [1, 4]. For this purpose Prime Number Generators are needed.

2.3 Prime Number Generator

In practice it is common to work with pseudo-prime numbers which fulfill most basic requirements for primes such as producing an odd number. Then a primality test, usually Miller-Rabin [5], is applied as depicted in Fig. 1.

The likelihood that a randomly picked or generated integer p is a prime is of further interest. In case of RSA, e.g., in order to generate a 1024-bit modulus n, the two primes p and q each should have a length of about 512 bits [4]. The chance that

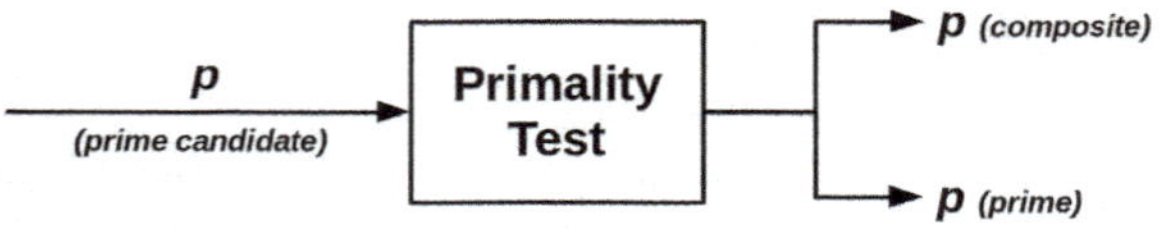

Fig. 1 Approach to generate primes

a random integer of that size is prime is still sufficiently high based on the prime number theorem and is approximately $\frac{1}{\ln(p)}$ as shown below [2, 6]:

$$p \text{ is prime } \approx \frac{2}{\ln(p)} = \frac{2}{\ln(2^{512})} = \frac{2}{512\ln(2)} \approx \frac{1}{177} \quad (5)$$

This means on average that 177 random numbers must be generated and tested before finding a prime. The density of primes, for even much larger bit numbers, is still adequate high [2].

This is relevant if similar prime numbers are generated. Often a value is then added to the result if it fails the primality test until a prime number is finally found. But this can lead to similar generated numbers.

Consequently, a prime used in RSA must be unpredictable. However, if at least one of the primes is easily obtained, RSA would be broken. This paper therefore analyses whether our current process of determining the quality of randomness for primes is still valid.

2.4 Evaluation of PRNG

Every published analysis dealing with PNGs or RNGs such as [3, 7–10] are just focusing on this unpredictability of subsequent sequences. As long as the produced output, e.g., a pair of two primes, is unique compared to the previous one, then the primes are sufficient enough for cryptographical use.

However, by generating more than one billion primes of specific bit lengths, (32, 64, 128, 256, and 512) and displaying the result into different statistics as presented in the following Sects. 3–7. The generated prime numbers show similar characteristics and seem related to each other.

A statistical analysis of the PNG used in LibreSSL was conducted. The results demonstrate suspicious behavior indicating that the numbers are not fully random as they seem.

3 Statistical Analysis

The statistical analysis is based on two essentials aspects:

(a) Prime Numbers
(b) Prime Distances

Each aspect itself is separated into sub-aspects again.

(a.1) Smallest prime number
(a.2) Largest prime number
(a.3) Mean prime number

The exact same statistical approach applies to the generated corresponding distances between prime numbers:

(b.1) Smallest distance
(b.2) Largest distance
(b.3) Maximum distance between (b.1) and (b.2)
(b.4) Mean distance

In general the statistics are showing the following relationship between two consecutive generated primes, as depicted in Fig. 2 and all generated primes in general:

Furthermore a variance analysis and standard deviation will be presented and explained in Sect. 3.3 for the prime numbers and the prime distances in Sect. 3.7. Due to the large amount of data and the large numbers involved the following subsections are going to present the most outstanding properties with regard to the above-mentioned listing in (a) and (b) and the specific bit lengths of 32, 64, 128, 256, and 512 bit because they are most commonly used.

3.1 *Largest and Smallest Prime Numbers*

The largest and smallest prime numbers that have ever occurred are listed in Tables 1 and 2.

Although these specific numbers did not occur very frequently they give a good reference point to search for boundaries and patterns within and among other primes. This includes, among other properties, the occurrences of numbers near threshold values (see Sects. 4 and 5), such as the largest and smallest prime.

Fig. 2 Distance between p and q

Table 1 Largest prime numbers

Bit	Value	Occurrence
32	4,294,967,291	13
64	18,446 … 876,649	2
128	340,282 … 715,813	2
256	115,792 … 191,919	1
512	13,407 … 162,089	1

Table 2 Smallest prime numbers

Bit	Value			Occurrence
32	3,221,225,473			3
64	13,835	...	607,023	2
128	255,211	...	346,557	1
256	86,844	...	462,243	1
512	100,558	...	779,483	1

Table 3 Mean value of all prime numbers

Bit	Mean value		
32	3,756,397,796		
64	16,138	...	636,565
128	297,724	...	539,463
256	101,315	...	281,316
512	11,731	...	587,394

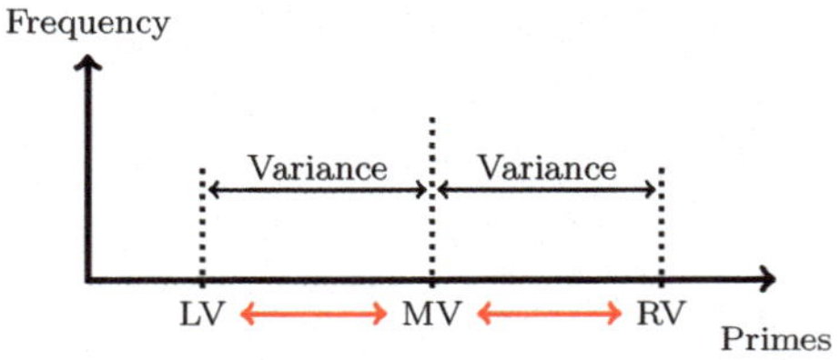

Fig. 3 Variance analysis

3.2 *Mean Value of Prime Numbers*

The mean value is calculated via all generated prime numbers which then was used to determine the standard deviation which will be explained in the following subsection. Table 3 shows the results.

3.3 *Standard Deviation of Prime Numbers*

The variance analysis as shown in Fig. 3 was conducted with regard to the standard deviation. This applies to both the distance and the prime analysis. The evaluation of the standard deviation provides good matches with the previous results.

The abbreviations used in the following tables and figures as well as in the above Fig. 3 are:

SP:	Smallest prime	MV:	Mean value
LP:	Largest prime	xDR:	x delta right
SD:	Smallest distance	xDL:	x delta left
LD:	Largest distance	Sx:	Sector x
LV:	Left value to MV	LD:	Last digit
RV:	Right value to MV		

Fig. 4 Variance analysis of prime numbers

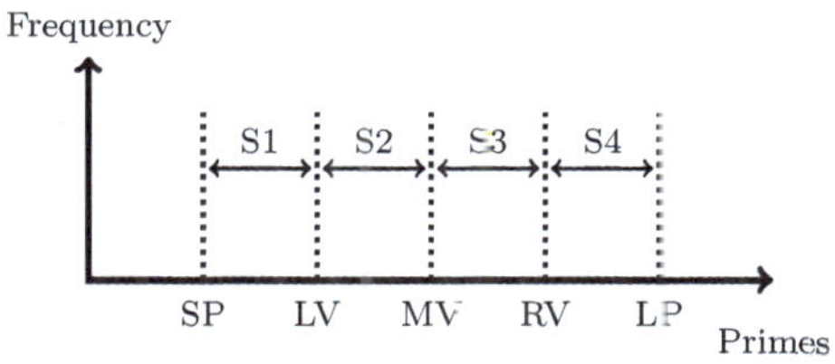

Table 4 Statistical distribution of all prime numbers

Bit	Sector 1 (S1)	Sector 2 (S2)	Sector 3 (S3)	Sector 4 (S4)
32	21.13%	28.95%	28.78%	21.13%
64	21.13%	28.90%	28.84%	21.13%
128	21.13%	28.88%	28.85%	21.13%
256	21.14%	28.87%	28.86%	21.13%
512	21.55%	28.72%	28.71%	21.02%

To get the first overview the huge number space was divided into four different sectors to analyze the distribution as shown in Fig. 4.

The largest and smallest prime numbers found marked the upper and lower boundaries while the mean value is marking the approximate center. The left and right value to the mean value helped again to separate the number range. Table 4 shows the proportional distribution of the primes for each sector and bit length.

First of all Table 4 reveals that the distribution for every bit length is almost identical. But it proved at the same time that for every bit length most of the generated primes, approximately 60%, are located between LV and RV.

3.4 *Largest and Smallest Prime Distances*

Looking at the distances is relevant due to the fact that they can provide information about the properties and dependencies between two consecutive generated primes. This can be very important, i.e., in relation to improve a prime factorization processes and knowing the approximate location of two primes. The largest and smallest prime distances that occurred are listed in Tables 5 and 6.

Table 5 Largest prime distance

Bit	Value	Occurrence
32	1,073,726,616	1
64	4611 … 887,704	1
128	85,069 … 365,142	1
256	28,947 … 658,100	1
512	335,188 … 410,182	1

Table 6 Smallest prime distance

Bit	Value	Occurrence
32	2	3
64	2,031,919,274	1
128	97,049 … 555,744	1
256	24,277 … 917,692	1
512	1982 … 318,524	1

Table 7 Distance between two generated primes

Bit	Maximum distance
32	1,073,726,608
64	4611 … 968,430
128	85,069 … 094,649
256	28,947 … 018,638
512	355,188 … 303,022

Table 8 Mean value of all prime distances

Bit	Mean value
32	35,794,328
64	1537 … 847,280
128	28,357 … 762,191
256	9649 … 455,788
512	1173 … 562,455

3.5 Maximum Distance

The maximum distance was computed via the results of the largest and smallest distance. The results are shown in Table 7.

3.6 Mean Value of Prime Distances

The result for the mean value of the primes distances is shown in Table 8.

3.7 *Standard Deviation of Prime Distances*

Table 9 shows the proportional distribution of the prime distances for each sector and bit length.

Both standard deviations show similar results. However, the deviation of the distances in sector 2 is prominent. Again, sector 2 and 3 together contain most of the generated distances. In this case approximately 63%.

4 Occurrence of Primes Near the Threshold Values

Given the fact that the generated integers are becoming increasingly larger with the increasing length of bits the analyzed scope was adjusted based on the discovered pattern of the largest prime. Thus, these patterns were used in order to locate other primes within a predefined range. The most important ones are shown in Table 10.

The first digits of a prime are mainly responsible for the specific pattern with regard to the threshold value. The delta (Δ) range was chosen based on the position of the last digit of the found pattern. The idea is illustrated using the 32 bit value in Table 11 where x is representing any digit. The last digit of the pattern ends in the fourth position to the right. So, this position is set to one filled with zeros in the delta value.

The classification of the number line is depicted in Fig. 5. The delta (Δ) between each sector represents the searching area for close related numbers in relation to the threshold value.

In case of the 32 bit pattern you can create a specific amount of groups of patterns ($\Delta_{Pattern-Range}$) computed via the difference between both patterns as shown in Eq. (6). This then represents the actual search range for these specific patterns. In

Table 9 Statistical distribution of all prime distances

Bit	Sector 1 (S1)	Sector 2 (S2)	Sector 3 (S3)	Sector 4 (S4)
32	18.58%	36.98%	25.78%	18.57%
64	18.57%	36.98%	25.87%	18.57%
128	18.57%	36.98%	25.87%	18.57%
256	18.57%	36.98%	25.87%	18.57%
512	18.57%	36.98%	25.87%	18.57%

Table 10 Patterns for LP, SP

Bit	Pattern LP	Pattern SP
32	4,294,967 ...	3,221,225 ...
64	18,446,744 ...	1,383,505,805 ...
128	34,028,236 ...	255,211,775 ...
256	115,792,089 ...	8,684,406 ...
512	1,340,780 ...	100,558,559 ...

Table 11 Patterns for LP, SP

Pattern	Chosen Δ
4,294,967,xxx	1000
3,221,225,xxx	1000

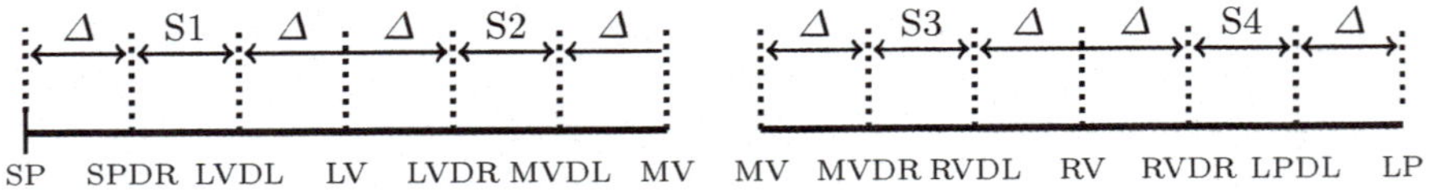

Fig. 5 Likelihood of occurrence near the threshold values

Fig. 6 Dissemination of patterns

this case it is computed as followed:

$$\begin{aligned} \Delta_{Pattern-Range} &= Pattern_{LP} - Pattern_{SP} \\ &= 4,294,967 - 3,221,225 \\ &= 1,073,742 \end{aligned} \quad (6)$$

Figure 6 depicts the dissemination of all patterns across the entire number range. In this respect only the 32 bit values are displayed again.

As can be seen the minimum amount of patterns is distributed approximately evenly above a certain threshold value indicated with the red line. On the other hand it shows a slowly but surely decreasing distribution approaching to the threshold value which can be traced back to the fact that the integers are getting larger.

Figure 7 is depicting a much smaller range showing that there are areas of a greater concentration of patterns as indicated with the red circles. In this view these accumulations can be seen as hills in the graph.

Table 12 shows the preliminary results of the first analysis of the percentage distribution of primes within certain areas. The numbers in brackets present the percentage of the amount of found located primes. As a matter of fact the frequency of occurrence will start to drop as greater the bit size becomes. But the relative frequency is still very high in relation to the bit size.

Fig. 7 Accumulations of patterns

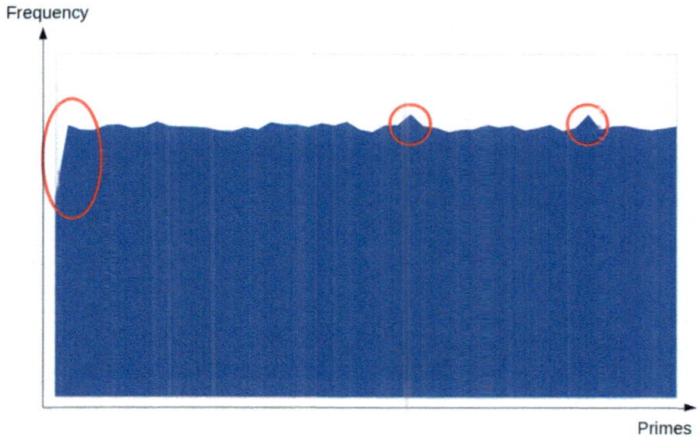

Table 12 Percentage distribution of primes

Section	Bit 32	Bit 64	Bit 128	Bit 256	Bit 512
SD:	0.09%	0.22%	0.12%	0.35%	0.56%
LV:	0.19%	0.43%	0.24%	0.69%	0.06%
MV:	0.19%	0.43%	0.24%	0.69%	0.06%
RV:	0.19%	0.22%	0.23%	0.69%	0.06%
LD:	0.09%	0.00%	0.12%	0.35%	0.03%
Sector-1:	20.95%	20.70%	20.90%	20.45%	20.96%
Sector-2:	28.76%	28.46%	28.65%	28.18%	28.66%
Sector-3:	28.60%	28.41%	28.62%	28.17%	28.65%
Sector-4:	20.95%	20.70%	20.90%	20.44%	20.96%

Table 13 Patterns for LD, SD

Bit	Pattern LP	Pattern SP
32	1073 ...	2 ...
64	46,116 ...	20,319 ...
128	8506 ...	97,049 ...
256	2894 ...	24,277 ...
512	3351 ...	1982 ...

5 Occurrence of Distances Near the Threshold Values

The exact same procedure as described with the primes in the previous section was conducted with the distances, too. Table 13 shows the patterns for the largest and smallest distances while Table 14 illustrated using the 32 bit value to find out the corresponding delta. In this respect, it should be noted that the delta for the smallest distance was set to the same value as for the largest distance due to the small factor for the smallest distance. Table 15 presents the final results.

In addition to Tables 12 and 15 a more detailed analysis is still underway but the first analysis already shows that the generated primes can be separated into groups of patterns. Then these patterns can be assigned to the different sectors

Table 14 Patterns for LD, SD

Pattern	Chosen Δ
1073 xxx xxx	1,000,000
2	1,000,000

Table 15 Percentage distribution of distances

Section	Bit 32	Bit 64	Bit 128	Bit 256	Bit 512
SD:	0.19%	0.04%	0.02%	0.07%	0.06%
LV:	0.34%	0.08%	0.04%	0.12%	0.11%
MV:	0.25%	0.06%	0.03%	0.09%	0.08%
RV:	0.16%	0.04%	0.02%	0.06%	0.05%
LD:	0.00%	0.00%	0.00%	0.00%	0.00%
Sector-1:	18.22%	18.49%	18.53%	18.44%	18.46%
Sector-2:	36.69%	36.91%	36.94%	36.87%	36.89%
Sector-3:	25.67%	25.82%	25.85%	25.80%	25.81%
Sector-4:	18.49%	18.55%	18.56%	18.55%	18.55%

and narrowed down the area of concentration where most of the primes a primarily located. This means that the actual located primes within that range is supposed to be much higher considering that some patterns only differ in one digit of the primary pattern. Currently it can be stated that both distances and primes have a noticeable behavior. While the distances seem to be more concentrated near the mean value the primes seem to have a close proximity to some threshold values with fluctuations in between. But this assumption still needs to be proven in detail.

What definitely can be said about this property is that both the smallest and largest prime numbers have some patterns to adjacent numbers that are close to the maximum and minimum values. In the case of the largest 512 bit prime the most noticeable property is the fact that all smaller numbers are closely related to each other and lap with the first six digits. The same applies to the smallest prime numbers and to all other bit sizes.

6 Patterns Within Primes

The search for patterns is very complex and difficult. However, first tests show that there are not only patterns at the beginning of a prime but also within a prime. This will be demonstrated with a small example of a 64 bit number as shown below:

17261221124532159023
17267101136670495809

The multiple patterns are marked in red. Patterns of a similar kind are very probable but, as already mentioned, are very difficult to find. The more such patterns

Table 16 Likelihood of occurrence of the last digit

Bit	1	3	7	9
32	25.50%	23.78%	26.07%	24.65%
64	25.52%	23.73%	26.10%	24.65%
128	0%	33.41%	34.06%	32.53%
256	0%	33.35%	34.02%	32.62%
512	0%	33.41%	34.06%	32.62%

are found within a prime number, the more advantageous it is to find other prime candidates. This has two decisive consequences. On the one hand the number space is automatically restricted and on the other hand the probability of finding possible prime numbers increases due to the classification into groups of patterns.

7 Searching for Last Digits

Another special feature of primes is the fact that the last digit of a prime candidate can either end with 1, 3, 7, or 9. The numbers 2 and 5 are excluded in this consideration because they are the only exception to this rule. For this reason it was analyzed how evenly the primes are distributed with regard to their last digit. The final result of this analysis is shown in Table 16.

As can be seen every last digit appears nearly with the same likelihood within the absolute values for every bit length. The last digit of 7 appears the most followed by 1, 9, and 3. However, this applies only to integers with a bit length of 32 and 64 bit. Much more prominent is the result shown in the second column representing the last digit of 1. Integers at a length of 128 bit and above do not show any single prime number ending with the digit of 1. 7 is still the most common followed again by 3 and 9. The analysis of this result is still ongoing.

8 Benefits from the Statistics

The first analysis found several aspects including block patterns of same digits within the prime number itself as well as among several primes. Furthermore, the generated primes seem to be generated within a certain undocumented upper and lower boundary, which means that the generated primes do not exceed or fall below a certain generated limit. But the most remarkable property is the fact the generated prime numbers do not differ much from each other as one could assume. Whether or not the primes were produced at once or in several runs and on different machines. All in all this knowledge already reveals a lot about the generation process and further reduces randomness as a consequence of this. This entire knowledge can be of great importance when it comes to factorizing a large integer and to accelerate and facilitate the factorization process. Knowing and visualizing the proportional distribution of primes as well as their related distances and their relationship of distance can be a tremendous help in cracking keys that are based on PNGs and used in RSA, for instance. This allows a faster prime factorization within a predefined range based on the properties and relationships. This knowledge can also be used to increase the forecast probability for a number being a potential prime key candidate to search for.

9 Conclusions

This paper outlines and indicates that the current implementations of LibreSSL and the one-sided approach to analyze PNGs are not sufficient enough to prove randomness. However, this still requires further analysis including sorting and analyzing the primes for these patterns as well as an intense code audit in order to find correlations between the available statistics and the generation process itself. This code audit is temporarily using LibreSSL as an example. A parallel analysis of GNU Multiple Precision Arithmetic Library is underway and being set up so that the results can be compared regarding their predictability and security. As a result, this work might demonstrate that the current status "symmetric encryption cannot be undone without the private key" might be wrong and thereby enable investigators to decipher encrypted data.

References

1. C. Eckert, *IT-Sicherheit: Konzepte - Verfahren - Protokolle*, 8th edn. (Oldenbourg, München, 2013), pp. 327–329, 431–433
2. C. Paar, J. Pelzl, *Understanding Cryptography - A Textbook for Students and Practitioners*, 2nd corrected printed (Springer, Berlin, 2010)
3. M. Blum, S. Micali, How to generate cryptographically strong sequences of pseudo-random bits. SIAM J. Comput. **13**(4), 850–864 (1984)
4. R.L. Rivest, A. Shamir, L. Adleman, *A Method for Obtaining Digital Signatures and Public-Key Cryptosystems*, vol. 21(2) (Communications of the ACM, New York, 1978), pp. 120–126
5. D. Knuth, *The Art of Computer Programming*, vols. I–III, 3rd edn. (Addison-Wesley Pub Co, Boston, 1997)
6. M. Schubert, *Mathematik für Informatiker - Ausführlich erklärt mit vielen Programmbeispielen und Aufgaben* (Vieweg+Teubner Verlag, Wiesbaden, 2009)
7. Bundesamt für Sicherheit in der Informationstechnik, *Quellcode-basierte Untersuchung von kryptographisch relevanten Aspekten der OpenSSL-Bibliothek*, Projekt 154, Bonn, Version 1.2.1, 2015-11-03
8. Bundesamt für Sicherheit in der Informationstechnik, Documentation and Analysis of the Linux Random Number Generator, Bonn, Version 3.5, 2019-12-13
9. L. Accardi, M. Gäbler, Statistical Analysis of Random Number Generators (2011), pp. 117–128. https://doi.org/10.1142/9789814343763_0009
10. R. Fujdiak, J. Misurec, P. Mlynek, Analysis of Random Number Generator from Texas Instrument in MSP430 x5xx Families (2014). https://doi.org/10.1109/TSP.2015.7296344

Secure Authentication Protocol for Drones in LTE Networks

Dayoung Kang, Gyuhong Lee, and Jin-Young Choi

1 Introduction

In December 2016, Amazon, the largest e-commerce company in the USA, succeeded in delivering a delivery service using a drone to a farm in Cambridge, England [15]. The time from order to delivery was only 13 min, and people began to notice the possibility of using commercial drones. Drones began to be developing for military use during World War I in the USA [16], but now they are in higher demand from the civilian sector. The drone applications are diverse, such as disaster management, search and rescue, agricultural use, and the arts, and the expected economic effect is excellent. In order to operate and control drones in vast areas, diverse methods of communication have been developed. Cellular networks have become highly attractive to assist with drone identification, authentication, and communication, and people started to get interested in drones [18].

Even though we are transitioning from LTE to 5G, more than 50% of GSM subscribers use LTE networks for communication in 2020. GSMA (GSM Association) announces that the number of 5G connections will reach 1.4 billion by 2025, 20% of total connection. Nevertheless, 4G (LTE) will continuously grow over this period, accounting for 56% of global connections by 2025 [11].

There are various vulnerabilities on LTE networks [6], but we are focusing on the security vulnerability that could reveal the unique identifier (IMSI) of mobile subscribers during mutual authentication and key exchange. If drone pilots used LTE

D. Kang (✉) · J.-Y. Choi
The Graduate School of Information Security, Korea University, Seoul, Republic of Korea
e-mail: dayokiki@korea.ac.kr; narnia@korea.ac.kr

G. Lee
The Department of Cyber-Warfare, Korea Army Academy at Yeongcheon, Yeongcheon, Gyeongsangbuk-do, Republic of Korea

K. Daimi et al. (eds.), *Advances in Security, Networks, and Internet of Things*, Transactions on Computational Science and Computational Intelligence,
https://doi.org/10.1007/978-3-030-71017-0_2

for drone communication and a malicious adversary obtained IMSI of the drone, the adversary could conduct passive or active attacks related to location privacy. If a drone leaks its location, a malicious adversary can predict the frequency, destination, starting point, or mission of the drone. Therefore the leakage of IMSI needs to be addressed for the security and privacy of LTE drones.

In this paper, we first review various studies [12, 19, 21, 22] on how to improve the LTE authentication protocol for mobile phones and show that these techniques are not fit to LTE drone, which means that we need an authentication protocol suitable for LTE drones. We describe the operation concept of a drone using LTE, a network concept based on LTE, and a general user authentication process of the LTE network. Later, we perform a security analysis to derive risks about leakage of IMSI. Next, we proposed a key authentication protocol suitable for LTE drones as a countermeasure. Finally, we specify and verify the proposed protocol using Scyther [9], an automated protocol verification tool.

2 Related Works

Since there does not exist studies related to the LTE authentication protocol specialized in LTE drones, we refer to a paper that poses the possibility of IMSI leakage during the authentication process of a mobile phone on an LTE network and proposes various security protocols.

Broek et al. [21], in 2015, proposed Dynamic ID-based authentication. The author insisted that Home Subscription Server (HSS) must authenticate the subscriber and the key exchanged by receiving the Pseudo IMSI (PMSI), instead of receiving the ISMI from the SIM user at the initial stage. After shared with the HSS, the subscriber stored two PSMIs in the SIM. The SIM subscriber receives an Authenticate from the server by throwing one of the PMSIs rather than ISMI at the initial phase. At this time, HSS discard the PMSI used for authentication and make a new PMSI, and the subscriber receives the new PMSI and use it for the next communication.

Norrman et al. [19] proposed that a SIM subscriber sends the IMSI encrypted with a public key of HSS to the HSS so the IMSI would not be leaked. Since the HSS's private key is required to decrypt the IMSI encrypted with the HSS's public key, a malicious adversary without the private key cannot decrypt the ciphertext and get the IMSI.

Hsieh et al. [12] provided an authentication protocol that used One-Time Passwords (OTPs) based on the time and location information to authenticate the subscriber securely. When the mobile subscriber transmits the location information to the HSS, the HSS can predict the next moveable distance compared with the previously received location information. The author claimed that if the mobile phone subscriber was outside the predictable travel distance, the HSS could not authenticate the subscriber.

Abdrabou et al. [5] proved that the EPS-AKA protocol used for LTE authentication transmitted an unencrypted IMSI and proposed an authentication protocol modified the EPS-AKA protocol. For this, the proposed authentication algorithm should be stored in USIM on subscriber-side and in the authentication center on the home subscriber server-side.

The above-discussed authentication protocols proposed an improvement of the current protocol and modification of the existing infrastructure, but these were studies on mobile phone subscribers, not drones. In this paper, assuming that the current infrastructure is not changed, we propose an authentication protocol for LTE drones that utilizes the secure channel of the drone operating system. In particular, we introduce an authentication protocol suitable for LTE drone operation using Asymmetric cryptography and 2-Factor Authentication.

3 LTE Drone Control System

3.1 General System Architecture

The drone control system is divided into a UAV (Drone) and a ground control station (GCS) that commands and controls the drone, and for mutual communication, there exist data link to transmit and receive information between the drone and the GCS.

We can divide the drone data link into two types [22], and one is a peer-to-peer link where the drone is directly connected and controlled to the GCS, another is a network-type link where the drone connects to the GCS using a high-speed wireless network such as LTE or IEEE 802.16. There is a message protocol (e.g., MAVLink [1]) used to exchange messages between the drone and GCS, and the drone transmits the MAVLink message to the GCS over the LTE networks. In this paper, we discuss a network-type link drone control system using LTE.

The drone performs through a base station (eNodeB), SGW (serving gateway), and PGW (packet gateway) to communicate with GCS. The mutual authentication between the drone and the LTE network is required at the initial phase. For this authentication, the required information is stored in the drone's SIM and the HHS/AuC (home subscriber server/authentication center) operated by the subscribed cellular carriers (e.g., Verizon, AT&T). Also, there exists MME (mobility management entity), which is responsible for initiating authentication of the drone device.

Figure 1 presents the architecture of the drone control system using LTE networks [7].

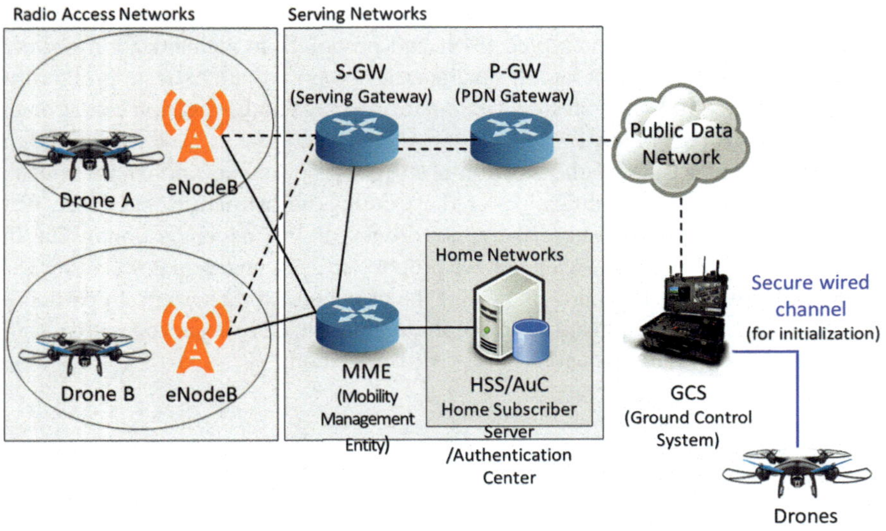

Fig. 1 Using LTE for drone control system

3.2 LTE Authentication Protocol

The drone needs to be authenticated by HSS/AuC to use LTE networks, but the drone and the HSS do not send and receive authentication information directly, but perform the authentication procedure through the MME [7, 22].

The drone sends an Attach Request with authentication information (IMSI, UE Network Capability) in plaintext to make a connection from the MME at the initial phase.

Upon receiving the Attach Request from the drone, the MME sends an Authentication Data Request (IMSI, SN id, Network Type) to the HSS to authenticate the drone.

The HSS puts the received Authentication Data into the EPS-AKA algorithm, obtains Authentication Vector (RAND, XRES, AUTN, and K_{asme}) as a result, and sends the AVs to the MME. After receiving the Authentication Vectors(AVs) from HSS, the MME stores the AVs and sends an Authentication Request (RAND, AUTN) to the drone to request authentication.

The drone that received the Authentication Request puts it into the EPS-AKA algorithm in SIM, acquires AUTN, RES, K_{asme}, and sends RES to MME.

On receiving RES from the drone, the MME compares it to the XRES in possession. Finally, the MME completes the authentication procedure if it is the same as the XRES it has.

After that, MME set the encryption method, and the Drone, MME, and HSS have secure communication performed using the K_{asme} owned by both sides.

Figure 2 shows how HSS authenticates drone through MME according to the LTE standard [2–4].

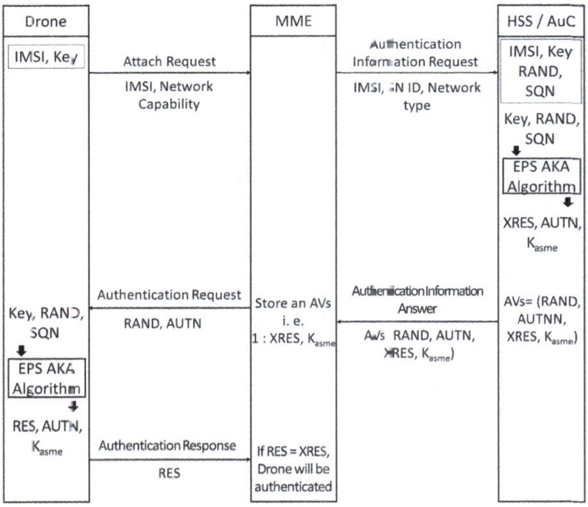

Fig. 2 Authentication and key agreement protocol

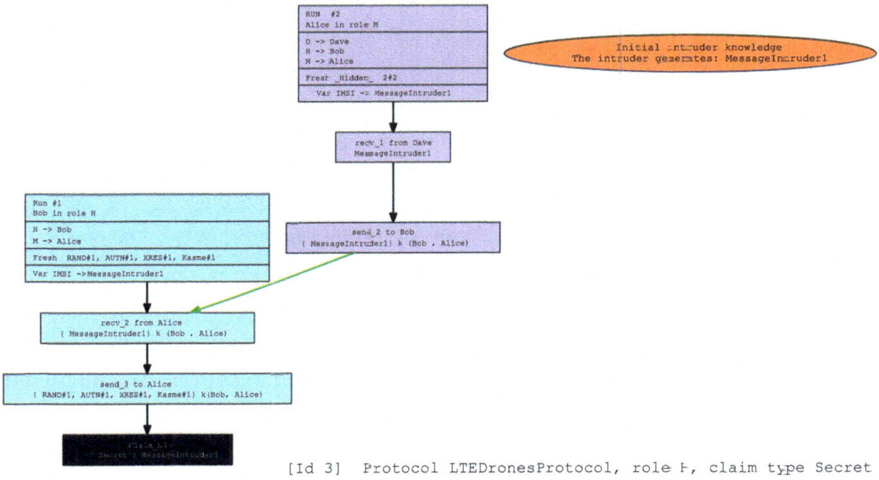

Fig. 3 Attacks for claim secret IMSI

3.3 Security Analysis

We specified the ESP-AKA protocol using Scyther, which is a protocol verification tool. We verified that a malicious intruder could sniff the IMSI over the LTE networks, and pretend to be the drone. We will introduce Scyther in Sect. 5,and the result of the security analysis is shown in Fig. 3.

According to Mjølsnes et al. [17], even if an adversary has not experienced high-level hacking, it is easy for him to reveal the IMSI in LTE. Once the adversary sniffs IMSI, the threats that will affect the drone missions are as follows :

Disclosure of the Drone Identity A drone user must be authenticated by HSS to access the LTE networks at the first access. During the authentication, the drone transmits his IMSI in plaintext. Thus, an adversary could intercept the IMSI through a sniffing attack. Once the adversary sniffed IMSI, the adversary could obtain subscriber information, location privacy, and conversation information. Furthermore, the attacker could hide the real drone user and commit other cyber attacks such as DoS by using the IMSI [5].

Man in the Middle (MITM) Attack at the Authentication Phase A malicious adversary can always intercept an Authentication Request sent to MME by disguised as a Rogue base station. If the adversary gets the drone's IMSI, it can hold both the IMSI and the Authentication Request, so a MITM attack is possible[5].

Denial of Service A malicious attacker attempts to access the network by sending a Fake Attach request continuously to the MME pretended to be a drone after intercepting the IMSI and Authentication Request of the Drone. As the adversary attacks, MME and HSS consume its computational powers, and normal drones cannot access the network. Therefore, the drone cannot communicate with the GCS over LTE networks, so it fails to operate the mission in the affected area [17].

Down Grade Attack Using IMSI If a malicious adversary steals IMSI of a target drone and launches a DoS attack at the stage of sending an Attach Request, the drone cannot use the LTE network. In such a situation, the drone uses 3G networks as an alternative. Then, the adversary can use the 3G network's weakness to attack the drone[7, 13].

Location Leakage If a malicious attacker launches a DoS Attack toward a drone to cause the drone to use a 3G network rather than an LTE network and has the IMSI of the drone, the adversary can reveal the location of the target drone running within a radius of 2 km^2 [13, 20].

4 Proposed Protocol

NIST proposed methods to mitigate various threats in LTE networks [7]. To prevent a hacker disguised as a rogue base station, wiretapping, or downgrade attack, NIST identified third party over-the-top solutions with the mitigation method.

Therefore, we propose an authentication protocol that uses the trusted 3rd party (GCS), subscriber information (signature value), and location information as authentication factors. We propose a secure authentication protocol that we protect the confidentiality of IMSI by encrypting with asymmetric keys of GCS and HSS at first attach request phase.

In this paper, we propose an authentication protocol with the goal of not leaking IMSI by using the EPS-AKA Algorithm of USIM and the existing infrastructure without modification.

4.1 Architecture of Proposed Protocol

When buying a SIM and a cellular plan, the drone provides IMSI, LTE K of the SIM, and user information to HSS. The HSS stores the received IMSI in the form of $h(i)$ using a hash function h and ISMS i, and stores the subscriber information in AuC as well.

The secret key for MAVLink protocol should be created on a GCS and shared with drones via secure channels (e.g., local USB cable or local wired Ethernet cable). GCS receives IMSI of the drone using this must-connect secure channel.

Once GCS gets IMSI from the drone through the wired secure channel, the GCS encrypts the $h(i)$ and the current location information (GPS) with the drone user's private key and HSS's public key to request authentication.

After receiving the $h(i)$ and GCS encrypted by the subscriber, the HSS decrypts the cyphertext with HSS's private key and subscriber's private key and checks whether the subscriber is valid or not. If the requester is a valid subscriber, the HSS creates an Authentication Vectors (RAND, AUTN, XRES, K_{asme}) using the EPS-AKA algorithm shared with the subscriber.

After the HSS select MME near drones based on the GPS information, HSS sends the AVs (XRES, K_{asme}) and the temporary IMSI (TMSI) to the MME and addresses the AVs (RAND, AUTN) and TMSI to the subscriber (GCS). The drone receives AVs (RAND, AUTN) and TMSI from the GCS via a secure channel.

The drone puts the AVs (RAND), IMSI, Key into the EPS-AKA algorithm in SIM, acquires AUTN, RES, K_{asme}, and compares the acquired AUTN and received AUTN from HSS.

If the acquired AUTN and received AUTN are the same, the drone considers that HSS sent the TMSI, and K_{asme} is the encryption key for the LTE network.

Since the MME has been holding the AVs (XRES, K_{asme}) received from the HSS, once receiving TMSI and AVs (RES) encrypted with K_{asme} from the Drone, the MME can decrypt the cyphertext. After comparing it with the RES and XRES, the MME authenticates the drone if the values are the same.

Finally, LTE drone completes the authentication procedure, and a data link between the Drone and GCS over the LTE network is connected.

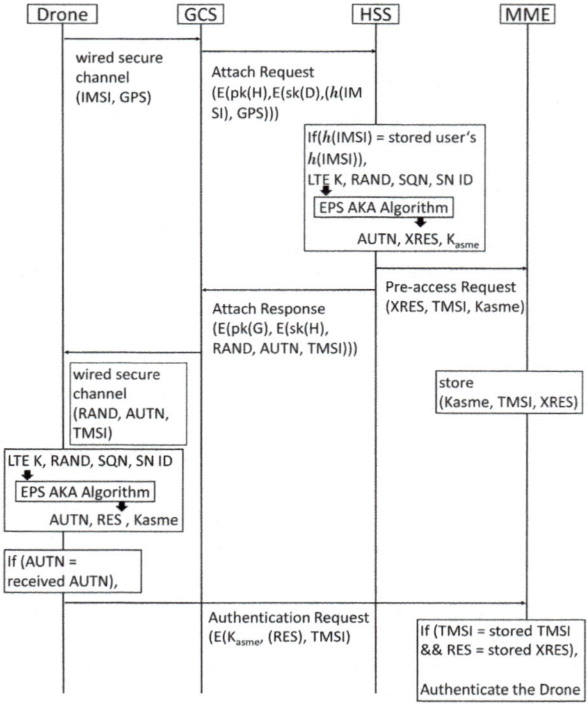

Fig. 4 LTE drone authentication phase

4.2 *Phase of Proposed Protocol*

The proposed protocol consists of the authentication phase between four components. We depict the authentication phases in Fig. 4, and list the used notations in Table 1.

4.3 *Security Analysis of Proposed Protocol*

The proposed protocol is secure against the leakage of IMSI, so it protects the drone from the adversary attacks related to the vulnerability.

Security Against Disclosure of Drone Identity In the proposed protocol, a drone transmits IMSI to its GCS over a secure wired channel, which is USB via so it is impossible to leak IMSI at this stage. The GCS encrypts IMSI with its private key, then encrypts it with HSS's public key and sends it to the HSS. Even if a malicious adversary sniffs the LTE air interface, he cannot obtain the IMSI.

Security Against MITM Attack at the Authentication Phase In our proposed protocol, a GCS sends an authentication request for the drone directly to the HSS

Table 1 Notations of the proposed protocol

Parameter	Meaning
Drone	LTE device (user)
GCS	Ground control station (command and control the drone)
HSS	Home subscriber server
MME	Mobility management entity
IMSI (TMSI)	Drone USIM number (temporarily ID)
GPS	Current location of drone
K_{asme}	Secret key for communication between drone, MME and HSS
RAND	Random number
$h(.)$	Hash function
pk(X)	Public key of X
sk(X)	Private (secret) key of X
E(a,b)	Encrypt message 'b' with a key 'a'
XRES	Expected response
RES	Response for authentication
AUTN	Authentication token

without going through the eNodeB and MME. Therefore, since the GCS sends IMSI as an encrypted message directly to the HSS, even if a malicious adversary masquerades as a rogue eNodeB, the adversary cannot receive the authentication request and obtain the IMSI.

Security Against Denial of Service and Down Grade Attack Using IMSI As the proposed protocol is secure against MITM attacks at the authentication request phase, the malicious attacker cannot interrupt the authentication messages between HSS and GCS. After successful drone authentication, the HSS allocates TMSI instead of IMSI and provides the TMSI to the GCS, so the malicious adversary cannot do DoS attacks using IMSI. Since TMSI is periodically changed, MITM attacks or downgrade attacks using them would be troublesome to the malevolent adversary.

Security Against Location Leakage If a malicious adversary fails to collect IMSIs of drones, the attacker will not be able to detect the location or movement of the drone over time.

5 Formal Analysis

5.1 Protocol Verification Tool : Scyther

Scyther is an automatic tool for the verification, falsification, and analysis of security protocol under the perfect cryptography assumption. We assumed that all

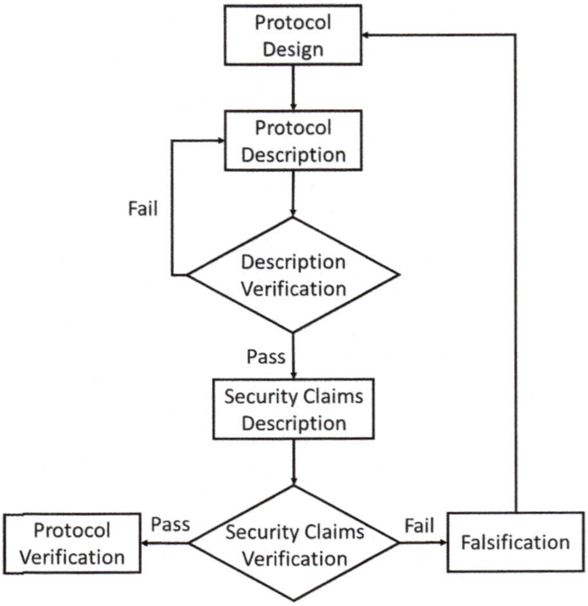

Fig. 5 Protocol verification using Scyther

cryptographic functions are perfect, which means any adversary learns nothing from an encrypted message unless he knows the key. We can use Scyther to find the potential security problems from the protocol development [8, 9].

Figure 5 shows an overview of a protocol verification process using Scyther. We could check the security properties for LTE Drone Authentication and to verify that certain values are confidential (secrecy) or certain properties hold for the communication partners (authentication). Scyther allows us to verify these properties or falsify them.

5.2 *Specification*

The descriptions of the proposed protocols and the claims are written in Security Protocol Description Language (SPDL), which allows Scyther to verify the properties. The descriptions of the LTE authentication protocol are as follows :

```
const pk : Function;
secret sk : Function;
inversekeys (pk, sk);
hashfunction h;
usertype Message;
protocol LTEDroneAuth(D, G, H, M){
  role D{
    fresh RES : Nonce;
```

```
        fresh TMSI2 : Message;
        send_4 (D, M, {RES}k(D,H,M),TMSI2); }
    role G{
        fresh Imsi, GPS : Nonce;
        var rand, autn : Nonce;
        var TMSI2 : Message;
        send_1 (G, H, {{h(Imsi),GPS}sk(G)}pk(H));
        recv_3 (H, G, {{rand, autn, h(Imsi),
        TMSI2}sk(H)}pk(G)); }
    role H{
        var Imsi, GPS : Nonce;
        fresh rand, autn, XRES : Nonce;
        fresh authenticateduser : Nonce;
        fresh TMSI1 : Message;
        fresh TMSI2 : Message;
        recv_1 (G, H, {{h(Imsi),GPS}sk(G)}pk(H));
        match (h(Imsi),h(authenticateduser));
        send_2 (H, M, {XRES, TMSI1,
        k(D,H,M)}k(H,M));
        send_3 (H, G, {{rand, autn, h(Imsi),
        TMSI2}sk(H)}pk(G)); }
    role M{
        var XRES, RES : Nonce;
        var TMSI1 : Message;
        var TMSI2 : Message;
        recv_2 (H, M, {XRES, TMSI1, k(D,H,M)}k(H,M));
        recv_4 (D, M, {RES}k(D,H,M), TMSI2);
        match(RES, XRES);
        match(TMSI1, TMSI2); }
}
```

5.3 *Analysis of the Verification Results*

Cas Cremers[10] and G. Lowe [14] suggest a methodology for the formal specification and verification of abstract security protocols. The author introduces security properties related to secrecy and several forms of authentication. Based on the approach, the security properties that we used to verify that our proposed authentication protocol are as follows:

Secrecy Secrecy property explains that certain information is not revealed to adversaries, even though the message is communicated over untrusted networks. Secrecy claim is written as a claim(A, secret, rt), where A is the role executing this event, and rt is the term that should be secret, which is not known to the adversary.

Aliveness Aliveness is a form of authentication that intends to authorize that an aimed communication partner has executed some events that means the partner is alive. Aliveness claim is written as a claim(A, alive), where A is the role executing this event and should be alive.

Non-injective Synchronization Non-injective synchronization is that everything we intended to happen in the protocol description also occurs in the trace. Non-injective synchronization claim is written as a claim(A, Nisynch), where A is the role executing this event.

Non-injective Agreement The definition of non-injective agreement expresses that for all claims in any trace, there exist runs for the other roles in the described protocol, such that all communication events causally preceding the claim must have occurred before the claim. Non-injective agreement claim is written as a claim(A, Niagree), where A is the role executing this event.

Weak Agreement Weak agreement is that if an initiator completes a run of the protocol, apparently with a responder, then the responder has previously been running the protocol, apparently with the initiator. Weak agreement claim is written as a claim(A, Weakagree), where A is the role executing this event.

We verify that the secrets, which required for authentication, such as IMSI and RES and XRES, are not leaked by using Scyther to verify the proposed authentication protocol. Also, we proved that the proposed protocol satisfies the other security properties. The verification result is as follows :

- LTE Drones : We verify that LTE drone satisfies Non-injective synchronization and Non-injective Agreement in the proposed protocol. While transmitting RES from drone to MME, they are secure against malignant adversaries. The verification result is shown in Fig. 6.
- MME : We prove that MME satisfies the security properties, which are Aliveness, Weak agreement, Non-injective synchronization, and Non-injective Agreement within its bound. MME keeps the secret, K_{asme}, and XRES from HSS and RES from LTE drone. It is proved that the proposed mechanism withstands all automatic attacks, and no attack was found within its bounds. The analysis result is shown in Fig. 7.

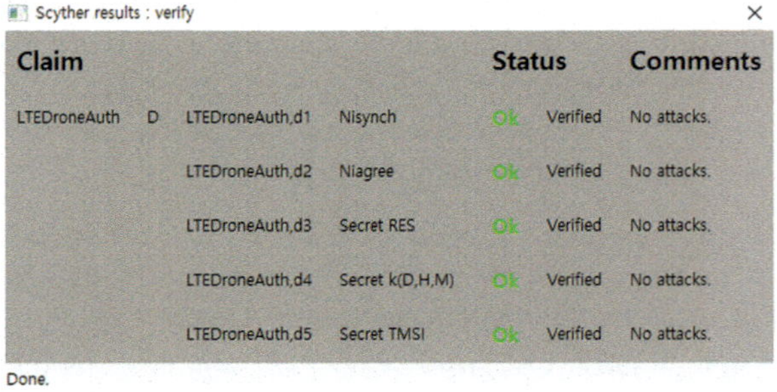
Scyther results : verify

Claim				Status		Comments
LTEDroneAuth	D	LTEDroneAuth,d1	Nisynch	Ok	Verified	No attacks.
		LTEDroneAuth,d2	Niagree	Ok	Verified	No attacks.
		LTEDroneAuth,d3	Secret RES	Ok	Verified	No attacks.
		LTEDroneAuth,d4	Secret k(D,H,M)	Ok	Verified	No attacks.
		LTEDroneAuth,d5	Secret TMSI	Ok	Verified	No attacks.

Done.

Fig. 6 Verification result of LTE drone

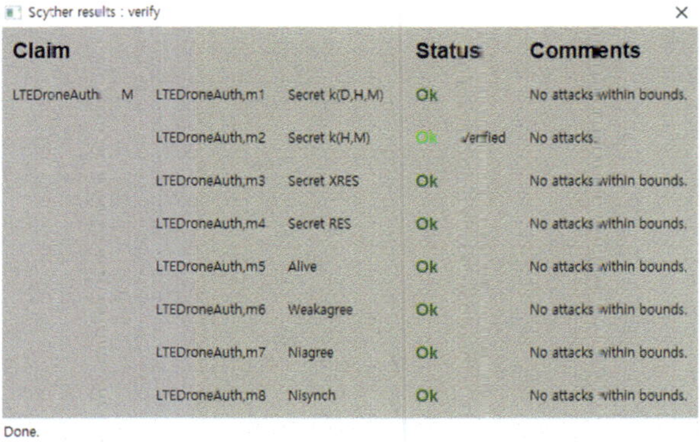

Fig. 7 Verification result of MME

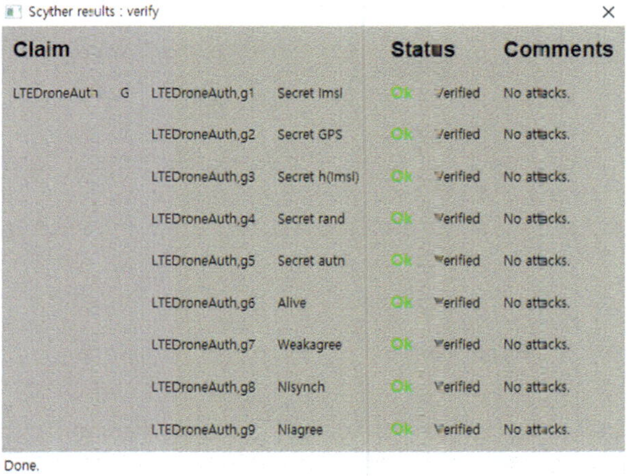

Fig. 8 Verification result of GCS

- GCS: We verify that GCS satisfies the security properties, which are Aliveness, Weak agreement, Non-injective synchronization, and Non-injective Agreement. While sending a message to HSS, GCS keeps the secrets, the location information of the drone, and IMSI by hiding with a hash function. While receiving a message from HSS, a malicious adversary cannot capture the secrets, RAND, AUTN, and K_{asme}. The verification result is shown in Fig. 8.
- HSS : We verify that HSS satisfies the security properties, which are Aliveness, Weak agreement, Non-injective synchronization, and Non-injective Agreement. While sending a message to MME, HSS keeps the secret, XRES and K_{asme}. While receiving a message from GCS, a malicious adversary cannot capture the secrets, IMSI, and the location information of the drone. While replying a message to GCS, HSS keeps the secrets, RAND and AUTN. The analysis result is shown in Fig. 9.

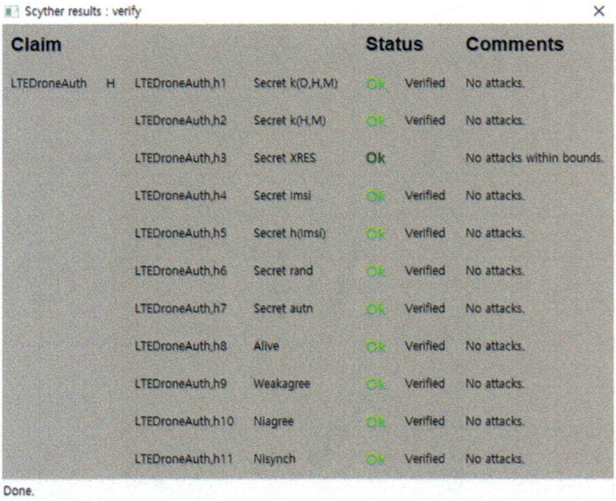

Fig. 9 Verification result of HSS

As shown in results, the proposed protocol proves a mutual authentication between LTE drone and HSS by using a trusted third party (GCS). We can claim that our proposed protocol is secure against the leakage of IMSI.

6 Conclusion

In this paper, we introduced LTE drones communication architectures, and a general authentication protocol, EPS-AKA protocol in LTE. Since there exist the security flaws in the LTE authentication protocol, it could be vulnerable to the location privacy and security of LTE drones. Once HSS authenticates LTE drones by using general LTE authentication protocol, adversaries could detect IMSI since it was not encrypted.

We proposed a secure authentication protocol that is appropriate for the LTE drone environment consisting of command and control systems such as GCS. The proposed authentication protocol for LTE drones uses signed and encrypted messages by a public key of GCS and a private key of HSS in order to hide the IMSI of drones from the wireless section. As a result, we could address the security and privacy risk of LTE drones by hiding IMSI.

Besides, we proved the formal analysis of the new proposed protocol against the frequent attacks and automated adversaries. The proposed authentication protocol has been proved secure by using Scyther, and we verified the protocol all the security property requirements specified in the abstract model.

We believe that this work represents a significant step toward secure LTE drones. In our future work, we will evaluate and discuss the performance of the proposed protocol.

Acknowledgments This work was supported by Institute for Information & communications Technology Promotion (IITP) grant funded by the Korea government (MSIT) (No.2018-0-00532, Development of High-Assurance (≥EAL6) Secure Microkernel).

References

1. Overview · mavlink developer guide (nd). https://mavlink.io/en/protocol/overview.html
2. 3GPP TS 23.003: Technical Specification Group Core Network and Terminals; Numbering, addressing and identification (Release 16) (2020). http://www.3gpp.org/dynareport/23003.htm
3. 3GPP TS 23.401: Technical Specification Group Services and System Aspects; General Packet Radio Service (GPRS) enhancements for Evolved Universal Terrestrial Radio Access Network (E-UTRAN) access (Release 16) (2020). http://www.3gpp.org/dynareport/23401.htm
4. 3GPP TS 23.402: Technical Specification Group Services and System Aspects; Architecture enhancements for non-3GPP accesses (Release 16) (2019). http://www.3gpp.org/dynareport/23402.htm
5. M.A. Abdrabou, A.D.E. Elbayoumy, E.A. El-Wanis, LTE authentication protocol (EPS-AKA) weaknesses solution, in *2015 IEEE Seventh International Conference on Intelligent Computing and Information Systems (ICICIS)* (IEEE, New York, 2015), pp. 434–441
6. J. Cao, M. Ma, H. Li, Y. Zhang, Z. Luo, A survey on security aspects for LTE and LTE-A networks. IEEE Commun Surv. Tutor. **16**(1), 283–302 (2013)
7. J. Cichonski, J. Franklin, M. Bartock, Guide to LTE security. Tech. rep., National Institute of Standards and Technology (2016)
8. C.J. Cremers, The scyther tool: verification, falsification, and analysis of security protocols, in *International Conference on Computer Aided Verification* (Springer, New York, 2008), pp. 414–418
9. C. Cremers, Scyther user manual. Department of Computer Science, University of Oxford, Oxford (2014)
10. C. Cremers, S. Mauw, Operational semantics, in *Operational Semantics and Verification of Security Protocols* (Springer, New York, 2012), pp. 13–35
11. GSMA: The mobile economy (2020). https://www.gsma.com/mobileeconomy/
12. W.B. Hsieh, J.S. Leu, Design of a time and location based one-time password authentication scheme, in *2011 7th International Wireless Communications and Mobile Computing Conference* (IEEE, New York, 2011), pp. 201–206
13. D.F. Kune, J. Koelndorfer, N. Hopper, Y. Kim, Location leaks on the GSM air interface. ISOC NDSS (Feb 2012) (2012)
14. G. Lowe, A hierarchy of authentication specifications, in *Proceedings 10th Computer Security Foundations Workshop* (IEEE, New York, 1997), pp. 31–43
15. M. McFarland, Amazon makes its first drone delivery in the U.K (2016). https://money.cnn.com/2016/12/14/technology/amazon-drone-delivery/index.html
16. S.L. McFarland, T.D. Biddle, ISIS-international review devoted to the history of science and its cultural influence, in *America's Pursuit of Precision Bombing, 1910–1945*, vol. 87(2) (Smithsonian Institution Press, Washington, 1996), pp. 389–389
17. S.F. Mjølsnes, R.F. Olimid, Easy 4G/LTE IMSI catchers for non-programmers, in *International Conference on Mathematical Methods, Models, and Architectures for Computer Network Security* (Springer, New York, 2017), pp. 235–246
18. H.C. Nguyen, R. Amorim, J. Wigard, I.Z. Kovacs, P. Mogensen, Using LTE networks for UAV command and control link: a rural-area coverage analysis, in *2017 IEEE 86th Vehicular Technology Conference (VTC-Fall)* (IEEE, New York, 2017), pp. 1–6
19. K. Norrman, M. Näslund, E. Dubrova, Protecting IMSI and user privacy in 5G networks, in *Proceedings of the 9th EAI International Conference on Mobile Multimedia Communications* (2016), pp. 159–166

20. A. Shaik, R. Borgaonkar, N. Asokan, V. Niemi, J.P. Seifert, Practical attacks against privacy and availability in 4G/LTE mobile communication systems (2015). Preprint. arXiv:1510.07563
21. F. Van Den Broek, R. Verdult, J. de Ruiter, Defeating IMSI catchers, in *Proceedings of the 22nd ACM SIGSAC Conference on Computer and Communications Security* (2015), pp. 340–351
22. G. Wang, B. Lee, K. Lim, J. Ahn, Technical trends on security of control and non-payload communications network for unmanned aircraft systems. Electron. Telecommun. Trends **32**(1), 82–92 (2017)

Memorable Password Generation with AES in ECB Mode

Timothy Hoang and Pablo Rivas

1 Introduction

Password security is a constant bother in the modern world. Today, it is possible to save passwords into Google's autofill and other password manager programs. These programs can generate passwords on the fly and store them on a computer or online account so that one would only have to remember a single password to access every other password they have. This is useful for people to have a multitude of different secure passwords for all of their accounts to make it more difficult for malicious entities to access them all. However, the downside of this technology is that all of the accounts are linked to one crucial location. Another downside is needing to first log into the master account if signing onto one of the accounts from another location, making access to the desired account difficult. The last downside is that many of these programs charge a subscription fee to securely store passwords. With ALP program, it is possible to create multiple randomly generated passwords that should be easy to remember given the user's parameters. This solves the one location and access from other computers issues since one can more easily remember each individual password since they are readable. Since the password will be readable, and therefore more memorable, one will have an easier time accessing their accounts from other locations without access to the password manager.

T. Hoang (✉) · P. Rivas
Marist College, Poughkeepsie, NY, USA
e-mail: timothy.hoang1@marist.edu

K. Daimi et al. (eds.), *Advances in Security, Networks, and Internet of Things*, Transactions on Computational Science and Computational Intelligence,
https://doi.org/10.1007/978-3-030-71017-0_3

2 Concerns

There is a concern for dictionary attacks with normal random word combinations for password generation [1]. That is why the lexicon was made completely customizable, as it will negate the effect of this type of brute force attack. With the customizable lexicon, the user is not limited to traditional words found in the dictionary. The user is able to add whatever string they deem memorable. This includes slang, names, made-up words, non-English words, sentimental numbers, and any bit of information. This is useful as it allows one to create passwords that are equally complex as a completely random string of the same length, while also being secure from traditional dictionary attacks since a stylistic touch is added to the lexicon. Therefore, unless the attackers know exactly what "words" were in the dictionary at the time of creating the password, they cannot perform a dictionary attack. This makes the customizable lexicon approach protected against brute force dictionary attacks.

In the lexicon, for example, the user puts in these strings, "4Plakilt3, WaR75atel8, M1zule, Laz4Apt, M0der0ck." To the user, these made up words are memorable for some reason (the numbers, capitalization, and segments of the words hold some significance to that specific user and no one else). Now, say the user requests that the password be 16 characters in length. From those words, the program can combine them to generate a password such as M1zuleWaR75atel8 or 4Plakilt3Laz4Apt. As long as the user has enough words and variety in word lengths within the dictionary, it is possible to create many completely unique passwords at any length. Since the "words" in the dictionary are deemed memorable according to the user, what may look like a bunch of random characters to someone can be easily remembered by the user. Unless the attacker knows every word within the lexicon that the user utilized at the time of generating their password, the only way to brute force it would be to brute force every single character. This gives the same time complexity as if they were to brute force a password where every character is random.

3 Methodology/Experimental Setup

The lexicon reading and password generation portion of the ALP application is complete. The function is for the program to read a lexicon that the user creates. From the lexicon of words, a random "readable" password will be generated. Each word in the .txt file is separated by line. Although any length word can be added to the lexicon, the program, as it stands, will only choose words that are 16 characters or less since that is what is required for the AES 128. This character count can be expanded easily in the code, but for the purpose of demonstrating its use in an AES 128 cipher, it was limited to 16. If the smallest word in the lexicon is too big to fill in the remaining characters needed, randomly generated characters will be chosen for the remainder of the password. In addition to this, the user is able to toggle between a "readable" and a completely random password. In the completely random passwords, each character is randomly generated and not read from the lexicon at all. These programs are only intended for English letters and numbers, so it may produce problems when introduced to other characters.

For the decryption portion, inverse programs were created for the ShiftRow(), NibbleSub(), and MixColumn() functions that are present in the AESencryption.java file which encrypts messages with AES 128. The changes to these functions include the change present in InvMixColumn(), where there is Galois multiplication in different fields. In addition to these inverse functions, the order of key operations has been reversed.

Figure 1 contains the order of operations to be done for encryption and decryption in AES. Since the encryption part was completed previously, here is an explanation of the right side of the picture from the bottom up. The program starts by adding the round key. Then, for the next nine keys, the following will take place: invShiftRow(), invNibbleSub(), add the round key, then InvMixColumn(). For the last key, the program will do the same process except it will leave out the InvMixColumn()

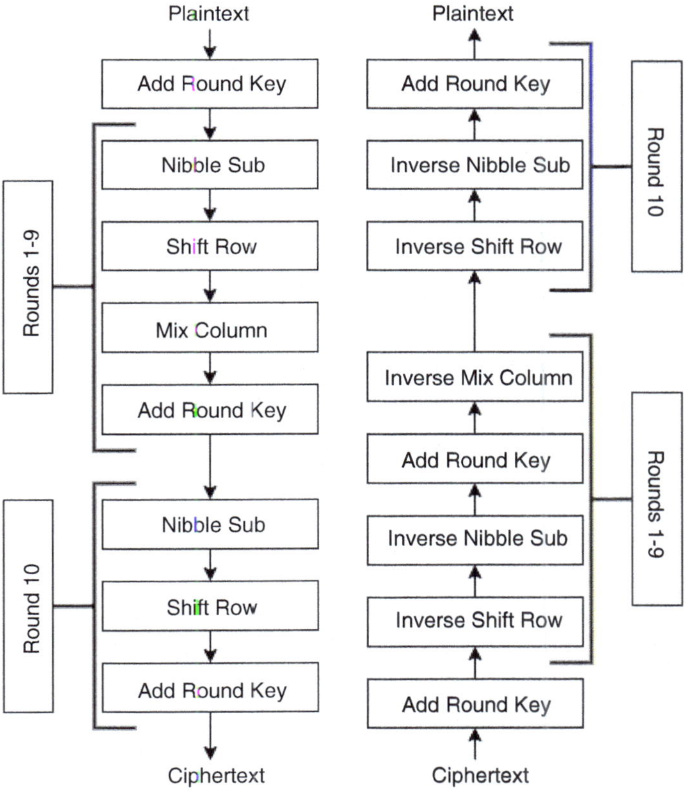

Fig. 1 Order of operations for AES encryption and decryption

function. The order of operations is achieved in the AESdecrypt() function, which is located in the AESdecipher.java file [2].

In InvShiftRow(), all the program does is change the rows opposite to how it was shifted in ShiftRow(). That is, instead of taking the last one, two, and three subjects of the bottom three rows in the matrix and bringing them to the front, the program takes the first three, two, and one subjects from those rows and moves them to the back. This is displayed visually in Fig. 2 [3].

For InvNibbleSub(), the program does the exact same operation as AESNibble-Sub() but uses the substitution values contained in the table shown in Fig. 3 [4].

For InvMixColumn(), the program will perform operations as shown in Fig. 4, where the numbers in the second matrix are the Galois field numbers [5].

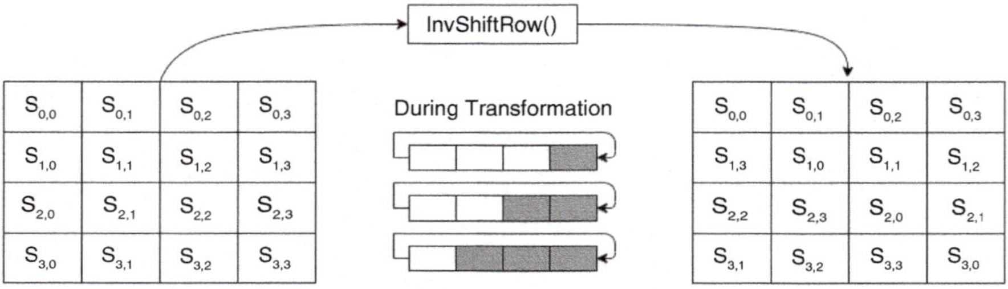

Fig. 2 Inverse Shift Row function

x \ y	0	1	2	3	4	5	6	7	8	9	a	b	c	d	e	f
0	52	09	6a	d5	30	36	a5	38	bf	40	a3	9e	81	f3	d7	fb
1	7c	e3	39	82	9b	2f	ff	87	34	8e	43	44	c4	de	e9	cb
2	54	7b	94	32	a6	c2	23	3d	ee	4c	95	0b	42	fa	c3	4e
3	08	2e	a1	66	28	d9	24	b2	76	5b	a2	49	6d	8b	d1	25
4	72	f8	f6	64	86	68	98	16	d4	a4	5c	cc	5d	65	b6	92
5	6c	70	48	50	fd	ed	b9	da	5e	15	46	57	a7	8d	9d	84
6	90	d8	ab	00	8c	bc	d3	0a	f7	e4	58	05	b8	b3	45	06
7	d0	2c	1e	8f	ca	3f	0f	02	c1	af	bd	03	01	13	8a	6b
8	3a	91	11	41	4f	67	dc	ea	97	f2	cf	ce	f0	b4	e6	73
9	96	ac	74	22	e7	ad	35	85	e2	f9	37	e8	1c	75	df	6e
a	47	f1	1a	71	1d	29	c5	89	6f	b7	62	0e	aa	18	be	1b
b	fc	56	3e	4b	c6	d2	79	20	9a	db	c0	fe	78	cd	5a	f4
c	1f	dd	a8	33	88	07	c7	31	b1	12	10	59	27	80	ec	5f
d	60	51	7f	a9	19	b5	4a	0d	2d	e5	7a	9f	93	c9	9c	ef
e	a0	e0	3b	4d	ae	2a	f5	b0	c8	eb	bb	3c	83	53	99	61
f	17	2b	04	7e	ba	77	d6	26	e1	69	14	63	55	21	0c	7d

Fig. 3 Inverse Nibble Substitution function substitution values

$$\begin{bmatrix} a_{0,0} & a_{0,1} & a_{0,2} & a_{0,3} \\ a_{1,0} & a_{1,1} & a_{1,2} & a_{1,3} \\ a_{2,0} & a_{2,1} & a_{2,2} & a_{2,3} \\ a_{3,0} & a_{3,1} & a_{3,2} & a_{3,3} \end{bmatrix} \times \begin{bmatrix} \text{galois}(14) & \text{galois}(11) & \text{galois}(13) & \text{galois}(9) \\ \text{galois}(9) & \text{galois}(14) & \text{galois}(11) & \text{galois}(13) \\ \text{galois}(13) & \text{galois}(9) & \text{galois}(14) & \text{galois}(11) \\ \text{galois}(11) & \text{galois}(13) & \text{galois}(9) & \text{galois}(14) \end{bmatrix} = \begin{bmatrix} b_{0,0} & b_{0,1} & b_{0,2} & b_{0,3} \\ b_{1,0} & b_{1,1} & b_{1,2} & b_{1,3} \\ b_{2,0} & b_{2,1} & b_{2,2} & b_{2,3} \\ b_{3,0} & b_{3,1} & b_{3,2} & b_{3,3} \end{bmatrix}$$

Fig. 4 Inverse Mix Column function

4 Experiment Results

The program will produce a memorable password using words provided in the lexicon of choice. By running the main program, it has produced passwords such as hic3cco7dinhl4mb and contr2buto9hedrd. These passwords are concatenations of the strings ["hic", "3cco7dinh", "l4mb"] and ["contr2buto9", "hedrd"], respectively. The test cases for each of the inverse functions can be performed by running test.java program. test.java will display the matrices passed into each of the functions as well as their resulting matrices. The output for these test cases can be found in the testcases.txt file located within the data folder of this project. All of the inverse functions perform as intended. Attempting to run the full decryption process does not decrypt the message properly. Despite the fact that several experiments were conducted to address this, the source of this issue has not been found.

5 Conclusion

In conclusion, the ALP program is able to create secure, memorable passwords for use with the lexicon of choice. However, the decryption process of the program is still being investigated. Work has been made on the individual processes and order methodology functioning properly as far as the test cases and results show, but there is something holding back the full decryption. The password generation portion of ALP has many use cases for individuals of all kinds, as everyone needs proper security for their many accounts in the modern world. For further improvement, figuring out and fixing the error concerning the order of operations in the AES decryption process is the most crucial. In addition to this, adding the ability to have special characters in the lexicon for password generation and being able to set requirements for the resulting password, such as requiring a certain number of numbers, special characters, and capitalization can be done.

The code to reproduce the experiments can be accessed under the MIT license in this repository: github.com/timhoangt/ALP

Acknowledgments This work was supported in part by the New York State Cloud Computing and Analytics Center, and by the VPAA's office at Marist College.

References

1. L. Bošnjak, J. Sres, B. Brumen, Brute-force and dictionary attack on hashed real-world passwords, 1161–1166 (2018). https://doi.org/10.23919/MIPRO.2018.8400211
2. A. Wadday, S. Wadi, H. Mohammed, A. Abdullah, Study of WiMAX based communication channel effects on the ciphered image using MAES algorithm. Int. J. Appl. Eng. Res. **13**, 6009–6018 (2018)
3. B. Bhattarai, N.K. Giri, FPGA Prototyping of the secured biometric based Identification system. (2015). https://doi.org/10.13140/RG.2.1.4067.2729
4. G. Selimis, A. Kakarountas, A. Fournaris, A. Milidonis, O. Koufopavlou, A low power design for Sbox cryptographic primitive of advanced encryption standard for mobile end-users. J. Low Power Electron. **3**, 327–336 (2007). https://doi.org/10.1166/jolpe.2007.139
5. L.M. Raju, S. Manickam, Secured high throughput of 128-Bit AES algorithm based on interleaving technique. Int. J. Appl. Eng. Res. **10**, 11047–11058 (2015)

A Comprehensive Survey on Fingerprint Liveness Detection Algorithms by Database and Scanner Model

Riley Kiefer and Ashokkumar Patel

1 Introduction

Fingerprint biometrics are of one of the most popular forms of biometric data for authentication. Many modern cell phones have some form of fingerprint recognition, and fingerprints are commonly used by the police and investigation services. While other biometric data like face recognition is gaining traction, it will take time for it to reach mainstream use. With a growing number of people using biometric authentication, it falls into the hands of researchers and companies to ensure the security of these systems. One of the biggest threats to biometric authentication is the ability for an attacker to spoof the biometric data. For a fingerprint, this is possible by making an artificial fingerprint from the residue on a surface. Some of the most common types of spoofing materials include gelatin, latex, and glue. While researchers have developed algorithms to counter the most common spoofing materials, novel or unseen materials can easily fool an algorithm. Alternatively, cadaver fingers also pose a threat to security. There is a need for robust algorithms to detect fingerprint spoofing of all types. The goal of this research is to compile the latest performance data on software-based anti-spoofing schemes for fingerprints to assist researchers in developing next-generation anti-spoofing measures.

This work is inspired by the E. Marasco and A. Ross' 2014 survey on Anti-spoofing Schemes for Fingerprint Recognition [1]. Their survey details almost all facets of fingerprint biometrics, and it provides a comparison of algorithms published on or before 2014. This survey focuses on a comparison of algorithms published in late 2014 to 2019. This work is an extension and continuation of our original brief survey on Spoofing Detection Systems for Fake Fingerprint

R. Kiefer · A. Patel (✉)
Florida Polytechnic University, Lakeland, FL, USA
e-mail: rkiefer@floridapoly.edu; apatel@floridapoly.edu

K. Daimi et al. (eds.), *Advances in Security, Networks, and Internet of Things*
Transactions on Computational Science and Computational Intelligence,
https://doi.org/10.1007/978-3-030-71017-0_4

Presentation Attacks [2]. This chapter expands upon our original work by including a summary of the LivDet competition results over time, adding algorithm variants, and including algorithms that were tested on non-LivDet datasets.

2 A Brief Review of the LivDet Competition Series

The LivDet competition is held once every 2 years. Competitions have been held in 2009 [3], 2011 [4], 2013 [5], 2015 [6], 2017 [7], and 2019 [8]. At the competition, biometric spoofing data of all types are used to test algorithms submitted by researchers. The fingerprint dataset is divided by images from three or four different scanner models. The primary metric of performance is the Average Classification Error (ACE), which is the average of the Type I and Type II statistical errors. Table 1 includes the various fingerprints scanners used in the LivDet datasets. Most scanners are optical; however, there is a recent shift to thermal images (from Orcanthus) to test algorithm performance on the latest hardware technologies.

Figure 1 shows the improvement of the average overall ACE for LivDet competitors over all competition years. A projection of the trendline shows the possible performance of LivDet-2021; however, this is highly dependent on the data itself. As seen in LivDet-2011, all submitted algorithms had over 20% ACE, alluding to the relative difficulty of the dataset. Figure 2 provides the best ACE metric of all algorithms submitted by competitors, for all scanner types across all competition years. This data is often used as a baseline performance for researchers testing their algorithm.

Table 1 A complete summary of the fingerprint scanner specifications used in the LivDet fingerprint competition series

Company	Years Used	Model	Resolution	Image Size	Format	Scanner Type
Biometrika	09, 11, 13	Fx2000	569	312x372	RAW	Optical
CrossMatch	09	Verfier 300 LC	500	480x640	RAW	Optical
Identix	09	DFR2100	686	720x720	RAW	Optical
Digital Persona	11	4000B	500	355x391	RAW	Optical
Italdata	11, 13	ET10	500	640x480	RAW	Optical
Sagem	11	MSO300	500	352x384	RAW	Optical
CrossMatch	13, 15	L Scan Guardian	500	800x750	-	-
Swipe	13	-	96	208x1500	-	-
Biometrika	15	HiScan-PRO	1000	1000x1000	BMP	Optical
GreenBit	15	DactyScan26	500	500x500	PNG	Optical
Green Bit	17, 19	DactyScan84C	500	500x500	PNG	Optical
Orcanthus	17, 19	Certis2 Image	500	300xn	PNG	Thermal Swipe
Digital Persona	15, 17, 19	U .are.U 5160	500	252x324	PNG	Optical

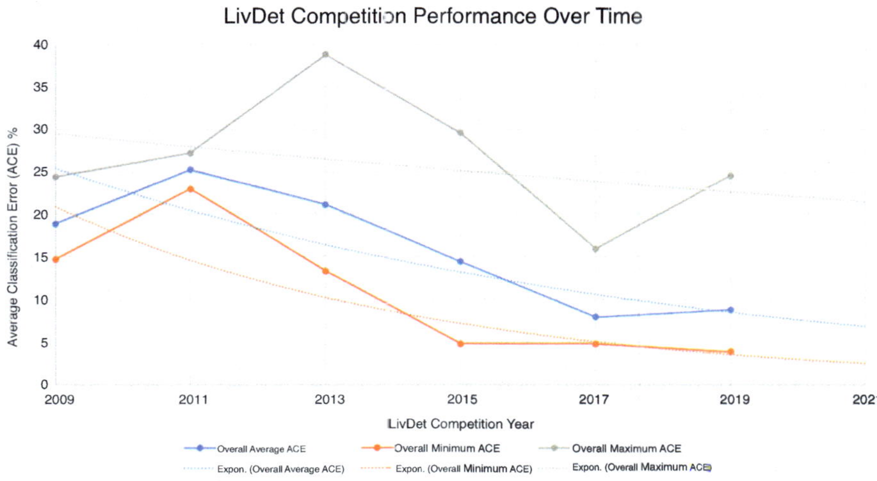

Fig. 1 A graph of the performance of LivDet competitions over time in terms of average, minimum, and maximum overall ACE on all scanners and projections for LivDet-2021

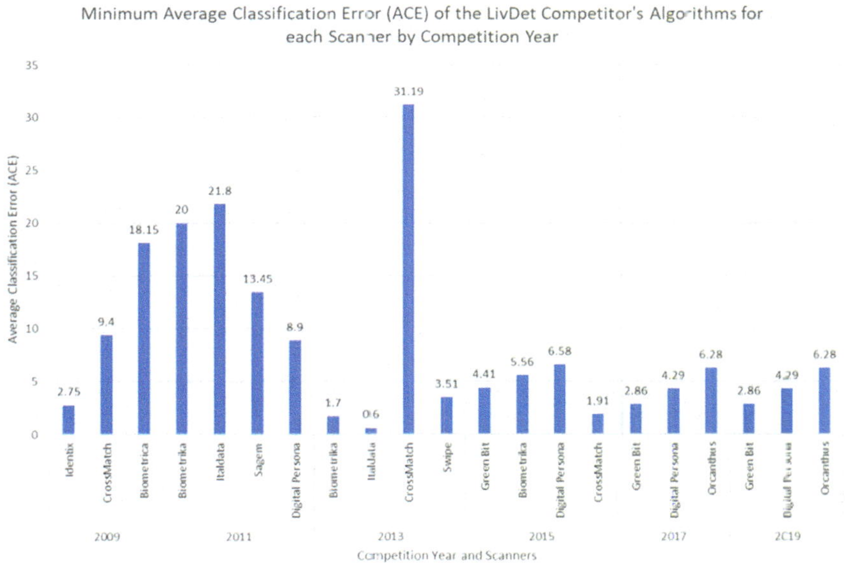

Fig. 2 A graph of the best competitor's algorithms by scanner type for each competition year

3 The LivDet-2009 Competition Dataset

The LivDet-2009 competition dataset [3] includes three scanner models: Biometrika, CrossMatch, and Identix. All fingerprint images were collected in a consensual manner, and there were three spoof materials used: gelatin, silicone, and Play-Doh. Table 2 ranks the algorithms that were submitted between 2014 and 2019 and tested on the LivDet-2009 dataset in terms of the ACE metric.

4 The LivDet-2011 Competition Dataset

The LivDet-2011 competition dataset [4] includes four scanner models: Biometrika, Digital Persona, Italdata, and Sagem. All fingerprint images were collected in a consensual manner and there were six spoof materials used: latex, gelatin, silicone, Play-Doh, wood glue, and Eco-flex. Table 3 ranks the algorithms that were submitted between 2014 and 2019 and tested on the LivDet-2011 dataset in terms of the ACE metric, except for [9], which used the ER metric.

5 The LivDet-2013 Competition Dataset

The LivDet-2013 competition dataset [5] includes four scanner models: Biometrika, CrossMatch, Italdata, and Swipe. Half of the fingerprint images were collected in a consensual manner (CrossMatch and Swipe) and there were six spoof materials used: Body double, latex, Play-Doh, wood glue, gelatin, Eco-Flex, and Mudsill. Table 4 ranks the algorithms that were submitted between 2014 and 2019 and tested on the LivDet-2013 dataset in terms of the ACE metric, except for [9], which used the ER metric.

6 The LivDet-2015 Competition Dataset

The LivDet-2015 competition dataset [6] includes four scanner models: Biometrika, CrossMatch, Digital Persona, and Green Bit. All fingerprint images were collected in a consensual manner and there were ten spoof materials used: Eco-Flex, gelatin, latex, wood glue, liquid Eco-Flex, RTV, Play-Doh, Body Double, OOMOO, and a special gelatin. Table 5 ranks the algorithms that were submitted between 2014 and 2019 and tested on the LivDet-2015 dataset in terms of the ACE metric.

Table 2 A complete summary of all algorithms and their variants ranked by the average ACE of all scanners of the LivDet-2009 dataset, with green representing the best for that scanner

Reference	Algorithm Name and/or Brief Description	Average	Biometrika	Cross Match	Identix
16	DCNN and SVM Trained with RBF Kernel	0	0	-	-
16	DCNN and SVM Trained with Polynomial Kernel Order 2	0.3837	0.3837	-	-
16	DCNN and SVM Trained with Polynomial Kernel Order 3	0.5756	0.5756	-	-
18	MvDA: G1- SID RICLBP LCPD DSIFT LPQ-3 + LPQ-5, M =4424	0.73	1.3	0.9	0
18	MvDA: G5- SID RICLBP LCPD DSIFT RILPQ-3, M=4168	0.73	1.2	1	0
22	Deep Triplet Embedding (TNet)	0.77	0.71	1.57	0.044
18	MvDA: G6- SID RICLBP LCPD DSIFT, M=3912	0.8	1.1	1.3	0
18	MvDA: G4- SID RICLBP LCPD DSIFT WLD, M=6793	0.9	1.6	1.1	0
18	MvDA: G2- SID RICLBP LCPD DSIFT WLD LPQ-5 + LPQ-7, M=7304	0.93	1.7	1.1	0
18	MvDA: G3- SID RICLBP LCPD DSIFT WLD BSIF-5, M= 7826	1	1.9	1.1	0
18	Spidernet: G2- SID RICLBP LCPD DSIFT WLD LPQ-5 + LPQ-7, M=7304	1.23	1.6	1.8	0.3
18	Spidernet: G3- SID RICLBP LCPD DSIFT WLD BSIF-5, M= 7826	1.23	1.6	1.8	0.3
18	Spidernet: G5- SID RICLBP LCPD DSIFT RILPQ-3, M=4168	1.27	1.4	2	0.4
18	Spidernet: G1- SID RICLBP LCPD DSIFT LPQ-3 + LPQ-5, M =4424	1.3	1.5	2.1	0.3
18	Spidernet: G6- SID RICLBP LCPD DSIFT, M=3912	1.3	1.6	2	0.3
18	Spidernet: G4- SID RICLBP LCPD DSIFT WLD, M=6793	1.33	1.9	1.8	0.3
18	AdaBoost: G1- SID RICLBP LCPD DSIFT LPQ-3 + LPQ-5, M =4424	1.37	1.7	2.4	0
18	AdaBoost: G2- SID RICLBP LCPD DSIFT WLD LPQ-5 + LPQ-7, M=7304	1.37	2.1	2	0
18	Linear SVM: G2- SID RICLBP LCPD DSIFT WLD LPQ-5 + LPQ-7, M=7304	1.4	1.6	1.6	1
18	Linear SVM: G3- SID RICLBP LCPD DSIFT WLD BSIF-5, M= 7826	1.4	1.7	1.5	1
18	AdaBoost: G3- SID RICLBP LCPD DSIFT WLD BSIF-5, M= 7826	1.43	2.2	2.1	0
18	AdaBoost: G4- SID RICLBP LCPD DSIFT WLD, M=6793	1.47	2.3	2.1	0
18	Linear SVM: G1- SID RICLBP LCPD DSIFT LPQ-3 + LPQ-5, M =4424	1.47	1.7	1.8	0.9
18	Linear SVM: G4- SID RICLBP LCPD DSIFT WLD, M=6793	1.47	1.8	1.6	1
18	Linear SVM: G5- SID RICLBP LCPD DSIFT RILPQ-3, M=4168	1.5	1.9	1.7	0.9
18	AdaBoost: G6- SID RICLBP LCPD DSIFT, M=3912	1.53	2.3	2.3	0
18	AdaBoost: G5- SID RICLBP LCPD DSIFT RILPQ-3, M=4168	1.57	2.3	2.4	0
18	Linear SVM: G6- SID RICLBP LCPD DSIFT. M=3912	1.57	2.1	1.7	0.9
13	Model 1- CNN-VGG- 227x227	1.63	4.1	0.6	0.2
13	Model 2- CNN-Alexnet- 224x224	2.73	5.6	1.1	0.4
43	Wavelet-Markov	2.83	5.4	2.8	0.3
41	Local Uniform Comparison Image Descriptor (LUCID)	2.86	0.14	7.94	0.49
20	Binarised Statistical Image Features (BSIF)	3.03	3.48	4.6	1.02
11	CNN, Random Sample Patch	3.42	-	-	3.42
16	DCNN and SVM Trained with linear Kernel	3.84	3.84	-	-
13	Model 3- CNN-Random	3.9	9.2	1.7	0.8
38	Augmented Convolutional Network PCA SVM	3.94	9.23	1.78	0.8
29	Local Quality Features (LQF)	5.3	8.5	5.6	2
38	Convolutional Network PCA SVM	5.36	9.49	3.76	2.82
13	Model 4- Local Binary Patterns (LBP)	5.53	10.4	3.6	2.6
38	Augmented Local Binary Patterns (LBP) PCA SVM	5.58	10.44	3.65	2.64
16	DCNN and LR Classifier	8.06	8.06	-	-
42	Image Quality Assesment (IQA)-based	8.23	12.8	10.7	1.2
48	Local Accumulated Smoothing Pattern (LASP)	11.51	9.99	12.28	12.24
19	Local Coherence Patterns and SVM	13.17	13.21	15.58	10.71
38	Local Binary Patterns (LBP) PCA SVM	19.25	50	6.81	0.95

7 The LivDet-2017 Competition Dataset

The LivDet-2017 competition dataset [7] includes three scanner models: Green Bit, Digital Persona, and Orcanthus. All fingerprint images were collected in a consensual manner and there were six spoof materials used: wood glue, Eco-flex, Body Double, gelatin, latex, and liquid Eco-Flex. Due to the LivDet-2017 dataset

Table 3 A complete summary of all algorithms and their variants ranked by the average ACE of all scanners of the LivDet-2011 dataset, with green representing the best for that scanner

Reference	Algorithm Name and/or Brief Description	Average	Biometrika	Digital Persona	Italdata	Sagem
10	LBP and Discrete Shearlet Transform	0	-	0	-	-
26	CNN-MobieNet-v1 and Munitae-based Local Patches	1.67	1.24	1.61	2.45	1.39
18	Spidernet: G4- SID RICLBP LCPD DSIFT WLD, M=6793	1.68	2	0.7	2.8	1.2
18	Spidernet: G5- SID RICLBP LCPD DSIFT RILPQ-3, M=4168	1.9	1.7	1.3	2.9	1.7
34	Fully Convolutional Network 64 x 64	1.95	1.55	0.8	4.1	1.34
18	Spidernet: G1- SID RICLBP LCPD DSIFT LPQ-3 + LPQ-5, M =4424	1.95	2.3	1.4	2.1	2
18	Spidernet: G2- SID RICLBP LCPD DSIFT WLD LPQ-5 + LPQ-7, M=7304	2	2.4	1	3.1	1.5
18	Spidernet: G3- SID RICLBP LCPD DSIFT WLD BSIF-5, M= 7826	2	2.3	1	2.6	2.1
18	MvDA: G1- SID RICLBP LCPD DSIFT LPQ-3 + LPQ-5, M =4424	2.08	0.7	0.7	4.6	2.3
34	Fully Convolutional Network 48 x 48	2.13	1.1	1.1	4.75	1.56
18	MvDA: G5- SID RICLBP LCPD DSIFT RILPQ-3, M=4168	2.13	0.7	0.8	4.7	2.3
49	Fisher Vector	2.13	3.45	0.2	3.1	1.75
18	Spidernet: G6- SID RICLBP LCPD DSIFT, M=3912	2.15	2.2	1.1	3.1	2.2
18	Linear SVM: G4- SID RICLBP LCPD DSIFT WLD, M=6793	2.25	2	1.3	3.5	2.2
18	MvDA: G2- SID RICLBP LCPD DSIFT WLD LPQ-5 + LPQ-7, M=7304	2.28	0.9	0.7	5.3	2.2
18	MvDA: G6- SID RICLBP LCPD DSIFT, M=3912	2.28	0.9	0.8	4.9	2.5
18	Linear SVM: G2- SID RICLBP LCPD DSIFT WLD LPQ-5 + LPQ-7, M=7304	2.33	2	1.4	3.8	2.1
18	Linear SVM: G3- SID RICLBP LCPD DSIFT WLD BSIF-5, M= 7826	2.33	1.8	1.2	4	2.3
18	Linear SVM: G5- SID RICLBP LCPD DSIFT RILPQ-3, M=4168	2.35	1.5	1.4	4	2.5
18	Linear SVM: G6- SID RICLBP LCPD DSIFT, M=3912	2.38	1.8	1.5	3.8	2.4
18	Linear SVM: G1- SID RICLBP LCPD DSIFT LPQ-3 + LPQ-5, M =4424	2.43	1.8	1.4	4.3	2.2
34	Fully Convolutional Network 32 x 32	2.44	2.35	0.9	5.4	1.09
28	Gram-128 Model	2.45	2.75	0.55	5	1.5
18	MvDA: G4- SID RICLBP LCPD DSIFT WLD, M=6793	2.45	0.5	0.8	5.9	2.6
21	DCNN-Inception v3 + Minutiae-based local patches	2.59	2.6	2.7	3.25	1.8
18	MvDA: G3- SID RICLBP LCPD DSIFT WLD BSIF-5, M= 7826	2.7	1.2	0.8	6.7	2.1
49	Vector Locally Aggregated Descriptors	2.88	3.9	0.1	6.5	1
17	CNN-VGG 227x227 With Dataset Augmentation	2.9	-	-	-	-
22	Deep Triplet Embedding (TNet)	3.33	5.15	1.85	5.1	1.23
18	AdaBoost: G3- SID RICLBP LCPD DSIFT WLD BSIF-5, M= 7826	3.55	3.3	2.9	5.9	2.1
28	Gram-128 Model with Augmentation	3.58	4.95	2	4.8	2.56
18	AdaBoost: G2- SID RICLBP LCPD DSIFT WLD LPQ-5 + LPQ-7, M=7304	3.58	3	3.3	5.8	2.2
18	AdaBoost: G4- SID RICLBP LCPD DSIFT WLD, M=6793	3.58	3.2	3.1	5.8	2.2
18	AdaBoost: G1- SID RICLBP LCPD DSIFT LPQ-3 + LPQ-5, M =4424	3.68	3.1	3.2	5.8	2.6
17	CNN-Alexnet With Dataset Augmentation	3.7	-	-	-	-
18	AdaBoost: G5- SID RICLBP LCPD DSIFT RILPQ-3, M=4168	3.98	3.5	3.5	6	2.9
18	AdaBoost: G6- SID RICLBP LCPD DSIFT, M=3912	4.03	3.7	3.5	6.1	2.8
17	CNN-VGG 227x227 Without Dataset Augmentation	4.2	-	-	-	-
13	Model 1- CNN-VGG- 227x227	4.53	5.2	3.2	8	1.7
17	CNN- Random With Dataset Augmentation	4.7	-	-	-	-
17	CNN-Alexnet Without Dataset Augmentation	5	-	-	-	-
13	Model 2- CNN-Alexnet- 224x224	5.6	5.6	4.6	9.1	3.1
31	Deep Residual Network- ROI	5.65	7.6	2.1	11	2.5
23	Weber Local Binary Descriptor (WLBPD)	5.96	5.65	4.1	11.85	2.25
38	Convolutional Network- PCA SVM	6.19	9.9	1.9	5.09	7.86
13	Model 3- CNN-Random	6.4	8.2	3.6	9.2	4.6
33	DCNN with image scale equalization	6.45	9.2	1.35	12.35	2.9
38	Augmented Convolutional Network- PCA SVM	6.45	8.25	3.65	9.27	4.64
31	Deep Residual Network- ROI+LGP	6.68	9.6	1.9	13.5	1.72
49	Bag of Words	6.7	8.15	3.15	11.15	4.35
12	Low Level Features and Shape Analysis: SURF+PHOG+Gabor	6.9	7.89	6.25	8.1	5.36
20	Binarised Statistical Image Features (BSIF)	7.17	6.8	3.55	13.65	4.68
12	Low Level Features and Shape Analysis: SURF+PHOG	7.32	8.76	6.9	7.4	6.23
12	Low Level Features and Shape Analysis: SURF	8.04	9.12	7.95	8.35	6.77
13	Model 4- Local Binary Patterns	8.18	8.8	4.1	12.3	7.5
38	Augmented Local Binary Patterns (LBP) PCA SVM	8.22	8.85	4.15	12.34	7.54
41	Local Uniform Comparison Image Descriptor (LUCID)	8.54	-	-	-	8.54
46	Local Binary Patterns and Principle Component Analysis	8.625	7.1	9.7	10.5	7.2
50	Bayesian Belief Network-MLQc	8.89	9.45	7.1	12.6	6.4
50	Bayesian Belief Network-MLQ	9.09	9.45	7.1	13.4	6.4
17	CNN- Random Without Dataset Augmentation	9.4	-	-	-	-
12	Low Level Features and Shape Analysis: Gabor	9.46	11.21	7.85	12.5	6.28
50	Direct Modeling-Gaussian Mixture Model	9.75	9.5	8.35	13.95	7.2
40	Local Binary Patterns (LBP) with image denoise	10.2	10.2	-	-	-
38	Local Binary Patterns (LBP) PCA SVM	10.32	8.2	3.85	23.68	5.56
39	Pore Characteristics: Fusion	12	18.4	7.8	15.2	6.7
40	LBP without image denoise	12.17	12.17	-	-	-
39	Pore Characteristics: Baseline	12.9	20.6	8.4	14	8.4
50	Bayesian Belief Network-ML	13.5	14.85	9.3	21.1	8.75
50	Spoof Detector	14.08	15	11	21.55	8.75
12	Low Level Features and Shape Analysis: PHOG	17.92	22.45	13.07	20.05	16.1
48	Local Accumulated Smoothing Pattern (LASP)	21.22	22.6	27.1	17.6	17.58
39	Pore Characteristics: Pore Analysis	25.9	26.6	23.4	31.4	22
19	Local Coherence Patterns and SVM	33.21	-	-	-	-
45	Pyramid Hisogram of Gradients (PHOG)- QDA	ER = 5.0	ER = 4.5	ER = 5.7	ER = 5.4	ER = 4.3
45	Pyramid Hisogram of Gradients (PHOG)-GMM	ER = 5.1	ER = 4.8	ER = 7.5	ER = 4.7	ER = 3.4
45	Pyramid Hisogram of Gradients (PHOG)-GC	ER = 5.1	ER = 5.1	ER = 6.1	ER = 4.6	ER = 4.4
45	Local Binary Patterns (LBP)-GMM	ER = 5.4	ER = 4.4	ER = 8.7	ER = 4.8	ER = 3.6
45	Local Phase Quantization (LPQ)-QDA	ER = 6.1	ER = 3.8	ER = 10.7	ER = 3.7	ER = 6.1
45	Local Binary Patterns (LBP)-GC	ER = 6.2	ER = 5.1	ER = 9.1	ER = 7.8	ER = 2.9
45	Local Phase Quantization (LPQ)-GMM	ER = 7.0	ER = 3.9	ER = 11.7	ER = 4.6	ER = 7.9
45	Local Binary Patterns (LBP)-QDA	ER = 7.1	ER = 4.2	ER = 8.3	ER = 12.3	ER = 3.4
45	Local Phase Quantization (LPQ)-GC	ER = 7.7	ER = 3.7	ER = 14.9	ER = 5.3	ER = 6.7

Table 4 A complete summary of all algorithms and their variants ranked by the average ACE of all scanners of the LivDet-2013 dataset, with green representing the best for that scanner

Reference	Algorithm Name and/or Brief Description	Average	Biometrika	CrossMatch	Italdata	Swipe
26	CNN-MobieNet-v1 and Munitae-based Local Patches	0.25	0.2	-	0.3	-
34	Fully Convolutional Network- 32 x 32	0.28	0.15	-	0.4	-
34	Fully Convolutional Network- 48 x 48	0.38	0.35	-	0.4	-
34	Fully Convolutional Network- 64 x 64	0.43	0.2	-	0.65	-
21	DCNN-Inception v3 + Minutiae-based local patches	0.5	0.6	-	0.4	-
22	Deep Triplet Embedding (TNet)	0.57	0.55	-	0.5	0.66
44	Filter: Optimized spoofnet	0.72	0.15	1.77	0.05	0.92
28	Gram-128 Model with Augmentation	0.8	0.7	-	0.9	-
44	Filter: Optimized Architecture Optimization (AO)	1.03	0.7	1.96	0.55	0.92
28	Gram-128 Model	1.05	0.85	-	1.25	-
18	MvDA: G2- SID RICLBP LCPD DSIFT WLD LPQ-5 + LPQ-7, M=7304	1.58	0.4	3.9	0.4	1.6
18	MvDA: G3- SID RICLBP LCPD DSIFT WLD BSIF-5, M= 7826	1.63	0.4	4.7	0.4	1
18	Spidernet: G1- SID RICLBP LCPD DSIFT LPQ-3 + LPQ-5, M =4424	1.65	0.7	3.4	0.9	1.6
18	Spidernet: G3- SID RICLBP LCPD DSIFT WLD BSIF-5, M= 7826	1.68	0.9	3.3	0.7	1.8
18	MvDA: G1- SID RICLBP LCPD DSIFT LPQ-3 + LPQ-5, M =4424	1.68	0.5	4.4	0.5	1.3
35	Slim-ResNet- Convolutional Nueral Network (thres)	1.74	0.47	-	3.01	-
18	MvDA: G5- SID RICLBP LCPD DSIFT RILPQ-3, M=4168	1.8	0.5	4.3	0.6	1.8
18	Spidernet: G2- SID RICLBP LCPD DSIFT WLD LPQ-5 + LPQ-7, M=7304	1.83	1	3.7	1	1.6
18	Spidernet: G4- SID RICLBP LCPD DSIFT WLD, M=6793	1.83	1.1	3.3	1.1	1.8
18	Spidernet: G5- SID RICLBP LCPD DSIFT RILPQ-3, M=4168	1.85	1	3.6	0.9	1.9
18	Spidernet: G6- SID RICLBP LCPD DSIFT, M=3912	1.85	1.1	3.5	0.7	2.1
49	Fisher Vector	1.88	1.3	3.7	0.6	1.9
23	Weber Local Binary Descriptor (WLBPD)	1.89	0.4	-	0.95	4.31
20	Binarised Statistical Image Features (BSIF)	1.9	0.55	-	0.55	4.61
18	MvDA: G4- SID RICLBP LCPD DSIFT WLD, M=6793	2	0.5	4.6	0.5	2.4
18	MvDA: G6- SID RICLBP LCPD DSIFT, M=3912	2.03	0.5	5.2	0.4	2
44	Filter: Optimized cf10-11	2.03	1.5	2.67	2.65	1.3
18	Linear SVM: G1- SID RICLBP LCPD DSIFT LPQ-3 + LPQ-5, M =4424	2.15	0.7	3.9	1.3	2.7
18	Linear SVM: G5- SID RICLBP LCPD DSIFT RILPQ-3, M=4168	2.25	0.7	4.4	1.3	2.6
18	Linear SVM: G2- SID RICLBP LCPD DSIFT WLD LPQ-5 + LPQ-7, M=7304	2.3	0.7	3.9	1.7	2.9
13	Model 1- Convolutional Nueral Network-VGG- 227x227	2.33	1.8	3.4	0.4	3.7
18	Linear SVM: G3- SID RICLBP LCPD DSIFT WLD BSIF-5, M= 7826	2.33	0.6	4.3	1.7	2.7
18	Linear SVM: G6- SID RICLBP LCPD DSIFT, M=3912	2.35	0.7	4.8	1.3	2.6
18	Linear SVM: G4- SID RICLBP LCPD DSIFT WLD, M=6793	2.5	0.7	4.7	1.8	2.8
12	Low Level Features and Shape Analysis: SURF+PHOG+Gabor	2.61	2.27	2.5	2.17	3.5
49	Vector Locally Aggregated Descriptors	2.68	1.7	4.3	0.7	4
35	Slim-ResNet- Convolutional Nueral Network	2.84	0.47	-	5.21	-
13	Model 2- Convolutional Nueral Network-Alexnet- 224x224	2.85	1.9	4.7	0.5	4.3
31	Deep Residual Network- ROI+LGP	2.96	-	-	-	-
31	Deep Residual Network- ROI	2.99	-	-	-	-
12	Low Level Features and Shape Analysis: SURF+PHOG	3.27	3.42	2.96	2.85	3.85
18	AdaBoost: G4- SID RICLBP LCPD DSIFT WLD, M=6793	3.3	1	5.6	1.3	5.3
18	AdaBoost: G3- SID RICLBP LCPD DSIFT WLD BSIF-5, M= 7826	3.4	1.1	5.6	1.6	5.3
47	GoogLeNet	3.4	3.4	-	-	-
18	AdaBoost: G6- SID RICLBP LCPD DSIFT, M=3912	3.45	1.2	6.3	1.3	5
13	Model 3- Convolutional Nueral Network-Random	3.5	0.8	3.2	2.4	7.6
18	AdaBoost: G5- SID RICLBP LCPD DSIFT RILPQ-3, M=4168	3.53	1.3	6.3	1.4	5.1
38	Augmented Convolutional Network- PCA SVM	3.55	0.8	3.29	2.45	7.67
47	CaffeNet	3.55	3.55	-	-	-
18	AdaBoost: G1- SID RICLBP LCPD DSIFT LPQ-3 + LPQ-5, M =4424	3.6	1	6.4	1.5	5.5
18	AdaBoost: G2- SID RICLBP LCPD DSIFT WLD LPQ-5 + LPQ-7, M=7304	3.65	0.9	5.8	1.5	6.4
33	DCNN with image scale equalization	3.7	4.35	7	1.4	2.05
12	Low Level Features and Shape Analysis: Gabor	3.72	2.7	4.67	4.02	3.5
12	Low Level Features and Shape Analysis: SURF	5.26	5.75	6.08	4.6	4.6
44	Filter: Random Architecture Optimization (AO)	6.26	3.5	7.91	2.55	11.06
47	Siamese	6.95	6.95	-	-	-
12	Low Level Features and Shape Analysis: PHOG	7.24	3.87	9.32	6.7	9.05
49	Bag of Words	7.26	4.95	5.5	12.25	6.35
44	Filter: Random spoofnet	7.42	5.3	12.18	8.95	3.25
35	ResNet- Convolutional Nueral Network	10.63	4.09	-	17.16	-
37	Histogram of Invariant Gradients (HIG): Minutiae Circle Combined	12.2	3.9	28.76	1.7	14.44
39	Local Uniform Comparison Image Descriptor (LUCID)	12.24	-	-	-	12.24
39	Pore Characteristics: Pore Analysis	12.7	2.2	34.9	1	-
13	Model 4- Local Binary Patterns	14.18	1.7	49.4	2.3	3.3
38	Augmented Local Binary Patterns- PCA SVM	14.2	1.7	49.45	2.3	3.34
12	Low Level Features and Shape Analysis: WLD+LPQ	14.31	1.4	45.95	3.4	6.51
39	Pore Characteristics: Baseline	14.4	2.4	38.4	2.5	-
39	Pore Characteristics: Fusion	14.4	2	39.6	1.6	-
37	Histogram of Invariant Gradients (HIG): Dense Block Packing Extended	14.87	10.9	28.76	1.7	18.11
37	Histogram of Invariant Gradients (HIG): Dense Block Packing	15.19	3.9	34.13	8.3	14.44
38	Convolutional Network PCA SVM	15.84	4.55	5.2	47.65	5.97
44	Filter: Random cf10-11	18.85	22.55	16.89	23.55	12.4
37	Histogram of Invariant Gradients (HIG): Minutiae Circle	21.82	4.3	39.96	10.6	32.41
38	Local Binary Patterns- PCA SVM	33.75	25.65	49.87	55.45	4.02
45	Local Binary Patterns-GC	ER = 13.9	ER = 2.4	ER = 1.2	ER = 48.8	ER = 3
45	Pyramid History of Invariant Gradients (PHOG)-GMM	ER = 6	ER = 3.2	ER = 4.6	ER = 11.8	ER = 4.5
45	Pyramid History of Invariant Gradients (PHOG)-GC	ER = 7.2	ER = 3.9	ER = 6.3	ER = 12.1	ER = 6.6
45	Pyramid History of Invariant Gradients (PHOG)- QDA	ER = 7.5	ER = 3.9	ER = 6.4	ER = 12.9	ER = 6.7
45	Local Phase Quantization (LPQ)-GMM	ER = 7.6	ER = 6.8	ER = 5.5	ER = 14.3	ER = 2.6
45	Local Binary Patterns-QDA	ER = 7.8	ER = 2	ER = 1.9	ER = 22.3	ER = 4.9
45	Local Phase Quantization (LPQ)-QDA	ER = 8.2	ER = 7.9	ER = 6.3	ER = 15.1	ER = 3.3
45	Local Phase Quantization (LPQ)-GC	ER = 8.4	ER = 7.7	ER = 7.2	ER = 14.7	ER = 3.8
45	Local Binary Patterns-GMM	ER = 8.5	ER = 1.7	ER = 3.2	ER = 24	ER = 5.1

Table 5 A complete summary of all algorithms and their variants ranked by the average ACE of all scanners of the LivDet-2015 dataset, with green representing the best for that scanner

Reference	Algorithm Name and/or Brief Description	Average	Biometrika	CrossMatch	Digital Persona	GreenBit
26	CNN-MobieNet-v1 and Munitae-based Local Patches	0.97	1.12	0.64	1.48	0.68
34	Fully Convolutional Network- 48 x 48	1.26	0.35	1.09	3.4	0.2
34	Fully Convolutional Network- 32 x 32	1.34	1.25	0.82	3	0.3
21	DCNN-Inception v3 + Minutiae-based local patches	1.39	1.76	0.81	1.08	2
34	Fully Convolutional Network- 64 x 64	2.01	0.6	1.44	5.45	0.55
52	Electrocardiogram Fingerprint (ECGFP)-30	2.6	3.5	-	2.6	1.8
35	Slim-ResNetCNN: Model 2- 7 improved residual block_bs	3.07	2.77	2.82	4.53	2.17
35	Slim-ResNetCNN: Structure 1- padding channel layer stride =1	3.08	2.78	2.82	4.53	2.17
35	Slim-ResNet Convolutional Nueral Network	3.11	2.78	3.03	4.48	2.14
35	Slim-ResNet Convolutional Nueral Network (thres)	3.11	3.1	4.32	2.37	2.64
49	Fisher Vector	3.2	3.2	3.56	4.75	1.3
32	Template Probe-CNN	3.21	2.08	0.44	5.88	3.64
52	Electrocardiogram Fingerprint (ECGFP)-10	3.3	4.3	-	3.4	2.1
28	Gram-128 Model	3.56	4.1	0.27	8.5	1.35
35	Slim-ResNetCNN: Model 1- 4 improved residual block_bs	3.58	4.51	3.12	4.88	1.8
52	Electrocardiogram Fingerprint (ECGFP)-7.5	3.7	4.7	-	4.2	2.2
35	Slim-ResNetCNN: Structure 3- 1x1 convolution layer stride =1	3.78	4.23	3.65	4.81	2.41
35	Slim-ResNetCNN: Model 3- 10 improved residual block_bs	3.79	3.8	4.1	4.53	2.71
32	Liveness Map-CNN	4.13	3.28	0.85	8.08	3.04
28	Gram-128 Model with Augmentation	4.15	3.75	3.4	7	2.46
49	Vector Locally Aggregated Descriptors	4.16	4.2	4.85	5.2	2.4
52	Electrocardiogram Fingerprint (ECGFP)-5	4.2	4.9	-	5	2.6
29	Local Quality Features (LQF)	4.22	4.78	1.93	5.84	4.33
35	Slim-ResNetCNN: Structure 2- padding channel layer stride =2	4.49	3.79	3.51	6.4	4.27
25	Liveness Map-CNNp	4.72	4.2	1.4	9.5	3.8
31	Deep Residual Network- ROI	5.32	-	-	-	-
31	Deep Residual Network- ROI+LGP	5.86	-	-	-	-
23	Weber Local Binary Descriptor (WLBPD)	9.6775	9.64	10.82	13.72	4.53
49	Bag of Words	10.67	11.15	10.38	14.1	7.05
35	ResNet Convolutional Nueral Network	24.25	32.06	9.59	40.61	14.74

Table 6 A complete summary of all algorithms and their variants ranked by the average ACE of all scanners of the LivDet-2017 dataset, with green representing the best for that scanner

Reference	Algorithm	Average	Green Bit	Digital Persona	Orcanthus
35	Slim-ResNetCNN	1.01	0.48	0.97	1.57
57	Spoof Buster with UMG Wrapper	4.12	2.58	4.8	4.99
56	Spoof Buster	4.56	3.32	4.88	5.49

recently opening to the public, there are several publications that have tested on this dataset as shown in Table 6.

8 Traditional Machine Learning Algorithms

Several researchers used traditional machine learning algorithms with modified parameters. Research on Fractional Energy of Cosine Transformed Fingerprint Images [10] showed incredible success on the FVC2000 and ATVS dataset with correctly tuned hyperparameters of size and age. Other research on Directional Ridge Frequency [11] showed the importance of using a combination of horizontal, vertical, and diagonal ridge orientations, compared to using them separately. Table 7 ranks the best traditional machine learning algorithms variations on the ATVS and

Table 7 A summary of traditional machine learning algorithms and their variants ranked by average Accuracy Rate on the FVC2000 and ATVS dataset, with green representing the best for that dataset

Reference	Algorithm/Classifier	Average	ATVS AR (%)	FVC2000 AR (%)
53	MLP-- CTFC- Size-8, Age 0.097%	98.41	97.12	99.69
53	Random Forest- CTFC- Size-8, Age 0.097%	98.28	97.49	99.06
53	MLP-- CTFC- Size-4, Age 0.024%	97.76	96.51	99
53	Random Forest- CTFC- Size-4, Age 0.024%	97.64	95.89	99.38
53	MLP-- CTFC- Size-16, Age 0.39%	96.18	92.72	99.64
53	Random Forest- CTFC- Size-16, Age 0.39%	96.1	94.35	97.84
53	Random Forest- CTFC- Size-2, Age 0.006%	95.77	94.35	97.19
53	J48- CTFC- Size-4, Age 0.024%	95.63	93.44	97.81
53	Random Forest- CTFC- Size-32, Age 1.56%	95.1	92.85	97.34
53	SVM- CTFC- Size-8, Age 0.097%	95.01	90.32	99.69
53	J48- CTFC- Size-8, Age 0.097%	94.25	91.3	97.19
53	Naïve Bayes- CTFC- Size-8, Age 0.097%	93.43	88.11	98.75
53	SVM- CTFC- Size-16, Age 0.39%	93.29	87.58	99
24	Random Forest- All Fusion/Combination	93.1	94.86	91.33
53	MLP-- CTFC- Size-2, Age 0.006%	93.07	86.58	99.56
24	SVM-All Fusion/Combination	92.54	93.87	91.21
53	J48- CTFC- Size-16, Age 0.39%	91.41	85.7	97.12
53	J48- CTFC- Size-32, Age 1.56%	91.35	85.57	97.12
53	J48- CTFC- Size-2, Age 0.006%	91.13	85.7	96.56
53	SVM- CTFC- Size-32, Age 1.56%	91.11	82.94	99.28
53	SVM- CTFC- Size-4, Age 0.024%	90.9	82.11	99.69
53	Naïve Bayes- CTFC- Size-4, Age 0.024%	90.84	82.29	99.38
53	MLP-- CTFC- Size-32, Age 1.56%	90.78	82.55	99
24	J48- All Fusion/Combination	90.12	89.13	91.1
53	Naïve Bayes- CTFC- Size-2, Age 0.006%	89.73	88.83	90.63
53	Naïve Bayes- CTFC- Size-16, Age 0.39%	88.66	88.83	88.49
53	Naïve Bayes- CTFC- Size-32, Age 1.56%	88.65	88.58	88.71
24	Random Forest- Diagonal Ridges	88.59	88.43	88.75
24	MLP- All Fusion/Combination	87.56	89.42	85.7
53	Naïve Bayes- CTFC- Size-64, Age 6.25%	87.32	87.82	86.82
53	Random Forest- CTFC- Size-64, Age 6.25%	87.2	87.2	87.2
53	Random Forest- CTFC- Size-128, Age 25%	87.03	87.08	86.98
53	Random Forest- CTFC- Size-256, Age 100%	86.28	86.28	86.28
24	Random Forest- Horivertical Ridges	86.19	78.52	93.86
53	Naïve Bayes- CTFC- Size-128, Age 25%	86.13	87.13	85.13
53	SVM- CTFC- Size-2, Age 0.006%	85.04	87.58	82.5
53	J48- CTFC- Size-64, Age 6.25%	84.94	84.94	84.94
53	J48- CTFC- Size-128, Age 25%	84.15	83.65	84.65
53	Naïve Bayes- CTFC- Size-256, Age 100%	83.82	83.82	83.82
24	SVM- Horivertical Ridges	82.48	79.49	85.46
24	MLP- Diagonal Ridges	81.47	81.65	81.29
24	J48- Horivertical Ridges	80.5	76.76	84.24
53	J48- CTFC- Size-256, Age 100%	80.48	79.48	81.48
53	MLP-- CTFC- Size-64, Age 6.25%	78.54	78.54	78.54
53	SVM- CTFC- Size-64, Age 6.25%	76.58	76.78	76.38
24	J48- Diagonal Ridges	75.99	78.21	73.76
53	SVM- CTFC- Size-128, Age 25%	75.09	74.49	75.69
24	MLP- Horivertical Ridges	74.99	78.52	71.46
53	SVM- CTFC- Size-256, Age 100%	74.04	73.89	74.19
24	SVM- Diagonal Ridges	72.94	82.27	63.6
53	MLP-- CTFC- Size-128, Age 25%	71.61	70.61	72.61
24	Naïve Bayes-All Fusion/Combination	71.1	56.42	85.78
53	MLP- CTFC- Size-256, Age 100%	70.52	70.02	71.02
24	Naïve Bayes- Diagonal Ridges	67.68	60.27	75.09
24	Naïve Bayes- Horivertical Ridges	53.1	53.71	52.48

FVC2000 dataset using the average of the Accuracy Rate (AR) performance metric for each scanner.

9 Performance on Other Datasets

While the LivDet dataset is the most popular dataset to test a liveness detection algorithm, other datasets still play an important role in testing an algorithm's robustness. The datasets tested in this section include MSU-FPAD, ATVS, PBSKD, and a custom dataset.

9.1 Performance on ATVS Data by Scanner Type (Capacitive, Optical, and Thermal)

The ATVS dataset is unique because of the variety of scanner types. With a capacitive, optical, and thermal sensor, algorithm cross-sensor robustness is easily identifiable if the performance metrics are relatively similar. Table 8 summarizes several publications that tested all three types of scanners using the Error Rate metric. As illustrated by [12], the method of a Gaussian Filter with pore extraction has a strong optical sensor performance, but a relatively low capacitive performance, which shows improvements must be made on cross-sensor algorithm robustness.

9.2 Performance on Miscellaneous Datasets Using ACE, FAR, and FRR

Table 9 shows the performance of various algorithms on the various datasets by the ACE metric. The false acceptance rate (FAR) and false rejection rate (FRR) are included if available. It is important to note that some of the datasets in this table are private, such as the MSU-FPAD and the custom dataset of [13]. While performance tests on private datasets provide a quantifiable metric of how the well the algorithm

Table 8 A ranked summary of algorithm performance on the ATVS dataset by scanner type in terms of Error Rate

Reference	Algorithm	Database	All Average	Capacitive Sensor Error Rate (%)	Optical Sensor Error Rate (%)	Thermal Sensor Error Rate (%)
9	Multi-Scale Center Symmetric Local Binary Patterns 1 (MSLBP-1)	ATVS	4.13	4.1	5.2	3.1
9	Multi-Scale Center Symmetric Local Binary Patterns 2 (MSLBP-2)	ATVS	5.13	5.2	6.1	4.1
9	First Proposed Method (MS-CS-LBP)	ATVS	3.43	3.1	4.1	3.1
9	Second Proposed Method (CS_LBP and MSLBP-2)	ATVS	4.77	4.1	6.1	4.1
19	Local Coherence Patterns and SVM	ATVS	6.51	9.05	3.58	6.9
51	Gaussian filter and extracted pores	ATVS-FFp	7.62	13.03	2.05	7.79

Table 9 A summary of algorithm performance ranked by ACE on various datasets

Reference	Brief Algorithm Description or Name	Database and or Sensor	ACE (%)	FAR (%)	FRR (%)
26	CNN-MobieNet-v1 and Munitae-based Local Patches	MSU-FPAD- CrossMatch Guardian 200	0.11	0.11	0.1
54	Security Level=High	ATVS- FFp- All Sesnors	0.25	-	-
26	CNN-MobieNet-v1 and Munitae-based Local Patches	MSU-FPAD- CrossMatch Guardian 200	0.5	0	1
26	CNN-MobieNet-v1 and Munitae-based Local Patches	PBSKD- CrossMatch Guardian 200	0.67	0.33	1
55	ANN with texture descriptors- 4 Principal Components	Custom Dataset	0.74	0	1.47
26	CNN-MobieNet-v1 and Munitae-based Local Patches	PBSKD- CrossMatch Guardian 200	0.83	0.65	1
54	Security Level=Medium	ATVS- FFp- All Sesnors	0.98	-	-
26	CNN-MobieNet-v1 and Munitae-based Local Patches	MSU-FPAD- Lumidigm Venus 302	1.15	1.3	1
26	CNN-MobieNet-v1 and Munitae-based Local Patches	PBSKD- Lumidigm Venus 302	1.97	3.84	0.1
54	Pattern of Oreitned Edge Magnitudes (POEM)	ATVS- FFp- All Sesnors	2.2	-	-
55	ANN with texture descriptors- 2 Principal Components	Custom Dataset	2.21	2.21	2.21
26	CNN-MobieNet-v1 and Munitae-based Local Patches	PBSKD- CrossMatch Guardian 200	2.71	5.32	0.1
9	Method 1- (MS-CS-LBP)	ATVS- All Sensors	3.43	-	-
54	Census Transform Histogram (CENTRIST)	ATVS- FFp- All Sesnors	3.84	-	-
9	Multi-Scale Center Symmetric Local Binary Patterns 1 (MSLBP-1)	ATVS- All Sensors	4.13	-	-
9	Method 2- (CS_LBP and MSLBP-2)	ATVS- All Sensors	4.77	-	-
26	CNN-MobieNet-v1 and Munitae-based Local Patches	MSU-FPAD- Lumidigm Venus 302	5.07	10.03	0.1
9	Multi-Scale Center Symmetric Local Binary Patterns 2 (MSLBP-2)	ATVS- All Sensors	5.13	-	-
19	Local Coherence Patterns and SVM	ATVS- All Sensors	6.51	-	-
54	Local Uniform Comparison Image Descriptor (LUCID) (SL=Low)	ATVS- FFp- All Sesnors	7.17	-	-
51	Gaussian filter and extracted pores	ATVS-FFp- All Sensors	7.62	-	-

performed, it is difficult to compare the performance of two algorithms on separate datasets.

10 Conclusion

In summary, our survey reviews the Fingerprint LivDet competition's growth over the years, the many algorithms published between 2014 and 2019 and their performance on the LivDet competition datasets, the performance of traditional machine learning algorithms with varying hyperparameters and inputs, and the performance of published algorithms on non-LivDet competition sets. As expected, the performances observed at the LivDet competitions continue to improve in terms of ACE with more powerful and robust algorithms. However, with more sophisticated datasets and different scanner hardware, it may pose a significant challenge to the algorithm performance. With the numerous published algorithms and their tests on the popular LivDet datasets, it is easy to compare algorithm performance.

With all the data collected in this survey, the state-of-the-art algorithm performance for each scanner type is easily identifiable and accessible for other researchers to study to improve their own model. The LivDet performance data of Tables 2, 3, 4, and 5 also gave some useful insights when plotted on a graph, which could not be included in this chapter due to page constraints. While the data from these graphs need further analysis to gain additional insight, the graph's scanner-based trendlines show the relative difficulty of each dataset. A scatter plot of the data presented in Table 2 (LivDet-2009) shows that the Identix dataset is typically easier for most algorithms to classify compared to CrossMatch and Biometrika. A scatter plot of the data presented in Table 3 (LivDet-2011) shows that that algorithm performance on the Digital Persona images typically had a lower ACE compared to Italdata images, which typically had a much higher ACE compared to the other model's images. A scatter plot of the data presented in Table 4 (LivDet-2013) shows that images from Biometrika and Italdata typically had the lowest ACE metrics, while CrossMatch images typically had the highest ACE metric. Finally, a scatter plot of the data presented in Table 5 (LivDet-2015) shows that images from GreenBit and CrossMatch typically had a lower ACE metric, while the Digital Persona dataset had a much higher ACE. This data pinpoints the scanner datasets that researchers have mastered, like the LivDet-2009 Identix dataset, while also highlighting the scanner datasets that need additional research, like the LivDet-2013 CrossMatch dataset.

The traditional machine learning algorithms with modified parameters and inputs also provided powerful solutions to anti-spoofing with the correct tuning. The research from [10] reveals that a size parameter of 64–128 and an age parameter of 6.25–25% yields the best results on the FVC2000 dataset across all tested traditional machine learning algorithms. On the ATVS dataset, with a size parameter of 64 and an age parameter of 6.25%, the Random Forest and Naïve Bayes performed slightly worse compared to the FVC2000 dataset, but significantly better than the other tested machine learning algorithms on the ATVS dataset. With the right parameters, the research from [10] shows impressive results on the FVC2000 dataset but will need some tweaking to have an equal performance on the ATVS dataset. Research from [11] also pinpoints which traditional machine learning algorithm works the best with varying ridge orientations. The data provided also highlights the importance of using a combination of ridge angles with the significant increase in performance for most algorithms on both datasets. The miscellaneousdatasets discussed at the end of this chapter, while the algorithm comparability is low, still provides a means for additional data to train for model robustness, and it reveals promising models [14, 15] that should also be tested on the LivDet datasets for comparability. Some datasets like the ATVS can provide a powerful measure of cross-scanner-type robustness. By using optical, thermal, and capacitive scanners, performance results can show how dependent an algorithm is on a certain type of input.

With the introduction of more powerful and unique hardware, algorithms need to adapt. While this survey reviews most of the data that compares a model's performance by scanner type, there is still a need to research, survey, and analyze the cross-spoof material data. Attempts have been made to learn the characteristics of varying spoof materials, but with novel spoofing materials research seems to be outpaced. Perhaps, the further development on the One-Class Classifier [16] will combat the rapid number of novel spoof materials or research on Optical Coherence Tomography (OCT) [17] will inhibit the effectiveness of many spoof materials by altering the approach to liveness detection solutions. In conclusion, as the data in this survey suggests, there is a continual need for more advanced generalization in terms of cross-sensor and cross-material robustness to ensure the security of fingerprint biometrics.

References

1. E. Marasco, A. Ross, A survey on anti-spoofing schemes for fingerprint recognition systems. ACM Comput. Surv **47**(2) (2014)., Article 28, 36 Pages
2. R. Kiefer, J. Stevens, A. Patel, M. Patel, A survey on spoofing detection systems for fake fingerprint presentation attacks, in *Fourth International Conference on ICT for Intelligent Systems (ICTIS – 2020)*. Accepted and pending publication
3. G. Marcialis et al., First international fingerprint liveness detection competition—LivDet 2009, Clarkson.edu, 2009
4. D. Yambay, L. Ghiani, P. Denti, G. Marcialis, F. Roli, S. Schuckers, LivDet 2011 – fingerprint liveness detection competition 2011, Clarkson.edu, 2011
5. L. Ghiani et al., LivDet 2013 fingerprint liveness detection competition 2013, in *2013 International Conference on Biometrics (ICB)*, (Madrid, 2013), pp. 1–6
6. V. Mura, L. Ghiani, G. Marcialis, F. Roli, LivDet 2015 fingerprint liveness detection competition 2015, Clarkson.edu, 2015
7. V. Mura et al., arXiv:1803.05210v1 [cs.CV] 14 Mar 2018 LivDet 2017 fingerprint liveness detection competition 2017, Arxiv.org, 2019
8. G. Orru et al., LIVDET inaction- fingerprint liveness detection competition 2019, Arxiv.org, 2019
9. Z. Akhtar, C. Micheloni, G.L. Foresti, Correlation based fingerprint liveness detection, in *2015 International Conference on Biometrics (ICB)*, (Phuket, 2015), pp. 305–310
10. S. Khade, S.D. Thepade, Novel fingerprint liveness detection with fractional energy of cosine transformed fingerprint images and machine learning classifiers, in *2018 IEEE Punecon*, (Pune, India, 2018), pp. 1–7
11. S. Khade, S.D. Thepade, A. Ambedkar, Fingerprint liveness detection using directional ridge frequency with machine learning classifiers, in *2018 Fourth International Conference on Computing Communication Control and Automation (ICCUBEA)*, (Pune, India, 2018), pp. 1–5
12. M. Lu, Z. Chen, W. Sheng, A pore-based method for fingerprint liveness detection, in *2015 International Conference on Computer Science and Applications (CSA)*, (Wuhan, 2015), pp. 77–81
13. C. Zaghetto, M. Mendelson, A. Zaghetto, F.D.B. Vidal, Liveness detection on touch-less fingerprint devices using texture descriptors and artificial neural networks, in *2017 IEEE International Joint Conference on Biometrics (IJCB)*, (Denver, CO, 2017), pp. 406–412
14. F. Pala, B. Bhanu, On the accuracy and robustness of deep triplet embedding for fingerprint liveness detection, in *2017 IEEE International Conference on Image Processing (ICIP)*, (Beijing, 2017), pp. 116–120
15. T. Chugh, A.K. Jain, Fingerprint presentation attack detection: Generalization and efficiency, in *ICB*, (2019)
16. J.J. Engelsma, A.K. Jain, Generalizing fingerprint spoof detector: Learning a one-class classifier. arXiv:1901.03918 (2019)
17. T. Chugh, A.K. Jain, OCT fingerprints: Resilience to presentation attacks. arXiv:1908.00102 (2019)

Suitability of Voice Recognition Within the IoT Environment

Salahaldeen Duraibi, Fahad Alqahtani, Frederick Sheldon, and Wasim Alhamdani

1 Introduction

User authentication refers to the process in which a user submits his/her identity credential (often represented by paired username and password) to an information system in order to validate the person who he/she claims to be. In general, within the context of IoT, three factors of authentication are usually employed: (i) something a user knows (e.g., a password); (ii) something a user has (e.g., a secure token); and (iii) something a user is (e.g., biometric characteristics). Passwords are the most common authentication mechanism (i.e., single factor). However, password- and token-based authentications have many security issues [1, 2] and are not suitable for smart devices because of the unattended nature of IoT smart devices. A biometric

S. Duraibi (✉)
Computer Science Department, University of Idaho, Moscow, ID, USA

Computer Science Department, Jazan University, Jazan, Saudi Arabia
e-mail: dura6540@vandals.uidaho.edu

F. Alqahtani
Computer Science Department, University of Idaho, Moscow, ID, USA

Computer Science Department, Jazan University, Jazan, Saudi Arabia

Computer Science Department, Prince Sattam Bin Abdulaziz University, Al-Kharj, Saudi Arabia
e-mail: Alqa0199@vandals.uidaho.edu

F. Sheldon
Computer Science Department, University of Idaho, Moscow, ID, USA
e-mail: sheldon@uidaho.edu

W. Alhamdani
Computer Science Department, University of the Cumberlands, Williamsburg, KY, USA
e-mail: wasim.alhamdani@ucmberlands.edu

K. Daimi et al. (eds.), *Advances in Security, Networks, and Internet of Things*,
Transactions on Computational Science and Computational Intelligence,
https://doi.org/10.1007/978-3-030-71017-0_5

authentication system verifies the identity of a person based on either their unique physiological traits [3, 4] or their unique behavioral biometrics [5, 6]. Biometric authentication is more user friendly in nature than the approaches that rely on passwords and secure tokens. While physiological traits can achieve high accuracy in user authentication, they are subject to a variety of attacks [7, 8] and also raise privacy concerns [9]. Moreover, the accuracy of physiology-based mechanisms may be substantially degraded by environmental factors, such as the viewing angle, illumination, and background noise [10, 11]. In contrast, behavioral biometrics (i.e., key stroke, voice, or gait analysis) appear less sensitive to ambient light or noise [12, 13].

There have been a few studies on the security and usability of behavioral biometric authentication for the IoT ecosystem. To the best of our knowledge, there are only two studies that adopted voice biometrics as an authentication mechanism for the IoT ecosystem. Shin and Jun [14] implemented voice recognition technology to verify authorized users for controlling and monitoring a smart home environment. Shin et al. proposed a voice recognition system that is divided into server and device parts. The role of the server part of the system is for user preregistration, user recognition, and command control analysis. The role of the device is command reception, device control, and then response. The type of models and techniques employed in their research is not discussed. Likewise, the implementation of their [1] model is this chapter.

The rest of the chapter is laid out as follows: Sect. 2 is the background, Sect. 3 is the related work, Sect. 4 discusses the motivation, Sect. 5 our proposed model is presented, Sect. 6 is the implementation, and Sect. 7 concludes the chapter.

2 Background

Automatic speaker verification [ASV] (Fig. 1) is a pattern recognition problem that predominantly works on speech signals. ASV systems intend to acquire different information from voice data and combine them for each speaker. For example, *idiolectal* and *prosody* identify [6] high-level attributes of a speaker's voice, while *short-term spectral* identifies low-level attributes of the speech signals. The latter is the main source of individuality in speech [5]. Low-level attributes are easy to extract and are most common when applied to ASV systems.

As can be seen in Fig. 1, there are three processes in any ASV system. These are *voice feature extraction*, *speaker modeling*, and *decision making*. The *feature extraction* process is the same in the enrollment and verification stages where the voice signals are converted into a sequence of frames. Each frame is a short window of the waveform with overlapping adjacent windows [15]. Considering the unique resource-constrained characteristics of IoT devices, our discussion will focus only on *short-term spectral qualities* [16] as it requires less computational resources. Additionally, the selection of appropriate feature extraction methods is crucial in this process because they influence the performance of the system. The two most popular

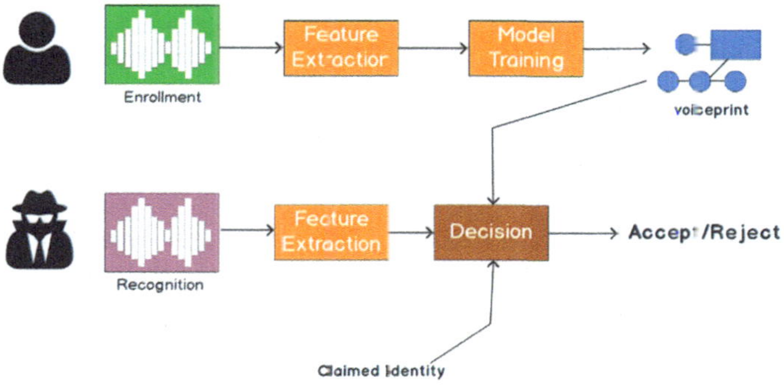

Fig. 1 ASV

spectral feature extraction methods, *filter bank analysis* and *linear predictive coding*, are discussed in the following sections [9, 17, 18].

2.1 Filter Bank Analysis

In filter bank analysis, the voice signals are expected to pass through a bank of band-pass filters that cover a range of frequencies consistent with the transmission characteristics of the signal. The spacing of the filters can be either be uniform or nonuniform, based on the perceptual criteria such as linear frequency cepstral coefficient [LFCC] or mel-frequency cepstral coefficient [MFCC]; the latter provides a linear spacing [9, 19].

2.2 Linear Predictive Coding [LPC]

In the LPC, the speech signal can be modeled by a linear process prediction. Signals at each time step use unique and specific periods of preceding samples that capture the temporal evolution of the features from one speech segment to another [9, 19].

1. In the ASV systems, 2. The speaker modeling step comes, 3. After features of the voice are extracted. ASV has the ability to construct a model λ_s for each user where "s" is user and λ is the specific model. Such a modeling depends on, for example, whether it is used for applications that use fixed words (text-dependent), or applications that use phonemes not seen in the enrollment data (text-independent).

Speaker modeling methods can be of two categories: nonparametric or parametric. The nonparametric approaches include templates which are suitable for a text-dependent verification system [9].

The parametric speaker modeling includes *vector quantization (VQ), Gaussian mixture models (GMM), and hidden Markov models (HMM).* In VQ, a set of representative samples of the user's enrollment voice is constructed by clustering the feature vectors. *GMM* has been proven to be very effective in the cases of a text-independent speaker recognition [15], also referred to as a refinement of vector quantization. HMM is suitable for text-dependent ASVs and is used for access control of personal information or bank accounts. Among the three techniques, GMM has been proven very effective for phones, and may be the best candidate for IoT devices [20]. There are also some other nonparametric and parametric approaches in the literature. However, those presented in this chapter are the most common techniques implemented in ASVs. The final process is decision making. In this process, an "accept or "reject" decision is delivered based on the verification models discussed above.

3 Related Work

This section reviews different biometric authentication mechanisms employed in the IoT ecosystem. Generally, there are two main types of IoT authentication methods including centralized and distributed architectures as shown in Figs. 2 and 3, respectively. The centralized architecture uses a centralized server to manage the credentials used in the authentication, whereas in the distributed architecture the authentication is accomplished point-to-point between the communicating parties [21]. Biometric authentication is achieved based on these two architectures. There are four basic biometric authentication performance measuring strategies" they include accuracy, scale, security, and privacy. Elements such as enrollment, biometric reference, comparison, networking, and personal biometric criteria are common in biometric authentication systems. Biometric-based authentication systems use two other factors: physical and behavioral [21]. The physical factors include fingerprint, face, iris, hand geometry, and palm print recognition, while the behavioral factors may include, but are not limited to, voice, signature, and gait recognitions [11]. Biometric systems have some advantages over conventional identity-based methods (password and ID); they cannot be transferred, stolen, lost, broken, or easily guessed [12, 22]. The acceptance and performance of the biometric

Fig. 2 Centralized authentication architecture

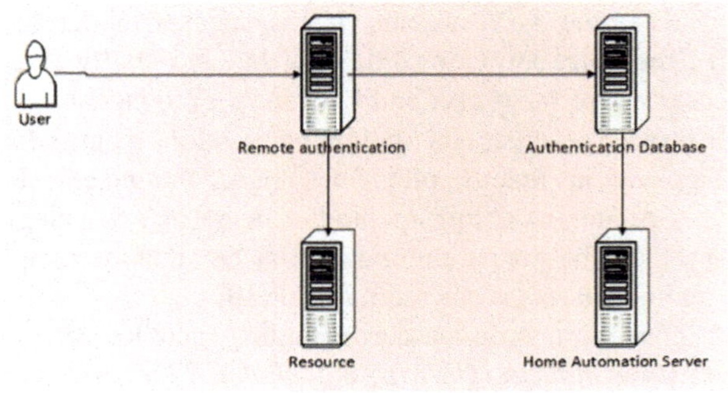

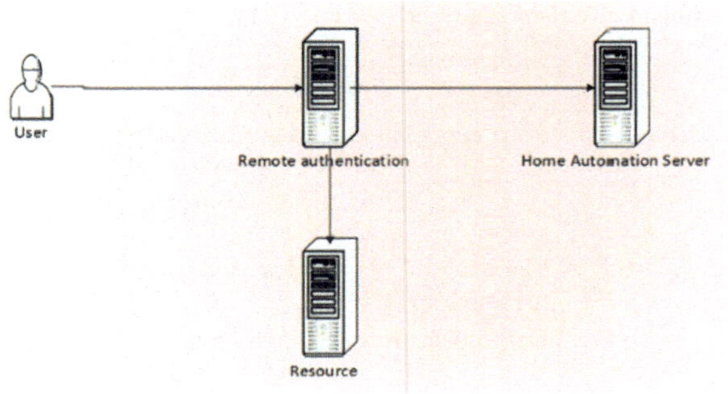

Fig. 3 Distributed authentication architecture

systems are presented in Table 1. From Table 1, the voice biometric systems are more suitable compared to other biometric systems. The voice biometric systems use voice rather than complicated input methods, like fingerprints which need special hardware for input. Hence, voice is appropriate for IoT where convenience is important. Some researchers have investigated and adopted voice based-biometric systems into the IoT ecosystem.

For example, Kim and Hong [23] used MFCC and pitch as voice features and the GMM in the voice authentication process for speaker recognition. Likewise, Chen et al. [24] proposed an authentication and authorization scheme that uses voice rhythmic pattern [11] for mobile IoT devices.

4 Motivation

Entity authentication refers to a process by which an agent in a distributed system gains confidence in the identity of a communication partner. In other words [25], defines authentication as a provision for ensuring the correctness of the claimed identity of an entity. Most of the time authentication is mistaken with authorization, which is concerned with the level of access or privilege an entity may possess.

In this light, security throughout the process of authentication has to be maintained. For example, if an attacker steals the credentials of a user and gains access to a smart-door lock of a house or the health-monitoring smart device of a patient, this could be life-threatening. Hence, security before, during, and after the authentication of a smart device is of the highest importance. However, biometric authentication suffers from the public nature of some biometrics, including the facial-feature method that uses the face as a biometric, which is easy to be replicate. Fingerprint biometrics are commonly left everywhere and can be reproduced. Likewise, voice biometrics also suffer from the issue of recording and replaying for authentication. Storing biometric data on servers also raises concern. For example, if a perpetrator gains access to the server where the biometrics are stored, the attacker may take those biometrics and access anything the biometric is used to protect. This poses a major problem.

Table 1 Comparison of biometric system

Factor type	Biometric systems	Weakness	Strength
Behavioral	Voice recognition	Has a relatively low accuracy, inefficiencies in certain circumstances	Needs no hardware, ease of use, widespread usage, can be used for remote authentication
	Signature recognition	Has a relatively low accuracy	Wide acceptance, non-rigging
	Detect behavior	Shows nonperformance in certain conditions	Continuous authentication
	Tough dynamics	Inconsistent accuracy, lack of efficiency under certain conditions	Continuous authentication, does not require specific hardware
	Keystroke dynamics	Inconsistent accuracy, lack of efficiency under certain conditions	Continuous authentication, does not require specific hardware
Physical	Fingerprint recognition	The need for additional hardware, the difficulty of obtaining high-quality images, the lack of efficiency in certain circumstances	Use and wide acceptance, low cost, good accuracy
	Face recognition	The need for additional hardware, lack of efficiency in certain circumstances	Use and wide acceptance, good accuracy and less fraud
	Iris recognition	The need for additional hardware, high cost, time-taking authentication	High precision, non-rigging
	Hand geometry recognition	The need for additional hardware, precision	Easy to use, less fraud
	Palm detection	The need for additional hardware, high cost	Public acceptance, high precision

There are also issues that are concerned with remote authentication. Normally, personal non-attended smart devices ask a user to remotely authenticate himself to his devices. However, there is an issue of trust with remote verification. Because the user sends his biometrics remotely for authentication, he cannot ensure that his biometrics data will not be hijacked and potentially be misused or mishandled. This raises an issue of trust or privacy. Because of this, our research focuses on a secure voice biometric-based authentication. This chapter specifically focuses on steps taken towards the primary testing of an IoT user-authentication model that uses voice biometrics. The security and resilience aspects of the model are ongoing.

5 Our Proposed Model

A voice biometric-based IoT user authentication model was proposed in one of our previous papers as presented in Fig. 4 [26]. The model has two phases, the enrollment phase when the user of the smart device speaks his voice for registration. The other phase is verification when the system checks whether the identity claimer is the real user by comparing the previously enrolled voice with the voice of the identity claimer. Subsequently, if the similarity of the two voices reaches a certain predefined threshold, then access is granted to the claimer; otherwise the claimer is rejected. The design criteria of the model is given in Table 2.

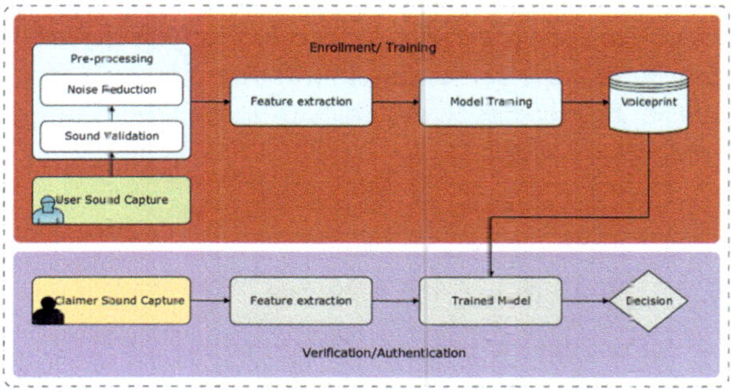

Fig. 4 The proposed IoT voice biometric model

Table 2 Design criteria of the model

Design criteria	Description
Universality	A very high percentage of the population should have the characteristic. For example, virtually everyone has recognizable fingerprints, but there are rare exceptions.
Distinctiveness	No two people should have identical characteristics. For some otherwise acceptable characteristics, identical twins share virtually the same patterns, such as facial features and DNA, but not other features, such as fingerprints and iris patterns.
Permanence	The characteristic should not change with time. For otherwise acceptable characteristics, such as facial features and signatures, periodic re-enrollment of the individual may be required.
Collectability	Obtaining and measuring the biometric feature(s) should be easy, nonintrusive, reliable, and robust, as well as cost-effective for the application.
Performance	The system must meet a required level of accuracy, perform properly in the required range of environments, and be cost-effective.
Circumvention	The difficulty of circumventing the system must meet a required threshold. This is particularly important in an unattended environment where it would be easier to use such countermeasures and a fingerprint prosthetic or a photograph of a face.
Acceptability	The system must have high acceptance among all classes of users. Systems that are uncomfortable to the user, appear threatening, require contact that raises hygienic issues, or are nonintuitive are unlikely to be acceptable to the general population

6 Implementation

6.1 Description of the Implementation

The normal process of voice biometric implementation usually includes creation of models from audio data, generation of tables, and self-authentication to the IoT manager. We, therefore, used a client-server model for the initial implementation. A user mobile device simulates running [1] a client, where the server simulates the IoT manager and handles the verification requests. By contrast, the IoT devices receive the command from the IoT manager and automatically respond to the command.

First, the user inputs his own voice command using the smartphone. Subsequently, the system on the server side determines whether the connection is for enrollment or verification and accordingly performs the process in each phase. New connections are considered first for enrollment, while the returning connections are considered by the server for verification by prompting the identity claimer with challenging words.

6.2 Open Source Software

To accomplish the initial test, we used a software package called Mistral/Alize. Mistral is tested by the National Institute of Standards and Technology [NIST] and with 0.5 error rate. Figure 5 shows the process of Mistral package.

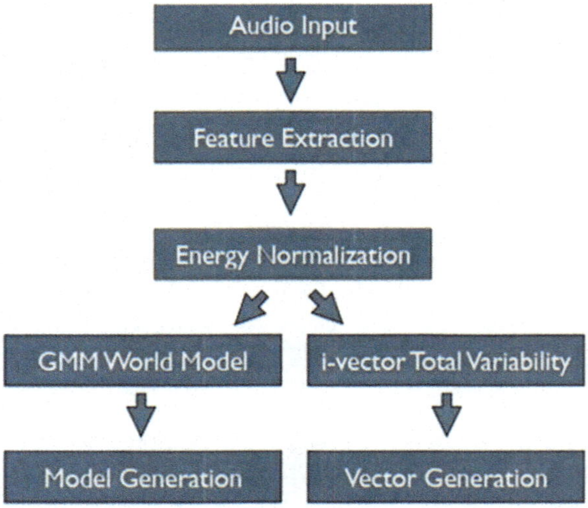

Fig. 5 The processing of the Mistral software package

6.3 *Dataset*

There are a number of datasets for audio dataset selection, but we selected the MIT dataset, which appears to be recent and is considered to be good for initial results. MIT is designed for mobile environments. The dataset consists of 48 speakers including 22 female and 26 male. The dataset is used for the universal background model (UBM) training.

6.4 *Preprocessing*

In this step, after the user inputs his voice using his own smartphone, the voice is validated, and noise is removed from the voice signal. For example, Fig. 6 shows the raw voice signal without noise reduction, while Fig. 7 illustrates that the noise (silent part) is removed before it is submitted for feature extraction. In this process, using SPro [27], which is supported in the Mistral program, frame selection is performed by excluding silent frames longer than 100 ms. In addition, if a sample is different from 16 KHz, SPro performs resampling by default. Mistral requires the voice to be more compact. The most important aspect of this process is that the voice is converted into the binary format.

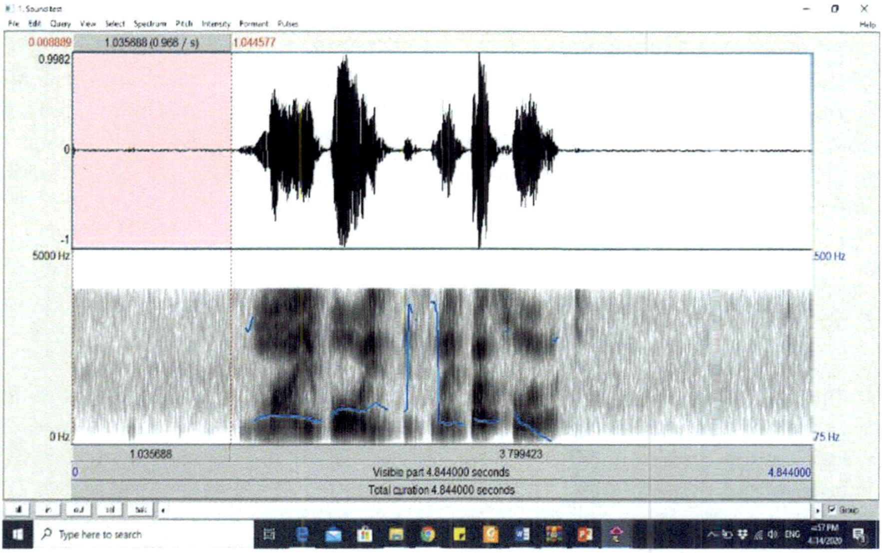

Fig. 6 Voice waves before noise reduction

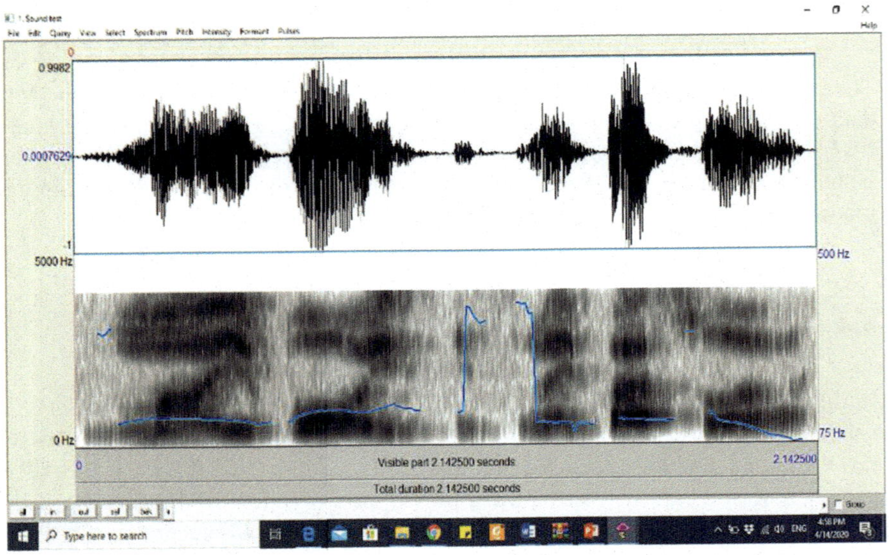

Fig. 7 Voice waves after noise reduction

6.5 Feature Extraction

In this step, using SPro, the feature extraction process is conducted. The voice data, preprocessed in the preceding step, is subjected to feature extraction. Feature extraction is conducted through 12 MFCCs, as shown in Fig. 8, and is subsequently stored in a parameter file. The front end of the system actually stops at this step and, according to our work, is accomplished at the user's smartphone. At this point of the testing, we used a virtualized Android OS to represent the mobile part of the implementation.

6.6 Voice Model Training/Server Side

In the Mistral toolkit, the next step is training the model. This step controls the list of users of the smart IoT device. Using the Train World process, we used the Gauss mixture model [GMM] to conduct this part of the training with the MIT dataset for the UBM training. This procedure requires an already trained UBM. The remaining 12 were kept for testing. For this part of the implementation, we used a virtualized Linux server to host the Android Things operating system, which acts as the manager of the IoT devices.

```
frame [1]:
    numberOfCoefficients = 12
    c0 = 999.8048601455131
    c []:
      c [1] = -44.22161952301026
      c [2] = -65.33400756773091
      c [3] = -68.47878265501046
      c [4] = -77.75213262672578
      c [5] = -39.118073561448305
      c [6] = -27.97100271788985
      c [7] = -16.280815060716662
      c [8] = -28.29389137258743
      c [9] = -48.891519014221736
      c [10] = -21.636367646690204
      c [11] = -41.99618380453937
      c [12] = -33.930970331179246

frame [2]:
    numberOfCoefficients = 12
    c0 = 981.7050655028498
    c []:
      c [1] = -34.444664351555556
      c [2] = -60.544813681496144
      c [3] = -56.850407405322389
      c [4] = -106.549861142603 3
      c [5] = -13.863919139306429
      c [6] = -52.19559700536588
      c [7] = -28.201681724169667
      c [8] = -40.525337031971048
      c [9] = -78.32348751266473
      c [10] = -38.554244958442941
      c [11] = -51.863737275336856
      c [12] = -32.592918613884919
```

Fig. 8 12 MFCCs features

6.7 *Verification*

After both parts of the model have been trained, the data is to against the world UBM model to create verification. The tests were run twice; the first test included using enrolled speakers. The second test did not include enrolled speakers. After the testing phase started, scores were calculated for each test speech segment verification based on the Mistral toolkit. By using a decision threshold speaker model, verification could either be accepted or rejected. In commercial verification systems, users are required to test and decide on the threshold. However, in our implementation, we only speculated on the possible use of the threshold.

6.8 *Result and Discussion*

We had no problem with the enrollment of the speakers in the dataset. In the first test, only the intended target speakers were used to train at the UBM before they were enrolled. Figure 9 shows the detection error trade-off curve of the MIT voice dataset.

In the future, we will collect enrollment segments from many users anonymously and use different devices in regular daily life settings for a significantly improved result; this will build a more reliable UBM to be distributed for application. In addition, a secure remote authentication mechanism will be investigated. Voiceprint security, once stored in the voice database, will be included in the future research.

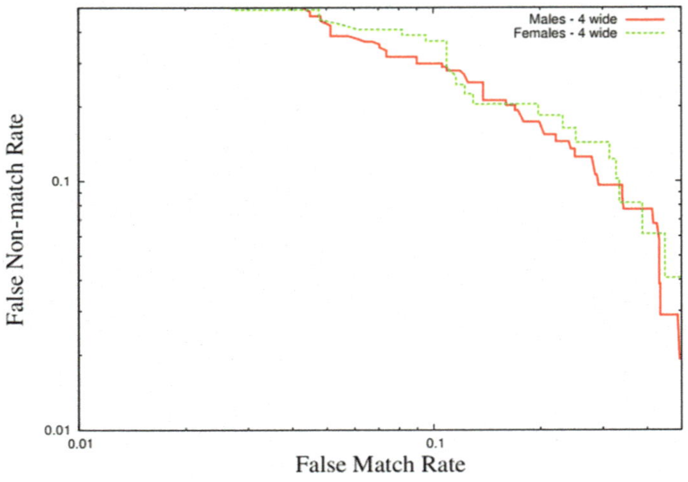

Fig. 9 DET plots for a male-only trial and female-only trial

7 Conclusion

Our research demonstrates that voice biometric-based authentication models can be used in IoT ecosystems. We proposed and implemented a model that is intended to be used by IoT technology owners to remotely and securely authenticate themselves to smart technologies. We tested the model for its usability and have shown a promising result. It is worth mentioning again that we did not include any security measures in our implementation, leaving this for future work.

References

1. O. Olazabal et al., Multimodal biometrics for enhanced IoT security, in *2019 IEEE 9th Annual Computing and Communication Workshop and Conference (CCWC)*, (IEEE, 2019)
2. Y. Ren et al., Replay attack detection based on distortion by loudspeaker for voice authentication. Multimed. Tools Appl. **78**(7), 8383–8396 (2019)
3. S. Nainan, V. Kulkarni, Performance evaluation of text independent automatic speaker recognition using VQ and GMM, in *Proceedings of the Second International Conference on Information and Communication Technology for Competitive Strategies*, (ACM, 2016)
4. M. McLaren, Automatic speaker recognition for authenticating users in the Internet of Things, 26 August 2016
5. S. Gupta, S. Chatterjee, Text dependent voice based biometric authentication system using spectrum analysis and image acquisition, in *Advances in Computer Science, Engineering & Applications*, (Springer, Berlin, Heidelberg, 2012), pp. 61–70
6. C. Kolkata, About voice biometric and speaker recognition, 2014
7. A.S. Thakur, N. Sahayam, Speech recognition using Euclidean distance. Int. J. Emerg. Technol. Adv. Eng. **3**(3), 587–590 (2013)

8. D. Petrovska-Delacrétaz, A. El Hannani, G. Chollet, Text-independent speaker verification: state of the art and challenges, in *Progress in Nonlinear Speech Processing*, (Springer, Berlin, Heidelberg, 2007), pp. 135–169
9. A.E. Rosenberg, F. Bimbot, S. Parthasarathy, Overview of speaker recognition, in *Springer Handbook of Speech Processing*, (Springer, Berlin, 2008), pp. 725–742
10. A. Yassine et al., IoT big data analytics for smart homes with fog and cloud computing. Futur. Gener. Comput. Syst. **91**, 563–573 (2019)
11. M.A. Ferrag, L. Maglaras, A. Derhab, Authentication and authorization for mobile IoT devices using biofeatures: Recent advances and future trends. Secur. Commun. Netw. **2019**, 1–20 (2019)
12. H. Hamidi, An approach to develop the smart health using Internet of Things and authentication based on biometric technology. Futur. Gener. Comput. Syst. **91**, 434–449 (2019)
13. A.N. Moussa, N. Ithnin, A. Zainal, CFaaS: Bilaterally agreed evidence collection. J. Cloud Comput. **7**(1), 1 (2018)
14. K. Brunet et al., Speaker recognition for mobile user authentication: An android solution, 2013
15. F. Thullier, B. Bouchard, B.-A. Menelas, A text-independent speaker authentication system for mobile devices. Cryptography **1**(3), 16 (2017)
16. X. Zhang et al., Voice biometric identity authentication system based on android smart phone, in *2018 IEEE 4th International Conference on Computer and Communications (ICCC)*, (IEEE, 2018)
17. A. Khitrov, K. Simonchik, System for text-dependent speaker recognition and method thereof. Google Patents, 2019
18. A.N. Moussa et al., A consumer-oriented cloud forensic process model, in *2019 IEEE 10th Control and System Graduate Research Colloquium (ICSGRC)*, (IEEE, 2019)
19. A. El Hannani et al., Text-independent speaker verification, in *Guide to Biometric Reference Systems and Performance Evaluation*, (Springer, London, 2009), pp. 167–211
20. A.E. Rosenberg, S. Parthasarathy, Speaker background models for connected digit password speaker verification, in *1996 IEEE International Conference on Acoustics, Speech, and Signal Processing Conference Proceedings*, (IEEE, 1996)
21. M. El-hajj et al., A survey of Internet of Things (IoT) authentication schemes. Sensors **19**(5), 1141 (2019)
22. A.N. Moussa, N.B. Ithnin, O.A. Miaikil, Conceptual forensic readiness framework for infrastructure as a service consumers, in *2014 IEEE Conference on Systems, Process and Control (ICSPC 2014)*, (IEEE, 2014)
23. D.-S. Kim, K.-S. Hong, Multimodal biometric authentication using teeth image and voice in mobile environment. IEEE Trans. Consum. Electron. **54**(4), 1790–1797 (2008)
24. Y. Chen et al., Your song your way: Rhythm-based two-factor authentication for multi-touch mobile devices, in *2015 IEEE Conference on Computer Communications (INFOCOM)*, (IEEE, 2015)
25. M. Bhatia et al., *IS-IS Generic Cryptographic Authentication* (Network Working Group, 2009)
26. S. Duraibi, F.T. Sheldon, W. Alhamdani, Voice biometric identity authentication model for IoT devices. Int J Secur Privacy Trust Manag. (IJSPTM) **9**(2), 1–10 (2020)
27. G. Guillaume, SPro: Speech signal processing toolkit (2004). Software available at http://gforge.inria.fr/projects/spro

Chor-Rivest Knapsack Cryptosystem in a Post-quantum World

Raúl Durán Díaz, Luis Hernández-Álvarez, Luis Hernández Encinas, and Araceli Queiruga-Dios

1 Introduction

It is known that the quantum algorithms proposed by Shor [1] will break the main asymmetric cryptosystems used today, if a quantum computer with sufficient computing capacity is developed (it is estimated that with more than 10^7 qubits). Indeed, the mathematical problems on which their security is based (integer factorization and discrete logarithms) could be solved in just a few hours. In fact, if a current PC needs $\mathcal{O}\left(2^{\sqrt[3]{\log n}}\right)$ bit operations to break an algorithm, a quantum computer using Shor algorithm could reduce such number of bit operations to $\mathcal{O}\left(\log^3 n\right)$ while requiring a memory storage of $\mathcal{O}(\log n)$ bits.

As to symmetric cryptography, Grover [2] and Simon [3] algorithms could reduce the computing time required to break the main symmetric cryptosystems to the square root of the current time. That is, if a quantum computer with the sufficient computing capacity is developed, the security of current symmetric cryptosystems would be equivalent to that of the same cryptosystems with half-length keys.

Due to these quantum threats to the security of the information when it is protected with the current cryptosystems, the National Institute of Standards and

R. Durán Díaz
Departamento de Automática, Universidad de Alcalá, Alcalá de Henares, Spain
e-mail: raul.duran@uah.es

L. Hernández-Álvarez · L. Hernández Encinas (✉)
Instituto de Tecnologías Físicas y de la Información (ITEFI), Consejo Superior de Investigaciones Científicas (CSIC), Madrid, Spain
e-mail: luis.hdez.alvarez@iec.csic.es; luis@iec.csic.es

A. Queiruga-Dios
Departamento de Matemática Aplicada, Universidad de Salamanca, Salamanca, Spain
e-mail: queirugadios@usal.es

K. Daimi et al. (eds.), *Advances in Security, Networks, and Internet of Things*, Transactions on Computational Science and Computational Intelligence,
https://doi.org/10.1007/978-3-030-71017-0_6

Technology (NIST) has launched an International Call to select new quantum-resistant cryptographic algorithms [4] for asymmetric encryption, digital signatures, and key encapsulated mechanisms. In January 2019, NIST published the list of the algorithms that had been promoted to the second round [5]. The security of such algorithms is based on specific mathematical problems founded in lattices, error correction codes, quadratic multivariate polynomials, hash functions, and elliptic curve isogenies. The objective of NIST is to select new cryptographic systems invulnerable to quantum computers (or quantum resistant), no matter how much computing power they have.

On the other hand, traditionally, cryptosystems based on knapsack problems have had bad press because most of them have been completely broken. In this sense, it is important to note that Chor-Rivest cryptosystem [6, 7] was one of the systems only broken for the original parameters proposed [8]; whereas it has been proved that it is possible to select safer parameters [9, 10] which could permit to implement such cryptosystem in a secure way.

In this paper, we present the state of art of some of the mathematical problems proposed to define quantum-resistant asymmetric cryptosystems [11] and moreover, we want to focus on another problem, not considered so far, that could offer alternatives of interest for post-quantum security. This is a knapsack-type problem, which uses the arithmetic of finite fields and needs to compute discrete logarithms in order to determine the keys of the system. The interesting point is that the security of such cryptosystem depends on the knapsack problem but not on the discrete logarithm problem (DLP). In fact, if the DLP becomes tractable, then the Chor-Rivest cryptosystem is easier to implement, but not easier to break.

The rest of this paper is organized as follows: in Sect. 2 we present the most important mathematical foundations upon which the new cryptographic proposals are based. Section 3 shows an alternative cryptographic primitive, not considered in the NIST call, that could be added to the list of quantum-resistant algorithms. Finally, Sect. 4 presents a discussion about the possible benefits that such proposal could offer to the quantum scenario and some future works.

2 Post-quantum Proposals

We focus our attention over asymmetric cryptography since as we have mentioned before, it seems to be the most affected by quantum algorithms. The most important proposals, which will be dealt with in turn, are: (1) hash-based, (2) code-based, (3) lattice-based, (4) multivariate-quadratic-equations, and (5) elliptic-curve-isogeny-based.

2.1 Hash-Based Cryptography

This proposal stems from an old one-time signature scheme due to Lamport and Diffie [12] and provides a quantum-resistant signature scheme. The key ingredient is a hash function H, which is required to be simply *pre-image resistant* (this means that an adversary provided with x, y, with $y = H(x)$, is not able to determine $x' \neq x$ such that $y = H(x')$). To fix ideas, let us assume that H outputs n-bit hashes for any input message, i.e.,

$$H\colon \{0, 1\}^* \to \{0, 1\}^n.$$

The first step is for the user to generate her public/private key pair. In order to do so, she selects uniformly at random $2n$ n-bit strings, which make up the secret key:

$$SK = (sk_{n-1}[0], sk_{n-1}[1], \dots, sk_0[0], sk_0[1]),$$

and publishes the corresponding public key, namely

$$PK = (pk_{n-1}[0], pk_{n-1}[1], \dots, pk_0[0], pk_0[1]),$$

where $pk_i[j] = H(sk_i[j])$, for $0 \leq i \leq n-1$, $j = 0, 1$.

Given a message $M \in \{0, 1\}^*$ to be signed, the user first computes $h = H(M)$, which is an n-bit length string, so that $h = (h_{n-1}, \dots, h_0)$. The signature process yields the signature σ for M by setting

$$\sigma = (\sigma_{n-1}, \dots, \sigma_0) = (sk_{n-1}[h_{n-1}], \dots, sk_0[h_0]).$$

The verification of the pair (M, σ) is accomplished as follows: the verifier computes $h = H(M)$, and then sets $h = (h_{n-1}, \dots, h_0)$. The signature is deemed correct if

$$H(\sigma_i) = pk_i[h_i], \; 0 \leq i \leq n-1.$$

Several remarks are in order here First, it is clear that the public/private key pair can be used safely only once, which accounts for the term "one-time signature." Both private and public keys have $2n^2$ bits, and the signature needs n^2 bits. For example, if one chooses SHA-256 as a hash primitive, then $n = 256$, so the signature has a length of $n^2 = 65{,}536$ bits, and the key pair $2n^2 = 131{,}072$ bits.

However, the key point is that the signature scheme just described remains safe as long as the underlying hash functions remain hard to invert. And the good piece of news is that this last statement is true even in the presence of quantum algorithms (for example, Grover's). Public key cryptography is saved at least for the time being.

2.2 Merkle Trees

Merkle trees are a device intended to open the possibility of signing several messages and builds upon the one-time signature scheme above.

The user chooses a positive integer ℓ, and generates 2^ℓ key pairs (X_i, Y_i), $0 \leq i \leq 2^\ell - 1$, following Lamport's one-time signature scheme. Equipped with these ℓ key pairs, a signer uses them up to exhaustion as long as she commits the public keys to some kind of "global public key." The latter can be accomplished by means of the so-called Merkle trees, which are binary trees such as the one shown in Fig. 1.

Let us denote the nodes as $n_h[k]$, where h is the "floor number" and k is the ordering inside a floor, from left to right, so that $0 \leq k \leq 2^{\ell-h} - 1$ for floor number h. Merkle tree will be filled in as follows:

$$\begin{aligned} n_0[k] &= H(Y_k),\ \ 0 \leq k \leq 2^\ell - 1, \\ n_h[k] &= H(n_{h-1}[2k] \mid\mid n_{h-1}[2k+1]), \end{aligned} \tag{1}$$

for $1 \leq h \leq \ell, 0 \leq k \leq 2^{\ell-h} - 1$, where $||$ is the "concatenation" operator. Once this operation is complete, the user publishes the root node value, $n_\ell[0]$, as the "global public key" and serves as a commitment of the whole set Y_i of public keys.

To sign messages, the user takes the secret keys X one at a time, until exhaustion. Assuming it is the time for the secret key X_s to be used, the user signs a message using Lamport's one-time scheme with the pair (X_s, Y_s), yielding the signature, say, σ_s. The user publishes this one-time signature along with the verification key Y_s.

In order to verify the signature, the verifier performs a standard one-time signature verification using the verification key Y_s, and then uses Merkle tree in order to verify Y_s. To assist the verifier in this process, the signer publishes the *verification path*, $V = (a_0, \ldots, a_{\ell-1})$, where each a_h corresponds to the sibling of the height-h node along the path from leaf $H(Y_s)$ to the Merkle tree root node. With this information at hand, the verifier can reproduce the computations in Eq. (1), traversing the tree upwards to the root; the verification of Y_s is successful if the final result matches the Merkle tree public key, $n_\ell[0]$.

Fig. 1 Merkle tree for $\ell = 3$

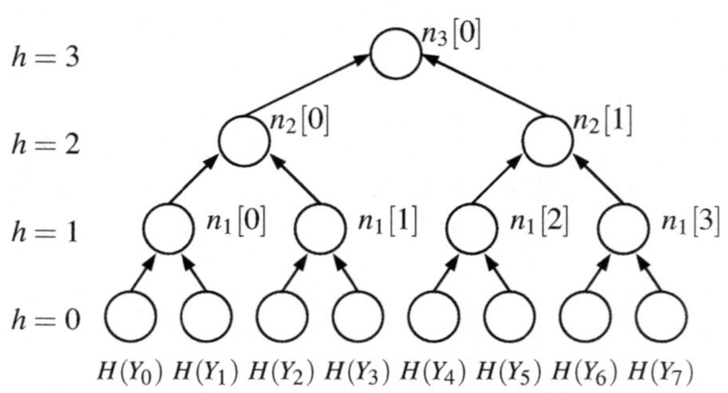

2.3 Code-Based Cryptography

The first code-based proposals can be traced back to McEliece's in 1978 [13]. The idea behind this and other similar encryption schemes stems from the world of electronic transmissions over noisy channels: most of the time the receiver is expected to receive a distorted signal, so the sender adds some extra information that depends on the data transmitted, allowing the receiver to reconstruct the original data even in the presence of transmission errors (up to a certain level).

Assuming that we transmit bits, this strategy ensures that not any bit pattern but only specific ones, termed *codewords*, are valid. These codewords are a certain "distance" apart in the code space so that any received bit pattern is either a valid codeword or can be "corrected" to a valid one, as long as the error remains below a certain bound. Often, Hamming distance is chosen, whereby two n-bit strings are at a Hamming distance of t if they differ precisely in t bits.

Translating these ideas to the cryptographic world, one can think of an encryption scheme such that a sender "disguises" the information by "adding" some noise, which only the legitimate receiver is able to prune since she is in possession of the "key", namely the right "correcting code."

The "workhorse" for a McEliece-like cryptosystem is, therefore, a (linear) correction code, $\mathcal{C}$, able to correct up to t bits in error, and possessing an efficient decoding algorithm. We assume in the sequel that we are provided with such a code, represented by a $k \times n$ matrix G, and the base field is $\mathbb{F}_2$.

The user A generates her public/private key pair as follows:

1. A chooses a matrix $S \in GL(n, \mathbb{F}_2)$ with entries at random.
2. A chooses at random a $k \times k$ permutation matrix P.
3. A computes the $k \times n$ matrix $H = PGS$.
4. A publishes (H, t) as her public key. She keeps (S, G, P) as her private key.

If a user B is willing to send a message m to A, she takes the following steps:

1. B encodes the message m as an n-bit string.
2. B computes $c^* = Hm$.
3. B generates at random $z \in \mathbb{F}_2^k$ with weight t (i.e., with exactly t bits set to 1).
4. B computes and sends to A the ciphertext $c = c^* + z$.

Remark that z plays the role of "adding errors" along the transmission, transforming c^* into a "distorted" value c.

Upon receipt of c, A takes the following steps to recover the message:

1. A computes $c' = P^{-1}c$.
2. A uses the decoding algorithm to decode c' to m'.
3. A recovers $m = S^{-1}m'$.

We can check that the deciphering is correct since

$$c' = P^{-1}c = P^{-1}c^* + P^{-1}z = P^{-1}Hm + P^{-1}z$$

$$= P^{-1}PGSm + P^{-1}z = GSm + P^{-1}z.$$

The last term means that the decoding algorithm returns $m' = Sm$, discarding the "noise" represented by $P^{-1}z$, whence $m = S^{-1}m'$.

2.4 Lattice-Based Cryptography

Given a set $\mathbf{B} = \{\mathbf{b}_1, \ldots, \mathbf{b}_n\}$, $\mathbf{b}_i \in \mathbb{R}^n$, of linearly independent vectors, a *lattice* $\mathcal{L}$, is defined as the set

$$\mathcal{L}(\mathbf{b}_1, \ldots, \mathbf{b}_n) = \left\{ \sum_{i=1}^{n} z_i \mathbf{b}_i : z_i \in \mathbb{Z} \right\}.$$

Under this definition, $\mathcal{L} \subset \mathbb{R}^n$ has the structure of a discrete additive subgroup of $\mathbb{R}^n$. We consider the standard inner product in $\mathbb{R}^n$ and the norm of any vector $\mathbf{x}$ as the standard L_2 Euclidean norm, $||\mathbf{x}||^2 = \langle \mathbf{x}, \mathbf{x} \rangle$. The set $\mathbf{B}$ is called a basis for the lattice $\mathcal{L}$, though any lattice admits multiple bases: for example, if $\mathbf{U}$ is a unimodular matrix, then $\mathcal{L}(\mathbf{B}) = \mathcal{L}(\mathbf{UB})$. In particular, the determinant of the matrix representing the base $\mathbf{B}$ is independent of the choice of $\mathbf{B}$. The seminal work of Ajtai [14] set the foundations of lattice-based cryptography defining the first two problems of those described below.

Among the best-known lattice-related problems that give rise to cryptographic applications, we can find:

1. SVP (shortest-vector problem): given a lattice, $\mathcal{L}(\mathbf{B})$, find the shortest (non-zero) vector in it.
2. CVP (closest-vector problem): given a lattice, $\mathcal{L}(\mathbf{B})$, and a target vector $\mathbf{t} \in \mathbb{R}^n$, find the vector $\mathbf{v} \in \mathcal{L}(\mathbf{B})$ closest to $\mathbf{t}$ according to the selected L_2 norm.
3. SIS (short-integer-solution problem): Let n and q be integers (n is a security parameter, and usually $q = \text{poly}(n)$), and let $\beta > 0$. Given a uniformly random matrix $A \in \mathbb{Z}_q^{n \times m}$ for some $m = \text{poly}(n)$, find a (non-zero) integer vector $\mathbf{z} \in \mathbb{Z}^m$, such that $A\mathbf{z} = \mathbf{0} \pmod{q}$ and $||\mathbf{z}|| \leq \beta$, using L_2 norm. Remark that β must be large enough to ensure that a solution does exist, but $\beta > q$ makes the problem trivially easy to solve. Ajtai [14] showed that for appropriate parameters, solving SIS on the average is (with non-negligible probability) at least as hard as approximating several lattice problems on n-dimensional lattices in the worst case (up to poly(n) factors).

4. LWE (learning-with-errors problem): Given parameters n and q as before, and a "noise rate," α, we choose $\mathbf{s} \in \mathbb{Z}_q^n$ and form pairs $(\mathbf{a}_i, b_i = \langle \mathbf{a}_i, \mathbf{s} \rangle + e_i) \in \mathbb{Z}_q^n \times \mathbb{Z}_q$, where each $\mathbf{a}_i$ is uniformly randomly selected, and e_i is picked from $\mathbb{Z}_q^m$ following a Gaussian-like distribution $\chi : \mathbb{Z}_q \to \mathbb{R}/\mathbb{Z}$ with standard deviation roughly equal to αq. The LWE problem (in its "search" version) consists in finding the secret vector $\mathbf{s} \in \mathbb{Z}_q^n$. This problem was posed by Regev [15] who showed that for certain choices of q and χ, solving LWE on the average is (with non-negligible probability) at least as hard as approximating lattice problems in the worst case to within $\tilde{O}(n/\alpha)$ factors using a quantum algorithm. Later on, similar results were proved for classical algorithms [16].

   There also exists the "decision" version of the LWE problem, whereby given pairs $(\mathbf{a}_i, b_i)$, the target is deciding whether they have been generated as above or they are truly random.

   The problem can be used to define a public key cryptosystem. The user A selects a vector $\mathbf{s} \in \mathbb{Z}_q^n$ as private key. She also picks random m vectors $\mathbf{a}_i \in \mathbb{Z}_q^n$, and m values $e_i \in \mathbb{Z}_q$ according to error distribution χ. She publishes the m pairs $(\mathbf{a}_i, b_i)$, with $b_i = \langle \mathbf{a}_i, b_i \rangle + e_i$, as her public key. If a user B is willing to send a ciphered bit to A, first selects a random subset S from the power set of $\{1, \ldots, m\}$. The encryption is defined as $(\bar{\mathbf{a}}, \bar{b}) = (\sum_{i \in S} \mathbf{a}_i, \sum_{i \in S} b_i)$ if the bit to be encrypted is 0; and $(\bar{\mathbf{a}}, \bar{b}) = (\sum_{i \in S} \mathbf{a}_i, \lfloor q/2 \rfloor + \sum_{i \in S} b_i)$ otherwise.

   To decipher, user A simply computes $\bar{b} - \langle \bar{\mathbf{a}}, \mathbf{s} \rangle$, and checks whether it is nearer to 0 (then the decrypted bit is 0) or to $\lfloor q/2 \rfloor$ (the decrypted bit is 1).

   These descriptions follow closely [17].

5. R-LWE (ring-learning-with-errors problem): It is a variant of the previous LWE aiming at optimizing the efficiency in terms of key length and speed of operation. We give hereafter a simplified version of this problem, following [18]. Let $f(x) = x^n + 1 \in \mathbb{Z}[x]$, with n a power of 2, an irreducible polynomial over the rationals. We consider the ring $R = \mathbb{Z}[x]/\langle f(x) \rangle$, the ring of integer polynomials modulo the ideal generated by $f(x)$. Technically, given $m = 2^k$, then $f(x)$ is the m-th cyclotomic polynomial, which is of degree $\varphi(m) = 2^{k-1} = n$, and R can be seen as the ring of integers of the algebraic number field $\mathbb{Q}(\zeta_m)$, where ζ_m is a primitive m-root of the unity (ζ_m and its powers co-prime to m are precisely the roots of $f(x)$). Let q be a sufficiently large prime number such that $q \equiv 1 \pmod{m}$. We now define $R_q = R/\langle q \rangle$, namely the finite ring of polynomials with integer coefficients modulo q and degree less than n.

   Fixing now a certain probability (error) distribution χ over R concentrated on "small" elements of R (informally, those with small integer coefficients) and randomly selecting an element $s \in R_q$ form a number of pairs $(a_i, b_i = a_i \cdot s + e_i)$, where each a_i has been uniformly randomly selected from R_q and each e_i has been drawn according to the error distribution χ. Given the pairs (a_i, b_i), the "search" version of the R-LWE problem consists in finding the secret element s. The "decision" problem consists in distinguishing pairs generated as above from truly random uniform pairs.

   The gain in efficiency with the variant R-LWE is twofold. First, the product $b = a \cdot s + e$ gives simultaneously n values over $\mathbb{Z}_q$, where the LWE gives only a scalar, while the cost is much smaller by using FFT multiplication techniques. Second, each pair $(a_i, b_i) \in R_q \times R_q$ can replace n samples $(\mathbf{a}_i, b_i) \in \mathbb{Z}_q^n \times \mathbb{Z}_q$, which results in an n-fold saving in the size of the public key.

To encrypt a message $m \in \{0, 1\}^n$, we embed it in R via "coefficient embedding." Then we draw "small" elements $r, e_1, e_2 \in R$ with the error distribution and consider the ciphertext as the pair $(u, v) \in R \times R$ given by

$$u = a \cdot r + e_1 \quad \text{and} \quad v = b \cdot r + e_2 + \lfloor q/2 \rceil \cdot m.$$

To decipher, the legitimate user computes

$$v - u \cdot s = (r \cdot e - s \cdot e_1 + e_2) + \lfloor q/2 \rceil \cdot m.$$

If the parameters of the system are wisely selected, the coefficients of $r \cdot e - s \cdot e_1 + e_2$ are smaller than $q/4$ so the bits of m can be recovered rounding each coefficient in $v - u \cdot s$ to either 0 or $\lfloor q/2 \rceil$, whichever is closer.

The interesting point is that all the previous described problems are proved to be $\mathcal{NP}$-hard.

2.5 *Multivariate-Quadratic-Equations Cryptography*

Another vein of hard problems stems from the world of Algebraic Geometry and gives rise to the so-called Multivariate Quadratic Cryptography. Introduced by Matsumoto and Imai in 1988 [19], it is based on the difficulty of the $\mathcal{MQ}$ problem.

The $\mathcal{MQ}$ problem over a finite field of q elements $\mathbb{F}_q$ consists in finding a solution $\mathbf{x} \in \mathbb{F}_q^n$ to a given system of m quadratic polynomial equations $\mathbf{p} = (p_1, \ldots, p_m)$ over $\mathbb{F}_q$ in n indeterminates, where each polynomial p_k is of the form:

$$p_k(x_1, \ldots, x_n) = \sum_{i \geq j} a_{ijk} x_i x_j + \sum_i b_{ik} x_i + c_k,$$

for $1 \leq k \leq m$, and all the coefficients a_{ijk}, b_{ik}, c_k, in $\mathbb{F}_q$.

This problem is $\mathcal{NP}$-complete over any finite field for a randomly selected polynomial vector $\mathbf{p} = (p_1, \ldots, p_m)$ ([20, p.251]; [21, section 2.5]; [22]). However, a real-life cryptosystem needs some kind of structure in order to implement the "trap-door" but even in this case, current research seemingly suggests that $\mathcal{MQ}$ problem is also hard on average (see [23], [24]).

$\mathcal{MQ}$ problem gives rise both to encryption and signature schemes. A simple encryption scheme works based upon a *central map*, an easy-to-invert quadratic function $\mathcal{Q}\colon \mathbb{F}_q^n \to \mathbb{F}_q^m$, which plays the role of a private key. In order to hide the central map, we choose two (usually affine) invertible transformations, $S\colon \mathbb{F}_q^n \to \mathbb{F}_q^n$

Fig. 2 The (blue) public and the (red) private "routes"

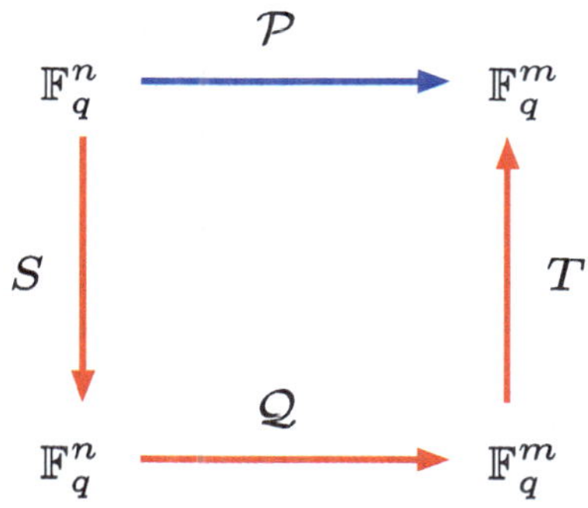

and $T: \mathbb{F}_q^m \to \mathbb{F}_q^m$. The public key is the composition of those maps, namely $\mathcal{P} = T \circ \mathcal{Q} \circ S$. Figure 2 depicts the composition.

The ciphering process consists in evaluating the polynomial $\mathcal{P}$ over a vector $\mathbf{m}$ that represents the message in order to get the ciphertext $\mathbf{c} = \mathcal{P}(\mathbf{m})$. An adversary willing to break the system must needs solve an instance of the $\mathcal{MQ}$ problem to recover the message. For the legitimate receiver this task is very easy, since she knows the secret key (the decomposition of $\mathcal{P}$) so she can compute S^{-1}, T^{-1}, and $\mathcal{Q}^{-1}$, and carry out the deciphering process:

$$\left(S^{-1} \circ \mathcal{Q}^{-1} \circ T^{-1}\right)(\mathbf{c}) = \left(S^{-1} \circ \mathcal{Q}^{-1} \circ T^{-1} \circ \mathcal{P}\right)(\mathbf{m})$$
$$= \left(S^{-1} \circ \mathcal{Q}^{-1} \circ T^{-1} \circ T \circ \mathcal{Q} \circ S\right)(\mathbf{m}) = \mathbf{m}.$$

A simple signature scheme is also possible: Consider a hash function $H: \{0, 1\}^* \to \mathbb{F}_q^m$, and let A be a user with public key $\mathcal{P}$ and corresponding private key $(S, \mathcal{Q}, T)$. Assume that A is willing to sign a message $\mathbf{m}$: Then A first computes $h = H(\mathbf{m}) \in \mathbb{F}_q^m$, and obtains the signature σ by applying the composition $\left(S^{-1} \circ \mathcal{Q}^{-1} \circ T^{-1}\right)(h) = \sigma \in \mathbb{F}_q^n$.

Suppose a verifier is given the pair $(\mathbf{m}, \sigma)$. The verification process consists of applying the public key to σ, thus obtaining $\bar{h} = \mathcal{P}(\sigma)$. The verifier considers the signature valid if and only if $\bar{h} = H(\mathbf{m})$.

2.6 *Elliptic-Curve-Isogeny-Based Cryptography*

The core of this problem and its application to cryptography is rather involved, and its description would require too much space. We content ourselves with citing it in the context of the class of problems that are seemingly quantum-resistant.

3 Chor-Rivest Knapsack Cryptosystem

Chor and Rivest proposed a cryptosystem based on knapsack problem and on arithmetic in finite fields [7]. It needs to compute discrete logarithms in order to determine the keys of the system. One of the advantages of this cryptosystem is that the Chor-Rivest knapsack density is greater than 1. This made difficult low density-based attacks [25]. Although the system was broken for some specific parameters [26] enabling the recovery of the private key from the public one, it remains unbroken for some other parameters. The knapsack problem is considered a "decision" $\mathcal{NP}$-complete problem [27].

For the sake of brevity, we will denote Chor-Rivest cryptosystem simply as C-RC.

3.1 *Key Generation System*

To define the C-RC, the following parameters must be considered (see [7]):

- The power of a primer number, $q = p^{\lambda}$, and a positive integer $h \leq q$ such that the calculation of logarithms in $\mathbb{F}_{q^h}$ can be efficiently carried out.
- A root $t \in \mathbb{F}_{q^h}$ from an irreducible monic polynomial of degree h, $F(x) \in \mathbb{F}_q[x]$. Any element of $\mathbb{F}_{q^h}$ can be represented as a polynomial in t with degree $\leq h$ and coefficients in $\mathbb{F}_q$.
- A generator, g, of the multiplicative group $\mathbb{F}^*_{q^h}$. Observe, in passing, that the order of $\mathbb{F}^*_{q^h}$ is $n = q^h - 1$.
- The following q logarithms that are to be computed:

$$a_i = \log_g (t + \alpha_i),$$

  for all $\alpha_i \in \mathbb{F}_q$, $0 \leq i \leq q - 1$.
- The elements a_i are reordered by a random permutation

$$\pi : \{0, 1, \ldots, q-1\} \to \{0, 1, \ldots, q-1\},$$

  in such a way that $b_i = a_{\pi(i)}$.
- A noise is added, considering a randomly chosen integer number $0 \leq r \leq q^h - 2$, and then

$$c_i \equiv (b_i + r) \pmod{q^h - 1}, \quad i = 0, 1, \ldots, q - 1.$$

- The user's *public key* is the set $\{c_0, c_1, \ldots, c_{q-1}, q, h\}$.
- The user's *private key* is the set $\{t, g, \pi, r\}$.

Thus, the key size in this cryptosystem is dependent on the size of the parameters originally considered.

3.2 Encryption Process

To cipher a message, M, it must first be converted into a vector of q bits with weight h, namely with exactly h bits set to 1,

$$M = \left(x_0, x_1, \ldots, x_{q-1}\right), \quad x_i \in \{0, 1\}, \quad i = 0, \ldots, q-1.$$

The ciphertext, C, corresponding to M is calculated by adding the values c_i for which $x_i = 1$, i.e.,

$$C = \sum_{i=0}^{q-1} x_i \cdot c_i \pmod{q^h - 1}.$$

3.3 Decryption Process

In order to decrypt a ciphertext C, assuming that it comes from a valid q-bit message with weight h, the following steps are considered:

1. Compute $C' \equiv C - h \cdot r \pmod{q^h - 1}$.
2. Obtain $g^{C'}$ written as a polynomial in x. There exists a unique polynomial $Q(x) \in \mathbb{F}_{q^h}[x]$ with degree $\leq h - 1$, such that

$$Q(x) \equiv g^{C'} \pmod{F(x)}.$$

   Naming as $I = \{i_1 < \cdots < i_h\}$ the set of indices with corresponding bits equal to 1, i.e., $x_{i_1} = \cdots = x_{i_h} = 1$, then:

$$g^{C'} = g^C \cdot g^{-h \cdot r} = g^{\sum_{i=0}^{q-1} x_i \cdot c_i} \cdot g^{-h \cdot r} = \prod_{i \in I} g^{c_i - r}$$

$$= \prod_{i \in I} g^{b_i} = \prod_{i \in I} g^{a_{\pi(i)}} = \prod_{i \in I} \left(t + \alpha_{\pi(i)}\right).$$

3. As the polynomial with degree h, $F(x) + Q(x)$ factorizes linearly in the field $\mathbb{F}_q$, then

$$F(x) + Q(x) = \prod_{i \in I} \left(x + \alpha_{\pi(i)}\right). \tag{2}$$

Replacing all values $\alpha_0, \alpha_1, \ldots, \alpha_{q-1} \in \mathbb{F}_q$, the h roots of that polynomial are obtained. If those roots are $\alpha_{j_1}, \ldots, \alpha_{j_h}$, then applying the inverse permutation π^{-1} to those roots indices, $\pi^{-1}(j_l) = i_l$, subindices of the message with terms equal to 1 are obtained.

3.4 Safe Parameters

Several attacks are known to break C-RC but, in practice, they succeed only for the system parameters originally proposed by the authors in 1988. We mention some of them hereafter.

Chor and Rivest [7] describe both *specialized attacks*, where the attacker knows parts of the recipient's secret key, such as the one designed by Goldreich and Odlyzko; and *general attacks*, where only the public key is available. Among the latter, we can count the brute-force attack designed by Brickell [28], low-density knapsack attacks such as the one performed by Lagarias and Odlyzko [25], and others such as the one by Schnorr and Hörner [29].

The pairs of parameters (q, h) originally proposed for C-RC by Chor and Rivest were $(197, 24)$, $(211, 24)$, $(3^5, 24)$, and $(2^8, 25)$, whose number of digits are, respectively, 56, 56, 58, and 60. If the number of digits is small enough, as is the case for the pairs $(103, 12)$ and $(151, 16)$, the Schnorr-Hörner attack is capable of breaking the cryptosystem [29].

In 2001, Vaudenay [8] was the first to really break C-RC under a much broader range of parameters (including those proposed by Chor and Rivest). Vaudenay considers the finite field $\mathbb{F}_{q^h}$, q being a prime or the power of a prime number, and h an integer. His attack is remarkable since he exploits the fact that this C-RC has *equivalent* secret keys, and he is able to recover one of them. An *equivalent* key is (in general) different than the original secret key, but it works the same as it were the right one. All users can use the same values for q and h because the risk of collisions (that is, that two users have the same password) is extremely small. As the private key is made up of the set $\{t, g, \pi, r\}$, there are

$$h \cdot \phi(q^h - 1) \cdot (q^h - 2) \cdot q!$$

different private keys, from which $h \cdot q(q-1)$ are equivalent, where ϕ is the Euler's indicator. Then, the number of non-equivalent keys is

$$n_{q,h} = \phi(q^h - 1) \cdot (q^h - 2) \cdot (q-2)!$$

Hence, for a set of k users the number of possible collisions will be

$$\frac{(n_{q,h}-1)(n_{q,h}-2)\cdots(n_{q,h}-k+1)}{n_{q,h}^k},$$

which is a small number, given that $1/n_{q,h}$ is an upper bound of that value. As an example, for the pair of parameters $q = 197$, $h = 24$, the bound could be calculated as

$$\frac{(n_{q,h}-1)(n_{q,h}-2)\cdots(n_{q,h}-k+1)}{n_{q,h}^k} < 0.16155 \cdot 10^{-472}.$$

However, all the attacks just described are able to break the C-RC *only* if the original parameters (remember they date back to 1988!) or parameters showing certain properties (in the case of Vaudenay's) are used. This opens the vein of exploring either parameters lying in broader ranges or parameters selected to circumvent the known attacks.

In [9, 10] it is proven that several pairs of parameters (q, h) could be found verifying the conditions established by C-RC and making the cryptosystem strong against the known attacks, namely:

1. $2 \leq h \leq q$, with q a prime number, and h a prime number or the square of a prime number,
2. the number of digits of n is $t(q, h) \geq 36$,
3. the bitlength of the public key satisfies $l(q, h) < 15,000$,
4. the density of the knapsack, $d(q, h)$, is greater than 1, and
5. the smoothness of n, $u(n)$, has at most 18 decimal digits.

The pairs verifying these conditions are included in Table 1. Remark that the pair (1123, 13) exhibits the longest public key size, which is on the order of 150 kilobits. Such a size could be deemed too large in 2008 or 2009, but this is no longer the case: It suffices to observe that the key sizes of the proposals presented to the second round of the NIST call [5] are on the order of several megabytes.

3.5 Experimental Results

We have conducted an experiment in order to assess the computation time required to perform an encryption and a subsequent decryption of messages with various representative bit sizes, and system parameters.

We implemented the protocol with Magma computer algebra system [30], in particular using its language with version Magma V2.20-10. The Magma programs were run over an Intel Core i7/860 at 2.80 GHz with 12 GB of RAM under Windows 10 Pro using a single running thread.

We chose three message sizes: a "small-sized" message, with about 1 KB; a "medium-sized" message, with about 200 KB; and finally a "large-sized" message,

Table 1 Values of the pairs (q, h) verifying the conditions (1)–(5)

h	q	$t(q,h)$	$l(q,h)$	$d(q,h)$	$u(n)$	h	q	$t(q,h)$	$l(q,h)$	$d(q,h)$	$u(n)$
13	547	36	64, 546	4.626	14	13	1123	40	147, 113	8.524	16
13	571	36	67, 949	4.796	13	17	127	36	14, 986	1.068	15
13	577	36	68, 663	4.838	14	17	131	36	15, 589	1.095	15
13	599	37	71, 281	4.994	13	17	167	38	20, 875	1.330	14
13	601	37	72, 120	5.008	11	17	193	39	24, 897	1.495	17
13	613	37	73, 560	5.092	15	17	233	41	30, 989	1.742	15
13	631	37	75, 720	5.218	11	17	263	42	35, 768	1.924	15
13	641	37	77, 561	5.288	17	17	277	42	37, 949	2.008	15
13	659	37	79, 739	5.413	16	17	317	43	44, 697	2.244	17
13	683	37	83, 326	5.579	14	17	331	43	47, 002	2.326	17
13	719	38	88, 437	5.828	15	17	409	45	60, 123	2.773	9
13	757	38	93, 868	6.088	16	17	433	45	64, 084	2.908	18
13	787	38	98, 375	6.292	15	17	587	48	91, 572	3.754	14
13	797	38	99, 625	6.360	13	17	643	48	101, 594	4.054	17
13	839	39	105, 714	6.644	11	17	661	48	105, 099	4.150	16
13	877	39	111, 379	6.900	16	23	173	52	29, 410	1.011	18
13	887	39	112, 649	6.967	16	23	191	53	33, 234	1.095	18
13	941	39	120, 448	7.327	14	23	199	53	34, 825	1.132	15
13	953	39	121, 984	7.407	16	23	283	57	52, 921	1.510	16
13	967	39	123, 776	7.500	14	23	563	64	118, 230	2.679	14
13	977	39	126, 033	7.566	16	25	601	70	138, 230	2.604	17
13	1093	40	143, 183	8.329	13	25	613	70	141, 603	2.648	16

Table 2 Encryption/decryption times for various (q, h) and message sizes

Size	Action	(167, 17)	(409, 17)	(631, 13)	(839, 13)
1 KB	Enc time	0.042	0.075	0.090	0.120
1 KB	Dec time	0.153	0.138	0.106	0.109
200 KB	Enc time	8.636	12.547	16.975	23.655
200 KB	Dec time	30.858	25.048	23.019	24.878
1 MB	Enc time	52.895	68.667	106.403	147.969
1 MB	Dec time	185.998	153.794	147.488	158.383

with some 1 MB. In most common situations, one would use a "small-sized" message in order to setup a key exchange (for example, to implement a key encapsulation mechanism). We selected four different configuration for the system parameters (q, h).

Table 2 summarizes the results: The computation times are measured in seconds. Each pair of lines represents the encryption time (Enc time) and decryption time (Dec time) for a particular message size and four different system parameters: (167, 17), (409, 17), (631, 13), and (839, 13). The interesting point is that for the "small-sized" message, both encryption and decryption times are on the order of one tenth of a second, whereas the system remains of little use for "large-sized" messages.

4 Discussion and Future Work

Modern advances in quantum computers pose a looming threat to current cryptosystems, those based on (classically) intractable problems, since quantum algorithms do exist that solve in polynomial time (= fast) those problems. This fact has triggered the interest and research towards quantum-resistant problems that could act as a "drop-in replacement" to safeguard the security and strength of cryptography.

In this paper, we have introduced some of the problems currently considered as quantum-resistant, which make up the building blocks of the "Post-Quantum Cryptography."

However, we feel that in this new scenario C-RC has not received proper attention and its security has not been analyzed in due detail. For one thing, the mathematical problems in which C-RC are based upon have not been considered by the NIST Call in [4]. It is true that the majority, if not all, knapsack-type cryptosystems has been broken at some point, a fact that may account for the distrust currently elicited by this type of cryptosystems.

The interesting point is that C-RC is a knapsack-type cryptosystem whose security is not based upon discrete logarithms, even though it needs to compute them in order to set up the system. Actually, any reduction in the computation time of such logarithms means an advantage for C-RC, since both the public and the private keys could be obtained in a much shorter time-frame. Still better, we are allowed to use much longer (thus safer) key sizes if the advent of the universal quantum computer is ever realized: Shor's algorithm [1] would permit users to reach key sizes not dreamed of when the system was originally proposed. In fact, one of its drawbacks was precisely the necessity of computing such discrete logarithms, for which the only known "efficient" algorithms were "Baby step-Giant step," or Pohlig-Hellman's [27]. Since these algorithms exhibit a subexponential running time, the use of long and safe key lengths was precluded in practice. All these facts may very well make C-RC deserve a proper place at the post-quantum candidates dinner table. In view of the latter, we suggest several research lines to be considered:

- To increase the key sizes in knapsack-like cryptosystems so as to be safe in the face of quantum computation while keeping efficiency in the key generation process.
- Analyze the security of this cryptosystem endowed with new, revamped parameter sizes in front of the threat posed by quantum computers.
- Improve the key generation process so as to thwart attacks based on equivalent keys (such as Vaudenay's) even in the presence of quantum computers.

Acknowledgments This work has been partially supported by Ministerio de Economía, Industria y Competitividad (MINECO), Agencia Estatal de Investigación (AEI), and European Regional Development Fund (ERDF), through project COPCIS, grant no. TIN2017-84844-C2-1-R, and by Comunidad de Madrid (Spain) through project CYNAMON, grant no. P2018/TCS-4566-CM, co-funded with ERDF.

References

1. P. Shor, Polynomial-time algorithms for prime factorization and discrete logarithms on a quantum computer. SIAM J. Comput. **26**, 1484–1509 (1997)
2. L. Grover, Quantum mechanics helps in searching for a needle in a haystack. Phys. Rev. Lett. **79**, 325–328 (1997)
3. D. Simon, On the power of quantum computation. SIAM J. Comput. **26**(5), 1474–1483 (1997)
4. NIST, Post-quantum cryptography. On-line publication. 2017. https://csrc.nist.gov/Projects/Post-Quantum-Cryptography
5. NIST, Post-quantum cryptography, 2nd round. On-line publication, 2019. https://csrc.nist.gov/projects/post-quantum-cryptography/round-2-submissions
6. B.-Z. Chor, Two Issues in Public Key Cryptography. RSA Bit Security and a New Knapsack Type System. ACM Distinguished Dissertation. The MIT Press, Cambridge, MS (1986)
7. B. Chor, R. Rivest, A knapsack-type public key cryptosystem based on arithmetic in finite fields. IEEE Trans. Inform. Theor. **34**(5), 901–909 (1988)
8. S. Vaudenay, Cryptanalysis of the Chor-Rivest cryptosystem. J. Cryptol. **14**, 87–100 (2001)
9. L. Hernández Encinas, J. Muñoz Masqué, A. Queiruga Dios, Safer parameters for the Chor-Rivest cryptosystem. Comput. Math. Appl. 56, 2883–2886 (2008)
10. L. Hernández Encinas, J. Muñoz Masqué, A. Queiruga Dios, Analysis of the efficiency of the Chor-Rivest cryptosystem implementation in a safe-parameter range. Inf. Sci. **179**, 4219–4226 (2009)
11. D. Bernstein, J. Buchmann, E. Dahmen (eds.), *Post-quantum Cryptography* (Springer, Berlin, Heidelberg, 2009)
12. L. Lamport, Constructing digital signatures from a one way function. SRI International Computer Science Laboratory, Technical Report SRI-CSL-98 (1979)
13. R.J. McEliece, A public-key cryptosystem based on algebraic coding theory. Jet Propulsion Laboratory, Technical Report 42–44 (1978)
14. M. Ajtai, Generating hard instances of lattice problems (extended abstract), in *Proceedings of the Twenty-Eighth Annual ACM Symposium on Theory of Computing, STOC '96* (Association for Computing Machinery, New York, NY, 1996), pp. 99–108
15. O. Regev, On lattices, learning with errors, random linear codes, and cryptography. J. ACM 56(6), 1–40 (2009). Art. no. 34
16. C. Peikert, Public-key cryptosystems from the worst-case shortest vector problem: extended abstract, in *Proceedings of the Forty-First Annual ACM Symposium on Theory of Computing, STOC '09* (Association for Computing Machinery, New York, NY, 2009), pp. 333–342
17. D. Micciancio, C. Peikert, Hardness of SIS and LWE with small parameters, in *Advances in Cryptology – CRYPTO 2013*, ed. by R. Canetti, J.A. Garay (Springer, Berlin, Heidelberg, 2013), pp. 21–39
18. V. Lyubashevsky, C. Peikert, O. Regev, On ideal lattices and learning with errors over rings. J. ACM **60**(6), 1–35 (2013)
19. T. Matsumoto, H. Imai, Public quadratic polynomial-tuples for efficient signature-verification and message-encryption. Lect. Notes Comput. Sci. **330**, 19–453 (1988)
20. M.R. Garey, D.S. Johnson, *Computer and Intractability: A Guide to the Theory of NP-Completeness* (W. H. Freeman & Co, New York, 1990)

21. C. Wolf, Multivariate quadratic polynomials in public key cryptography, Ph.D. dissertation, Katholieke Universiteit Leuven, November (2005)
22. J. Patarin, Cryptanalysis of the Matsumoto and Imai public key scheme of Eurocrypt '88. Lect. Notes Comput. Sci. **963**, 248–261 (1995)
23. N. Courtois, A. Klimov, J. Patarin, A. Shamir, Efficient algorithms for solving overdefined systems of multivariate polynomial equations. Lect. Notes Comput. Sci. **1807**, 392–407 (2000)
24. N. Courtois, L. Goubin, W. Meier, J.-D. Tacier, Solving underdefined systems of multivariate quadratic equations. Lect. Notes Comput. Sci. **2274**, 211–227 (2002)
25. J.C. Lagarias, A.M. Odlyzko, Solving low-density subset sum problems. J. ACM **32**, 229–246 (1985)
26. A.M. Youssef, Cryptanalysis of a knapsack-based probabilistic encryption scheme. Inf. Sci. **179**(18), 3116–3121 (2009)
27. A.J. Menezes, P.C. Van Oorschot, S.A. Vanstone, *Handbook of Applied Cryptograpay* (CRC Press, Boca Raton, 1996)
28. E. Brickell, Solving low density knapsacks, in *Proceedings of Crypto '83* (Plenum Press, New York, 1984), pp. 25–37
29. C.-P. Schnorr, H. H. Hörner, Attacking the Chor-Rivest cryptosystem by improved lattice reduction, in *International Conference on the Theory and Applications of Cryptographic Techniques* (Springer, New York, 1995), pp. 1–12
30. W. Bosma, J.J. Cannon, C. Playoust, The Magma Algebra System I: the user language. J. Symb. Comput. **24**(3–4), 235–265 (1997)

An Effective Tool for Assessing the Composite Vulnerability of Multifactor Authentication Technologies

Adam English and Yanzhen Qu

1 Introduction

Information systems have become centric to many business and government sectors, crossing industries and maintaining a growing adoption level rate [1]. A central concept in cyber security and information security is the Confidentiality, Integrity, and Availability (CIA) triad [2]. The National Institute of Standards and Technology (NIST) now has been continuously evolving and enhancing cyber security principles to include characterization of risk. Authentication is critical in the protection of information systems, with username-and-password-based authentication representing a large array of successful breaches into systems [3].

Multifactor authentication was the latest evolution in authentication technology for information systems, reducing the capability for credential compromises [4]. Multifactor authentication with recent studies and technologies present novel and potentially enhanced means in providing secure information systems and cyber infrastructure [5]. For example, the smartcard-based password authentication, the technology set implemented through Public Key Infrastructure (PKI) smartcard tokens combined with a passphrase or password, has been widely considered as a material evolution from username and password authentication. However, along with the implementations of the multiple technology sets the complicity and anecdotal cases indicate user dissatisfaction and an array of human actions resulting in compromise of one or more of the authentication factors. Multifactor authentication is neither limited by a two-factor implementation nor is it confined by a singular implementation, rather that it is comprised of a wide range of technologies with each of them addressing one or more specific vulnerabilities. The superset of

A. English · Y. Qu (✉)
Colorado Technical University, Colorado Springs, CO, USA
e-mail: adam.english5@student.ctuonline.edu; yqu@coloradotech.edu

K. Daimi et al. (eds.), *Advances in Security, Networks, and Internet of Things*, Transactions on Computational Science and Computational Intelligence,
https://doi.org/10.1007/978-3-030-71017-0_7

all the vulnerabilities relating to the multiple factors in multifactor authentication is considered as the composite vulnerability throughout this chapter. That is semantically the aggregation of all the vulnerabilities generated from multiple factors in a specific technical architecture, and implementation of a multifactor authentication is called the composite vulnerability of that specific multifactor authentication technology.

Authentication is a prerequisite in role-based access control (RBAC) and attribute-based access control (ABAC) technologies, representing a core cybersecurity principle. Contemporary research and cyber practitioners argue in favor of multiple authentication factors beyond the commonly implemented two-factor authentication solutions [6] with a movement toward continuous authentication integration within systems and inclusion of several concurrent authentication modalities [7]. Therefore, a concise and grounded comparative analysis of authentication factor multiplicity is needed in order to address concerns regarding the relevancy of authentication factor technologies to meet or exceed NIST governance and standards.

To achieve this purpose, we have made a novel extension to the Common Vulnerability Scoring System (CVSS) calculator to enable it naturally applicable to any multifactor authentication technologies. CVSS is an industry standard vulnerability impact rating mechanism, providing an objective evaluation of technologies and threat vectors resulting in relative vulnerability scores and determination of the vulnerability impacts. This extension work has provided a tool to effectively determine composite vulnerability scores when the authentication factor multiplicity is scaled within systems, in both theoretical and real-world scenarios.

This chapter is structured as follows. The second section explores the contemporary-related research. The third section provides the problem statement for the research and presents the hypothesis and research question. The fourth section describes the methodology for the work. The fifth section shows an application example of conducting a comparative study by using the extension to calculate the base composite vulnerability scores of a two-factor authentication and a three-factor authentication. The final section presents the conclusion and future work.

2 Related Work

2.1 Prior Research on CVSS

The common vulnerability scoring system (CVSS) is a standardized mechanism for identifying and assessing vulnerabilities within an open framework [8]. The Forum of Incident Response and Security Teams (FIRST) provides and maintains the CVSS toolkit [9]. CVSS v2.0 was comprised of the base vulnerability metrics translated into a score and assigned to a vector, with the recommended addition of temporal and environmental metrics providing enhancement to contextual data and accuracy of the CVSS scores [8]. FIRST, cyber researchers, and industry experts recognized gaps within CVSS 2.0, resulting in the publication and release of CVSS 3.0.

In CVSS v3 calculations, the base metric of access vector became attack vector, maintaining similar characteristics, though incorporating in network distance as a factor in the metric [10, 11]. Attack complexity became two separate metrics; the new access complexity metric comprised of the condition in components outside of adversarial control necessitated for a successful exploitation and the necessity for human interaction, resulting in a new user interaction metric [10]. CVSS v3.0 further altered the authentication metric to the privileges required metric, defining the level of access adversaries required for exploitation rather than the number of authentication attempts [11]. Since CVSS v3.0 represented enhancements over the CVSS v2.0 release, importance on contextual metrics diminished, materially reducing the influence of temporal metrics while the additional environment metrics introduced mitigating and compensating control factors. The evolution from CVSS v2.0 to CVSS v3.0 increased the average base metric scoring from a 6.5 to 7.4, representing an increased awareness of cyber security adversarial advancements [12]. However, the number of critical-, medium-, and low-rated vulnerabilities diminished when evaluated with CVSS v3.0 compared with CVSS v2.0, while high vulnerability categorization experienced a significant increase [12].

Cyber professionals and researchers have provided empirical-based studies regarding the scaling difficulties with CVSS application as network complexity increases, noting the need for extension of CVSS calculations [13]. Researchers applied CVSS scoring systems in theoretical determinations for modern authentication and confidentiality operations such as block chaining, demonstrating a high resistance to common vulnerabilities in the technology sets, demonstrating benefits in vulnerability reductions with modern and advanced authentication modalities [14]. Contemporary researchers determined CVSS scoring presents inadequacies in the incorporation of emerging exploitations, noting that CVSS v3 variants provided greater mitigation definition capabilities than previous versions and recommending the inclusion of Jaccard similarity metrics against the CVSS vector and the associated National Vulnerability Database (NVD) values [15]. The industry perspective and reputation of CVSS continues to advance as CVSS scoring accounts for an increased array of relevant variables with a demonstrated need for CVSS v3 variants to provide a means of evaluating technologies, effectively facilitating comparative analysis.

2.2 *Prior Research on Multifactor Authentication*

Authentication provides an initial line of defense in cyber security, servicing the scheme of identity verification to the information system prior to authorization to resources within that information system [16]. Authentication operations begin with an entity's digital identity; the unique aspects of the subject accessing digital resources constrained in the defined authenticator assurance level [17].

Authentication factors are characterized as essential data points used to validate user identities within the boundaries of an information system or organization, commonly grouped into three categories: knowledge-based being what the user knows, such as passwords; possession-based being what the user owns, such as smartcards or RSA tokens; and inherence-based being what or who the user is, commonly referred to as biometrics [18].

Multifactor authentication has demonstrated capability to provide heightened security, resiliency, and robustness to access control methods, validating legitimate users of an information system while increasing mitigations of unauthorized access attacks and vulnerability exploitations [19]. A common implementation of multifactor authentication involves password authenticated key exchange schemes, where a level of efficiency equivalent to single-factor authentication methods exists while ensuring a high level of security where if one authentication factor is successfully compromised, the user identity has not been compromised and authentication credentials do not require modification in a single factor compromise scenario [20]. The advancements in multifactor authentication with multiplicity of factors provide a safer and more secure computing environment for users and organizations, though at the expense of other organizational considerations.

3 Problem Statement, Hypothesis, and Research Question

3.1 Problem Statement

The standard CVSS v3 equations do not provide a mechanism for evaluating composite scores of full-spectrum authentication mechanisms exemplified through multiplicity in multifactor authentication technologies, preventing an accurate characterization of risk aiding in selection of authentication modalities for information systems.

3.2 Hypothesis

If we can extend the CVSS v3 calculator through introduction of a mathematical formula, enabling objectively and determining the composite vulnerability scores of multifactor authentication technologies, we will be able to accurately predict and characterize risk to systems as multiplicity of authentication factors increases.

3.3 Research Question

What will be a simple mathematical formula that can be applied to any metric of CVSS v3 calculator to calculate the aggregated value of any type of multifactor authentication technology?

4 Methodology

In this section, we will present our extension to the CVSS v3 calculator. We will first present the definitions of relevant concepts through set theory. Then we will provide the relevant mathematical formula and show an example to demonstrate the usage of the formula.

4.1 Definitions

One of the key concepts used in the CVSS v3 calculator is the attack vector, which defines the context that a vulnerability exhibits exploitation capability. Attack vector (AV) is one metric within the CVSS v3 calculator within the Base Score category. While AV itself does not comprise the CVSS score itself, it represents a metric with multiple metric value options, and the equation demonstrated is applicable throughout all metrics in the CVSS v3 calculator. Therefore, if we can extend the calculation of AV to cover the impact of multiple factors of an authentication mechanism, a similar approach can be applied to all other parameters used by CVSS v3 calculator.

Definition 1
Let us assume we have the following:

Attack Vector Set $A = \{a\_1, \ldots, a\_i, \ldots, a\_n\}$, where a_i is the ith element in the A, and $i = 1, 2, \ldots\ldots, n$, such that n is an integer.

Numerical Value Set $V = \{v\_1, \ldots, v\_i, \ldots, v\_n\}$, where v_i is the ith numerical value for the ith element in A, and $i = 1, 2, \ldots\ldots, n$, such that n is an integer, and $0 < v\_i < 1$.

One example of Attack Vector is shown in Table 1. Based on the table, we can have the following sets:

$$A = \{\text{Network, Adjacent, Local, Physical}\} \quad (1)$$

$$V = \{0.85, 0.62, 0.55, 0.2\} \quad (2)$$

Table 1 An example of Attack Vector

Attack Vector metric components	
Metric value	*Numerical value*
Network	0.85
Adjacent	0.62
Local	0.55
Physical	0.2

Definition 2

The existing CVSS v3 calculator allows users to select a singular classification in the metric value for each metric; this has limited specification for complex attack vectors and combination technologies. Therefore, to work around this limitation, we have proposed a concept called "Impact by the Factor" to allow the impact of each factor in a multifactor authentication technology to be considered for each element of an attack metric such as an AV, so that an aggregated vulnerability impact score generated as the singular attack metric value can be provided per metric of CVSS calculator. To avoid the definition of a multitude of metric and numerical values, we introduce the "Factor Set F". This allows for optimization of the calculation. That is, we will define the following:

Factor Set $F = \{f\_1, \ldots, f\_j, \ldots, f\_m\}$, where f_j represents the jth factor in a multifactor authentication technology, and $j = 1, 2, \ldots, m$, and m is an integer.

Factor's Impact Set $Qj = \{q\_j1, ..., q\_jn\}$, where q_ji represents the jth factor's impact to the ith optional value of the parameter, and $\sum(i = 1 \text{ to } n)\ q\_ji = 1$.

For example, if we have a three-factor authentication mechanism, we denote the three factors simply as Factor1, Factor2, and Factor3. We assume the factors make the following impacts to the metric as shown in Table 1:

- Factor1: made 50% impact to Network, 50% to Local
- Factor2: made 30% impact to Adjacent, 30% impact to Local, and 40% impact to Physical
- Factor3: made 30% impact to Network, 10% to Adjacent, 20% to Local, and 40% to Physical

Based on the above, we can have

$$F = \{\text{Factor1}, \text{Factor2}, \text{Factor3}\} \tag{3}$$

$$\text{Q1} = \{50\%, 0, 50\%, 0\} \tag{4}$$

$$\text{Q2} = \{0, 30\%, 30\%, 40\%\} \tag{5}$$

$$\text{Q3} = \{30\%, 10\%, 20\%, 40\%\} \tag{6}$$

4.2 Math Formula for the Aggregation

By using the two definitions above, we can have the following two math formula for the aggregation of the impact made by all the factors in a multifactor authentication technology.

First, if we use T to represent the aggregated impact of all factors toward the ith optional value v_i, where $i = 1, 2, \ldots, n$, and n is an integer, then we have

$$T = \sum (j = 1 \text{ to } m)\ q\,ji\Big)/m \tag{7}$$

where q_ji is the jth element of Factor Impact Set Qj, m is an integer representing the number of authentication factors.

Then, if we use X to represent the "composition value" for the entire Attack Vector Set AV, then we have:

$$X = \sum (i = 1 \text{ to } n)\ \Big\{vi * \Big[\Big(\sum (j = 1 \text{ to } m)\ q\,ji\Big)/m\Big]\Big\} \tag{8}$$

where v_i is the ith element of Optional Value Set V, q_ji is the jth element of Factor Impact Set Qj, n and m are both an integer.

To demonstrate how these formulas can be used, let us apply the values of (1), (2), (4), (5), and (6) into (7), we can get

$$\text{T (Network)} = [(50\% + 0 + 30\%)/3] = 0.267 \tag{9}$$

$$\text{T (Adjacent)} = [(0 + 30\% + 10\%)/3] = 0.133 \tag{10}$$

$$\text{T (Local)} = [(50\% + 30\% + 20\%)/3] = 0.333 \tag{11}$$

$$\text{T (Physical)} = [(0 + 40\% + 40\%)/3] = 0.267 \tag{12}$$

Then, if we apply the values of (2) and the results of (9), (10), (11), and (12) into (8), we can get

$$X = (0.85 * 0267) + (0.62 * 0.133) + (0.55 * 0.333) + (0.2 * 0.267) = 0.545 \tag{13}$$

From (13), we demonstrate a singular composite value of 0.545 on the AV metric under the base score in the CVSS v3 calculator. Thus, we have demonstrated our extension work to CVSS v3 calculator. This work will enable the researcher and cybersecurity professional to use CVSS v3 calculator as it is but also cover the needs of assessing the vulnerabilities of any multifactor authentication technology.

5 Application Example

In this section, as an application example, we will provide a comparative study between a two-factor biometric fingerprint- and password-based authentication and a three-factor smartcard-, password-, and biometric fingerprint-based authentication by using CVSS v3 calculator to calculate their composite vulnerability scores based on the extension work we have shown in the Sect. 4. A subset of published Common Vulnerabilities and Exposures (CVE) under each authentication factor technology was selected through pseudo-random sampling within the CVE database.

The base score composite metric within the CVSS v3 calculator provided the foundation in our casual-comparative analysis of the multifactor authentication strategies. We focused on application of our formula to the metrics of access vector (AV), access complexity (AC), and privileges required (PR) through purposive sampling, keeping user interaction (UI) at "none," scope (S) at "changed," and the CIA triad metrics at "low." As the experiment considers the exploitation of authentication, static assignments of values in UI, S, and CIA triad metrics provide a level of control while maintaining simplicity to our experiment.

5.1 Authentication Factor Vulnerability Summary

Password or PIN factors continue to exhibit novel vulnerabilities in brute force attacks, exploitable via varying attack vectors [21]. Password vulnerabilities remain existent for internal user operations and failures in technologies providing secure password increment, storage, or updating capabilities [22]. Smartcard authentication is a theoretically strong authentication mechanism. However, the implementations for smartcard authentication and supporting PKI architectures can be complex, resulting in introduction of high-criticality exploitation mechanisms [23]. Buffer overflow attacks from various attack vectors also present exploitations on smartcard technologies [24]. Biometric fingerprint authentication demonstrates known weaknesses in firmware errors and capability for actors to retrieve user fingerprint images through exploiting encryption through various attack vectors and complexity levels [25]. The set of technologies supporting biometric fingerprint authentication remains vulnerable to advanced threats capable of interception or successful soliciting of protected user authentication data [26]. Encryption implementation errors or employment of weak cryptographic mechanisms increases the potential for compromise in biometric authentication mechanisms [27].

Table 2 Fingerprint and password AV impacts

Attack Vector metric components				
Factor				
Metric value	*Numeric value*	*Fingerprint*	*Password*	*Combined*
Network	0.85	25%	70%	95/200
Adjacent	0.62	25%	15%	40/200
Local	0.55	30%	10%	40/200
Physical	0.2	20%	5%	25/200

Table 3 Fingerprint and password AC impacts

Attack Vector metric components				
Factor				
Metric value	*Numeric value*	*Fingerprint*	*Password*	*Combined*
Low	0.77	45%	70%	115/200
High	0.44	55%	30%	85/200

Table 4 Fingerprint and password PR impacts

Attack Vector metric components				
Factor				
Metric value	*Numeric value*	*Fingerprint*	*Password*	*Combined*
None	0.85	40%	60%	100/200
Low	0.62	30%	25%	55/200
High	0.27	30%	15%	45/200

5.2 CVSS Metrics Calculations

Our first set of metric calculations consider the two-factor biometric fingerprint- and password-based authentication. Tables 2, 3, and 4 provide the source values input into our formula for the AV, AC, and PR metrics. Equations (14), (15), and (16) provide the output composite scores from the metrics following our proposed formula.

$$\text{X1 (AV)} = (0.85 * 0.475) + (0.62 * 0.2) + (0.55 * 0.2) + (0.2 * 0.125) = 0.663 \tag{14}$$

$$\text{X1 (AC)} = (0.77 * 0.575) + (0.44 * 0.425) = 0.63 \tag{15}$$

$$\text{X1 (PR)} = (0.85 * 0.5) + (0.62 * 0.275) + (0.27 * 0.225) = 0.656 \tag{16}$$

Table 5 Smartcard, fingerprint, and password AV impacts

Attack Vector metric components					
Factor					
Metric value	*Numeric value*	*Smartcard*	*Password*	*Fingerprint*	*Combined*
Network	0.85	30%	70%	25%	125/300
Adjacent	0.62	35%	15%	25%	75/300
Local	0.55	15%	10%	30%	55/300
Physical	0.2	20%	5%	20%	45/300

Table 6 Smartcard, fingerprint, and password AC impacts

Attack Vector metric components					
Factor					
Metric value	*Numeric value*	*Smartcard*	*Password*	*Fingerprint*	*Combined*
Low	0.77	15%	70%	45%	125/300
High	0.44	85%	30%	55%	170/300

Table 7 Smartcard, fingerprint, and password PR impacts

Attack Vector metric components					
Factor					
Metric value	*Numeric value*	*Smartcard*	*Password*	*Fingerprint*	*Combined*
None	0.85	35%	60%	40%	135/300
Low	0.62	50%	25%	30%	105/300
High	0.27	15%	15%	30%	60/300

Our second set of metric calculations consider the three-factor smartcard-, password-, and biometric fingerprint-based authentication. Tables 5, 6, and 7 provide the source values input into our formula for the AV, AC, and PR metrics. Equations (17), (18), and (19) provide the output composite scores from the metrics following our proposed formula.

$$\mathrm{X2}\,(\mathrm{AV}) = \left(0.85 * 0.417\right) + \left(0.62 * 0.25\right) + \left(0.55 * 0.183\right) + \left(0.2 * 0.15\right) = 0.64 \tag{17}$$

$$\mathrm{X2}\,(\mathrm{AC}) = (0.77 * 0.417) + (0.44 * 0.567) = 0.57 \tag{18}$$

$$\mathrm{X2}\,(\mathrm{PR}) = (0.85 * 0.45) + (0.62 * 0.35) + (0.27 * 0.2) = 0.654 \tag{19}$$

Table 8 Base score outputs

Base score results for two- and multifactor authentication technologies	
Authentication factors	*CVSS v3 score*
X1. Fingerprint and password	6.10
X2. Smartcard, password, and fingerprint	5.86

5.3 CVSS Composite Vulnerability Scores Calculations

We now calculate the CVSS base score utilizing the inputs derived from the work demonstrated in Sect. 5.2. Equation (20) represents the CVSS calculator equation where the scope metric is "changed."

$$\text{Roundup}\left(\text{Minimum}\left[1.08 * (\text{Impact} + \text{Exploitability}), 10\right]\right) \tag{20}$$

The impact sub score (ISC) is not altered in the experiment, rather the exploitability score only, which is calculated via Eq. (21). The *Impact* subscore utilized in the comparative analysis, summarized in Table 8, is 3.733.

$$8.22 * AV * AC * PR * UI \tag{21}$$

where UI is defined as "none" and calculated at 0.85.

From Table 8, we can see that based on the results of the base CVSS composite vulnerability scores, the three-factor authentication technology demonstrates a lower composite vulnerability score by 0.24 in compared to the two-factor authentication technology. This is an indication that extending the multiplicity of authentication factors may reduce vulnerability profile, although more evaluation examples are needed to confirm this discovery.

6 Conclusion

In this chapter, we have presented an extension to the CVSS v3 calculator, demonstrating the capability of the equation in determining the composite metric. The approach provides organizations and researchers an effective tool to rapidly and objectively assess and characterize the composite vulnerability impact of adding authentication factors in multifactor authentication technology sets. The extent we propose to the CVSS v3 calculator is intended to provide enhancements to the current CVSS scheme in the representation of composite vulnerabilities. The chapter is limited in providing minimal sample CVE sources, rather than derived from complete common vulnerability databases. Future work will focus on calculations utilizing real vulnerability sets and expansion of the equation application to all metrics within the CVSS v3 calculator. Additionally, we will apply the proposed CVSS v3 extension on assessing the composite vulnerability of an entire IT system including more sophisticated authentication factor solutions to determine points of diminishing returns and optimal multiplicity in multifactor authentication.

References

1. D. Lipaj, V. Davidavičienė, Influence of information systems on business performance. Mokslas – Lietuvos Ateitis **5**(1), 38–45 (2013). https://doi.org/10.3846/mla.2013.06
2. L. Henderson, *Multi-Factor Authentication Fingerprinting Device Using Biometrics* (Villanova University, 2019)
3. A. Tang, Two-factor authentication: The death of the password? [Blog] (2020). Retrieved from https://www.itproportal.com/2015/01/15/two-factor-authentication-death-p4ssw0rd/
4. T. Webb, *An Architecture for Implementing Enterprise Multifactor Authentication with Open Source Tools* (SANS Institute Reading Room, 2013)
5. Y. Choi, Security weakness of efficient and secure smart card-based password authentication scheme. Int. J. Appl. Eng. Res. **12**(7), 1222–1226 (2017)
6. S. Carberry, DOD pushes toward CAC replacement. FCW (2017). Retrieved from https://fcw.com/articles/2017/08/14/dod-cac-replacement-carberry.aspx
7. J. Couretas, M. Ucal, Organizational adoption of innovation: Background, programs & a descriptive modeling approach, in *Military Modeling & Simulation Symposium*, (Society for Computer Simulation International, Boston, MA, 2011), pp. 44–52
8. P. Mell, K. Scarfone, S. Romanosky, A complete guide to the common vulnerability scoring system version 2.0. CVSS (2007). Retrieved from https://www.first.org/cvss/v2/guide
9. E. Kovacs, FIRST announces CVSS version 3.1 [Blog] (2019). Retrieved from https://www.securityweek.com/first-announces-cvss-version-31
10. Common Vulnerability Scoring System v3.0: User Guide. Retrieved from https://www.first.org/cvss/v3.0/user-guide (Accessed May 25, 2021)
11. Common Vulnerability Scoring System v3.0: Specification Document. Retrieved from https://www.first.org/cvss/specification-document (Accessed May 25, 2021)
12. O. Santos, The evolution of scoring security vulnerabilities: The sequel (2016). Retrieved from https://blogs.cisco.com/security/cvssv3-study
13. L. Allodi, S. Biagioni, B. Crispo, K. Labunets, F. Massacci, W. Santos, Estimating the assessment difficulty of CVSS environmental metrics: An experiment. Future Data Secur. Eng., 23–39 (2017). https://doi.org/10.1007/978-3-319-70004-5_2
14. D. Nguyen, D. Nguyen-Duc, N. Huynh-Tuong, H. Pham, CVSS, in *Proceedings of the Ninth International Symposium on Information and Communication Technology – SoICT 2018*, (2018). https://doi.org/10.1145/3287921.3287968
15. Latent Feature Vulnerability Ranking of CVSS Vectors, in *Summer Computer Simulation Conference (SCSC)* (2017). https://doi.org/10.22360/summersim.2017.scsc.019
16. S. Hazari, Challenges of implementing public key infrastructure in Netcentric enterprises. Logist. Inf. Manag. **15**(5/6), 385–392 (2002). https://doi.org/10.1108/09576050210447073
17. National Institute of Standards and Technology, *Security and Privacy Controls for Information Systems and Organizations* (National Institute of Standards and Technology, Gaithersburg, MD, 2017), pp. 6–17
18. I. Velásquez, A. Caro, A. Rodríguez, Authentication schemes and methods: A systematic literature review. Inf. Softw. Technol. **94**, 30–37 (2018). https://doi.org/10.1016/j.infsof.2017.09.012
19. D. Dasgupta, A. Roy, A. Nag, Toward the design of adaptive selection strategies for multi-factor authentication. Comput. Secur. **63**, 85–116 (2016). https://doi.org/10.1016/j.cose.2016.09.004

20. D. Stebila, P. Udupi, S. Chang, Multi-factor password-authenticated key exchange, in *Eighth Australasian Conference on Information Security*, (Australian Computer Society, Inc., Brisbane, Australia, 2010), pp. 56–66
21. NVD - CVE-2020-11052, (2020). Retrieved from https://nvd.nist.gov/vuln/detail/CVE-2020-11052
22. NVD - CVE-2019-14833, (2020). Retrieved from https://nvd.nist.gov/vuln/detail/CVE-2019-14833
23. NVD - CVE-2019-3980, (2020). Retrieved from https://nvd.nist.gov/vuln/detail/CVE-2019-3980
24. NVD - CVE-2018-16393, (2020). Retrieved from https://nvd.nist.gov/vuln/detail/CVE-2018-1639
25. NVD - CVE-2019-13603, (2020). Retrieved from https://nvd.nist.gov/vuln/detail/CVE-2019-13603
26. NVD - CVE-2020-7958, (2020). Retrieved from https://nvd.nist.gov/vuln/detail/CVE-2020-7958
27. NVD - CVE-2019-12813, (2020). Retrieved from https://nvd.nist.gov/vuln/detail/CVE-2019-12813

Part II
Computer and Network Security and Related Issues

Phishing Prevention Using Defense in Depth

Joel Williams, Job King, Byron Smith, Seyedamin Pouriyeh, Hossain Shahriar, and Lei Li

1 Introduction

A social engineering attack is when an attacker uses human interaction to obtain information about an organization or its computer systems. The attacker usually assumes the role of an employee, vendor, or customer and offers details to support their identity. By asking questions they piece together information that allows them to compromise the controls in place to protect the network. Typically an attacker will take the information found from one source and continue to contact additional sources within the organization to gain more knowledge and add to their credibility.

Global events impacting countries worldwide can make organizations more vulnerable to social engineering attacks. For example, the worldwide COVID-19 pandemic has resulted in a dramatic increase in teleworking which has led to a rise in exploitation attempts. According to the CISA Alert (AA20-099A) [37], cybercriminals are using the pandemic to lure victims using the following techniques:

- Phishing via emails with the subject of coronavirus or COVID-19.
- Malware distribution, using coronavirus or COVID-19 themes to lure people into clicking links.
- DNS name registration using related wording techniques similar to coronavirus or COVID-19 to create fake sites.

Phishing is one of the prominent social engineering attacks that occurs when the target(s) is contacted by someone posing as a legitimate source to trick individuals

J. Williams (✉) · J. King · B. Smith · S. Pouriyeh · H. Shahriar · L. Li
Information Technology Department, Kennesaw State University, Marietta, GA, USA
e-mail: jwil1240@students.kennesaw.edu; jking189@students.kennesaw.edu; bsmit513@students.kennesaw.edu; spouriye@kennesaw.edu; hshahria@kennesaw.edu; lli13@kennesaw.edu

K. Daimi et al. (eds.), *Advances in Security, Networks, and Internet of Things*, Transactions on Computational Science and Computational Intelligence,
https://doi.org/10.1007/978-3-030-71017-0_8

into providing sensitive data such as personally identifiable information, banking details, credit card numbers, and passwords. Typically this is done by email, telephone, or text message. They then take this information to access essential accounts so they can gain access to critical information within the organization.

Phishing consequences on organizations can be severe, ranging from loss of data, compromised credentials, business disruption, and reputation damage. While it is desirable to make an organization bulletproof to all security threats, it is not financially feasible to reduce all risks an organization faces. Considering the facts that 94% of malware is delivered by email and 33% of all breaches involve some form of social attack [35], it is clear that phishing prevention has the potential for a high return on investment. By leveraging a defense-in-depth approach to phishing prevention, we can address this issue not only from a technology perspective but also from the most at-risk factor, the human element.

To help address the challenges created by phishing-based attacks, we investigate a defense-in-depth strategy from three different perspectives consisting of the following:

- Web Security Gateway
- Email Security Gateway
- Security Awareness Training

By using this strategy plan, we have layered protection starting at the perimeter of the user, confirmed the validity of emails, and strengthened the final layer of defense, the human firewall. In this scenario, we do not rely on a single-layer protection instead the combination of all three layers to cover a comprehensive security protection. In this paper we investigate the aforementioned security layers and explore the cutting edge models with respect to each layer.

The remainder of the paper is organized as follows. In Sect. 2, we review recent work in phishing prevention from web security gateway stand point. Section 3 covers the importance and recent studies of email security gateway to protect enterprise. Security awareness training is discussed in Sect. 4 and finally we discuss, conclude, and outline some future work in Sects. 5 and 6, respectively.

2 Web Security Gateway

2.1 Web Security Functionality

Defense-in-depth strategies to protect enterprises from phishing attacks must start at the network's perimeter. Perimeter phishing defense can have two components: preventing as many phishing attempts as possible from entering the corporate network (inbound) and attempts to reach phishing links from leaving the network (outbound). Incorporating intrusion detection systems (IDS) is a common way to accomplish this objective.

Intrusion detection systems are passive automated systems that monitor and analyze network traffic and generate "alerts" in response to activity that either match known patterns of malicious activities or is unusual. There are several types of IDS based on methods of detection employed. In this section we briefly cover the general functionality of the common types of IDS and focus more on their roles in phishing prevention (Table 1 shows the role of different IDS in phishing prevention concisely).

Network Intrusion Detection System (NIDS) Intrusion detection systems rely on signature-based and anomaly-based detection systems and come in two forms including host-based (HIDS) and network-based (NIDS) [31]. HIDS is generally software installed on end user clients and other devices where NIDS is deployed at

Table 1 Summary: the role of IDSs in phishing prevention

Technique	Pros	Cons	Reference
Anomaly-based IDS	Effective against new threats	High false positives	[30]
		Machine learning algorithms may impact performance	[25]
		Expensive hardware	[3]
		State-based protocols less effective in modern wireless networks	[26]
Signature-based IDS	Low false positives	Ineffective against zero day threats	[34]
	Fast and inexpensive	State-based protocols less effective in modern wireless networks	[31]
Honeypots IDS	Enhance anomaly, signature and collaborative detection	Increased management complexity	[2, 33]
	Divert some phishing attacks	Deployment in unprotected network region increases risk and may violate policy	[21]
	Provide some early warning/zero day functionality		[20]
Collaborative IDS	Effective at more sophisticated phishing methods designed to evade anomaly and signature	Deployment and management complexity	[18, 42]
		Security and policy issues raised by network connections to security devices at other	[15]

the network boundary, usually between the corporate network on one side and the edge router and firewall devices that connect to the Internet on the other. Figure 1 shows a typical placement of an NIDS.

Signature-based IDS compares network information with datasets of known threats and offers low false positive rates at the expense of being unable to detect zero day and other unknown attacks [34]. For example, the Snort IDS which is a widely used is known as signature-based NIDS [14]. Anomaly-based IDS, on the other hand, uses different machine learning approaches to build a normal model and trustworthy behavior of the system. As a result, any deviation of that model or unusual behavior will be flagged as anomalous or suspicious. In general, anomaly-based IDS has a high rate of false alarms (false positive). Additionally, they cannot provide details about the detected attacks [1, 19]. Zeek, formerly known as Bro, is an example of an anomaly IDS [14]. A hybrid approach where an anomaly-based IDS is used to update datasets used by a signature-based IDS in order to detect both known and unknown attacks with low false positive rates is recommended [12]. The hybrid architecture prioritizes low false positive rates (Fig. 2).

According to Sharifi et al. [30] an NIDS is placed in a network where it can monitor traffic traversing network interfaces. When applied to the phishing use case, an NIDS can monitor inbound SMTP traffic in a SPAN—switched port analyzer—port configuration described by Edwards [5] that allows it to read a copy of all packets directed to an email server. This is commonly referred to as port mirroring, which can be configured on both Ethernet and wireless switches that offer the capability. An alternative is to use a dedicated Ethernet or wireless network TAP device to copy inbound email traffic and direct it to the NIDS.

In this strategy, signature-based detection monitors the SMTP traffic for the presence of known URLs, DNS names, and other data to detect likely phishing emails while anomaly-based ID would use machine learning and other tools at its disposal to do the same. Upon detecting a likely phishing email candidate, a range of actions can be configured including alerting a notification system, withholding email for administrative review before permitting their delivery and automatically blocking all emails matching phishing profiles according to whatever detection, prevention, and/or response functionality is available.

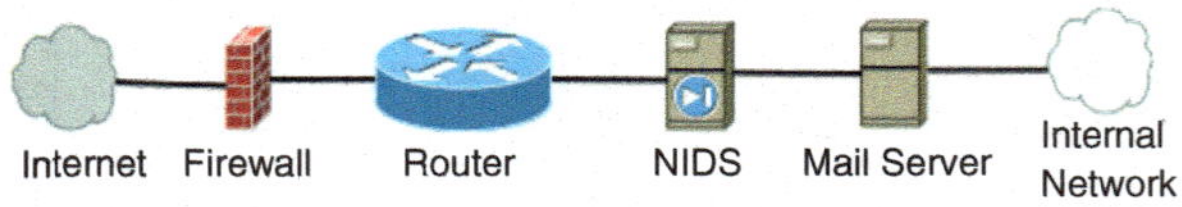

Fig. 1 A typical placement of NIDS

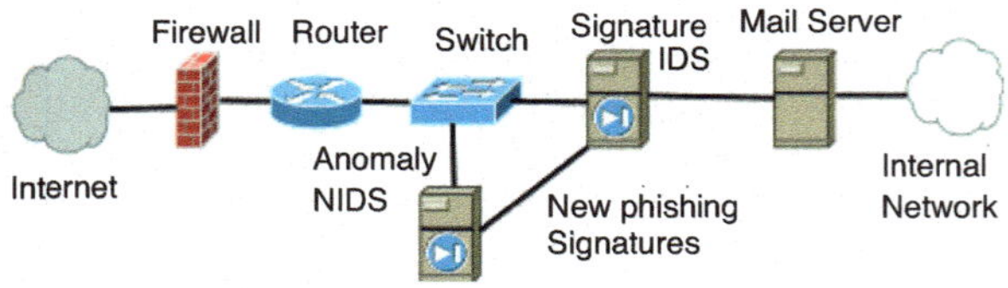

Fig. 2 Network with anomaly and signature IDS

Knowledge-Based Intrusion Detection In Knowledge-Based Intrusion Detection Systems [6], different knowledge-based techniques such as finite state machine, description language, and expert system techniques are utilized to extract the knowledge from the attacks and system vulnerabilities [31]. Those types of IDS maintain information on how specific attacks affect network activities and use them to detect the intrusions or attacks happening in the network or system [31]. In the estimation of Vacas et al. description language intrusion detection can be very effective if the IDS receives regular knowledge updates on new attacks from Open Source Intelligence (OSINT) data feeds [36]. OSINT receives information from public and private sources through feeds such as phish tank and malware domain list. OSINT-enabled IDS can also be updated by local sources such as honeypots [36], a concept that will be discussed later. Zhou et al. [42] proposed an approach where knowledge-based IDs from multiple organizations are interconnected in a peer-to-peer network so that whenever a member IDS detects a new phishing threat it notifies the rest of the network with the data from the threat. This could potentially achieve faster updates than OSINT data feeds and can also be a method of providing new information for OSINT feeds.

Honeypots for Knowledge-Based and Machine Learning Intrusion Detection System Chang et al. [2] proposed the use of virtual machines as honeypots for an intrusion detection strategy, with HoneyMonkey being the specific implementation. HoneyMonkey involves a virtual machine serving as "honey" to attract phishing attacks with the "monkey" being automated processes that emulate user activity running inside the virtual machine.

The goal of the honeypot is to attract as many phishing emails as possible for the purposes of generating data. For this reason, the honeypot solution would be located prior to the firewall. Please see Fig. 3 for an example network configuration.

To serve as the monkey, Moradpoor et al. [21] proposed using Python email parsing scripts capable of identifying every hyperlink in each email message received and applying various anomaly-based algorithms to the hyperlinks to detect and predict unknown and zero day phishing links, a strategy also endorsed by Uğurlu and Doğru [34]. Pham et al. [25] and Moon et al. [20] propose dynamic URL analysis, where URLs are accessed via HTTP GET requests in order to explicitly test and record whether URLs are phishing links. Web browsers, scripts, programs, or some combination may be used to access the URLs. In addition Pham et al. [25] propose that downloaded malware and all other outcomes that result from accessing the phishing link be recorded using State-Transition Analysis Technique(STAT) with Moon et al. advocating similarly [20]. This approach can provide a significant amount of contextual data for not only internal use for IDS solutions but also can be provided to the Open Source Intelligence phishing feeds [20] and to peer-to-peer IDS solutions [42]. The downside is that it can be very challenging to implement. However, this can be accomplished in practice by connecting honeypots to an IDS [2] and also by deploying host-based IDS solutions to the honeypots to perform the STAT analysis [20].

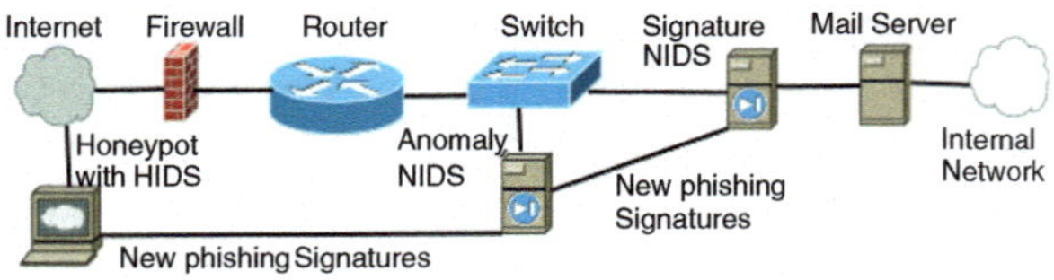

Fig. 3 Network with anomaly and signature IDS and Honeypot

Collaborative Intrusion Detection System The method proposed by [15, 36, 42] is a form of Collaborative Intrusion Detection System (CIDS), where anti-phishing IDS data is exchanged and analyzed by multiple IDS systems. CIDS can make signature-based IDS more effective against phishing, and in particular the fast flux issue [15, 42] where phishing sites either change their TCP/IP address and domain name rapidly or utilize botnet devices as front-end proxies to obfuscate them. CIDS turns this very concept against phishing sites by creating its own interconnected network of IDS devices that share and rapidly update each other with DNS record and other information about these sites so that they can be tracked as they change. In addition, this data sharing allows traffic to be traced back through the botnet proxies and load balancers to their true origin hosts [42]. This is possible because the large number of data sharing IDS devices can operate logically as a single IDS device that tracks multiple emails received from a single origin source. However, the same phishing email received by multiple IDS devices from different botnets would allow a CIDS scheme to ignore the botnet address, "search further down the chain" for the true originating phishing host controlling the botnets and update the CIDS dataset with information about the phishing host and the botnet, a process referred to by Zhou et al. [42] as trace-back. Vacas et al. [36] dealt primarily with the datasets that CIDS use where Zhou et al. [42] investigated CIDS architecture. Meanwhile Khonji et al. [15] had how CIDS operates in practice for an area of emphasis.

2.2 *Web Security Challenges*

NIDS as the most commonly used IDS has its limitations. Despite various strategies to limit and mitigate it, false positives linger as an issue [30]. The opposite is also true: false negatives can be a problem and furthermore NIDS cannot detect all phishing attempts and other forms of intrusion [30]. A bigger issue of being reliant on state-based protocols [26] and algorithms [39] in network design is the assumption that it remains static. Many things cause this not to be true, like mobile devices that are presumed to be static. As a result, changing the network design or topology as done through mobile devices [21] make NIDS techniques less effective. The NIDS devices themselves can also be exploited with denial-of-service attacks being a common method [30]. Furthermore, NIDS is more effective at stopping the further propagation of attacks through a network than at preventing the attacks entirely [30]. Another issue is that some forms of NIDS are better than others in addressing specific attacks, yet deploying multiple NIDS solutions may impact the cost, performance, and management complexity of the network [41].

Shayat et al. promoted cloud NIDS solutions to mitigate some of these limitations, particularly in the area of mobile devices [21] as well as cost. Host-based IDS is another mitigation technique to be considered, also for mobile devices [39] and as HIDS is generally implemented in software, mass deployment of open source or even commercially licensed—if at bulk rate—solutions may both lower the number of NIDS hardware devices needed as well as limit the scope of what is required of them to goals more attainable. For example, consider the work of Hurtado et al. [10] on the practicality of combining commodity hardware with open source software to build lower cost NIDS solutions.

3 Email Security Gateway

An important layer of protection within a layered infrastructure defense is a secure email gateway. Email is used by billions of users everyday which makes it a prime target for an attacker to infiltrate an organization. Phishing attacks by email can result in disaster when an organization relies solely on the human firewall. According to Om et al. [22] secure email gateways can defend more than 99% of spam and antivirus attacks along with additional detection benefits. The goal of an email gateway is to protect intellectual property by preventing attacks in the form of spam, malware and phishing to ensure safe and secure email correspondence.

A crucial tool utilized by email security gateways is the process of content filtering. Content filtering is the process of blocking certain emails or websites that are considered dangerous or distracting for those on a network. This process utilizes keyword matching and URL filtering to specify what is stopped by the filter along with allowing legitimate content to pass through [16].

A secure email gateway is a tool or software that is used to scan emails that are being sent and received on an email server. The goal is to filter out all unwanted spam, malware, and phishing attempts before they can be read by the end user. Content filtering can be configured to perform the following:

- Virus and Malware Filtering
- Spam Filter
- DNS Authentication to Combat Spoofing
- Block Impersonation Attempts
- Stop Malicious URLs
- Scan Attachments for Malicious Code
- Outbound Scan (preventing sensitive information from leaving the organization)

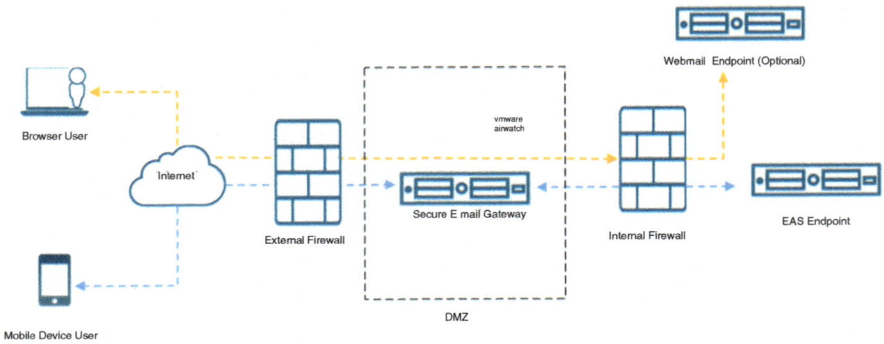

Fig. 4 Secure email gateway with DMZ [38]

3.1 *Gateway Functionality*

For a secure email gateway to function optimally a server must be placed within the email pipeline (Fig. 4). First, a virtual demilitarized zone (DMZ) will be established to hold the gateway server within its own subnet [13]. The gateway scanning server is then created within the DMZ all of which must be located between the message transport agent (MTA) server entering the network and the MTA that is directing mail to the internal email server [9]. This allows the scanning process to take place in isolation.

The scanning server is configured to detect spam, viruses, malware, run attachment tests and domain name real-time blackhole list scans. If a message is flagged for any of the above it is either deleted immediately or it may be quarantined on the server for a given amount of time to determine whether it is indeed a threat. The quarantined message may then be forwarded to the recipient after being cleared. All clean messages are immediately forwarded along to the MTA directing traffic to the internal user account server where the end user may access their messages via internet message access protocol (IMAP).

In addition to scanning for vulnerabilities, email gateways encrypt all internal correspondence as well as external emails leaving the network. Data must be encrypted at rest and in transit so that an attacker cannot make sense of anything they may intercept. Additional functions may be implemented such as an integrity check to guarantee that an email was not edited during the transmission process. A signature function will also support non repudiation [7].

3.2 *Email Gateway Challenges*

A secure email gateway cannot be relied upon solely by itself as there are additional threats that it cannot solve. Internal communication within an organization by email can lead to lost IP when documents are mishandled or lost via misplaced hardware [22]. An additional concern when configuring an email gateway is to notify users explicitly that emails are being monitored. Without user agreement and clearly stated compliance documentation regarding monitoring in place legal issues may arise [22]. Like most network feature additions an email gateway will increase costs for the organization in hardware, maintenance, and manpower. False positives can also be distracting to IT staff and even hinder productivity by blocking employee email correspondence. With these challenges properly addressed email security gateways are most effective when part of a layered security approach and should not be considered a standalone solution.

4 Security Awareness Training

Security awareness training is a process that provides training for employees about cybersecurity, IT best practices, and even regulatory compliance. A comprehensive security awareness program should train users on a variety of IT, security, and other business-related topics. These may include:

- Avoid phishing and different types of social engineering cyberattacks.
- Spot potential malware behaviors.
- Report security threats
- Understand and follow IT policies and best practices.
- Understand and adhere to any applicable data privacy and compliance regulations (GDPR, PCI DSS, HIPAA, etc.)

Despite businesses thinking their employees are too smart to be fooled by a phishing scam, cybercriminals use this method because it continues to be successful. The 2019 Verizon Data Breach Investigations Report [37] revealed that 96% of successful security breaches start with a phishing attack. By using regular training that includes phishing simulations, courses on IT and security best practices, and data protection and compliance training, companies can reduce risk, decrease infections and related help desk costs, protect their reputation by experiencing fewer breaches, and secure their overall cybersecurity investment. The key to having an effective Security Awareness program consists of two key steps: Building a Training Plan and turning employees into a Human Firewall. There are challenges in creating an effective Security Awareness Training program, but the benefits in my opinion far out weight those difficulties.

4.1 Build a Training Plan

A key to this process is building an effective training plan [26]. This ideally integrates with a system that allows the tracking of user training and support testing employees accordingly using various social engineering attacks. Only by tracking employee progress in training and reactions to testing methods will one be able to identify users who may need additional or specific training to address their weaknesses. Also, this process is on a continuous cycle and needs to adapt as threats change. An effective training plan should consist of the following and required by the entire workforce of the company:

- Monthly installations of training material highlighting various topics. Each month the subject should be rotated so that over time all the topics are covered [27]. Ideally, this would include:
  - Enrolling employees to complete short videos covering a cybersecurity topic about two to five minutes in length. Completing all these training modules must be mandatory and enforced.
  - Email newsletters that provide information on cybersecurity best practices and compliance.
  - Posting and rotating posters in common areas that provide information on cybersecurity best practices and compliance.
- Send monthly phishing simulations to employees to verify if they are following training steps to confirm if the email is safe. These emails need to adapt and change to avoid users learning past examples to spot them. They also need to be deployed at different days and times in the month to avoid being predictable based on a schedule.
- Perform monthly vishing and texting simulations against employees to verify if they are following training steps to confirm legitimacy. Again, creating variety allows more accurate results and prevents employees from learning patterns [24].
- Each month place test USB devices around different departments or locations to tempt employees to use them [40].

Again for companies to achieve optimal long-term results, the training plan must:

1. Require the total workforce to be a part of the program.
2. Training needs to cover the entire span of the year and not follow a once a year model. Cybersecurity must be something discussed frequently to remain fresh on employee's minds.
3. Training needs to be data-driven, adjusting with the changing threats impacting the company.
4. Employee awareness testing should be done on a continuous cycle to reduce predictability.

Employee satisfaction with security training increases as simulations and training content are considered to be relevant and worthwhile. By introducing more challenging attacks based on employees' past performances, employers can prevent sophisticated hacker attempts from tricking their staff, further reinforcing your security program's enduring relevance. Most importantly, the right training can support companies in transforming employees' behavior toward potential email attacks for the long term, which represents a significant competitive advantage in any industry that relies heavily on digital communication.

4.2 The Human Firewall

For companies to build their employees into human firewalls, it requires extensive training and a lot of practice. To accomplish this requires an effective training plan, as described above. Employees must train to spot and know how to react to these top threat activities:

- Phishing Attacks
- Malware
- Theft/Loss

By training employees to know how to react and respond effectively with testing they are better prepared to know what to do in real world cases that will arise. It is also important that employees know how to report incidents effectively in the organization so security professionals in the organization can determine if additional steps may be needed.

4.3 Security Awareness Training Challenges

Protecting a company from social engineering attacks is not only an IT challenge; it is an organizational challenge [39]. In short, all employees are responsible for protecting the company from these threats by avoiding these pitfalls and reporting when events occur so the organization can react to reduce impact. By employees working together with their IT departments, organizations effectively create a last line of defense against threats that may have penetrated physical and logical controls.

However, with all solutions there are challenges to an effective Security Awareness Training program. First, there is the cost for the solution, which is both financial in nature and time for the IT department to setup and end-users to use. Second, it is difficult to keep users invested in training and participating as they need to learn new threats that are emerging. Lastly, the time investment to develop and adopt the program to reflect the current challenges that are emerging is great.

5 Discussion

Please see Table 2 for a phishing defense-in-depth summary. More work does need to be done to increase the effectiveness of network intrusion and email

Table 2 Summary: phishing prevention using three defense layers

Method	Pros	Cons
Web security gateway	• Protects users and systems from malicious web sites and links	• Cost
	• Regulates access to approved content for the organization	• Configuration Complexity
	• Improved load times for web sites	• False positives impacting productivity
	• Reduce internet bandwidth consumption	• Maintenance of hardware and software
Email security gateway	• Protects users from malicious attachments and links via email	• Cost
	• Filters out inappropriate emails	• Configuration complexity
	• Ensures emails sent are safe for the recipients	• False positives impacting productivity
	• Improves users productivity by removing clutter	• Maintenance of hardware and software
Security awareness training	• Users are the last line of defense and educating them on how to spot dangers is key	• Cost
	• Users learn they are responsible for the organizations data and safeguarding against misuse	• Difficult to keep users invested in training
	• Timely delivery of best practices as they change and evolve	• Maintaining current content to reflect current challenges and vulnerabilities

security gateway techniques. Das et al. [4] propose machine learning algorithms based on feature selection and improved phishing datasets in order to improve automatic detection of zero day phishing websites and URLs. In the latter scenario, the computational ability offered by cloud solutions [11] provides the potential of machine learning and other artificial intelligence techniques to be applied to email security gateways [29] and perhaps would be more effective for this purpose than would be cloud-based IDPS. For human firewalls, further study on holistic approaches that synergistically combine the roles of individuals, organizations as well as such technologies as host-based intrusion detection is needed [8]. In addition a four-layer human-based approach that implements combinations of blocking features, user warnings, educational messages to users, and user reporting proposed by Stembert et al. [32] should be investigated.

While such an approach may seem expensive to implement and complex to manage, the costs of being intimidated by these into inaction may be much higher. Sadique et al. [28] relate that the financial toll of cybercrime can exceed $600 million per year. Equally damaging are the psychological tolls that organizations should seek to protect its employees from. For example, global COVID-19 epidemic saw successful phishing attacks that were used to spread misinformation and foment fear [17] that can be potentially exploited by criminal, terrorist, and hostile state actors to incite crimes and social unrest. These serve as examples of why creation of effective tools and strategies to combat phishing is an urgent responsibility of information technology and cybersecurity professionals.

6 Conclusion

The primary security issue of most organizations is now criminal activities carried out by means of computers or the Internet [36]. Social engineering, which relies on tricks designed to exploit the intellectual tendencies of human decisions, consistently ranks as a chief cybersecurity concern [2]. Phishing combines two very effective cybersecurity techniques, email spam and social engineering, to create a threat that is pervasive in scope and quite challenging to surmount [43]. Despite these obstacles, the potential for significant harm to finances and reputations of organizations and individuals requires an approach based on proven philosophies and technologies which provides a high likelihood of success. In order to contribute to this worthy goal, this article presents a strategy that incorporates three layers to defend organizations against phishing:

1. Protecting the perimeter using network intrusion detection.
2. Protecting the interior using email security gateway.
3. Protecting the endpoints using security awareness training.

While it is very appropriate to compare and contrast them in order to identify their various relative and absolute strengths and weaknesses, it should not be for the purposes of choosing one over the other. Instead such analysis should be done in order to determine how best to combine these methods as part of a single defense-in-depth security strategy against phishing. If this is done, these methods become complementary layers in a single defense plan. Similar to the layered interactivity of the OSI Network model [23] and other similar conceptual frameworks, each layer will need to be designed, planned, implemented, and maintained well enough to operate according to specification independently. However, it is only through the combined effect of the three layers working together in a manner where they benefit from each other's unique strengths while mitigating their individual weaknesses that they can achieve maximum effectiveness in combating the phishing threat.

An example implementation would be a 90/9/0.9 strategy. This would position the NIDS as the designated defender against known or well-defined threats that make up the vast majority of the traffic. The email gateway's role would be to handle a smaller number of zero day threats and more sophisticated attacks. This leaves the well-trained and highly motivated end user responsible for the minute number of phishing emails that evade technological detection. In this way such a layered approach can create an effective plan to defend an organization against phishing email attacks.

In conclusion, this survey paper intends to illustrate the value of approaching the issue of phishing from three different perspectives. Each creates value to address the issue. However, combined we believe a layered defense is created. Ensuring better results than using an individual option or without consideration of how each can be used together.

References

1. D.J. Brown, B. Suckow, T. Wang, A survey of intrusion detection systems. Department of Computer Science, University of California, San Diego (2002)
2. J. Chang, K.K. Venkatasubramanian, A.G. West, I. Lee, Analyzing and defending against web-based malware. ACM Comput. Surv. **45**(4), 1–35 (2013)
3. M.E. Cueva Hurtado, G. Gutierrez, C.R. Narvaéz Guillen, F.J. Álvarez Pineda, M.D.C. Ruilova Sanchez, Systematic literature review: open source tools for intrusion detection in wired and wireless networks, in *2019 International Conference on Information Systems and Computer Science (INCISCOS)* (2019), pp. 208–215
4. A. Das, S. Baki, A.E. Aassal, R. Verma, A. Dunbar, SOK: a comprehensive reexamination of phishing research from the security perspective. IEEE Commun. Surv. Tutor. **22**(1), 671–708 (2019)
5. S. Edwards, Network Intrusion Detection Systems: Important IDS Network Security Vulnerabilities (2002)
6. J.M. Estevez-Tapiador, P. Garcia-Teodoro, J.E. Diaz-Verdejo, Anomaly detection methods in wired networks: a survey and taxonomy. Comput. Commun. **27**(16), 1569–1584 (2004)
7. L. Fan, Y. Ma, W. Kou, D. Kang, T. Wang, Mail security gateway mechanism for email security, in *2015 International Symposium on Computers & Informatics* (Atlantis Press, Paris, 2015)
8. E.D. Frauenstein, R. von Solms, Combatting phishing: a holistic human approach, in *2014 Information Security for South Africa* (IEEE, New York, 2014), pp. 1–10
9. C. Gaboret, The successful low-cost deployment of a secure email gateway. Available at psu.edu, Citeseer database.
10. J.L. García-Dorado, P.M.S. del Río, J. Ramos, D. Muelas, V. Moreno, J.E.L. de Vergara, J. Aracil, Low-cost and high-performance: VoIP monitoring and full-data retention at multi-gb/s rates using commodity hardware. Int. J. Netw. Manage. **24**(3), 181–199 (2014)
11. E.S. Grant, A.F. Mohammad, Cloud computing security gateway proposed service architecture, in *Computer Games, Multimedia and Allied Technology (CGAT 2012)* (2012), p. 124
12. Z. Inayat, A. Gani, N.B. Anuar, S. Anwar, M.K. Khan, Cloud-based intrusion detection and response system: open research issues, and solutions. Arab. J. Sci. Eng. **42**(2), 399–423 (2017)
13. N. Jiang, H. Lin, Z. Yin, L. Zheng, Performance research on industrial demilitarized zone in defense-in-depth architecture. in *2018 IEEE International Conference on Information and Automation (ICIA)* (2018), pp. 534–537
14. G.B. Joe Schreiber, R. Langston, List of open source IDS tools (2020)
15. M. Khonji, Y. Iraqi, A. Jones, Phishing detection: a literature survey. IEEE Commun. Surv. Tutor. **15**(4), 2091–2121 (2013)
16. S. Khurshid, S. Khan, S. Bashir, Text-based intelligent content filtering on social platforms, in *2014 12th International Conference on Frontiers of Information Technology* (2014), pp. 232–237
17. M. Lanterman, Cybersecurity in pandemic times (2020). Accessed 03 Jun 2020
18. D. Laufenberg, L. Li, H. Shahriar, M. Han, Developing a blockchain-enabled collaborative intrusion detection system: an exploratory study, in *Future of Information and Communication Conference* (Springer, New York, 2020), pp. 172–183

19. A. Lazarevic, L. Ertoz, V. Kumar, A. Ozgur, J. Srivastava, A comparative study of anomaly detection schemes in network intrusion detection, in *Proceedings of the 2003 SIAM International Conference on Data Mining* (SIAM, New York, 2003), pp. 25–36
20. D. Moon, H. Im, I. Kim, J.H. Park, DTB-IDS: an intrusion detection system based on decision tree using behavior analysis for preventing apt attacks. J. Supercomput. **73**(7), 2831–2895 (2017)
21. N. Moradpoor, B. Clavie, B. Buchanan, Employing machine learning techniques for detection and classification of phishing emails, in *2017 Computing Conference* (IEEE, New York, 2017), pp. 149–156
22. K. Om, Secure email gateway, in *2017 IEEE International Conference on Smart Technologies and Management for Computing, Communication, Controls, Energy and Materials (ICSTM)* (2017), pp. 49–53
23. S.-N. Orzen, Interaction understanding in the OSI model functionality of networks with case studies, in *2014 IEEE 9th IEEE International Symposium on Applied Computational Intelligence and Informatics (SACI)* (IEEE, New York, 2014), pp. 327–330
24. K. Parsons, A. McCormac, M. Pattinson, M. Butavicius, C. Jerram, Phishing for the truth: a scenario-based experiment of users' behavioural response to emails, in *IFIP International Information Security Conference* (Springer, New York, 2013), pp. 366–378
25. T.S. Pham, T.H. Hoang et al., Machine learning techniques for web intrusion detection—a comparison, in *2016 Eighth International Conference on Knowledge and Systems Engineering (KSE)* (IEEE, New York, 2016), pp. 291–297
26. P. Puhakainen, M. Siponen, Improving employees' compliance through information systems security training: an action research study. MIS Quart. **34**(4), 757–778 (2010)
27. K. Renaud, Cooking up security awareness & training. Netw. Secur. **2018**(5), 20–20 (2018)
28. F. Sadique, R. Kaul, S. Badsha, S. Sengupta, An automated framework for real-time phishing URL detection, in *2020 10th Annual Computing and Communication Workshop and Conference (CCWC)* (2020), pp. 0335–0341
29. D. Sarabia-Jacome, I. Lacalle, C.E. Palau, M. Esteve, Efficient deployment of predictive analytics in edge gateways: fall detection scenario, in *2019 IEEE 5th World Forum on Internet of Things (WF-IoT)* (IEEE, New York, 2019), pp. 41–46
30. A.A. Sharifi, B. Akram Noorollahi, F. Farokhmanesh, Intrusion Detection and Prevention Systems (IDPS) and security issues. Int. J. Comput. Sci. Netw. Secur. **14**(11), 80 (2014)
31. S. Soniya, S. Maria Celestin Vigila, Intrusion detection system: classification and techniques, in *2016 International Conference on Circuit, Power and Computing Technologies (ICCPCT)* (2016), pp. 1–7
32. N. Stembert, A. Padmos, M.S. Bargh, S. Choenni, F. Jansen, A study of preventing email (spear) phishing by enabling human intelligence, in *2015 European Intelligence and Security Informatics Conference* (IEEE, New York, 2015), pp. 113–120
33. A. Umamaheswari, B. Kalaavathi, Honeypot TB-IDS: trace back model based intrusion detection system using knowledge based honeypot construction model. Cluster Comput. **22**(6), 14027–14034 (2019)
34. M. Uğurlu, I.A. Dogru, A survey on deep learning based intrusion detection system, in *2019 4th International Conference on Computer Science and Engineering (UBMK)* (2019), pp. 223–228
35. U.S. Department of Homeland Security. Covid-19 exploited by malicious cyber actors
36. I. Vacas, I. Medeiros, N. Neves, Detecting network threats using OSINT knowledge-based IDS, in *2018 14th European Dependable Computing Conference (EDCC)* (IEEE, New York, 2018), pp. 128–135
37. Verizon Wireless, 2019 Data Breach Investigations Report (2019). Accessed 3 Jun 2020
38. VMware, The Secure Email Gateway Architecture (2020)
39. M.E. Whitman, P. Fendler, J. Caylor, D. Baker, Rebuilding the human firewall, in *Proceedings of the 2nd Annual Conference on Information Security Curriculum Development* (2005), pp. 104–106

40. B. Wilson, Introducing cyber security by designing mock social engineering attacks. J. Comput. Sci. Coll. **34**(1), 235–241 (2018)
41. L.H. Yeo, X. Che, S. Lakkaraju, Understanding modern intrusion detection systems: a survey (2017). Reprint arXiv:1708.07174
42. C.V. Zhou, C. Leckie, S. Karunasekera, T. Peng, A self-healing, self-protecting collaborative intrusion detection architecture to trace-back fast-flux phishing domains, in *NOMS Workshops 2008-IEEE Network Operations and Management Symposium Workshops* (IEEE, New York, 2008), pp. 321–327
43. E. Zhu, Y. Chen, C. Ye, X. Li, F. Liu, OFS-NN: an effective phishing websites detection model based on optimal feature selection and neural network. IEEE Access **7**, 73271–73284 (2019)

Phishing Detection using Deep Learning

Beatrice M. Cerda, Shengli Yuan, and Lei Chen

1 Introduction

With rapid advancement in technology comes complex security challenges. One such security challenge that leaves users exposed is phishing attacks. Attackers set up fake websites to trick users into believing the website is legitimate, and is safe to enter sensitive information such as their passwords [1]. Anti-phishing frameworks have been developed in different forms. The most recent implementation involves datasets used to train machines in detecting phishing sites [2]. This chapter discusses implementation of a Deep Feedforward Artificial Neural Network using supervised learning to detect malicious URLs.

In this study, we developed a python program that extracted up to 30 features from thousands of Uniform Resource Locators, that is, URLs. These features were then used as dataset to train a feedforward artificial neural network to detect malicious URLs. During the training phase, several models were created using only one feature at a time. This allowed for all features to be evaluated individually. Efforts were made to see how effective each feature was in detecting phishing URLs. Groups of features were later used to train models. Models using only one feature yielded poor accuracies, while models trained with groups of features produced higher accuracy ratings. Initially, the hypothesis was that the models with more features yielded better accuracies because they contained more occurrences

B. M. Cerda · S. Yuan (✉)
Department of Computer Science and Engineering Technology, University of Houston – Downtown, Houston, TX, USA
e-mail: cerdab1@gator.uhd.edu; yuans@uhd.edu

L. Chen
Department of Information Technology, Georgia Southern University, Statesboro, GA, USA
e-mail: LChen@georgiasouthern.edu

K. Daimi et al. (eds.), *Advances in Security, Networks, and Internet of Things*,
Transactions on Computational Science and Computational Intelligence,
https://doi.org/10.1007/978-3-030-71017-0_9

of said features within the dataset. After sorting features by highest occurrence and training new models using the top 20 features, it was discovered that the number of occurrences of a feature within a dataset had no effect on accuracy. The reason a group of features performed better was simply together they were the better choice of features to use in yielding the highest accuracy.

The chapter is organized in the following manner: Sect. 2 introduces the background and reviews related works, Sect. 3 discusses the models in detail, Sect. 4 explains the results, and Sect. 5 discusses further research, together with closing statements.

2 Background

2.1 Definition of Phishing

Phishing as a form of security attack has been around for nearly 25 years [3]. There are different types of phishing, including deceptive phishing, spear phishing, pharming, and whaling, among others [4, 5]. Deceptive phishing is considered the most common scam. The idea behind deceptive phishing is replication of legitimate websites to steal sensitive information. Users open emails and receive a link that they believe leads to a legitimate site. The truth is the anchor shown on the email is linked to a site that looks very similar to the legitimate site, but is actually a fake website set up to steal sensitive data such as login credentials. Deceptive phishing is the most common type seen by users. Spear phishing is another form, and it targets a specific person or group of people. Personalizing the email makes it looks credible, luring users to click on the malicious link. Pharming is a form of redirection that modifies the IP address through the DNS server. By going through the server, it prevents the users from visualizing the malicious URL because they are entering the valid URL but are being redirected to the malicious website. Whaling is a scam directed at executives. It gives a professional look pretending to be official business. The scam is termed whaling because they are targeting the "big fish," or rather corporate giants. In this chapter, we focus on deceptive phishing.

2.2 Related Work

Authors of [1] reintroduce a framework for phishing detection using a layered approach. CANTINA+ architecture has a feature extractor that identifies 15 features used in the training and testing phases in a machine learning model. It is noted that the results of false positives and false negatives are affected by individual features while combining all features gives the best results. Overall performance of CANTINA+ was competitive with its layered approach.

Authors of [2] is a survey that discusses the different machine learning techniques which have occurred in recent years. Sahoo et al. saw the malicious URL detection as a "machine learning task" that uses a set of URLs as the training dataset to classify the URLs as malicious or legitimate. One key discovery was that feature representations have different distributions when it comes to malicious or benign URLs.

Authors of [6] give a brief description on how effective neural networks are in detecting phishing websites. Some important phishing characteristics that are extracted as features and used in machine learning are URL domain identity, security encryption, source code with JavaScript, page style with contents, web address bar, and social human factor. The authors extracted a total of 27 features to train and test the model.

Authors of [7] discuss methods to detect phishing emails by extracting email headers and the HTML of the email body. They propose a "multi-layer feedforward artificial neural network with back propagation" to train a model. The neural network uses 18 features. To train the artificial neural network, they used one hidden layer using the sigmoid activation function for the output layer. The number of nodes in the hidden layer ranged from 1 node to 18 nodes. A total of 9100 emails were used with 6000 of them training the network and 3100 testing the network. Results yielded a high accuracy of 98.72% when the node count reached 5.

Authors of [8] investigate the different machine learning models and their accuracies. Thirty features were extracted, and more than 11,000 URLs were used from PhishTank and Millersmiles repositories. There were eight machine learning algorithms that were compared by their effectiveness. These eight learning algorithms are: Bayes Nte, C4.5, SVM, AdaBoost, eDRI, OneRule, Conjunctive Rule, and RIDOR. C4.5 yielded the highest accuracy, but was not considered because it overfit the dataset. Ridor and eDRI algorithms worked because they produced high-quality models and have rules that are simple to understand. Also, the most effective set of features to use in a machine learning environment were the URL of Anchor and the SSL Final State. During training, measures were taken to avoid overfitting, by stopping at five epochs.

Authors of [9] detail an approach for stochastic optimization called Adam, a first-order gradient-based optimization algorithm that works well with large data and large parameters. This algorithm works well with nonstationary objectives and/or sparse gradient. It can control "exponential decay rates." Bias-corrected estimates can be used to prevent bias towards zero. The algorithm was used on logistic regression and multilayer neural networks and found that Adam converges as fast as Adagrad and shows it is best at convergence as opposed to other methods.

Authors of [10] discuss in detail the limitations that deep neural networks have when using standard gradient descent. Attention was paid to activation functions and their relations to initializations, in that initialization may affect how activation functions behave. A feedforward neural network was set up with five hidden layers, with 1000 hidden units in each layer. Stochastic back-propagation was used with ten mini-batches on ten consecutive pairs with varied nonlinear activation functions: sigmoid, hyperbolic tangent, and softsign. They concluded sigmoid or hyperbolic tangent units with standard initialization do not perform well. Networks using hyperbolic tangent performed well with normalized initialization since the layered transformation kept activations and gradients.

Authors of [11] test the performance of sigmoid function up against the hyperbolic tangent function in a series of experiments. They found by using a small training set overfitting occurred when many units were used, while underfitting happened with a minimal number of nodes. When a good size training set was used, the number of nodes did not affect output. Four features were used with a simple feedforward neural network that had one input layer, one hidden layer, and one output layer. Different machine learning classifiers were compared: Decision Tree, K-nearest, Naive Bayes, Neural Network, Support Vector Machine1, Support Vector Machine 2. It was discovered that the neural network was the best performer in identifying phishing emails, yielding the highest recall with high accuracy.

Authors of [12] seek to find if lexical features are sufficient in detecting malicious URLs. Anh Le et al. compared two types of features: lexical features that are visible on the URL, and external features that must be searched by remote servers. Using only lexical features is enough in locating malicious URLs, but when using all features, the results still yield highest accuracy. The paper discusses implementation of the AROW algorithm using only lexical features to classify URLs.

Authors of [13] implement an approach that divides URLs into three types of websites: benign, phishing, and malware. Benign websites are legitimate and safe; phishing websites pose as legitimate websites to steal sensitive information; malware websites are designed to hack into victims' computer systems. Using URLs spares the "run-time latency," avoiding any exposure to web content. Three learning algorithms were used: Decision Tree, Support Vector Machine, and Logistic Regression, to compare three groups of features: lexical, popularity, and host-based.

Authors of [14] study issues with phishing detection when it is discovered that up to 80% of malicious sites are shut down just hours of being used. For this reason, URLs were inspected using two kinds of features: lexical features and domain features. These features can be detected without having to access the website. A classification model was trained using extracted features; a total of 18 features were used. Each feature was evaluated individually, yielding accuracies ranging from 61.89% to 88.44%. Weibo Chu et al. implemented an algorithm that selected features to be grouped together yielding highest accuracy.

Authors of [15] use a purely lexical approach in classifying malicious URLs producing a "lightweight" approach using machine learning. A linear classification model with logistic regression was created to guarantee high performance. Functional features must be tuned to reduce false positives. By balancing the ratio of malicious versus benign URLs, the classifier can be tuned. Datasets can be valid by removing URLs with duplicate tokens. Using a large dataset's distinct URLs increases the probability of high accuracy. Measures were taken on our part to balance datasets with exactly half malicious and half benign URLs to ensure better accuracy.

Authors of [16] discover the importance of datamining in detecting phishing sites. Several features were extracted to create the dataset: Address Bar-Based

Features, Abnormal-Based Features, HTML and JavaScript-Based Features, and Domain-Based Features. A table displays 17 features by frequency of use in URLs, with request URL being at 100%. Efforts were made to reduce the number of features that must be used for efficiency. The top nine features that can be used to produce the highest accuracy were: request URL, age of domain, HTTPS and SSL, website traffic, long URL, sub-domain and multi-sub-domain, adding prefix or suffix separated by (−) to domain, URL of anchor, and using the IP address.

3 Phishing Detector

Our phishing detector has two parts: the Feature Extractor and the Deep Feedforward Artificial Neural Network (Fig. 1). First, a list of URLs is fed into the feature extractor that takes a defined checklist of features and inspects each URL to see what features it has [17]. Features were classified into four groups: Address Bar-Based Features, Abnormal-Based Features, HTML and JavaScript-Based Features, and Domain-Based Features and General Features [18]. Once the feature extractor identifies all the features, it outputs a value for each of the features onto a spreadsheet. This is the dataset that will be used to train the Deep Feedforward Artificial Neural Network.

When creating Dataset 1, the URLs were found in two different websites. The malicious URLs were taken from a trusted source that specializes in phishing URLs, PhishTank [19]. The legitimate URLs were found through Kaggle [20]. In total, there were 37,000 verified phishing URLs, and 37,000 legitimate URLs. The feature extractor was able to extract 25 features for each of the 74,000 URLs. In addition, we also obtain Dataset 2 from [21] that has 9769 URLs, and 30 slightly different features already extracted. Both datasets were split into 80% training and 20% testing.

For the machine learning model, a Deep Feedforward Artificial Neural Network was chosen as the network architecture [22]. To find the best model, we must determine the optimal network architecture and the best feature(s) to use. The network architecture relies on three parameters:

- Number of hidden layers
- Number of epochs
- Batch size

Fig. 1 Flowchart of the phishing detector

Table 1 This table shows the different combinations used in creating 13 training models

Machine Learning Models
2 Hidden Layers; 15 Epochs; Batch Size 4
1 Hidden Layer; 50 Epochs; Batch Size 1
1 Hidden Layer; 50 Epochs; Batch Size 4
1 Hidden Layer; 50 Epochs; Batch Size 10
1 Hidden Layer; 100 Epochs; Batch Size 1
1 Hidden Layer; 100 Epochs; Batch Size 4
1 Hidden Layer; 100 Epochs; Batch Size 10
5 Hidden Layers; 50 Epochs; Batch Size 1
5 Hidden Layers; 50 Epochs; Batch Size 4
5 Hidden Layers; 50 Epochs; Batch Size 10
5 Hidden Layers; 100 Epochs; Batch Size 1
5 Hidden Layers; 100 Epochs; Batch Size 4
5 Hidden Layers; 100 Epochs; Batch Size 10

The model that produced the highest accuracy with the lowest error was chosen to test the accuracy of various features

We built 13 machines with different combinations (Table 1). Each row in the table represents one Deep Feedforward Artificial Neural Network. The neural networks were built using advanced calculation libraries. These libraries included: Theano, Tensorflow, and Keras.

To find the feature or group of features that work the best, more models were built, each with a different set of features that were used for training. The following are the groups of features in Dataset 1 that were used to train a total of 25 models:

- Individual Features
- Grouped Features
  - Address Bar-Based Features
  - Lexical-Based Features
  - General Features
- All Features

The following are the groups of features in Dataset 2 that were used to train a total of 34 models:

- Individual Features
- Grouped Features
  - Address Bar-Based Features
  - Abnormal-Based Features
  - HTML and JavaScript-Based Features
  - Domain-Based Features
- All Features

4 Results and Analysis

When creating Dataset 1, we noticed that most of the URLs from Kaggle shared the same domain names, which might limit its effectiveness in training the models. Additionally, the feature extractor was not able to extract Abnormal-Based Features, Domain-Based Features, nor HTML and JavaScript-Based Features, thus providing an unbalanced dataset with limited features selections. Therefore, we decided to train the models with Dataset 2.

After training 13 models using different combination of parameters, one of them generated the highest accuracy of 96.99% (Table 2). This model used the following parameters (Fig. 2):

- 5 hidden layers
- 100 epochs
- Batch size 4

After the preferred model had been selected, it was used to determine the feature(s). During training with Dataset 1, we noticed the model stopped improving after 5 epochs. For this reason, the number of epochs was reduced from 100 to 5 epochs.

After the preferred model was tuned, the neural network took one feature at a time and tried to identify all the URLs. This was first done for all 25 features of Dataset 1. The results are shown in Table 3. The highest accuracy was 86.27%. We believed the unbalanced data in Dataset 1 might be a factor for this mediocre performance. For comparison, we trained the model with each feature in Dataset 2. The results are shown in the top portion of Table 4. The highest accuracy was 88.88%, an improvement over Dataset 1 but not very significantly.

Table 2 13 Deep Feedforward Artificial Neural Networks were trained on different combinations of parameters to find the best model yielding highest accuracy

Machine Learning Models	TN	FN	FP	TP	Accuracy
2 Hidden Layers; 15 Epochs; Batch Size 4	939	50	55	916	94.64%
1 Hidden Layer; 50 Epochs; Batch Size 1	951	36	43	930	95.97%
1 Hidden Layer; 50 Epochs; Batch Size 4	944	30	50	936	95.92%
1 Hidden Layer; 50 Epochs; Batch Size 10	951	41	43	925	95.71%
1 Hidden Layer; 100 Epochs; Batch Size 1	960	42	34	924	96.12%
1 Hidden Layer; 100 Epochs; Batch Size 4	963	30	31	936	96.89%
1 Hidden Layer; 100 Epochs; Batch Size 10	961	39	33	927	96.33%
5 Hidden Layers; 50 Epochs; Batch Size 1	958	32	36	934	96.53%
5 Hidden Layers; 50 Epochs; Batch Size 4	936	26	58	940	95.71%
5 Hidden Layers; 50 Epochs; Batch Size 10	936	20	58	946	96.02%
5 Hidden Layers; 100 Epochs; Batch Size 1	950	27	44	939	96.38%
5 Hidden Layers; 100 Epochs; Batch Size 4	963	28	31	938	96.99%
5 Hidden Layers; 100 Epochs; Batch Size 10	958	38	36	928	96.22%

TN True Negative, *FN* False Negative *TP* True Positive, *FP* False Positive

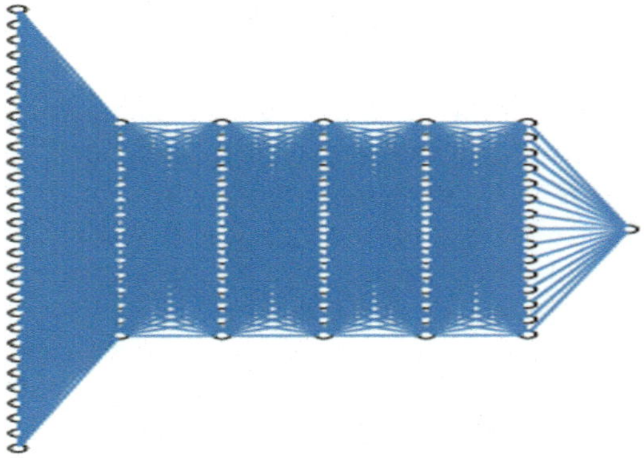

Fig. 2 Deep Artificial Neural Network composed of 30 inputs, 5 hidden layers with 15 nodes each, and one output layer

Table 3 Models were trained using 25 different sets of features from Dataset 1

Feature	FN	TN	TP	FP	Accuracy
Exe Present	49.33%	50.67%	0.00%	0.00%	50.67%
Mail To	49.33%	50.67%	0.00%	0.00%	50.67%
IP Present	48.93%	50.66%	0.41%	0.01%	51.07%
At Present	46.59%	50.67%	2.74%	0.00%	53.41%
Hyphen Present	41.78%	48.55%	7.55%	2.12%	56.10%
Sensi Present	37.50%	49.61%	11.83%	1.05%	61.45%
Number of Dots Url	31.92%	41.07%	17.41%	9.60%	58.48%
Length of Host	30.99%	42.78%	18.34%	7.89%	61.12%
Domain Token Count	26.36%	43.23%	22.97%	7.44%	66.20%
Number of Dots Host	25.72%	43.26%	23.61%	7.41%	66.87%
Avg Domain Token Length	25.22%	42.07%	24.11%	8.60%	66.18%
Average Path Token Length	24.22%	40.51%	25.11%	10.16%	65.62%
Large Domain Token Length	23.82%	37.28%	25.51%	13.39%	62.80%
Large Path Token Length	18.42%	40.99%	30.91%	9.68%	71.91%
AddressBarBased	18.16%	43.68%	31.17%	6.99%	74.84%
Length of URL	15.54%	38.16%	33.79%	12.51%	71.95%
Lexical	**9.93%**	**42.14%**	**39.40%**	**8.53%**	**81.54%**
General	**8.58%**	**45.52%**	**40.75%**	**5.15%**	**86.27%**
Length of Path	7.89%	37.78%	41.45%	12.89%	79.23%
Path Token Count	**5.13%**	**38.80%**	**44.20%**	**11.87%**	**83.00%**
Short URL	0.00%	0.00%	49.33%	50.67%	49.33%
Double Slash Position	0.00%	0.00%	49.33%	50.67%	49.33%
Favicon	0.00%	0.00%	49.33%	50.67%	49.33%
HTTPS Present	0.00%	0.00%	49.33%	50.67%	49.33%
Redirect	0.00%	0.00%	49.33%	50.67%	49.33%

The highest accuracy is 86.27%

Next, groups of features by category were used to train the neural network. This was done for all four groups in Dataset 2:

Table 4 35 models were trained using different sets of features from Dataset 2

Feature	FN	TN	TP	FP	Accuracy
HTTPS Token in Domain	0.00%	0.00%	49.29%	50.71%	49.29%
Submitting to Email	0.00%	0.00%	49.29%	50.71%	49.29%
Website Forwarding (Redirect)	5.41%	6.02%	43.88%	44.69%	49.90%
Disabling Right Click	47.30%	48.01%	1.99%	2.70%	50.00%
iFrame	44.44%	45.56%	4.85%	5.15%	50.41%
Port	49.29%	50.71%	0.00%	0.00%	50.71%
Pop Up Window	49.29%	50.71%	0.00%	0.00%	50.71%
Favicon	0.00%	0.00%	50.71%	49.29%	50.71%
DNS Record	16.17%	17.76%	33.11%	32.96%	50.87%
Double Slash	42.24%	44.34%	7.04%	6.38%	51.38%
On Mouse Over	5.26%	7.55%	44.03%	43.16%	51.58%
Long URL	39.59%	42.40%	9.69%	8.32%	52.09%
Statistical Report	6.12%	8.93%	43.16%	41.79%	52.09%
Abnormal URL	41.12%	43.98%	8.16%	6.73%	52.14%
Short Service	41.73%	44.90%	7.55%	5.82%	52.45%
At Symbol	5.97%	9.29%	43.32%	41.43%	52.60%
Age of Domain	23.21%	26.73%	26.07%	23.98%	52.81%
Links Pointing to Page	25.82%	30.31%	23.47%	20.41%	53.78%
Google Index	4.64%	9.95%	44.64%	40.77%	54.59%
Using IP	13.27%	19.64%	36.02%	31.07%	55.66%
Page Rank	33.52%	40.10%	15.77%	10.61%	55.87%
SFH	35.71%	44.08%	13.57%	6.63%	57.65%
Links In Tags	12.65%	23.72%	36.63%	26.99%	60.36%
Domain Registration Length	11.63%	23.27%	37.65%	27.45%	60.92%
Prefix Suffix	37.40%	50.71%	11.89%	0.00%	62.60%
Request URL	14.34%	28.83%	34.95%	21.89%	63.78%
Sub Domain	25.41%	42.60%	23.88%	8.11%	66.48%
Web Traffic	11.48%	35.00%	37.81%	15.71%	72.81%
URL Anchor	0.46%	35.15%	48.83%	15.55%	83.98%
SSL Final State	3.88%	43.47%	45.41%	7.24%	88.88%
HTMLJavaScriptBasedFeatures	42.70%	44.49%	6.58%	6.22%	51.07%
DomainBasedFeatures	13.93%	38.83%	35.36%	11.89%	74.18%
AbnormalBasedFeatures	5.66%	42.09%	43.62%	8.62%	85.71%
AddressBarBasedFeature	3.32%	43.72%	45.97%	6.99%	89.69%
All Features	**2.70%**	**47.60%**	**46.58%**	**3.11%**	**94.18%**

- HTML and JavaScript-Based Features
- Domain-Based Features
- Abnormal-Based Features
- Address Bar-Based Features

This time the highest accuracy was 89.69% when using the Address Bar-Based Features (Table 4). Features in this category were mainly lexical features.

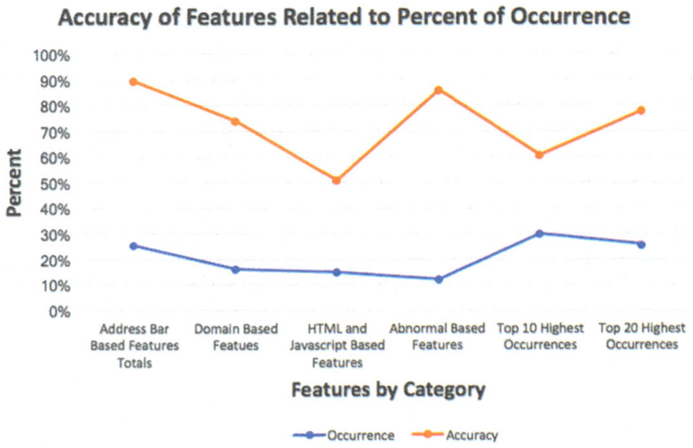

Fig. 3 The accuracies yielded from features did not depend on their frequency or occurrence within the dataset

Finally, we used all features in Dataset 2 to train the model which yielded the highest accuracy of 94.18% (Table 4).

Here we list the top five models using said features from Dataset 2:

1. All features	**94.18%**
2. Address bar-based features	**89.69%**
3. SSL final state	**88.88%**
4. Abnormal-based features	**85.71%**
5. URL anchor	**83.98%**

We questioned why there were variants among the accuracies from individual features, groups of features, and all features. Intuitively, the more data we used, the better performance we should receive. However, SSL Final State and URL Anchor were both individual features, and they generated better accuracies than certain groups of features. To find out whether occurrences of certain individual feature have an impact on their performance, each feature was then taken and the number of occurrences was recorded for all URLs in Dataset 2. Once the top 10 features with the highest occurrences were identified, they were grouped and used to train the model. The same process was done for the top 20 features. Results in Fig. 3 show that accuracies yielded from features did not depend on their frequency or occurrence within the dataset.

5 Conclusion

Phishing attacks are serious security threats. Finding ways to detect phishing URLs before a user clicks on it is important. Even more important is stopping such URLs from spreading via emails. This study verified machine learning can provide viable solutions toward achieving this goal. Out of the 60 models trained using various features, several models produced satisfactory accuracies. For best practice, when training a Deep Feedforward Artificial Neural Network, it is recommended to use as many features as possible. Future research will focus on improving the algorithms in the hidden layers of the deep artificial neural networks, and fine-tuning the machines with better parameters, as well as including more contextual information in addition to the URLs.

References

1. G. Xiang, J. Hong, C.P. Rose, L. Crancr, Cantina. ACM Trans. Inf. Syst. Secur. **14**(2), 1–28 (2011). Web
2. D. Sahoo et al, Malicious URL detection using machine learning: A Survey [1701.07179], 16 March 2017, arxiv.org/pdf/1701.07179.pdf
3. KnowBe4, History of phishing. Phishing, www.phishing.org/history-of-phishing
4. DMBisson, David BissonFollow, 6 common phishing attacks and how to protect against them. The State of Security, 3 June 2016, www.tripwire.com/state-of-security/security-awareness/6-common-phishing-attacks-and-how-to-protect-against-them/
5. J. Chen, C. Guo, Online detection and prevention of phishing attacks, in *2006 First International Conference on Communications and Networking in China*, (2006) n. pag. Web
6. C.J. Chandan et al., A Machine learning approach for detection of phished websites using neural networks. Https://Pdfs.semanticscholar.org. Int. J. Recent Innov Trends Comput Commun. 2014. pdfs.semanticscholar.org/7e3f/4613751db651f3cbf43836fa783b8433_8bd.pdf
7. N.G.M. Jameel, E.G. Loay, Detection of phishing emails using feedforward neural network. Https://Pdfs.semanticscholar.org. Int. J.Comput Appl. (0975–8887) **77**(7) (2013), pdfs.semanticscholar.org/33fa/22cc24349b4a27872f269baf424badd41db1.pdf
8. N. Abdelhamid et al., Phishing detection: A recent intelligent machine learning comparison based on models content and features. IEEE Xplore, 2017, ieeexplore.ieee.org/stamp/stamp.jsp?tp=&arnumber=8004877. 978-1-5090-6727-5/17/ ©2017 IEEE
9. D.P. Kingma, J.L. Ba, Adam: A method for stochastic optimization. Https://Arxiv.org, ICLR 2015, 23 July 2015, arxiv.org/abs/1412.6980
10. X. Glorot, Y. Bengio, Understanding the difficulty of training deep feedforward neural networks, in *Proceeding.MLR.Press, 13th International Conference on Artficial Intelligence and Statistics*, 2010, proceedings.mlr.press/v9/glorot10a/glorot10a.pdf. Chia La-guna Resort, Sardinia, Italy. Volume 9 of JMLR: W&CP 9
11. N. Zhang, Y. Yuan, Phishing detection using neural network – cs229.Stanford.edu, cs229.stanford.edu/proj2012/ZhangYuan-PhishingDetectionUsingNeuralNetwork.pdf
12. A. Le et al., PhishDef: URL names say it all. ArXiv.org, 12 September 2010, arxiv.org/pdf/1009.2275.pdf. arXiv:1009.2275v1
13. H. Liu et al., Learning based malicious web sites detection using suspicious URLs. https://pdfs.sematicscholar.orghttp://pdfs.semanticscholar.org/bf5e/84bc5572c4dc535cc01da87a49d79ba5bf46.pdf

14. W. Chu et al., protect sensitive sites from phishing attacks using features extractable from inaccessible phishing URLs, in *Protect Sensitive Sites from Phishing Attacks Using Features Extractable from Inaccessible Phishing URLs – IEEE Conference Publication*, pdfs.semanticscholar.org/dad5/a7b4d1eb0e51805090e3289955de38275991.pdf
15. M. Darling, A lexical approach for classifying malicious URLs. Digitalrepository.unm.edu, Engineering ETDs at UNM digital repository, 2015, digitalrepository.unm.edu/cgi/viewcontent.cgi?article=1062&context=ece_etds
16. R.M. Mohammad et al., Intelligent rule-based phishing websites classification. Https://Pdfs.sematicscholar.org,http://pdfs.semanticscholar.org/17fe/6bb5a2248c1524dda71bdbfbc8346c479224.pdf
17. Fancyarora, Fancyarora/URL-feature-extraction. GitHub, github.com/fancyarora/URL-Feature-Extraction
18. R. Mohammad et al., Phishing websites features, pp. 1–7
19. PhishTank > Developer Information, PhishTank – out of the net, into the tank, www.phishtank.com/developer_info.php
20. Rohk, One week of global news feeds | Kaggle, 29 September 2017, www.kaggle.com/therohk/global-news-week/data
21. A. Kumar, Phishing website dataset. Kaggle, 12 January 2018, www.kaggle.com/akashkr/phishing-website-dataset/data
22. S. Xu, L. Chen, A novel approach for determining the optimal number of hidden layer neurons for FNN's and its application in data mining, in *Semantic Scholar, 5th International Conference on Information Technology and Applications (ICITA2008)*, (2008)., pdfs.semanticscholar.org/254f/3f0fce2905675445d48ca8ab61e3761b1e9b.pdf

Enhancing Data Security in the User Layer of Mobile Cloud Computing Environment: A Novel Approach

Noah Oghenfego Ogwara, Krassie Petrova, Mee Loong (Bobby) Yang, and Stephen MacDonell

1 Introduction

Mobile Cloud Computing (MCC) is a distributed computing technology that improves on the resource limitations posed by the mobile device (MD). This technology leverages the benefits of the powerful computing resources offered by Cloud Computing (CC) infrastructure [1]. MCC integrates CC services into the Mobile Computing (MC) environment by outsourcing data to the cloud and allowing MD to utilize less memory, but the security of user data during outsourcing may be questionable [2]. Extensive research and development effort have been carried out, with the aim to provide a more secure MCC environment and attract more consumers to use these services. However, security and privacy have been reported as a major challenge that hinders the MCC technology [3–14].

Data security issues cut across the different models that make up the MCC architecture, including the User Layer (UL), the Communication Layer (CL), and the Cloud Service Provider Layer (CSPL) [6]. The sensitivity of the information stored or accessible at the UL makes it vulnerable to cybercrime attacks. Studies have reported that most MD users do not install protective applications such as antivirus, antimalware, antispyware, among others making them more vulnerable in the MCC environment [11–13]. Google's simple publishing policy about official Android applications (apps) makes their users a prime cybercrime target. Malicious apps have been uploaded successfully to the Google Play Store despite the protection of its inbuilt security model [14]. Furthermore, existing defensive solutions such as

N. O. Ogwara (✉) · K. Petrova · M. L. (B.) Yang · S. MacDonell
School of Engineering, Computer and Mathematical Sciences, Auckland University of Technology, Auckland, New Zealand
e-mail: fego.ogwara@aut.ac.nz; krassie.petrova@aut.ac.nz; bobby.yang@aut.ac.nz; stephen.macdonell@aut.ac.nz

K. Daimi et al. (eds.), *Advances in Security, Networks, and Internet of Things*, Transactions on Computational Science and Computational Intelligence,
https://doi.org/10.1007/978-3-030-71017-0_10

antivirus for MDs are not efficient at eliminating the ever-increasing mobile malware threat. This is as a result of their total dependence on signature-based approach for detection. Furthermore, MDs are not suited for constant malware scanning due to their known resource limitations [13].

This chapter presents the current state-of-the-art IDS solutions in the cloud and mobile environment and proposes a model that provides a protection to the highly exposed UL. The chapter builds on our earlier review which explored the MCC security solution landscape environment and suggested that Intrusion Detection Systems (IDS) provided a feasible approach towards addressing a wide spectrum of MCC security threats [15]. The rest of the chapter is organized as follows. First, it provides a brief discussion about the nature of intrusion types in the MCC environment, followed by a comprehensive review of existing IDS solutions. Next, the chapter identifies security issues that need to be addressed and proposes a model that may be used to build a security-enhancing solution for the UL of MCC. Directions for further research and development are also highlighted.

2 Intrusion and Intrusion Detection System

IDS monitors the activities that take place within the system or network in order to identify intrusions activities that violate the security policy of the systems. In MCC, intrusion can occur both at the MD, and in the cloud infrastructure.

2.1 Intrusion in Mobile Devices

In the UL of the MCC infrastructure, MDs are used to access the resources of this environment. As the MD architecture is like the architecture of personal computers, MDs are vulnerable to the same types of intrusions and malicious activities, for example, viruses, Trojan horses, spyware [16]. The following are brief description of some attacks that affects the UL of the MCC architecture.

(a) Virus: In this type of intrusion, the attacker tries to replicate itself and infect the MD by using malware, which affects the operating system.
(b) Spyware: In this type of intrusion, the attacker infects the user devices without the user permission and steal information from the devices unknown to the user.
(c) Bot Process: In this type of intrusion, the attacker executes commands remotely and takes control of the infected device.
(d) Worm: In this type of intrusion, the attacker creates copies of itself and spreads their copies without any user interference.
(e) Phishing apps: In this type of intrusion, the attackers try to reveal the destructive application as a reliable site but containing mobile phishing application that steals users' sensitive information.

2.2 *Intrusion in the Cloud Infrastructure*

MCC environments are also vulnerable to intrusion that targets the security of technology itself. Confidential information that may be stored by users in cloud resources may become the target of an attack. By obtaining unauthorized access, attackers may violate the privacy and confidentiality of the data stored in the cloud by cloud users. Among others, the type of intrusion attacks that are prevalent in the CC infrastructure include insider attacks, flooding attacks, Denial of Services (DoS) attacks, user to root attacks, port scanning, Virtual Machine (VM) attacks, covert-channel attacks. Such intrusion attacks are dangerous since they affect both the MD users and the cloud service provider (CSP). Moreover, it is the responsibility of CSPs to provide adequate security protection of user information [16–18]. The following are brief description of some attacks that affect the CSPL of the MCC architecture.

(a) Insider attacks: This type of attacks allows a legitimate user of the cloud infrastructure to misuse unauthorized privileges by performing malicious activities in the cloud environment, thereby accessing or modifying information of another user without authorization.
(b) Flooding attacks: In this attack, intruders send many packets from an innocent host in the network, thereby making them not to respond to legitimate traffic. In MCC, virtual machines can be accessed by the user when connected to the Internet, which an attacker can use to cause DoS via the innocent host. Flooding attacks affect the availability of services to an authorized user when intruder attacks servers that provide services to a user; it affects the availability of services offers by such servers in MCC environment.
(c) User to root attacks: In this type of intrusion, attackers hijack user credentials and gain root-level access and explore cloud server as a root user, which gives unlawful access to data stored in such server.
(d) Port scanning: In this type of intrusion, the intruder attempts to locate and access open port on the Internet for launching attacks on the cloud.
(e) Attacks on Virtual Machine (VM): The intruder attempts to compromise installed hypervisor, which hosts all the VMs in a specific server thereby gaining control on all installed VMs.
(f) Covert-channel attacks: In this type of attacks, the intruder exploits the weakness in the isolation of shared resources and uses hidden part to steal confidential information.

2.3 *Types of IDS*

There are mainly four different types of IDS used in Cloud: Host-based Intrusion Detection Systems (HIDS), Network-based Intrusion Detection Systems (NIDS),

Hypervisor-based IDS (Hy-IDS), and Distributed Intrusion Detection Systems (DIDS).

HIDS These IDS are installed in the host machine and detect intrusion by analyzing the information they receive. Information sources include, for example, the file system, system calls, and log files.

NIDS These IDS detect intrusion by analyzing the network packets in order to discern malicious activities in the network. NIDS compare the current network behavior with previously observed behavior to identify suspicious activities.

Hy-IDS These IDS allow users to monitor the communication channels across VMs and analyze communication patterns in order to detect possible intrusion.

DIDS These IDS may comprise several HIDS and NIDS over a large network. As DIDS deploy the detection techniques used by both NIDS and HIDS, DIDS inherit the benefits of both types of systems [16, 17].

2.4 IDS Detection Method

Detection methods (DM) used in IDS can be categorized according to the following approach. Signature-based detection (SB) attempts to define a set of rules (signatures), which are used to detect (and predict consequentially) the appearance of known intrusion attack patterns. In a cloud-based system, this method can be used to identify a known attack [19].

Anomaly-based detection (AB) is concerned with identifying and labeling a "malicious" event that deviates from the normal cloud network behavior. The method has the advantage of identifying attacks that may have not been found previously [17]. Finally, hybrid detection (HB) is used to enhance the capabilities of an existing IDS by combining SB and AB in order to enable the detection of both known attacks and unknown attacks [17].

3 Methodology and Results

A total number of 371 papers were obtained from our search across four different electronic databases (IEEE, Science Direct, ACM, and Springer Link), published between 2010 and 2020. Only peer-reviewed journal articles and conferences papers that were written in English language were used in this study. The 58 papers that were selected for a detailed review proposed either solutions or solution frameworks for intrusion and/or malware detection either in the MD, or in the CC/MCC environment. A summary of the review results is shown in Table 1. Each article is assigned an identifier and is referenced to the source paper (columns 1, 2, 3). The IDS type, its detection method and scope are shown in columns 4, 5, 6. The last three

Table 1 Results of the analysis of the selected IDS frameworks and solutions

ID	Source	Year	IDS type	Detection method	Targeted environment	Prevention component	Machine learning	Performance analysis
F1	[20]	2011	DIDS	AB	CC	✗	✓	✗
F2	[21]	2011	HIDS	SB	CC	✓	✗	✗
F3	[22]	2011	HIDS	AB	MD	✗	✗	✗
F4	[23]	2012	NIDS	HB	CC	✗	✓	✓
F5	[24]	2012	HIDS	SB	CC	✓	✗	✗
F6	[25]	2012	NIDS	AB	CC	✗	✓	✓
F7	[26]	2012	NIDS	SB	CC	✗	✗	✗
F8	[27]	2012	DIDS	SB	CC	✗	✗	✗
F9	[28]	2012	NIDS	SB	CC	✗	✗	✗
F10	[29]	2012	NIDS	HB	CC	✓	✓	✗
F11	[30]	2013	DIDS	HB	MD	✓	✓	✗
F12	[31]	2013	NIDS	AB	CC	✗	✗	✗
F13	[32]	2014	HIDS	AB	MD	✓	✓	✓
F14	[33]	2014	HIDS	AB	MD	✗	✓	✗
F15	[34]	2014	NIDS	AB	MD	✗	✓	✓
F16	[35]	2014	NIDS	AB	CC	✗	✓	✓
F17	[36]	2014	DIDS	HB	CC	✗	✓	✗
F18	[37]	2014	NIDS	AB	CC	✓	✓	✓
F19	[38]	2014	NIDS	AB	MD	✗	✓	✓
F20	[39]	2015	NIDS	SB	CC	✗	✗	✗
F21	[40]	2015	NIDS	SB	CC	✓	✗	✗
F22	[10]	2015	Hy-IDS	AB	MCC	✓	✗	✓
F23	[11]	2015	Hy-IDS	SB	CC	✗	✗	✓
F24	[42]	2015	DIDS	AB	CC	✗	✓	✗
F25	[43]	2015	Hy-IDS	AB	CC	✗	✓	✗
F26	[44]	2015	NIDS	HB	CC	✓	✓	✓
F27	[45]	2016	NIDS	AB	CC	✗	✗	✗
F28	[46]	2016	HIDS	AB	MD	✗	✓	✓
F29	[47]	2016	HIDS	AB	MD	✗	✓	✓

(continued)

Table 1 (continued)

ID	Source	Year	IDS type	Detection method	Targeted environment	Prevention component	Machine learning	Performance analysis
F30	[48]	2016	NIDS	SB	CC	✗	✗	✓
F31	[49]	2016	DIDS	HB	CC	✓	✗	✓
F32	[50]	2016	Hy-IDS	AB	CC	✗	✓	✓
F33	[51]	2017	DIDS	HB	CC	✓	✗	✗
F34	[52]	2017	HIDS	HB	MD	✗	✗	✓
F35	[53]	2017	Hy-IDS	AB	CC	✗	✗	✗
F36	[54]	2017	NIDS	SB	CC	✓	✗	✗
F37	[55]	2019	NIDS	AB	CC	✓	✗	✓
F38	[56]	2017	DIDS	SB	CC	✓	✗	✓
F39	[57]	2017	Hy-IDS	AB	CC	✗	✗	✗
F40	[58]	2017	NIDS	AB	CC	✗	✓	✓
F41	[59]	2019	NIDS	AB	CC	✗	✓	✓
F42	[60]	2018	NIDS	AB	CC	✓	✓	✓
F43	[61]	2018	NIDS	AB	CC	✓	✗	✓
F44	[62]	2018	NIDS	AB	CC	✗	✗	✓
F45	[63]	2018	NIDS	AB	MCC	✗	✓	✓
F46	[64]	2018	NIDS	HB	CC	✓	✗	✓
F47	[65]	2018	HIDS	AB	CC	✗	✓	✓
F48	[66]	2018	DIDS	HB	CC	✗	✗	✓
F49	[67]	2018	HIDS	AB	CC	✗	✓	✓
F50	[68]	2019	NIDS	AB	CC	✗	✓	✓
F51	[69]	2018	NIDS	HB	CC	✗	✓	✓
F52	[70]	2018	NIDS	AB	MCC	✗	✓	✓
F53	[71]	2019	NIDS	AB	CC	✗	✓	✓
F54	[72]	2019	HIDS	AB	MD	✗	✓	✓
F55	[73]	2019	DIDS	AB	MCC	✗	✓	✓
F56	[74]	2019	NIDS	AB	CC	✗	✗	✓
F57	[75]	2020	HIDS	AB	MD	✗	✓	✓
F58	[76]	2019	HIDS	AB	MD	✗	✗	✓

columns describe the characteristics, or dimensions, of the proposed solution or framework. Several criteria have been used in prior research to investigate existing solutions and identify weakness [5, 6, 17, 73]. Extracted from the literature, the solution dimensions described below were used to analyze the IDS frameworks included in the review. The ✓ indicates that the framework includes the dimension in its architecture while ✗ indicates that the framework does not include it.

IDS Type This dimension specifies the kind of IDS that was proposed in each framework or solution such as HIDS, NIDS, Hy-IDS, and DIDS.

Detection Method This identifies the method of intrusion detection used in each proposed solution or framework, such as SB, AB, and HB.

Prevention Component This is used to indicate whether the framework or the solution provides a prevention technique whenever intrusion is detected.

Machine Learning (ML) Component This specifies whether the framework incorporates any ML process or algorithm as part of the intrusion detection process.

Performance Analysis This indicates whether the research discusses the performance and performance analysis methods associated with the proposed framework or solution.

3.1 CC-Based IDS Solutions and Frameworks

The solutions proposed in F2, F5, F47, and F49 are of the HIDS type, and target the CC environment. Frameworks F2 and F5 use an SB detection approach and proxy servers for in-depth forensic analysis of files stored locally at the device. The frameworks proposed in F47 and F49 use the AB detection approach with ML techniques. F47 uses a mobile agent to automatically collect data from each host for its detection process. However, F49 focuses on the protection of the VM in the CC environment. In this review, the NIDS proposed solution that uses the SB detection method have their detection engine located at the cloud server. Only F21 and F36 contain attack prevention modules. The solution presented in F9 uses correlated alerts for its detection process. Some of the solutions that target the CC environments have shifted the detection target to the MD node as evidenced in F20 and in F30. The solution presented in F7 enhances the SB detection method by updating new signatures automatically. The NIDS frameworks that use the AB detection method for analyzing network traffic applies various ML techniques in detecting intrusions in the CC environment. F41 and F53 are concerned with detecting DoS attacks in the CC infrastructure. Framework F56 provides a novel technique for anomaly detection using the statistical features of time series. F50 applies both supervised and unsupervised ML to improve detection and classification of attacks in the CC environment. The security approach in F43 incorporates the uses of immune mobile agents while F44 uses correlation of alerts for a more effective detection of malicious activities in a network.

The solutions in F27, F40, and F42 also uses ML techniques. The solution in F42 is concerned with the prediction of the malicious device in the cloud network, while F27 relies on the identification of a network path where disruption occurs as a result of longer transmission time and reduced speed in transmission, in order to detect intrusion.

The NIDS solutions presented in F37 and F18 address security issues concerning communication using a cloudlet controller and a Virtual Private Network (VPN), respectively. However, the frameworks presented in F12, F6and F16 analyze system calls and model the device behavior to enable the IDS to identify attacks. The NIDS frameworks presented in F4, F10, F26, F46, and F51 apply the HB detection method to identify intrusions in the CC environment. Most of these solutions combine SNORT with ML algorithms. However, the solution in F46 uses honeypot technology to produce an early warning about possible threats and attacks.

The solutions presented in F23, F25, F32, F35, and F39 are Hy-IDS. Only F23 uses the SB detection method while F25, F32, F35, and F39 use the AB detection in their detection engines, located at the CC infrastructure. However, none of the Hy-IDS uses a hybrid detection method. The use of mobile agents is common to these solutions. These mobile agents carry intrusion alerts from each VM in the cloud to a management server for analysis, in order to detect distributed intrusion at the hypervisor layer. The solutions presented in F38, F8, F1, F24, F17, F31, F33, and F48 are DIDS. F8 and F38 use the SB detection method. The AB detection method was used in F1 and F24. The HB detection method was applied in F17, F31, F33, and F48.

3.2 *MD-Based IDS Solutions and Frameworks*

The solutions proposed in F3, F13, F14, F28, F29, F34, F54, F57, and F58 are of the HIDS type and target the MD environment. In these frameworks, the detection engines are normally located at the device level except for F13 and F14. The solution presented in F13 focuses on device resource optimization and places its detection engine at the cloud. In contract, in F14, some parts of the detection engine are located at the device level while others reside at the cloud server. The HB detection method was applied in the solution presented in F34 while the other frameworks have adopted the AB detection method. However, none of the HIDS type that targets the MD environment apply the SB detection method. In F34, the dynamic and static analysis of malicious applications in the MD node using system calls is the adopted approach. In a similar fashion, F28 extracts system calls from the applications that reside on the devices, constructs a weighted directed graph, and applies a deep learning algorithm in order to detect new attacks. Framework F54 features an autonomous detection of malicious activities related to both known and unknown attacks, using ML techniques.

The frameworks presented in F3 and F13 use location-based services for detecting intrusion at the MD node. F14 runs a local malware detection algorithm at the MD node to check for a known malware family. The security solutions presented in F58 and F57 analyze system calls and system log files, respectively, to determine if a given app is malicious or not. The security techniques in F29 revolve around the Google Cloud Messaging service for malware detection. In this review, NIDS frameworks that target the MD environment were found in F15 and F19. Both frameworks use the AB detection method. The detection engine in F15 resides at the cloud while that of F19 resides at the device. The framework presented in F11 uses DIDS, with an HB detection approach. The detection engine in F11 resides at both the device and the cloud.

3.3 MCC-Based IDS Solutions and Frameworks

The frameworks in F22, F45, F52, and F5 target the MCC environment. F22 and F55 are of the DIDS type while F45 and F52 are of the NIDS type. All the frameworks that target the MCC environment presented in this review apply an AB detection approach, using ML techniques. Only F22 has an attack prevention module. In the framework presented in F45, the attack detection module analyzes incoming requests and classifies each request as normal or suspicious based on a trained deep learning model. The security techniques used in F52 involve the application of principal component analysis and simulation in order to identify intrusion events in the MCC environment. The framework in F55 presents an ML-based IDS that secures data collection and data fusion in a distributed environment The framework analyzes network traffic from different VM using a multilayer intrusion detection approach. In the event of detected malicious activities, decisions are made about restricting access to a specific VM.

4 Issues Identified in the Review

The detailed analysis of the results of the review of the selected IDS solutions and frameworks shows that most of the existing IDS solutions locate their detection engine at the cloud server. Most of the existing solutions gather and correlate alerts from different nodes, and forward network traffic to the cloud for analysis using a proxy server. Some require the duplication of real-life device into an emulator at the cloud server for an in-depth analysis. These common approaches, as seen in the existing solutions, may cause additional security issues.

The first issue is the leakage of sensitive information during the forwarding of network traffic to a proxy server. The second issue is the relatively low effectiveness of the approach to centralize the correlation of alerts received from various nodes; this process may consume significant time before an attack is detected. In addition, this process also requires a constant connection to the server which might be disrupted at the device layer, for example, due to lack mobile network coverage. Third, most of the existing solutions have adopted AB detection methods using ML techniques. This approach is associated with a high rate of false alarms. Nevertheless, the issue of insider attacks in the CC environment remains a serious issue; applying ML and deep learning techniques may help combat this threat. Furthermore, mitigation process was seen in few works, to manage intrusion events. Finally, the security issues in the MCC environment affecting the UL and the communication channel have not received much attention. However, attackers focus their attention on the UL that is relatively more exposed due to the lack of security awareness among its users. Attackers may target the UL using malicious applications, code obfuscation, and repackaging of popular and legitimate apps with a malicious payload that is difficult to detect by the existing defensive techniques. The model proposed below addresses the security issues identified, especially at the UL in the MCC environment. It adopts a hybrid IDS approach.

5 The Proposed Model

The model proposed in this chapter is called MINDPRES (*M*obile-Cloud *In*trusion *D*etection and *Pre*vention *S*ystem). It aims to enhance data security at the UL in the MCC environment using dynamic analysis of the device behavior and applying ML technique at run time (Fig. 1). It is described below.

5.1 Functional Description of the Model

The model comprises three major components, namely, Application Evaluator, Detection Engine, and Prevention Engine. The *application evaluator* is responsible for the preliminary assessment of each app that resides at the MD node, in order to ascertain the risk level of each user-installed application. The application evaluator extracts detailed information about the app from the manifest file of the application at run time, including the list of requested permissions, intent, and hardware required. The extracted information is offloaded to the cloud where a risk assessment is conducted using an ML model. The ML model would have been trained previously using preprocessed application data collected from different sources, with different ML classification algorithms such as Support Vector Machine, LightGbmClassifier, StochasticDualCoordinateAscentClassifier, Naive Bayes classifier, and K-means classifier. As a result of the risk assessment, each user-installed app is classified as either of high or medium or low-risk level; this result is sent as feedback to the device and the suspicious app watch list is updated accordingly. The watch list enables MINDPRES to monitor only the user-installed app included in the watch list database rather than all user-installed apps.

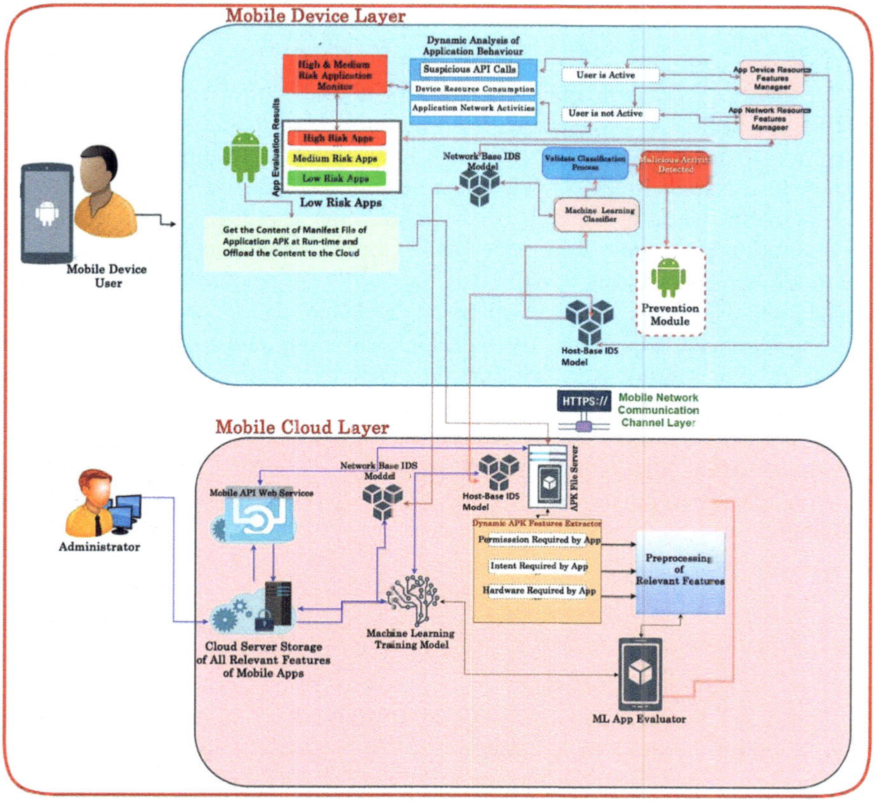

Fig. 1 *M*obile-Cloud *In*trusion *D*etection and *Pre*vention *S*ystem (MINDPRES)

The *detection engine* uses a combination of an IDS and an Intrusion Prevention System (IPS) to safeguard the device. The HIDS dynamically monitors suspicious API calls to the device (e.g., device root access, installation of new application), and the device resource consumption (e.g., CPU usage, memory usage)) by the apps in the watch list. The network activities of the apps in the watch list are monitored by a NIDS, including Internet usage, requested URLs, upBytes, DownBytes, and other network activities. The device activities are monitored both when the device is active (i.e., used by its owner) and when the device is idle. The *prevention engine* in MINDPRES aims to provide an automatic mitigation process once an intrusion is detected. This will stop the execution of the detected application or block all malicious network traffic from a specific host. However, MINDPRES gives the user the flexibility to either allow the execution of the application if the user feels the application is safe for the device due to the possibility of false alarms.

5.2 System Design Procedures

In this phase, we intend to collect data from Android applications by downloading benign and malicious applications from different known sources such as Google Play Store and AppInChina Store. We will submit each application to the VirusTotal engine to ascertain if the application is benign or malicious. The manifest file will be extracted from the apk of the application using the Android apk easy tool. Information from the manifest file of each application contains details of permissions requested, intent requested, and hardware resources required by each application. The collected data will be preprocessed, and a corresponding dataset will be generated. Thereafter, we will apply different data mining techniques to extract relevant features for distinguishing malicious activities in the MCC environment, instead of using all the features extracted. These relevant features will be stored in a central cloud server to build the ML model using different ML classifier algorithms. These ML algorithms are Support Vector Machine, LightGbmClassifier, StochasticDualCoordinateAscentClassifier, Naive Bayes classifier, and K-means classifier.

5.3 Implementation and Evaluation Procedures

The proposed model will be implemented using C#.net programming language with Xamarin Android in the Microsoft Visual Studio Integrated development environment. After the implementation of the proposed model, the effectiveness of the proposed model will be verified with a testbed of over 1000 mobile applications that will be collected from different sources. The data collected will consists of both benign and malicious applications. These data will be preprocessed, and the data will be divided into two parts. Eighty percent of the data will be used for training and 20% of the data will be used for testing. The preprocessed data will be used to build an ML model for both the application evaluator and the detection engines. The evaluation procedures of the proposed model will be in two stages. The first procedure is a pre-evaluation plan for the application evaluator using the confusion matrix. The classification accuracy and false alarm rate will be used as our validation metrics. The best ML classifier will be selected to build the final ML model. Second, the evaluation of the prototype system will involve real-life experiment by installing the prototype system (MINDPRES) in the three devices. The apps that will be installed in each of the devices will be tested on VirusTotal to ascertain whether the app is benign or malicious. The performance overhead will be determined using a standard benchmark tool named quadrant standard edition app that is available at Google Play Store.

6 Conclusion and Future Works

In this chapter, we highlight some security issues that MCC infrastructure face. The issues were identified after a comprehensive review of literature of existing IDS solutions proposed in the CC, MD, and MCC environments. One of the key issues identified is the vulnerabilities of the UL. Hence a novel approach for enhancing the security of the UL in the MCC architecture named MINDPRES was proposed. MINDPRES has an application evaluator that is trained with the ML classifiers. This evaluator ascertains the risk level of all user-installed applications that resides at the UL in the MCC environment. MINDPRES also contains a hybrid IDS that detect intrusion at the device level by constantly monitoring applications activities while the device is being used or idle and an IPS for managing intrusions detected. The proposed approach is the first of its kind in the MCC domain to the best of our knowledge. Only few works have been done using IDS and IPS in the MCC environment. Most of such work has not really focused on the UL. Also, no works in MCC has combined dynamic analysis of device behavior at run time with user activities using ML techniques for protection of MCC resources against attacks. In our proposed model, the detection engine resides locally at the device level. This eliminates the need for constant connection to a remote cloud for protection as proposed by most of the existing solution.

In the future, we intend to complete the development and implementation of the proposed model across different mobile platforms to test its effectiveness in solving security issues in the MCC environment.

References

1. M. Alizadeh, S. Abolfazli, M. Zamani, S. Baaaharun, K. Sakurai, Authentication in mobile cloud computing: A survey. J. Netw. Comput. Appl. (2016). https://doi.org/10.1016/j.jnca.2015.10.005
2. Vishal, B. Kaur, S. Jangra, Assessment of different security issues, threats with their detection and prevention security models in Mobile Cloud Computing (MCC), in *Communications in Computer and Information Science*, (2019). https://doi.org/10.1007/978-981-13-3143-5_27
3. T. Bhatia, A.K. Verma, Data security in mobile cloud computing paradigm: A survey, taxonomy and open research issues. J. Supercomput. (2017). https://doi.org/10.1007/s11227-016-1945-y
4. K. Zkik, G. Orhanou, S. El Hajji, Secure mobile multi cloud architecture for authentication and data storage. Int. J. Cloud Appl. Comput. (2017). https://doi.org/10.4018/ijcac.2017040105
5. M.B. Mollah, M.A.K. Azad, A. Vasilakos, Security and privacy challenges in mobile cloud computing: Survey and way ahead. J. Netw. Comput. Appl. (2017). https://doi.org/10.1016/j.jnca.2017.02.001
6. T.H. Noor, S. Zeadally, A. Alfazi, Q.Z. Sheng, Mobile cloud computing: Challenges and future research directions. J. Netw. Comput. Appl. (2018). https://doi.org/10.1016/j.jnca.2018.04.018
7. S.K. Khatri, Monica, V.R. Vadi, Biometrie based authentication and access control techniques to secure mobile cloud computing, in *2nd International Conference on Telecommunication and Networks, TEL-NET 2017*, (2018). https://doi.org/10.1109/TEL-NET.2017.8343558

8. L.T. Chean, V. Ponnusamy, S.M. Fati, Authentication scheme using unique identification method with homomorphic encryption in Mobile Cloud Computing, in *ISCAIE 2018–2018 IEEE Symposium on Computer Applications and Industrial Electronics*, (2018). https://doi.org/10.1109/ISCAIE.2018.8405469
9. N. Agrawal, S. Tapaswi, A trustworthy agent-based encrypted access control method for mobile cloud computing environment. Pervasive Mob. Comput. (2019). https://doi.org/10.1016/j.pmcj.2018.11.003
10. Y. Shi, S. Abhilash, K. Hwang, Cloudlet mesh for securing mobile clouds from intrusions and network attacks, in *Proceedings – 2015 3rd IEEE International Conference on Mobile Cloud Computing, Services, and Engineering, MobileCloud 2015*, p. 2015. https://doi.org/10.1109/MobileCloud.2015.15
11. J. Walls, K.K.R. Choo, A review of free cloud-based anti-malware apps for android, in *Proceedings – 14th IEEE International Conference on Trust, Security and Privacy in Computing and Communications, TrustCom 2015*, (2015). https://doi.org/10.1109/Trustcom.2015.482
12. R. Kumar, R. Goyal, On cloud security requirements, threats, vulnerabilities and countermeasures: A survey. Comput. Sci. Rev. (2019). https://doi.org/10.1016/j.cosrev.2019.05.002
13. F. Idrees, M. Rajarajan, M. Conti, T.M. Chen, Y. Rahulamathavan, PIndroid: A novel Android malware detection system using ensemble learning methods. Comput. Secur. (2017). https://doi.org/10.1016/j.cose.2017.03.011
14. Gartner Inc., Gartner says global smartphone sales to only grow 7 per cent in 2016, 2016
15. N.O. Ogwara, K. Petrova, M.L.B. Yang, Data security frameworks for mobile cloud computing, in *2019 29th International Telecommunication Networks and Applications Conference (ITNAC) IEEE*, pp. 1–4
16. Z. Inayat, A. Gani, N.B. Anuar, S. Anwar, M.K. Khan, Cloud-based intrusion detection and response system: Open research issues, and solutions. Arab. J. Sci. Eng. (2017). https://doi.org/10.1007/s13369-016-2400-3
17. C. Modi, D. Patel, B. Borisaniya, H. Patel, A. Patel, M. Rajarajan, A survey of intrusion detection techniques in Cloud. J. Netw. Comput. Appl. (2013). https://doi.org/10.1016/j.jnca.2012.05.003
18. J. Wu, L. Ding, Y. Wu, N. Min-Allah, S.U. Khan, Y. Wang, C2Detector: A covert channel detection framework in cloud computing. Secur. Commun. Netw. (2014). https://doi.org/10.1002/sec.754
19. A. Patel, M. Taghavi, K. Bakhtiyari, J. Celestino Júnior, An intrusion detection and prevention system in cloud computing: A systematic review. J. Netw. Comput. Appl. (2013). https://doi.org/10.1016/j.jnca.2012.08.007
20. S.N. Dhage, B.B. Meshram, R. Rawat, S. Padawe, M. Paingaokar, A. Misra, Intrusion detection system in cloud computing environment, in *International Conference and Workshop on Emerging Trends in Technology 2011, ICWET 2011 – Conference Proceedings*, (2011). https://doi.org/10.1145/1980022.1980076
21. A. Houmansadr, S.A. Zonouz, R. Berthier, A cloud-based intrusion detection and response system for mobile phones, in *Proceedings of the International Conference on Dependable Systems and Networks*, (2011). https://doi.org/10.1109/DSNW.2011.5958860
22. N. Ulltveit-Moe, V.A. Oleshchuk, G.M. Køien, Location-aware mobile intrusion detection with enhanced privacy in a 5G context. Wirel. Pers. Commun. (2011). https://doi.org/10.1007/s11277-010-0069-6
23. C. Modi, D. Patel, B. Borisanya, A. Patel, M. Rajarajan, A novel framework for intrusion detection in cloud, in *Proceedings of the 5th International Conference on Security of Information and Networks, SIN'12*, (2012). https://doi.org/10.1145/2388576.2388585
24. R.S. Khune, J. Thangakumar, A cloud-based intrusion detection system for Android smartphones, in *2012 International Conference on Radar, Communication and Computing, ICRCC 2012*, p. 2012. https://doi.org/10.1109/ICRCC.2012.6450572
25. W. Yan, CAS: A framework of online detecting advance malware families for cloud-based security, in *2012 1st IEEE International Conference on Communications in China, ICCC 2012*, (2012). https://doi.org/10.1109/ICCChina.2012.6356881

26. W. Yassin, N.I. Udzir, Z. Muda, A. Abdullah, M.T. Abdullah, A cloud-based intrusion detection service framework, in *Proceedings 2012 International Conference on Cyber Security, Cyber Warfare and Digital Forensic, CyberSec 2012*, (2012). https://doi.org/10.1109/CyberSec.2012.6246098
27. M. Ficco, S. Venticinque, B. Di Martino, mOSAIC-based intrusion detection framework for cloud computing, in *Lecture Notes in Computer Science (including subseries Lecture Notes in Artificial Intelligence and Lecture Notes in Bioinformatics)*, (2012). https://doi.org/10.1007/978-3-642-33615-7_12
28. N.D. Man, E.N. Huh, A collaborative intrusion detection system framework for cloud computing, in *Lecture Notes in Electrical Engineering*, (2012). https://doi.org/10.1007/978-94-007-2911-7_8
29. A. Patel, M. Taghavi, K. Bakhtiyari, J. Celestino, Taxonomy and proposed architecture of intrusion detection and prevention systems for cloud computing, in *Lecture Notes in Computer Science (including subseries Lecture Notes in Artificial Intelligence and Lecture Notes in Bioinformatics)*, (2012). https://doi.org/10.1007/978-3-642-35362-8_33
30. R. Roshandel, P. Arabshahi, R. Poovendran, LIDAR: A layered intrusion detection and remediationframework for smartphones, in *ISARCS 2013 – Proceedings of the 4th ACM Sigsoft International Symposium on Architecting Critical Systems*, (2013). https://doi.org/10.1145/2465470.2465475
31. A. Dolgikh, Z. Birnbaum, Y. Chen, V. Skormin, Behavioral modeling for suspicious process detection in cloud computing environments, in *Proceedings – IEEE International Conference on Mobile Data Management*, (2013). https://doi.org/10.1109/MDM.2013.90
32. S. Yazji, P. Scheuermann, R.P. Dick, G. Trajcevski, R. Jin, Efficient location aware intrusion detection to protect mobile devices, in *Personal and Ubiquitous Computing*, (2014). https://doi.org/10.1007/s00779-012-0628-9
33. J. Milosevic, A. Dittrich, A. Ferrante, M. Malek, A resource-optimized approach to efficient early detection of mobile malware, in *Proceedings – 9th International Conference on Availability, Reliability and Security, ARES 2014*, (2014). https://doi.org/10.1109/ARES.2014.51
34. J. Li, L. Zhai, X. Zhang, D. Quan, Research of android malware detection based on network traffic monitoring, in *Proceedings of the 2014 9th IEEE Conference on Industrial Electronics and Applications, ICIEA 2014*, (2014). https://doi.org/10.1109/ICIEA.2014.6931449
35. F. Idrees, R. Muttukrishnan, War against mobile malware with cloud computing and machine learning forces, in *2014 IEEE 3rd International Conference on Cloud Networking, CloudNet 2014*, (2014). https://doi.org/10.1109/CloudNet.2014.6969008
36. S. Manthira Moorthy, M. Roberts Masillamani, Intrusion detection in cloud computing implementation of (SAAS and IAAS) using grid environment, in *Advances in Intelligent Systems and Computing*, (2014). https://doi.org/10.1007/978-81-322-1299-7_6
37. V.A. Pandian, T.G. Kumar, A novel cloud based NIDPS for smartphones, in *Communications in Computer and Information Science*, (2014). https://doi.org/10.1007/978-3-642-54525-2_42
38. Y. Qi, M. Cao, C. Zhang, R. Wu, A design of network behavior-based malware detection system for android, in *Lecture Notes in Computer Science (including subseries Lecture Notes in Artificial Intelligence and Lecture Notes in Bioinformatics)*, (2014). https://doi.org/10.1007/978-3-319-11194-0_52
39. M. Kumar, M. Hanumanthappa, Cloud based intrusion detection architecture for smartphones, in *ICIIECS 2015–2015 IEEE International Conference on Innovations in Information, Embedded and Communication Systems*, (2015). https://doi.org/10.1109/ICIIECS.2015.7193252
40. T.M. Marengereke, K. Sornalakshmi, Cloud based security solution for android smartphones, in *IEEE International Conference on Circuit, Power and Computing Technologies, ICCPCT 2015*, (2015). https://doi.org/10.1109/ICCPCT.2015.7159512
41. Y. Mehmood, M.A. Shibli, A. Kanwal, R. Masood, Distributed intrusion detection system using mobile agents in cloud computing environment, in *Proceedings – 2015 Conference on Information Assurance and Cyber Security, CIACS 2015*, (2016). https://doi.org/10.1109/CIACS.2015.7395559

42. H. Toumi, M. Talea, K. Sabiri, A. Eddaoui, Toward a trusted framework for cloud computing, in *Proceedings of 2015 International Conference on Cloud Computing Technologies and Applications, CloudTech 2015*, (2015). https://doi.org/10.1109/CloudTech.2015.7337013
43. A. Fischer et al., CloudIDEA: A malware defense architecture for cloud data centers, in *Lecture Notes in Computer Science (including subseries Lecture Notes in Artificial Intelligence and Lecture Notes in Bioinformatics)*, (2015). https://doi.org/10.1007/978-3-319-26148-5_40
44. C.N. Modi, network intrusion detection in cloud computing, in *Emerging Research in Computing, Information, Communication and Applications*, (2015)
45. T. Singh, S. Verma, V. Kulshrestha, S. Katiyar, Intrusion detection system using genetic algorithm for cloud, in *ACM International Conference Proceeding Series*, (2016). https://doi.org/10.1145/2905055.2905175
46. S. Hou, A. Saas, L. Chen, Y. Ye, Deep4MalDroid: A deep learning framework for android malware detection based on Linux kernel system call graphs, in *Proceedings – 2016 IEEE/WIC/ACM International Conference on Web Intelligence Workshops, WIW 2016*, (2017). https://doi.org/10.1109/WIW.2016.15
47. W.G. Hatcher, D. Maloney, W. Yu, Machine learning-based mobile threat monitoring and detection, in *2016 IEEE/ACIS 14th International Conference on Software Engineering Research, Management and Applications, SERA 2016*, (2016). https://doi.org/10.1109/SERA.2016.7516130
48. T. Dbouk, A. Mourad, H. Otrok, C. Talhi, Towards ad-hoc cloud based approach for mobile intrusion detection, in *International Conference on Wireless and Mobile Computing, Networking and Communications*, (2016). https://doi.org/10.1109/WiMOB.2016.7763251
49. H.A. Kholidy, A. Erradi, S. Abdelwahed, F. Baiardi, A risk mitigation approach for autonomous cloud intrusion response system. Computing (2016). https://doi.org/10.1007/s00607-016-0495-8
50. N. Pandeeswari, G. Kumar, Anomaly detection system in cloud environment using fuzzy clustering based ANN. Mob. Networks Appl. (2016). https://doi.org/10.1007/s11036-015-0644-x
51. U. Nagar, X. He, P. Nanda, Z. Tan, A framework for data security in cloud using collaborative intrusion detection scheme, in *ACM International Conference Proceeding Series*, (2017). https://doi.org/10.1145/3136825.3136905
52. F. Tong, Z. Yan, A hybrid approach of mobile malware detection in android. J. Parallel Distrib. Comput. (2017). https://doi.org/10.1016/j.jpdc.2016.10.012
53. A. Nezarat, A game theoretic method for VM-To-hypervisor attacks detection in cloud environment, in *Proceedings – 2017 17th IEEE/ACM International Symposium on Cluster, Cloud and Grid Computing, CCGRID 2017*, (2017). https://doi.org/10.1109/CCGRID.2017.138
54. D. Moloja, N. Mpekoa, Towards a cloud intrusion detection and prevention system for M-voting in South Africa, in *International Conference on Information Society, i-Society 2017*, (2018). https://doi.org/10.23919/i-Society.2017.8354666
55. V. Balamurugan, R. Saravanan, Enhanced intrusion detection and prevention system on cloud environment using hybrid classification and OTS generation. Cluster Comput. (2019). https://doi.org/10.1007/s10586-017-1187-7
56. H. Idrissi, M. Ennahbaoui, S. El Hajji, E.M. Souidi, A secure cloud-based IDPS using cryptographic traces and revocation protocol, in *Lecture Notes in Computer Science (including subseries Lecture Notes in Artificial Intelligence and Lecture Notes in Bioinformatics)*, (2017). https://doi.org/10.1007/978-3-319-55589-8_24
57. A. Nezarat, Y. Shams, A game theoretic-based distributed detection method for VM-to-hypervisor attacks in cloud environment. J. Supercomput. (2017). https://doi.org/10.1007/s11227-017-2025-7
58. S. Raja, S. Ramaiah, An efficient fuzzy-based hybrid system to cloud intrusion detection. Int. J. Fuzzy Syst. (2017). https://doi.org/10.1007/s40815-016-0147-3
59. S. Velliangiri, J. Premalatha, Intrusion detection of distributed denial of service attack in cloud. Cluster Comput. (2019). https://doi.org/10.1007/s10586-017-1149-0

60. A.S. Sohal, R. Sandhu, S.K. Sood, V. Chang, A cybersecurity framework to identify malicious edge device in fog computing and cloud-of-things environments. Comput. Secur. (2018). https://doi.org/10.1016/j.cose.2017.08.016
61. Y. Li, M. Du, J. Xu, A new distributed intrusion detection method based on immune mobile agent, in *Proceedings – 2018 6th International Conference on Advanced Cloud and Big Data, CBD 2018*, (2018). https://doi.org/10.1109/CBD.2018.00046
62. S. Ghribi, A.M. Makhlouf, F. Zarai, C-DIDS: A Cooperative and Distributed Intrusion Detection System in Cloud environment, in *2018 14th International Wireless Communications and Mobile Computing Conference, IWCMC 2018*, (2018). https://doi.org/10.1109/IWCMC.2018.8450478
63. K.K. Nguyen, D.T. Hoang, D. Niyato, P. Wang, D. Nguyen, E. Dutkiewicz, Cyberattack detection in mobile cloud computing: A deep learning approach, in *IEEE Wireless Communications and Networking Conference, WCNC*, (2018). https://doi.org/10.1109/WCNC.2018.8376973
64. S. Ravji, M. Ali, Integrated intrusion detection and prevention system with honeypot in cloud computing, in *Proceedings – 2018 International Conference on Computing, Electronics and Communications Engineering, iCCECE 2018*, (2019). https://doi.org/10.1109/iCCECOME.2018.8658593
65. T. Qin, R. Chen, L. Wang, C. He, LMHADC: Lightweight method for host based anomaly detection in cloud using mobile agents, in *2018 IEEE Conference on Communications and Network Security, CNS 2018*, (2018). https://doi.org/10.1109/CNS.2018.8433208
66. O. Achbarou, M.A. El Kiram, O. Bourkoukou, S. Elbouanani, A multi-agent system-based distributed intrusion detection system for a cloud computing, in *Communications in Computer and Information Science*, (2018). https://doi.org/10.1007/978-3-030-02852-7_9
67. E. Besharati, M. Naderan, and E. Namjoo, "LR-HIDS: logistic regression host-based intrusion detection system for cloud environments," J. Ambient Intell. Humaniz. Comput., 2018, doi: https://doi.org/10.1007/s12652-018-1093-8
68. H. Kim, J. Kim, Y. Kim, I. Kim, K.J. Kim, Design of network threat detection and classification based on machine learning on cloud computing. Cluster Comput. (2019). https://doi.org/10.1007/s10586-018-1841-8
69. C. Modi, D. Patel, A feasible approach to intrusion detection in virtual network layer of Cloud computing. Sadhana – Acad. Proc. Eng. Sci. (2018). https://doi.org/10.1007/s12046-018-0910-2
70. K. Peng, L. Zheng, X. Xu, T. Lin, V.C.M. Leung, Balanced iterative reducing and clustering using hierarchies with principal component analysis (PBirch) for intrusion detection over big data in mobile cloud environment, in *Lecture Notes in Computer Science (including subseries Lecture Notes in Artificial Intelligence and Lecture Notes in Bioinformatics)*, (2018) https://doi.org/10.1007/978-3-030-05345-1_14
71. R. Rajendran, S.V.N. Santhosh Kumar, Y. Palanichamy, K. Arputharaj, Detection of DoS attacks in cloud networks using intelligent rule based classification system. Cluster Comput. (2019). https://doi.org/10.1007/s10586-018-2181-4
72. J. Ribeiro, G. Mantas, F.B. Saghezchi, J. Rodriguez, S.J. Shepherd, R.A. Abd-Alhameed, Towards an autonomous host-based intrusion detection system for android mobile devices, in *Lecture Notes of the Institute for Computer Sciences, Social-Informatics and Telecommunications Engineering, LNICST*, (2019). https://doi.org/10.1007/978-3-030-05195-2_14
73. S. Dey, Q. Ye, S. Sampalli, A machine learning based intrusion detection scheme for data fusion in mobile clouds involving heterogeneous client networks. Inf. Fusion (2019) https://doi.org/10.1016/j.inffus.2019.01.002
74. Y. Weng, L. Liu, A collective anomaly detection approach for multidimensional streams in mobile service security. IEEE Access (2019). https://doi.org/10.1109/ACCESS.2019.2909750
75. J. Ribeiro, F.B. Saghezchi, G. Mantas, J. Rodriguez, S.J. Shepherd, R.A. Abd-Alhameed, An autonomous host-based intrusion detection system for android mobile devices. Mob. Netw. Appl. (2020). https://doi.org/10.1007/s11036-019-01220-y
76. Q. Zhou, F. Feng, Z. Shen, R. Zhou, M.Y. Hsieh, K.C. Li, A novel approach for mobile malware classification and detection in Android systems. Multimed. Tools Appl. (2019). https://doi.org/10.1007/s11042-018-6498-z

Vulnerability of Virtual Private Networks to Web Fingerprinting Attack

Khaleque Md Aashiq Kamal and Sultan Almuhammadi

1 Introduction

Privacy in the internet is a challenge for both users and service providers. Every step of internet activities can be tracked by the eavesdropper. An internet user location might be disclosed easily. If a user accesses his own residence server remotely, an eavesdropper can easily snip the traffic between server of that residence and user. Therefore, presently accessing private resources securely is observed as most crucial need. Even though maintaining privacy is not an easy task, Virtual Private Networks (VPNs) are one of the most effective ways to achieve privacy in the internet [1]. The main objectives of the VPNs are to evade the sniffing attack and to maintain the data integrity in the untrusted network of the internet [2].

In VPN communication, all the traffic is end-to-end encrypted between the VPN client user and the VPN server. The IP of the destination web address is kept hidden to the hops in the VPN tunnel. In this case, any eavesdropper will fail to find out which web page is being actually accessed by the VPN user as the IP is hidden. As a result, sometimes people might misuse the VPN service. For example, companies may not allow their employees to use social media, video streaming site, playing online games, etc. during office hours and sometimes schools block some website for their students for containing adult content [3]. This censorship can be easily avoided by using VPN services.

According to the renowned market research company Global Web Index [4], around 410 million people all over the world use anonymous software, like: VPN, Tor browser, Proxy Servers, etc. to hide their identity. Among them, 166 million

K. Md. Aashiq Kamal · S. Almuhammadi (✉)
College of Computer Sciences and Engineering, King Fahd University of Petroleum and Minerals, Dhahran, Saudi Arabia
e-mail: sultan@almuhammadi.com

K. Daimi et al. (eds.), *Advances in Security, Networks, and Internet of Things*, Transactions on Computational Science and Computational Intelligence,
https://doi.org/10.1007/978-3-030-71017-0_11

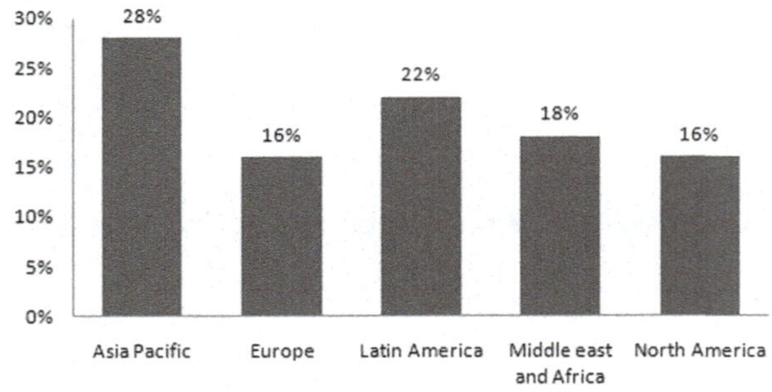

Fig. 1 Frequency of VPN users in different region

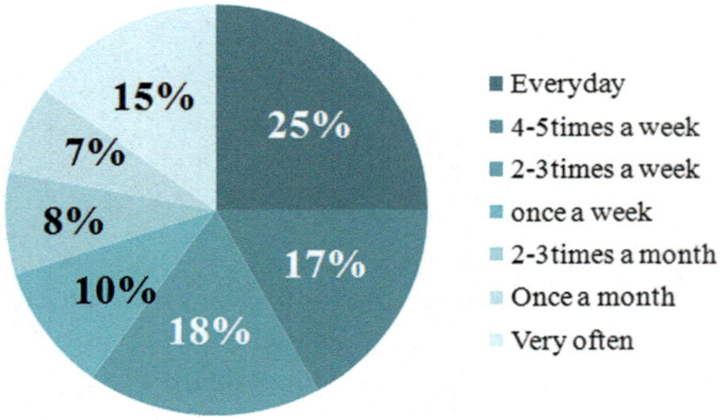

Fig. 2 Frequency of uses of VPNs

people use VPNs. Figure 1 shows the region-wise VPN users in the world. It shows that users in Asia pacific and Latin America region use VPN more than other regions in the world. They have 50% of the total VPN users. Middle East come next with 18% of VPN users. While Europe and North America have 16% each. The frequency of the VPN usage varies from daily to once a month as shown in Fig. 2. One fourth of the VPN users need it every day. While 7% of them use it once a month.

Moreover, the statistics in [4] show that most of the VPNs usage is not limited to security purposes. In fact, 72% of the VPNs users need it to access blocked websites, access blocked content at work, or hide identity from government. Therefore, it is very important to analyze VPN traffic to find out which web service is accessed by any given user.

This paper presents a comparative analysis of five different VPN services based on the website fingerprinting attack given by Cai et al. [5], to find out which VPN service is more vulnerable to this attack. Moreover, it estimates the efficiency of the web traffic classification through Psiphon VPN for four different web services, namely: video call communication, video streaming, online gaming site, and peer-to-peer file sharing.

The remaining of this paper is as follows: Sect. 2 gives a general background on VPN. While Sect. 3 discussed related work of the website fingerprinting attack in the literature. Section 4 explains the main goals of this study. Section 5 highlights the main phases of our methodology. The data collection process and the fingerprinting techniques are discussed in Sects. 6 and 7, respectively. The results are presented in Sect. 8, and further discussed in Sect. 9. Finally, the conclusion comes in Sect. 10 with useful recommendations based on our results.

2 Background

VPNs transfer sensitive information through public networks while minimizing the risk of information exposure. The goal is achieved by encrypting traffic payload at the source before sending it through public network while decrypting at the destination. An usual VPN tunnel communication is shown in Fig. 3. VPNs permit user of internet to transmit and receive data over open networks transversely, as if they were connected to a private network. This also helps avoiding blockade and overcoming geo-restrictions.

2.1 Security Protocols in VPN

There are many variations of VPN in terms of the tunneling protocol used to create a secure tunnel. Two of the most popular protocols in use are the Internet Protocol Security (IPsec) and the Secure Socket Layer/Transport Layer Security (SSL/TLS).

- **IPSec**: IPsec is standard network protocol suite that is used to provide a security at the network layer of the Internet model. It contains three protocols: Authentication Header (AH) Protocol, Encapsulating Security Payload (ESP) Protocol, and Internet Key Exchange (IKE) Protocol. IPsec operates in one of two different modes: transport mode, which is typically used for host-to-host communication, and tunnel mode, which is used when one or both ends of a security association are a security gateway [6].
- **SSL/TLS**: SSL/TLS is one of the most widely used security protocols at the transport layer of the Internet model. It is a general-purpose service implemented

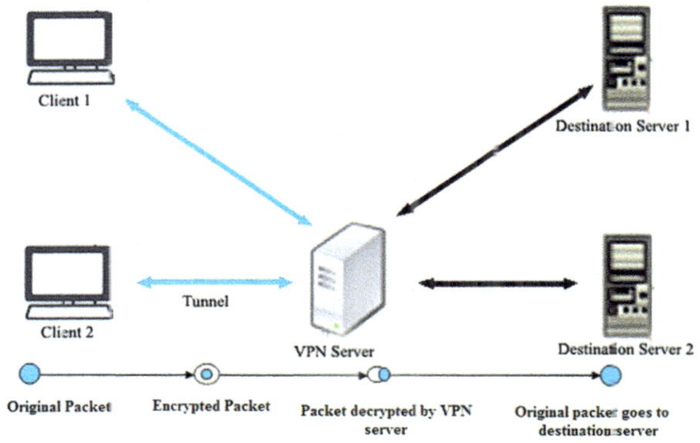

Fig. 3 Encryption configuration through a VPN tunnel

as a set of protocols that rely on TCP/SCTP. It provides five services which are fragmentation, compression, authentication, confidentiality, and framing. It can be implemented either as a part of the underlying protocol suite or embedded in specific packages. It accomplishes its tasks by four protocols (Record, Alert, Change Cipher Spec, and Handshake protocols) in two layers [6].

Since IPsec works at the network layer and SSL/TLS works at the transport layer, they are both capable of protecting the traffic payload, but in different ways. SSL/TLS does not need any extra header to hide the metadata, but IPSec makes a new header to protect the metadata.

IPSec protects metadata of the payload, by making a new header. On the other hand, SSL/TLS does not protect the metadata, and therefore, it does not need any extra header to hide the metadata. Typical structures of an IPsec packet and an SSL/TLS packet are shown in Figs. 4 and 5.

Fig. 4 IPSec VPN tunnel

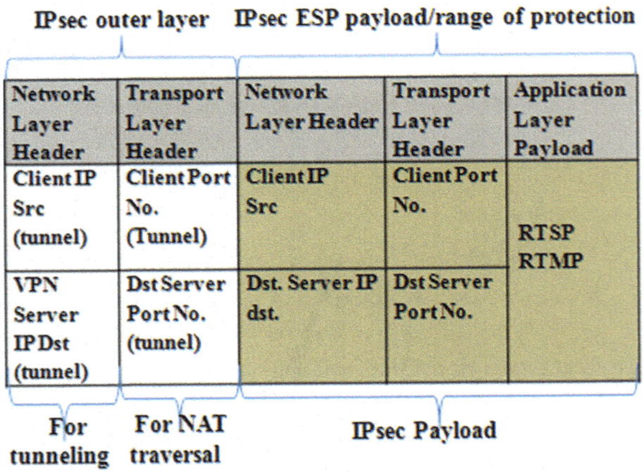

Fig. 5 SSL/TLS VPN tunnel

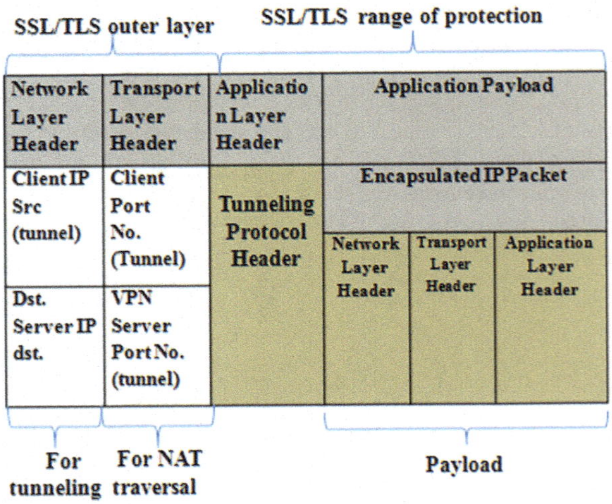

2.2 VPN Services

There are many different VPN services. We choose five of them in this work. Table 1 shows a technical summary of the VPN services used in this work. For each VPN service, it shows the communication and security protocols. It also indicates whether or not the VPN service changes the IP address (IP Change) or the port number (Port No. Change).

2.3 Fingerprinting Approach

Traffic analysis or fingerprinting is technological process that allows capturing the traffic activities even though its content is concealed or encrypted. For the web applications or services, it is attained by observing specific patterns inside the traffic packets. Those features can direct to the web services accessed by the users. These features or patterns are size of network packet, and direction of the packet.

To build a fingerprinting approach, a number of sub-processes needs to be achieved. Figure 6 shows the complete scenario of the fingerprinting approach. First, inward/outward traffic of internet should be observed by using a network analyzer or snipping tool, like *Wireshark*. Then, a number of traces of web access is recorded. Next, selected packet features are collected to signify the fingerprint for particular web services. Those features are classified through a specific Machine Learning algorithm. After the training stage, a testing stage is executed and the accuracy of the classifier algorithm is observed and further fine-tuned accordingly.

3 Related Work

There is a good number of works on encrypted and non-encrypted traffic analysis. This section highlights the related work in both types of traffic analysis, with more focus on the analysis of encrypted traffic, more particularly on Tor and VPN traffic analysis.

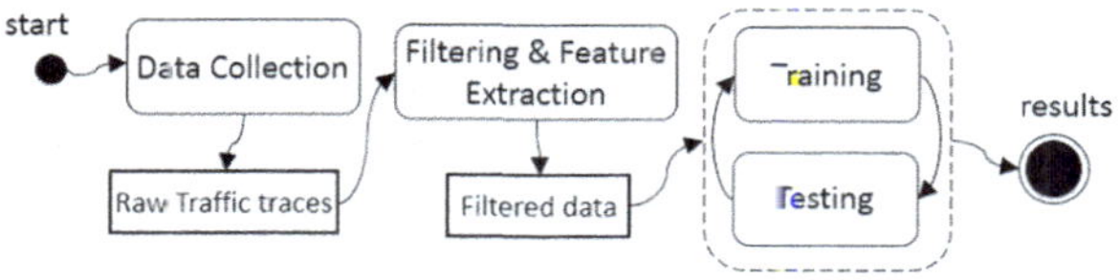

Fig. 6 Fingerprinting procedure model

Table 1 Technical summary of VPNs

Name	Protocol	Change of IP	Changes of port no. (during session)	Changes of port no. (new session)	Security protocol
Softether	UDP	Yes	No	Yes	IPSec/SSL
Open VPN	UDP	Yes	No	Yes	Own protocol based on SSL
Avira	TCP	Yes	No	Yes	Own protocol based on SSL
Psiphon	TCP	Yes	Yes	Yes	IPSec/SSL
Hotspot shield	TCP	Yes	No	Yes	SSL
Tunnel bear	TCP	Yes	No	Yes	SSL
Betternet	UDP	Yes	No	Yes	SSL

3.1 Web Traffic Classification

Li and Moore [7] used a supervised method named C4.5 to categorize the internet traffic. They classified browsing, peer to peer application, e-mail, FTP, and services with good accuracy rate of 99%, but it was not on encrypted traffic. Moor and Zaev [8] also classified non-encrypted internet traffic based on port number, inter arrival time, flow length, etc. They used Naive Bayes classifier to distinguish among peer to peer, email, web browsing, and multimedia and achieved an accuracy level of 95%.

MCgregor et al. [9] proposed a method to distinguish different traffic pattern including: SMTP, HTTP, DNS, FTP, etc. They clustered different trafic flows based on their pattern by employing Expectation Maximization algorithm. Zander et al. [10] also proposed a method to cluster different traffic flow which includes: Telnet, FTP, HTTP, SMTP etc. They used AutoClass (which is based on Bayesian algorithm) for clustering the traffic.

Another method based on clustering approach was developed by Bernaille et al. [11], in which K-means algorithm was used to cluster the traffic flow. The k-means uses distance vector scheme to cluster the given data. Their method was only capable to classify TCP based traffic. Their study includes: HTTP, POP3, SMTP, SSH , HTTPS, etc. K-means algorithm was also used in Erman et al. [12] to distinguish peer to peer traffic from normal web traffic including FTP. They assumed that traffic flow can be distinguished using payload and information of header. Another approach was studied by Junior et al. [13] to identify P2P traffic from other application type traffic.

Perenyi and Molnar [14] and Freire et al. [15] independently proposed methods to identify Skype traffic flows from normal web traffic. However, Freire et al. achieved a relatively better success rate with 5% false positive.

According to the literature, we found that some researchers focused on only one application type, while others focused on number of applications. HTTP, SMTP, P2P, web traffic protocols are the main focus in most of these studies. : Moreover, encrypted traffic classification is mostly related to either SSL or SSH. However, analysis or classification of encrypted traffic faces more challenges and difficulties than others.

Wright et al. [16] proposed a model to identify traffic flow in encrypted communication using Hidden Markov Models (HMM). They considered as features: packet sizes, direction, and timing information and set the main focus on HTTPS communication. They identified 20% applications in their work.

Alshammari and Zincir-Heywood [17] worked on SSH traffic and used machine learning algorithms, like RIPPER and AdaBoost, to classify the SSH traffic without knowing the payload (IP address and port numbers). They succeed to classify applications such as DNS, FTP, and telnet with high accuracy of 99%. They extended their work in [18], to identify traffic of Skype as a P2P VoIP. They successfully distinguished between SSH traffic from non-SSH and Skype from non-Skype. They used five types of machine learning algorithms, namely: SVM, Naive

Bayes, RIPPER, AdaBoost, and C4.5 to deploy their work. Among these algorithms, C4.5 has given the best result of 97% accuracy.

Leroux et al. [19] also developed a fingerprinting attack based on machine learning techniques. This attack targeted traffic through IPsec and Tor tunnel. They distinguished between four types of traffic: web browsing, voip, video streaming, and P2P. But, they only considered single application from every type of traffic. Naive Bayes, logistic regression, and random forest were used as classifier algorithm. They considered the timing and size of the packets as features to train the classifier.

3.2 Web Fingerprinting Through Encrypted Communication

In [20], the author presented an analysis based on weakness in web proxy named SafeWeb. This weakness exposes to eavesdroppers the website being browsed by the user. To analyze the test dataset, the author used the client's port number with the size and the direction of the traffic. Their program can decide how to distinguish two fingerprints of different websites.

On the other hand, Sun et al. [21] identified web traffic in a SSL communication from a large sample. Their traffic signature was based on website's requested object number and size of objects. Using this approach, they achieved an accuracy level of 75% in identifying web traffic.

Liberatore and Levine [22] showed how unique packet lengths are a powerful WF feature. They made two attacks using the Jaccard coefficient and the Naive Bayes classifier. Each packet sequence was mapped to its set of unique packet length, and Jaccard coefficient attack was used to measure it. It discarded packet ordering and packet frequency. After that, the Jaccard coefficient has been used to measure the distance of the sequences of two packets. The classifier of Naive Bayes also used packet lengths and their frequencies of occurrence, but the packet ordering and timing are discarded. The Naive Bayes assumption is that the probabilities of occurrence of different packet lengths are independent of each other.

Later, Herrmann et al. [23] presented a fingerprinting approach in different encrypted traffic that uses text mining techniques. They used Multinomial Naive Bayes as a machine learning classifier. Single hop and multihop systems have been used in their approach. They considered OpenSSH, OpenVPN, CiscoVPN, Stunnel as single hop and Tor as multihop systems. Their work can correctly classify 97% of unique website from encrypted communication.

Zhou et al. [24] developed a website fingerprinting attack based on Profile Hidden Markov Model (PHMM). They conducted experiments of both closed world and open world scenario by collecting web dataset via SSH. They used packet size and direction as feature. They achieved more than 95% accuracy rate in every case.

3.3 Web Fingerprinting Through Tor Communication

In 2012, Cai et al. [5] also used SVM to classify web fingerprinting attack on Tor communication. For post-processing data, they used Damerau–Levenshtein edit distance algorithm for calculating distance between packet sizes of two different traffic traces of web browsing, which is described in more detail in Section VIII. They achieved an accuracy of 87%.

In 2013, Tao Wang and Ian Goldberg [25] improved the attack for tor by modifying the distance based algorithm of Cai et al. In the closed world experiments, their accuracy is 91%, as compared to 87% from the best previous classifier on the same data.

Jahani and Jalili [26] introduced a technique based on the Fast Fourier Transform (FFT) to estimate similarity distance between two different instances from traffic flows.

Tobias Pulls and Rasmus Dahlberg [27] introduced a robust technique based on the Website Oracle (WO). It gives a website fingerprinting attacker the ability to find out whether a particular website was among the websites visited by Tor users during the victim's trace. Their work showed that combining of website oracle and website fingerprinting significantly decreases false positive (FP) for about half of the visited websites. They also used packet size and direction as feature to train the classifier. They achieved 95% in tor network.

3.4 Web Fingerprinting Through VPN

Shi and Biswas [3] worked on traffic analysis to detect web traffic in encrypted tunnel named Juniper VPN. Their target was to detect video streaming from encrypted tunnel. They designed a signature based on the packet size and timing of each packet. Finally, they achieved good results for BayesNet Classifier with lower false positive rate. In their related work [28], packet size distribution has been used as a feature vector to analyze the traffic. J4.8 tree classifier has been used in this work to recognize the video stream from other web traffic. They achieved 90% accuracy in JuniperVPN, which is the same accuracy achieved by Herrman et al. in [23].

Shi and Biswas also used traffic analysis to detect encrypted video traffic [29] where packet arrival interval (PAI) was used as a classification feature to detect video streaming traffic from encrypted OpenVPN tunnel. They also used J4.8, SVM, and 1-NN as classifier in this work. The 1-NN classifier gave the best results with 94% accuracy. Table 2 summarizes the details of these web fingerprinting techniques through VPN.

Feghhi et al. [30] introduced a traffic analysis attack on VPN traffic. They used timing information of packet as feature for analyzing the encrypted traffic. Their success rate is 90%. This attack is suitable for wired and wireless network.

Table 2 Summary of the web fingerprinting through VPN

Study	Techniques	Feature	Domain	Result
Herrman et al. [23]	Multinomial Naïve Bayes	Packet size and direction	Open VPN and Cisco VPN	90%
Shi et al. [3]	BayesNet classifier	Packet sizes, timing info	Video traffic in Juniper VPN	80%
Shi et al. [28]	Decision tree classifier	Packet Size distribution	Video traffic in Juniper VPN	90%
Shi et al. [29]	k-NN	Packet arrival information	Video traffic in Open VPN	94%

According to our literature review no work has been conducted on different VPN services to find out which web services were accessed by the user. In this work, different top most visited dynamic web pages, social networking sites, video streaming sites, online games, and video communications will be considered as web services. Moreover, no work has been found related to traffic classification with diverse and mixed traffic dataset through VPN.

4 Objective

The main goals of this paper are as follows:

- ***Website fingerprinting of different VPNs***: There are several VPN services today, and the most commonly used are listed in Table 1 with brief descriptions. The listed VPN services have been installed and configured for our experiment. Different VPN services use different security protocols to encapsulate their network traffic before sending it to the public network. We performed a comparative analysis of these different VPN services based on website fingerprinting attacks. The comparative analysis gives a clear idea about the vulnerability of these VPNs to web fingerprint attack. The goal here is to assess how vulnerable the different VPNs are to this type of attack.
- ***Website traffic Classification through VPN***: For the second goal, we use one VPN service to access different web services. Then we record the encrypted traffic, and analyze it using fingerprint to retrieve high level information, namely the web service has been accessed through VPN. The goal here is to identify which web service is accessed given only the recorded encrypted traffic.

5 Methodology

The proposed experiment builds an environment for different VPN services and uses the real world web through these VPNs to study the web traffic applying website

Table 3 Experimental setup

Operating system	Windows (64 bit)
CPU	CPU 2020M 2.40GHz
Physical memory	2048 MB
Browser	Google Chrome version 57.0.2987.133
Network protocol analyzer	*tshark* 2.2.1

fingerprinting attack on these VPNs, in order to find out the identity of the accessed website. This work will be carried out in several phases as follows:

- ***Installing and Configuring VPN Services***: This phase starts after selecting five most commonly used VPN services. Selected VPN tools will be installed and configured to real world uses for different web services.
- ***Analyzing VPN Services***: In this phase, installed VPN services are analyzed according to their distinctiveness. More specifically, the protocol (TCP or UDP) which is being used by any specific VPN tools is studied.
- ***Data Collection of Different Web Services Traffic***: This phase involves collecting data of different types of web traffic through five VPN services. Top twenty most visited websites according to Alexa [31] have been used as a data source for website fingerprinting attack. In order to do the, traffic classification, we collected data of video calls, video streaming, online gaming, and P2P sites. Every access for all applications has been repeated for forty times. Then the individual incoming and outgoing web traffic for each access has been captured at the client side by using *tshark* tool.
- ***Data Setup***: After collecting the required data, we need to process them to make appropriate format for the Cai et al. [5] fingerprinting techniques The details are given in Sect. 6.
- ***Website fingerprinting attack***: We apply Cai et al. [5] fingerprinting techniques on collected data of five different VPNs.
- ***Web traffic Classification***: We investigate the recorded data in offline mode to identify which specific web service is accessed by a given user based on the captured traffic data.

6 Data Collection and Setup

In this experiment, we used a fresh windows based computer. Table 3 shows the details of the computer and installed software. We have chosen Google Chrome as it is more secure than other browsers for fingerprinting attacks [32]. Then we installed and configured five most commonly used VPN client in the client machine. As discussed in Sect. 2.

6.1 Data Collection for Website Fingerprinting Through VPN

In order to apply the fingerprinting attack on the five different VPN services, twenty most popular websites have been chosen according to Alexa [31]. The list of these websites with their type is given in Table 4. An encrypted VPN communication has been initiated from VPN client to the VPN server. Website traffic traces are collected by visiting every website forty times. The visited traffic has been logged using network protocol analyzer *tshark*.

To automate the browsing and traffic capture process we used a windows batch program. The script follows the following steps: (1) It opens the Chrome browser and enables the *tshark*, (2) it reads the file containing website names and requests the browser to open it, (3) it waits for 30 s to load the website, (4) it captures and stores the individual traffic, and (5) it waits 10 s to ensure a delay between two visits of websites. The whole process is repeated forty times for twenty websites, one for each of the five VPNs. In order to avoid the noise in the traffic the browser cache has been completely cleared after every iteration. The data collection process is done during morning, evening, and night to simulate the real network traffic scenario. We ended up having $20 \times 40 \times 5 = 4000$ *pcap* files of web traffic at the end of the capture process.

Table 4 List of top 20 website

Website name	Type of website
www.google.com	Search engine
www.youtube.com	Video sharing
www.facebook.com	Social network
www.baidu.com	Search engine
www.wikipedia.com	Encyclopedia
www.yahoo.com	Portal media
www.amazon.com	E-commerce
www.qq.com	Portal media
www.live.com	Software services
www.taobao.com	E-commerce
www.vk.com	Social network
www.twitter.com	Social network
www.instagram.com	Social network
www.hao123.com	Web directories
www.sohu.com	Portal
www.sina.com.cn	Portal
www.reddit.com	Entertainment
www.linkedin.com	Social network
www.tmall.com	E-commerce
www.weibo.com	Social network

6.2 Data Collection for Web Traffic Classification Through VPN

For applying the fingerprinting attack on different web services, we selected four common web services: video call communication, video streaming website, online gaming sites, and peer to peer file sharing. From every services, we selected three different service providers. The list is given in Table 5. An encrypted VPN communication has been initiated from VPN client to the VPN server. To collect the video call communication data, 40 times video call has been initiated using three different service providers mentioned in Table 5. The same process is repeated for other services for only one VPN to collect data.

6.3 Data Setup for Cai Classifier

We have used Cai et al. [5] classifier to evaluate the vulnerability of the different VPNs to website fingerprinting attack. After collecting both types of data (for website fingerprinting and web traffic classification), we process them to make appropriate format for the Cai et al. fingerprinting techniques. We used *tsahrk* command to filter out all the packet sizes from stored *pcap* files of each and every iteration of web browsing and web services. The packet sequences of every website visit are stored in different text files of the form *X_N.txt*, where *X* is the number of website, and *N* is the trace number of that website. For example, the file 5_1.txt contains processed packets with their corresponding packet sizes in the first attempt web trace of the fifth website (which is wikipedia).

To automate the process, we used a batch script which reads all forty captured *pcap* files of one individual website from a folder, then filter out the packet sizes and sequences from that file and save it on text files. The process is repeated for 20 websites for each VPN. Then the data process is also repeated on captured data of traffic classification through VPN.

Table 5 List of different services

Services types	Name of the service provider
Video call communication	Facebook, Hangout, Skype
Video streaming	Youtube, Metacafe, Vimeo
Online gaming	Dota2, Solitaire, Patterns
Peer2Peer file sharing	Thepiratebay, Extratorrent, Torrentz2

7 Fingerprinting Techniques

In this experiment, we used Cai et al. [5] approach to assess the accuracy of website fingerprinting attack on five VPNs. Moreover, we used it for classifying different web services through VPN. Cai et al. developed an efficient website fingerprinting attack based on SVM. This approach takes specific types of input data files, named: XN.txt where X is the website and N is the trial number. All combinations of X and N, where $1 \leq X \leq$ WebsiteNum, $1 \leq N \leq$ TrialNum, must exist in the folder, or the process cannot continue. The user can set website number and trial number. Other combinations are ignored. Each such file represents a traffic instance and it is a list of integers separated by newline.

Once the data file is read, it calculates the Damerau–Levenshtein edit distance [33], which is a string metric for measuring the difference between two sequences. It is the minimum number of single character edits (insertions, deletions, substitutions, transpositions) required to change one word into the other. The bigger the return value is, the less similar the two text are, because different words take more edits than similar words.

In Cai et al. work, the costs of insertion, deletion, and substitution are the same, but they assign a lower cost to transpositions than others. After post-processing the input data, the training and testing are done based on SVM classifier. Finally it gives an output result with percentage accuracy.

8 Results and Analysis

In this section, we show the results of our experiment and analyze them. Further discussion is given in Sect. 9.

8.1 Website Fingerprinting Through VPN

Popular VPNs differ in several aspects. In particular, they use different protocols and different packet sizes for same websites. The types of the protocols and packet sizes influence the encrypted communication traffic. In order to compare the resistance of popular VPNs to website fingerprinting attacks, traces of web visits are collected using different VPNs and converted to the setup of Cai's technique as mentioned in Sect. 7.

We evaluated the accuracy of website fingerprinting attack using the data of most popular five VPNs as shown in Table 6. The accuracy results of the experiment are illustrated in histogram shown in Fig. 7. Avira Phantom has the best resistance to the attack with almost only 32% of website visit traffic recognized appropriately. The next best VPN in resisting this attack is Psiphon, with the accuracy of about 42%.

Table 6 Website fingerprinting using Cai classifier

Name of the VPNs	Classification accuracy (%)
Psiphon	41.125%
Softether	49.75%
Hotspot Shield	65.00%
OpenVPN	72.00%
Avira Phantom	31.50%

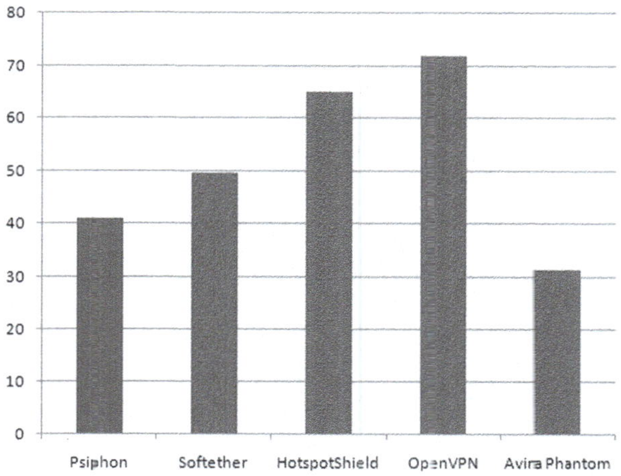

Fig. 7 Accuracy of the website fingerprinting attack on the five VPNs using Cai et al.

Next come Softether VPN, with accuracy of around 50%. Finally, Hotspot Shield and OpenVPN have the highest levels of accuracy of 65% and 72% respectively, which implies that they are very vulnerable to Cai's technique. It states that almost three-fourth of the website visits through OpenVPN can be detected successfully by this fingerprinting attack.

8.2 *Web Traffic Classification Through VPN*

In order to classify between different web services, uses of different web service mentioned in Table 5 are collected using Psiphon VPN, and the traces of traffic are converted to the setup essential for the Cai's technique as mentioned in Section VIII. Different video Call communication, video Streaming, online gaming, and Peer2Peer file sharing have been used through Psiphon VPN to collect data. We assess the accuracy of web services classification attack using the Psiphon VPN's data.

Table 7 depicts the accuracy findings of the experiment on our diverse collected data. The average accuracy of traffic classification through VPN is around **82%**. It implies that 82% of different web services have been correctly classified. This traffic classification method has a great impact in the field of censorship. If

Table 7 Efficiency of the web traffic classification estimated with Cai et al.

Results	Classification accuracy (%)
	68.75%
	81.25%
	84.375%
	84.375%
	87.50%
Average	81.25%

any organization wants to block video streaming sites for their employees, this classification technique can be used to distinguish real web traffic and apply censorship on the contents. Although the user depends on VPN communications, our technique can successfully intercept and correctly detect with the high accuracy rate (over 80%).

9 Discussion

The purpose of this section is to discuss how different factors may affect the accuracy rate of our experiments.

9.1 Challenges of Data Storage

When the analysis is done in offline mode, maintaining the data storage of different websites is a real challenge. To maintain the huge signatures of each visited websites is both time and space consuming. On the other hand, discovering which web services are targeted is a complex assessment as it is not known to the internet service provider which service is suspicious.

9.2 Web Browsers Variety

Web browsers are numerous nowadays with various editions. Every browser can have special strategy about transmission of packets, which can take part in defense of consumer's privacy. Particularly, traces created by using a web service can have a different mark or signature, depending on the browser which was employed. This matter deserves a study on how security mechanisms are engaged by browsers activity. This study can facilitate in fingerprinting the browsers in the early period of investigation, and subsequently, investigate web services based on the resulting browser. During our investigation, we have observed that most fingerprinting techniques have been estimated under data collected from the Firefox web browser. Based on [32], the accuracy of fingerprinting techniques might be change if evaluated under data collected from different web browsers, given that browsers might contribute in preserving user's privacy. In our work, we used Google chrome which is the most secure [32]. The less secure browser may increase the vulnerability rates.

9.3 *Diversity of Data Collection Ways*

The majority of web fingerprinting techniques are performed on recorded traffic, in offline mode, in several phases. First, huge amount of data is collected. Then, data is and assessed in offline mode. In contrast, different fingerprinting techniques assess the traffic in online mode. Definitely, a fingerprinting approach will not generate the similar results if performed in these different traditions. In our work, we analyzed in offline mode. The traffic analysis in online mode may affect the results since the online mode is an unpredictable open world.

10 Conclusion

The web fingerprinting attack is an effective analysis attack on encrypted traffic. It is based on monitoring the manners of traffic for the purpose of finding out valuable patterns in the packets flow. Although Web fingerprinting attack is regarded as a harmful action, it can sometimes be beneficial and be used in beneficial ways, e.g., when government agencies and organizations may require to carry out a Web fingerprinting attack for homeland security reasons and cyber-crimes prevention.

This paper investigates the vulnerability of five different VPN services through website fingerprinting attack using Cai et al. approach. It shows that Open VPN is vulnerable with 72% accuracy rate. This means that if someone browses any websites through Open VPN, 72% websites can be traced successfully by the fingerprinting attack. Although the VPN communication is totally encrypted, it leaves traceable metadata that can be further analyzed to carry out the attack. The Open VPN has the highest vulnerability as compared to other studied VPNs. On the other hand, Avira phantom VPN is the least vulnerable with accuracy rate of 32%.

Moreover, we estimated efficiency of the web traffic classification through VPN for four different web services, namely: video streaming, online games, video call communications, and peer to peer application with Cai et al. classifier. The result shows an accuracy level of almost 82%.

According to these results, we recommend avoiding the use of VPN whenever the service type is critical, since VPN is vulnerable to fingerprinting attack. However, other usages of VPN, like hiding identity of users and content of media are still protected. Moreover, government agencies need to develop more sophisticated attacks to retrieve high level information other than the types of the web services.

Acknowledgments The authors would like to thank King Fahd University of Petroleum and Minerals, Dhahran, Saudi Arabia, for supporting this research.

References

1. A. Thomas, G. Kelley, Cost-effective VPN-based remote network connectivity over the internet, Department of Computer Science, University of Massachusetts, vol. 100 (2002)
2. R. Venkateswaran, Virtual private networks. IEEE Potentials **20**(1), 11–15 (2001)
3. Y. Shi, S. Biswas, Detecting tunneled video streams using traffic analysis, in *7th International Conference on Communication Systems and Networks (COMSNETS)* (IEEE, 2015), pp. 1–8
4. Globalwebindex, www.globalwebindex.net
5. X. Cai, X.C. Zhang, B. Joshi, R. Johnson, Touching from a distance: Website fingerprinting attacks and defenses, in *Proceedings of the 2012 ACM Conference on Computer and Communications Security* (ACM, 2012), pp. 605–616
6. R. Stanton, Securing VPNs: Comparing SSL and IPsec. Comput. Fraud Secur. **2005**(9), 17–19 (2005)
7. W. Li, A.W. Moore, A machine learning approach for efficient traffic classification, in *15th International Symposium on Modeling, Analysis, and Simulation of Computer and Telecommunication Systems* (IEEE, 2007), pp. 310–317
8. A.W. Moore, D. Zuev, Internet traffic classification using Bayesian analysis techniques, in *ACM SIGMETRICS Performance Evaluation Review*, vol. 33(1) (ACM, 2005), pp. 50–60
9. A. McGregor, M. Hall, P. Lorier, J. Brunskill, Flow clustering using machine learning techniques, in *International Workshop on Passive and Active Network Measurement* (Springer, 2004), pp. 205–214
10. S. Zander, T. Nguyen, G. Armitage, Automated traffic classification and application identification using machine learning, in *The IEEE Conference on Local Computer Networks, 2005, 30th Anniversary* (IEEE, 2005), pp. 250–257
11. L. Bernaille, R. Teixeira, I. Akodkenou, A. Soule, K. Salamatian, Traffic classification on the fly. ACM SIGCOMM Comput. Commun. Rev. **36**(2), 23–26 (2006)
12. J. Erman, A. Mahanti, M. Arlitt, C. Williamson, Identifying and discriminating between web and peer-to-peer traffic in the network core, in *Proceedings of the 16th International Conference on World Wide Web* (ACM, 2007), pp. 883–892
13. G.P.S. Junior, J.E.B. Maia, R. Holanda, J.N. de Sousa, P2p traffic identification using cluster analysis, in *First International Global Information Infrastructure Symposium, 2007, GIIS 2007* (IEEE, 2007), pp. 128–133
14. M. Perényi, A. Gefferth, T.D. Dang, S. Molnár, Skype traffic identification, in *IEEE Global Telecommunications Conference, 2007, GLOBECOM'07* (IEEE, 2007), pp. 399–404
15. E.P. Freire, A. Ziviani, R.M. Salles, Detecting Skype flows in web traffic, in *IEEE Network Operations and Management Symposium, 2008, NOMS 2008* (IEEE, 2008), pp. 89–96
16. C.V. Wright, F. Monrose, G.M. Masson, On inferring application protocol behaviors in encrypted network traffic. J. Mach. Learn. Res. **7**, 2745–2769 (2006)
17. R. Alshammari, A.N. Zincir-Heywood, A flow based approach for SSH traffic detection, in *IEEE International Conference on Systems, Man and Cybernetics* (IEEE, 2007), pp. 296–301
18. R. Alshammari, A.N. Zincir-Heywood, Machine learning based encrypted traffic classification: Identifying SSH and Skype. CISDA **9**, 289–296 (2009)
19. S. Leroux, S. Bohez, P.-J. Maenhaut, N. Meheus, P. Simoens, B. Dhoedt, Fingerprinting encrypted network traffic types using machine learning, in *NOMS 2018-2018 IEEE/IFIP Network Operations and Management Symposium* (IEEE, 2018), pp. 1–5
20. A. Hintz, Fingerprinting websites using traffic analysis, in *International Workshop on Privacy Enhancing Technologies* (Springer, 2002), pp. 171–178

21. Q. Sun, D.R. Simon, Y.-M. Wang, W. Russell, V.N. Padmanabhan, L. Qiu, Statistical identification of encrypted web browsing traffic, in *IEEE Symposium on Security and Privacy* (IEEE, 2002), pp. 19–30
22. M. Liberatore, B.N. Levine, Inferring the source of encrypted http connections, in *Proceedings of the 13th ACM Conference on Computer and Communications Security* (ACM, 2006), pp. 255–263
23. D. Herrmann, R. Wendolsky, H. Federrath, Website fingerprinting: attacking popular privacy enhancing technologies with the multinomial Naïve-Bayes classifier, in *ACM Workshop on Cloud Computing Security* (ACM, 2009), pp. 31–42
24. Z. Zhuo, Y. Zhang, Z.-L. Zhang, X. Zhang, J. Zhang, Website fingerprinting attack on anonymity networks based on profile hidden Markov model. IEEE Trans. Inf. Forensics Secur. **13**(5), 1081–1095 (2017)
25. T. Wang, I. Goldberg, Improved website fingerprinting on Tor, in *Proceedings of the 12th ACM Workshop on Privacy in the Electronic Society* (ACM, 2013), pp. 201–212
26. H. Jahani, S. Jalili, A novel passive website fingerprinting attack on tor using fast Fourier transform. Computer Communications **96**, 43–51 (2016)
27. T. Pulls, R. Dahlberg, Website fingerprinting with website oracles. Proc. Privacy Enhancing Technol. **2020**(1), 235–255 (2020)
28. Y. Shi, S. Biswas, Characterization of traffic analysis based video stream source identification, in *IEEE International Conference on Advanced Networks and Telecommuncations Systems (ANTS)* (IEEE, 2015), pp. 1–6
29. Y. Shi, S. Biswas, Protocol-independent identification of encrypted video traffic sources using traffic analysis, in *IEEE International Conference on Communications (ICC)* (IEEE, 2016), pp. 1–6
30. S. Feghhi, D.J. Leith, A web traffic analysis attack using only timing information. IEEE Trans. Inf. Forensics Secur. **11**(8), 1747–1759 (2016)
31. Top visited website, http://www.alexa.com/topsites
32. S. Zhioua, M. Langar, Traffic analysis of web browsers. in *FMS@ Petri Nets* (Citeseer, 2014), pp. 20–33
33. V.I. Levenshtein, Binary codes capable of correcting deletions, insertions, and reversals. Sov. Phys. Doklady **10**(8), 707–710 (1966)

Intrusion Detection Through Gradient in Digraphs

S. S. Varre, Muhammad Aurangzeb, and Mais Nijim

1 Introduction

Intrusion in networks is not new. Intrusion within networks is as old as the concept of networks by themselves Numerous schemes have been proposed for intrusion detection. Generally, the intrusion detection techniques are based on neural networks (NN), machine learning (ML), and artificial intelligence (AI). In their paper [1], Li et al. have proposed an intrusion detection scheme with the help of intrusion sensitivity-based trust management model where trustworthiness of nodes is determined through supervised machine learning techniques [1]. Similarly, in [2], Yin et al. have presented an intrusion detection system based on ML and recurrent NN. In [3], Hodo et al. have presented a hazard analysis of the Internet of Things and used a supervised NN to prevent Distributed Denial of Service attacks. In their paper [4], Alom et al. have studied the intrusion detection abilities of supervised NN through a series of trials. In [5], Li et al. presented a sensitivity-based mechanism for the detection of intrusion due to pollution attacks in networks. The work is based on the study that each intrusion discovery has different levels of sensitivity in discovering certain categories of invasions. For all these schemes, there is an AI learning involved and training time is involved [2].

This chapter poses and proposes a solution to the problem of having a malicious agent within a cooperative network [6, 7]. In such network, the nodes intend to reach a consensus about a parameter of interest, like headings, size of the target, or any other pivotal information. Whenever two or more nodes collaborate with each other to pursue common goals, they form networks. Graph theory is the area

S. S. Varre · M. Aurangzeb · M. Nijim (✉)
Department of Electrical Engineering and Computer Science, Texas A&M University-Kingsville, Kingsville, TX, USA
e-mail: Muhammad.aurangzeb@tamuk.edu; mais.nijim@tamuk.edu

K. Daimi et al. (eds.), *Advances in Security, Networks, and Internet of Things*, Transactions on Computational Science and Computational Intelligence,
https://doi.org/10.1007/978-3-030-71017-0_12

of mathematics which deals with the analyses of networks [8]. In graph theory, a network is mathematically expressed by a graph; background knowledge on graphs is presented in Sect. 2. The proposed solution is proposed by extending the notions of scalar point functions, level surfaces, and gradient to the graphs. The notions of gradient, level surfaces, and scalar fields are not new to the scientific community [9, 10]. By extending the notions of scalar fields, level surfaces, and gradient to the graphs, this chapter proposes a mechanism for the detection of intrusion and a linear time algorithm to reach the intruding or malicious node.

The chapter is arranged in the following sections. Section 2 presents background knowledge about the graph theory, gradient, level surfaces, and scalar fields, to be used further. Section 3 presents the mathematical model to extend the notions of scalar point function, level surfaces, and gradient. Moreover, this section presents a conjecture to find the min path to the intruder/malicious node within the network. Section 4 introduces the simulation results supporting the mathematical model presented in Sect. 3. At the end, the conclusions are provided in Sect. 5.

2 Background

A network consists of various entities like sensor nodes, autonomous vehicles, routers, and switches. When it comes to formal mathematical analyses of the networks, these individual entries are called nodes or vertices. In a network, these nodes interact with each other. The nature of their interaction depends upon the functionality of the network. Some examples of the interaction include communication between the nodes, consensus within the nodes, exchange of information about the state of the network, etc. The nodes' interaction is the vital component which transforms individual nodes into a network. The interaction between the nodes can be bidirectional or unidirectional according to the nature of the flow of information. Whether the flow is bidirectional or unidirectional, it can always be modeled as an arc (a, b). The arc (a, b) denotes a flow of information from node a to node b. The case of bidirectional flow of information is formally expressed with a pair of arcs (a, b) and (b, a). Depending upon the nature of the network, a weight can also be associated with an arc. In such cases, an arc from node a to node b with a weight w is expressed as (a, b, w). For example, a network where the network operation directly depends upon the physical distance in between the vertices, the weights are proportional to physical distance between the nodes. For a routing network, the weight on arcs are proportional to the time delay on the arc etc. In certain cases where all the arcs are essentially same, each one of them is assigned with a constant weight of *1*. A network is thus composed of a collection V of vertices or nodes and a collection E of arcs, and it is formally studied under the mathematical structure known as a graph $G(V, E)$. If the direction of arcs matters in the analyses, then such graphs are called digraphs. Otherwise, they are referred to as simple graphs. In this proposed research, we want to focus on sensor networks and networks of autonomous vehicles.

Based on the above discussion, in this research, any sensor network or network of autonomous vehicles is represented as a graph $G(V, E)$. In G, V is the collection of all the agents, nodes, or vertices, and E is the set of all the weighted arcs. Each node in the network is expressed as a vertex within the set V and each direct interaction between the nodes is expressed as an arc within the set E. The research uses the tools of Graph theory and Algebraic Graph Theory [8] to investigate the networks for intrusion detection and isolation. The fundamental concepts of Graph Theory are elaborated in the following two paragraphs for the sake of completion of this discussion.

A digraph is a pair $G\ (V, E)$. In G, V is a fixed set and $E \subseteq V \times V \times W$ where $V \times V$ is the set of all the ordered pairs of the elements of V, and W is the collection of all the weights. The cardinality of $V \times V$ is represented as $|V \times V|$ and equal to $|V|^2$ where $|V|$ is the number of elements of V called the order of the graph, and in our discussion, represented by N. Here V is called the collection of nodes, agents, or vertices, and E is called the collection of arcs. Consequently, the elements of V are called vertices of the digraph and usually denoted by v, and those of E are knowns as arcs of the digraph and denoted by e. If $e = (v_1, v_2, w)$ is an arc with v_1 and v_2 as vertices, then e is said to connect v_1 to v_2. It is also said that v_1 dominates v_2, or in other words, v_1 is at the tail of e, and v_2 is at the head of e. The number of arcs with heads at v is the in-degree of v and denoted as $d^-(v)$. Similarly, the number of arcs with tails at the v is the out-degree of v and is denoted as $d^+(v)$. Here v refers to a vertex. The sum of in-degree and out-degree of a node v is called degree of v and is denoted as $d(v)$. A graph can also be represented with the help of an adjacency matrix A. An adjacency matrix is matrix with equal number of rows and columns with the entry in its ith row and jth column as the weight w of the arc from the vertex v_i to the vertex v_j. Adjacency matrix plays a fundamental role in the analyses of a graph. A sequence of explicit vertices $(u =)\ v_0,\ v_1,\ \ldots,\ v_m\ (= v)$ such that starting from a vertex u to another vertex v such that there exists an arc from vertex v_i to the vertex v_{i+1} for all i goes from $0,1,2,\ldots,m-1$ is called a directed path from u to v within a digraph G. A vertex u is said to be connected to a vertex v, if there occurs a dipath from u to v. Dipath stands for directed path. The sum of the weights associated with each of the arc along the path makes the path weight. The minimum path weight, or minimum hop count for graphs with uniform weights, over the paths from node u to node v is termed as distance between u to v, and such path is called shortest path. A digraph is said to have a root node r if r is connected to every vertex v. A root node with no in-degree is known as a pinning node. In this research, a pinning node which is assigned in this role by the design of the networks is called a leader node, whereas a node taking up the role of pinning node due to some foreign intervention is called a malicious node. In a network where information about a certain parameter like heading of a target flows through the network, the pinning nodes are the sources of the information, whereas the rest of the nodes reach on stable values by averaging the information they received from the dominating nodes. Consider the scenario of a flock of army autonomous vehicles that are heading for a mission in the direction of their target. The direction is being propagated to all the members in the flock through a leader node. To undermine the operation, an enemy can hijack one of the nodes or insert one of its own which will start giving out a wrong direction to the flock; such a node is named as malicious node. In this situation, it is vital for the successful completion of the mission that not only the malicious node be tracked down, but it can also be isolated from the rest of the flock. A rigorous mathematical model is developed based on graph theory. Several simulations were run under such scenarios, and it was found that the detection and isolation of malicious node is possible by incorporating the ideas of scalar field, level surfaces, and gradient [9, 10] within the graph theoretic model.

In a network of nodes where the information about a parameter is spreading from a leader, and possibly also from a malicious node as discussed above, the state of the network at a discrete time k is written as $\boldsymbol{x}(k)$, and it is a vector having the parameter values of all the nodes at that discrete time slot. It is already established in the literature [7] that the state $\boldsymbol{x}(k)$ at time slot k is related to the next state as $x(k+1) = (D+I)^{-1}(A+I)x(k)$, where D is the matrix with entries in the diagonal equal to the in-degree of each node, and the rest of the entries 0, I denotes the identity matrix and A denotes the adjacency matrix. This relation is elaborated in the next section. After a few iterations, the nodes reach their stable values. This chapter establishes that the stable values of all the nodes exist between the values of the pinning nodes.

The concepts of gradient, level surfaces, and scalar field are familiar concepts in several fields of engineering and science [9, 10]. Scalar field is defined as a function such that it assigns to each point in the field of interest a unique scalar value. Scalar field is seen as heat level at several spots around a hot flame. Temperature in a spot in a vicinity of flame varies upon the position of the spots with respect to the flame. Moreover, numerous simulations, infrared scans, and scientific models support that incessant surfaces exist around the flame where the heat levels are identical at their entire points. These surfaces are known as level surfaces [10]. Gradient points to the direction of max change of scalar field relative to the change in scalar field. Gradient thus offers the direction of min displacement to the flame at a spot. These concepts are not still undefined in the area of network analyses. This research extends these vital concepts of gradient, level surfaces, and scalar field to the networks. A malicious node spreading deception can be taken as a hot flame. The mathematical model in this chapter and simulation results show that as the malicious information spreads through the network, it forms a scalar field with level surfaces around the malicious node. A gradient can be defined within the network, which according to our simulation results gives the direction of minimal distance to the malicious node.

3 Mathematical Model

As discussed in the last section, networks can be expressed as their adjacency matrix $A = [a_{ij}]$ where a_{ij} is the weight of the arc from i to j, where both i and j are the nodes. As mentioned earlier that this chapter focuses on the cooperative networks like that of autonomous vehicles and sensor networks, where all the nodes intend to reach a consensus about a scalar function like their heading in accordance with a leader node. In the further discussion in this chapter, the scalar function and heading are used interchangeably. There is also a possibility of presence of a malicious node within the network. In such network, nodes receive headings of their dominating nodes and iteratively adjust their headings by averaging their own heading with that of their dominating nodes. If node i is a dominating node of node j, then an arc from i to j exists.

In the further discussion, for the sake of simplicity all the arcs, if exist, are considered equally weighted, that is,

$a_{ij} = 1$ if an arc exists from i to j
$a_{ij} = 0$ if no arc exists from i to j

The heading of each node j at a discrete time k is x_{k}^{j} which is the average of dominating nodes heading values with that of node itself.

$$x_{\mathrm{k}+1}^{j} = \frac{x_{k}^{j} + \sum_{i=1}^{n} a_{ij}\, x_{k}^{i}}{1 + \sum_{i=1}^{n} a_{ij}} \ldots \quad (1)$$

Combining the above equation for the entire nodes in the matrix form yields

$$\boldsymbol{x}_{k+1} = (D + I)^{-1}\,(A + I)\,.\boldsymbol{x}_{k} \ldots \quad (2)$$

In this equation, $\boldsymbol{x}_{k+1}$ is a vector having the headings of all the nodes at a discrete time $k + 1$, D is the matrix of in-degrees for all the nodes in the diagonal, and A is the adjacency matrix for the network.

Rest of the section builds up on top of the mathematical model explained above. A lemma establishes that the steady-state values of the heading are linear interpolation of the headings of the leader and malicious nodes. Based on the lemma, the notions of gradient, level surfaces, and scalar field are extended to the networks. Finally, a conjecture is suggested about the shortest distance to the malicious node starting from the leader node.

Lemma In a network where a leader node has one heading value while there is a presence of a malicious node with another heading value, with all the rest of the nodes iteratively adjusting their headings by averaging their own heading with that of their dominating nodes, all the nodes get steady-state values interpolated between the heading values of the leader node and malicious node.

Proof For a network in which all the nodes other than the malicious and leader node nodes iteratively adjusting their headings by averaging their own heading with that of their dominating nodes, the headings of the entire nodes is given by (2) above

$$\boldsymbol{x}_{k+1} = (D + I)^{-1}\,(A + I)\,.\boldsymbol{x}_{k} \ldots \quad (2)$$

Here D is the matrix with diagonal values as in degrees of nodes with remaining entries as 0, and A is the adjacency matrix for the network. Please note that the leader and malicious nodes do not have any in–degrees, thus corresponding in-degrees are 0. Applying one side z-transform to (2) [11]

$$X^{+}(z) = (D + I)^{-1}(A + I)\left(x(-1) + z^{-1}X^{+}(z)\right)$$

Rearrangement of the above equation gives

$$\left(I - z^{-1}(D + I)^{-1}(A + I)\right)X^{+}(z) = (D + I)^{-1}(A + I)\,x(-1)$$

Since $(D + I)^{-1}(A + I)x(-1) = x(0)$, the above equation becomes

$$\left(I - z^{-1}(D + I)^{-1}(A + I)\right)X^{+}(z) = x(0)$$

$$X^{+}(z) = z(z(D + I) - A - I)^{-1}(D + I)\;x(0)$$

$$X^{+}(z) = \frac{z\,\mathrm{adj}\,(z(D + I) - A - I)}{\det(z(D + I) - A - I)}(D + I)\,x(0)\ldots \tag{3}$$

From the construction of $(D + I)$, the first two elements of $(D + I)x(0)$ are identical to those of (0), now

$$A + I = \begin{bmatrix} I_{2\times 2} & O_{2\times N-2} \\ L_{N-2\times 2} & R_{N-2\times N-2} \end{bmatrix}$$

The aggregate of nodes is N.

$$D + I = \begin{bmatrix} \begin{bmatrix} 1\,0 \\ 0\,1 \end{bmatrix} & O_{2\times N-2} \\ O_{N-2\times 2} & D\prime_{(N-2)\times(N-2)} \end{bmatrix}$$

which means

$$(z(D + I) - A - I) = \begin{bmatrix} \begin{bmatrix} (z-1) & 0 \\ 0 & (z-1) \end{bmatrix} & & O & \\ & zd_3 - 1 & \ddots & -a_{ij} \\ -L_{(N-2)\times 2} & & & \\ & -a_{ji} & & \ddots\, zd_N - 1 \end{bmatrix}$$

or

$$(z(D + I) - A - I) = \begin{bmatrix} \begin{bmatrix} (z-1) & 0 \\ 0 & (z-1) \end{bmatrix} & O \\ -L_{(N-2)\times 2} & zD' - R \end{bmatrix}$$

Suppose that

$$E = \operatorname{adj}\left(z\left(D+1\right) - A - I\right)$$

Let E in the elements form be

$$E = \begin{bmatrix} E_{11} & \dots & E_{1n} \\ \vdots & \ddots & \vdots \\ E_{n1} & \cdots & E_{nn} \end{bmatrix}$$

where

$$E_{11} = E_{22} = (z-1)\det\left(zD' - R\right)$$

$$E_{2k} = 0 = E_{1j} \text{for } j \neq 1 \text{ for } k \neq 2 \text{ as it has a zero row}$$

$$E_{i1} = (z-1)\det\left(zD' - R\right)^{i1}$$

i1 in the superscript amounts to substituting the ith location of (zD'-R) with first location of ($-L_{(N\text{-}2)\times 2}$). For the rest of the cofactors,

$$E_{ij} = (z-1)^2 . e_{ij}$$

The denominator term in (3) is given by

$$\det\left(zD - A - I\right) = (z-1)^2 . \det\left(zD' - R\right)$$

The Eq. (3) leads to

$$(z-1)\,X^{+}(z) = z\frac{\begin{bmatrix} \begin{bmatrix} \det\left(zD'-R\right) & 0 \\ 0 & \det\left(zD'-R\right) \end{bmatrix} & \begin{bmatrix} 0\cdots\cdots 0 \\ 0\cdots\cdots 0 \end{bmatrix} \\ \vdots & \begin{bmatrix} (z-1)\,e & \ddots & (z-1)\,e \\ (z-1)\,e & \ddots & (z-1)\,e \end{bmatrix} \end{bmatrix}}{\det\left(zD'-R\right)}(D+I)\,x(0)$$

By final value theorem [11] in digital signal processing $\lim_{n\to\infty} x(n) = \lim_{z\to 1} (z-1)\,X^{+}(z)$

$$\lim_{n\to\infty} x(n) = \begin{bmatrix} \begin{bmatrix} 1 & 0 \\ 0 & 1 \end{bmatrix} & \begin{bmatrix} 0 \cdots\cdots 0 \\ 0 \cdots\cdots 0 \end{bmatrix} \\ \begin{bmatrix} \det\left(D' - R\right)^{31} & \vdots \\ \vdots & \vdots \end{bmatrix} & \begin{bmatrix} 0 & 0 \\ 0 & 0 \end{bmatrix} \end{bmatrix} x(0)$$

$$\lim_{n\to\infty} x(n) = \frac{\begin{bmatrix} \begin{bmatrix} 1 & 0 \\ 0 & 1 \end{bmatrix} & 0_{2\times(N-2)} \\ \begin{bmatrix} \det\left(D' - R\right)^{31} & \det\left(D' - R\right)^{32} \\ \det\left(D' - R\right)^{41} & \det\left(D' - R\right)^{32} \\ \vdots & \vdots \\ \det\left(D' - R\right)^{N1} & \det\left(D' - R\right)^{N2} \end{bmatrix} & 0_{(N-2)\times(N-2)} \end{bmatrix}}{\det\left(D' - R\right)} x(0) \ldots \tag{4}$$

Since $A + I$ is dominant diagonally, $\det(D' - R) \neq 0$. Moreover, it can be seen from (4) that the steady-state value of all nodes is interpolated between the heading values of the leader and malicious nodes. □

The above lemma and the simulation results in the next section lead to the following definition.

Definition (Scalar Field) In a network of nodes where there is a malicious and a leader node while the rest of the nodes are iteratively adjusting their headings by averaging their own heading with that of their dominating nodes, the steady-state value of the headings for any node is a function f assigning each node a unique heading value which is a linear interpolation of the headings of the leader node and malicious node. This function is called a scalar field.

Definition (Level Surface) In a network of nodes where there is a malicious and a leader node while the rest of the nodes are iteratively adjusting their headings by averaging their own heading with that of their dominating nodes, the scalar field values of the nodes form virtual level surfaces by putting nodes with same values on a surface and extrapolating the surfaces through network arcs by interpolation.

Definition (Gradient or Grad) In a network of nodes where there is (are) leader and malicious node(s) while the rest of the nodes are iteratively adjusting their headings by averaging their own heading with that of their dominating nodes, the gradient of the scalar point function at a node j is the maximum of the ratio of change in the scalar point function in the direction of an arc to its weight. Mathematically, if there is a node j and f is the scalar field, then grad $f(j) = \max_{\text{arc } ij} \frac{f(i)-f(j)}{w_{ij}}$

On the bases of the above lemma, the definitions and the simulation results in the next section, the following conjecture is proposed. The conjecture needs to be established into a theorem in future research.

Conjecture (Minimum Distance to Malicious Node) In a network of nodes where there are leader and malicious node each, while the rest of the nodes are iteratively adjusting their headings by averaging their own heading with that of their dominating nodes, if the network is iteratively traversed starting from the leader node in the direction of gradient, the iterative traversal leads to the malicious node in minimum distance.

The simulation results in the next section support the above conjecture.

4 Simulation Results

The theoretical model developed in this chapter is extensively tested through simulation results. For this purpose, a grid graph structure of 25 nodes is selected. The selection of a grid of size 5 × 5 is made owing to two reasons: first, the grid structure makes it possible to visually observe the concepts of level surfaces and gradients and second the size of 5 × 5 makes it possible to depict/show the simulation results for all the nodes with acceptable clarity.

The entire simulation is based on nodes in the grid network having some random initial value of the heading, Fig. 1. There is a leader node whose initial heading is 0, which it wants to propagate to the entire network by developing a consensus within

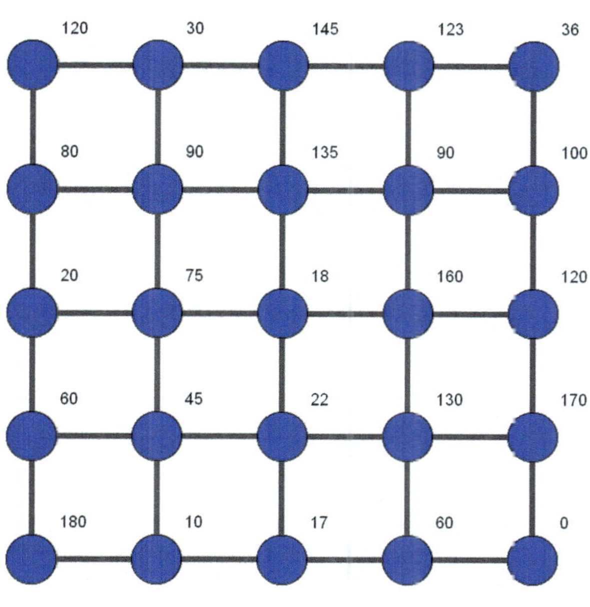

Fig. 1 Grid network used for simulation. All the nodes are assigned with random initial headings with the leader node in the bottom-right corner with a heading value of 0

the network. The simulation results show that in the absence of a malicious node it is possible for the entire nodes to reach the consensus in some finite number of transactions, all the nodes were having a heading value of 0, Fig. 2.

In the next phase of simulation, a malicious node is introduced within the network. In the following figures of the grid network, the malicious node can be identified as the one with a heading value of 180. In all these various network settings, the malicious node is placed at various locations relative to the node in leader role. In the presence of a malicious node who wants to pull the entire network to a heading of 180, the entire nodes get their steady-state values which are interpolation of heading values of the leader node and malicious node. Figure 3 depicts the steady-state values of the headings of entire nodes within the grid network. Figure 4 shows the convergence to steady-state values by the entire nodes.

It can be observed that various nodes within the networks have the same steady-state values. The nodes with the same steady-state values are joined through continuous contours. The corresponding contour is said to have the corresponding heading. In order to draw a continuous contour, linear interpolation is used to draw contour lines through the arcs of the grid networks. These contour lines are depicted

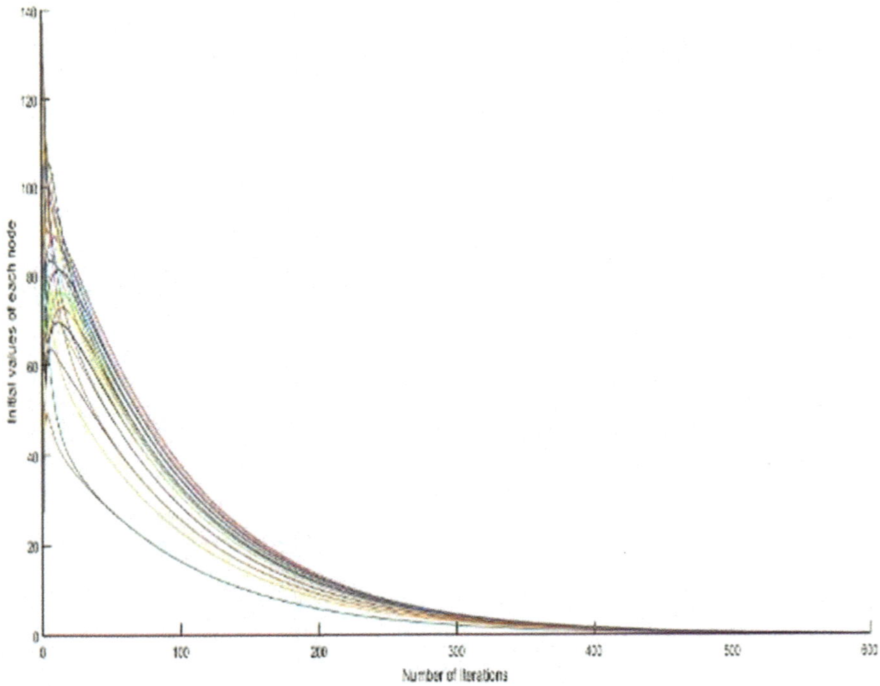

Fig. 2 Simulation results with one leader node with a heading of 0, while rest of the nodes with some random initial heading values. Simulation results show that in the presence of a leader node the entire nodes within the network reach a consensus about their heading with a value of 0

in Fig. 5. The simulation results show that in all cases without any exception, level surfaces were formed, and the notion of the gradient, which is maximum rate of change of malicious information along the arcs, always give us direction to the shortest distance towards the malicious node, Fig. 6.

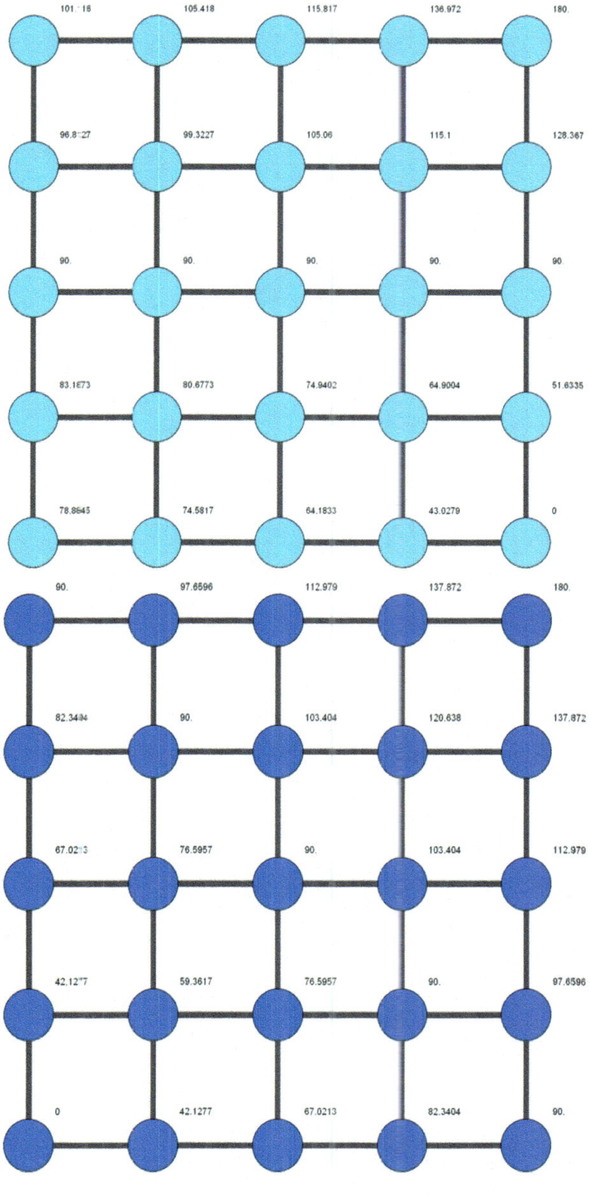

Fig. 3 Stable values of heading for all the nodes in the grid network

Fig. 3 (continued)

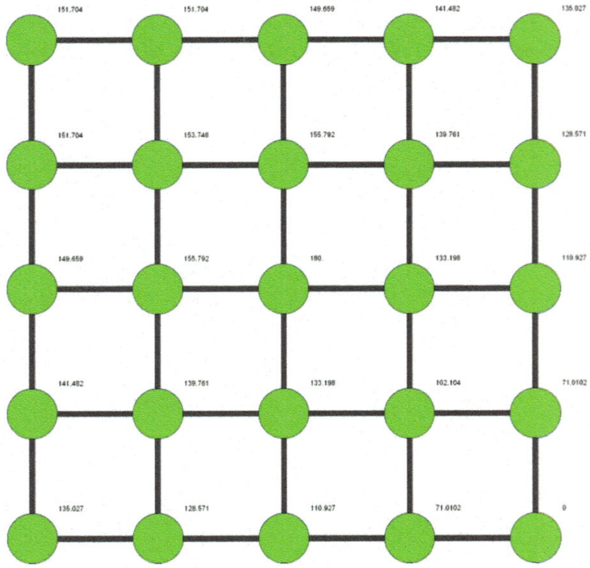

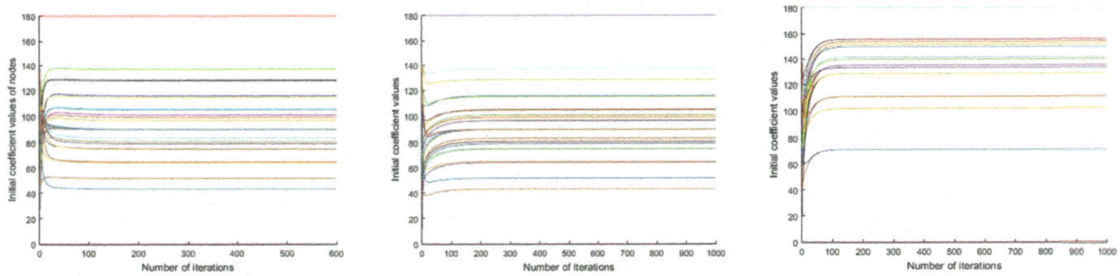

Fig. 4 Headings vs time show the convergence of headings of all the nodes in the network

Since the contour lines pass through the nodes with the same steady-state heading values, these are the level surfaces [9, 10] within the network. The gradient, as defined earlier, is the maximum value of the change of the scalar function relative to the arc weight. In this simulation, the leader node has a heading of 0, making any nonzero value of the heading an error in the heading. The error as distributed across the networks in the simulation form a scalar field. It can be seen further that starting from the leader node one can follow the path of maximum rate of change to reach the malicious node.

5 Conclusion

The notions of gradient, level surfaces, and scalar field are not new to the scientific community; these notions are extensively used in areas of study like thermodynamics and electrostatics to determine the center of a scalar field. This chapter extends these concepts to the networks and utilizes these concepts for intrusion detection within the network. A mathematical model introduced with a conjecture is proposed about the shortest path to the malicious node, starting from the leader. This research is to be further extended to establish a mathematical proof of the proposed conjecture in Sect. 3. Moreover, the effect of having multiple malicious nodes is also to be studied.

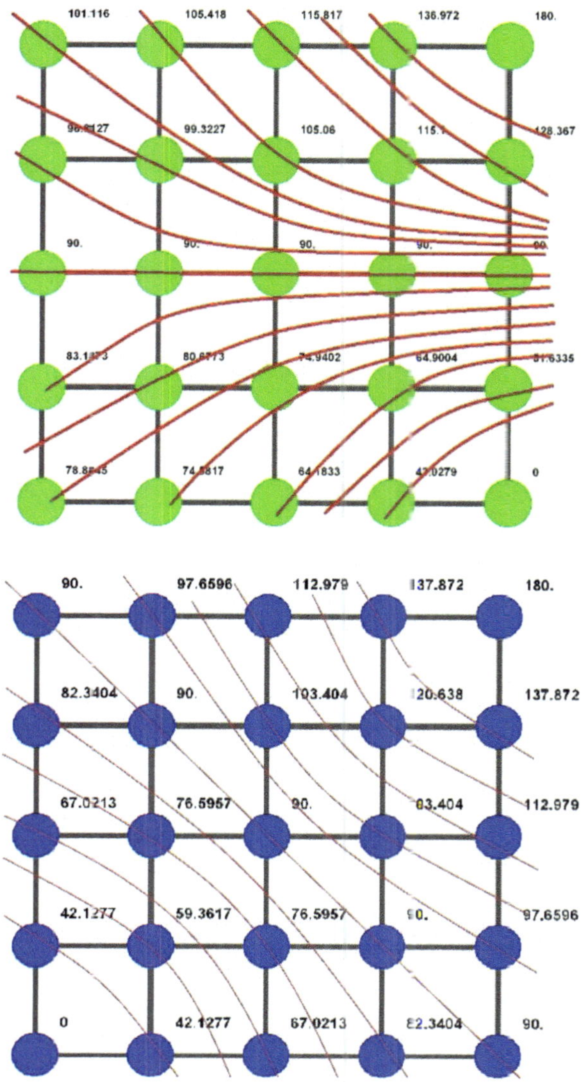

Fig. 5 The contour lines drawn through nodes with equal steady-state values of the headings for the nodes. The continuous contour lines are drawn by using interpolation as the contour lines cross the edges of the grid network

Fig. 5 (continued)

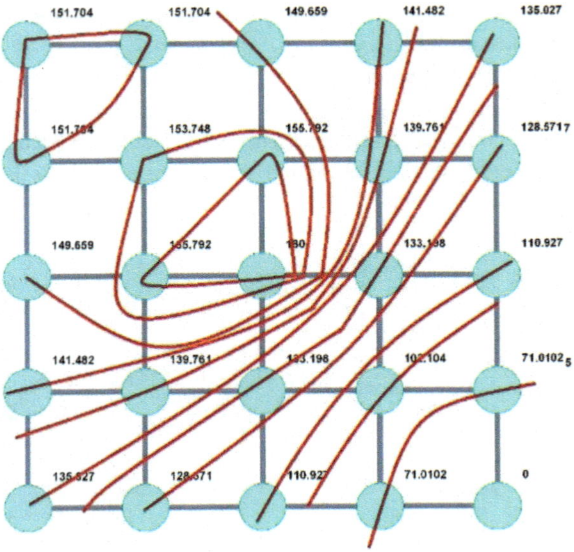

Fig. 6 Path from the leader node by following the gradient, max change of scalar function relative to arc weight, is the shortest path to the malicious mode starting from the leader node. The path is shown in the figure by thick line

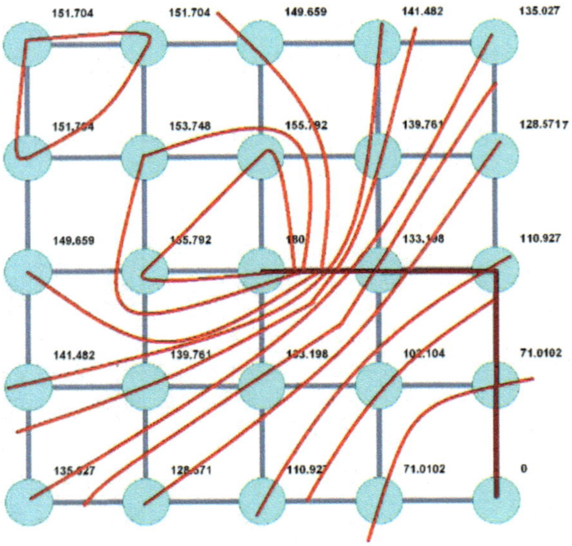

References

1. W. Li, W. Meng, L.-F. Kwok, H.S. Horace, Enhancing collaborative intrusion detection networks against insider attacks using supervised intrusion sensitivity-based trust management model. J. Netw. Comput. Appl. **77**, 135–145 (2017)
2. C. Yin, Y. Zhu, J. Fei, X. He, A deep learning approach for intrusion detection using recurrent neural networks. IEEE Access **5**, 21954–21961 (2017)

3. E. Hodo, X. Bellekens, A. Hamilton, P.-L. Dubouilh, E. Iorkyase, C. Tachtatzis, R. Atkinson, Threat analysis of IoT networks using artificial neural network intrusion detection system, in *2016 International Symposium on Networks, Computers and Communications (ISNCC)*, (IEEE, 2016), pp. 1–6
4. M.Z. Alom, V.R. Bontupalli, T.M. Taha, Intrusion detection using deep belief networks, in *2015 National Aerospace and Electronics Conference (NAECON)*, (IEEE, 2015), pp. 339–344
5. W. Li, W. Meng, Enhancing collaborative intrusion detection networks using intrusion sensitivity in detecting pollution attacks. Inf. Comput. Secur **24**, 265 (2016)
6. M. Aurangzeb, F.L. Lewis, Internal structure of coalitions in competitive and altruistic graphical coalitional games. Automatica **50**(2), 335–348 (2014)
7. F.L. Lewis, H. Zhang, K. Hengster-Movric, A. Das, *Cooperative Control of Multi-Agent Systems: Optimal and Adaptive Design Approaches* (Springer Science & Business Media, 2013)
8. C. Godsil, G.F. Royle, *Algebraic Graph Theory*, vol 207 (Springer Science & Business Media, 2013)
9. W.H. Hayt, *Engineering Electromagnetics* (McGraw-Hill Companies, 1974)
10. M.E. Gurtin, E. Fried, L. Anand, *The Mechanics and Thermodynamics of Continua* (Cambridge University Press, 2010)
11. J.G. Proakis, *Digital Signal Processing: Principles Algorithms and Applications* (Pearson Education India, 2001)

A Practice of Detecting Insider Threats within a Network

Jeong Yang, David Velez, Harry Staley, Navin Mathew, and Daniel De Leon

1 Introduction

Every attack starts with reconnaissance and has a systemic pattern which it must follow in order to be successful. This methodology consists of the following steps: reconnaissance, gaining access, escalating privileges, creating backdoors to maintain access, and covering the tracks of the attack. The first step, reconnaissance, the focal point of our research, consists of scanning and enumeration. Scanning (usually done via tools such as Nmap) is necessary to map out the network, find open ports, and figure out what service is running on those ports, leading to the next step which is enumeration, which is not possible without scanning in the first place.

Enumeration is the process of finding all possible attack vectors. These potential attack vectors are things like network shares, SNMP data, IP tables, usernames, and passwords. Various tools exist for enumeration, but the ones most frequently used are NTP Suite, enum4linux, and SMTP-user-enum. After enumeration, the attacker uses these attack vectors, or vulnerabilities, to gain access to the network. The attacker will then use these "entry-level" credentials to escalate their privileges until they have gained super admin, or super user, permissions. Using entry-level credentials will allow the attacker to create backdoors for future ease of access, as well as the ability to delete all logs containing the information of the attack [1].

There are two distinct methodologies of reconnaissance: passive and active. Passive reconnaissance is stealthy and undetectable. It is this stealth factor, however, which does not yield much in terms of results and limits the amount of information

J. Yang (✉) · D. Velez · H. Staley · N. Mathew · Daniel De Leon
Department of Computing and Cyber Security, Texas A&M University-San Antonio, San Antonio, TX, USA
e-mail: jyang@tamusa.edu; dvele01@jaguar.tamu.edu; harrya.staley@jaguar.tamu.edu; nmath0@jaguar.tamu.edu; ddele01@jaguar.tamu.edu

K. Daimi et al. (eds.), *Advances in Security, Networks, and Internet of Things*, Transactions on Computational Science and Computational Intelligence, https://doi.org/10.1007/978-3-030-71017-0_13

gained. Passive reconnaissance is done via tools such as Wireshark, Google, and various websites. These websites (such as FindSubDomains, VirusTotal, and Shodan) offer various passive reconnaissance utilities [2]. Active reconnaissance, however, requires actual interaction with the network, which acquires much more information at a significantly faster rate at the cost of detection via an Intrusion Detection Service (IDS). Some active reconnaissance tools are Nmap, Nessus, OpenVAS, Nikto, and Metasploit [2]. Each of these tools analyzes potential attack vectors or vulnerabilities. These range from port scanners (Nmap) to vulnerable application detectors (Nessus and OpenVAS) [3]. Active reconnaissance consists of scanning open ports and services that are running on those open ports. This can then allow the hostile user to discover weaknesses of the network's internal security.

This research investigates the gravity of threats within a network and the notion that it is a large attack surface on a small scale. Cybersecurity is typically focused on external threats such as the threat of nation-states, terrorists, hackers, anonymous black hats, and thieves. However, external threats are no longer the only danger to a network.

Inside the comfort and security provided by the IT (Information Technology) professionals that shelter a network from the would-be attackers outside lies numerous potential vulnerabilities. Behind layers of protection consisting of firewalls, security policies, access-list controls, and alerts, the average user is oblivious to internal threats and oftentimes completely unaware of the existence of many potential dangers that lurk within the internal network.

The research intends to explore the common network vulnerabilities that exist in typical workspaces and reveal how the insider threat can be more of a problem than those from external sources. Using a typical infrastructure design, we built a network to emulate a small business as a baseline model for a typical network environment. This network contains numerous resources and standards that generally exist in a shared network, such as a file, email, chat, and web servers.

From the emulated network, we collect network traffic from port scans to correlate, and "fingerprint" reconnaissance scans to understand the nuances of network attack surfaces. With this, we can distinguish malicious activity from regular network traffic, thereby assisting cybersecurity engineers in detecting the recon tools used for network host/port scans.

2 Related Work

External network security threats are familiar to most people, but within a network, security tends to be more laxed because of the obstruction of productivity that security protocols often perpetuate. This friction of user productivity vs security is a common problem, and when it comes to resistance from employees, it is typical to see management trade security for comfort and flexibility.

Current research suggests that internal threats are on the rise in recent years. According to Shein, "while organizations are the victims of external hackers, many are finally realizing an uncomfortable truth: insider threats are and have long been an authentic problem" [4]. Nevertheless, the issue of internal threats is not exactly new. Recently, because of the infamous leaks from Edward Snowden and Chelsea Manning, more companies are looking for ways to detect the insider threat and find classification systems that help distinguish a level of threat amongst their employees. "Companies are seeking out technologies to develop employee profiles of what is 'normal' behavior versus actions that should send up a red flag" [4].

To classify behavior requires vast amounts of data collection. Data must be collected from computers, servers, login sessions, system logs, weblogs, database logs, SSH (Secure Shell) logins, network activity, and much more. If a company is small, collecting logs on a day-to-day basis can still be a daunting task. Getting information from logs and visualizing it, even from a little bit of packet data, is an even more enormous task as we demonstrate in our research. Senator et al.'s research on user activity found that in a medium-sized company, an "approximate 5.5 million actions per day from approximately 5500 users" were collected, which is a thousand actions per user [5]. Being able to parse through all that data requires automated parsing, collection, and visualization. However, even with visualization tools to sift through logs, without filtering to hone in on bad traffic vs. regular traffic, it still poses an immense challenge.

According to Senator et al., "IT detection is more difficult than many other anomaly detection (AD) problems not only because insiders are knowledgeable about an organization's computer systems and procedures and authorized to use these systems, but also, and more important, because malicious activity by insiders is a small but critical portion of overall activity on such systems," [5]. An insider's first step in exploitation is scoping, and because insiders are already inside the network, many of them are familiar with the network and its structure as well as being thoroughly aware of where big payloads may reside. Furthermore, a network has so much going on internally, such as email activity, file transfers, ongoing chats, and intranets being scoured, that detecting nefarious activity is nearly impossible. The small activities that could lead to an internal attack are challenging to detect without the appropriate software and infrastructure. Determining a pattern of abuse is paramount in detection, which leads to the investigation of standard methods that attackers use to perform attacks.

According to Alsaleh et al., "scanning is an effective way to search for potential weaknesses in dedicated servers." It means that the preliminary approach to any kind of threat starts with a level of reconnaissance that relates to scoping out targets and their vulnerabilities [6].

3 Research Design and Methodology

Our research set out to answer three research questions. First, can our chosen Intrusion Detection System (IDS) detect active reconnaissance activities such as port scanning on a network, and if not, at what intensity level or levels will it fail?

Second, if the IDS fails to detect the active reconnaissance, can we analyze the data generated by the scans in a white box test and deduce a pattern for scans at all levels paying close attention to those intensities that previously failed to be detected? Third, if we can deduce a pattern, can we write software to patch the system through a rules-based approach or an artificial intelligence-based approach?

The contributions that our research makes towards the improved detection of insider threats is an expansion of research conducted by research teams from the Navy Postgraduate School, the SANS Institute, and others. Our expansion focuses explicitly on the detection of standard and custom implementations of Nmap. The approach of our research was to design a network that simulates an ordinary, small business environment. Deviating from this design would suggest false or inaccurate conclusions about the risk of internal threats. Emulation of a real-world environment using products, software, and hardware was paramount in our research in order to allow relevant and pressing concerns yielded from our analysis and results collected. Our model, however, cannot, and will not accurately depict every small business environment.

The internal network is open to the outside world, but is not affected by any other systems on other networks, which necessitated segregation from them. Segregating the network required equipment that supports VLANs (Virtual Local Area Networks) and firewall rules to deny and drop packets not originating and ending in the research network.

The test environment was not accessible by all researchers, so to allow researchers to access the system, we set up a secure VPN (Virtual Private Network) service that enabled access to the network from the outside. Each user account had a distinct VPN configuration file with a password to authenticate into the system to ensure a closed environment for testing.

We set up two physical hosts on the network: an Asus Tinker Board hosting DietPi (headless Raspbian) and a Windows 10 Professional Server running VMware Workstation 14 to run the virtual machines that make up the remaining nodes. The VMware Workstation Server's memory was increased to 32 GB (Gigabytes) of RAM (Random Access Memory), and a 500 GB Hard Drive was dedicated as the datastore to handle the various guest machines. The eight-core desktop contains seven Linux Servers, three Windows 10 Professional workstations, and one Windows Server 2012.

The Windows Server serves as the internal network's domain: *research.local*, and is the DNS (Domain Name Service) server of the network to keep track of the hostnames of each of the machines. The Linux servers maintain various services that are standard in business networks such as email, chat, websites, and shared file services. A couple of these services are directly tied to the domain via LDAP (Lightweight Directory Access Protocol), forcing the use of domain accounts.

Many services were configured insecure and open, used shared passwords for service accounts, and installed software that leaves ports exposed to mimic a typical internal network. Many assume that proper network security for external threats translates internally. Thus, the assumption of strong network security posturing often leads to poor configuration of interior services because of the high level of trust from an employee.

The last bit of the infrastructure required is the collection of network activity within the network; this level of system resources needed the most effort, resources, and research. We decided to use RockNSM, an open-source network security-monitoring platform. RockNSM helps us collect, interpret, and visualize, using an ELK stack, to harness the data, which includes the filters and analytics needed for monitoring. Also, this all-in-one software supported custom visualizations for a better examination of pertinent intelligence.

To enable data collection, RockNSM required that we:

1. Added a second NIC (Network Interface Card) to the VMware Workstation Server.
2. Connect the second NIC into a mirror port on a Layer-3 Switch to copy the inbound traffic.
3. The Ubiquiti EdgeSwitch 8–150 W enabled the use of VLAN (Virtual Local Area Network) Tagging and Port Mirroring to duplicate the network traffic coming in and out of the research network.
4. The virtual machine hosting RockNSM also required two vNICs (Virtual Network Interface Card): one for management and one for monitoring/sensing.

Following RockNSM configuration, we loaded the Kibana Web Interface (the data visualization module of RockNSM) and enabled the Suricata and Bro modules to allow capturing, analyzing, and graphing of network data. Figure 1 shows a high-level perspective of the Internal Research Network infrastructure.

Once on the Kibana Data Visualization GUI (Graphical User Interface), we configured basic data visualizations integrated with RockNSM, such as Suricata. Suricata is an IDS, Intrusion Prevention System, and NSM (Network Monitoring System) that is free and open source. NSM tools are used to inspect traffic using rule sets and signatures for the detection of sophisticated threats. We utilized a predefined SHELL script set to run a CRON job hourly to generate our network discovery shown in Fig. 2.

These are the relevant configurations of NMAP that were used to scan ports from inside the network. We would compare the time stamps generated with the logs and visualizations on Kibana to form an initial starting point for this research. This helped us test our internal network and review the capabilities of Kibana in real time. The different scan types utilized in our testing will be discussed in more detail later in this chapter. It was necessary to customize our Suricata Dashboard to capture various forms of traffic in order to completely understand how to recognize the appearance of internal threats. To visualize this, we needed network and host-level data, including SSH detection and failure. This required the addition of a Filebeat so that we could capture the /Syslog directory, where SSH logging is stored.

A Filebeat is a lightweight agent for forwarding logs. With it, the Filebeat actively monitors the host. It is installed on the remote host. It listens to log files and forwards them to a centralized server. With RockNSM, it uses Elasticsearch and Logstash in the backend for data collection; the researchers configured Filebeats

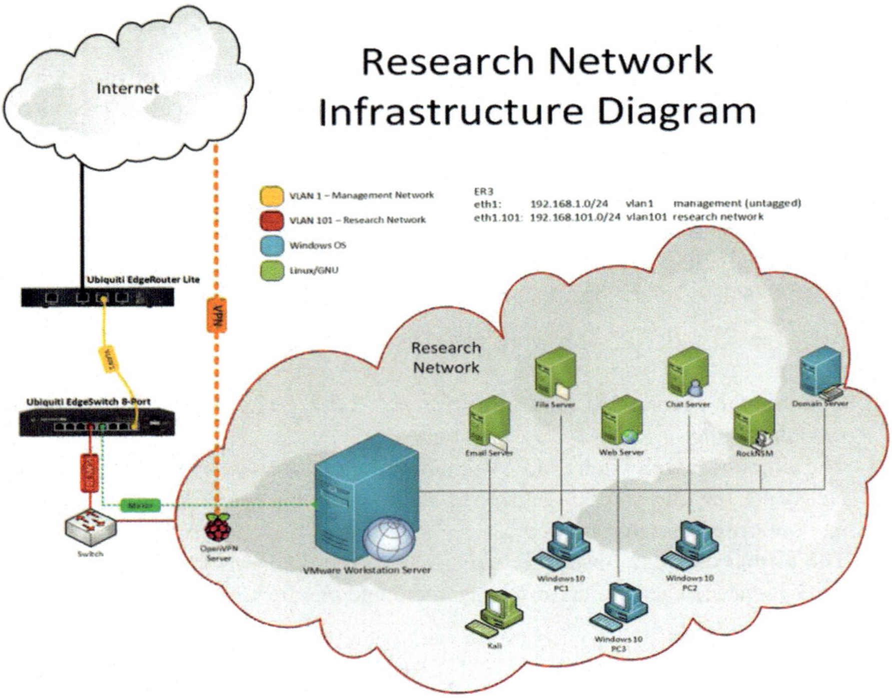

Fig. 1 Network infrastructure

```
#!/bin/bash

echo "starting NMAP scans"
FLAGS=("-sS" "-sT" "-sU" "-sY" "-sN" "-sF" "-sX" "-sA" "-sW" "-sM")

declare -A name
name["-sS"]="tcpSynScan"
name["-sT"]="tcpConnectScan"
name["-sU"]="udpScan"
name["-sY"]="sctpInitScan"
name["-sN"]="tcpNullScan"
name["-sF"]="tcpFinScan"
name["-sX"]="tcpXmasScan"
name["-sA"]="tcpAckScan"
name["-sW"]="tcpWindowScan"
name["-sM"]="tcpMaimonScan"
name["-O --fuzzy"]="fuzzyOSScan"

for i in ${FLAGS[@]}; do
  echo "starting ${name[$i]} at $(date)."
  echo $($SCAN 192.168.101.1-254 $i > nmap/${name[$i]}.txt)
done
echo "Done with nmap scans at $(date)"
```

Fig. 2 Nmap configurations for port scanning

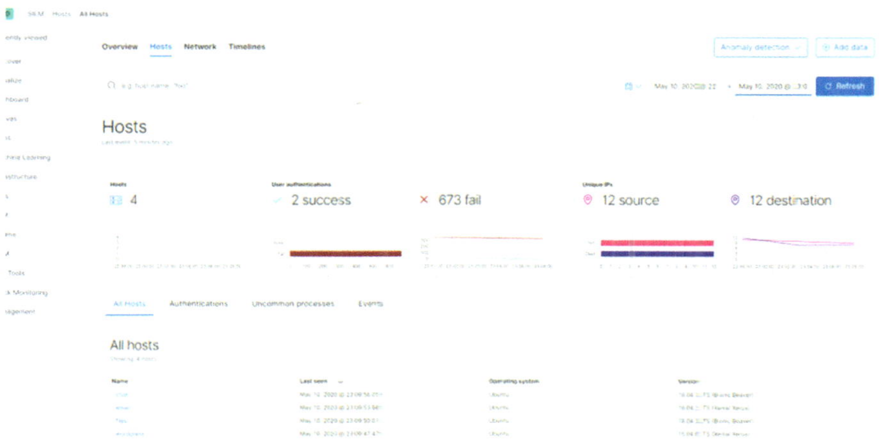

Fig. 3 Logs information with Filebeat

to collect system logs using a Filebeat module called "system." With Syslogs forwarded to the RockNSM Server, we visualized logging information from each of our servers, including SSH login attempts and failures.

SIEM (Security Information and Event Management) is a module built into RockNSM that centralizes the workflow and management of logs for our target machines. The SIEM module breaks down data on the target machines into two levels of logging: host level and network level. At the host level, these are general logs, kernel logs, SSH authentication and failures, source and destination IP (Internet Protocol) addresses, and a list of known entities (hosts). The visual in Fig. 3 demonstrates this capability.

To visualize SSH login failures further, we created a custom dashboard, as displayed in Fig. 4, with a list of SSH failures, how many attempts, and the responsible party. The custom dashboard offers insight into detecting the insider threat, which can prove useful in the development of a tool used to identify a consistent pattern of abuse, along with the help of historical data. This system is using the Filebeat "System" module mentioned above, with a timeline of SSH failures and successes.

We use Bro for packet capturing, enabling real-time packet stream analysis and granular network data collection to dissect network traffic further. Kibana then parses the network packets into various categories such as source, destination IP and port, TCP (Transmission Control Protocol)/UDP (User Datagram Protocol) identification, protocols in use, network packet count, and more. With Syslog forwarding through a Filebeat from each Linux server, we visualized other information, such as commands used and kernel errors, along with the origin of the host with the respective errors, shown below.

However, to detect network reconnaissance, we opted to test Nmap as the first scanner. Nmap is a port scanning tool that is a free and open-source utility used for

Fig. 4 List of SSH failures

security auditing and discovery. This tool is commonly used by system engineers, system administrators, and network administrators in the IT field. It is also a popular tool in networking classes used to scan networks and provide students with an understanding of network posture and cybersecurity. The capability of the port scanner extends beyond retrieving the status of open ports to perform operating system identification, and software versions of services in use by the targets. This is particularly handy for potential hackers who can compare the versions of software and look for common exploits associated with said versions of an operating system or software service.

The other benefit of using Nmap is that it does not require a GUI and can run from a console (terminal). Nmap does have a GUI version of itself, known as Zenmap, which is more commonly used in Windows Operating Systems, although they are effectively the same tool. Nmap is also capable of bypassing ICMP (Internet Control Message Protocol) blocks by utilizing TCP or UDP packet discovery, allowing the scanner to go around firewalls that commonly block pings for various reasons, including DoS (Denial of Service)/DDoS (Distributed Denial of Service) attacks. For all these reasons, Nmap was our choice of scanning for the purposes of our research.

4 Results and Findings

In the beginning, it was essential to test network captures and visualizations with specific dashboards using the Suricata module. As shown in Fig. 5, a regular Nmap scan across the network showed spikes in network activity along with alerts classified as "Generic Protocol Command Decode," which is the tampering of

Fig. 5 Regular Nmap scan across the network

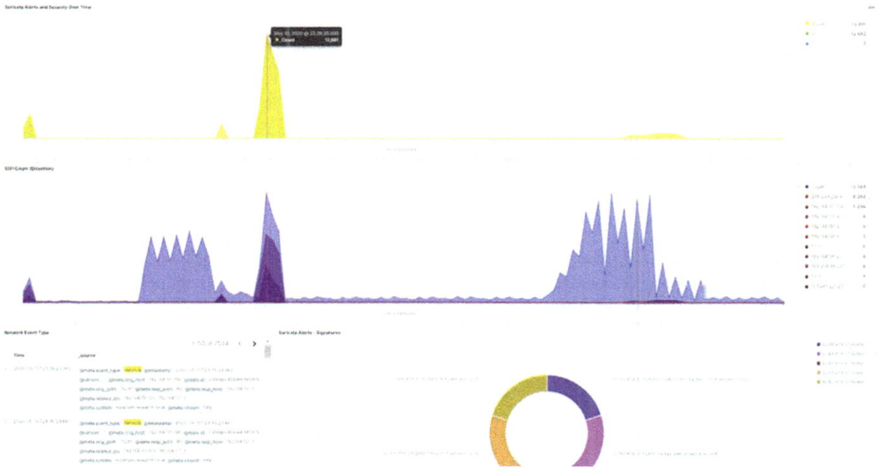

Fig. 6 Aggressive Nmap scan across the network

network packets, as well as "Attempted User Privilege Gain," or a process to gain unauthorized access to a system deliberately.

Moving to more aggressive scans by upping the intensity level of Nmap port scanning shows spikes of network counts and brought the IP Addresses of the top offenders to the top of the list, as well as attempted SSH scans, a frequent port of entry into a server (Fig. 6).

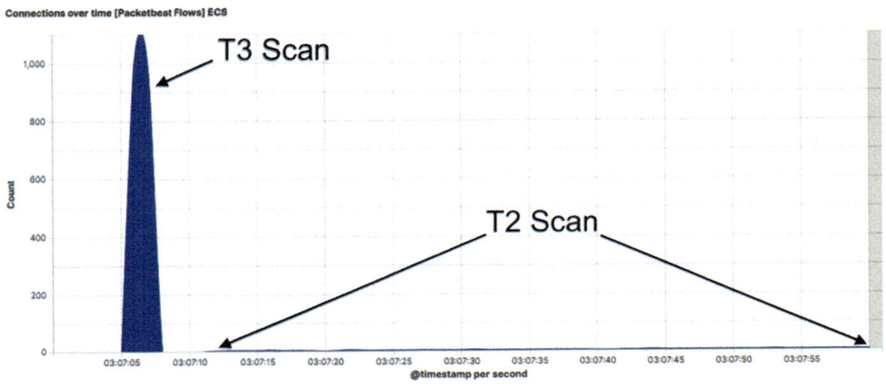

Fig. 7 Traffic pattern of standard T3 and polite T2 scans

However, when using Nmap to scan standard ports on a single host at every intensity level, Suricata failed to detect any threat at all! With a popular IDS failing to detect even T5 scans from Nmap, the question was whether another module could potentially capture only the Nmap traffic which could then serve as a starting point of classifying suspicious network traffic. This would answer our first question: can our chosen Intrusion Detection System (IDS) detect active reconnaissance activities such as port scanning on a network, and if they fail at what intensity level or levels will it fail?

To attempt to detect a traffic pattern in this methodology of active reconnaissance, we looked into other Filebeat modules such as Packetbeat, a network monitor that has gauges to detect a spike in response times, a slowdown in network transactions, and network flow of all network packets. Using this Packetbeat module, we were able to filter and isolate network traffic, which allowed us to discover the Nmap traffic and even compare the various intensity levels. The results below satisfy our second research question: if the IDS fails to detect the active reconnaissance, can we analyze the data generated by the scans in a white box test and deduce a pattern for scans at all levels paying close attention to those intensities that previously failed to be detected?

For example, as seen in Fig. 7, we were able to see a standard T3 scan, the default level of an Nmap scan, and the polite T2 scan. Upon closer analysis, we were able to determine through Packetbeat that a T2 port scan uses approximately five packets per second for reconnaissance scanning. This was significantly less than a T3 scan with over 1100 packets in the span of 1 second, which is over 200 times greater in frequency per second.

Moving down to a T1 scan, a *stealthy* Nmap scan, the data shows that a single packet is sent every 15 seconds, while the *paranoid* T1 Nmap scan sends a single packet every 5 minutes, as shown in Fig. 8. Using Packetbeat, we can identify the pattern of Nmap scans to distinguish them from benign traffic activity, and in theory, customize rules on Suricata to detect such traffic.

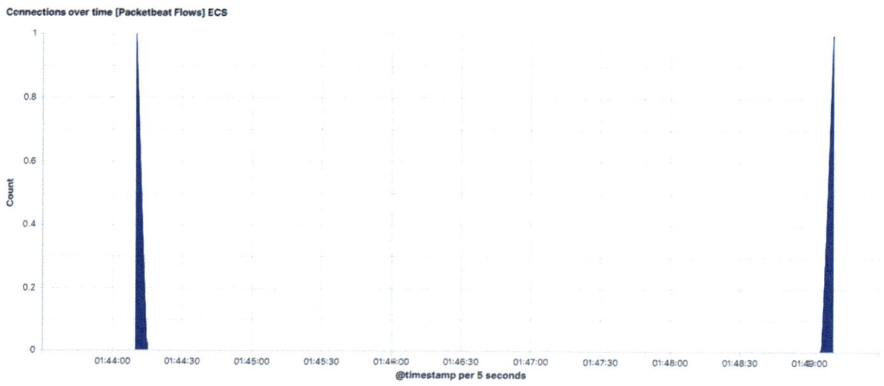

Fig. 8 Traffic pattern of T1 scan

5 Limitations

Though the research covers numerous services seen in small businesses, it will not cover all instances of networks. Some of these limitations include but are not limited to: endpoint protection software, comprehensive firewalls (on the host and network level), application firewalls, IPS systems, and other software that may change the nature of the results seen in our research. Another factor is the equipment used, which can also affect the outcome of the results found in our research network. The software used in our collection and visualization, RockNSM, requires a system that has 4 CPU cores, a minimum of 8 GB of RAM, 256 GB of storage, and 2 NICs (one for management and one for monitoring/sensing).

6 Conclusion and Future Work

It is true that insider threat has become one of the more common methods of attack in the digital age. We have seen that no one is safe from the insider threat and that those threats can come in the form of accidental, negligent, or malicious threat actors. However, every attack begins with reconnaissance. Fingerprinting and detecting this reconnaissance can allow networks to stay on top of potential threats. The most ubiquitous tool for scanning and information gathering on a network for systems administrators and malicious actors alike is Nmap.

This research has shown that Suricata failed to detect Nmap scans at all levels except for scans targeting ports outside of the normal range. On the other hand, deeper traffic inspection has successfully detected the patterns inherent in Nmap scans through the use of Packetbeat. This leads us to our further research in creating rules for Suricata to detect scans more efficiently and alert responsible parties about the state of the network, thus allowing the user to make an informed decision on what action to take. Additionally, we will be building an Artificial Intelligence tool to detect patterns that were previously undetected with our current detected patterns and respond in a similar manner to the scripts that we are developing.

The future work will focus on our third research question: if we can deduce a pattern, can we write software to patch the system through a rules-based approach or an artificial intelligence-based approach?

References

1. What is privilege escalation and why is it important? (2019). Retrieved from https://www.netsparker.com/blog/web-security/privilege-escalation/
2. Introduction to reconnaissance. (2019). Retrieved from https://www.hackingloops.com/introduction-to-reconnaissance-part-1-terms-and-methods/
3. INFOSEC: Top 10 Network Recon Tools. (2019). Retrieved from https://resources.infosecinstitute.com/category/certifications-training/ethical-hacking/network-recon/#gref
4. E. Shein, Companies proactively seek out internal threats. Commun. ACM **58**(11), 15–17 (2015). https://doi.org/10.1145/2820423
5. T. Senator, E. Chow, I. Essa, J. Jones, V. Bettadapura, D.H. Chau, O. Green, O. Kaya, A. Zakrzewska, E. Briscoe, R. Mappus, H. Goldberg, R. Mccoll, L. Weiss, T. Dietterich, A. Fern, W.–K. Wong, S. Das, A. Emmott, D. Bader, Detecting insider threats in a real corporate database of computer usage activity, 1393 (2013). https://doi.org/10.1145/2487575.2488213
6. M. Alsaleh, P.C.V. Oorschot, Network scan detection with LQS: A Lightweight, Quick and Stateful Algorithm. Proceedings of the 6th ACM Symposium on Information, Computer, and Communications Security - ASIACCS 11, 102–113, 2011 https://doi.org/10.1145/1966913.1966928

Toward Home Area Network Hygiene: Device Classification and Intrusion Detection for Encrypted Communications

Blake A. Holman, Joy Hauser, and George T. Amariucai

1 Introduction

As technology advances, the threats to our networks and data are continuously increasing. New devices and new forms of technology are produced on a greater scale than ever before, and with these new devices come new threats and new attack surfaces. One main technological advancement that also poses a great security concern is the development of Internet of Things (IoT) devices. These devices are made to be cheap and small and often sacrifice security in favor of fast installation, ease of use, and low cost. The Internet of Things spans a broad range of devices, from the more traditional general-purpose computation devices like laptop and desktop computers and mobile phones, to home area network-specific devices like smart meters, smart appliances, thermostats, security cameras, etc., and to very simple sensors and actuators like temperature or humidity sensors and smart power switches and outlets. Devices like these are currently connected to most private local area networks and pose a major security threat to their owners. The threat comes mainly in the form of malware that can cause a device to secretly collect data and share it with unauthorized parties, to manipulate actuators (e.g., unlock doors, etc.) for malicious purposes, or to become bots for various types of distributed denial of service (DDoS) attacks.

The use of artificial intelligence (AI), and in particular artificial neural network (ANN), is currently changing the technological landscape. Traditionally, ANNs are used for pattern recognition and prediction. ANNs are embedded into many different applications in IoT devices and form the foundation of a promising technological field that has the potential to expand the uses of technology in many different areas

B. A. Holman (✉) · J. Hauser · G. T. Amariucai
Kansas State University, Manhattan, KS, USA
e-mail: baholman@ksu.edu; jhauser@ksu.edu; amariucai@ksu.edu

K. Daimi et al. (eds.), *Advances in Security, Networks, and Internet of Things*, Transactions on Computational Science and Computational Intelligence, https://doi.org/10.1007/978-3-030-71017-0_14

including analyzing traffic data, distilling relevant consumer information, or, in the context of IoT security, determining intrusions on a private network. The latter application has the ability to prevent malicious actors from entering private networks or stop them for using the private networks for malicious purposes.

ANN-based network intrusion detection is mainly focused around anomaly detection in the network statistics, such as packet counts, average packet size, duration of connection, packet types, etc. [1–4]. More recent work [5, 6] uses recurrent neural networks (RNNs) to capture the temporal patterns in sequences of packets associated with a single communication flow.

In this chapter, the concern is with a problem similar to intrusion detection, but more specialized to the home area network (HAN). Specifically, the chapter aims to classify the devices communicating over the HAN solely by their communication patterns while completely disregarding the packet contents. The appeal of the proposed system is that no assumptions are made about the environment in which it is deployed, other than that enough time is provided for learning, in the absence of any attacks. The proposed system thus automatically clusters the devices into categories, prompts the user to select labels (or enter brief descriptions) of all the identified categories, and continues to monitor the network for new devices or new behaviors. Upon detecting a new device (or atypical device behavior), the system raises an alarm and asks the user whether a new device was added, or whether an old device was intentionally repurposed. Because IP addresses are often allocated dynamically, no IP address information is used during classification, except to distinguish between different conversations carried out by different pairs of devices. Instead, only sequences of packet lengths and interpacket pauses are used to characterize each device on the HAN. To capture the patterns in these sequences, we construct a classifier in the form of a recurrent neural network (RNN).

The remainder of the chapter is structured as follows. The most relevant related work is presented in Sect. 2, and the problem is formulated in Sect. 3. The technical details of the proposed solution are described in Sect. 4, while the experimental design and results are given in Sect. 5. Finally, conclusions are drawn in Sect. 6.

2 Related Work

The existing ANN-based network intrusion detection literature focuses on different intrusions, different neural networks, and different parameters and features extracted from network traffic to be used as the input of the respective ANNs. Many of the existing works focus exclusively on the detection of predetermined attacks—like the Distributed Denial of Service (DDoS) attacks. In this case, the traffic features used for classification are often chosen according to the authors' expertise and expectations regarding those specific attacks. By contrast, the mechanism proposed in this chapter has to work in the absence of any predetermined beliefs on the devices' behavioral traits.

In [7], the authors explain how an ANN can be used for packet classification to determine if a DDoS attack is being conducted on a network. First, they collect normal data from the network in the absence of an attacker and then attack data from a network undergoing a DDoS attack. This data is in a PCAP file format. Next, the authors extract "average packet size, number of packets, time interval variance, packet size variance, packet rate, and number of bytes." Then, the authors use the extracted data to train an ANN. Lastly, they use the ANN to classify the packet data into normal network activity or DDoS network activity. They claim that the best accuracy rate was 99.6% [7]. Such high accuracy rates are expected when the attack is known to differ significantly from normal network operation. In fact, DDoS attacks are by definition characterized by unusually high volumes of traffic.

Similarly, [8] discusses the use of neural networks to identify denial of service attacks, distributed denial of service attacks, and port scans. In this article, the authors use network traffic data as input into an ANN to determine whether a network is undergoing an attack. In this case, the authors use Self-Organizing Maps (SOMs) to cluster data together to provide more accurate anomaly and intrusion detection. The authors claim that "Neural Networks using both supervised and unsupervised learning have many advantages in analyzing network traffic" [8].

In [9], Devikrishna and Ramakrishna similarly discuss the use of an ANN in determining intrusions on a network. Specific features are extracted from the PCAP files for use in an ANN. These features consist of "duration of the connection, protocol type, service type, status flag, total bytes sent to destination host, total bytes sent to source host, whether source and destination addresses are the same or not, number of wrong fragments, and number of urgent packets" [9]. The authors then input this data into an ANN to determine whether there is an intrusion on the network.

The use of a Recurrent Neural Network in determining network intrusions is first discussed in [5]. In this article, the authors use 90 GB of network traffic data for training. An unsupervised approach is also used along with the RNN. The work in [6] analyzes how Recurrent Neural Networks compare to other methods of intrusion detection. First, the authors explain how the current ways neural networks are used to identify network intrusions will often miss some specific intrusions. Then they explain the potential use of Sequence Forward Selection algorithm and the Decision Tree model to enhance the ability to identify network intrusions and anomalies. Next, the authors explain the uses of different types of neural networks to enhance network intrusion identification. They conclude that the use of Sequence Forward Selection algorithm and the Decision Tree models incorporated into a Long Short-Term Memory (LSTM) RNN "not only reduced execution time and the amount of required memory but also significantly improved the performance of conventional LSTM model" [6].

To conclude, the works outlined above show the benefits of an ANN for intrusion detection. Specifically, they show how features can be extracted from the PCAP files representing captured network traffic to be used to train an ANN for intrusion detection. The reasonably high accuracy rates achieved for the proposed intrusion detection mechanisms based on ANNs are particularly encouraging for the problem discussed in this chapter.

3 Problem Description

The problem of identifying intrusions in a network is one of great importance for both current and future HAN architectures. There are many different ways to tackle it. This chapter addresses using a Long Short-Term Memory (LSTM) Artificial Neural Network. This type of ANN "remembers" some of the previous data input to help in classification.

The first question to answer is what type of data should be used to identify an intrusion. The articles discussed previously extract various features from PCAP files to input into the ANN. In this chapter, the interest is in extracting only the simplest, and as such the least privacy-invasive, metadata for each device on a network: packet lengths and interpacket times. For training purposes, the extracted information is split into classes based on the device type.

Next, the separated data from the PCAP files is used to train the LSTM RNN to classify devices based on type. By doing so, when a new device is introduced to the network, its type can be identified and one can determine if it belongs on the network. By determining devices that do not belong on a network (or equivalently, and indiscriminately, misbehaving devices), intrusions can in turn be identified.

To accomplish these tasks, a set of Python scripts were created to automatically identify all the devices on a network and then use the LSTM RNN to provide each device with a label of the type of device. For example, a security camera would be labeled as a camera. This suite of scripts is called the *Device Detector* for the remainder of this chapter. If the Device Detector cannot determine the type of a device, it will label it as unknown and raise an alarm. An alarm is similarly raised when a new device (although correctly classified into one of the available types) is observed.

The overall goal can be broken down into three different tasks:

1. Classify a device when the LSTM RNN has been trained on some of the PCAP data from the same device.
2. Classify a device when the LSTM RNN has been trained on PCAP data from the same type of device, but not from the exact same device.
3. Determine when a new device type (previously not represented in any training data) has been introduced to the network.

The remainder of this chapter explains how to go about accomplishing each of the previous tasks. But to be able to identify intrusions, first the appropriate setup for the application to run must be completed. Below are the items required in order to run this application:

- A wireless network with IoT devices connected to it.

- A software tool for capturing PCAP traffic files from the wireless network.

These required items will allow the LSTM to be trained on the properly extracted information. This is explained in the next section.

4 Proposed Solution

The solution is broken into several different parts. First, a large data set of PCAP files is obtained. Next, the PCAP data is broken down into an understandable format, so initially the PCAP files are broken down based on device. Next, these device files are changed to "conversations," which is explained later. Then, the LSTM RNN is trained to recognize a device class from one of its conversations. Next, boundaries are set in the RNN's output space to classify unknown devices. Lastly, the program is evaluated. Throughout this section, each step of the process is explained.

4.1 PCAP Content Extraction

The first thing the program must accomplish is the extraction of the contents in a PCAP file. To do this, a module from Scapy called rdpcap is used. The rdpcap module assisted in collecting the packets from the PCAP files and converting them into plain text. By extracting the text, the program is able to collect a variety of information about the packets, such as the packet timestamp, the packet type, the header of the packet, and the body of the packet.

Within the header, the Ethernet, ARP, IP, TCP, UDP, ICMP, and DNS headers for each packet were extracted if they existed. Additionally, the text for the body for the packets that have a body was also extracted. In some packets, the packet body was very useful and provided valuable information about the device. However, many of the packets either did not have a body in their packets or the body was encrypted. Therefore, the packet timestamp and packet length were the main focus to determine the type of the device for labeling the training and test data.

The extraction began by separating each device by MAC address. Each device has extracted PCAP information based on the devices it was sending packets to.

4.2 Conversation Extraction

For this project, each device classification was strictly based on the packet length and packet timestamp. However, the packet timestamp on its own provides no unique information on the device. To solve this issue, the use of "conversations"

Table 1 An example of a conversation

Packet length	Packet time difference
6.80e+01	0.00e+00
2.99e+02	5.81e−04
2.99e+02	8.37e−02
3.75e+02	8.50e−05
2.99e+02	1.02e−01

was implemented. A conversation is a period of time, in this case a minute, where a device is sending and receiving packets from another device.

A conversation starts when the first packet is sent. The length of the first packet is recorded with initial time set to zero. The next packet received in the sixty-second time period will have its length and time recorded in seconds between it and the last packet sent. This will continue until the next conversation begins. Each device will have conversations recorded from every device it sends packets to. An example of a conversation can be seen in Table 1.

Now that the data is set up, each device type must have the same number of conversations. If they do not, it creates an imbalanced data set and the LSTM RNN is bound to learn a bias.

4.3 SMOTE

Not every device type has the same number of conversations associated with it. This means that the LSTM neural network that is used will produce more accurate results for the device types with more data. Also, it will classify devices with less data incorrectly more often. There are two main ways to eliminate this bias: undersampling and oversampling. Undersampling would force the least amount of conversation data to be used to train the neural network. But since the device types with the least amount of data do not contain nearly enough information, the use of an oversampling technique was selected instead.

Oversampling is a technique that produces more data for the classes with too little data. One method of doing this is to reuse the conversation data from classes with too little data associated with it. However, this can cause the neural network to overfit for the device types reusing too much data. So, synthetic data was created to mitigate the risk of overfitting.

The method used to create our synthetic data is called *Synthetic Minority Oversampling Technique* (SMOTE). SMOTE is used to balance imbalanced data sets by creating synthetic data [10]. For the conversations, two files from a device type that needed more data were randomly selected. Next, the differences of the packet length and time between packets were calculated. Then, a random "gap" between zero and one was generated. Next, the first file's packet length and time were

Table 2 Sample conversation 1

Packet length	Packet time difference
6.80e+01	0.00e+00
2.99e+02	5.81e−04
2.99e+02	8.33e−02
3.75e+02	8.50e−02
2.99e+02	1.02e−01

Table 3 Sample conversation 2

Packet length	Packet time difference
3.75e+02	0.00e+00
6.80e+01	5.34e−05
2.99e+02	5.68e−04
6.80e+01	5.32e−02
2.99e+02	1.80e−04

Table 4 Sample synthetic conversation

Packet length	Packet time difference
2.20e+01	0.00e+00
3.34e+02	5.97e−04
2.99e+02	9.62e−02
3.29e+01	8.98e−02
2.99e+02	1.17e−04

selected, and this was added to the difference multiplied by the gap. Below is a visual representation of this calculation:

- length difference = (conversation 1 length) − (conversation 2 length)
- time difference = (conversation 1 time) − (conversation 2 time)
- gap = random number between 0 and 1, inclusive
- synthetic length = (conversation 1 length) + gap * (length difference)
- synthetic time = (conversation 1 time) + gap * (time difference)

This calculation was done repeatedly until all device types contained the same number of conversations. We understand that this produces a bias. However, we determined that the bias is useful for preventing overfitting and adding some noise to our limited data set.

In Tables 2 and 3, an example of two conversations that were randomly selected is given.

After the conversations from Tables 2 and 3 are obtained, we randomly calculate a gap between zero and one. In this case, let us assume that the gap is 0.15. Table 4 will be the synthetic data calculated from the conversations in Tables 2 and 3.

4.4 The Artificial Neural Network

For the classification of devices, an Artificial Neural Network (ANN) is used. Specifically, a Long Short-Term Memory (LSTM) ANN is used. An LSTM is used so the network can "remember" the sequences of the conversations.

To begin, an input size of two is used for the LSTM. The first input is the packet length, and the second input is the time between packets in a conversation. The neural network will process each time–length pair in the conversation before moving to the next conversation.

For this experiment, two hidden layers with eighteen nodes each are created. Eighteen nodes and two hidden layers produced the highest accuracy when tested. Next, the output size is equal to the number of device types. This allows each time–length sequence (conversation) to be classified into a type of device because each output node is associated with a device type. The outputs are normalized so that all outputs sum up to one. This means that the output vectors live on a probability simplex, and each specific output can be interpreted as the probability that the analyzed device falls under the corresponding type. The output node with the greatest value then corresponds to the device type a conversation will be classified as. The label for each device type will be a similar vector, only a sparse one, containing all zeros except for the component identifying the device type, which will equal one. This is, in essence, a *one-hot* representation of the device type, consistent with a probability-one assignment to the correct device class. This means that the distance between the label vectors and the output vectors can be calculated over the probability simplex.

4.5 Boundaries

For each conversation for a device type, the distance the output is from the label is calculated. Then, the distance at the 90-th percentile for a particular device type is found. This gives the "boundaries" associated with each device type.

The boundaries' purpose is to determine when unknown device types are introduced to the network. This prevents a device from being classified incorrectly. For example, if the ANN has not been trained to recognize a smart thermostat, then the ANN will incorrectly classify it as another device type. However, if the boundary has been set, one can recognize when the ANN is struggling to classify a device. When the distances to all device types' labels are abnormally large and outside the corresponding boundaries, one can assume that the analyzed device does not belong to any known device type. Then, one can classify it as a new and unknown device type.

Let us take the example in Fig. 1. For this example, we have two device types initially. The point (0,1) represents the first device label (blue marks), and the point (1,0) represents the second device label (red marks).

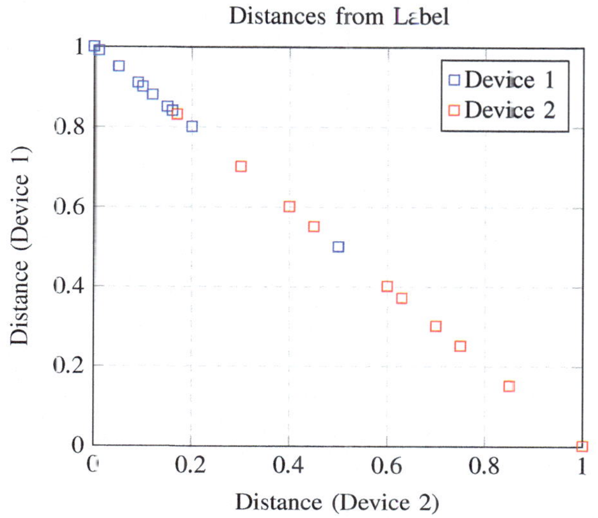

Fig. 1 Distance of two devices from label

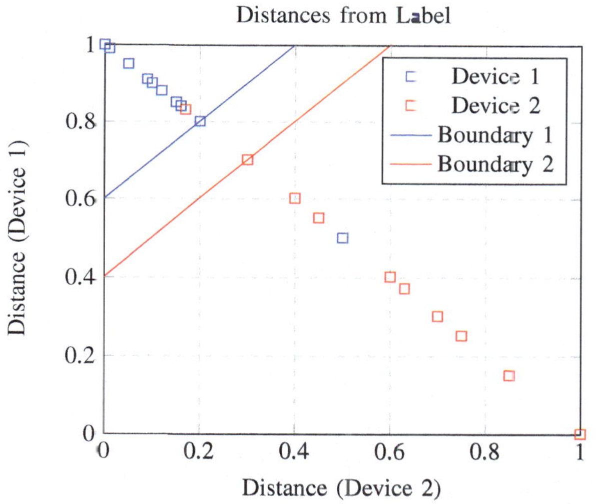

Fig. 2 Distance of two devices from label with boundary line

Now that the distances are plotted, the boundary at the 90-th percentile can be calculated. In Fig. 2, we added lines to represent where these boundaries would be for the same situation.

Boundary 1, in this case (Fig. 2), is much closer to its label than boundary 2. This is because the distances for boundary 1 are closer to the label. Also, one may note that many points for device 2 are incorrectly classified in this example. However, this does not affect the positioning of the boundary.

Now let us examine how an unknown device type (green marks) might look in this example (Fig. 3).

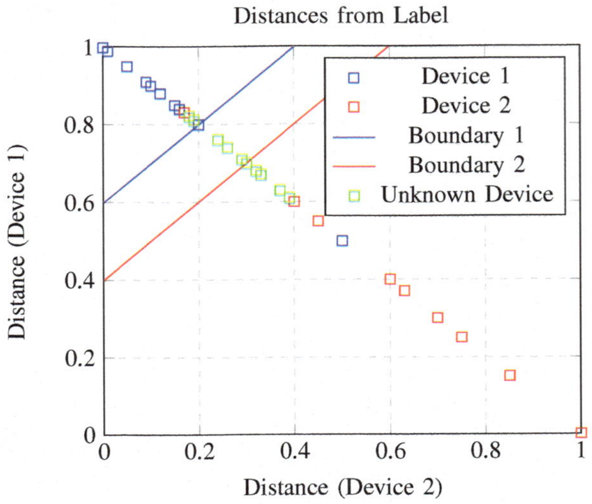

Fig. 3 Distances of unknown devices compared to known device

Because the new, unknown device is classified the same as device 1, device 2's marks and lines can be ignored (red). In the new graph (Fig. 3), one can see that the new device lies mostly outside the device 1's boundary. From this, one can infer that the new device, although classified as device 1 type, does not belong to that type. Now the neural network can be retrained to include the new device type.

5 Experiment Design and Results

5.1 The Data

The experiment began with approximately 6 GB of PCAP data from many different devices. This data was provided by the Johns Hopkins Physics Laboratory. The data set contains three cameras, two Google Assistants, two Amazon Echos, one Apple Assistant, one mobile device, two smart outlets, one speaker, one clock, and one smart home hub. With these devices, we made nine device-type classes, as follows:

- Camera
- Google Assistant
- Amazon Echo
- Apple Assistant
- Mobile Devices
- Smart Outlets
- Speaker
- Clock
- Smart Home Hub

The largest number of conversations for any device type was 133,459. Using SMOTE, as explained previously, we created synthetic data for each of the device types to match the largest number of conversations. Therefore, each device type ended with 133,459 conversations.

5.2 *The Neural Network*

Since there were nine device types, we used nine output nodes for our Long Short-Term Memory Artificial Neural Network. Each output node was correlated with a device type. For the hidden layers, two layers of eighteen nodes each were used. Furthermore, the Neural Network train was trained on 80% of the conversation data leaving 20% for testing.

After testing a few different numbers of hidden layers and hidden layer nodes, we determined that two hidden layers with eighteen nodes each provided the most accurate classification. For testing, we tested one, two, three, and four hidden layers with five, eleven, and eighteen nodes. Five, eleven, and eighteen were selected based on our input and output sizes. Five is just over double the input size, eleven is the summation of the output and input sizes, and eighteen is double the output size. Architectures with just one hidden layer produced poor results. There was almost no difference between architectures using three and four hidden layers, but the results were still relatively inaccurate. Two hidden layers with five nodes each still produced poor results, but two layers with eleven nodes each classified much more accurately. The best results came from a two-layer system with eighteen nodes each.

5.3 *The Results*

In the training rotations, we started with a 32.44% accuracy rate and ended with a 69.25% accuracy rate. For testing rotations, each device type had 30,000 conversations tested. The results are shown in Table 5. Note that this table is read vertically; that is, each column represents the number of conversations from a single device type that were classified as each available device type. For example, the first column in Table 5 states that out of 10,000 conversations carried out by Google Assistants, 3512 were classified correctly, 1570 were classified as Amazon Echo, 365 were classified as Apple Assistant, and so on. The table is helpful in identifying which device types cause the most errors, which impact the overall accuracy, etc.

Table 5 Test results Table 1

Device (horizontally) classified as (vertically)	Google Assistant device	Amazon Echo device	Apple Assistant device	Camera device	Smart outlet device	Mobile device	Speaker device	Smart home hub device	Clock device
Google Assistant class	16,492	2090	3938	203	1475	1817	534	17	443
Amazon Echo class	2930	19,291	3590	5140	1886	2103	124	0	167
Apple Assistant class	99	206	12,007	0	0	19	1	0	0
Camera class	3637	514	864	20,675	225	4952	7567	157	70
Smart outlet class	2300	2704	847	353	23,043	1490	167	193	424
Mobile device class	3267	3511	8220	524	914	18,418	2416	274	731
Speaker class	423	84	25	270	8	296	18,905	0	25
Smart home hub class	193	47	66	1	97	107	72	29,359	2
Clock class	659	1553	443	2834	2352	798	214	0	28,138
Total	30,000	30,000	30,000	30,000	30,000	30,000	30,000	30,000	30,000

5.4 Data Error

After running many different tests and analyzing the results, we can assume that a few different devices produce similar conversations. This will cause our neural network to improperly classify one or more devices with similar conversations and is mainly responsible for the relatively low accuracy. For example, in Table 5 under the device type Apple Assistant, we can see the Neural Network correctly classifies a smaller portion of the total conversations tested. It classifies some of the Apple Assistant conversations as a Mobile Device or another assistant. Therefore, one can assume that these two device types can act similarly and produce similar conversation data. Knowing that both an Apple Assistant and a Mobile Device can be connected and conduct similar operations if they both were doing similar actions while the data was captured, both device types would have similar looking conversations.

We also noticed that devices that are more often interconnected with other devices, such as a mobile device, speaker, and the assistants, are more likely to produce less accurate results. This may be because of the similarity in the conversations between interconnected devices. We can also notice that devices that are less likely connected to other devices produce the best results. For example, the smart clock is a more accurate device, and it is the least interconnected device.

5.5 Task 1

The first task we intended to solve was to classify a device when the LSTM has been trained on some of the PCAP data from that device. The Apple Assistant, Mobile Device, Speaker, Smart Home Hub, and Clock device types all contained only one device per device type. As one can see, the Smart Hub device had the highest accuracy rate with 97.86% of the conversations classified correctly. The next highest accuracy was the Smart Clock with 93.79% accuracy. Next, the Speaker had an overall accuracy of 63.02%. This was followed by the Mobile Device with 61.39% accuracy. The worst accuracy from Task 1 and overall was the Apple Assistant, which classified with an accuracy of 40.02%. Most of the devices classified the majority of the conversations in the correct device type. However, the Apple Assistant device type did not follow this pattern and had the worst accuracy rate of all device types. This Task had the most diverse results.

5.6 Task 2

The next task we set out to accomplish was to classify a device when the LSTM has been trained on PCAP data from the same type of device. The Google Assistant, Amazon Echo, Camera, and Smart Outlet device types all contained more than one device per device type. This means that other devices of the same type were used to train the LSTM, and these other devices were also used to classify a different device of the same type. For this task, one can see that Smart Outlet device had the highest accuracy rate with 76.81% of the conversations classified correctly. The second highest accuracy rate from this task was the Camera with 68.92% accuracy. This was followed by the Amazon Echo with an accuracy of 64.30%. Lastly, the most inaccurate device type for this task was the Google Assistant that had an accuracy rate of 54.97%. Each of these device types classified most of the conversations correctly. One can conclude that the more devices per device type are available for training, the more reliable the results are. Overall, this task produced better and more consistent results than device types in Task 1.

5.7 Task 3

The last task was to determine when a new device type has been introduced to the LSTM. For this task, "boundaries" were created, as explained in the previous section. These boundaries allow us to determine whether a new device type is introduced to the LSTM. Unfortunately, for this to work properly, a higher accuracy than what was obtained from this LSTM RNN must be produced.

6 Conclusions and Future Work

We introduced a new mechanism for classifying devices on a home area network that relies solely on the sequences of packet lengths and interpacket pauses that define *conversations*. Given the nature of the data, these experiments show encouraging results, although far from ideal. For this reason, we propose that such classification mechanism be used in conjunction with other classification methods that use richer data, with more attributes mined from the network traffic.

For future implementations on this project, first the overall accuracy must be improved. There are a few different ways to do this. First, one can attempt to improve the synthetic data production. Another option is to use different devices and PCAP data to train the LSTM. One issue might have been the lack of diversity in the training set, which could have caused less accurate data. Also, the simplest solution would be to adjust the hidden layers and iterations of the LSTM. Changing the number of nodes in the hidden layer, number of hidden layers, and the number of iterations through the LSTM could potentially improve the accuracy. These are all options we might pursue to produce more accurate results in the future.

As stated earlier, one potential issue with the neural network producing inaccurate results could be the interconnectivity of the data. By using data that we know will not interfere by having similar conversations as other devices, we may be able to produce a more accurate result. One can reasonably assume this because the device types that are interconnected with other device types tend to produce more inaccurate results. By eliminating these bad results, we hope to be able to achieve a better overall accuracy.

Another possible future development would be device fingerprinting. Through this project, we realized that not only does each device type produce distinct patterns but devices of the same type can produce quite different patterns as well. This means that, given the right neural network setup, one could potentially identify each individual device, rather than device type, on a network.

Acknowledgments This work was supported in part by the US National Science Foundation under grant numbers 1527579 and 1619201.

References

1. Z. Zhang, J. Li, C. Manikopoulos, J. Jorgenson, J. Ucles, Hide: a hierarchical network intrusion detection system using statistical preprocessing and neural network classification, in *Proceedings of the IEEE Workshop on Information Assurance and Security* (2001), pp. 85–90
2. J.Z. Lei, A. Ghorbani, Network intrusion detection using an improved competitive learning neural network, in *Proceedings. Second Annual Conference on Communication Networks and Services Research, 2004* (IEEE, New York, 2004), pp. 190–197
3. S.M. Botros, T.A. Diep, M.D. Izenson, Method and apparatus for training a neural network model for use in computer network intrusion detection, Jul. 27 2004, US Patent 6,769,066
4. E. Hodo, X. Bellekens, A. Hamilton, P.-L. Dubouilh, E. Iorkyase, C. Tachtatzis, R. Atkinson, Threat analysis of IoT networks using artificial neural network intrusion detection system, in *2016 International Symposium on Networks, Computers and Communications (ISNCC)* (IEEE, New York, 2016), pp. 1–6
5. B. Radford, L. Apalonio, A. Trias, J. Simpson, Network traffic anomaly detection using recurrent neural networks (2018). https://arxiv.org/abs/1803.10769 (Accessed: May 25, 2021)
6. T. Le, Y. Kim, H. Kim, Network intrusion detection based on novel feature selection model and various recurrent neural networks. Appl. Sci. **9**, 1392 (2019)
7. I. Riadi, A.W. Muhammad, Network packet classification using neural network based on training function and hidden layer neuron number variation. Network **8**(6) (2017). https://doi.org/10.14569/IJACSA.2017.080631
8. A. Bivens, C. Palagiri, R. Smith, B. Szymanski, M. Embrechts, Network-based intrusion detection using neural networks, in *Intelligent Engineering Systems Through Artificial Neural Networks*, vol. 12 (2002)
9. K.S. Devikrishna, B. Ramakrishna, An artificial neural network based intrusion detection system and classification of attacks. Int. J. Eng. Res. Appl. **3**, 1959–1964 (2013)
10. N. Chawla, K. Bowyer, L. Hall, W. Kegelmeyer, SMOTE: synthetic minority over-sampling technique. J. Artif. Intell. Res. **16**, 321–357 (2002)

Part III
Security Education, Training, and Related Tools

The Impact of Twenty-first Century Skills and Computing Cognition Cyber Skills on Graduates' Work Readiness in Cyber Security

Anna J. Griffin, Nicola F. Johnson, Craig Valli, and Lyn Vernon

1 Background

By 2026, it is expected that the global demand for cyber security will increase by 86 percent, and by 2026 Australia will be short of 17,000 cyber security employees [1]. To meet the demand, higher education (HE) is increasing the availability of courses in cyber security, with an expected 400% increase to 2000 per year, by 2026 [1]. However, little research exists about the quality of the courses and whether graduates are work ready for careers in cyber security.

This research project is the first stage of a four-part PhD which will evaluate the usefulness of the Bachelor of Science (cyber security) program in optimizing students to be effective cyber security workers. This research will examine how students' perceived competencies in twenty-first century skills and cyber security skills impact their perceived work readiness?

Graduate quality has been a common employer concern in many industries, as employers feel graduates often do not have the required skills [2]. There is a common criticism that HE struggles to produce graduates who are work ready [3]. Work ready is a concept that many researchers have investigated [4–6] particularly in areas like business [2] and some areas in the ICT industry [7] but not specifically in cyber security. The concept of work readiness is that graduates are ready to enter the workforce and are adequately prepared to succeed by having the required skills [8]. However, these are not necessarily specific technical cyber security skills [3] and often employers are seeking graduates that possess a variety of soft skills [1, 9–11] that are transferable between roles, rather than role-specific. This transferability is particularly important in cyber security, as the industry is dynamic and requirements

A. J. Griffin · N. F. Johnson · C. Valli (✉) · L. Vernon
Security Research Institute, Edith Cowan University, Joondalup, WA, Australia
e-mail: c.valli@ecu.edu.au

K. Daimi et al. (eds.), *Advances in Security, Networks, and Internet of Things*, Transactions on Computational Science and Computational Intelligence,
https://doi.org/10.1007/978-3-030-71017-0_15

within roles are changing quickly. Therefore, to ensure HE are producing the quality graduates in cyber security, there needs to be a balance between the development of soft skills and technical skills.

The growth of cyber security in HE has experienced challenges including fragmented curriculum due to a lack of national framework, a lack of supply of quality teachers and costs associated with setting up effective infrastructure, for example, cyber labs [1]. In Australia, there is currently no nationally accepted, accredited framework to base course curriculum upon [1]. This is unlike other countries where frameworks and accreditations have been developed and supported [12].

In the USA, the NICE framework was established in 2012 and provides a taxonomy of the field of cyber security [12]. The NICE framework identifies 33 specialist areas within cyber security and then maps the identified key knowledge, skills, and abilities (KSA) required for each of the different specialisations [12, 13]. In contrast, in Australia, the diversity within the profession has led to a fragmented curriculum in cyber security within universities [12, 14] as it is impossible for a 3-year university program to cover all specialty areas. It has been argued that universities must take a more holistic approach to cyber security education [13].

A holistic approach considers the development of skills that are referred to as soft skills or twenty-first century skills. A study of Australian graduates' work readiness by Prikshat et al. [15] identified the following deficits in graduates: interpersonal, self-management skills, communication (written and oral), leadership, teamwork, cognitive abilities (critical thinking and problem solving), lack of innovation and creativity. These skills develop within practical activities, for example, war games and real work settings, rather than theory-based or classroom-based learning [9]. It is believed twenty-first century skills are essential in preparing graduates work ready, as they are applicable to all work situations and are transferable. The Australian Computer Society (ACS), which provides professional accreditation in IT including cyber security, recognises the significance of 'soft skills' by including accreditation of courses on 'soft skills' [16]. However, these soft skills are not demonstrated in outcomes.

Therefore, the question is: what skills are essential for developing students' work readiness? In particular, are the technical skills within the field of cyber security more important than the twenty-first century skills that can be transferable between roles in the industry, contributing more to a graduate's work readiness? The purpose of this study is to examine student's perception of their work readiness through a survey and examine whether twenty-first century skills and cyber security skills contribute to the perception of their work readiness. In later stages of the PhD study, employers' perception of work readiness will be explored. However, the main aim of this project is to carry out a pilot study to develop a model of work readiness which can be used as a reference throughout the PhD project.

2 Methodology

2.1 Research Design

Edith Cowan University is one of six universities that are part of the Cyber Security Cooperative Research Centre (CSCRC). The PhD will complete research within all six CSCRC universities to understand the students' work readiness and will include surveying students, interviewing graduates, and industry stakeholders to further evaluate the Bachelor of Science (cyber security) programs

A quantitative methodology was chosen due to its ability to gather a high number of responses through surveying which will facilitate a broader capacity for generalisation [17]. Specifically, the survey design has the capacity to explore the phenomena of work readiness, usually investigated through interviews and focus groups (qualitative), via quantitative data that can be statistically analysed [18].

Data collection through survey was chosen because of its proficiency to gather demographic information and perceptions and attitudes of students [19] towards their work readiness. By surveying a sample of a population, responses can be collated and statistically analysed with associations being conveyed that, depending on sample limitations, allow for generalisations that may be cautiously applied to the whole population [20]. A survey also allows for standardised measurement across the sample, by asking the same questions of all participants [17].

The research examines how students perceived competencies in twenty-first century skills and cyber security skills are associated with their perceived work readiness within the Bachelor of Science (cyber security) program at Edith Cowan University. The dependent variable, which is the variable affected by a cause and what is being measured [17], is perceived work readiness. The independent variables, which are what makes something happen [17], are cyber security skills and twenty-first century skills. The covariates (e.g. gender, age) will be used to explain some of the sample variances in these three constructs. The four hypotheses are outlined in Fig. 1.

This pilot study will examine work readiness from the students' point of view. The PhD will be an explanatory sequential mixed method study that will involve interviews of graduates and industry stakeholders to develop an in-depth picture of the program and develop rubrics to demonstrate skill development for graduates' work readiness in cyber security.

2.2 Survey

Using the online platform Qualtrics, a survey was developed which is easily circulated to a large audience, and cost-effective [17]. The survey questions were developed through adapting and combining established survey questions from existing scales that measure work readiness and twenty-first century skills, with

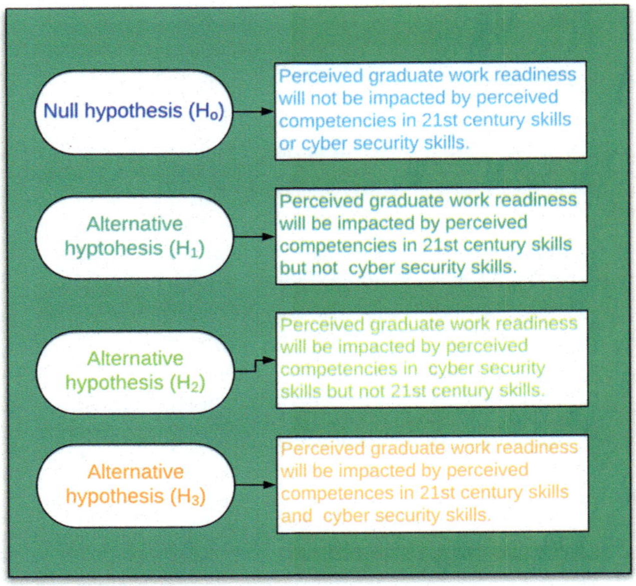

Fig. 1 Hypothesis for mini-project

original development of cyber security skills questions using the NICE Framework. Previous research scales were used as validity and reliability have been established [17]. The work readiness scale, developed by Caballero et al. [5], was used as it had previously assessed college students' perceived work readiness. The twenty-first century skills were divided into problem-solving [21], teamwork [22] and critical thinking [23], as these were deemed important to cyber security and previously used by Prikshat et al. [15]. The cyber security skills were developed from the NICE Framework [24] as there were no available research scales and were arranged using higher-order thinking skills using Bloom's taxonomy [25].

The benefit of completing a pilot study is that it will highlight any possible risks and allow for changes to the survey [17] before data collection commences in the PhD. The aim is to develop the basis for a model measuring students' perceived work readiness which will be used in follow-up PhD studies. Apart from the demographic questions, the survey questions will use a four-point Likert scale [26]. It was decided to use a four-point Likert Scale as this avoids bias towards the centre score [26]. The survey data will then be combined to develop a summated scale [27] from questions about the same variables. The summated scale will provide overall scores for each of the x variables (See Fig. 2).

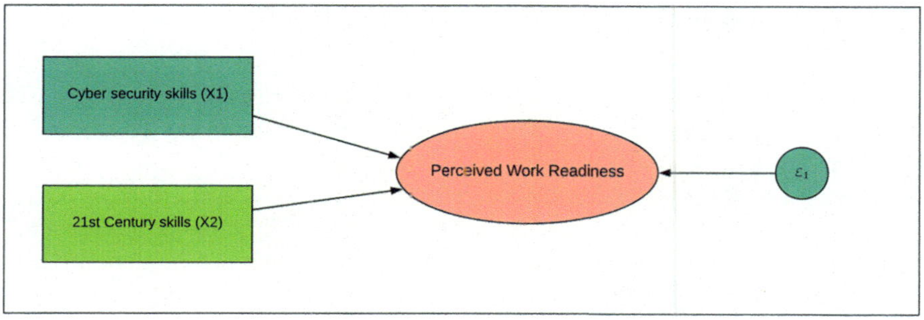

Fig. 2 SEM model of work readiness

2.3 *Sample and Recruitment*

The population size of 3-year students in the Bachelor of Science (cyber security) is approximately 80. The target sample size is 40, as this represents 50 percent of the population.

Convenience sampling will be used, as participants will opt in and volunteer to complete the survey. This may not provide a representative sample, as it only takes the participants who are available and willing to participate, and they may have a particular interest in the survey [17]. However, due to ethical considerations around recruitment within the university, this is the best method to ensure students do not feel pressure to participate. There will be two methods of recruitment, due to on-campus and off-campus learning.

First, the chief investigator, who has no connection to the teaching of the program, will go to tutorials of selected third-year classes at the start of the second semester. The project will be explained to the students and they will be asked to sign up by providing their email address. There will then be an anonymous link for the survey emailed to the volunteers, using the facility that is available in Qualtrics.

The other method of recruitment will be electronic; this will be via advertising on the blackboard site for the relevant Computer Science unit to capture all student modalities and will provide an anonymous link. If there are not enough numbers via these methods, then the unit coordinators will be asked to email a request to the students asking if they would be prepared to participate. All identifying features will be removed from the survey data, and participants will be informed about this via the participation letter (Appendix 2), which will be electronic and at the start of the survey.

2.4 Data Analysis

Qualtrics will provide raw data for analysis after the survey is completed. The raw data will be transferred to SPSS to complete detailed analysis. The basic tests of normality will be run using a p-value of less than 0.05. The Likert Scale can be considered either an ordinal or interval data type; however, Roni et al. [17] suggests that interval is common. Part of the analysis will be to check the validity and reliability of the combined scales in the survey, which can be done by using Cronbach's alpha [28].

Descriptive statistics (frequencies, correlations, means and standard deviations for each item and for the summated scales with reliability of scales also reported – Cronbach's alpha) will be determined using the statistical package *SPSS24*. Structural equation modelling (SEM) using the statistical package *AMOS* will assess the associations between the constructs. A confirmatory factor analysis can be modelled for the latent variable 'perceived work readiness' and the model fit will be assessed. This method is used when attempting to measure concepts within social sciences, such as education, where there is no accepted measurement instrument that exists [27], as with work readiness. SEM will then be used to model the associations between cyber security skills and perceived work readiness and twenty-first century skills and perceived work readiness. They will be placed in the one model to account for each other. Covariates will also be added to the model to explain the variance in the constructs. See Fig. 2 which demonstrates the proposed SEM for assessing the connections between the dependent variable work readiness with the independent cyber security skills and twenty-first century skills variables. Mediation (explaining why two variables are linked with a third variable) and moderation (differences between groups, e.g. gender) [29] techniques will explore the relationship each variable has on one another.

3 Significance

This pilot study will contribute directly to the PhD project that will be evaluating the Bachelor of Science (Cyber security) throughout the six CSCRC universities, to understand if they are optimizing students to be effective cyber security workers. The project provides the opportunity to test the survey instrument. It will also contribute to the development of a model that will be utilised in the mixed method PhD research design.

The PhD will include interviews of graduates and industry stakeholders about the work readiness of graduates who have completed the program. The study will review the curriculum within the CSCRC universities and compare it with the data collected from interviews and the survey. This information will contribute to developing skills rubrics which can be used within HEs to assess students' work readiness within cyber security.

The goal is to produce original research into ensuring students within the undergraduate degrees in cyber security are work ready when graduating, by providing guidance to universities about what skills students must develop during the program. This will be important information for the CSCRC universities directly, as they can adapt their curriculum and use the suggested skills rubrics to assess students' work readiness.

On a larger scale, not only will universities within the CSCRC use this study to develop or re-sculpture their courses in cyber security but so will other universities within Australia. With an industry that is facing worker shortages, it is important that we are training a skilled workforce for the future security of businesses within Australia.

4 Conclusion

Within HE institutes in Australia, there has been an increased demand for cyber security courses due to the increasing skills shortage within the cyber security industry. Currently, there is no accredited framework for these courses to develop their curriculum upon, curriculum has become fragmented across HE. The increasing demand from employers is that graduates should be work ready, which includes the development of twenty-first century skills such as teamwork, critical thinking and problem solving. There is also the need for graduates to be technically competent for an array of different cyber security roles.

Therefore, this pilot project aims to assess the connection between the perceived competencies in twenty-first century and the cyber security skills of third-year students in the Bachelor of Science (cyber security) program at Edith Cowan University with their perceived work readiness. A survey will be conducted, and with SEM analysis, this project aims to answer the research question. This pilot study is a significant step towards the larger PhD project which will examine the usefulness of the Bachelor of Science (cyber security) program in optimizing cyber security workers.

References

1. AustCyber, Australia's Cyber Security Sector Competitiveness Plan 2019 Update, AustCyber, Australia, 2019. [Online]. Available: https://www.austcyber.com/resource/australias-cyber-security-sector-competitiveness-plan-2019
2. J. Cavanagh, M. Burston, A. Southcombe, T. Bartram, Contributing to a graduate-centred understanding of work readiness: An exploratory study of Australian undergraduate students' perceptions of their employability. Int J. Manag. Educ **13**(3), 278–288 (2015). https://doi.org/10.1016/j.ijme.2015.07.002
3. D. Jackson, Modelling graduate skill transfer from university to the workplace. J. Educ. Work. **29**(2), 199–231 (2016)

4. J. Borg, C.M. Scott-Young, Priming the project talent pipeline: Examining work readiness in undergraduate project management degree programs. Proj. Manag. J. **51**(2), 165–180 (2020). https://doi.org/10.1177/8756972820904220
5. C.L. Caballero, A. Walker, M. Fuller-Tyszkiewicz, The work readiness scale (WRS): Developing a measure to assess work readiness in college graduates. J. Teach. Learn. Graduate Employab **2**(1), 41–54 (2011)
6. D. Jackson, Student perceptions of the development of work readiness in Australian undergraduate programs. J. Coll. Stud. Dev. **60**(2), 219–239 (2019). https://doi.org/10.1353/csd.2019.0020
7. S. Chillas, A. Marks, L. Galloway, Learning to labour: An evaluation of internships and employability in the ICT sector. N. Technol. Work. Employ. **30**(1), 1–15 (2015). https://doi.org/10.1111/ntwe.12041
8. C.L. Caballero, A. Walker, Work readiness in graduate recruitment and selection : A review of current assessment methods. J. Teach. Learn. Graduate Employab **1**(1), 13–25 (2010)
9. V. Gore, 21st century skills and prospective job challenges. IUP J. Soft Skills **7**(4), 7–14 (2013)
10. S. Greiff, C. Niepel, S. Wüstenberg, 21st century skills: International advancements and recent developments. Think. Skills Creat. **18**, 1–3 (2015). https://doi.org/10.1016/j.tsc.2015.04.007
11. K.A. Sliter, Assessing 21st century skills: Competency modeling to the rescue. Ind. Organ. Psychol. **8**(2), 284–289 (2015). https://doi.org/10.1017/iop.2015.35
12. E.L. McDuffie, V.P. Piotrowski, The future of cybersecurity education, (in eng). Computer **47**(8), 67 (2014)
13. C. Paulsen, E. McDuffe, W. Newhouse, P. Toth, NICE: Creating a cybersecurity workforce and aware public, (in eng). IEEE Secur. Priv. **10**(3), 76 (2012)
14. A. Rashid et al., Scoping the cyber security body of knowledge. IEEE Secur. Priv. **16**(3), 96–102 (2018). https://doi.org/10.1109/MSP.2018.2701150
15. V. Prikshat, A. Montague, J. Connell, J. Burgess, Australian graduates' work readiness – Deficiencies, causes and potential solutions. Higher Educ. Skills Work-Based Learn **10**(2), 369–386 (2019). https://doi.org/10.1108/HESWBL-02-2019-0025
16. ACS. ACS. https://www.acs.org.au/home.html. Accessed 29 May 2020
17. S.M. Roni, M.K. Merga, J.E. Morris, *Conducting Quantitative Research in Education* (Springer, Singapore, 2019) [Online]. Available: https://public.ebookcentral.proquest.com/choice/publicfullrecord.aspx?p=5850781
18. D. Muijs, *Doing Quantitative Research in Education with SPSS*, 2nd edn. (Sage Publications, Los Angeles, 2011) [Online]. Available: http://srmo.sagepub.com/view/doing-quantitative-research-in-education-with-spss-2e/SAGE.xml
19. L. Andres, *Designing & Doing Survey Research* (SAGE, London, 2012), p. 197
20. R.M. Groves, F.J. Fowler, M. Couper, J.M. Lepkowski, E. Singer, R. Tourangeau, *Survey Methodology*, 2nd edn. (Wiley, Hoboken, 2009) [Online]. Available: https://public.ebookcentral.proquest.com/choice/publicfullrecord.aspx?p=819140
21. D.V. Tesone, M.J. Ross, R. Upchurch, An analysis of event managers' problem-solving propensity: Applying the Problem-Solving Inventory (PSI) to the field of event management. Event Manag **14**(1), 83–89 (2010). https://doi.org/10.3727/152599510X12724735767633
22. M.L. Leeann, A.-B. Dawn, J.N. Tarkington, Validity and reliability of the teamwork scale for youth. Res. Soc. Work. Pract. **27**(6), 716–725 (2017). https://doi.org/10.1177/1049731515589614
23. E.J.N. Stupple, F.A. Maratos, J. Elander, T.E. Hunt, K.Y.F. Cheung, A.V. Aubeeluck, Development of the Critical Thinking Toolkit (CriTT): A measure of student attitudes and beliefs about critical thinking. Think. Skills Creat. **23**, 91–100 (2017). https://doi.org/10.1016/j.tsc.2016.11.007
24. National Initiative for Cybersecurity Education (NICE) Cybersecurity Workforce Framework (2017).
25. D.R. Krathwohl, A revision of Bloom's taxonomy: An overview. Theory Pract. **41**(4), 212–218 (2002)

26. S.S.S. Aznal et al., Validation of a 'Work Readiness Scale' for health professional (HP) graduates. Med. Teach., 1–6 (2019). https://doi.org/10.1080/0142159X.2019.1697434
27. N.J. Blunch, *Introduction to Structural Equation Modeling Using IBM SPSS Statistics and AMOS*, 2nd edn. (SAGE, Los Angeles, 2013) [Online]. Available: http://methods.sagepub.com/book/introduction-to-structural-equation-modeling-using-ibm-spss-statistics-and-amos/
28. J.J. Vaske, J. Beaman, C.C. Sponarski, Rethinking internal consistency in Cronbach's Alpha. Leis. Sci. **39**(2), 163–173 (2017) Online]. Available: http://ezproxy.ecu.edu.au/login?url=http://search.ebscohost.com/login.aspx?direct=true&db=s3h&AN=121166347&site=ehost-live&scope=site
29. P. Jose, *Doing Statistical Mediation and Moderation* (The Guildford Press, New York, 2013)

Enhancing the Cybersecurity Education Curricula Through Quantum Computation

Hisham Albataineh and Mais Nijim

1 Introduction

Cybersecurity is a global multidimensional area that involves governmental and private sectors at the national and international levels. In recent years, the cyberspace threats have increased rapidly as the use of the Internet has grown remarkably as well. Networks, governmental and corporate, are increasingly at risk, malicious acts are escalating sharply as attackers and organized terrorist groups have improved their cyber skills and capabilities [1]. As the number of cyberspace threats has proliferated, all effected parties, individuals, private sectors, governments, etc., struggled to keep their emerging liabilities safe and under control [2]. Government networks, military defenses, political organizations, private companies have been the main targets of organized terrorist groups. An educated cybersecurity workforce is instantly needed as stated recently in a study by the Centre for Strategic and International Studies: "We not only have a shortage of the highly technically skilled people required to operate and support systems already deployed, but also an even more desperate shortage of people who can design secure systems, write safe computer code, and create the ever more sophisticated tools needed to prevent, detect, mitigate and reconstitute from damage due to system failures and malicious acts" [2, 3].

H. Albataineh (✉)
Department of Physics and Geoscience, Texas A&M University-Kingsville, Kingsville, TX, USA
e-mail: hisham.albataineh@tamuk.edu

M. Nijim
Department of Electrical Engineering and Computer Science, Texas A&M University-Kingsville, Kingsville, TX, USA
e-mail: mais.nijim@tamuk.edu

K. Daimi et al. (eds.), *Advances in Security, Networks, and Internet of Things*, Transactions on Computational Science and Computational Intelligence,
https://doi.org/10.1007/978-3-030-71017-0_16

Cyber defense is an extraordinary challenging mission, as all automated defense systems cover significant threats and major weaknesses [2]. Furthermore, the setback is more prominent due to the lack of a qualified workforce. I.T. experts, in general, groups of personnel who have the applicable proficiency, are highly needed because of the diverse skills they acquire. Cybersecurity experts, personnel require a set of multidisciplinary skills of numerous domains and technical fields, are even more demanded because of their extraordinary talents and their tiny numbers of the whole I.T. professional available candidates. Consequently, employing skilled experts in the I.T. and Cybersecurity is challenging worldwide while having to choose from a small fraction of the I.T. pool [4]. The world is undergoing increased cyberspace attacks because of the fast growth of various technologies, various mobile devices such as automatous cars and drones, and the growing usage of the Internet. Therefore, the ever-increasing trend in technologies and devices increases the risks and concerns associated with hackers, and terrorist groups attempt to gain access to sensitive data and crucial information. Nowadays, we see the young generation is becoming more addicted to their phones, tablets, and other personal devices. They have access to several social media applications that may pose risks to their personal or work-related information of being hacked. Cyber-attacks are on the rise, for consumers, corporations, governments, etc., whether they are using the Internet, sensors, the cloud to control and manage their products from pipelines, powerlines, etc..

Quantum computation is a revolutionary technology for Cybersecurity education. The traditional digital computers run on encoded data based on binary system 0 or 1 [5]. With the emerging of quantum computation, quantum computers run on qubits or quantum bits data rather than 0 and 1. Quantum computation is vital for Cybersecurity education due to several factors [6]. The first factor is the high speed of solving complex mathematical problems, and specifically, quantum computation excels in solving cryptography algorithms. The second factor is security, where quantum computation provides a tremendous productive tool for distributing cryptographic keys remotely to other parties with a very high level of implicit security using distribution functions. The distribution functions use quantum effects, such as superposition, entanglement, and randomness, when processing information. The third factor quantum computation offering is safety, as current quantum computers require specific conditions for running, near-zero absolute temperature space, and need to be isolated from the surroundings to prevent interference of different waves and noise. Therefore, it is improbable that hackers and cybercriminals could afford such powerful machines. However, such powerful computers with significant computing capabilities will be a massive challenge for cryptography in the future as hackers and cybercriminals will expand and develop their skills to be able to deploy very sophisticated quantum attacks. To prevent such events from happening, specialists in the field are currently developing resistant algorithms. The fourth factor is resistance, as quantum computing offers a random number of generators or randomness effect. This kind of encryption is unbreakable as it creates a one-time encrypted key for single use.

Our goal in this chapter is to point out the necessity and the significant role of quantum computation in Cybersecurity education by offering a cybersecurity curriculum that universities and colleges can adopt in the form of a minor degree in Cybersecurity or a certificate.

The rest of this chapter is organized as follows. Section 2 summarizes the necessary background to comprehend the cybersecurity field. In Sect. 3, we present the proposed educational model for preparing skilled cybersecurity personnel. In Sect. 4, we conclude the chapter.

2 Cybersecurity Overview

Cybersecurity is one of the most important primary goals to achieve at national and global levels. It is one of the many challenges all over the world. by governments, societies, international economics, etc. The main goal of Cybersecurity is protecting everything, such as computer networks, electronic objects, programs, and data, etc., against any threats. A plan can be achieved by using a combination of different technologies and information processing [7]. Professionals in the field of Cybersecurity defined it as an information-technology discipline that involves information, technology, individuals, and processes in an interdisciplinary way [8, 9]. Furthermore, an expert in Cybersecurity is an individual who acquires a set of multidisciplinary skills, like information security, networks, programming, etc.

The academic community, nationally and globally, has a long history of meeting the demands of governments, societies, and industry. Professionals in the society of Cybersecurity have declared the areas and the concepts which form the core of Cybersecurity that experts in the field must require. These areas are: (1) data security which addresses all aspects of data protection, (2) software security which focuses on development and usage of software, (3) system security which focuses on maintaining a robust combination of all components, infrastructure, computers, and related networks, (4) human security which addresses human aspects and privacy, (5) organizational security which aims to minimize all types of cybersecurity risks, (6) societal security that addresses issues like intellectual properties, codes of conduct, and legislations, and (7) connection security which addresses the connection between the different areas using crosscutting concepts, such as confidentiality, integrity, risk, availability, adversarial thinking, and system thinking [10].

In the last two decades, the world witnessed a real digital revolution and transformation in companies and organizations. There has been a confluence of technologies: social networks, mobile, Internet of Things (IoT), cloud computing, big data analysis, etc. Moreover, another revolution has appeared as a result of a combination of nanotechnology, biotechnology, genomics, and quantum computing. The basis for this revolution was set by many exceptional scientists, such as Einstein, Schrödinger, Dirac, Heisenberg, Pauli, and many more. They settled the basis for quantum mechanics, which describes the behavior of nature at the subatomic levels, behavior that classical mechanics cannot explain.

In 1982, Feynman initiated the second quantum revolution when he asked: What kind of computer are we going to use to simulate physics? [11]. Feynman's question was the spark that initiated the idea of the quantum computer, and so the birth of quantum computer science. Our knowledge and understanding of quantum computers has grown drastically over the last two decades. Using various quantum computing effects, such as superposition and entanglement, quantum computers will provide faster computing speed, offering high value in diverse applications. There are thousands of exciting applications for quantum computing in various areas: economics and finance services; chemistry; medicine and health; supply chain and logistics; energy; agriculture etc. [12]. Nowadays, the projections of quantum computing are fascinating, and extraordinary expectations are now driving a large-scale, worldwide, effort that aims to perfect quantum computing [13].

Quantum computers enable parallel computation of an assembly of states simultaneously as they work on particles that are in superposition. If we can build such a machine that performs to a scale, then all current encryption will be cracked. Encryption in modern systems is meant to ensure resilience when under attack. The hope is that quantum cryptography will be our assurance against cybersecurity attacks. Quantum communication is employed through quantum effects; in particular, entanglement occurs when two particles interact physically even when distanced. Such communications were demonstrated in China across 1200 kilometers. Transformation of information across such distance provided an extremely secure communication network, a space-based quantum Internet [5].

Quantum computation offers homomorphic encryption, a technology to transmit arbitrary computation on encrypted data. This technology will keep all data encrypted and anonymous. Unfortunately, full homographic encryption is not ready yet. In partial homomorphic encryption, some computational functions can be employed on distributed encrypted data across multiple subsystems. This combination ensures resilience under attack and adds another layer of protection [14].

3 Cybersecurity Educational Model

This section presents the proposed educational model. We present the Cybersecurity topics needed to be included in the curricula. We have considered several guidelines [8, 9] while preparing the Cybersecurity curriculum. At the same time, we also thought of the urgent needs and demands that industry seeks in their future employees, such as in-depth knowledge, training, and skills in the field of Cybersecurity [4].

The overall goal is to provide a new general curriculum that can be adopted by universities. This will give the universities the ability to produce Cybersecurity professionals and workforce to meet the needs of Cybersecurity professionals. The proposed curriculum will aim at increasing the number of underrepresented

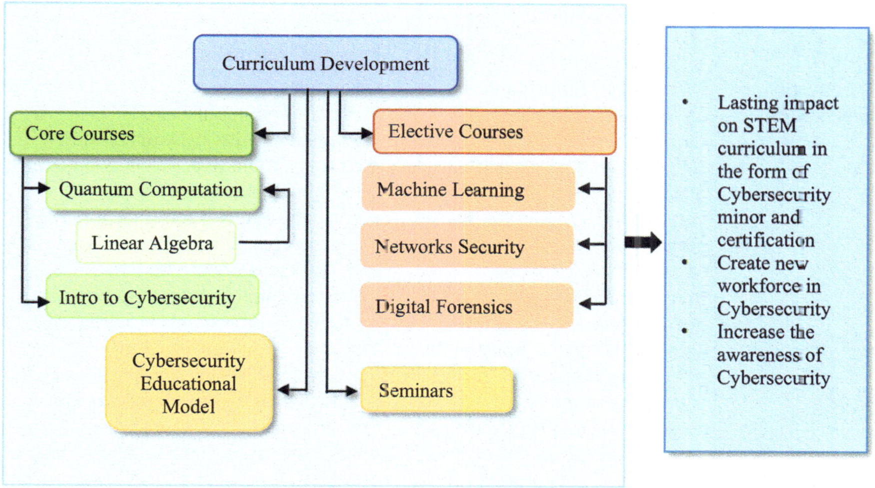

Fig. 1 Cybersecurity minor/certificate track for undergraduate students

minorities working in Cybersecurity and available as trained professionals for the need of the Department of Energy, Department of Homeland Security, and others. The multidisciplinary Cybersecurity curriculum offers courses within the departments of Computer Science and Physics. This interdisciplinary program will be open to all STEM fields. Figure 1 shows the cybersecurity model with the proposed courses. For the students pursuing a certificate or a minor in Cybersecurity, they will be required to complete four of the offered cybersecurity courses, two core cybersecurity courses and choose two electives cybersecurity courses out of three offered courses, to be selected from the list of approved elective courses. For the students pursuing a minor in Cybersecurity, they will be required to complete five of the offered cybersecurity courses, three core cybersecurity courses and two elective cybersecurity courses, to be chosen from the list of approved elective courses. These lists of the core and elective cybersecurity courses are given and described in detail in Table 1.

The minor in cybersecurity degree or certificate would require two core courses and two elective courses. The students will be required to maintain a minimum required GPA of 2.5. They must obtain a grade of C or higher in each course to be used towards the certificate or minor. The first core course is quantum computation; the linear algebra math course is a required prerequisite for the taking the Quantum Computing course. The Quantum Computing course is a mandatory course that will cover the fundamentals of quantum mechanics from the practical perspective of computer science while introducing the exciting and fast-developing field of quantum computing. It will explore the concept of a quantum computer, including algorithms that outperform classical computation and methods for performing quantum computation. As this is a multidisciplinary subject, the course will cover basic concepts in theoretical computer science and physics in

Table 1 Overview of courses included in the education model

Course	Course Objectives	Learning Outcomes
Introduction to cybersecurity	Discuss the origin and nature of Cyberspace and Cybersecurity. Introduce the fundamental concepts of Cybersecurity. Introduce the threats, vulnerabilities, and risks that pertain to national, commercial, or personal information. Introduce resiliency and risk management. Understand how to adjust system protections and responsive actions over time. Understand complex legal, political, social, technical, and regulatory requirements. Discuss the basic requirements for developing, implementing, and managing organizational cybersecurity programs. Discuss computer network basics and the relation to Cybersecurity. Developing, implementing, and managing organizational cybersecurity programs.	Understand Cybersecurity and the way it is defined and implemented. Recognize the threats, vulnerabilities, and risks that pertain to national, commercial, or personal information. Realize the range of tools, technologies, and procedures that help to mitigate risks and foster resilience. Understand how to adjust system protections and responsive actions over time. Identify complex legal, political, social, technical, and regulatory requirements. Comprehend the basic requirements for developing, implementing, and managing organizational cybersecurity program.
Introduction to quantum computation	Review the basic mathematics of quantum mechanics for application to quantum computation. Discuss quantum entanglement and non-locality Discuss the classical physical computation (not with digital circuits), in particular reference to computational complexity. Develop computation with quantum qubit circuits Cover Glover's search algorithm and its optimality. Cover Shor's factorization algorithms. Discuss quantum error correction and fault tolerance. Discuss the more recent Adiabatic Quantum Computation, together with a general perspective on quantum and physical computation in general	The theoretical foundation of quantum information and computation Basic understanding of the quantum mechanical framework, why quantum computation can outperform classical computation. Familiarity with the main results of quantum computation field. Develop an intuition for quantum physics and its interface with compDuter science. Learn about the research frontier of one specific topic via the course project. Understand engineering challenges currently faced by developers of real quantum computers. Evaluate one essential technology requirement for quantum computers to be able to function correctly. Utilize various simulators of relevant experimental setups available on the web and explore quantum circuits with the help of quantum computer simulator.

(continued)

Table 1 (continued)

Course	Course Objectives	Learning Outcomes
Digital forensics	Discuss a systematic methodology for computer investigations. Utilize numerous forensic kits for collecting digital evidence. Discuss digital forensics analysis via different operating systems, i.e., Windows, MAC, Linux, iOS, and Android. Practice email investigations. Discuss anti-forensics, methods, tools, and their usage. Discuss the implications of anti-forensics to the digital forensics investigator. Analyze various image files, logical and physical ones. Explain the guidelines for successful investigation reporting. Research various topics in Forensics and relate to Cybersecurity.	Understand the fundamentals of Computer Forensics. Prepare for computer investigations. Comprehend and apply various forensic tools to collect digital evidence. Practice digital forensics analysis in different cases. Demonstrate an awareness of current approaches of decreasing efficiencies of anti-forensics. Experimenting with different forensic algorithms. Explore new topics in Forensics and applying that to Cybersecurity.
Machine learning	Code in Python and R to create machine-learning models from training part of a dataset and evaluate them on the testing part of the dataset. Evaluate the accuracy and other statistical measures of the models created from the data using the different machine-learning algorithms. Gain an understanding of how to create models from data by modeling and regression. Learn to use a gamut of machine-learning algorithms. Apply machine-learning algorithms to build predictive models for decision-making in real-world problems.	Build predictive models from big data or large datasets and develop decision-making systems. Develop an algorithm that can receive input parameters and use statistical models based on the data to predict an accurate output parameter. Use machine-learning algorithms to learn information from data or to train a model without depending on a predetermined equation. Adaptively enhance model performance with the increase in learning samples using machine-learning algorithms Utilize different categories of methods, such as supervised learning in which labeled training data can be used and classification or regression can be performed, and unsupervised learning in which unlabeled training data can be used and clustering and pattern recognition can be achieved. Create computational models in programs written in Python and R to process large volumes of data efficiently to make predictions or decisions automatically. Apply machine-learning techniques for industrial and research applications.

addition to introducing core quantum computing topics. The second core course is Introduction to Cybersecurity. This course introduces students to the interdisciplinary field of Cybersecurity by discussing the evolution of information security into Cybersecurity, cybersecurity theory, and relating cybersecurity applications to nations, businesses, society, and people. Students will be exposed to multiple cybersecurity technologies, environments, processes, and procedures to learn how to analyze threats, vulnerabilities, and risks present in these environments, and develop appropriate strategies to mitigate potential cybersecurity problems and impacts in the modern information environment. Additionally, this course will cover some computer network-related basics such as security domains, packets, filters, access control, and proxies.

The first elective course is machine learning. This course introduces machine-learning algorithms and methods such as decision trees, Bayesian networks, support vector machines, k-nearest neighbors, neural networks, logistic regression, discriminant analysis, etc. It covers the modeling and development of models from datasets for prediction and decision-making. This course also covers machine-learning algorithms and methods to create models from data and evaluate the models for prediction and decision-making. The course includes supervised and unsupervised learning methods, optimization, computational statistics, and logistic regression. Also, it covers machine-learning methods such as decision trees, Bayesian networks, support vector machines, k-nearest neighbors, neural networks, logistic regression, discriminant analysis, etc. Machine-learning algorithms are essential tools for engineers nowadays working in the industry in fields such as big data, data mining, mathematical modeling, etc. There are several applications such as robotics, data science, self-driving automatic vehicles, bioinformatics, pattern recognition for image processing, and signal processing. Currently, machine learning is also used in multiple applications such as in bioinformatics, diagnosis of biomedical images, or signals to predict diseases, stock markets, retail markets to predict products customers would potentially buy, media, and entertainment industry to predict favorite media, and customer relationship management.

Digital Forensics is also an elective course. This course will go over the fundamentals of computer forensics and investigations. Topics included in this course will cover previous and current computer forensic cases and inspective security issues; a systematic methodology to computer investigations; digital forensics, electronic mail (email), and imaging file analysis; and important guidelines for successful, effective reporting. Numerous forensic tools will be introduced during the laboratory associated with the class. Development of computer forensics laboratory will be covered thoroughly, including software and hardware development. Introduce creating marks (e.g., using watermarking), then reconstruct these marks after going through different changes such as compression, resizing, etc.

4 Conclusion

In light of the urgent need and the importance of the Cybersecurity domain, and the emphasis of quantum computation in Cybersecurity, we have addressed the knowledge areas and interdisciplinary skills cybersecurity experts should possess. We have considered the aspect of interdisciplinary domains as presented by The National Institute of Standard and Technology (NIST). Moreover, we presented an overview of a cybersecurity education model that has core courses and elective courses that can be adopted by the university wishing to propose a cybersecurity field. The presented education model is established to meet the need of the industry in Cybersecurity, taking into account all required training, knowledge, and skills their future employees should have.

References

1. M. Nijim, H. Albataineh, M. Khan, FastDetict: A Data Mining Engine for predicting and preventing DDoS attacks. IEEE international symposium on technologies for homeland security, 2017
2. K. Evans, F. Reeder, Human capital crisis in cybersecurity technical proficiency matters, A report of the CSIS Commission on Cybersecurity for the 44th Presidency, Center for Strategic and International Studies, Nov 2010
3. E. Beidel, S. Magnuson, Government, military face severe shortage of cybersecurity experts, National Defense (National Defense Industrial Association), August 2011. http://www.nationaldefensemagazine.org/archive/2011/August/Pages/Government, MilitaryFaceSevereShortageOfCyberseurityExperts.aspx
4. M. Turkanovic, M. Holbl, An example of a cybersecurity education model, 29th annual conference of the European Association for Education in Electrical and Information Engineering, 2019
5. P. Gabriel, China's quantum satellite achieves 'spooky action' at record distance. http://www.sciencemag.org/news/2017/06/china-s-quantum-satellite-achieves-spooky-action-record-distance, Jun 15, 2017. Retrieved 15 Sept 2018
6. B. Arslan, M. Ulkter, et al., A Study on the use of quantum computers, risk assessments, and security problems, IEEE international symposium on digital forensic and security, 2018
7. B. Caulkins, T. Marlowe, A. Reardon, Cybersecurity skills to address today's threats, in *Advances in Human Factors in Cybersecurity*, (2018), pp. 187–192
8. S. von Solms, A. Marnewick, Towards educational guidelines for the security systems engineer, in *IFIP Advances in Information and Communication Technology*, vol. 531, (2018), pp. 57–68
9. M. Bishop et al., Cybersecurity curricular guidelines, in *IFIP Advances in Information and Communication Technology*, vol. 503, (2017), pp. 3–13
10. V. Solms, L. Futcher, Identifying the cybersecurity body of knowledge for a postgraduate module in systems engineering, in *IFIP Advances in Information and Communication Technology*, vol. 531, (2018), pp. 121–132
11. R.P. Feynman, Simulating physics with computers. Int. J. Theor. Phys. **21**, 467–488 (1982) https://doi.org/10.1007/BF02650179
12. Allende López, Marcos; Da Silva, Marcelo Madeira. Quantum Technologies. Digital Transformation, Social Impact, and Cross-Sector Disruption. Interamerican Development Bank (2019). https://doi.org/10.18235/0001613
13. T.S. Humble, E.P. DeBenedictis, Quantum realism. IEEE Comput **52**(6), 13–17 (2019)
14. P. Koola, Cybersecurity- a system perspective, MTS DP conference, 2018

CyberCheck.me: A Review of a Small to Medium Enterprise Cyber Security Awareness Program

Craig Valli, Ian Martinus, Jayne Stanley, and Michelle Kirby

1 Introduction

The CyberCheck.me initiative is a local government and academic engagement project trying to improve cyber security awareness and skills of Small to Medium Enterprise (SME). The initiative utilizes staff from Edith Cowan University (ECU), City of Joondalup (CoJ) and City of Wanneroo (CoW) to manage and run the engagement activities with additional funding from the Australian Centre of Cyber Security Excellence (ACCSE) [1] and WA AustCyber Innovation Hub (WAACIH) [2]. Students from Edith Cowan University and North Metro TAFE (NMTAFE) engage at various business events and forums to disseminate knowledge about cyber security to SMEs. In addition, the SMEs are offered the opportunity to attend a personalized one-on-one cyber consultation to provide them an overview of the potential cyber risks for their business as a result of a detailed cyber health check questionnaire they must complete prior to their consultation. This organized face-to-face engagement is further supported by a public-facing website https://cybercheck.me that has the latest news, and simple guides for securing mobile devices and computers. This chapter will outline the processes undertaken and observed benefits of the CyberCheck.me program, which secured significant funding to increase interactions from 2017.

C. Valli (✉) · J. Stanley · M. Kirby
Security Research Institute, Edith Cowan University, Perth, WA, Australia
e-mail: c.valli@ecu.edu.au; sri@ecu.edu.au

I. Martinus
Western Australian Cyber Innovation Hub, Joondalup, WA, Australia
e-mail: ian@wacyberhub.org

K. Daimi et al. (eds.), *Advances in Security, Networks, and Internet of Things*, Transactions on Computational Science and Computational Intelligence,
https://doi.org/10.1007/978-3-030-71017-0_17

1.1 The Background

In 2014, ECU, CoJ and CoW commissioned and ran a survey of SME businesses in the respective catchments [3]. The survey was the first of its kind conducted in Australia and possibly the world. The research was action-based and the results from the survey informed the basis for skills intervention workshops that were run in the local government catchments, post survey and publications resulted. The workshop attendances exceeded expectations and a model to further extend the engagement was planned in 2015 which resulted in the CyberCheck.me initiative being launched in early 2016. In 2017, additional funding allowed the program to increase its scope and performance indicators. At present, the program is further consolidating and codifying processes with the objective of continual improvement.

1.2 CyberCheck.me 2019 and Beyond

Since August 2019, six CyberCheck.me pop-ups have been held at various business events and locations with more than 200 SMEs visiting the stands and speaking to the cyber students from ECU and NMTAFE. In addition to this, more than 50 businesses registered their interest with CyberCheck.me to attend a one-on-one free cyber consultation. The consultations have increased in number in 2020. Small businesses can now book a session online after completing the pre-consultation survey. Promotion of the consultations occurs through various online channels. Feedback from the initial consultations has been positive with one business stating, "I have learned a lot. My understanding and vigilance toward cyber security has changed dramatically."

In addition to CoW and CoJ, a number of other local councils are in the pipeline to roll out the initiative taking available outreach to over 700,000 people.

2 The CyberCheck.me Engagement Model

A secure ecosystem for the conduct of cyber-enabled business is an essential underpinning of any digitally enabled economy. One of the issues for SMEs is the ability to access expert help at a reasonable cost. The CyberCheck.me initiative addresses this in that ECU, CoJ and CoW, respectively, have co-funded the provision of expert advice to the city catchments. The model also allows for targeted expert advice and training to be given to SMEs who opt in to deal with their self-identified cyber security issues.

Dissemination of cyber security material for CyberCheck.me is achieved on multiple levels and channels of engagement. The activities for CyberCheck.me are published through the cybercheck.me website, CoJ and CoW Facebook and

Twitter channels, and various ECU twitter channels. Additionally, local government pushes notifications to their respective business associations, such as the Wanneroo and Joondalup Business Associations. Their combined membership exceeds 700 businesses, representing every major industry group and ANZSIC code. The interactive information sessions run at civic facilities such as community libraries as well as at local business events within those primary and secondary centres.

Relevant and easy-to-understand fact sheets have been produced for dissemination to the community. The first is a fan fold brochure that outlines the eight key things people should do to protect themselves. It contains general information that can be applied to a business or anyone in the community. The tips include:

1. Use strong passwords
2. Protect your computer
3. Use email and the web safely
4. Use your mobile devices securely
5. Keep up to date
6. Install AV software
7. Back up your data
8. Use encryption

The brochure also contains information regarding other available online resources to get further advice and information. In addition to this, an animation book was created with cartoon graphics to make the message clear to a wider variety of audiences. The messaging still had a small business focus. The same eight cyber security tips were highlighted, but applied to a business scenario in an easily digestible format. These animation books have proved to be very popular with SMEs who visit, and engaged with, the students at the CyberCheck.Me stand during the events.

Complementing the two guides are single sheet A4 information sheets about securing mobiles through use of pin or pattern, enabling file encryption on computer operating systems and how to use encryption to protect files in transit. These are used also in the face-to-face engagements with people. These sheets use freely available either embedded in the operating system features or freeware/opensource solutions, for example, VeraCyrpt [4] for USB disk encryption.

2.1 *Outreach Sessions Uncover Serious Attacks*

In 2019, we saw an increase in businesses self-identifying as suffering a significant cyber-attack. Some of these businesses have gone further and told their story in an effort to make others realize they are vulnerable to cyber-attack and the follow-on consequences of weak cyber defence. The attacks have resulted in significant damage to their businesses.

A national insurance company reported a $2 million loss due to a crypto-locker attack that wiped out their entire laptop fleet and servers. This attack further resulted in the business being offline for 18 days and having to replace over 400 laptops. As the owner of this business said, "Like a burglary, it's not if it happens, it's when, and a Disaster Recovery Plan is paramount".

A small catering and event management business that primarily used a web page as a channel for promoting and securing new business almost never recovered from the cyber attack. As a result of the attack, they had lost view of all customer bookings for the next 12 months, their website completely disappeared from the Internet and the business had to be closed for 2 months and all staff laid off. A common thread in these stories is a failure to patch, backup, antiviral protections and firewalling. The effects of a weak cyber defence were swift and costly.

3 Survey Results Section

Preliminary results (n = 30) from the survey this year would indicate there are still some very troubling patterns and trends being observed in the responses from the SMEs. The survey covered basic demographics, technologies in use to access the Internet, technologies used in the SME, cyber security countermeasures applied to those technologies and devices in the SME, implementation of cyber policies and plans. This following section outlines and discusses some of the key preliminary trends and results uncovered in the survey so far.

3.1 Survey Demographics

The survey had the following demographics. The age of respondents were as follows 18–35 18%, 35–49 50%, 50–64 32%, 65 or older 0%. The following Tables 1, 2, and 3 also outline basic demographics for the SME businesses that have participated in the survey so far.

Table 1 Which industry sector(s) does your business operate in?

Construction	7%	Transport, postal, warehousing	7%
Professional, scientific and technical Services	30%	Rental, hiring and real estate	3%
Financial and insurance Services	10%	Manufacturing	3%
Retail trade	0	Other	40%

Table 2 Business size

Sole proprietor/partnership (non-employing)	60%	Microbusiness (less than 5 employees)	11%
Small business (5–19 employees)	11%	Medium business (20–199 employees)	11%
Large business (200+ employees)	7%		

Table 3 Where they conduct business

Commercial premises	33%	Home	37%
Mobile (e.g. vehicle-mobile tradesperson)	18%	Anywhere/completely virtual	12%

Table 4 Internet access methods (%)

NBN	100	Mobile phone	61	WiFi/wireless	61%
4G/Wireless hotspot	7	NBN satellite	4		

3.2 *Business Devices and Access*

The devices used to access the Internet by the businesses are desktop computer 68%, smartphone 79%, tablet 46% and laptop 75%. It should be noted that some organisations have multiple devices in use. The following Table 4 outlines how the businesses access the Internet and what technology they use to access it again; some businesses use multiple technologies.

NBN in an Australian context means the Australian National Broadband Network (NBN) which is a high-speed network developed by the Australian government to provide network infrastructure to the nation. It should be of little surprise that NBN is at 100%. It is interesting to note in this survey that all of the respondents are technically in a metropolitan area with some respondents from rural areas using satellite because of physical cable distance issues. This outcome may be as a result of the universal service guarantee to overcome network blackspots which occur with some frequency in certain parts of the metropolitan area. For the first time as well, the use of 4G/wireless hotspots has been registered.

With respect to countermeasures being deployed on devices by device type, there is some stark differences. The following table shows each category of device as per the questionnaire which is PC/laptop, tablet and smartphone.

Authenticating to Devices

To authenticate to computers/laptops, the following percentages were observed in survey response: password 86%, multifactor or two-factor authentication 43%, fingerprint 19%, face recognition 10% and hardware token 4%. For tablets and mobile phones, the following authentication methods were observed: password/PIN 100%, facial recognition 31%, multifactor or two-factor authentication 15% and fingerprint 61%.

Survey respondents answered a question about password policy setting and enforcement thereof, the responses were Yes 45%, Partial 33% and No 22%. When asked if they used a password manager to manage passwords, the responses were Yes 33%, No 41% and "do not know what it is" 26%. This points to a significant gap that needs to be addressed with one third of respondents using a password manager. This type of tool is an effective countermeasure against poor passwords and password strength when used correctly. As this program is tied educating SMEs about cyber security, this is an area for us to focus on. We already have some materials available and have delivered presentations on why they should be used. This outcome is an indicator that this work needs to continue to educate users on the importance of a password manager for password security.

What was of significant concern is that 73% have shared accounts and 27% do not have shared accounts. From a cyber security and auditing perspective, this is dangerous and needs to be addressed. If this is then coupled with frequent staff turnover and infrequent changing of passwords, this becomes a significant vector for insider attack.

There was an incongruity in the survey in that when asked 80% restrict access to sensitive and critical data which is in direct contrast to 73% of people having shared accounts. This response does not make a lot of sense. The survey also returned that 50% provide basic information security training but this result could be a self-selection effect in the survey; we would not posit that this is not normal.

3.3 Maturity of Wi-Fi Security

Wi-Fi is actively used by 95% of the SMEs surveyed. Of these respondents, only 63% have implemented WPA2 that they can confirm, 8% confirmed No and 29% were unsure. Furthermore, 59% of respondents that had used WPA2 had reset their password to a complex password. Alarmingly, 35% still are using default passwords either provided by the manufacturer or the Internet service provider. Further study will reveal if this is a matter of laziness or a lack of awareness of the dangers of keeping default settings on devices. A strong indication that Wi-Fi is starting to be seen as a significant risk for the SME businesses has seen 38% of the respondents place their Wi-Fi on a separate NBN connection that does not connect to their internal business systems. This question will be modified to further investigate the need for this separation of networks in future surveys.

Approximately 40% of respondents use VPN to protect connections. However, 32% answered No to the use of VPN and 28% were unsure. It could be reasonably argued that 60% of businesses are not using VPN technologies to protect their data; this needs to be addressed. Combining this observed response with 30% of all businesses operating in a mobile or virtual fashion and they are not using VPN. This is a significant risk.

3.4 Applied Countermeasures

One of the simplest things that SMEs can do is install standard countermeasures such as a firewall, some form of antivirus or malware protection, and possibly a spam killer. It should be noted that firewalls are now routinely provided free with all operating systems that are in use in SMEs including but not limited to computers, laptops, PCs, tablets and phones. The following Table 5 outlines a percentage of

Table 5 Applied countermeasures by technology type

Countermeasure	AV	Spam killer	Firewall	Malware
PC/laptop	95%	32%	73%	50%
Phone	34%	5%	53%	21%
Tablet	15%	0%	15%	7%

Table 6 Application of operating systems and vendor patches frequency by hardware platform

Vendor	Auto	Weekly	2–3 a month	Month	Less than monthly	Does not know
PC/laptop	62%	8%	4%	4%	8%	12%
Phone	79%				21%	
Tablet	79%				21%	

Table 7 Updating of antiviral countermeasures frequency by hardware platform

AV	Auto	Week	2–3 a month	Month	Less than month	Never	Does not know
PC	62%	8%	4%	4%	8%		12%
Phone	100%						
Tablet	62%					13%	25

software installed on each given hardware platform of the identified cyber security countermeasures.

As can be seen in Table 5, in the PC/laptop technology 95% of SMEs have now applied antivirus. Microsoft Defender Antivirus with Microsoft Windows 10 [5] has now made its entry into the market and has a good effect, as most SMEs use Microsoft operating systems on their PCs and laptops. The enablement of firewalls is also increasing based on previous similar surveys [3, 6], which is encouraging. What is also trending similar to previous surveys is the lack of antivirus software that is installed on both phones and tablets. Commercial offerings and competent-free versions of antivirus software are available. However, there still seems to be a disconnection on cyber risk when it comes to these devices.

3.5 *Applying Updates*

The application of updates and patching of operating systems, applications and countermeasures is a critical function in any cyber security program. The figures expressed in Tables 6 and 7 gives us an insight into what SMEs are doing with respect to their software patching of their critical business devices and systems.

It is alarming to see that the automatic update of operating system patches and applications is still not occurring. The percentages in phones and tablets while presenting as high given that phones and tablets by default automatically update, it means that someone has potentially altered this away from good practice. In the case of smart phones and tablets, it could be that the update processes are consuming scant space on memory cards and so users turn them off to save space, a bad idea.

With antivirus being installed on phones at the rate of 34% and tablets at the rate of 15%, it is encouraging to see, however, that those have taken the time to install have left it on to automatically update the antivirus signatures. PCs and laptops have 62% of respondents who have enabled automatic updates. The remaining 38% either do not know or are doing it at such a frequency that they are putting themselves at a higher risk of compromise. This trend is of major concern given the growth of crypto locker-based attacks on the business sector in the past few years. This is only expected to increase.

3.6 Backups Often the Last Great Refuge of Clean Data

In the survey only 50% of respondents indicated that they backup regularly, the remaining 50% backup sometimes, rarely or do not even know about status of backup. This is a very alarming trend in the age of crypto locker-based attacks. A regular routine of backing up important company data is one of the few significant countermeasures that businesses can rely on to recover from these type of cyber attacks.

When we delve into the more granular statistics, the picture is worse for all technologies: computers/laptops 46%, phones 21% and tablets only 15% are backed up. Surprisingly, the backup of network accessible storage (NAS) is higher at 18% than for phones. In, this survey, we asked where they backup their critical data. The use of a cloud service was actually the highest at 37%. This is unsurprising given the tight coupling and bundling offers that modern vendors are placing in their operating systems and the cloud. The use of local USB drives was only 18%, with backup tapes now becoming rapidly arcane with only 3% of businesses using this as a backup media. The use of network-attached storage has grown to replace or supplant tapes with 11% of the respondents backing up to these locations.

These trends and statistics, however, point to a critical problem or failure point in SMEs currently, which is insufficient backup of critical data. Further, the complete lack of any good frequency of backup with only 50% of respondents regularly backing up is of significant concern and needs to be addressed.

3.7 Cyber Security Responses, Planning, Maintenance and Insurance for the Lack Thereof

The authors surveyed the businesses to ascertain if they had sustained a cyber attack 54% of respondents claim they have never suffered a cyber attack. When they identified as being cyber attacked, we asked what remedies they undertook to address and recover from the attack. Thirteen percent (13%) of respondents fixed the problem themselves, a further 25% had their IT support person fix the problem.

Further study on the nature, cost and time taken to remedy the attack might be useful in that it might give some insight into the reliance of micro and small businesses on outsourced IT services.

In terms of reporting the incident to the authorities, no one reported anything to police or law enforcement agencies at all, simply astounding. A mere 8% indicated they reported it to ACORN (Australian CyberCrime Online Reporting Network) [7] or the new replacement site ReportCyber [8] functionality in Australia. This is a significant problem in that crimes go unreported and that government and others cannot get statistics and signals about current state of impacts of cyber attacks on the Australian economy.

A question was asked as to who in the small business was responsible for the cyber security function. Not surprisingly, but nonetheless alarming, 11% of SMEs have no one to carry out this function. Other patterns identified were 4% use a trusted family member, 11% use a specialist cyber security service, 25% use their existing IT support team and 50% look after it themselves. Also, in terms of a cyber response plan only 22% had a cyber response plan, 64% had no plan and 14% indicated they were working on a plan. Of those with plans and processes, 64% did not review any of these processes after they had been developed. This was expected, and in many ways unsurprising.

The marketing of cyber insurance to small business has increased in recent times, with many large global firms entering this potentially lucrative source of new insurance product revenue. In this survey, 20% of respondents had some form of cyber insurance, 55% no insurance and 25% were considering it. Given that most of them did not have information security processes mapped and plans, it would be questionable as to whether they could get cyber insurance as most insurance require some basic policy and process for cyber security to be able to claim Cyber insurance is quickly moving and evolving component of cyber security, and one that signals the opportunity for a greater awareness of small business governance, risk and compliance (GRC) requirements in the future.

4 Conclusion

The survey is indicating yet again that SMEs need access to easily digestible cyber education. There is a strong role that local government and its local business association and business chamber partners can play to assist a greater awareness of the vital role cyber security plays. As businesses digitally store and transmit data on an increasing basis over a wide range of networks and platforms, their understanding of basic countermeasures and available protections can make them more resilient to cyber-attack and cyber-criminal exploits. There are many concerning trends in this initial analysis, some of which has some relatively simple remedies. Implementation of the simple remedies will significantly increase the cyber security posture and resilience of any small business at little or no cost. The CyberCheck.me platform allows easy access to the basic cyber security countermeasures. The program is gaining national and international recognition given its blend of face-to-face and online consultations. The success of the program is due to its practical and immediate suggestions to a small business where instant action is possible.

In this age of a wide variety of disruptive and deadly distributed cyber threats to business, the lack of backup being applied to vital business data is of extreme concern. A key focus of the CyberCheck.ME small business awareness training is how to backup critical business data with adequate frequency and assurance. Equally, patching and updating systems and applications are techniques that small businesses need to be aware.

As the practical value of programs customised to suit a small business audience increases, SMEs need to realise in greater numbers that their smart phone and tablets are a critical point of vulnerability. The way in which they use and protect them needs to change in step with the increasing level of risk associated with attacks, whether they be motivated by profit, disruption, fun or all those things.

References

1. Anonymous, Academic Centres of Cyber Security Excellence (ACCSE), May 11, 2020; https://www.education.gov.au/academic-centres-cyber-security-excellence-accse
2. Anonymous, AustCyber, Australian Cyber Security Growth Network Limited, 2020
3. C. Valli, I.C. Martinus, M.N. Johnstone, Small to medium enterprise cyber security awareness: An initial survey of Western Australian business, in *SAM2014 International Conference on Security and Management, Las Vegas, USA*, (2014), pp. 71–75
4. IDRIX, Veracrypt., IDRIX, 2016
5. Microsoft, Microsoft Defender Antivirus, Microsoft, 2019
6. C. Valli, M.N. Johnstone, R. Fleming, A survey of lawyers' cyber security practises in Western Australia, in *2018 ADFSL Conference on Digital Forensics, Security and Law, University of Texas, San Antonia, USA*, (2018), pp. 219–226
7. ACORN, Australian Cybercrime Online Reporting Network (ACORN), 29/09/2017, 2017; https://www.acorn.gov.au/
8. Anonymous, Welcome to ReportCyber, 10th May, 2020; https://www.cyber.gov.au/report

Part IV
Security, Forensics, Management and Applications

A Hybrid AI and Simulation-Based Optimization DSS for Post-Disaster Logistics

Gonzalo Barbeito, Dieter Budde, Maximilian Moll, Stefan Pickl, and Benni Thiebes

1 Introduction

From earthquakes to terror attacks, the concept of disaster gathers a wide variety of events, both natural and human-made, covering a diverse set of characteristics. During the last twenty years, the field of Disaster Management (DM) has gained significant presence in the scientific community, as well as in the political dimension. Some authors correlate this increased interest with the 2004 Tsunami in the Indian Ocean, where the need for humanitarian operations became clear to the world [1, 6, 10]. Studies also suggest a growing need for this field due to an ever-increasing frequency of disasters [6, 10], with too expensive and long-lasting consequences to be ignored by the research community.

Despite its growth in popularity, there are still diverse challenges at the intersection of emergency management and safety and security. This work focusses particularly on the uncertainty associated with matching the physical properties of a disaster with the human component associated with humanitarian operations [6]. It does so by developing an Emergency Response Decision Support System (ERDSS) framework combining well-known mathematical models, simulation, and artificial intelligence (AI) to support decision making post-disaster.

In order to address stakeholder needs, experienced practitioners in the field of DM were consulted regarding particular requirements for this framework to be helpful. This resulted in three main requirements to be considered:

G. Barbeito (✉) · D. Budde · M. Moll · S. Pickl
Fakultät für Informatik, Universität der Bundeswehr München, Neubiberg, Germany
e-mail: gonzalo.barbeito@unibw.de

B. Thiebes
German Committee for Disaster Reduction (DKKV), Bonn, Germany

K. Daimi et al. (eds.), *Advances in Security, Networks, and Internet of Things*, Transactions on Computational Science and Computational Intelligence,
https://doi.org/10.1007/978-3-030-71017-0_18

1. **Usability:** This framework targets experienced practitioners in the field of DM. There are no assumptions, however, on the required knowledge a practitioner should possess on theoretical or applied concepts supporting this work. It is also expected that, by avoiding unnecessary complexity, engagement of potential practitioners will be easier to achieve.
2. **Flexibility:** As already mentioned, disasters are normally characterized by their unpredictability. This requires a framework that can be quickly adapted to the ongoing and developing situation both in model structure as well as in parameterization. In addition, a modular approach is required in order to simplify improvements to the system or its adaptation to a new scenario.
3. **Transparency:** An AI is used to estimate certain simulation variables. The approach used for this task involves the automatic training and sequential querying of a Bayesian Network, which results in a clear mapping of the variable interdependencies involved in the decision making process.

This work attempts to address all three requirements while partially addressing the uncertainty challenge in the context of safety and security. The concrete focus is on relief distribution, particularly resource management as described in [17], in the aftermath of a disaster. The distribution of relief goods is not only a question of theoretical analysis, of the flow of relief and information, but also a challenge for practical implementation. The coordination of the different agents is of crucial importance for a successful process. The different agents from politics, military, aid organizations, and volunteers must be brought together in coordinated relief operations. Only in this way it is possible to provide aid in a timely manner. In the following, however, it is not a question of practical implementation, but of a model for decision making.

In order to test the framework with meaningful parameters, a conceptual case study is presented, introducing a disaster requiring the relocation of civilian population. The current implementation uses shelters to receive and protect the displaced population. Beyond the parameterization of this case study, the framework offers the option to apply alternative relief distribution strategies, re-purposing the shelters as distribution centers, in line with national structures of civil protection.

This paper is organized as follows: After the introduction, Sect. 2 presents some concepts that will be used throughout the paper, while Sect. 3 describes the proposed ERDSS framework and its associated concepts.

In Sect. 4, the case study used to validate the approach as well as the general mathematical problem are formulated. First results of the framework are presented in Sect. 5, and the conclusion and future work on this research are outlined in Sect. 6.

2 Preliminaries

This section attempts to give context to this work by describing some of the core ideas that will be used later on in this paper.

2.1 Humanitarian Logistics

Within the context of Humanitarian Supply Chain (HSC), the following classification of operations is established in literature [10]:

- **Pre-Disaster**:
  - *Mitigation*: reduce vulnerability to impact.
  - *Preparedness*: develop community awareness to adopt a proactive approach.
- **Post-Disaster**:
  - *Response*: address immediate threats to minimize losses.
  - *Recovery*: support restoration of damage.

Unlike traditional supply chains, in HSC, delays in relief not only have financial consequences but can also cost lives. This puts responsiveness over cost as the key goal in humanitarian operations, particularly in the first 72 hours after the disaster occurrence [10, 12]. Thus, to increase the responsiveness and efficiency of the HSC, Humanitarian Logistics (HL) acts as the bridge between disaster preparedness (pre-disaster phase) and response (post-disaster phase) [9]. The focus in HL is on managing and controlling the flow of relief and information between origin and destination to meet the needs of the affected [17]. An important problem in HL is the distribution of relief goods from a depot hub to its final delivery destination, sometimes referred to as *last-mile distribution*. In this problem, the focus is to deliver these goods to the demand locations as fast and as efficiently as possible. The latency between demand request and order fulfillment is called *lead time*, and its formulation is similar to its counterpart in commercial supply chains, although with an associated uncertainty introduced by the disaster situation [17].

2.2 Information and Decision Making: ERDSS

In humanitarian actions, decision-makers follow, in the generality, two main principles [11, 18]:

1. *human need principle*: assigns priority on saving lives and in alleviating suffering.
2. *impartiality principle*: the decision should be taken based solely on the need, without consideration of any other features among the population.

Literature has widely condensed both principles down to the *priority* approach, which impartially decides on allocation of relief resources according to a priority list based on the needs of the considered group of individuals [12].

In order for a good decision to be made, decision-makers can be assisted by an Emergency Response Decision Support System (ERDSS). An ERDSS is a tool to assess emergency plans and select the best strategy for the given conditions, using the available information [16]. Due to the complexities of disaster scenarios, however, useful and timely information is not easy to obtain [1, 8]. Moreover, for operations that precede the disaster (mitigation and preparedness), the information is often incomplete or does not exist yet and needs to be estimated [18].

Next to information considerations, another relevant aspect of designing an ERDSS Framework is the characterization of the potential user or stakeholder for the solution [16]. This work is currently targeted to practitioners in the field of DM and HL, and aims to satisfy their particular requirements, as explained in Sect. 1, from an Operations Research perspective.

2.3 Simulation and Data-Driven Modeling

A distinction needs to be made between simulation modeling and data-driven modeling to understand the modeling concepts used in this work. The former is an in-depth representation of a complex system, describing the trajectory of state changes over time [7]. The latter is an analytical, data-driven reconstruction of a specific process (in this particular case, a process within the simulation model), describing the variable interdependency observed in that process [5].

Regarding the particular task of **simulation modeling**, three modeling techniques were used simultaneously [7]:

1. *System Dynamics (SD)* uses a set of simple building blocks and entities to describe how systems change over time with a high level of abstraction and is normally considered a strategic modeling methodology.
2. *Discrete Event Simulation (DES)* is a method that requires the modeler to divide the studied system into a sequence of operations performed across entities, over discrete time (i.e., the model clock only advances when something significant happens in the model). It is generally considered to be a low abstraction modeling technique and used to model processes in-depth.
3. *Agent-Based Modeling (ABM)* is a more recent modeling approach, focusing on the behavior of individual interacting entities (namely, agents) to create emergent behavior (bottom up approach), instead of the process affecting those entities (top down).

A geographic information system (GIS) is used as a canvas embedded in the simulation model to show a simple map of the considered region and to calculate the routes and distances between shelters and depots.

Data-driven modeling, on the other hand, primarily takes data as input and trains a model to represent the data generation process. In this particular research, the approach used is Bayesian Networks (BNs). BNs represent a general Artificial Intelligence (AI) framework for describing and encoding complex probability distributions using directed graphs and the concept of conditional probability [13]. BNs provide a complete framework for system analysis and query formulation, thanks to its white box approach. Furthermore, they achieve a balanced combination of the objectivity found on data and the expertise and previous knowledge that domain experts can contribute.

A combination of both simulation and BN modeling approaches, as described in [4, 5], can be used to support decision-makers. On the one hand, the simulation model is used to represent the physical constraints and in-depth behavior of the system, and create environmental data. The BN, on the other hand, receives these data, structures the variable interdependency, and predicts a highly stochastic variable, which is then fed back into the simulation model.

3 ERDSS: Concept and Framework

This section describes the ERDSS and the framework supporting it. There are five main components in this ERDSS:

A. a *disaster scenario* characterized by quali-quantitative variables describing an expected regional impact,
B. a *simulation model*, used as a testbed for different scenarios,
C. a *control tower* supporting the crisis manager decision process. It observes the environment and the predictions simultaneously and makes decisions accordingly, using an online assignment problem as optimization basis,
D. a *prediction algorithm* (data-driven model) capable of forecasting lead times according to environmental variables, and
E. a *management cockpit* to provide the crisis manager with a fast overview of the system situation.

The left side of Fig. 1 shows a conceptual description of the architecture used for the ERDSS. Compared to other architectures for similar problems, the key differentiating element of this approach is the introduction of a stochastic, white box

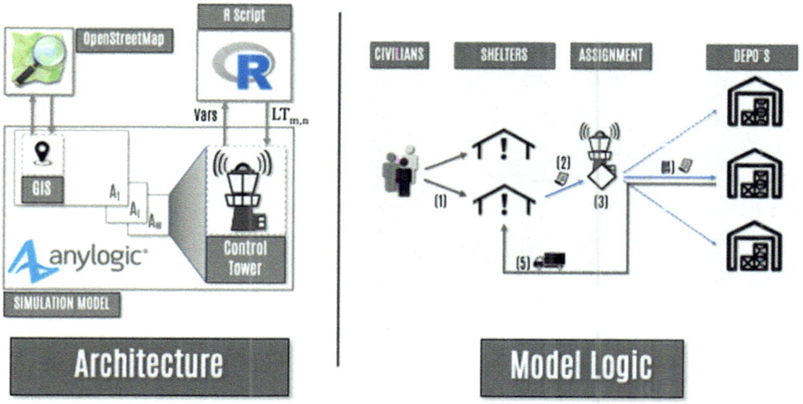

Fig. 1 Left: Proposed architecture for the ERDSS. Right: Conceptual description of the simulation distribution logic

AI algorithm (i.e., a BN) that is continually observing the system and improving the prediction rate as more information becomes available. The following lines will go more in-depth on each one of those elements.

3.1 Disaster Scenario

The framework is built around the quali-quantitative description of a *Disaster Scenario*, relying on the assumptions of experts. A loose definition of the disaster and its particularities is a desired feature of this framework, to account for the gap between the uncertainty in the pre-disaster phase and the flexibility required to achieve a fast response post-disaster. In its current state, the framework relies on a set of static parameters to characterize the disaster, although it allows the combination with a disaster model, influencing these parameters dynamically.

3.2 Simulation Model

The *Simulation Model* was used to represent and observe the temporal evolution of a Last-Mile distribution system, based on particular instances of an online assignment problem. A version of this problem using forecasts in a similar way to this current work has been introduced in [19]. In that paper, the demand is known, and the supply nodes arrive one at a time, assuming that an arrival forecast is available. In the present paper, it is the demand nodes, instead of the supply, which arrive one by one in the form of orders placed with the system. Furthermore, due to the uncertainty of a disaster scenario, no forecast can be made on the arrival orders. The reason to consider this an assignment problem is the assumption that demand is excessively volatile, and a robust routing plan cannot be achieved. The proposed distribution model, based on the direct delivery model conceptually described in [22], connects various depots with various shelters using on-demand (online) orders, instead of one depot with multiple shelters and a vehicle visiting them all.

The model was programmed in Anylogic, using a combination of ABM, DES, and SD. There are a total of eight agent types in the model, plus the `main` agent (`shelters`, `depots`, `control tower`, `crew`, `civilians`, `resources`, `fleet`, and `orders`). Each agent type was developed as an individual module carrying its own logic and serving a specific purpose. Within each type, however, populations share the same behavior but different parameters (e.g., all `shelters` have the same behavior, but different location and occupancy parameters).

The logic of the model, shown in the right side of Fig. 1, tries to represent the real world. It does so by capturing how logistics are organized by civil protection agencies and HL as follows:

[1] Relocated `civilian` agents arrive to the `shelters` and consume `resources`, starting by the stock available at each `shelter`. These agents keep track of the average resource consumption and estimate the number of remaining days of stock.
[2] Once a reorder point (ROP) is reached, an `order` is placed to the assignment module embedded within a `control tower` agent.
[3] The `control tower` observes the system state and assigns a utility for sending the order to each depot, considering the full system.
[4] The `order` is sent to the `depot` that minimizes supply time while having the least negative impact in other `shelters`; this behavior is explained in detail in the next section.
[5] When a `depot` receives an `order`, it enters a queue that preserves the arrival sequence. If enough `vehicles` are available to satisfy an `order`, one truck will be assigned, prepared, and shipped to the `shelter` in need.

3.3 Control Tower

Decision-makers in the context of humanitarian operations require a precise, quick, and up to date overview of the system interdependencies in order to understand the situation's dynamics and be able to predict how the system will evolve [6]. In the context of this model, this overview is handled by a centralized control agent, which oversees the system state, predicts lead times, and assigns resources in a close to optimal solution for the system as a whole. A comprehensive review considering centralized and decentralized control schemes is given in [20]. The authors conclude that the majority of contributions in multi-sourcing models assume centralized control. This stems from better coordination and information sharing possibilities, allowing more complex assignment strategies (e.g., by comparing each shelter's situation with the system's situation).

The left side of Fig. 2 shows the actual modeling blocks used by the control tower; the right side shows a schematic of the workflow to simplify the visualization. There are seven important attributes to this module:

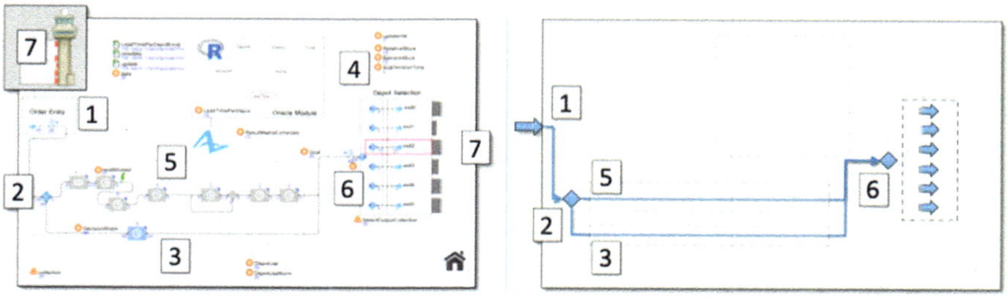

Fig. 2 Control tower—Detail (left) and Schematic (right)

1. This agent's involvement in the solution starts with an `order` agent entering the module.
2. The order can then follow two different paths: a random process assigning orders to depots arbitrarily (step 3) or using the prediction algorithm (step 5). A user-controlled switch changes the operation mode of the model between training (step 3) and querying (step 5).
3. This path contains a random assignment of orders to depots. It is only used to retrieve bulk training data for structural and parameter learning for the BN, and has no prediction functionalities. The random assignment function is corrected by the number of orders in queue per depot, in order to balance the data samples.
4. The calculations conducted in the next step rely on a prediction algorithm to forecast the instant lead times associated with each shelter–depot pair. This framework contemplates two different ways to forecast an order lead time:

   i. An explicit formulation of the simulation model, adapted from the discrete event processes described in [15]. Within this environment, the lead time can be calculated in the following way:

   $$\begin{cases} T_p^* + T_{d,s}^* & \text{if } V_A \geq Q \\ T_p^* + T_{d,s}^* + T_r^* + k(T_p^* + T_{d,s}^* + T_{s,d}^*) & \text{if } V_A < Q \end{cases} \quad (1)$$

   with:
   T_p^*: Preparation time of the vehicle.
   $T_{d,s}^*$: Travel time from depot to shelter.
   $T_{s,d}^*$: Return time from shelter to depot.
   T_r^*: Remaining time of the ongoing order.
   V_A: Available vehicle number.
   Q: Orders in queue.
   $k := \lfloor Q/V_A \rfloor$

   It should be noted that variables marked with an asterisk * represent stochastic variables, whose values cannot be exactly determined in the model. In this case, their expected value is used as an approximation. Furthermore, this approach can only be used when the underlying process is fully known, and the variables are sufficiently characterized. This makes this direct approach hard to implement at best, and unfeasible at worst, and is introduced here only as performance baseline of the BN approach proposed next.

   ii. An external variable interdependency reconstruction using a BN (explained in step 5). This approach only assumes that the lead time can be approximated as function of the distance between depot and shelter pairs, the number of orders in queue at the depot, the population in each shelter, and the number of trucks available to satisfy the orders.

5. This step uses a BN external to the model to forecast the lead time. When the order travels this path, a series of operations are performed in the following order:
   - Environmental variables `queue` (per depot) and `Population` (per barrack) are sent to an R script as observed variables, acting as a proxy for the current load on the system.
   - For m shelters and n depots, the R script will return the matrix $LT \in \mathbb{R}^{m \times n}$ with the lead times of each `shelter-depot` pair.
   - A results matrix $R \in \mathbb{R}^{m \times n}$ is calculated by subtracting the lead time of each decision from each shelter's remaining days of stock:

$$r_{i,j} = s_i - lt_{i,j} \tag{2}$$

   - In order to account for the principle of impartiality in humanitarian actions [11], the results matrix R is corrected by the stock level of each shelter relative to the average system stock resulting in a new matrix $RC \in \mathbb{R}^{m \times n}$:

$$rc_{i,j} = \frac{r_{i,j}}{s_i / \left(\frac{1}{m} \sum_{k=1}^{m} s_k\right)} \tag{3}$$

   - In the last step, the utility for each decision is calculated:

$$Utility_j = rc_{i_0,j} - \alpha \sum_{i=1}^{m} s_i rc_{i,j} \tag{4}$$

   with $s_k = \mathbb{1}_{k \neq i_0}$ and α a global optimization parameter.
6. The order is sent to $depot_j$ according to $argmax_j\ (Utility_j)$.
7. A real time histogram is used as visual aid of the `depot` assignments performed by the control tower.

3.4 Prediction Algorithm

The prediction algorithm is a BN, which is trained from the model's output data to understand the variable interdependency in the system. This algorithm is connected to the ERDSS as oracle [3, 14], predicting the value of a single variable without further involvement in the decision process [3].

The variable dependency structure learned from data can be seen in Fig. 3 and represents the variable dependency of the lead time for each shelter–depot pair according to the remaining assessed variables. The most important aspect of this

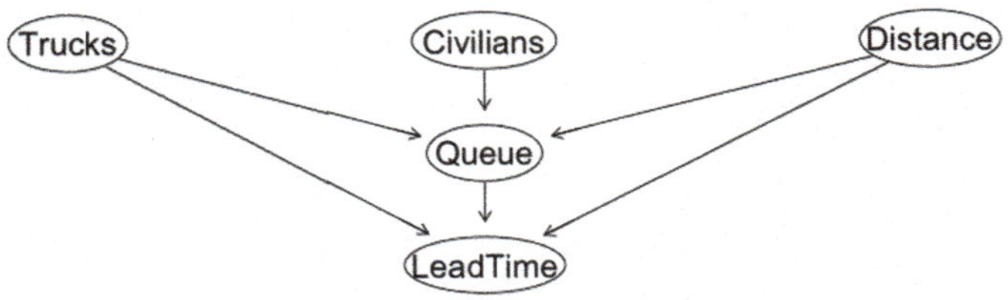

Fig. 3 BN trained to predict lead times in the model

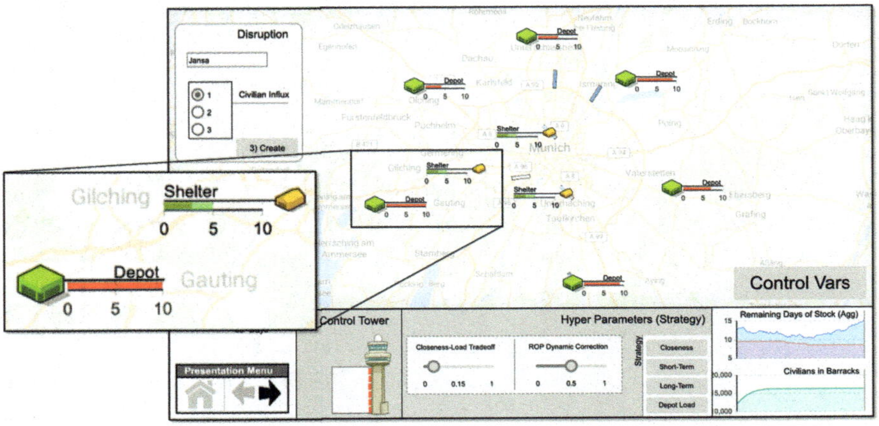

Fig. 4 Proposed management cockpit, integrating GIS map with real time feed update, plots, and control options

result is that the variable interdependence is created automatically from data without user intervention. Furthermore, it is aligned with the stakeholders' request for transparency, using a clear graphical representation.

3.5 Management Cockpit

This component aims to provide a general perspective of the system situation visually and interactively. Figure 4 shows the current implementation of this dashboard: The main area contains a GIS map with the position of shelters and depots and their corresponding attributes. The detail shows a shelter with a green bar, representing the number of days worth of stock remaining. The red bar in the Depot indicates the load on that particular depot, by measuring the number of orders in the queue.

The lower section of Fig. 4 shows the control options for the model, used to characterize the supplying strategy that the control tower will follow. The two plots on the right show a live feed of the aggregated system state, keeping track of the number of civilians in all shelters and the number of days before the three shelters run out of stock if not supplied.

4 Case Study: Long-Term Disaster Scenario

As a demonstrator of the proposed framework, a generic long-term disaster scenario was conceptualized. Despite its weak characterization, the disaster parameters were selected to encompass a wide variety of hazards, such as a long duration blackout or a disease outbreak forcing groups of the population to require protection or quarantine.

During the post-disaster phase, the immediate response strategy involves civilian relocation in shelters, and supply assignments from regional depots.

The simulation model is used to assist in developing and assessing strategies for the aftermath, particularly the response phase. The goal is to prevent **shortages** and **stockouts** to guarantee relief resources to the civilian shelters for as long as the crisis lasts, but without considerations of system recovery, to avoid the complexity layers brought by diverse stakeholders [8, 9]. Due to the focus being on the response phase, the utility function favored by the stakeholders should be the survival and well-being of the population, and not the cost of operations [9, 10].

4.1 Assumptions of This Work

1. The analysis encompasses only a single region affected by the disaster.
2. No infrastructure damage is considered.
3. Civilians are to be relocated in shelters, with a fixed location and stock capacity determined in the pre-disaster phase. The arrival rate of civilians to shelters is stochastic.
4. Relief is supplied to shelters by depots. They also have a fixed location, but infinite stock capacity.
5. Information flows uni-directional, from shelter to depot, in the form of order placement.
6. The fleets are operated by the depots.

4.2 Model Selected

Using the classification described in [10], the model used for this problem has the following properties:

Category: Relief Distribution.
Type: Last-Mile Relief Distribution.
Disaster Phase: Response.
Relief Commodity: Single Product (water).
Objective Function: See section III, item c.5.
Constraints: Vehicle capacity, fleet composition, number of shelters and depots, shelter capacities.
Variables: Stochastic.
Stages considered: 2-stages (Data generation and Query).
Model Formulation: Simulation [8].
Mathematical Model: Online Assignment Problem.
Decision Making Approach: Centralized

5 First Results

The results focus on assessing the performance of the predictive algorithm and the supplying strategies to the shelters. The goal is to extend the number of days before shelters run out of stock.

Figure 5 shows, in green, the performance of the BN when forecasting lead times. The *a posteriori* (observed) lead time is represented by a blue line. Finally, the red line indicates the values obtained when predicting the lead times by explicit knowledge of the system, within the simulation model. These results show a very good performance of the BN when predicting average and peak values, but a tendency to under-represent shorter lead times. The simulation prediction (red line), on the other hand, is more conservative when predicting peak values. This behavior is, in the context of HL, less desirable than the deviations observed with the BN due to the expensive risk of stockout.

Finally, in order to test different strategies, a set of simulations was performed, where the only variation was in the supplying strategy adopted by the control tower, summarized in the value of the variable *alpha*. As explained in Sect. 3, this is a global variable that controls the assignment strategy. In essence, a low value of *alpha* indicates a "greedy" strategy, where orders are sent to the depot that minimizes lead time for the requesting shelter. As the value of *alpha* increases, the strategy becomes cooperative. It considers the impact of each decision on the whole system, penalizing each decision by the opportunity cost of supplying another shelter, see Eq. 4.

Figure 6 shows a sensitivity analysis for the system's survival time (i.e., minutes before system stockout) according to different strategies. The decision to send orders to random depots was considered the baseline to improve (blue dotted line in Fig. 6), with an average value of 15200 min until all shelters are depleted. The left side shows *alpha* values in the range between (0, 0.5), while the right side shows a detail in the range (0.45, 0.5). These results indicate that a strategy with an *alpha* ≈ 0.46

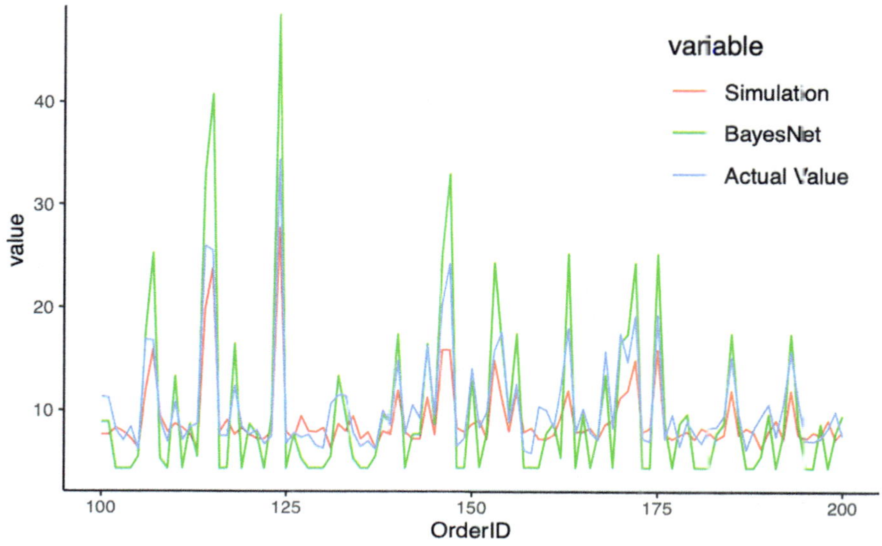

Fig. 5 Predicted and observed values for the lead time of each order placed

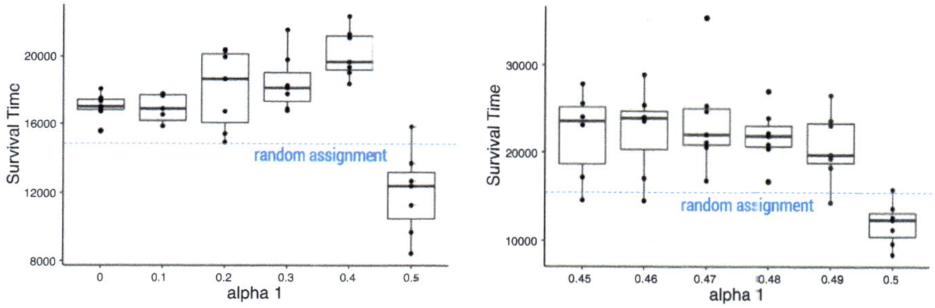

Fig. 6 Sensitivity analysis results. Survival time vs alpha values

may yield, on average, almost a sixty percent increase in the survival time of the system.

6 Conclusions and Future Work

This paper introduced a novel framework to combine BNs, a Probabilistic AI method, with Modeling and Simulation for decision support in the context of humanitarian actions. The first results of this approach seem promising concerning the three particular requirements set by stakeholders (usability, flexibility, and transparency). However, the actual value of this framework remains to be tested in a real scenario.

As future work in this framework, several ideas are being considered. The first extension of this research is the characterization and modeling of specific disasters and their properties, using the outputs of such a model as dynamic inputs for the current response simulation model. Another important consideration is the refinement of the simulation model, and the introduction of order splitting between multiple suppliers, resulting in the vehicle routing problem [2, 21]. An improvement is also required in the calculation of each shelter's remaining days of stock by accounting for the civilian's arrival rate. Finally, the model will be enriched by increasing the number of resources considered as well as the fleet composition and operation by different actors to account for extra transportation capacity when the system is stressed.

References

1. H. Abidi, S. de Leeuw, M. Klumpp, Humanitarian supply chain performance management: a systematic literature review. Supply Chain Manag. Int. J. (2014)
2. W.K. Anuar, M. Moll, L. Lee, S. Pickl, H. Seow, Vehicle routing optimization for humanitarian logistics in disaster recovery: A survey, in *Proceedings of the International Conference on Security and Management (SAM)*, pp. 161–167. The Steering Committee of The World Congress in Computer Science, Computer (2019)
3. S. Armstrong, A. Sandberg, N. Bostrom, Thinking inside the box: Controlling and using an Oracle AI. Minds Mach. **22**(4), 299–324 (2012)
4. G. Barbeito, In-depth behavior modeling of transportation networks: Description and preliminary results of a subway network model, in *Proceedings of the 52nd Hawaii International Conference on System Sciences* (2019)
5. G. Barbeito, M. Moll, W. Bein, S. Pickl, Deterministic and stochastic simulation: A combined approach to passenger routing in public transportation, in *Operations Research Proceedings 2019* (Springer, 2019)
6. A. Behl, P. Dutta, Humanitarian supply chain management: a thematic literature review and future directions of research. Ann. Oper. Res. **283**(1-2), 1001–1044 (2019). https://doi.org/10.1007/s10479-018-2806-2
7. A. Borshchev, *The Big Book of Simulation Modeling: Multimethod Modeling with AnyLogic 6* (AnyLogic North America, 2013)
8. A.M. Caunhye, X. Nie, S. Pokharel, Optimization models in emergency logistics: A literature review. Socio Econ. Plann. Sci. **46**(1), 4–13 (2012). https://doi.org/10.1016/j.seps.2011.04.004
9. S.R.A. da Costa, V.B.G. Campos, R.A.d.M. Bandeira, Supply chains in humanitarian operations: Cases and analysis. Procedia Social Behav. Sci. **54**, 598–607 (2012). https://doi.org/10.1016/j.sbspro.2012.09.777
10. M.S. Habib, Y.H. Lee, M.S. Memon, Mathematical models in humanitarian supply chain management: A systematic literature review. Math. Probl. Eng. **2016** (2016). https://doi.org/10.1155/2016/3212095
11. D. Hilhorst, Dead letter or living document? ten years of the code of conduct for disaster relief. Disasters **29**(4), 351–369 (2005)

12. K. Huang, Y. Jiang, Y. Yuan, L. Zhao, Modeling multiple humanitarian objectives in emergency response to large-scale disasters. Transp. Res. E Logist. Transp. Rev. **75**, 1–17 (2015). https://doi.org/10.1016/j.tre.2014.11.007
13. D. Koller, N. Friedman, *Probabilistic Graphical Models: Principles and Techniques* (MIT Press, 2009). https://doi.org/10.1016/j.ccl.2010.07.006
14. J. Pearl, *Causality* (Cambridge University Press, 2009)
15. A. Rushton, P. Croucher, P. Baker, *The Handbook of Logistics and Distribution Management: Understanding the Supply Chain* (Kogan Page Publishers, 2014)
16. S. Shan, L. Wang, L. Li, Y. Chen, An emergency response decision support system framework for application in e-government. Inf. Technol. Manag. **13**(4), 411–427 (2012)
17. J.B. Sheu, Challenges of emergency logistics management (2007). https://doi.org/10.1016/j.tre.2007.01.001
18. G.H. Tzeng, H.J. Cheng, T.D. Huang, Multi-objective optimal planning for designing relief delivery systems. Transp. Res. E Logist. Transp. Rev. **43**(6), 673–686 (2007)
19. E. Vee, S. Vassilvitskii, J. Shanmugasundaram, Optimal online assignment with forecasts, in *Proceedings of the 11th ACM Conference on Electronic Commerce*, pp. 109–118 (2010)
20. M. Yao, S. Minner, Review of multi-supplier inventory models in supply chain management: An update. Available at SSRN 2995134 (2017)
21. Y. Zhou, L. Zhao, X. Zhao, J. Jiang, A supplier selection and order allocation problem with stochastic demands. Int. J. Syst. Sci. **42**(8), 1323–1338 (2011)
22. H. Zijm, M. Klumpp, A. Regattieri, S. Heragu, *Operations, Logistics and Supply Chain Management* (Springer, 2019)

A Posteriori Access Control with an Administrative Policy

Farah Dernaika, Nora Cuppens-Boulahia, Frédéric Cuppens, and Olivier Raynaud

1 Introduction

The a posteriori access control is usually deployed in trustworthy environments, where actions of users are checked after granting them access. Generally, this mechanism of auditing is based on analyzing logs, where all the traces are being registered. The primary reference of this analysis is the security policy, as the goal is to detect potential violations of this policy. Thus, for security concerns, organizations need to analyze and review their logs from one time to another to make sure that their deployed security policy is being respected.

Many pieces of research have devoted some attention to this type of access control and proposed solutions to understand what is happening in the logs and correlate them with the security policy to detect violations such as [1]. However, the fact that the investigations are done a posteriori introduces many vital aspects that should be taken into account. The first consideration is *time* since many changes can take place between the time of the access and the time of the investigation. For example, subject attributes, object attributes, and circumstances might change, in addition to the policy itself that can evolve over time, including more or less rules.

F. Dernaika (✉)
IMT Atlantique, Rennes, France
Be-ys Research, Geneva, Switzerland
e-mail: farah.dernaika@imt-atlantique.fr

N. Cuppens-Boulahia · F. Cuppens
Polytechnique Montreal, Montreal, QC, Canada
e-mail: nora.boulahia-cuppens@polymtl.ca; frederic.cuppens@polymtl.ca

O. Raynaud
Be-ys Research, Geneva, Switzerland
e-mail: oraynaud.ext@almerys.com

K. Daimi et al. (eds.), *Advances in Security, Networks, and Internet of Things*, Transactions on Computational Science and Computational Intelligence,
https://doi.org/10.1007/978-3-030-71017-0_19

These types of modification can be done using an administrative security policy. Indeed, to have a complete access control model, an administration model should be provided. For instance, RBAC was associated with ARBAC97 [2], OrBAC was affiliated with AdOrBAC [3], and ABAC was recently paired with AMABAC [4]. These models control who is permitted to modify attributes and/or permissions as well as rules in the security policy.

Another thing to be pointed out is that administrators themselves should have the right privileges in order to modify the security policy. In consequence, administrative actions should also be monitored to check that the applied modifications were also allowed.

All previous works related to the a posteriori access control considered a static security policy and did not take into account its time-dependent evolution.

In this chapter, we propose a novel approach, based on the Event Calculus (EC), that treats the a posteriori access control in case of an administrative security policy. The Event Calculus was chosen as an appropriate basis to formalize our problematic since both logs and policy modifications are event-driven.

Our main contributions are the following:

- We propose a framework that takes into account the policy changes made by an administrator when performing an a posteriori access control. As the time of the investigation and the time of the access are different, our proposal permits to get the security rules that were legitimate at the time when an event was logged.
- We enhance the expressiveness of the AMABAC administrative policy and formalize it using the EC. We give expressions that show the relation between a logged event and the rules that held at the time of the access to detect violations, as well as the inter-dependency between administrative actions and the valid security rules.
- We consider the violations that can be caused by both the users and the administrators.
- We formalize the violation detection as a recursive process and show its termination.

Motivating Example The *"Stay Alive"* hospital has deployed a "break-glass" mechanism, where access authorizations outside the standard case can be given explicitly, on a case-by-case basis, by the administrator. *Mary* and *Jeanne* are two nurses who work, respectively, in the cardiology and the neurology departments of the hospital. At the beginning of spring, *Mary* took a vacation for two weeks. Since nurses are only allowed to access medical records that belong to the same department in which they work, *Jeanne*, who replaced *Mary* during that period of time, asked the administrator to grant her the necessary accesses to complete the job. During the same period, *Jeanne* has viewed the medical record of a patient in the oncology department. Did *Jeanne* have the right to do so? Was the created rule correct? Did the administrator abuse his privileges? Etc. The goal of this chapter is to answer these questions.

Chapter Organization Section 2 recalls the Event Calculus, Sect. 3 overviews AMABAC, Sect. 4 presents our proposal, Sect. 5 illustrates an example, Sect. 6 presents the implementation, Sect. 7 discusses related work, and finally Sect. 8 concludes.

2 Event Calculus

The Event Calculus is a logical language for representing and reasoning about events and their effects. The authors in [5] described it as "a logical mechanism that infers what's true when given what happens when and what actions do."

Besides, the language of the EC consists of (1) a set of event types or actions, (2) a set of fluents, that is a set of properties which values can change over time (and can be true or false), and (3) a set of time points. These three elements are essential to feed the basic predicates that constitute the language, and that is represented in Table 1.

Furthermore, relating the various predicates together can form domain-independent axioms that formalize the correct evolution of a fluent. We provide this set of axioms:

$$Happens(e, t_1) \wedge Initiates(e, f, t_1) \wedge (t_1 < t) \wedge \neg Clipped(t_1, f, t) \\ \rightarrow HoldsAt(f, t) \quad (1)$$

$$\exists e, t[Happens(e, t) \wedge (t_1 \leq t < t_2) \wedge Terminates(e, f, t)] \\ \longleftrightarrow Clipped(t_1, f, t_2) \quad (2)$$

Expression (1) indicates that a fluent is true at time t if it has been made true in the past and has not been made false in the meantime. The predicate *Initiates* introduces the event, which activates the fluent, at the time of its execution. For instance, assigning the role Doctor to a user leads to the user having the role Doctor. This can be expressed using Initiates as *Initiates(setRole(user, Doctor), Role(user,Doctor), t)*.

Table 1 Event Calculus basic predicates

Predicate	Meaning
Initiates(e,f,t)	If event *e* is executed at time *t*, fluent *f* is true after *t*
Terminates(e,f,t)	If event *e* is executed at time t, fluent *f* is false after *t*
Happens(e, t)	Event *e* occurs at time *t*
HoldsAt(f,t)	Fluent *f* holds at time *t*
Clipped(t1,f,t2)	Fluent *f* is terminated between times *t1* and *t2*

Similarly, *Terminates(removeRole(user,Doctor), Role(user,Doctor), t)* indicates that removing the role Doctor of a user terminates the fact of that user being a Doctor.

Moreover, the *Clipped* predicate presented in (2) states that an event's occurrence terminates a fluent during an interval of time.

Therefore, a fluent f holds for an open–closed interval $I =]t_1, t_2]$ as follows:

$$\begin{aligned} &Happens(e_1, t_1) \wedge Initiates(e_1, f, t_1) \wedge Happens(e_2, t_2) \wedge \\ &Terminates(e_2, f, t_2) \wedge \neg Clipped(t_1, f, t_2) \rightarrow HoldsFor(f, t_1, t_2) \end{aligned} \quad (3)$$

It must be noted that the interval is open–closed since an event has an effect on a fluent right after its occurrence. For instance, when an initiating event happens at t, the corresponding fluent will start holding right after t (at $t + 1$), and similarly for the terminating event.

In particular, a fluent can be started by an event and not terminated yet. In this case, the fluent will hold until *now*:

$$\begin{aligned} &Happens(e, t_1) \wedge Initiates(e, f, t_1) \wedge \neg \exists t_2 [t_1 < t_2 \wedge Clipped(t_1, f, t_2)] \\ &\rightarrow HoldsFor(f, t_1, now) \end{aligned} \quad (4)$$

In contrast, a fluent's value may remain the same over time. Thus, we introduce the predicate *Always(f)*, to express that a fluent f is always true if and only if it holds at any time t, as follows:

$$Always(f) \longleftrightarrow \forall\, t, HoldsAt(f, t) \quad (5)$$

It is worth mentioning that the events in EC can be natural events like lightning or accidental crash of a hard disk. Since we are dealing with access control, we shall consider, in the following, events that are caused by the execution of an action by a subject on an object.

3 AMABAC

In ABAC, access is granted according to user attributes, resource attributes, action attributes, and context attributes. Using multiple attributes makes ABAC a multi-dimensional access control system that is capable of supporting any access control model. Thus, its support of making fine-grained access decisions made it successful, the reason why we chose to model the security policy according to the ABAC model. As any access control model, ABAC needs to have an administrative representation. Several administrative models were proposed for ABAC [4, 6]. However, we chose AMABAC [4] for being the most recent and with less limitations.

AMABAC is an administrative model for ABAC, where a set of authorized administrative users *U*, who have a set of administrative attributes *A*, that can acquire possible values R_a, and a set of administrative relations *AP*, are defined. The set of attribute name–value pairs associated with an administrative user *a* is given by the expression *attr(a)*. Each administrative relation $Re_i \in AP$ is of the form <*ac*,*Par*>, where *ac* is an administrative attribute condition, that is a set of administrative attribute name–value pairs, and *Par* is an optional set of parameters passed to the relation Re_i. The role of these relations is to define the set of attributes that an administrative user must have to be able to modify a specific component in an ABAC system.

Furthermore, there are 20 administrative relations and commands in AMABAC, which are meant to modify subject, object, and environmental attribute-related components, as well as authorization rules-related components. In this chapter, we are interested in the rules modifications as a whole. Two administrative relations were defined for authorization rules modification in AMABAC. The first relation is to add a rule, and the second one is to remove a rule from the policy. Moreover, each relation is associated with an administrative command that is required to be executed to perform the actual modification. In this connection, some preconditions need to be satisfied prior to the execution of a command for this latter to take effect on the policy. For instance, a new rule *r* can be added (removed) to the policy *P* if there is an administrative user *a* who has the administrative attribute *ac* ($ac \subseteq$ *attr(a)*) that allows the insertion (removal) of a new rule (*can_add_rule(ac)*) and if the same rule *r* is not already in (is already in) the policy *P* ($r \in P$).

Yet, the fact of verifying if the rule belongs to the security policy or not before adding it is not enough. The verification should go further than that to include redundancy checking [7], as well as conflict resolution [8]; hence, the lack of these types of verification represents a limitation in AMABAC. Therefore, to simplify the problem, we will not take into account this requirement as a necessary precondition and will not consider the verification of the preconditions $r \notin P$ and $r \in P$, when, respectively, adding and removing a rule.

4 Our Proposal

In the a posteriori access control, security rules defined in the policy need to be checked to assure that they are being respected. This verification process starts by analyzing logs, where all the access attempts of the system are registered. Moreover, to have a good log analysis, useful information relating to the security policy should be extracted from logs. Therefore, as a log event usually contains an operation *op* that was performed by a user *u* on an object *o* at a specific time *t*, we denote it as e=<*u, op, o*>, and we consider that we retrieved it as well as its time of occurrence using a semantic mediator, as described in [9].

Conversely, an ABAC rule presumes that an action is permitted if the subject, object, and environment that are involved in it fulfill certain values. In consequence, we denote a security rule as *r* = <*SA, OA, EA, op*>, where *SA, OA,* and *EA* are the required attribute values of the subject, object, and environment, respectively, and *op* is a permitted action. Moreover, ABAC rules contain an *if-then* statement, in which the satisfiability of a condition (*the required attribute values*) leads to a permitted action; hence, a security rule can be expressed as *Condition* ⟶ *is-permitted(u,op,o)*.

In case of static security rules, which remain the same over time, the expression defining *Always* is applied to each rule as in (5):

$$Always(Condition \longrightarrow is\text{-}permitted(u, op, o). \tag{6}$$

Therefore, since the rules are always true, we only have to check if *SA, OA,* and *EA* were satisfied at the time the action was executed. Nevertheless, we consider the case of an administrative security policy, where an administrator can change the deployed security rules over time. The analysis consists then not only in checking if the condition holds at the time of the access but also in verifying which rule that relates the condition to the permitted action was valid at that time. Therefore, *Always* cannot be applied anymore.

We suppose that at the time of the investigation t_{invest} (now), the policy is in its last updated state and that the logged event to be analyzed happened at $t < t_{invest}$. Thus, we distinguish two types of verification: (1) check if the logged action at the past time *t* was permitted or not, by fetching the rules that were in the policy at that same time *t*, and collect the user, object, and context attributes defined in the holding rules for verification, and (2) check if the rules corresponding to the deployed policy at time *t* were created by administrators that had the right to create them.

That being said, we consider having two log databases: one that registers all the actions executed in the application domain by regular users, and one that records administrative actions. It must be pointed out that the separation between the two databases is purely conceptual. We could consider that there is only one database, but we made that choice to distinguish between regular and administrative actions.

4.1 Getting the Rules that Held at Access Time

Considering the administrative security policy according to the AMABAC model, a permitted action will hold at the time of its execution *t*, if the *postconditions* triggered by an administrative command hold at that time *t*. In consequence, an action done by a user on an object is permitted if there is a rule that assures its permission at the time of its execution as follows:

$$\begin{aligned} &Happens(e, t) \wedge e =< u, op, o > \wedge \exists r[HoldsFor(r \in P, t_1, t_2) \wedge t_1 < t <= t_2 \wedge \\ &HoldsAt(matches(A(e), A(r)), t)] \longrightarrow HoldsAt(is\text{-}permitted(u, op, o), t) \end{aligned} \tag{7}$$

We recall that $r = < SA, OA, EA, op >$, and t_1 and t_2 are the times when the rule r was added and removed from P, respectively. We also define accordingly *A(e)* and *A(r)* as the set of attributes concerned in the event and defined in the rule, including operations. The *matches* predicate returns true if the attributes values and operation in *A(e)* are the same as the ones defined in *A(r)*, and false otherwise. It is worth mentioning that if a rule was added and never removed before t_{invest} (now), then the fluent $r \in P$ will hold from t_1 to t_{invest} like in (4).

According to (3), *"a rule r is in the security policy"* holds for an interval *]t₁,t₂]*, if it was added at t_1, removed at t_2, and not removed in the meantime. Thus, from the administrative log, we should search for the activating and deactivating events of the fluent $r \in P$ (*add_rule and remove_rule*), if any. However, as mentioned earlier, in AMABAC, a set of preconditions should be satisfied for a successful execution of an administrative command. Therefore, we consider that when an administrator *a* has an administrative attribute that allows him/her to perform a certain action (*e.g., can_add_rule(ac) & ac* $\subseteq$ *attr(a)*), the action provided by that attribute becomes permitted (*e.g., is-permitted(a,add_rule,r)*). Thus, a rule is successfully added to the policy as follows:

$$HoldsFor(is\text{-}permitted(a, add\_rule, r), t_1, t_2) \wedge Happens(< a, add\_rule, r >, t)$$
$$\wedge\, t_1 < t <= t_2 \longrightarrow Initiates(< a, add\_rule, r >, r \in P, t) \tag{8}$$

Similarly, we can express the case of removing a rule from the policy:

$$HoldsFor(is\text{-}permitted(a, remove\_rule, r), t_1, t_2) \wedge t_1 < t <= t_2$$
$$Happens(< a, remove\_rule, r >, t) \longrightarrow Terminates(< a, remove\_rule, r >, r \in P, t) \tag{9}$$

4.2 Monitoring Administrative Actions

As stated in (8) and (9), the effect of an administrative action on the security policy depends on the administrative attributes and rights that the administrator who is performing the action has. Thus, with no doubt, we still need to check if that administrator has the right to perform a modification action (*e.g., HoldsAt(is-permitted(a,add_rule,r), t)*). In consequence, we need to apply (3) again for the fluents *is-permitted(a,add_rule,r)* and *is-permitted(a,remove_rule,r)*, by getting the appropriate initiating and terminating events from the administrative log. In the same way, we need to verify if the administrative user who is modifying the administrators' attributes also has the right attributes to do so. Therefore, a recursive process is introduced as shown in Fig. 1.

To put an end to this verification loop, we consider that there is only one administrator, whom we call *super administrator sad*, who can assign/remove

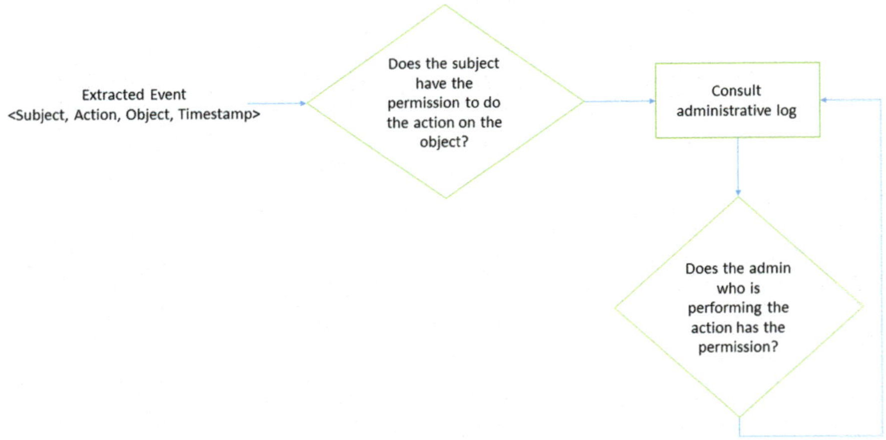

Fig. 1 Recursive aspect of administrative attributes verification

administrative rights to the rest of the administrators. This proposition will define the initial states, since at the time of the conception of the application only one administrator will delegate permissions to other administrators; hence, this will guarantee the end of the recursion. Consequently, we extend AMABAC to include two new administrative commands, *assign_admin_perm* and *remove_admin_perm*, that can be executed without any precondition. Therefore, we obtain the following:

$$Always(is\text{-}permitted(sad, assign\_admin\_perm, perm)). \tag{10}$$

Similarly, all the other administrative actions such as *remove_admin_perm*, *add_rule*, *remove_rule*, etc. are always permitted for *sad*. Moreover, we define a permission *perm* as a tuple *<owner, action, condition>*, where *owner* is the owner (administrator) to whom the permission is assigned, *action* is the operation that is allowed by the permission, and *condition* is the condition that should be satisfied by the object on which the permission is applicable. For instance, *perm*=< a_1,*add_rule,rule.SA=(role=doctor)*> is a permission where the administrator a_1 can add rules in which the subject has the role Doctor. It is worth mentioning that as in AMABAC, the administrative actions are general (e.g., *add_rule* permits creating any rule), and we added the *condition* element to have more expressivity and preciseness in the actions that an administrator can do.

As a result, the loop stops once it gets to the super administrator who has, for sure, the right to perform any action. This idea is depicted in Fig. 2.

Now that *sad* is assigning/removing the administrative permissions of other administrative users, the corresponding actions will be permitted for the defined user without any dependency of other actions, as follows:

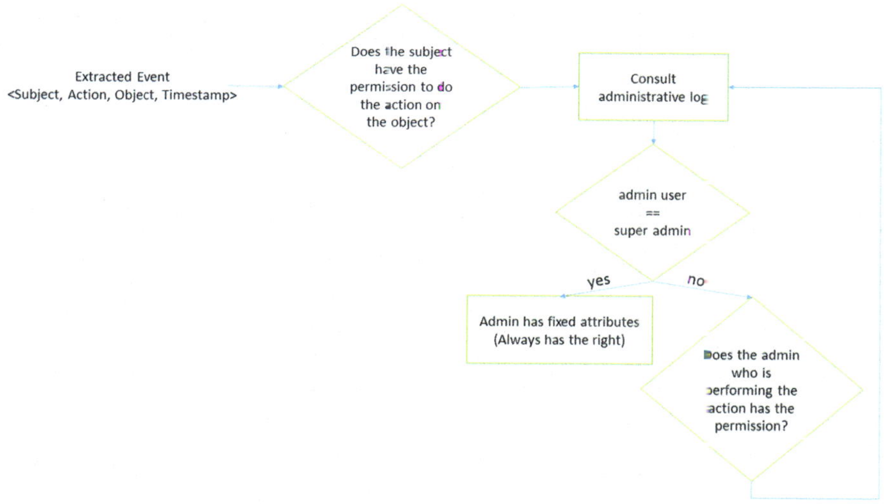

Fig. 2 Recursive aspect ends when *sad* is detected

$$Happens(< sad, assign\_admin\_perm, perm >, t) \wedge \exists o[satisfies(o, perm.condition)] \longrightarrow Initiates(< sad, assign\_admin\_perm, perm >, is\text{-}permitted(perm.owner, perm.action, o), t) \quad (11)$$

$$Happens(< sad, remove\_admin\_perm, perm >, t) \wedge \exists o[satisfies(o, perm.condition)] \longrightarrow Terminates(< sad, remove\_admin\_perm, perm >, is\text{-}permitted(perm.owner, perm.action, o), t) \quad (12)$$

4.3 Detecting Violations

Once all the necessary administrative events are obtained and after resolving expressions (8)–(12), we can get the rules that were in the policy at the time when the log event, to be checked, was executed. In this respect, we obtain from these rules all the required attributes that should be satisfied by the user and object at the same time; hence, we can start searching for these attributes in different databases to contextualize the extracted event and verify if they are compliant with the ones defined in the policy rules (*HoldsAt(matches(A(e),A(r)),t)*) as in (7). It must be pointed out that searching for the user, object, and environmental attributes requires also consulting the administrative log to see when an administrator assigned/removed an attribute to a regular user. Thus, the other administrative relations, and commands of AMABAC, that permit modifying subject, object, and environmental attribute-related components are solicited. Additionally, the verification of administrative privileges is also required in this step. Finally, if the executed action was not permitted at the time of the access, we can deduce a violation as follows:

$$Happens(< u, op, o >, t) \wedge \neg HoldsAt(is\text{-}permitted(u, op, o), t) \rightarrow violation \tag{13}$$

5 Use Case

In this section, we illustrate our proposal with a concrete example. Note that we use discrete time values for simplicity.

Consider an Electronic Health Record (EHR) application with two administrators a_1 and a_2 and one super administrator *sad*. At the time of the creation of the application ($t = 0$), *sad* assigned the privileges of adding and removing rules in which the subject is a doctor to a_1 and adding and removing rules concerning all the users except doctors to a_2. Figure 3 shows an excerpt of the corresponding administrative log.

We consider three ABAC rules, which were added and/or removed by a_1 and a_2 at different times as follows:

r_1: *"A lab technician can create a lab procedure."*

r_2: *"A doctor can create a prescription during an office visit."*

r_3: *"A nurse can view a health record that is in the same department in which he/she works in."*

Moreover, we suppose that we extracted from the application log the following event, which we want to check if it is a violation or not: $e\ =<$

sad	assign_admin_perm	add_rule	a1	r.SA = (role = doctor)	t=0
sad	assign_admin_perm	add_rule	a2	r.SA = (role != doctor)	t=0
sad	assign_admin_perm	remove_rule	a1	r.SA = (role = doctor)	t=0
sad	assign_admin_perm	remove_rule	a2	r.SA = (role != doctor)	t=0
	.				
	.				
	.				
a2	add_rule	r1			t=16
a1	add_rule	r2			t=18
a1	add_rule	r3			t=24
a2	remove_rule	r1			t=33
a2	remove_rule	r3			t=40

Fig. 3 Excerpt of the administrative log

$9003, CREATE, PRE35876, 35\ \ >$, and we consider having one policy named p_1. Furthermore, the investigation is done at $t = 45$ (now). We also define the conditions c_i and permissions $perm_i$ as follows: $c_1 =< r.SA = (role = doctor) >$, c_2=<r.SA=(role!=doctor)>, $perm_1$=<c_1,add_rule,c_1>, $perm_2$=<a_2,add_rule,c_2>, $perm_3$=<a_1,remove_rule, c_1>, and $perm_4$=<a_2, remove_rule,c_2>.

By expressing the administrative log events in the EC, we obtain the following: ***Happens(<sad, assign_admin_perm, $perm_1$>, 0), Happens(<sad, assign_admin_perm, $perm_2$>, 0), Happens(<sad, assign_admin_perm, $perm_3$>, 0), Happens(<sad, assign_admin_perm, $perm_4$>, 0), Happens(<a_2, add_rule, r_1>, 16), Happens(<a_1, add_rule, r_2>, 18), Happens(<a_1, add_rule, r_3>, 24), Happens(<a_2, remove_rule, r_1>, 33), Happens(<a_2, remove_rule, r_3>, 40)***.

Next, by applying (11) and (12), we get ***Initiates(<sad,assign_admin_perm, $perm_1$>,is-permitted(a_1,add_rule, r_2), 0)***, and ***Initiates(<sad, assign_admin_perm, $perm_3$>, is-permitted(a_1, remove_rule,r_2), 0)***. Similarly, the fluents ***is-permitted(a_2, add_rule,r_1), is-permitted(a_2, add_rule, r_3), is-permitted(a_2, remove_rule,r_1), and is-permitted(a_2, remove_rule,r_3)*** are initiated at *t=0*.

Since the fluents *is-permitted(a_i,add_rule,r_j)* and *is-permitted(a_i,remove_rule,r_j)* were never terminated, they hold until now: ***HoldsFor(is-permitted(a_i,add_ rule, r_j), 0, now), HoldsFor(is-permitted(a_i, remove_rule, r_j), 0, now)***.

In addition, (8) and (9) lead to having the following: ***HoldsFor($r_1 \in p_1$, 16, 33), HoldsFor($r_2 \in p_1$, 18, now)***.

In consequence, the only rule that held at *t = 35* was r_2.

Now that we know which rule was valid at the time of the access, we can start searching for the defined attributes (e.g., the subject's role, the subject's department, the type of object, etc.) to see if they verify expression (7), and whether there was a violation or not according to (13). Supposedly that *9003* is the identifier of a doctor, and the object *PRE35876* has the type prescription, but the condition does not respect an office visit, then a violation of r_2 is induced (*c.f. (7) and (13)*). Therefore, the user *9003* should be held accountable for not respecting the security rule that was in place when he/she created the prescription.

Now, we assume that we want to investigate the event *e=<7005,VIEW, HR8853, 37>*. The administrative log and the investigation time remain the same. Besides, we consider that all the requirements of r_3 are fulfilled by the occurring event (*7005* is a nurse who viewed the electronic health *HR8853* that is in the same department she works in). In this respect, when searching for the rules that were holding at $t = 37$, the answer will include only rule r_2, and two violations will be deduced. This can be explained with the following: according to the administration log, a_1 was assigned the permission to add rules concerning doctors only; hence, following (11), the fluent *is-permitted(a_1, add_rule, r_3)* was never initiated. Thus, a violation is produced according to (13). Continuing with (8), a rule is added successfully to the policy if the administrator who is performing the action has the right to do so. Since it is not the case, the fluent $r_3 \in p_1$ is not initiated. As a result, (3) leads to r_3 not being in the policy, which justifies why we will only get r_2 as holding rule at $t = 37$. Moreover, despite the fact that all the attributes are conform with the ones defined in r_3 and that the user technically did not violate the rule as it was added by a_1, a second violation will be provoked by resolving (7) and (13), since only r_2 was *legally* in place at the time of the access and the attributes defined in it are not respected.

At this point where violations are detected, responsibilities should be fixed to account the users and administrators. When only regular users violate the rules that were in place, the process is simpler because only they should justify their actions. Nevertheless, the accountability becomes more complex when the administrators themselves violate the administrative policy, the case in which not only administrators but also regular users should prove the legitimacy of their operations. As the administrators should be sanctioned for sure, the decision concerning the users should be looked into. This is where collateral damage comes around, to fix the responsibility of users when their accesses were authorized by rules that were created by someone who had not the permission to do so. We hereby leave this problematic for a future work.

6 Implementation

We used Semantic Web technologies to implement our approach. First, we represented the modified AMABAC administrative policy in the Web Ontology Language OWL [10]. In an OWL ontology, the concepts are named as classes and relationships as properties. Therefore, the *administrative users, administrative actions, rules, policy, etc.* are considered as classes, and their predicates *hasSubject, hasObject, etc.* are represented as properties. Moreover, the Semantic Web provides the Semantic Web Rule Language (SWRL) [11] to express rules of the form *antecedent* → *consequent*. Thus, we used this latter to express security rules. On the other hand, to express the EC in OWL, we adapted the ontology proposed in [12], for a simplified version of the EC that deals with discrete time points, to suit our case of the a posteriori access control. Furthermore, to justify their modelization, we recall the *"reification"* approach [13].

Therefore, expression (8) can be expressed in SWRL:

Happens(?happens) ∧ hasEvent(?happens,?e) ∧ hasTime(?happens,?t) ∧ add_rule(?e) ∧ hasSubject(?e,?a) ∧ hasObject(?e,?r) ∧ hasCond(?r,?cond) ∧ hasSubjAtt(?cond,?sa) ∧ hasObjAtt(?cond,?oa) ∧ hasEnvAtt(?cond,?ea) ∧ hasOperation(?cond,?op) ∧ hasConsequence(?r,?op) ∧ PermittedAction(?op) ∧ isPermitted(?e) ∧ HoldsFor(?holdsFor) ∧ hasFluent(?holdsFor,?e) ∧ hasStartTime(?holdsFor,?t1) ∧ hasEndTime(?holdsFor,?t2) ∧ swrlb:lessThan(?t1,?t) ∧ swrlb:lessThanOrEqual(?t,?t2) ∧ swrlx:makeOWLThing(?f,?happens) ∧ swrlx:makeOWLThing(?initiates,?happens) → Initiates(?initiates) ∧ isIn(?f) ∧ hasRule(?f,?r) ∧ hasPolicy(?f,p1) ∧ hasFluent(?initiates,?f) ∧ hasEvent(?initiates,?e) ∧ hasTime(?initiates,?t)

In contrast, the Clipped and NotClipped predicates were introduced in EC to deal with causal constraints; hence, it is necessary to support existential quantification and two-way implication to translate these axioms into rules. Nevertheless, SWRL lacks existential quantifiers which makes it impossible to express (3) and (7), for example, in SWRL alone. Therefore, we couple SWRL with an algorithm to have a correct implementation of the axioms. The algorithm is not provided in this chapter due to space limitation. Moreover, we used the Semantic Query Enhanced Web Rule Language (SQWRL) [14], an extension of SWRL, to query the OWL ontology so that the different predicate statements can be separated and resolved appropriately. Besides, SQWRL queries can only work on known individuals (instances) in an ontology, but they do not permit any alterations to the information that they might extract from the ontology.

For example, gathering up all the rules that were initiated by an *add_rule* event at a time point is done using the following SQWRL query:

Initiates(?initiates) ∧ isIn(?f) ∧ hasRule(?f,?r) ∧ hasPolicy(?f,p1) ∧ hasFluent(?initiates,?f) ∧ hasEvent(?initiates,?e) ∧ hasTime(?initiates,?t) → sqwrl:select(?initiates,?h,?e,?r,t)

As OWL and SWRL are based on the open-world assumption, where everything is assumed possible unless explicitly stated otherwise, SQWRL queries implement negation as failure on top of purely open-world systems as stated in [15]. We used Protege 5.2.0 as an environment for developing the administrative policy ontology according to the modified AMABAC, as well as the EC ontology. We also used the OWL and SWRL API, in a Java environment to infer and reason over our ontology, as well as executing SQWRL queries. Furthermore, we used the Java APIs for Date and Time to convert timestamps to discrete time points.

7 Related Work

Many attempts have been made to adapt to changes in traditional access control [16, 17]. Moreover, the authors in [18] addressed the problem of maintaining consistency through occasional changes in a collaborative environment. As they treated the a priori access control, and policy changes may occur while queries are actively being processed, these changes were accommodated online to synchronize and modify query planning.

Moving on to the a posteriori access control, a logical framework, based on logs, was introduced in [1] to check if the actions executed in a system are authorized or not. Furthermore, [19] presented a method to verify if the actions in an audit log adhere to a policy implemented in first-order logic. Their work is similar to previous approaches to validate compliance with a policy but assumes that the policy is correct and static. In contrast, the a posteriori access control had success in the healthcare domain. For instance, [20] and [21] provided a solution to perform an a posteriori analysis of security rules using ontologies and proposed a framework that transforms IHE-ATNA logs into a compliant format with an access and usage analysis led by an OrBAC policy [22]. Nevertheless, almost none of them treated log analysis in case of an expressive security policy and considered its possible changes. The best effort was [20], but it still does not treat the evolution of the security policy over time.

Besides, some works found that policy reconciliation can be done by reconstructing the policy from logs. For example, in [23], the authors proposed an approach to verify the enforcement of security policies and the usage of permissions. Their method was based on analytics and attempts to ensure the consistency of the used permissions with the configured policy. The consistency was provided by mining roles from usage logs and checking their correspondence with the actual policy. Leitner and Rinderle-Ma [24] identified anomalies in RBAC models that may indicate insider threats by comparing a prescriptive RBAC model to a generative RBAC model that can be derived from event logs. In both of these works, usage mining was used to compare logs to the security policy. However, roles, for example, are mined as general behavior, making it impossible to distinguish which role was used by a user at a specific time point.

As discussed above, related works on reconciling policy consider a static security policy. We considered that they are the closely related ones to our work, even if they are somewhat different. In order to make right decisions for accountability purposes, it is important to check which rules were in place when an access was done and to monitor administrators' actions as they can also be accountable. Our literature review has shown that this problem was not treated before.

8 Conclusion and Future Perspectives

In this chapter, we introduced a novel approach, based on the Event Calculus and SWRL, that considers an administrative policy in case of the a posteriori access control. To the best of our knowledge, it is the first work that takes into account the evolution of security rules over time to check policy compliance. Our proposal also permits the assurance of the conformity of administrative actions. Moreover, it has been shown that the Event Calculus is expressive enough to model the a posteriori access control. One drawback can be the translation of all the logs into SWRL facts, which can be costing in terms of time and memory. However, this can be solved by deploying a semantic mediator and a multi-agent system dedicated to collect the necessary events when needed.

As future perspectives, we would like to extend this model to include an accountability mechanism that handles the quantification of the violation level, as well as making decisions when the logs are incomplete. We have discussed briefly that the user can be held accountable in case there was a rule that permits his/her access, but that rule was created by an administrator who has not the right to do so. Thus, we wish to study the impact of users' actions to decide whether they should be held accountable in that case or not. Finally, it was shown that our model considers that there is no violation if at least one security rule is matched. Thus, we would like to take into account, in the future, the resolution of the conflicts and redundancies that may exist between the rules for a better violation detection.

References

1. S. Etalle, W.H. Winsborough, A Posteriori Compliance Control Categories and Subject Descriptors pp. 11–20
2. R. Sandhu, V. Bhamidipati, Q. Munawer, The arbac97 model for role-based administration of roles. ACM Trans. Inf. Syst. Secur. **2**(1), 105–135 (1999)
3. F. Cuppens, A. Miège, Administration model for Or-BAC, in *OTM Confederated International Conferences "On the Move to Meaningful Internet Systems"* (Springer, Berlin, 2003), pp. 754–768
4. S. Jha, S. Sural, V. Atluri, J. Vaidya, Security analysis of ABAC under an administrative model. IET Inf. Secur. **13**(2), 96–103 (2018)
5. M. Shanahan, The event calculus explained, in *Artificial Intelligence Today* (Springer, Berlin, 1999), pp. 409–430
6. S. Jha, S. Sural, V. Atluri, J. Vaidya, An administrative model for collaborative management of ABAC systems and its security analysis, in *2016 IEEE 2nd International Conference on Collaboration and Internet Computing (CIC)* (IEEE, Piscataway, 2016), pp. 64–73
7. M. Guarnieri, M. Arrigoni Neri, E. Magri, S. Mutti, On the notion of redundancy in access control policies, in *Proceedings of the 18th ACM Symposium on Access Control Models and Technologies* (2013), pp. 161–172
8. Mohan, A., Blough, D.M.: An attribute-based authorization policy framework with dynamic conflict resolution, in *Proceedings of the 9th Symposium on Identity and Trust on the Internet* (2010), pp. 37–50
9. F. Dernaika, N. Cuppens-Boulahia, F. Cuppens, O. Raynaud, Semantic mediation for a posteriori log analysis, in *Proceedings of the 14th International Conference on Availability, Reliability and Security* (2019), pp. 1–10
10. D.L. McGuinness, F. Van Harmelen et al.: Owl web ontology language overview. W3C Recomm. **10**(10), 2004 (2004)
11. I. Horrocks, P.F. Patel-Schneider, H. Boley, S. Tabet, B. Grosof, M. Dean et al.: SWRL: a semantic web rule language combining OWL and ruleML. W3C Member Submission **21**(79), 1–31 (2004)
12. W. Mepham, S. Gardner, Implementing discrete event calculus with semantic web technologies, in *2009 Fifth International Conference on Next Generation Web Services Practices* (IEEE, Piscataway, 2009), pp. 90–93
13. M. Dahchour, A. Pirotte, The semantics of reifying n-ary relationships as classes, in *International Conference on ICEIS*, vol. 2 (2002), pp. 580–586
14. M.J. O'Connor, A.K. Das, SQWRL: a query language for OWL, in *International Workshop on OWLED*, vol. 529 (2009)
15. G. Ng, Open vs closed world, rules vs queries: use cases from industry, in *International Workshop on OWLED* (2005)
16. L. Giuri, P. Iglio, Role templates for content-based access control, in *Proceedings of the Second ACM Workshop on Role-Based Access Control* (1997), pp. 153–159
17. G. Zhang, M. Parashar, Context-aware dynamic access control for pervasive applications, in *Proceedings of the Communication Networks and Distributed Systems Modeling and Simulation Conference* (2004), pp. 21–30
18. M. Le, K. Kant, S. Jajodia, Access rule consistency in cooperative data access environment, in *8th International Conference on Collaborative Computing: Networking, Applications and Worksharing (CollaborateCom)* (IEEE, Piscataway, 2012), pp. 11–20

19. D. Garg, L. Jia, A. Datta, A logical method for policy enforcement over evolving audit logs (2011). Preprint, arXiv:1102.2521
20. H. Azkia, N. Cuppens-Boulahia, F. Cuppens, G. Coatrieux, Reconciling IHE-ATNA profile with a posteriori contextual access and usage control policy in healthcare environment, in *2010 6th International Conference on Information Assurance and Security, IAS 2010* (2010), pp. 197–203. https://doi.org/10.1109/ISIAS.2010.5604060
21. H. Azkia, N. Cuppens-Boulahia, F. Cuppens, G. Coatrieux, Log content extraction engine based on ontology for the purpose of a posteriori access control. Int. J. Knowl. Learn. **9**(1–2), 23–42 (2014)
22. A.A. El Kalam, R. El Baida, P. Balbiani, S. Benferhat, F. Cuppens, Y. Deswarte, A. Miege, C. Saurel, G. Trouessin, Or-bac: un modèle de contrôle d'accès basé sur les organisations. Cah. Francoph. Rech. Sécur. Inf. **1**, 30–43 (2003)
23. S. Chari, I. Molloy, Y. Park, W. Teiken, Ensuring continuous compliance through reconciling policy with usage, in *Proceedings of the 18th ACM Symposium on Access Control Models and Technologies, SACMAT '13* (Association for Computing Machinery, New York, 2013), pp. 49–60. https://doi.org/10.1145/2462410.2462417
24. M. Leitner, S. Rinderle-Ma, Anomaly detection and visualization in generative rbac models, in *Proceedings of the 19th ACM Symposium on Access Control Models and Technologies*, pp. 41–52 (2014)

An Analysis of Applying STIR/SHAKEN to Prevent Robocalls

James Yu

1 Introduction

According to the Federal Trade Commission (FTC) website, "*If you answer the phone and hear a recorded message instead of a live person, it's a robocall.*" A news article reports an estimate of 4.7 million robocalls in the month of May 2019 [1], and many of those robocalls are phone scams. Another news article estimates that phone scams caused $10.5 billion of loss in 2019 [2]. The severity of the robocall issue prompts the Federal Communication Commission (FCC) to step up its action and requires all major Internet Telephony Service Providers (ITSP)[1] to authenticate their customers entering into their phone networks. The US Congress, with the bipartisan supports, passed the bill of Telephone Robocall Abuse Criminal Enforcement and Defense (TRACED) Act, which later signed into law at the end of 2019. FCC followed up with the law and mandated all phone companies to implement caller ID authentication by June 30, 2021 [3]. The mandate is to use STIR/SHAKEN to authenticate each incoming call and to attach an *attestation* on the call. The purpose of this chapter is to provide a thorough analysis of the effectiveness of STIR/SHAKEN to protect end users from being victims of phone scams.

In the taxonomy of VoIP security, researchers classify security requirements as follows [4]:

[1]ITSP is also known as phone companies. Almost all phone companies are offering Internet telephony service. Large ITSPs are also referenced as carriers.

J. Yu (✉)
DePaul University, Chicago, IL, USA
e-mail: jyu@cdm.depaul.edu

K. Daimi et al. (eds.), *Advances in Security, Networks, and Internet of Things*, Transactions on Computational Science and Computational Intelligence,
https://doi.org/10.1007/978-3-030-71017-0_20

- *Confidentiality* – The communication is restricted between the sender and the intended receiver. The security measure is to protect and prevent eavesdropping of the communication.
- *Integrity* – The content of the communication does not change during the communication. The security measure is to protect both signaling traffic and bearer traffic from being tampered by hackers. The attack is also known as man-in-the-middle attack.
- *Authentication* –Both the caller and the callee are authentic users as they are claimed. The security measure is to prevent unauthorized access to the VoIP system.
- *Availability* – The serviced is assured (guaranteed) to the users as specified by the service-level agreement (SLA). The security measure is to identify intrusions and defend against Denial of Service (DoS) attacks.

The issue of robocalls is in the category of user authentication where most of robocalls (and almost 100% of phone scams) have *faked* caller ID. On the legacy Public Switch Telephone Network (PSTN), telephone numbers follow the E.164 standard and are managed by the telephone service providers. Users, including Private Branch eXchange (PBX) users, do not have the ability to manipulate the phone numbers assigned by service providers. In addition, high-volume call generators were expensive devices and used exclusively by manufacturers developing large-scale telephone switches. Therefore, it is extremely difficult and very expensive to launch large-scale robcalls on the legacy telephone network.

With the popularity of Internet telephony, hackers can use the Session Initiation Protocol (SIP) to automatically generate calls (robocalls) from their computer, as illustrated in Fig. 1.

The SIP protocol, which is a text-based protocol, makes it relatively easy to generate calls automatically. For example, **sipp** (sipp.sourcforge.net) is a commonly used tool by researchers for stress testing of a VoIP server. The tool uses eXtensible Markup Language (XML) to specify the content of SIP messages and send these messages to a target SIP server. An example of the SIP INVITE message in XML is shown in Fig. 2 where the caller ID (highlighted) can be easily set in the **From** and **Contact** fields of the XML file.

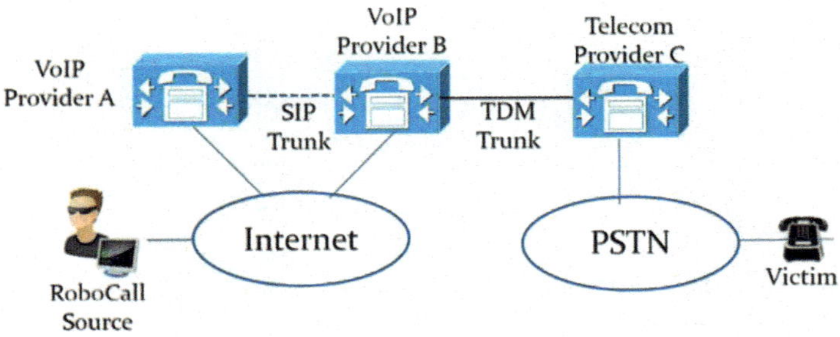

Fig. 1 Robocall generation

```
<send retrans="500">
  <![CDATA[

    INVITE sip:8002@140.192.40.20 SIP/2.0
    Via: SIP/2.0/[transport] [local_ip]:[local_port];branch=[branch]
    From: 8001 <sip:8001@140.192.40.20:[local_port]>;tag=[pid]SIPpTag07b[call_number]
    To: 8002 <sip:8002@140.192.40.20:[remote_port]>
    Call-ID: [call_id]
    CSeq: 1 INVITE
    Contact: sip:8001@[local_ip]:[local_port]
    Max-Forwards: 70
    Subject: Performance Test
    Content-Type: application/sdp
    Content-Length: [len]
    v=0
    o=user1 53655765 2353687637 IN IP[local_ip_type] [local_ip]
    s=-
    c=IN IP[media_ip_type] [media_ip]
    t=0 0
    m=audio [media_port] RTP/AVP 0
    a=rtpmap:0 PCMU/8000

  ]]>
</send>
```

Fig. 2 XML for SIP message creation

In our lab test, we can easily generate 1000 calls per second from a desktop machine. The **sipp** tool is also capable of creating different call scenarios and conducting interactive responses with the receiver (callee). For example, a hacker can send hundreds of calls to potential victims. If anyone answers the robocall, the tool follows up with a greeting, collects basic user information, and then routes the call to a real person to conduct a potential fraudulent scam.

2 Authentication of Call Identity

The major cause of robocall epidemic is a lack of user authentication, and caller IDs of most robocalls are fake. This section explores the user authentication schemes proposed in the SIP standards.

2.1 *Session Initiation Protocol (SIP)*

The original SIP protocol (RFC3261) [5] has a simple authentication process as illustrated in Fig. 3. The message highlighted can be authenticated. The authentication is one-way (server-side) only, where the server authenticates the client but not the other way.

If a SIP call involves multiple SIP servers, only the first SIP server authenticates the client. A SIP server can authenticate the trunk with its neighboring SIP server, but cannot authenticate the clients of its neighbors (Fig. 4).

Because of this deficiency, the data in the SIP header can be easily manipulated by a hacker. Figure 5 shows an example of faked IP addresses and phone numbers captured in our VoIP lab server where the source IP address (163.172.207.104), the SIP **Contact** IP address (212.10.129.158), and the SIP **From** IP address (140.192.40.4) are all different. It is also likely that the phone number of the **From** field and the **Contact** field are fake.

2.2 Call Identity

The issue of caller ID authentication is well known, and there are multiple research papers trying to address this issue [6, 7]. IETF proposes a companion standard, RFC3325 [8], which adds a new structure in the SIP header, **identity**. RFC3325 applies the privacy mechanism specified in RFC3323 [9]. It includes two data fields in the header:

- *P-Asserted Identity* is used among trusted SIP entities. A SIP header may have one or two P-Asserted identity values, and one of them must be the SIP address (in the URI format) of the caller. However, it should be noted that a SIP proxy may remove this header field.

Fig. 3 SIP call (message) flow

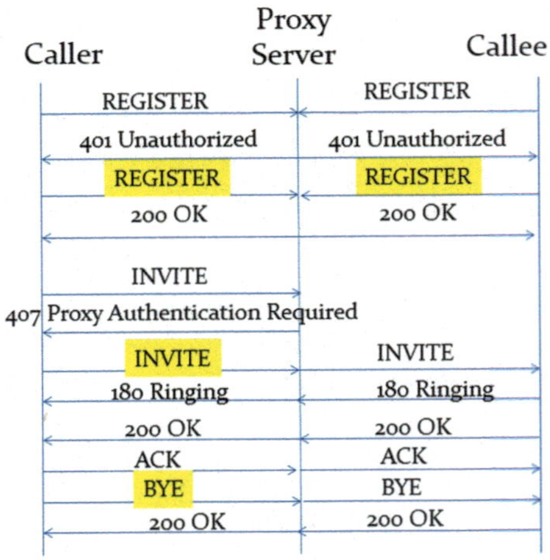

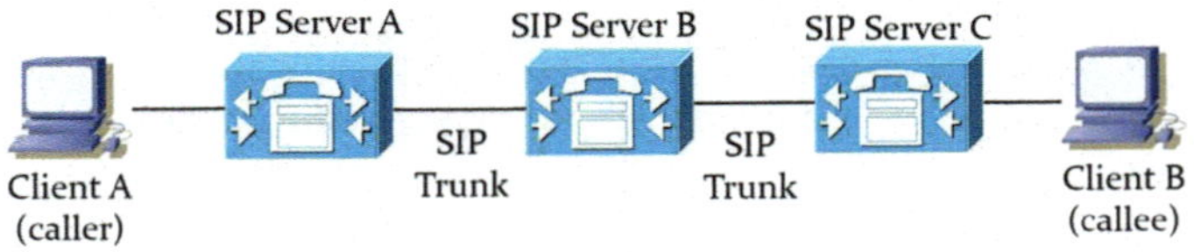

Fig. 4 Authentication of SIP trunks

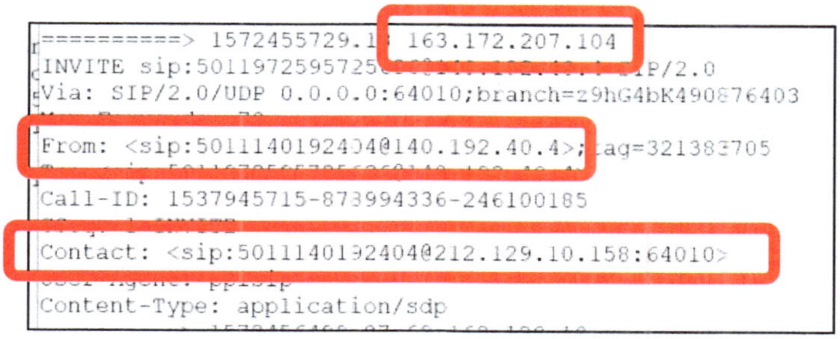

Fig. 5 Captured hacking traffic with fake ID

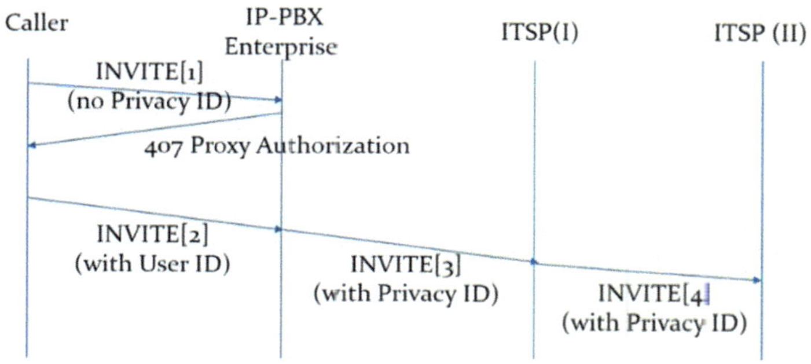

Fig. 6 SIP authentication with identity (RFC3325)

- *P-Preferred Identity* is for a user to carry the identity info to a trusted proxy indicating the identity the caller wishes to be used in the P-Asserted header field value.

An example of call flow with identity is illustrated in Fig. 6.

The first INVITE message from the caller has no user identity, and the second INVITE message includes the user identity.

```
Privacy: id
Proxy-Authorization: .... realm="sip.cisco.com" user="fluffy"
```

The enterprise IP-PBX then adds the real user information in the third INVITE message and sends it to the first ITSP:

```
P-Asserted-Identity: "Cullen Jennings" <sip:fluffy@cisco.com>
P-Asserted-Identity: tel:+14085264000
Privacy: id
```

The first ITSP then sends the fourth INVITE message, along with the identity information, to the next ITSP.

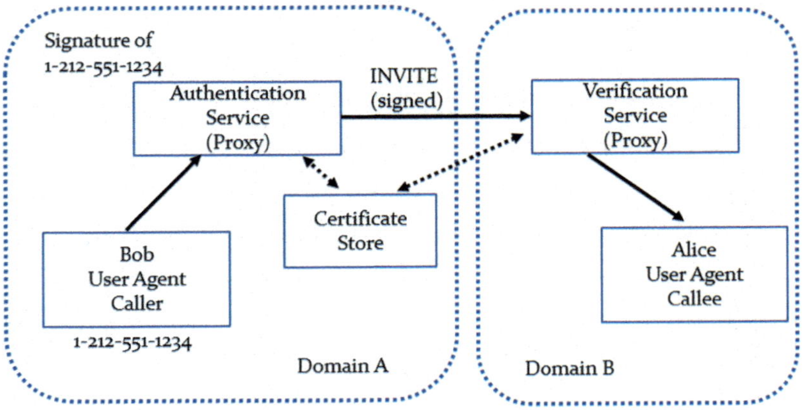

Fig. 7 Architecture of STIR

2.3 *Encryption of Call Identity*

A major issue with the identity (RFC3325) is a lack of authentication and protection of the identity information. RFC 4474 [10] was designed to address this issue. When SIP was used within an enterprise environment or within an ITSP network, there is no strong need for user authentication and protection. However, when Internet telephony becomes more popular and is used between ITSPs, there is a growing demand for user authentication and protection. Secure Telephony Identity Revisited (STIR) is an IETF working group that specifies the standards of STIR in RFC8824 [11], RFC8825 [12], and RFC8826 [13]. The key concept of STIR is to encrypt this identity with a public key encryption and the identity can be authenticated by the first ITSP and also be authenticated from one ITSP to the next ITSP [14]. A lightweight implementation of STIR is demonstrated in [15], which also claims its effective use in enterprise environment.

A key concept specified in RFC8824 is the use of a passport in the INVITE message header, called Passport Association Token (PASSporT) specified in RFC8825. The passport has the signature of user identity that can be authenticated from a Certificate Store, as illustrated in Fig. 7.

An example of an SIP header with STIR is given in Fig. 8. The data of *P-Asserted-Identity* is added by the proxy, and the signature is added by the first ITSP.

2.4 *SHAKEN and Attestation*

The Signature-based Handling of Asserted information using toKENs (SHAKEN) is specified in ATIS-1000074 [16] and RFC8588 [17]. It defines the framework of using STIR to implement the cryptographic validation of a call origination.

```
Contact: <sip:+12155551212@69.241.19.12:50207;rinstance=9da3088f36cc528e>
To: <sip:+12155551213@tel.example1.net>
From: "Alice"<sip:+12155551212@tel.example2.net>;tag=614bdb40
Call-ID: 79048YzkxNDA5NTI1MzA0OWFjOTFkMmFlODhiNTI2OWQ1ZTI

P-Asserted-Identity: "Alice"<sip:+12155551212@tel.example2.net>,<tel:+12155551212>
CSeq: 2 INVITE
Allow: SUBSCRIBE, NOTIFY, INVITE, ACK, CANCEL, BYE, REFER, INFO, MESSAGE, OPTIONS
Content-Type: application/sdp
Date: Tue, 16 Aug 2016 19:23:38 GMT
Identity:
eyJhbGciOiJFUzI1NiIsInR5cCI6InBhc3Nwb3J0IiwicHB0Ijoic2hha2VuIiwieDV1IjoiaHR0cDovL2N
lcnQtYXV0aC5wb2Muc3lzLmNvbWNhc3QubmV0L2V4YW1wbGUuY2VydCJ9eyJhdHRlc3QiOiJBIiwiZGVzdC
I6eyJ0biI6IisxMjE1NTU1MTIxMyJ9LCJpYXQiOiIxNDcxMzc1NDE4Iiwib3JpZyI6eyJ0biI6IisxMjE1NTUxMjEyIn0sIm9yaWdpZCI6IjEyM2U0NTY3LWU4OWItMTJkMy1hNDU2LTQyNjY1NTQ0MDAwMCJ9._28
kAwRWnheXyA6nY4MvmK5JKHZH9hSYkWI4g75mnq9Tj2lW4WPm0PlvudoGaj7wM5XujZUTb_3MA4modoDtCA
;info=<http://cert.example2.net/example.cert>;alg=ES256
```

Fig. 8 SIP message with STIR (signature highlighted)

Table 1 STIR/SHAKEN attestation

A Attestation	B Attestation	C Attestation
Caller on carrier's own network	Caller on carrier's own network	Caller **not** on carrier's network
Caller authenticated	Caller authenticated	Caller not authenticated
Phone number authorized	Phone number not authorized	Phone number not authorized

SHAKEN specifies three levels of attestation that can be assignedby ITSP Table 1.

It should be noted that STIR/SHAKEN does not make a decision about call screening. It simply adds information to the SIP header for authentication, validation, and attestation. It is up to the end user (callee) to determine whether to use this information to accept or reject an incoming call. It also makes it easier to develop an application to automatically filter out unwanted calls.

Another important capability of STIR/SHAKEN is the capability to track the call origination and identify the source of robocalls. For example, the US Department of Justice filed its first action to stop two telecom companies that facilitated fraudulent robocalls against American customers [18].

3 Interworking Between IP and PSTN

3.1 UNI and NNI

A telecom network interface is classified as either User-to-Network Interface (UNI) or Network-to-Network (NNI) interface, as illustrated in Fig. 9.

A UNI is an interface between a service provider and its customers. An NNI is an interface between service providers. There is also PNNI (private NNI) which is an interface within a service provider network.

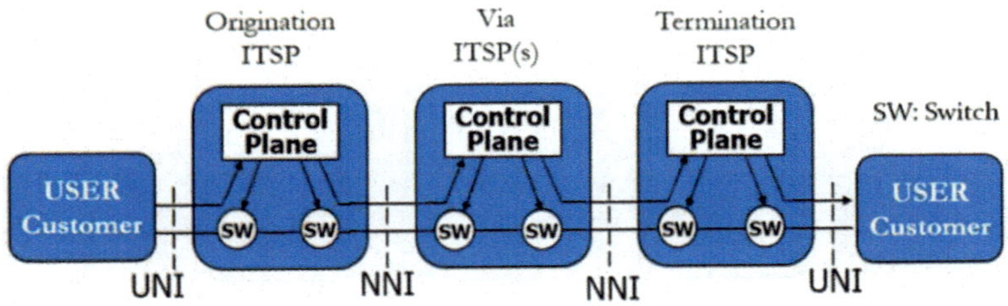

Fig. 9 UNI and NNI of network interworking

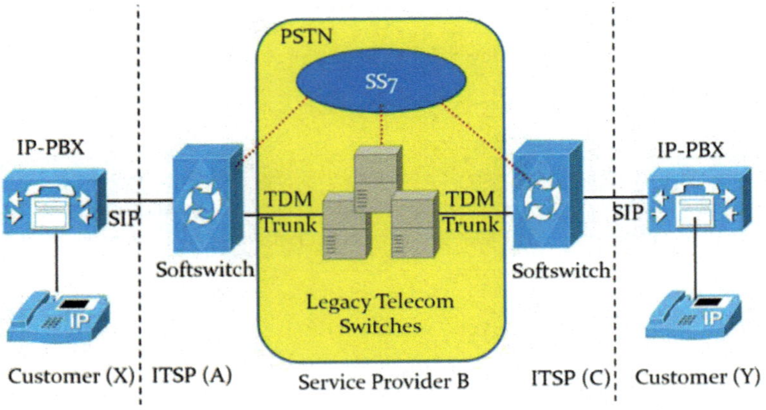

Fig. 10 VoIP and PSTN interworking network

The legacy telecom network is a circuit-switched network, and the establishment of a *circuit* assures the authenticity of call origination. For example, if the UNI side is an analog line, known as Plain Old Telephone Line (POTS), there would be no issue of user authentication as the voice channels determine the phone numbers. If the UNI side is IP, STIR/SHAKEN would be needed to create a user passport and embed the passport in the SIP header for authentication by the origination ITSP, the via ITSPs, and the termination ITSP.

3.2 *SIP and ISUP Interworking*

STIR/SHAKEN is based on SIP which is over IP. If the interface between two service providers is Time Division Multiplexing (TDM) trunk rather than IP, STIR/SHAKEN would not be applicable. The following network illustrates the VoIP traffic passing through the legacy Public Switch Telephone Network (PSTN) where the signaling is ISDN User Part (ISUP) of Signaling System 7 (SS7) (Fig. 10).

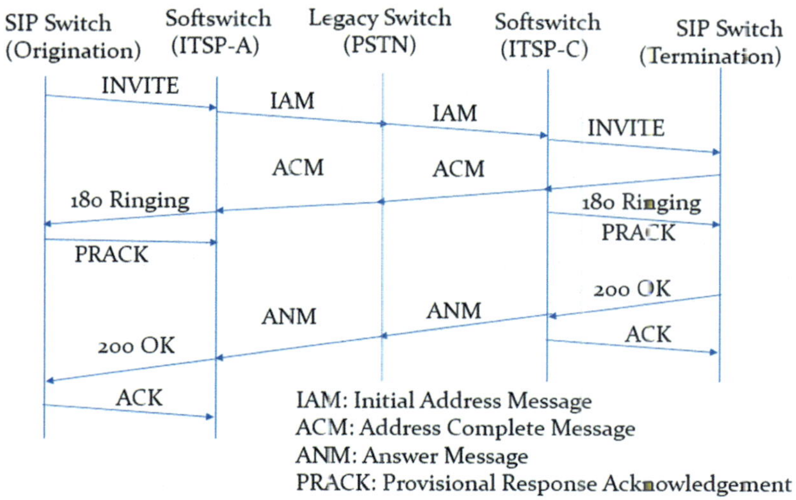

Fig. 11 SIP and ISUP call flow

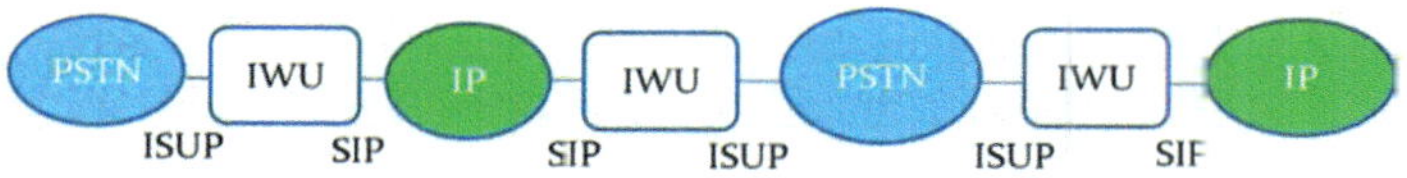

Fig. 12 Scope of SIP and ISUP interworking (Q.1912.5)

The call flow diagram of SIP and ISUP interworking (without caller and callee) is illustrated in Fig. 11.

A challenge for the above call flow diagram is tracking caller identity as there is no mechanism to carry the *passport* from the SIP header to the Initial Address Message (IAM) of the SS7 network. The legacy approach for SIP and SS7 interworking specified in RFC3398 [19] is to use the **From** data in the INVITE message mapping to the calling party number in the IAM message. As discussed before, the **From** data can be easily hacked, so it is would be a security concern of this approach.

To address this issue, a new standard, Q.1912.5 [20], from ITU-T is proposed to specify the interworking between SIP and ISUP. The scope of Q.1912.5 covers mapping from SIP to ISUP as well as ISUP to SIP, as illustrated in Fig. 12.

If an Interworking Unit (IWU) receives an IAM message, it shall encapsulate the IAM data in the SIP header. When an IWU receives an INVITE message, it shall check if it has encapsulated IAM message. If yes, IWU may simply copy the encapsulated IAM data from the INVITE message to the new IAM message. If not, Q.1912.5 provides a detailed procedure to map the data in the INVITE message to the IAM message, and an important provision is to use P-Asserted-Identity in the SIP INVITE message for the calling parting number in the SS7 IAM message. However, it does not reference how the *attestation* is transferred to the IAM message. It also does not reference if the SIP message does not have P-Asserted-Identity.

3.3 *SIP as NNI*

It is important to note that SIP is designed for UNI only, and SIP is not an NNI signaling. On the other hand, IP Multimedia Subsystem (IMS) is designed as an NNI signaling. Because STIR/SHAKEN is between ITSPs, several tutorials of STIR/SHAKEN use the IMS architecture to illustrate the framework of STIR/SHAKEN [21]. Therefore, an important question is whether IMS is required for STIR/SHAKEN deployment. A general response from the industry would be clearly **no** as STIR/SHAKEN has been demonstrated between ITSPs where IMS is not implemented.

The discussion of using *SIP as NNI* is likely more philosophical than technical. The Internet has multitier architecture, and the interface between two ITSP is either a peering agreement or a provider/customer relation, as illustrated in Fig. 13.

If the interface between two ITSPs is a provider/customer relation, it is really UNI rather than NNI. In practice, the interface between a top-tier ITSP (e.g., AT&T) and a lower tier ITSP (e.g., Comcast) is UNI, and there is no issue of using SIP trunks between them. As a result, the implementation of STIR/SHAKEN would be technically feasible.

On the other hand, the relationship between top-tier ITSPs (e.g., AT&T and Verizion) is a peering agreement, and their interface is NNI where SIP is not appropriate for NNI. As a result, the protocol between top-tier ITSPs would be either SS7 or IMS. The implementation of STIR/SHAKEN would be very challenging in this case of NNI.

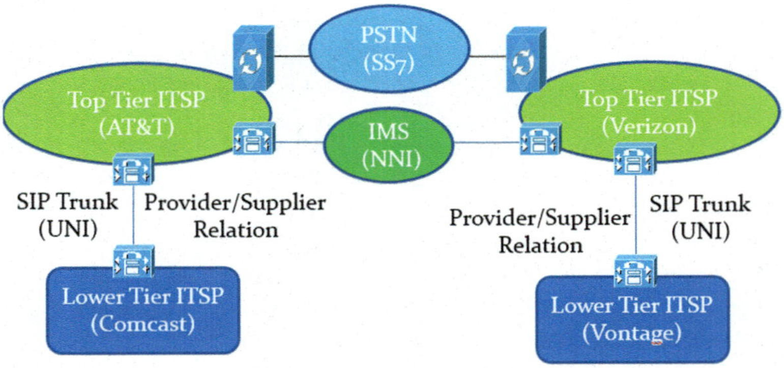

Fig. 13 UNI and NNI between service providers

In late 2018, FCC sent a letter to all major phone companies requesting their commitment to call authentication [22]. Although all major phone companies gave their commitment to support STIR/SHAKEN, their commitments are all at the UNI side, and none has a commitment at the NNI side. One top-tier company made it clear that [the company] *does not yet have VoIP **peers** or transit carrier partners who are prepared to accept signed calls from [the company].* Because other top-tier companies did not mention about their commitment with other VoIP *peers*, it is very likely that these companies are also in the similar situation. It is important to notice the difference between *peers* and *customers* in the context of UNI and NNI connections.

4 Challenges

Although STIR/SHAKEN is a well and thoroughly developed protocol to protect end users from robocalls, its effectiveness is limited to the UNI side and the IP network of ITSPs. We identified the following issues that would pose challenges for the successful deployment of STIR/SHAKEN.

4.1 Unprotected IP-PBX

If an enterprise IP-PBX is not protected or poorly protected from hacking, it could be easily compromised, as illustrated in Fig. 14 [23].

In this attacking scenario, a hacker can exploit a compromised IP-PBX and impersonate himself/herself as a legitimate user within the enterprise network. The hacker could then use the IP-PBX to launch robocalls. Because the calls are from a legitimate enterprise user, ITSP would not be able to tell the hacking scenarios.

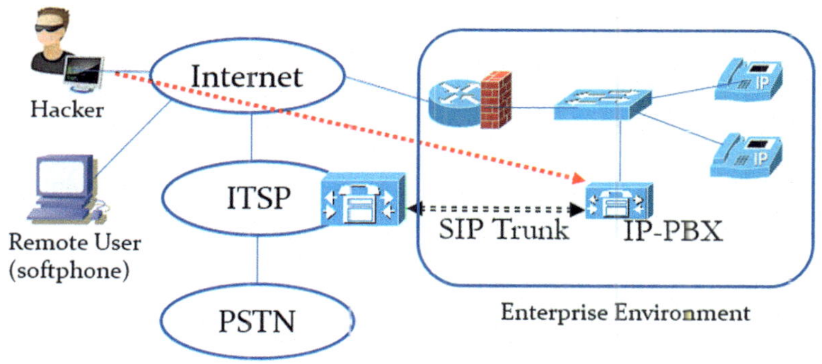

Fig. 14 Hacking unprotected IP-PBX

To address this issue, ITSP needs to be more proactive in helping their customers protecting their IP-PBX, and deployment monitoring programs to identify robocalls from enterprise IP-PBX.

4.2 Unscrupulous ITSP

An Internet Telephony Service Provider (ITSP) is able to provide legitimate telephone numbers (known as E.164 number) to its customers. When an ITSP receives an incoming call from its customers, the ITSP has the responsibility to verify the customer phone number and add **attestation** on it. If an ITSP (lower tier ITSP) does not provide attestation or provides incorrect attestation, it is not clear if and how its provider (upper tier ITSP) could check and verify the attestation.

STIR/SHAKEN also provides the traceability to the call origination. Therefore, it is feasible to trace back to the unscrupulous ITSPs that do not enforce call authentication or provide incorrect attestation. However, it would take nontrivial effort for the upper tier ITSPs and law enforcement to trace it back to those unscrupulous ITSPs.

4.3 Interworking Between SIP and ISUP

Because of a lack of peering agreement of SIP trunking between top-tier ITSPs, VoIP calls between them would be routed to PSTN, as illustrated in Fig. 10. In order to assure call identity and signature on the callee side, this information needs to be carried in the SS7 network. Therefore, the origination ITSP needs to follow the standard (Q.1912.5) to carry the *passport* information from the SIP header into the ISUP IAM message. The termination ITSP needs to retrieve this information from the IAM message and put it into the SIP header. It would be a challenge for ITSPs to follow the Q.1912.5 standard. However, none of the commitment letters from those phone companies to FCC mentioned about if and how they address the issue of interworking between the IP network and PSTN.

One approach to address this issue is to assure the authentication of call identity in PSTN. If a call identity cannot be authenticated, the call should be rejected from entering into PSTN. This is not an issue on the UNI side of PSTN. On the NNI side, it would require an ITSP to reject all C attestation calls. Such a policy would alleviate the interworking requirement between SIP and ISUP, but it is not clear if any ITSP would consider it unless there is another mandate by law.

5 Conclusions

This chapter provides an analysis of the effectiveness of STIR/SHAKEN to assure caller identity of VoIP calls. We conclude that STIR/SHAKEN is effective in an end-to-end IP calls with minor concerns of unprotected IP-PBX and unscrupulous ITSPs. The first one is the user responsibility but can be addressed with the help of more proactive actions by ITSP. The latter, unscrupulous ITSPs, could happen with lower tier ITSPs, and it requires higher tier ITSPs to check and verify the signature of incoming calls. It also requires law enforcement to be more diligent in tracking down those unscrupulous ITSPs.

The current FCC ruling mandates the *origination* ITSP and the *termination* ITSP to implement STIR/SHAKEN for caller ID authentication. However, there is no reference or mandate on via ITSP. Therefore, a major challenge for the STIR/SHAKEN deployment is the IP and PSTN interworking, and it requires commitment of ITSP to support and implement Q.1912.5. A better and long-term solution is to gradually retire circuit-switched PSTN to an all IP network. This vision would require the deployment of SIP/IMS between top-tier ITSPs which has been in waiting since 1999.

References

1. Katherine, Skiba, New, shocking statistics about robocalls, https://www.aarp.org/money/scams-fraud/info-2019/robocalls-statistics.html
2. M. Singh, Spam calls grew 18% in 2019, https://techcrunch.com/2019/12/03/truecaller-spam-call-robocall-report-2019/
3. FCC News, March 31, 2020. https://docs.fcc.gov/public/attachments/DOC-363399A1.pdf
4. D. Butcher, X. Li, J. Guo, Security challenge and defense in VoIP infrastructures. IEEE Trans. Syst. Man Cybern. **37-6**, 1152–1162 (2007)
5. J. Rosenberg, et al., SIP: Session Initiation Protocol, RFC3261, June 2002
6. H. Hakan Kilinc, T. Yanik, A survey of SIP authentication and key agreement schemes, IEEE communications surveys & tutorials, Vol 16, Issue 2, 2nd Quarter 2014, pp 1005–1023
7. Shao Bo, Li Cheng Shu, Identity-based SIP authentication and key agreement, 7th international conference on computational intelligence and security, Hainan, China, Dec 2011
8. C. Jennings, et al., Private extensions to the Session Initiation Protocol (SIP) for asserted identity within trusted networks, RFC3325, Nov 2002
9. J. Peterson, Neustar, A privacy mechanism for the session Initiation Protocol (SIP), RFC3323, Nov 2002
10. J. Peterson, et al., Enhancements for authenticated identity management in the Session Initiation Protocol (SIP), RFC4474, Aug 2006
11. J. Peterson, et al., Authenticated identity management in the Session Initiation Protocol (SIP), RFC8224, Obsolete 4474, Feb 2018
12. C. Wendt, et al., PASSporT: Personal Assertion Token, RFC8225, Feb 2018
13. J. Peterson, et al., Secure telephone identity credentials: Certificates, RFC8226, Feb 2018
14. J. McEachern, E. Burger, How to shut down robocallers, IEEE Spectrum, pp. 46–52, Dec 2019
15. M. Chiang, E. Burger, An affordable solution for authenticated communications for enterprise and personal use, IEEE 8th annual computing and communication workshop and conference (CCWC), Las Vegas, NV, USA, Feb 2018

16. Alliance for Telecommunications Industry Solutions, Signature-based Handling of Asserted information using toKENs (SHAKEN), ATIS-1000074, Jan 2017
17. C. Wendt, et al., Personal Assertion Token (PaSSporT) Extension for Signatured-based Handling of Asserted Information using toKENs (SHAKEN), RFC8588, May 2019
18. Justice News, The Department of Justice Files Actions to Stop Telecom Carriers Who Facilitated Hundreds of Millions of Fraudulent Robocalls to American Consumers, January 28, 2020., https://www.justice.gov/opa/pr/department-justice-files-actions-stop-telecom-carriers-who-facilitated-hundreds-millions
19. G. Camarillo, et al., Integrated Services Digital Network (ISDN) User Part (ISUP) to Session Initiation Protocol (SIP) Mapping, RFC3398, Dec 2002
20. ITU-T, Interworking between session initiation protocol (SIP) and bearer independent call control protocol or ISDN user part, Q.1912.5, January 2018
21. M. Dolly, An introduction and overview of the STIR/SHAKEN framework, the 8th SIPNOC Conference, Herndon, VA, USA, Dec 2018
22. FCC Site, Combating spoofed robocalls with caller ID authentication, https://www.fcc.gov/call-authentication
23. James Yu, Prevention of toll fraud against IP-PBX, Proceedings of 2015 international conference on security and management (SAM'15), Las Vegas, pp. 259–265, July 2015

Supervised Learning for Detecting Stealthy False Data Injection Attacks in the Smart Grid

Mohammad Ashrafuzzaman, Saikat Das, Yacine Chakhchoukh, Salahaldeen Duraibi, Sajjan Shiva, and Frederick T. Sheldon

1 Introduction

Today's smart grids, with generators, transmission systems, distribution systems, smart meters, distributed energy resources, and numerous other physical devices, are integrated with embedded computers, computation, networking, and other cyber technologies and therefore have been transformed into among the largest and the most complex cyber-physical systems (CPSs). With cyber capabilities, smart grids have inherited vulnerabilities and threats of cyber-attacks. Even though the industry has attempted to "air-gap" operational technology (OT) from information technology (IT) networks toward protecting valuable CPS assets critical to stable operations, OT networks are unfortunately still not fully insulated from the IT networks and are vulnerable to both internal and external threats [1].

The *false data injection* (FDI) attack is a new class of cyber-attacks against the state estimation process in the power grids [2]. The state estimation (SE) is a

M. Ashrafuzzaman (✉) · F. T. Sheldon
Department of Computer Science, University of Idaho, Moscow, ID, USA
e-mail: ashr3866@vandals.uidaho.edu; sheldon@uidaho.edu

S. Das · S. Shiva
Department of Computer Science, University of Memphis, Memphis, TN, USA
e-mail: sdas1@memphis.edu; sshiva@memphis.edu

Y. Chakhchoukh
Department of Electrical and Computer Engineering, University of Idaho, Moscow, ID, USA
e-mail: yacinec@uidaho.edu

S. Duraibi
Department of Computer Science, University of Idaho, Moscow, ID, USA

Department of Computer Science, Jazan University, Jazan, Saudi Arabia
e-mail: dura6540@vandals.uidaho.edu

K. Daimi et al. (eds.), *Advances in Security, Networks, and Internet of Things*
Transactions on Computational Science and Computational Intelligence,
https://doi.org/10.1007/978-3-030-71017-0_21

fundamental tool in the energy management system (EMS) at the power control center. It computes voltage magnitudes and phase angles at all of the different buses of the power system after collecting measurements that are communicated to the control center from remote terminal units (RTUs) [3]. In an FDI attack, an adversary modifies some of these measurement data with the intent of affecting the outcome of the SE and therefore reduces the control center operators' level of situational awareness [4] forcing the operators to take erroneous corrective actions. Stealthy FDI (SFDI) attacks are those that cannot be detected using traditional bad-data detection methods. An SFDI-attacked SE may disrupt the real-time operation of the grid by impacting tools such as contingency analysis, unit commitment, optimal power flow, and computation of locational marginal pricing for electricity markets. The SFDI is an important element of a coordinated attack on the power grid and represents an important class of attack on cyber-physical systems [5].

Investigations to devise detection methods for FDI attacks include traditional statistical approaches and approaches based on the physics of the state estimation [6]. In the recent years, data-driven machine learning-based approaches have been gaining popularity. The machine learning approaches treat the FDI attacks on the measurement data as anomalies compared to the normal data and use well-known machine learning algorithms or some improvisations to classify or cluster the FDI attacks as anomalies. It is well known that different classifiers may perform differently on the same data. Therefore, having an "ensemble" of classifiers may provide a wider coverage for detection of the FDI attacks. In an ensemble, the classification results given by the constituent classifiers are fed into another classifier for final decision [7]. Ensemble-based machine learning approaches have been shown to perform well in solving other problems [8–10].

In this chapter, we first use five well-known supervised models, namely logistic regression (LR), naïve Bayes with Gaussian function (NB), decision tree (DT), artificial neural networks (NN), and support vector machine (SVM) as classification models to detect SFDI attacks. Then, the outputs or decisions from these five classifiers are fed into seven models separately. This constructs seven ensemble models. The seven *ensemble classifiers* are majority voting (MV), logistic regression, support vector machine, naïve Bayes with Gaussian function, decision tree, artificial neural networks, and a model that performs an OR operation of all the stand-alone model outputs. Both the stand-alone and ensemble models are trained using historical data. The performances of all the 12 models are compared using standard evaluation metrics to determine the best performing model.

The major technical contributions of this chapter are summarized as follows:

- We design a scheme that consists of ensembles of supervised classification models. The ensembles are constructed using different classifiers with the goal to compare their performances and determine the best performing ensemble classifier.
- We simulate the standard IEEE 14-bus system using MATPOWER [11] and introduce stealthy FDI attacks to the measurement data generated by the simulation.

- We use this simulated data to test and evaluate our proposed scheme. We compare the performances of different stand-alone and ensemble models using standard evaluation metrics and find that the performance of stand-alone models and ensemble models is same.
- We reduce the feature set using random forest and run the models using this feature-reduced dataset. We compare the performances with feature-reduced dataset with those of full-feature dataset and find that the feature-reduced dataset runs much faster while training and provides the same performance as the performance of full-feature dataset.

The remainder of the chapter is organized as follows. Section 2 gives a brief review of related works. Section 3 describes the mathematical formulation for the static SE and the stealthy FDI attacks. Section 4 presents the ensemble-based machine learning scheme proposed in this chapter. A set of experiments with this scheme along with the results is presented in Sect. 5. Conclusions are presented in Sect. 6, followed by the references.

2 Related Works

Esmalifalak et al. [12] employed distributed support vector machine (SVM); Ozay et al. [13] used multi-layer perceptron, k-nearest neighbors (kNN), and SVM; He et al. [14] used conditional deep belief network; Wang et al. [15] used an algorithm based on the margin setting algorithm (MSA); Wang et al. [16] used kNN, neural network, SVM, naïve Bayes, and decision tree; Ashrafuzzaman et al. [17] used feed-forward neural networks, gradient boosting machines, generalized linear models, and distributed random forests; Ahmed et al. [18] proposed a Euclidean distance-based anomaly detection scheme; Niu et al. [19] used an LSTM-based convolutional neural network; Wang et al. [20] used stacked auto-encoders; Camana-Acosta et al. [21] used extremely randomized trees; and Mohammadpourfard et al. [22] used kNN to detect the FDI attacks. All of these models used above are supervised learning models.

In the unsupervised category, the models used are isolation forest by Ahmed et al. [23], sparse principal component analysis by Hao et al. [24], density ratio estimation by Chakhchoukh et al. [25], and sparse logistic regression and semi-supervised SVM by Ozay et al. [13]. Kurt et al. [26] used reinforcement learning algorithm SARSA.

Most of these works mention evaluation performance metrics in terms of model accuracy and precision. However, datasets used for detecting attacks, which are sparse compared to the non-attack or normal data, are imbalanced datasets, and for this kind of classifications, the more appropriate metrics are sensitivity or recall and false positive rate (FPR). These works do not discuss their results using these metrics. Also as shown above, many of the solutions proposed have used single classifiers, and a few used several classifiers as separate individual models, but none of the works used ensemble learning.

3 Stealthy False Data Injection Attacks on State Estimation

State estimation (SE) at the transmission system in electric power grids is a key function in supervisory control, operation, and planning of the system. It is used to provide the best estimate of the values of the system's unknown state variables, i.e., voltage magnitudes and phase angles of the system buses, from the measurements available from the network model and sent by the SCADA system to the control center. The functions of the state estimator include identifying and correcting contamination in the data, suppressing any bad data, and refining the measurements. Finally, it gives a set of state variables that is acceptable to the operator and as inputs to other computational programs of the energy management system (EMS) [27].

3.1 Formulation of State Estimation

The static state estimation is run after the SCADA units collect the measurements of power flows, power injections, and voltage magnitudes from the buses in the system. The static SE estimates the state vector $\boldsymbol{x} \in \mathbb{R}^n$ that contains phase angles and voltage magnitudes at the different buses, where $n = 2k - 1$ and k is the number of buses in the system. For AC static SE, the state vector $\boldsymbol{x}$ obeys the following nonlinear equation:

$$z = \boldsymbol{h}(\boldsymbol{x}) + \boldsymbol{e}. \tag{1}$$

In the above equation, the vector of measurements $z \in \mathbb{R}^m$ contains measurement readings from SCADA units, where m is the number of measurements. The nonlinear vector function $\boldsymbol{h}(\cdot)$ is computed from the grid topology and the transmission lines, transformers, and other grid devices parameters. The error vector $\boldsymbol{e} \in \mathbb{R}^m$ is assumed Gaussian with a covariance matrix R. The SE is executed to compute and estimate the state vector $\boldsymbol{x}$ using an iterative algorithm based on the weighted least squares (WLS).

$$\hat{\boldsymbol{x}}_k = \hat{\boldsymbol{x}}_{k-1} + H_k^{\sharp}\left(z_k - \boldsymbol{h}(\boldsymbol{x}_{k-1})\right), \tag{2}$$

where $H_k^{\sharp} = (H_k^{\top} R^{-1} H_k)^{-1} H_k^{\top} R^{-1}$ and H_k is the Jacobian matrix of $\boldsymbol{h}$ with respect to $\boldsymbol{x}$ at step k. The WLS algorithm is optimal under Gaussian noise.

After the algorithm converges, i.e., once $\|\hat{\boldsymbol{x}}_k - \hat{\boldsymbol{x}}_{k-1}\| < \delta$ for some chosen small threshold $\delta > 0$, the obtained residuals are analyzed for possible abnormal measurements by checking for residuals that do not obey the Gaussian assumption.

These abnormal or bad data could be due to natural failures such as sensor or communication error or due to FDI attacks.

3.2 Stealthy FDI Attacks

State estimation detects abnormal or bad data by analyzing the residual vector (i.e., the difference between the measurement vector z and the calculated value from the state estimation, i.e., $z - H\hat{\boldsymbol{x}}$). If the largest absolute value of the elements in normalized residual is greater than a predefined threshold $\alpha > 0$ (α is generally chosen to be 3), the corresponding measurement is identified as bad data and reported to system operators. Therefore, if the bad data are due to FDI attacks and are large enough, the conventional residual tests can detect them: these are called *non-stealthy FDI attacks* or simply FDI attacks. If the attackers have knowledge of the system topology or know the measurement matrix H, they can carefully and intelligently craft the false data in such a way that the residual $\boldsymbol{r}$ of the original measurement vector z remains the same as the residual $\boldsymbol{r}_a$ of the measurement vector z with the injected data z_a.

$$\boldsymbol{r}_a = z_a - H\hat{\boldsymbol{x}}_a = z - H\hat{\boldsymbol{x}} = \boldsymbol{r}. \tag{3}$$

These are *stealthy FDI attacks*, and they cannot be detected using the conventional methods based on residual analysis.

4 Machine Learning-Based Method

This section provides an overview of the proposed ensemble-based stealthy false data injection attack detection scheme. Figure 1 depicts the block diagram of the training process pipeline. It shows the processing phases, namely (1) data preprocessing, (2) feature selection, (3) classification using individual classifiers, (4) classification using ensemble methods, and (5) obtaining the best performing model. The training is performed offline with historical data, and the testing, when deployed, will be online in real time.

4.1 Data Preprocessing

Like in any machine learning (ML) training pipeline, data preprocessing phase removes unwanted and invalid data, imputes missing data, converts data suitable for the training, performs scaling, etc. If the dataset is imbalanced, additional steps are taken to balance the dataset.

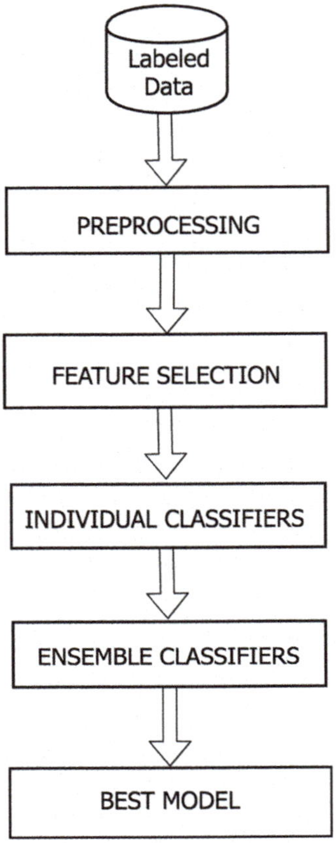

Fig. 1 The training pipeline with the ensemble-based ML framework

4.2 *Feature Selection*

Feature selection is used to eliminate the least important features from the dataset, thereby reducing the dimensionality without sacrificing much of the information. Dataset with reduced features often provides better performance and minimizes the running time. Our training pipeline currently supports random forest (RF) as a feature selection algorithm.

4.3 *Individual Classifiers*

The set of individual classifiers constitutes the first part of the two-part ensemble mechanism. The classifier models included in our framework are decision tree (DT), logistic regression (LR), naïve Bayes (NB), artificial neural network (NN),

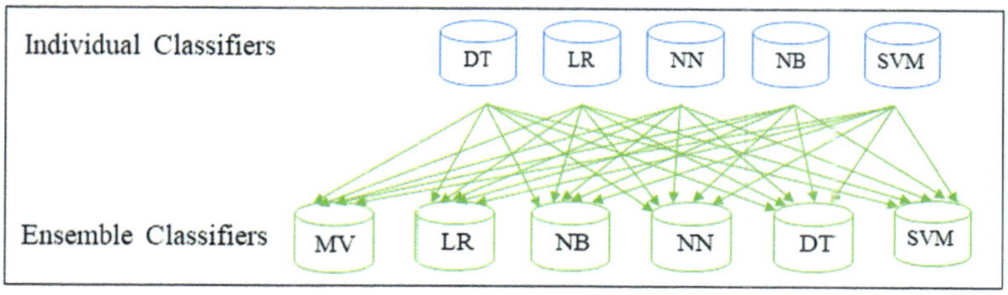

Fig. 2 Diagram showing the ensemble mechanism

and support vector machine (SVM). The classification decisions given by the five individual models are fed as input to the ensemble classifiers.

4.4 Ensemble Classifiers

Ensemble classifiers take the classifications decisions, a set of five 0s and 1s, from the five individual classifiers as input and classify these into 0 or 1, where 1 means attack and 0 means normal data. In order to find out the best performing ensemble classifier, we use six supervised classifiers in the pipeline. The ensemble classifiers are majority voting (Ens_MV), logistic regression (Ens_LR), naïve Bayes (Ens_NB), artificial neural network (Ens_NN), decision tree (Ens_DT), and support vector machine (Ens_SVM). Figure 2 shows the ensemble mechanism. In addition to these six models, we also use a model that performs OR operations on the outputs of the stand-alone models.

4.5 Best Performing Models

The datasets go through all of the individual and ensemble classifiers in the pipeline. The performance of all the classifiers is then compared using standard evaluation metrics. This comparison identifies the best performing model among the ensemble or individual models. The best model is to be deployed in the state estimation process for real-time detection of stealthy false data injection attacks.

5 Experiments and Results

This section presents an experiment with the framework proposed in this chapter and discusses the results.

5.1 Attack Model

In this chapter, SFDI attacks targeting the static AC state estimation of the transmission system are considered. The attacker is assumed to be capable of changing the communicated data such as voltages, currents, and power magnitudes. The adversary needs only selected partial knowledge of the network topology, which allows them to generate a stealthy attack on a single bus. The considered attack model assumes that only one fixed bus is targeted throughout the entire duration of an attack.

5.2 Simulation and Data Generation

Simulation of the standard IEEE 14-bus system is considered for generating data. The system has 5 generators and 11 loads [28], as shown in Fig. 3. The measurements are obtained from solving power flows using the MATPOWER toolbox [11] and adding Gaussian measurement noise. The measurements are 40 active power flows, 14 active power injections, 40 reactive power flows, 14 reactive power injections, and 14 voltage magnitudes giving a total of 122 measurements comprising the feature set. A new measurement vector z, corresponding to one set of data, is generated every 60 s. The dataset consists of 100,000 sets of measurement data.

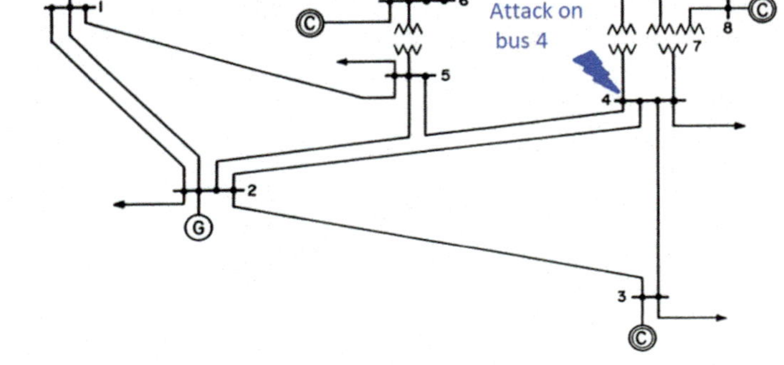

Fig. 3 Diagram of an IEEE 14-bus system (adapted from [28]) showing an attack that targets bus number 4

5.3 *Data Preprocessing*

In our dataset, 90% are "normal" data and 10% are "attack" data implying that the dataset is imbalanced. Classifiers perform poorly when trained with imbalanced datasets, especially for the minority class. In our case, the "attacks" are in the minority class, and our goal is to detect these precisely. In order to overcome this problem, we applied the synthetic minority oversampling technique (SMOTE) and the edited nearest neighbor (ENN) to oversample the "attack" sets of data and undersample the "normal" sets of data [29]. After this balancing act, the ratio of major and minor class samples in our dataset was 3:2.

The dataset did not have any missing data or invalid data; so, we did not need any data cleaning to perform. However, we applied standard scaling to the data. In standard scaling, the features are normalized by scaling the values in one feature to unit variance.

5.4 *Feature Reduction*

The random forest algorithm was used on the dataset to obtain an ordering of the features according to their importance. A plot showing the feature importance is given in Fig. 4. The figure shows that the first 21 features have the largest variances,

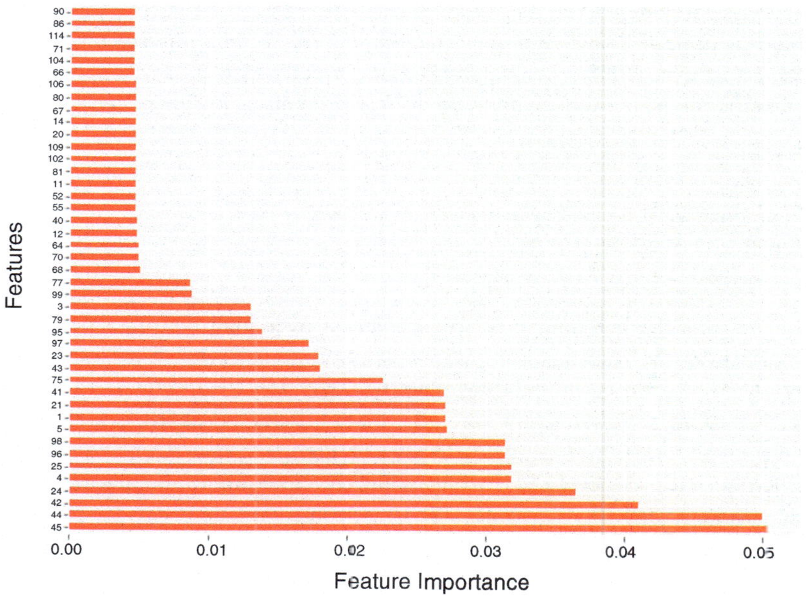

Fig. 4 Graph showing the features in order of importance

and therefore only these features were retained in the dataset as the predictor variables. The final feature set includes measurement numbers 45, 44, 42, 24, 4, 25, 96, 98, 5, 1, 21, 41, 75, 43, 23, 97, 95, 79, 3, 99, and 77 (listed here ordered by their importance).

5.5 Model Training

The experiment was conducted with individual classification first and then ensemble classification. The experiment ran the data through five individual models, and then the seven ensemble models were run with the outcomes of the individual models. We trained the models using grid search and retained the best values of the hyper-parameters. We split the dataset into two sets: 70% for training and 30% for testing. To avoid overfitting and to obtain robust models, we used 10-fold cross-validation over randomly divided training data during training of the models. Then, we used the test data for prediction and for measuring model performance.

5.6 Evaluation Metrics

A machine learning model for binary classification predicts class labels as output for a given input data as (1) true positive (TP), when the model correctly identifies an attack, (2) true negative (TN), when it correctly identifies a normal or non-attack, (3) false positive (FP), when a non-attack is incorrectly identified as an attack, and (4) false negatives (FN), when an attack is incorrectly identified as a non-attack. To evaluate the models in this chapter, the following metrics [30] are used:

1. Accuracy $= (TP + TN)/Total$,
2. Precision $= TP/(FP + TP)$,
3. False Positive Rate (FPR) $= FP/(FP + TN)$,
4. Recall $= TP/(FN + TP)$,
5. F1-Score $= 2TP/(2TN + FP + FN)$.

Accuracy is the percentage of true detection over total data instances. *Recall*, also known as the true positive rate, sensitivity, or detection rate, indicates how many of the attacks the model does identify. *Precision*, also known as the positive predictive value, represents how often the model correctly identifies an attack. The *F-measure* provides the harmonic average of precision and recall. In addition to these five metrics, the *ROC AUC score*, which is a measure of the diagnostic ability of binary classifier systems, is used. To demonstrate the detection performance of different models over all possible thresholds, the *ROC curves* are plotted. The ROC curve is a graph of false positive rate (FPR) versus true positive rate (TPR). The run times (i.e., elapsed times) were measured for comparing the speed of different training models.

5.7 Discussion of Results

In this section, we present and discuss the results from the experiment in terms of the evaluation metrics.

Table 1 shows the results, i.e., the values for the evaluation metrics, from running all the five supervised classifiers and seven ensemble classifiers on a feature-reduced dataset with 21 features. The values for individual classifiers and those for the ensemble classifiers are effectively the same for all the metrics. The table shows that precision values for the models are very close to 100%, whereas accuracy values are about 90%. The high F1 scores indicate that the models are quite robust. Therefore, it indicates that these models are very well suited for precisely and reliably detecting stealthy false data injection attacks. However, in a classification problem where the goal is to detect the minor class occurrences, the most important metrics are the recall or sensitivity which, in our case, measures the proportion of "attacks" that are correctly identified as such and the FPR which measures the proportion of "non-attacks" that are incorrectly identified as "attacks" raising a false alert. For supervised models, the sensitivity values for all the models are very similar, with the ensemble models having a little better number at 73.53%. This indicates that even the best model would be able to detect about 73% of the attacks and the rest 27% will go undetected. The FPR values for the models are 0.03% meaning that the models are able to identify a "non-attack" as such almost always and will seldom raise a false alert.

Figure 5 illustrates the ROC curves for all the models. It is not surprising that all the curves are the same within statistically negligible range.

Referring back to Table 1, we find that the elapsed time taken to train a model using the dataset having all the 122 features takes up to 400% more time than the corresponding time in the case of the feature-reduced dataset. The table also shows that not only the ensemble models do not perform any better, but they also take more time to run than the individual models. This is because the ensembles first run all the five individual models and then run the ensemble model, and the accumulated elapsed time, therefore, is higher.

As we have seen in Sect. 2, the SVM is a popular model among the researchers working on the problem of detecting false data injection attacks on the static estimation in the smart grid. However, our experiment shows that SVM performs the same as the other models. Moreover, SVM takes much more time to train. Whereas the other individual models take less than 2 s to train, SVM takes 2700 s or 45 min on the feature-reduced dataset. On the original dataset with 122 features, SVM takes an astounding 8900 s or 2.47 h. If we take SVM out as an individual model, then the times taken by the ensemble models reduce drastically without any reduction in performance. The last column in Table 1 shows times taken by the ensemble models when SVM is not included in the set of the individual models.

Table 1 Evaluation metrics values for supervised individual and ensemble models using the test dataset

Models	F1-score	Accuracy	Precision	Recall	FPR	ROC	Elapsed time (in seconds)		
						AUC	21 Features	122 Features	Without SVM
LR	0.8439	0.8931	0.9991	0.7304	0.0003	0.8639	0.56	1.02	–
NB	0.8439	0.8931	0.9991	0.7304	0.0003	0.8081	0.27	1.03	–
NN	0.8439	0.8931	0.9991	0.7304	0.0003	0.8650	0.57	0.83	–
DT	0.8438	0.8930	0.9991	0.7302	0.0003	0.8797	1.59	114.52	–
SVM	0.8439	0.8931	0.9991	0.7304	0.0003	0.8642	2713.82	8897.83	–
Ens_MV	0.8439	0.8931	0.9991	0.7304	0.0003	–	2718.96	9017.94	6.07
Ens_LR	0.8472	0.8961	0.9993	0.7353	0.0003	0.8675	2717.06	9015.59	4.15
En_NB	0.8472	0.8961	0.9993	0.7353	0.0003	0.8675	2717.01	9015.32	4.12
Ens_NN	0.8472	0.8961	0.9993	0.7353	0.0003	0.8675	2717.11	9015.91	4.33
Ens_DT	0.8472	0.8961	0.9993	0.7353	0.0003	0.8675	2717.02	9015.59	4.08
Ens_SVM	0.8472	0.8961	0.9993	0.7353	0.0003	0.8675	2733.31	9031.70	8.55
Ens_OR	0.8439	0.8931	0.9992	0.7304	0.0003	–	2716.21	9014.71	3.73

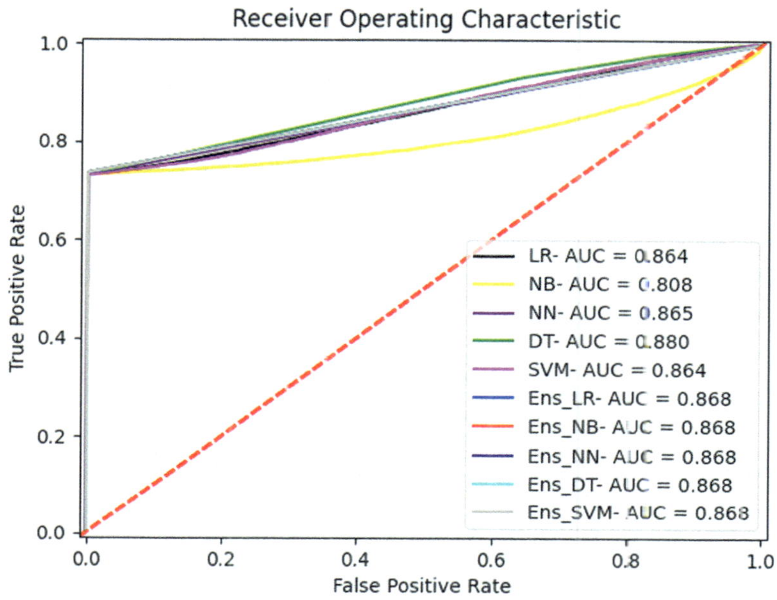

Fig. 5 ROC curves for the learning models. ROC curves predict probabilities for two-class problems

6 Conclusion

Stealthy false data injection attacks on the state estimation of a power grid can have severe consequences, and an early and accurate detection of these is critical to prevent economic loss or outages. In this chapter, we used both stand-alone supervised learning models and ensemble models for detecting stealthy FDI attacks. The ensembles are composed of five individual classifiers and seven ensemble classifiers. The scheme also includes random forest for dimension reduction. We implemented the scheme using the Python machine learning libraries and tested it using the standard IEEE 14-bus system simulated by MATPOWER. After training the models, we compared the performance of the individual and the ensemble classifiers. The test results demonstrate that the ensemble models do not perform any better than the individual classifiers. The models show 90% accuracy and 100% precision. However, these numbers may be misleading because we are dealing with imbalanced dataset. Looking into the recall and FPR numbers, we find that the models can detect about 73% of the attacks with very low false alerts.

Acknowledgments This research was partially supported by an Idaho Global Entrepreneurial Mission (IGEM) grant for Security Management of Cyber-Physical Control Systems, 2016 (Grant Number IGEM17-001).

References

1. S. McLaughlin, C. Konstantinou, X. Wang, L. Davi, A.-R. Sadeghi, M. Maniatakos, R. Karri, The cybersecurity landscape in industrial control systems. Proc. IEEE **104**(5), 1039–1057 (2016)
2. Y. Liu, P. Ning, M.K. Reiter, False data injection attacks against state estimation in electric power grids. ACM Trans. Inf. Syst. Secur. **14**(1), 13:1–13:33 (2011)
3. A. Abur, A. Gomez-Exposito, *Power System State Estimation: Theory and Implementation* (CRC Press, New York, 2004)
4. C. Alcaraz, J. Lopez, Wide-area situational awareness for critical infrastructure protection. Computer **46**(4), 30–37 (2013)
5. Y. Xiang, L. Wang, N. Liu, Coordinated attacks on electric power systems in a cyber-physical environment. Electr. Power Syst. Res. **149**, 156–168 (2017)
6. X. Liu, Z. Li, False data attack models, impact analyses and defense strategies in the electricity grid. Electr. J. **30**, 35–42 (2017)
7. R. Polikar, Ensemble learning in *Ensemble Machine Learning* (Springer, Berlin, 2012), pp. 1–34
8. N. Moustafa, B. Turnbull, K.-K.R. Choo, An ensemble intrusion detection technique based on proposed statistical flow features for protecting network traffic of internet of things. IEEE Internet Things J. **6**(3), 4815–4830 (2018)
9. X. Zhang, Z. Zhao, Y. Zheng, J. Li, Prediction of taxi destinations using a novel data embedding method and ensemble learning. IEEE Trans. Intell. Transp. Syst. **21**(1), 68–78 (2019)
10. S. Das, A.M. Mahfouz, D. Venugopal, S. Shiva, DDoS intrusion detection through machine learning ensemble, in *2019 IEEE 19th International Conference on Software Quality, Reliability and Security Companion (QRS-C)* (IEEE, Piscataway, 2019), pp. 471–477
11. R.D. Zimmerman, C.E. Murillo-Sánchez, R.J. Thomas, MATPOWER: steady-state operations, planning, and analysis tools for power systems research and education. IEEE Trans. Power Syst. **26**(1), 12–19 (2011)
12. M. Esmalifalak, L. Liu, N. Nguyen, R. Zheng, Z. Han, Detecting stealthy false data injection using machine learning in smart grid. IEEE Syst. J. **11**(3), 1644–1652 (2014)
13. M. Ozay, I. Esnaola, F.T.Y. Vural, S.R. Kulkarni, H.V. Poor, Machine learning methods for attack detection in the smart grid. IEEE Trans. Neural Netw. Learn. Syst. **27**(8), 1773–1786 (2015)
14. Y. He, G.J. Mendis, J. Wei, Real-time detection of false data injection attacks in smart grid: a deep learning-based intelligent mechanism. IEEE Trans. Smart Grid **8**(5), 2505–2516 (2017)
15. Y. Wang, M. Amin, J. Fu, H. Moussa, A novel data analytical approach for false data injection cyber-physical attack mitigation in smart grids. IEEE Access **5**, 26022–26033 (2017)
16. J. Wang, W. Tu, L.C. Hui, S.-M. Yiu, E.K. Wang, Detecting time synchronization attacks in cyber-physical systems with machine learning techniques, in *2017 IEEE 37th International Conference on Distributed Computing Systems (ICDCS)* (IEEE, Piscataway, 2017), pp. 2246–2251
17. M. Ashrafuzzaman, Y. Chakhchoukh, A. Jillepalli, P. Tosic, D. Conte de Leon, F. Sheldon, B. Johnson, Detecting stealthy false data injection attacks in power grids using deep learning, in *Wireless Communications and Mobile Computing Conference (IWCMC), 14th International* (IEEE, Piscataway, 2018), pp. 219–225
18. S. Ahmed, Y. Lee, S.-H. Hyun, I. Koo, Covert cyber assault detection in smart grid networks utilizing feature selection and Euclidean distance-based machine learning. Appl. Sci. **8**(5), 772–792 (2018)
19. X. Niu, J. Li, J. Sun, K. Tomsovic, Dynamic detection of false data injection attack in smart grid using deep learning, in *2019 IEEE Power & Energy Society Innovative Smart Grid Technologies Conference (ISGT)* (IEEE, Piscataway, 2019), pp. 1–6

20. H. Wang, J. Ruan, G. Wang, B. Zhou, Y. Liu, X. Fu, J.-C. Peng, Deep learning based interval state estimation of AC smart grids against sparse cyber attacks. IEEE Trans. Industr. Inf. **14**(11), 4766–4778 (2018)
21. M.R. Camana-Acosta, S. Ahmed, C.E. Garcia, I. Koo, Extremely randomized trees-based scheme for stealthy cyber-attack detection in smart grid networks. IEEE Access **8**, 19921–19933 (2020)
22. M. Mohammadpourfard, Y. Weng, M. Pechenizkiy, M. Tajdinian, B. Mohammadi-Ivatloo, Ensuring cybersecurity of smart grid against data integrity attacks under concept drift. Int. J. Electr. Power Energy Syst. **119**, 105947 (2020)
23. S. Ahmed, Y. Lee, S.-H. Hyun, I. Koo, Unsupervised machine learning-based detection of covert data integrity assault in smart grid networks utilizing isolation forest. IEEE Trans. Inf. Forensics Secur. **14**(10), 2765–2777 (2019)
24. J. Hao, R.J. Piechocki, D. Kaleshi, W.H. Chin, Z. Fan, Sparse malicious false data injection attacks and defense mechanisms in smart grids. IEEE Trans. Industr. Inf. **11**(5), 1–12 (2015)
25. Y. Chakhchoukh, S. Liu, M. Sugiyama, H. Ishii, Statistical outlier detection for diagnosis of cyber attacks in power state estimation, in *2016 IEEE Power and Energy Society General Meeting (PESGM)* (IEEE, Piscataway, 2016), pp. 1–5
26. M.N. Kurt, O. Ogundijo, C. Li, X. Wang, Online cyber-attack detection in smart grid: a reinforcement learning approach. IEEE Trans. Smart Grid **10**(5), 5174–5185 (2018)
27. M.S. Thomas, J.D. McDonald, *Power System SCADA and Smart Grids* (CRC Press, Boca Raton, 2015)
28. University of Washington, *Power System Test Case Archive* (2018). http://www.ee.washington.edu/research/pstca/
29. G.E. Batista, R.C. Prati, M.C. Monard, A study of the behavior of several methods for balancing machine learning training data. ACM SIGKDD Explor. Newslett. **6**(1), 20–29 (2004)
30. M. Sokolova, G. Lapalme, A systematic analysis of performance measures for classification tasks. Inf. Process. Manage. **45**(4), 427–437 (2009)

Vulnerability Analysis of 2500 Docker Hub Images

Katrine Wist, Malene Helsem, and Danilo Gligoroski

1 Introduction

Container technology has been known for a long time in Linux systems through Linux Containers (LXC), but it was not commonly used until a decade ago. The introduction of Docker in [1] made the popularity of containerization rise exponentially. Container technology has revolutionized how software is developed and is seen as a paradigm shift. More concretely, containerization is considered as a beneficial technique for Continuous Integration/Continuous Delivery (CI/CD) pipelines; it is providing an effective way of organizing microservices; it is making it easy to move an application between different environments; and in general, it is simplifying the whole system development life cycle.

Software containers got its name from the shipping industry since the concepts are fundamentally the same. A software container is code wrapped up with all its dependencies so that the code can run reliably and seamlessly in any computer environment isolated from other processes. Hence, containers are convenient, lightweight, and fast technology to achieve isolation, portability, and scalability.

Container technology is replacing virtual machines continuously, and the trend is that more companies are choosing to containerize their applications. Gartner predicts that more than 70% of global companies will have more than two containerized applications in production by 2023. This is an increase from less than

K. Wist · M. Helsem · D. Gligoroski (✉)
Department of Information Security and Communication Technology, Norwegian University of Science and Technology (NTNU), Trondheim, Norway
e-mail: danilog@ntnu.no

K. Daimi et al. (eds.), *Advances in Security, Networks, and Internet of Things*, Transactions on Computational Science and Computational Intelligence,
https://doi.org/10.1007/978-3-030-71017-0_22

20% in 2019.[1] With the advent of 5G communication technology, it seems that container technology, and particularly Docker, is finding new venue for application in the domain of network slicing, network management, orchestration, and in 5G testbeds [2].

Docker provides a popular registry service for the sharing of Docker images, called Docker Hub.[2] It currently hosts over 3.5 million container images, and the number keeps growing. Images could be uploaded and maintained by anyone, which creates an innovative environment for anyone to contribute and participate. However, on the downside, this makes it hard for Docker to ensure that packages and applications are up to date to avoid outdated and vulnerable software.

When looking at the security of Docker, two aspects need to be considered: the security of the Docker software at the host and the security of the Docker containers. Docker Inc. claims that "Docker containers are, by default, quite secure; especially if you run your processes as non-privileged users inside the container" [3]. However, it is a simple fact that Docker (the Docker daemon and container processes) runs with root privileges by default, which exposes a huge attack surface [4]. A single vulnerable container is enough for an adversary to achieve privilege escalation. Hence, the security of the whole Docker ecosystem is highly related to the vulnerability landscape in Docker images.

Related Work One of the first to explore the vulnerability landscape of Docker Hub was BanyanOps [5]. In 2015, they published a technical report revealing that 36% of official images on Docker Hub contained high priority vulnerabilities [5]. Further, they discovered that this number increases to 40% when community images (or general images as they call it in the report) are analyzed. BanyanOps built their own vulnerability scanner based on Common Vulnerabilities and Exposures (CVE) scores and analyzed all official images ($\approx$75 repositories with $\approx$960 unique images) and some randomly chosen community images. However, at that time, Docker Hub consisted of just $\approx$95,000 images.

In 2017, Shu et al. conducted a new vulnerability analysis of Docker Hub images [6]. With the aim of revealing the Docker Hub vulnerability landscape, they created their own analysis framework called DIVA (Docker image vulnerability analysis). The DIVA framework discovers, downloads, and analyses official and community images. It is based on the Clair scanner and uses random search strings to discover images on Docker Hub. The analysis revealed that, on average, an image (official and community) contains more than 180 vulnerabilities. They also found that many images had not been updated for hundreds of days, which is problematic from a security point of view. Further, it was observed that vulnerabilities propagate from parent to child images.

To our knowledge, the most recent vulnerability analysis of Docker Hub images was performed during Spring 2019 by Socchi and Luu [7]. They investigated

[1]Gartner: 3 Critical Mistakes That I&O Leaders Must Avoid With Containers.

[2]Docker Hub webpage: *https://hub.docker.com/*.

whether the security measures introduced by Docker Inc. (more precisely, the introduction of verified and certified image types) improved the security of Docker Hub. In addition, they inspected the distribution of vulnerabilities across repository types and whether vulnerabilities still are inherited from parent to child image. They implemented their own analyzing software using the Clair scanner and used the results from Shu et al. [6] from 2017 as a comparison. The data set they successfully analyzed consisted of 757 images in total. Of these, 128 were official, 500 were community, 98 were verified, and 31 were certified. They only analyzed the most recent images in each repository and skipped all Microsoft repositories. Their conclusion was that the security measures introduced by Docker Inc. do not improve the overall Docker Hub security. They stated that the number of inherited vulnerabilities had dropped since the analysis of Shu et al. However, they also found that the average number of new vulnerabilities in child images had increased significantly. Further, they found that the majority of official, community, and certified repositories contain up to 75 vulnerabilities and that the majority of verified images contain up to 180 vulnerabilities.

Our Contribution This is an extended summary of our longer and much more detailed work [8]. We scrutinized the vulnerability landscape in Docker Hub images at the beginning of 2020 within the following framework:

- Images on Docker Hub belong to one of the following four types: "official," "verified," "certified," or "community."
- We used a quantitative mapping of the Common Vulnerability Scoring System (CVSS) [9] (which is a numerical score indicating the severity of the vulnerability in a scale from 0.0 to 10.0) into five qualitative severity rating levels: "critical," "high," "medium," "low," or "none" plus one additional level "unknown."

For performing the analysis of a significant number of images, we used an open-source vulnerability scanner tool and developed our own scripts and tools. All our developed scripts and tools are available from [8] and from the GitHub repository.[3]

Our findings can be summarized as follows: (1) The median value (when omitting the negligible and unknown vulnerabilities) is 26 vulnerabilities per image. (2) Most of the vulnerabilities were found in the medium severity category. (3) Around 17.8% (430 images) do not contain any vulnerabilities, and if we are considering negligible and unknown vulnerabilities as no vulnerability, the number increase to as many as 21.6% (523 images). (4) As intuitively expected, when considering the average, community images are the most exposed. We found that 8 out of the top 10 most vulnerable images are community images. (5) However, to our surprise, the certified images are the most vulnerable when considering the median value. They had the most high rated vulnerabilities as well as the most vulnerabilities rated as low. As many as 82% of certified images contain at least either one high or critical vulnerability. (6) Official images come out as the most secure

[3]https://github.com/katrinewi/Docker-image-analyzing-tools.

Table 1 A summary comparison table of results reported in 2015[5], in 2017 [6] in 2019[7], and in our work (2020). The sub-columns "vuln" contain the percentage of images with at least one high rated vulnerability and the "avg" sub-columns contain the average number of vulnerabilities found in each image type

Image type	2015		2017		2019		**2020**	
	vuln	avg	vuln	avg	vuln	avg	**vuln**	**avg**
Official	36%	–	80%	75	–	170	**46%**	**70**
Community	40%	–	80%	180	–	150	**68%**	**150**
Verified	–	–	–	–	–	150	**57%**	**90**
Certified	–	–	–	–	–	30	**82%**	**90**

image type. Around 45.9% of them contain at least one critical or high rated vulnerability. (7) The median value of the number of critical vulnerabilities in images is almost identical for all four image types. (8) Verified and official images are the most updated, and community and certified images are the least updated. Approximately 30% of images have not been updated for the last 400 days. (9) There is no correlation between the number of vulnerabilities and the evaluated image features (i.e., the number of pulls, the number of stars, and the last update time). However, the images with many vulnerabilities generally have few pulls and stars. (10) Vulnerabilities in the Lodash library and vulnerabilities in Python packages are the most frequent and most severe. The top five most severe vulnerabilities are coming from two of the most popular scripting languages, JavaScript and Python. (11) Vulnerabilities related to execution of code and overflow are the most frequently found critical vulnerabilities. (12) The most vulnerable package is the `jackson-databind-2.4.0` package, with overwhelming 710 critical vulnerabilities, followed by `Python-2.7.5` with 520 critical vulnerabilities.

Last but not least, when put in comparison with the three previous similar studies [5–7], our results are summarized in Table 1. Note that some of the cells are empty due to differences in methodologies and types of images when the studies were performed.

2 Preliminaries

Virtualization is the technique of creating a virtual abstraction of some resources to make multiple instances run isolated from each other on the same hardware [10]. There are different approaches to achieve virtualization. One approach is using virtual machines (VMs). A VM is a virtualization of the hardware at the host. Hence, each VM has its own kernel, and in order to manage the different VMs, a software called hypervisor is required. The hypervisor emulates the central processing unit (CPU), storage, and random-access memory (RAM), among others, for each virtual machine. This allows multiple virtual machines to run as separate machines on a single physical machine.

In contrast to VMs, containers virtualize the operating system (OS) level. Every container running on the same machine shares the same underlying kernel, where only bins, libraries, and other run time components are executed exclusively for a single container. In short, a container is a standardized unit of software that contains all code and dependencies [11]. Thus, containers require less memory and achieve a higher level of portability than VMs. Container technology has simplified the software development process as the code is portable, and hence what is run in the development department will be the same as what is run in the production department [12].

On the Docker Hub, image repositories are divided into different categories. Repositories are either private or public and could further be either *official*, *community* or a *verified* repository. In addition, repositories could be certified, which is a subsection of the verified category. The official repositories are maintained and vetted by Docker. Docker vets the verified ones that are developed by third-party developers. Besides being verified, certified images are also fulfilling some other requirements related to quality, support, and best practices [13]. Community images could be uploaded and maintained by anyone. The distribution of the image repository types on Docker Hub can be seen in Table 2. The community repository category is by far the most dominant one and makes up to ≈99% of all Docker Hub repositories.

2.1 Vulnerability Databases and Categorization Method

The severity of vulnerabilities depends on a variety of different variables, and it is highly complex to compare them due to the diversity of different technologies and solutions. Already in 1997, the National Vulnerability Database (NVD) started working on a database that would contain publicly known software vulnerabilities to provide a means of understanding future trends and current patterns [14]. The database can be useful in the field of security management when deciding what software is safe to use and for predicting whether or not software contains vulnerabilities that have not yet been discovered.

Common Vulnerabilities and Exposures (CVE) National Vulnerability Database (NVD) contains Common Vulnerabilities and Exposures (CVE) entries and provides details about each vulnerability like vulnerability overview, Common Vulnerability

Table 2 Repository type distribution on Docker Hub (February 3rd, 2020)

Repository type	Quantity
Official	160
Verified	250
Certified	51
Community	3,064,454
Total	3,064,915

Scoring System (CVSS), references, Common Platform Enumeration (CPE), and Common Weakness Enumeration (CWE) [15].

CVE is widely used as a method for referencing security vulnerabilities that are publicly known in released software packages. At the time of writing, there were 130,094 entries in the CVE list.[4] The CVE list was created by MITRE Corporation[5] in 1999, whose role is to manage and maintain the list. They work as a neutral and unbiased part in order to serve in the interest of the public. Examples of vulnerabilities found in CVE are frequent errors, faults, flaws, and loopholes that can be exploited by a malicious user in order to get unauthorized access to a system or server. The loopholes can also be used as propagation channels for viruses and worms that contain malicious software [16]. Over the years, CVE has become a recognized building block for various vulnerability analysis and security information exchange systems, much because it is continuously maintained and updated, and because the information is stored with accurate enumeration and orderly naming.

Common Vulnerability Scoring System (CVSS) The Common Vulnerability Scoring System (CVSS) score is a numerical score indicating the severity of the vulnerability on a scale from zero to 10, based on a variety of metrics. The metrics are divided into three metric groups: Base Metric Group, Temporal Metric Group, and Environmental Metric Group. A *Base Score* is calculated by the metrics in the Base Metric Group and is independent of the user environment and does not change over time. The Temporal Metrics take in the base score and adjusts it according to factors that do change over time, such as the availability of exploit code [9]. Environmental Metrics adjust the score yet again, based on the type of computing environment. This allows organizations to adjust the score related to their IT assets, taking into account existing mitigations and security measures that are already in place in the organization.

In our analysis, it would not make sense to take into account the Temporal or Environmental Metrics as we wanted to discuss the vulnerability landscape independently of the exact time and environment. Therefore, only the Base Metric Group will be described in more detail. It is composed of two sets of metrics: the Exploitability metrics and the Impact metrics, as can be seen in Fig. 1 [9]. The first set takes into account *how* the vulnerable component can be exploited and includes attack vector and complexity, what privileges are required to perform the attack, and whether or not user interaction is required. The latter set reflects on the *consequence* of a successful exploit and what impact it has on the confidentiality, integrity, and availability of the system. The last metric is *scope*, which considers if the vulnerability can propagate outside the current security scope.

---

[4]The number of entries in the CVE list was retrieved on 28 Jan 2020 from the official website: https://cve.mitre.org.

[5]MITRE Corporation is a non-profit US organization with the vision to resolve problems for a safer world: https://www.mitre.org.

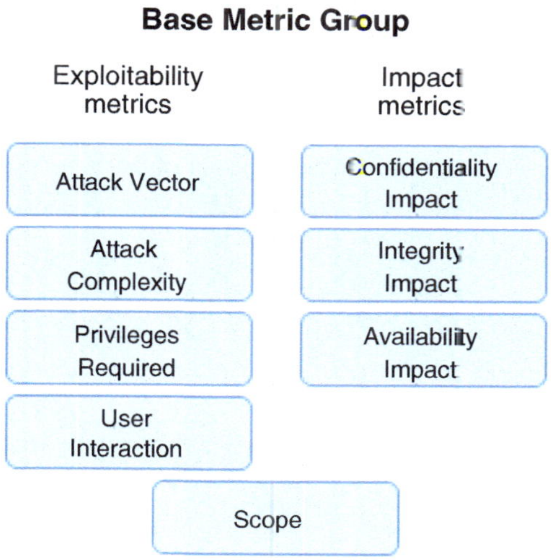

Fig. 1 Common Vulnerability Scoring System structure [9]

Table 3 CVSS severity scores

Rating	CVSS score
None	0.0
Low	0.1–3.9
Medium	4.0–6.9
High	7.0–8.9
Critical	9.0–10.0

When the Base Score of a vulnerability is calculated, the eight different metrics from Fig. 1 are being considered. Each metric is assigned one out of two to four different values, which is used to generate a vector string. The vector string is then used to calculate the Common Vulnerability Scoring System (CVSS) score, which is a numerical value between 0 and 10. In many cases, it is more beneficial to have a textual value than a numerical value. The CVSS score can be mapped to qualitative ratings where the severity is categorized as either critical, high, medium, low, or none, as can be seen in Table 3 [9].

3 Docker Hub Vulnerability Landscape

3.1 The Distribution of Vulnerabilities in Each Severity Category

To determine what the current vulnerability landscape is like in Docker Hub, the number of vulnerabilities found in each severity category is presented in Fig. 2. As it is interesting to see how many vulnerabilities that are found in total (Fig. 2a) and

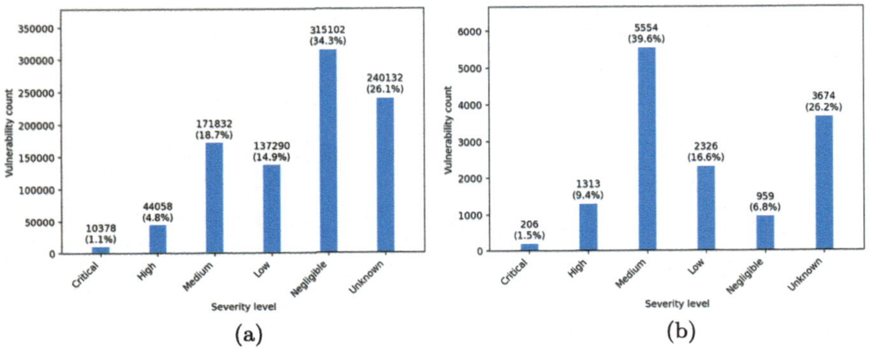

Fig. 2 Vulnerability distribution in severity levels. (**a**) Distribution of all 918,792 vulnerabilities. (**b**) Distribution of 14,032 unique vulnerabilities

how many unique vulnerabilities (Fig. 2b) there are, both these results are presented in this section.

In Fig. 2a, the results are based on vulnerability scanning of the complete data set, meaning that this result is based on all found vulnerabilities. The same vulnerability could potentially have multiple entries in the result. This is because a particular vulnerability could be found in multiple images and a single image could contain the same vulnerability in multiple packages. In Fig. 2b, only unique vulnerabilities are shown. However, some vulnerabilities are present in several severity categories, depending on which image it is found in. In cases like this, all versions of the vulnerability are included, which makes up a total of 14,031 vulnerabilities.

In Fig. 2a, the negligible and unknown categories clearly stand out, with a total of 315,102 and 240,132 vulnerabilities, respectively. When considering unique vulnerabilities (Fig. 2b), the medium category is the most dominant one with 5554 unique vulnerabilities. When examining the relation between Fig. 2a and b, one can observe the ratio of vulnerabilities between severity categories. It becomes clear that the negligible category contains a few number of unique vulnerabilities represented in many Docker images, whereas the medium category has many unique vulnerabilities represented at a lower ratio. The vulnerability ratio will be explained in detail in the next paragraph.

Table 4 shows the total number of vulnerabilities, the number of unique vulnerabilities, and the ratio measured as the total number of vulnerabilities divided by the number of unique vulnerabilities. So, for each unique vulnerability, there are a certain number of occurrences of the specific vulnerability in the data set. For example, for each unique vulnerability in the critical category, there are 50 occurrences of this vulnerability in the data set on average. For each unique negligible vulnerability, there are as many as 329 occurrences on average. This is significantly larger than the other values. Despite medium having the highest number of unique vulnerabilities, it has the lowest ratio.

Table 4 Vulnerability frequency in severity levels

Severity	Number of vulnerabilities (A)	Number of unique vulnerabilities (B)	Ratio (A/B)
Critical	10,378	206	50
High	44,058	1313	34
Medium	171,832	5554	31
Low	137,290	2326	59
Negligible	315,102	959	329
Unknown	240,132	3674	65
Total	918,792	14,031	65

Table 5 Statistical values for vulnerabilities per image type, disregarding negligible and unknown vulnerabilities

Image type	Number of analyzed images	Number of vulnerabilities	Average	Median	Max
Verified	60	6073	101.2	13	1128
Certified	22	1987	90.3	37	428
Official	157	11,489	73.2	9	1615
Community	2173	344,009	158.3	28	6509

3.2 Central Tendency of the Vulnerability Distribution

We have looked at the average and median values of the number of vulnerabilities in images when disregarding the vulnerabilities that are categorized as negligible and unknown. Looking at Table 4 from the previous section, one can see that negligible and unknown vulnerabilities together make up 555,234 out of the 918,792 vulnerabilities (around 60%). As vulnerabilities in these two categories are considered to contribute with little threat when investigating the current vulnerability landscape, it gives a more accurate result to exclude these. Therefore, we calculated the average and median number of vulnerabilities in images when disregarding negligible and unknown vulnerabilities (counting them as zero). The result was 151 for the average and 26 for the median.

To investigate the data when disregarding the negligible and unknown vulnerabilities further, we created Table 5 that shows statistical values of the number of vulnerabilities for each image type. The results show that community images have the highest average and maximum values (158 and 6509, respectively). The maximum value for community images is significantly larger than the average and the median, which is the case for the other three image types as well. The image type that is considered as the least vulnerable is official. It has the lowest average of 73 and the lowest median value of 9. Further, the maximum value for official images is the second lowest. The lowest maximum value belongs to certified and is only 428. Although certified has the lowest maximum value, it has the highest median value. This indicates that a larger portion of the images have many vulnerabilities. As a final note, all four image types contain at least one image with zero vulnerabilities.

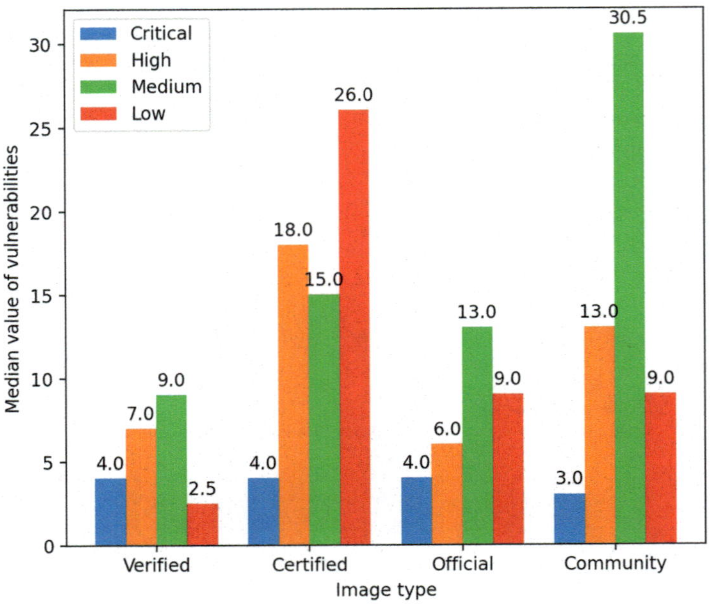

Fig. 3 Median values of vulnerabilities for each severity category and image type

3.3 Vulnerabilities in Each Image Type

Since the median describes the central tendency better than the average, when the data is skewed here we will work with the median values (given in Fig. 3). Note that only critical, high, medium, and low vulnerabilities are included in the figure. The negligible and unknown vulnerabilities are not included here because they do not usually pose as significant threats and therefore do not contribute with additional information when investigating the current vulnerability landscape.

The results show that the median of critical vulnerabilities is almost the same for all four image types (4.0 and 3.0). The other severity categories are more varied across the image types. The high severity category is the most represented in certified images, while the medium category is the most represented in the community images. For verified, official, and community images, the medium severity has the highest median, while the certified images have the most low vulnerabilities. Overall, it is the certified images that are the most vulnerable.

3.4 Images That Contain the Most Critical Vulnerabilities

Out of all 2412 successfully analyzed images, this section will present the most vulnerable ones. Table 6 displays the most vulnerable images based on the number

Table 6 The most vulnerable images sorted by critical count

	Image	Critical	High	Medium	Low	Number of pulls
1	pivotaldata/gpdb-pxf-dev	822	698	576	132	139,246,839
2	cloudera/quickstart	571	2155	1897	158	6,892,856
3	silverpeas	341	264	397	226	828,743
4	microsoft-mmlspark-release	184	428	264	252	1,509,541
5	anchorfree/hadoop-slave	168	636	797	107	5,375,424
6	saturnism/spring-boot-helloworld-ui	133	217	112	2	12,686,987
7	pantsel/konga	133	39	169	0	12,431,685
8	renaultdigital/runner-bigdata-int	127	335	691	103	4,787,745
9	springcloud/spring-pipeline-m2	125	293	2027	1357	8,359,973
10	raphacps/simpsons-maven-repo	122	271	399	2	36,136,733

of critical vulnerabilities in each image. In cases where the critical count is the same, the image with the highest number of high rated vulnerabilities is considered as the most vulnerable one. The *number of pulls* column denotes the total number of pulls (downloads) for each image. Out of the top 10 most vulnerable images, there are 8 community images, 1 official image (silverpeas), and 1 verified image (microsoft-mmlspark-release). There are big variations in the number of vulnerabilities in all presented severity levels. The most vulnerable image, *pivotaldata/gpdb-pxf-dev*, has ≈250 more critical vulnerabilities than the second most vulnerable image. However, the second most vulnerable image, *cloudera/quickstart*, contains as many as 2155 high rated vulnerabilities, which is ≈1500 more vulnerabilities than the one rated as the most vulnerable image. It was chosen to focus on the critical vulnerabilities in the ranking of the most vulnerable images. This is because it is the highest possible ranking, and hence the most severe vulnerabilities will be found in this category. The other severity categories are included in the table as extra information and to give a clear view on the distribution of vulnerabilities. From the number of pulls column, one can observe that the most vulnerable image is also the most downloaded one out of the top 10, with as many as 139,246,839 pulls. This is approximately 100 million more pulls compared to the second most pulled image on this list (the *raphacps/simpsons-maven-repo* image). There is no immediate correlation that could be observed between the number of pulls and the number of vulnerabilities in these images.

3.5 *Percentage of Images with Critical and High Vulnerabilities*

It is enough with a single vulnerability for a system to be compromised. Thus, we determine what percentage of images that contain at least one high or critical rated vulnerability for each image type, as shown in Fig. 4.

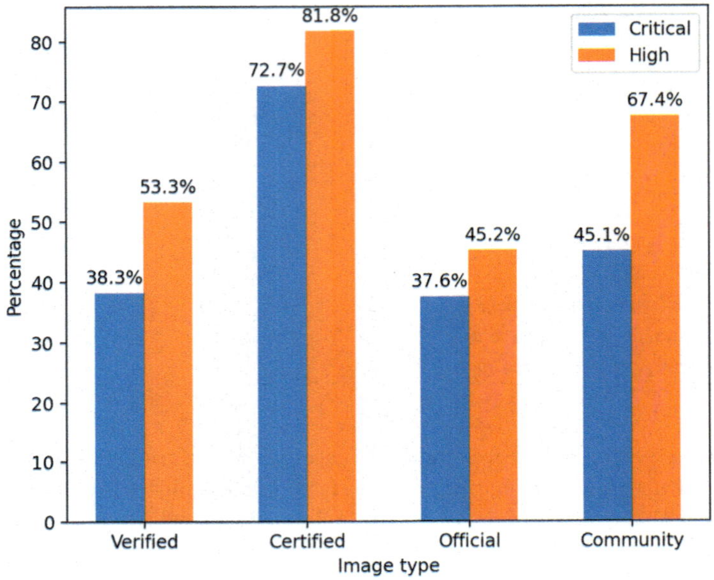

Fig. 4 The percentage of images that contain at least one high or critical rated vulnerability

Our results (Fig. 4) reveal that the certified image type, which is a subsection of the verified image type, is the most vulnerable by the means of this measure. 81.8% of all certified images contain at least one vulnerability with high severity level and 72.7% of them contain at least one critical vulnerability. Community images come out as the second most vulnerable image type. 67.4% have high vulnerabilities and 45.1% have critical vulnerabilities. The third most vulnerable image type is verified, followed by official.

When combining these results, to investigate what amount of the image types that contain *either* at least one critical or high rated vulnerability, the results are as follows: 81.8% for certified images, 68.4% for community images, 56.7% for verified images, and 45.9% for official images. This makes the official images the least vulnerable image type. However, it should be emphasized that still almost half of the official images contain critical or high rated vulnerabilities as presented in this section.

3.6 *The Trend in CVE Vulnerabilities*

This section will focus on the trend of all reported Common Vulnerabilities and Exposures (CVE) vulnerabilities each year compared to the number of unique CVE vulnerabilities found throughout our analysis. Data gathered from the CVE Details

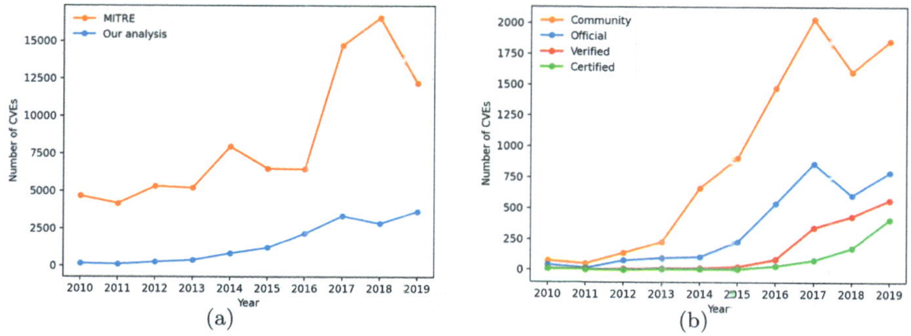

Fig. 5 CVE trend from 2010 to 2019: part (**a**) displays all reported CVEs and all found, unique CVEs in our analysis and part (**b**) displays the CVEs in the different image types from our analysis

database is used to display the number of new reported Common Vulnerabilities and Exposures (CVE) vulnerabilities each year.

In Fig. 5a, the reported Common Vulnerabilities and Exposures (CVE) vulnerabilities each year are presented together with the unique CVE vulnerabilities found in our analysis from 2010 to 2019. The orange line shows how the number of new discovered CVE vulnerabilities varies by a few thousand vulnerabilities each year. However, there is a significant increase in 2017. This increase is not reflected in the data from our analysis, which is following a steady increase in the years from 2014 to 2017. This increase can be explained by the introduction of Docker Hub in 2014, making new vulnerabilities more represented in images. As a final observation, the number of new reported vulnerabilities from MITRE between 2018 and 2019 is decreasing, while there is an increase in our results.

Figure 5b shows the number of unique vulnerabilities found in each image type (i.e., community, official, verified, and certified) in our analysis from 2010 to 2019. This figure gives an insight in how the overall changes are reflected in each image type. Verified and certified images have had an increase in the number of unique Common Vulnerabilities and Exposures (CVE) vulnerabilities each year from 2015. Community and official images, however, have had a significant decrease of unique vulnerabilities from 2017 to 2018. It is noteworthy to point out that the curves are affected by the time of introduction of the different image types. Official images were introduced in 2014, whereas verified and certified images were introduced in 2018.

3.7 *Days Since Last Update*

There is a high variation in how often Docker Hub images are updated. Intuitively, this affects the vulnerability landscape of Docker Hub. Hence, we have gathered

Table 7 The time since last update for all image types presented in percentage

Image type	More than 400 days	More than 200 days	Less than 14 days
Community	33.9%	47.0%	27.0%
Official	9.6%	14.7%	51.3%
Certified	18.2%	36.4%	13.6%
Verified	1.7%	5.0%	83.3%

data about when images were last updated, and calculated the number of days since the images were last updated, counting from February 25th, 2020.

A brief analysis of the numbers from our database revealed that 31.4% of images have not been updated in 400 days or longer and 43.8% have not been updated in 200 days or longer. The percentage of images that have been updated during the last 14 days is 29.8%. This implies that if these numbers are representative for all images on Docker Hub, a third of the images (31.4%) on Docker Hub have not been updated longer than 400 days.

To go into more detail, Table 7 presents how often images in each of the image types are updated. Community and certified images are the least updated image categories, where 47.0% of community images and 36.4% of certified images have not been updated for the last 200 days or more. The verified images are the most frequently updated category, where 83.3% of images have been updated during the last 14 days.

A handful of certified images are highly affecting the percentages from Table 7, because the overall number of certified images is small. Official images contain a high portion of images that have been updated recently (January 2020 to March 2020), and some more spread values with images that have not been updated since 2016. The verified images are the most updated image type, where there is only one image with the last updated time earlier than May 2019.

4 Correlation Between Image Features and Vulnerabilities

We investigate whether or not the number of vulnerabilities in an image is affected by a specific image feature, such as the number of times the image has been pulled, the number of stars an image has been given, or the number of days since the image was last updated. In order to find out whether there is a correlation, we used Spearman's r_s correlation coefficient [17]. Spearman's correlation was chosen because our data set contains skewed values and is not normally distributed. When handling entries that contained empty values, we opted for the approach of complete case analysis, which means omitting incomplete pairs. The alternative would be imputation of missing values, which means to create an estimated value based on the other data values. However, this approach was not chosen because the values of our data set are independent of each other.

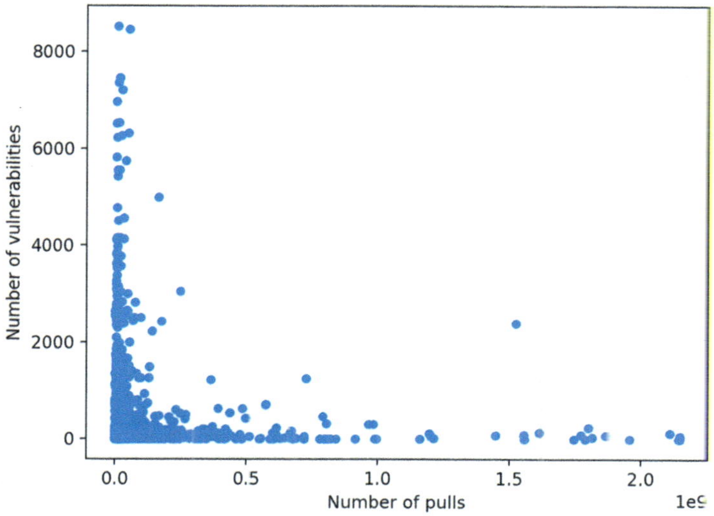

Fig. 6 Number of pulls and number of vulnerabilities for each image

Correlation Between Pulls and Vulnerabilities To check the folklore wisdom about the following correlation: *images with the most pulls generally have few vulnerabilities, and images with the most vulnerabilities generally have few pulls*, we created a scatter plot given in Fig. 6. However, after calculating the Spearman's correlation coefficient between the number of pulls and number of vulnerabilities for the whole set of investigated images we got $r_s = -0.1115$. This is considered as no particular correlation. To explain this, we refer to the meaning of having a high negative correlation: the markers would gather around a decreasing line (not necessarily linear), indicating that images with more pulls have less number of vulnerabilities. In the case of high positive correlation, the opposite would apply, i.e., the line would be increasing.

Correlation Between Stars and Vulnerabilities The correlation coefficient between the number of stars and number of vulnerabilities is $r_s = -0.0335$. The plot was similar to Fig. 6, but the correlation was even weaker.

Correlation Between Time Since Last Update and Vulnerabilities This correlation is calculated by computing the number of days since the last update counting from the day we gathered the data (which was February 25, 2020). The correlation was $r_s = 0.1075$, which shows a positive correlation as opposed to the other two. Figure 7 shows the scatter plot, and although the markers are approaching an increasing line a tiny bit, this is minimal. The value of 0.1075 is still not enough to state that there is a strong correlation between the number of vulnerabilities and time since the last update. The markers slightly approach an increasing line, indicating a weak tendency that there are more vulnerabilities in images that have not been updated for a long time. Still, the distribution of markers is relatively even along the

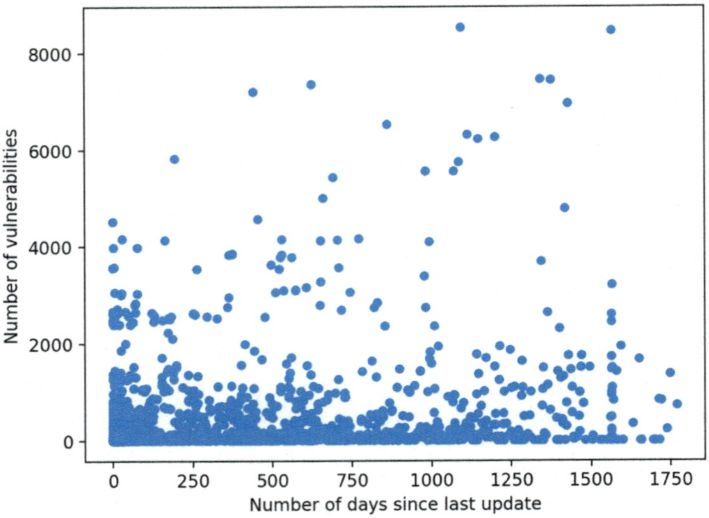

Fig. 7 Number of days since last update and number of vulnerabilities for each image

x-axis with the most markers in the lower part of the y-axis, supporting that there is no correlation.

5 The Most Severe Vulnerabilities

5.1 The Most Represented Critical Vulnerabilities

The most represented severe vulnerabilities are, intuitively, the ones having the highest impact on the vulnerability landscape. Table 8 presents the most represented critical rated vulnerabilities in descending order. The results are obtained by counting the number of occurrences for each vulnerability ID in the critical severity level. The critical count column is the number of occurrences for a specific vulnerability. Lastly, the type(s) column presents the vulnerability type of each of the vulnerabilities. This data is gathered from the CVE Details database [18].

5.2 Vulnerability Characteristics

We elaborate the top five most represented vulnerabilities presented in Table 8 regarding their characteristics and common features.The top five severe vulnerabilities are coming from two most popular script languages: JavaScript and Python.

Table 8 The most represented vulnerabilities (based on critical severity level)

	Vulnerability ID	Critical count	Type(s)
1	CVE-2019-10744	466	Improper input validation
2	CVE-2017-1000158	464	Execute code, overflow
3	CVE-2019-9948	378	Bypass a restriction or similar
4	CVE-2019-9636	374	Credentials management errors
5	CVE-2018-16487	365	Security features
6	CVE-2018-14718	354	Execute code
7	CVE-2018-11307	337	Deserialization of untrusted data
8	CVE-2018-7489	318	Execute code, bypass a restriction or similar
9	CVE-2016-5636	302	Overflow
10	CVE-2017-15095	295	Execute code

As a general observation, the execute code is the most common vulnerability type, followed by overflow.

The most represented critical vulnerability is found 466 times throughout our scanning. It has vulnerability ID *CVE-2019-10744*, and a base score of 9.8, which is in the upper range of the critical category (to examine how base scores are determined, see Sect. 2.1). The vulnerability is related to the JavaScript library lodash that is commonly used as a utility function provider in relation to functional programming. This particular vulnerability is related to improper input validation and makes the software vulnerable to prototype pollution. It is affecting versions of lodash lower than 4.17.12 [19]. In short, this means that it is possible for an adversary to execute arbitrary code by modifying the properties of the Object.prototype. This is possible as most JavaScript objects inherit the properties of the built-in Object.prototype object. The fifth vulnerability on the list, *CVE-2018-16487*, is also related to lodash and the prototype pollution vulnerability.

Further, the second, third, and fourth most represented critical vulnerabilities are related to Python vulnerabilities. The second vulnerability with vulnerability ID, *CVE-2017-1000158*, is related to versions of Python up to 2.7.13. The base score is rated 9.8, and the vulnerability enables arbitrary code execution to happen through an integer overflow leading to a heap-based buffer overflow [20]. Overflow vulnerabilities could be of different types, for instance, heap overflow, stack overflow, and integer overflow. Heap overflow and stack overflow are related to overflowing a buffer, whereas integer overflow could lead to a buffer overflow. A buffer overflow is related to overwriting a certain allocated buffer, causing adjacent memory locations to be overwritten. Any exploit of these kinds of vulnerabilities are typically related to the execution of arbitrary code, where the adversary is taking advantage of the buffer overflow vulnerability to run malicious code.

The third presented vulnerability with vulnerability ID *CVE-2019-9948* is affecting the Python module urllib in Python version 2.x up to 2.7.16. It is rated with 9.1 as base score. This vulnerability makes it easier to get around security mechanisms that blacklist the `file:URIs` syntax, which in turn could give an adversary access to local files such as the */etc/passwd* file [21]. The fourth vulnerability is found 374 times and has vulnerability ID *CVE-2019-9636*. It is affecting both the second and third versions of Python (versions 2.7.x up to 2.7.16, and 3.x up to 3.7.2). This vulnerability is also related to the urllib module, more precisely, incorrect handling of unicode encoding. The result is that information could be sent to different hosts than intended if it was parsed correctly [22]. It has a base score of 9.8.

6 Vulnerabilities in Packages

6.1 The Most Vulnerable Packages

Table 9 presents the packages that contain the most critical vulnerabilities. The critical count column is obtained by counting the total number of occurrences of critical vulnerabilities in each package, while the image count column is the number of images that uses each package.

There is a clear relation between the most vulnerable packages and the most represented vulnerabilities (Sect. 5), as expected. For example, vulnerabilities found in Python version 2.x packages and in the Lodash package are both presented in Sect. 5.

From Table 9, one can observe that the Python packages are by far the most used packages, and therefore they expose the biggest impact regarding the threat landscape. The lodash-3.10.1 package is found in 76 images. This package contains the prototype pollution vulnerability affecting JavaScript code, which is also the most represented vulnerability in Table 8. Further, the jackson-databind package is represented with four different versions in Table 9 (entry 1, 3, 8, and 9). This package is used to transform JSON objects to Java objects (Lists, Numbers, Strings, Booleans, etc.), and vice versa. In total, these packages are used by 44 images: a relatively low amount compared to the usage of the Python packages. Finally, the

Table 9 The most vulnerable packages (based on critical severity level)

	Package	Critical count	Image count
1	jackson-databind-2.4.0	710	15
2	Python-2.7.5	520	207
3	jackson-databind-2.9.4	354	4
4	lodash-3.10.1	312	76
5	silverpeas-6.0.2	280	1
6	Python-2.7.13	248	141
7	Python-2.7.16	224	117
8	jackson-databind-2.6.7.1	215	13
9	jackson-databind-2.9.6	192	12
10	Python-2.7.12	185	107

Table 10 Vulnerabilities in the most used packages

	Package	Critical	High	Medium	Low	Negligible	Unknown	Image count
1	tar-1.29b-1.1	0	0	0	0	482	0	241
2	coreutils-8.26-3	0	0	0	0	240	0	240
3	libpcre3-2:8.39-3	0	0	0	0	956	0	239
4	login-1:4.4-4.1	0	0	0	0	714	0	238
5	passwd-1:4.4-4.1	0	0	0	0	708	0	236
6	sensible-utils-0.0.9	0	0	103	0	0	111	214
7	libgcrypt20-1.7.6-2+deb9u3	0	0	0	0	211	0	211
8	libgssapi-krb5-2-1.15-1+deb9u1	0	0	0	0	621	0	207
9	libk5crypto3-1.15-1+deb9u1	0	0	0	0	621	0	207
10	libkrb5-3-1.15-1+deb9u1	0	0	0	0	621	0	207

silverpeas-6.0.2 package contains 280 critical vulnerabilities and is only used by a single image: the silverpeas image on Docker Hub.[6]

6.2 *Vulnerabilities in Popular Packages*

When considering the packages that have the most critical vulnerabilities (Table 9), some of the packages are only used by a few images (like the silverpeas package). Therefore, Table 10 is presented, as it is desirable to see what the vulnerability distribution is like in the most popular packages. The table shows the most used packages and the number of vulnerabilities that are present in them, considering all security levels. The image count column contains the number of images that use this package.

As observable from Table 10, the most used packages are not containing any critical, high, medium, or low vulnerabilities (except for one entry). However, they are containing a vast number of negligible vulnerabilities, which is of less significance from a security point of view, as mentioned in previous sections.

7 Conclusions and Future Work

This chapter summarizes the findings that we reported in a longer and much more detailed work [8]. We studied the vulnerability landscape in Docker Hub images

[6] *https://hub.docker.com/_/silverpeas.*

by analyzing 2500 Docker images of the four image repository categories: official, verified, certified images, and community. We found that as many as 82% of certified images contain at least one high or critical vulnerability and that they are the most vulnerable when considering the median value. Official images came out as the most secure image type with 45.9% of them containing at least one critical or high rated vulnerability. Only 17.8% of the images did not contain any vulnerabilities, and we found that the community images are the most exposed as 8 out of the top 10 most vulnerable images are community images.

Concerning the technical specifics about the vulnerabilities, we found that the top five most severe vulnerabilities are coming from two of the most popular scripting languages, JavaScript and Python. Vulnerabilities in the Lodash library and vulnerabilities in Python packages are the most frequent and most severe. Furthermore, the vulnerabilities related to execution of code and overflow are the most frequently found critical vulnerabilities. Our scripts and tools are available from [8] and from the GitHub repository.

For the future work, we first propose two improvements that are beyond our control and are mostly connected with the maintenance of all 3.5 million images at the Docker Hub website: (1) there is a need for a complete and well-documented endpoint for image data gathering and (2) there is a need for improvements on the Docker Hub webpages to make it possible to access all images through navigation.

Concerning improvements of this work, we consider a future analysis that will run over a more extended period. All previous studies conducted in this field, as well as ours, have only analyzed vulnerabilities in Docker Hub images captured from one single data gathering. Thus, changes in the data set over time are still not investigated. This type of analysis could reveal more in-depth details about the characteristics and evolution of the vulnerability landscape.

Lastly, we suggest future work to be targeting the false positives and false negatives in container scanners by integrating machine learning into container scanners.

References

1. A. Avram, Docker: Automated and consistent software deployments, in *InfoQ. Retrieved* (2013), pp. 08–09
2. A. Esmaeily, K. Kralevska, D. Gligoroski, A cloud-based SDN/NFV testbed for end-to-end network slicing in 4G/5G (2020). Preprint arXiv:2004.10455
3. Docker security, https://tinyurl.com/zhpa3dv. Accessed 15 April 2020
4. T. Micro, Why running a privileged container in Docker is a bad idea (2019). https://tinyurl.com/y9dwkqws. Accessed 15 April 2020
5. J. Gummaraju, T. Desikan, Y. Turner, Over 30% of official images in Docker hub contain high priority security vulnerabilities. Tech. Report, Banyan Ops, 2015
6. R. Shu, X. Gu, W. Enck, A study of security vulnerabilities on Docker Hub, in *Proceedings of the Seventh ACM on Conference on Data and Application Security and Privacy* (2017)
7. E. Socchi, J. Luu, A deep dive into Docker Hub's security landscape - a story of inheritance? (2019). https://tinyurl.com/y8axga4g. Accessed 5 Mar 2019

8. K. Wist, M. Helsem, An extensive analysis of the current vulnerability landscape in Docker Hub Images. Master's thesis, Norwegian University of Science and Technology (NTNU), 2020
9. FIRST, Common vulnerability scoring system version 3.1: Specification document (2019). https://tinyurl.com/ycbchlsr. Accessed 3 Feb 2020
10. D. Barrett, G. Kipper, Virtualization technique (2010). https://tinyurl.com/y6wx7577. Accessed 17 April 2020
11. D. Inc., What is a container? https://tinyurl.com/y9ca4jne. Accessed 31 Jan 2020
12. C. Anderson, Docker (2015). https://tinyurl.com/y8hvlkvw. Accessed 3 Feb 2020
13. J. Morgan, Introducing the new Docker Hub (2018). https://tinyurl.com/y9kewdsc. Accessed 3 Feb 2020
14. Z. Zhang, D. Caragea, X. Ou, An empirical study on using the national vulnerability database to predict software vulnerabilities (2011). https://tinyurl.com/ycqe7gz3. Accessed 27 Jan 2020
15. S. Na, T. Kim, H. Kim, A study on the classification of common vulnerabilities and exposures using Naïve Bayes (2017). https://tinyurl.com/y9reouf9. Accessed 28 Jan 2020
16. Z. Chen, Y. Zhang, Z. Chen, A categorization framework for common computer vulnerabilities and exposures (2009). https://tinyurl.com/ycyse3sa. Accessed 28 Jan 2020
17. A. Lehman, *JMP for Basic Univariate and Multivariate Statistics: A Step-By-Step Guide* (SAS Institute, Cary, 2005)
18. CVE details (2020). https://www.cvedetails.com/. Accessed 20 April 2020
19. Cve-2019-10744 detail (2019). https://tinyurl.com/y9lbjtat. Accessed 27 March 2020
20. Cve-2017-1000158 detail. https://tinyurl.com/ya7wml83. Accessed 27 March 2020
21. Cve-2019-9948 detail. https://tinyurl.com/yxthuynd. Accessed 27 March 2020
22. Cve-2019-9636 detail. https://tinyurl.com/ybcosfoe. Accessed 27 March 2020

Analysis of Conpot and Its BACnet Features for Cyber-Deception

Warren Z. Cabral, Craig Valli, Leslie F. Sikos, and Samuel G. Wakeling

1 Introduction

Conpot[1] is a mimicry ICS/SCADA honeypot, which supports the simulation of a number of protocols, such as HTTP, Modbus, SNMP, and BACnet. The functional version of the BACnet protocol was released in 1995 to facilitate communication of building automation and control systems for HVAC appliances, lighting control, access control, and fire detection systems and their corresponding equipment. This chapter presents the results of the research analysis of the cyber-deceptive potential of Conpot's built-in templates. Also, a brief literature review is provided regarding the vulnerabilities affecting the BACnet protocol, with subsequent analysis of the deceptive capabilities of the `bacnet.xml` protocol file.

[1] https://github.com/mushorg/conpot.

W. Z. Cabral (✉) · C. Valli · L. F. Sikos
Cyber Security Cooperative Research Centre, Joondalup, WA, Australia
Edith Cowan University, Joondalup, WA, Australia
e-mail: w.cabral@ecu.edu.au; c.valli@ecu.edu.au; l.sikos@ecu.edu.au

S. G. Wakeling
Edith Cowan University, Joondalup, WA, Australia
e-mail: s.wakeling@ecu.edu.au

K. Daimi et al. (eds.), *Advances in Security, Networks, and Internet of Things*, Transactions on Computational Science and Computational Intelligence,
https://doi.org/10.1007/978-3-030-71017-0_23

2 Literature Review

In the literature, attempts have been listed to change the default configuration of honeypots, including Conpot. For example, Scott successfully cloned Conpot's default template and configured it to mimic a Schneider Electric PowerLogic ION6200 smart meter with Syslog and Splunk monitoring and logging capabilities [1]. The honeypot was placed in the same subnet as the real ION6200 smart meter to act as another smart meter, and the logs were fed into a Syslog and Splunk server through a DMZ using appropriate port and firewall configurations. Their experiment was successful but did not incorporate a standard framework for understanding the function of each honeypot template, their subsequent variables, deception type, and how these can be configured to potentially increase the overall deceptiveness of Conpot. This section explores the use and security concerns around BACnet and how SCADA honeypots such as Conpot aim to mitigate them.

2.1 BACnet

BACnet is a data communication protocol designed for the Building Automation and Control Networks by the American Society of Heating, Refrigerating and Air-Conditioning Engineers (ASHRAE) [2]. Given the rise of data centers and their heating and cooling needs, such a protocol is becoming more and more critical to protect IT systems. Other systems using this protocol include, but are not limited to, food storage, perishable goods repositories, and industrial processes requiring strict temperature control via an environmental controller. The vulnerabilities of the BACnet protocol are well documented and well mapped, and their modus operandi and impacts are well known [3–6]. One of the major issues with BACnet controllers is that they are legacy deployments, and some of these deployments are several decades old. This presents a unique challenge for cyber-defenders in this increasingly interconnected world to protect critical systems that rely on BACnet controllers. SCADA honeypots aim to mitigate this challenge by mimicking devices simulating SCADA protocols such as BACnet, thereby baiting cybercriminals in order to study their attack methodology so that they can protect and reiterate their defenses and increase their cybersecurity posture.

2.2 SCADA Honeypots

Honeypots have been used not only to analyze how cybercriminals penetrate standard information security systems but also to notify cyber-defenders of a probable threat within their network [7, 8]. Similarly, SCADA honeypots are used for monitoring devices and networks for early threat detection within a SCADA environment. These honeypots reveal a new understanding of the threats concerning these legacy devices. SCADA honeypots such as Conpot utilize mimicry deception to bait cybercriminals by mimicking live SCADA environments. A conventional SCADA system consists of four parts: a central computer (host), a number of field-based remote measurement and control units known as RTUs, a wide area telecommunications system to connect them, and an operator interface to allow the operator to access the system [9, 10].

2.3 *Conpot*

Conpot[2] is an open-source low-interaction SCADA honeypot, which attempts to mimic an active SCADA system. It was released in May 2013 under the umbrella of the Honeynet Project and large industrial corporations [11]. Conpot is programmed to gather intelligence in real time about the motivations and approaches used by cybercriminals targeting industrial control systems. According to Jicha et al., Conpot is extremely efficient in implementation and supports the integration of PLC devices and the simulation of industrial protocols such as HTTP, Modbus, IPMI, SNMP, and BACnet [10].

3 Research Approach

The research was conducted within a virtual Ubuntu 18.04.4 (bridged network adapter) environment deployed using VMware Workstation Pro 15.5.2. Once all the necessary dependencies were installed, the Conpot 0.6.0 repository was acquired using the `git` command. The permissions of the Conpot installation directory had to be changed to prevent Conpot from throwing permission errors when running as a standard user. This is because, unlike previous versions of Conpot, templates would not load using the `sudo` command. Once installed, the default Conpot instance was executed using the following command:

```
conpot -f --template default
```

To verify whether the default template was loaded successfully, `localhost:8800` was loaded in Mozilla Firefox. The connection was successful, as shown in Fig. 1, and the correct timestamps displayed indeed matched the time the honeypot instance was executed.

Connection attempts or port scans using *NMAP*[3] against the honeypot were documented and stored in Conpot's log files.

[2] http://conpot.org.

[3] https://nmap.org.

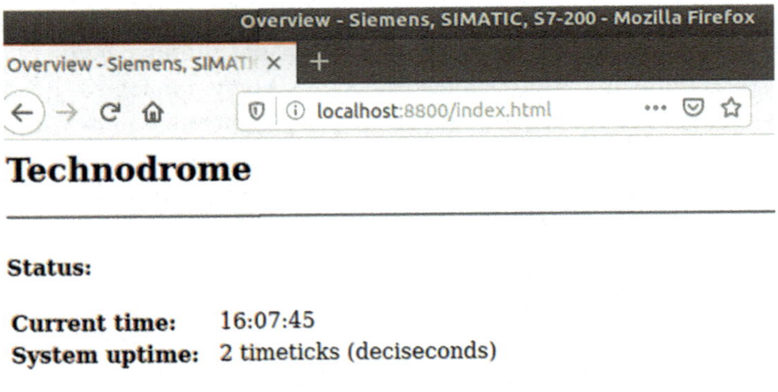

Fig. 1 Conpot's default Siemens S7-200 service webpage

The execution of the ipmi template resulted in *permission denied* errors. This is because this template uses port 623 by default, and nonroot users cannot bind to port numbers smaller than 1024. This error was fixed by using the following command and then loading the ipmi instance:

```
sudo setcap CAP_NET_BIND_SERVICE=+eip /usr/bin/python3.6
```

The above command allows any program that runs on Python 3.6 to open ports below 1024, although from a cybersecurity perspective, this is not a secure solution. Alternatively, the ipmi template can also be executed if the port number is configured to run on 6230 or any other subsequent ports above 1023.

All of Conpot templates were scanned using NMAP 7.80 within a Kali 2020.1 VM (bridged network adapter). Due to time restrictions, NMAP was the only tool used for scanning the templates, with the exception of the proxy template. The scans were performed using

```
nmap -sV -Pn -p- [IP Address]
```

where `sV` is used for daemon and service detection, `Pn` is used to suppress pings during scans to find out whether a host is up, and `p` is used to specify a port range. In our case, `-p-` will scan all ports between 1 and 65535.

While all templates were scanned, focus was placed on the default template, which mimics a Siemens S7-200 PLC and provides emulation for the BACnet protocol operating on port 47808. The reason for this is that Conpot provides basic but plausible BACnet device emulation. BACnet properties are populated from the `bacnet.xml` file[4] for the Conpot service. Therefore, any configuration for the `bacnet.xml` file has to be done in the file's directory to reflect the changes when a template is executed. At its core, Conpot uses the `bacpypes` Python

[4] `/home/[username]/.local/lib/python3.6/sitepackages/conpot-0.6.0-py3.6.egg/conpot/templates/default/bacnet/bacnet.xml`.

libraries, which provide reasonable emulation of the BACnet server components. Conpot implements a smart sensor, and this is why the - DM-RP-B (execute `ReadProperty`), - DM-DDB-B (execute `Who-Is`, initiate `I-Am`), and - DM-DOB-B (execute `Who-Has`, initiate `I-Have`) services are supported [12]. Whether NMAP could detect the BACnet service was checked using the command:

```
nmap -sV -Pn -p 47808 [IP Address]
```

The analysis of the `bacnet.xml` file was conducted, and its subsequent variables, functionality, and deception type populated (as shown in Table 1), using the research approach from a previous research study [13].

4 Observations and Results

The observations and results of the analysis of Conpot templates, NMAP scan, and the `bacnet.xml` file are described in the following sections.

4.1 Analysis of Conpot's Built-In Templates

Conpot 0.6.0 offers six honeypot templates, which are detailed in Table 2.

Their functionality and ports loaded on execution have also been documented. The NMAP scan was conducted to test how deceptive the analogous templates of Conpot appear to potential cybercriminals. (Note that SSH was disabled on the Ubuntu VM, and therefore SSH port 22 was not detected during the scans.)

4.2 NMAP Scan Discussion

The scan results indicate that the Conpot templates have a low deceptive capability against NMAP scans, because in most scenarios it correctly displayed the exact open ports loaded on honeypot execution. However, NMAP was unsure about the protocols associated with the open ports, except in the case of the IEC104 template, for which it correctly displayed protocol and port information. In contrast, the ipmi template did not return any search results. Furthermore, the NMAP BACnet scan using `nmap -sV -Pn -p- [IP Address]` could not detect the BACnet port. However, a targeted port scan using the `nmap -sV -Pn -p 47808 [IP Address]` command could successfully detect a closed BACnet port operating on port 47808, which belonged to the Conpot instance. These results can have two explanations: either the BACnet port has not been bound (despite Conpot indicating otherwise) or the NMAP scan gave incorrect results regarding the open BACnet

Table 1 Deceptive functionality of the variables in `bacnet.xml`

Variables	Usage	Deception
`enabled`	Enables the BACnet protocol for the default template. When disabled, BACnet will not function.	N/A
`host`	The IP address to which the Conpot instance connects to. By default, all hosts connect to 0.0.0.0.	Data masking
`port`	The port number on which the host IP address operates on. BACnet uses UDP port number 47808, but depending on the specification of the devices, it can be configured to listen on other ports.	Data masking
`device_name`	The name of the BACnet device.	Data mimicry
`device_identifier`	A number used to uniquely identify a BACnet instance within an interconnected network.	Data mimicry
`vendor_name`	The name of the vendor who supplies and maintains the BACnet device.	Data mimicry
`vendor_identifier`	A number used to uniquely identify a BACnet instance by the vendor who supplied the device.	Data mimicry
`model_name`	The model name of the BACnet device.	Data mimicry
`max_apdu_length_accepted`	The size of the largest message (in bytes) that the BACnet device can receive to prevent an internal memory buffer overflow when sending a packet.	Data mimicry
`protocol_version`	An integer used to check for incompatibilities between server versions.	Data masking
`object name`	Can be selected from the 60 standard object types defined by the ASHRAE 135-2016 standard, such as `objectanalog`, `objectaccesspoint`, and `objectbinary`.	Data mimicry
`segmentation_supported`	Segmentation is used to allocate BACnet messages that do not fit within a single packet. BACnet has four types of segmentation: `segmentedboth`, `segmentedtransmit`, `segmentedreceive`, and `nosegmentation`.	Data mimicry

Table 2 Functionality of Conpot's templates with ports loaded by the honeypot versus those detected by the NMAP scan

Template	Functionality	Ports loaded on Conpot execution	Ports detected by the NMAP scan
default	Conpot's default template created by the Conpot team, which provides mimicry deception for a basic Siemens S7-200 PLC with 2 slaves.	bacnet – port 47808 enip – port 44818 ftp – port 2121 http – port 8800 ipmi – port 6230 modbus – port 5020 s7comm – port 10201 snmp – port 16100 tftp – port 6969	ccproxy-ftp – port 2121 zenginkyo-1 – port 5020 sunwebadmin – port 8800 rsms – port 10201 enip – port 44818
guardian_ast	This template was also created by the Conpot team and is used to simulate a Guardian AST tank-monitoring system.	guardian_ast port 10001	scp config port 10001
IEC104	This template was released with Conpot 0.6.0. It was created by Patrick Reichenberger to mimic a simple IEC 60870-5-104 device.	iec104 – port 2404 snmp – port 16100	iec104 – port 2404
ipmi	This template was created by Lukas Rist to mimic the IPMI 371 device and provide basic deceptive capabilities. IPMI 371 provides an interface to monitor threats and faults for physical attributes of a server, such as temperature, voltage, fans, and power supplies [14].	ipmi – port 623	No open ports were detected. Even when configured to operate on port 6230, the NMAP scan returned no open ports.
kam-strup_382	Kamstrup 382 is an electricity meter, which allows for readings and registration of electric energy. This template was created by Johnny Vestergaard and registers a clone of an existing Kamstrup 382 smart meter.	kam-strup_management – port 50100 kamstrup_meter – port 1025	nfs or iis – port 1025 unknown service – port 50100
proxy	The proxy template was created by the Conpot team and provides proxy connection capabilities for built-in or custom templates. The proxy template can be configured to administer proxy connections.	N/A	N/A

port and the `bacnet.xml` file. Similar explanations can be stated for the other ports opened by Conpot yet not detected by the NMAP scan.

Furthermore, when executed, Conpot throws a number of unnecessary log traceback errors, which get stored in log files and potentially cause cyber-defenders to miss critical system errors. Furthermore, when interacting with a simulated environment, traceback errors can sometimes be used by cybercriminals to detect the honeypot service (even if they do not see the error) based on the traceback error and the way it is handled. Cybercriminals use commands and expect a certain output if the system is a production system. In this case, the traceback errors may result in a different output, or no output at all, inferring that they might be connected to a honeypot service.

As previously stated, the results showed us that Conpot has poor deceptive capabilities. In accordance with the ports and protocols detected by NMAP, we can modify the configuration file of the honeypot by theoretically inputting additional scripts, altering variables, and enabling or disabling features within the application. A NMAP scan will then be executed on the configured Conpot instance to test if the detected ports on this occasion differ from those seen in Table 2. This process will be repeated until a desired functionality is achieved. However, before we configure any templates or files, we must first attain an understanding of the numerous variables present and their deceptive functionality as seen in Table 2 for one such template, i.e., the `bacnet.xml` file.

4.3 Conpot's Default bacnet.xml File

Figure 2 displays the `bacnet.xml` template, and the results of its subsequent deception types are shown in Table 1, i.e., the function of each variable and their deceptive functionality.

The other templates of Conpot can be inspected and reviewed using a similar approach. This is done to study the function of the templates, their subsequent protocols, the extent of their functionality, and how they can be configured to mimic actual SCADA devices, thereby making them appear as legitimate systems. For example, the variables with a mimicry deception type (such as `device_name`) would need the appropriate `device_identifier` to go with them that would match the expected port for that particular device/model, coming from the correct `vendor_name` and `vendor_identifier`, etc. So basically, if a cybercriminal checked the manufacturer of the NIC based on the MAC address and found that it is a MAC of a real SCADA device, they would think that they are connecting to a production system rather than a honeypot. If it does not match, it would seem like the values are spoofed, implying a honeypot.

```
<bacnet enabled="True" host="0.0.0.0" port="47808">
    <device_info>
        <device_name>SystemName</device_name>
        <device_identifier>36113</device_identifier>
        <vendor_name>Alerton Technologies, Inc.</vendor_name>
        <vendor_identifier>15</vendor_identifier>
        <max_apdu_length_accepted>1024</max_apdu_length_accepted>
        <segmentation_supported>segmentedBoth</segmentation_supported>
        <model_name>VAV-DD Controller</model_name>
        <protocol_version>1</protocol_version>
    </device_info>
    <object_list>
        <object name="objectBinary">
            <properties>
                <object_identifier>12</object_identifier>
                <object_name>BI 01</object_name>
                <object_type>Binary Input</object_type>
            </properties>
        </object>
        <object name="objectAnalog">
            <properties>
                <object_identifier>14</object_identifier>
                <object_name>AI 01</object_name>
                <object_type>Analog Input</object_type>
                <present_value>68.0</present_value>
            </properties>
        </object>
        <object name="objectDoor">
            <properties>
                <object_identifier>16</object_identifier>
                <object_name>Door 01</object_name>
                <object_type>Access Door</object_type>
                <present_value>0</present_value>
                <out_of_service>True</out_of_service>
                <maintenance_required>2</maintenance_required>
            </properties>
        </object>
    </object_list>
</bacnet>
```

Fig. 2 Variables of `bacnet.xml`

5 Conclusions and Future Work

Considering that Conpot is an open-source honeypot and, as such, has its source code publicly released, thus not only implementers but also cybercriminals can analyze it. Honeypot deployments with default configuration settings are easier to exploit or more prone to attacks and therefore must be configured prior to deployment. Conpot's built-in templates and protocol support (including for BACnet) can be cloned and configured to mimic fully functional SCADA devices, provided that the configuration is a resemblance of a real SCADA service. Being coded in Python 3.5 and XML, the templates can be easily configured to include support for additional devices and protocols. Creating a custom template or cloning the default template and customizing it to add support for additional protocols, and simulations such as DNP3, AWL/STL, and OPC UA can significantly increase deceptive capability. This improves the honeypot's attack surface or artificially delays the service response time, thereby mimicking a server under constant load to appear more realistic.

As a future work, the Conpot honeypot will be configured when all the templates and their subsequent variables and deceptive functionality are addressed in accordance with the framework created. Ultimately, the honeypot will be deployed with configured recommendations for 2–3 months, and another one with default settings, and at the end of the experiment, a quantitative comparison will be made, making it clear how much better the configured deployment is in terms of deception.

6 Abbreviations

APDU	Application Layer Protocol Data Unit
ASHRAE	American Society of Heating, Refrigerating and Air-Conditioning Engineers
AWL	Anweisungsliste (German for STL)
BACnet	Building Automation and Control Network Protocol
DMZ	Demilitarized Zone
DNP3	Distributed Network Protocol 3
ENIP	EtherNet Industrial Protocol
FTP	File Transfer Protocol
HTTP	HyperText Transfer Protocol
HVAC	Heating Ventilation and Air Conditioning
ICS	Industrial Control System
IEC	International Electrotechnical Commission
IP	Internet Protocol
IPMI	Intelligent Platform Management Interface
MAC	Media Access Control
NIC	Network Interface Card
NMAP	Network Mapper
OPC UA	Open Platform Communications Unified Architecture
PLC	Programmable Logic Controller
RTU	Remote Terminal Unit
S7COMM	A proprietary Siemens communication protocol that runs between PLCs of the Siemens S7-300/400 family
SCADA	Supervisory Control and Data Acquisition
SNMP	Simple Network Management Protocol
SSH	Secure Shell
STL	Statement List Programming
TFTP	Trivial File Transfer Protocol
UDP	User Datagram Protocol
VAV	Variable Air Volume

Acknowledgments The work has been supported by the Cyber Security Research Centre, whose activities are partially funded by the Australian Government's Cooperative Research Centres program.

References

1. C. Scott, Designing and implementing a honeypot for a SCADA network (2014). SANS white paper. https://www.sans.org/reading-room/whitepapers/detection/designing-implementing-honeypot-scada-network-35252
2. ASHRAE: Standard 135-2016, BACnet™—a data communication protocol for building automation and control networks (2016). https://www.techstreet.com/ashrae/standards/ashrae-135-2016?product_id=1918140
3. D.G. Holmberg, BACnet wide area network security threat assessment (2003). https://doi.org/10.6028/NIST.IR.7009
4. M. Peacock, M.N. Johnstone, C. Valli, An exploration of some security issues within the BACnet protocol, in *Information Systems Security and Privacy*, ed. by P. Mori, S. Furnell, O. Camp (Springer, Cham, 2017), pp. 252–272. https://doi.org/10.1007/978-3-319-93354-2_12
5. L. Benbenishti, SCADA MODBUS protocol vulnerabilities (2017). https://www.cyberbit.com/blog/ot-security/scada-modbus-protocol-vulnerabilities/
6. B. Bowers, How to own a building: exploiting the physical world with BACnet and the BACnet attack framework (ShmooCon IX, Washington, 2013). https://infocondb.org/con/shmoocon/shmoocon-ix/how-to-own-a-building-exploiting-the-physical-world-with-bacnet-and-the-bacnet-attack-framework
7. L. Spitzner, *Honeypots: Tracking Hackers* (Addison-Wesley, Reading, 2002)
8. C. Valli, Honeypot technologies and their applicability as a strategic internal countermeasure. Int. J. Inf. Comput. Secur. **1**(4), 430–436 (2007). https://doi.org/10.1504/IJICS.2007 015503
9. S.M. Wade, SCADA Honeynets: the attractiveness of honeypots as critical infrastructure security tools for the detection and analysis of advanced threats. Master's Thesis, Iowa State University, Ames (2011)
10. A. Jicha, M. Patton, H. Chen, SCADA honeypots: an in-depth analysis of Conpot, in *2016 IEEE Conference on Intelligence and Security Informatics*, ed. by L. Zhou, L. Kaati, W. Mao, G.A. Wang (IEEE, Piscataway, 2016), pp. 196–198. https://doi.org/10.1109/ISI.2016.7745468
11. L. Rist, Introducing Conpot (2013). https://www.honeynet.org/2013/05/11/introducing-conpot/
12. MushMush Foundation, Welcome to Conpot's documentation!—Conpot 0.6.0 documentation (2020). https://conpot.readthedocs.io/en/latest/
13. W.Z. Cabral, C. Valli, L.F. Sikos, S.G. Wakeling, Review and analysis of Cowrie artefacts and their potential to be used deceptively, in *Proceedings of the 6th Annual Conference on Computational Science and Computational Intelligence* (IEEE, Piscataway, 2019), pp. 166–171. https://doi.org/10.1109/CSCI49370.2019.00035
14. R. Rajachandrasekar, X. Besseron, D.K. Panda, Computer Security Panda, Monitoring and predicting hardware failures in HPC clusters with FTB-IPMI, in *Proceedings of the 26th International Parallel and Distributed Processing Symposium Workshops* (IEEE Computer Society, Los Alamitos, 2012), pp. 1136–1143. https://doi.org/10.1109/IPDPSW.2012.139

Automotive Vehicle Security Metrics

Guillermo A. Francia, III

1 Introduction

Today's automobiles have more than 150 electronic control units (ECUs), which are embedded devices that control the actuators to ensure optimal engine performance. These vehicles have multiple wireless entry points, some connected to the Internet, that enable access convenience and online services [1].

A technical brief [2] by Trend Micro described the vulnerability found in modern vehicles' networks. This vulnerability enables a stealthy denial-of-service attack that practically works for every automotive vendor and had been disclosed and prompted an ICS-CERT alert: ICS-ALERT-17-209-01. Exploitable hardware design flaws in some capacitive micro-electromechanical system (MEMS) accelerometer sensors produced by prominent automobile parts manufacturers were reported in another ICS-CERT alert: ICS_ALERT-17-073-01A in early 2017.

Experience in securing traditional IT systems cannot simply be applied to vehicle systems due to their differing requirements and development; a special set of security metrics is needed for these systems. In this chapter, we present a literature review of various communication protocols, threats and vulnerability issues, and safety and security challenges on vehicle systems. We aggregate the information learned from the literature and devise a set of metrics for measuring the efficacy of security controls in vehicle systems.

The remainder of this chapter is organized as follows: Section 2 provides an overview of vehicle communication systems while Sect. 3 expounds on vehicle security and provides details on vehicle threats, vulnerabilities, and attacks. Section 4 presents automotive vehicle security metrics that were adapted from the Common

G. A. Francia, III (✉)
Center for Cybersecurity, University of West Florida, Pensacola, FL, USA
e-mail: gfranciaiii@uwf.edu

K. Daimi et al. (eds.), *Advances in Security, Networks, and Internet of Things*,
Transactions on Computational Science and Computational Intelligence,
https://doi.org/10.1007/978-3-030-71017-0_24

Vulnerability Scoring System (CVSS) and the Common Methodology for IT Security Evaluation. Finally, Sect. 5 provides concluding remarks and offers future research directions on automotive vehicle security metrics.

2 Vehicle Communication

The proliferation of electronic devices and the rapid advancement of communication technologies have ushered the steady progression of vehicular communication from an in-vehicle form to the far-reaching external variety.

Modern automotive vehicle communication can be classified into four main categories: in-vehicle communication, vehicle-to-device (V2D) communication, vehicle-to-vehicle communication (V2V), and vehicle-to-infrastructure (V2I) communication. An in-vehicle communication example could be a vehicle sensor transmitting operating signals to a controller connected to the vehicle network. The communication between the infotainment system and smart phone is an example of a V2D communication. An example of a V2V communication would be two or more vehicles connected through some form of ad hoc wireless network. This vehicular ad hoc network (VANET), first introduced at the turn of century, is an extension of the mobile ad hoc network (MANET). For the V2I communication category, a good example is the scenario wherein a vehicle captures and sends real-time data about the traffic conditions to the highway infrastructure management system through cellular communication. These captured data are then fed into an intelligent traffic system that manages and optimizes traffic control in that locality.

2.1 *Intra-vehicle Communication Protocols*

The intra-vehicle network communication protocol group consists of the three predominant communication protocols found in a modern automotive vehicle: Controller Area Network (CAN), Local Interconnect Network (LIN), and FlexRay. Recent advancements in in-vehicle protocol technology include the Automotive Ethernet.

The CAN [3] communication protocol works on a two-wired half-duplex high-speed serial network bus topology using the Carrier Sense Multiple Access (CSMA)/Collision Detection (CD) protocol. It implements most of the functions of the lower two layers of the International Standards Organization (ISO) Reference Model.

LIN [4] is an in-vehicle serial communication protocol that delivers a low-cost alternative to CAN and FlexRay [5] for vehicle network applications. However, it delivers a lower performance and less reliability. A LIN bus uses a single 12 V line and has a node that acts as a Master gateway for other LIN nodes. Up to 16 of these slave nodes can be connected to the LIN bus.

FlexRay [5] is an in-vehicle communication bus whose purpose is to meet the need for a fast, reliable, and greater bandwidth data communication system. National Instruments correctly pointed out that the optimization of cost and reduction of transition challenges can be accomplished by using FlexRay for high-end applications, CAN for powertrain communications, and LIN for low-cost body electronics .

Automotive Ethernet is an adaptation of the standard ethernet but works on two-wire instead of the four-wire configuration. It implements a physical network that is used to connect components within a car using a wired network. It is designed to meet the needs of the automotive market, including meeting electrical requirements and emissions, bandwidth requirements, latency requirements, synchronization, and network management requirements.

2.2 *Inter-vehicle Communication Protocols*

The inter-vehicle network communication protocol group consists of several short- and long-range communication protocols and standards that enable services necessary for a robust, secure, and efficient transportation infrastructure.

Dedicated Short Range Communications (DSRC), a variation of the Institute of Electrical and Electronics Engineers (IEEE) 802.11 Wi-Fi standard, is primarily intended for the automotive environment. It uses the IEEE 802.11p standard in the 5.9 GHz band. Additionally, this standard is a companion for the proposed IEEE 1609 Family of Standards for Wireless access in Vehicular Environments (WAVE).

The 802.11a is one of the earliest wireless standards operating on both the 2.4 GHz and 5.2 GHz Industrial, Scientific, and Medical (ISM) bands. The data rate for this standard ranges from 6 to 54 Mbps with an operating bandwidth of 20 MHz. Compared to DSRC, this standard operates on a limited distance of approximately 100 meters.

Vehicular Ad Hoc Network (VANET) is a form of a Mobile Ad Hoc Network (MANET) that utilizes the Wi-Fi (802.11 a/b/g), the Worldwide Interoperability for Microwave Access (WiMAX), a family of wireless broadband communication standards based on IEEE 802.16, or the Wireless Access in Vehicular Environments (WAVE) based on the IEEE 1609–12 standards. WAVE is a layered protocol architecture that includes the security of message exchange and operates on the Dedicated Short-Range Communication (DSRC) band.

3 Automotive Vehicle Security

There have been several initiatives towards the protection of a vehicle's electronic control units. Notable examples are the E-safety Vehicle Intrusion Protected Application (EVITA) Project [6], the Preparing Secure Vehicle-to-X Communication

Systems (PRESERVE) Project [7], Secure Vehicular Communication (SeVeCom) Project [8], and the Society of Automotive Engineers (SAE) J3061 Guidebook [9]. In a very recent work by Bauer and Schartner [10], a table depicting attack surfaces and the classification of attack potential according to common criteria is presented. The table includes information on the difficulty and the impact of a certain exploit to an asset. Further, the work introduced a novel solution towards a realistic assessment of the integration of specialized countermeasures into the design of vehicular cybersecurity concepts.

3.1 Automotive Vehicle Threats and Vulnerabilities

Leopold postulates that "as cybersecurity risks are not covered by existing safety norms for surface vehicles, new guidelines and standards for automotive cybersecurity need to be established [11]." Acknowledging this urgent need, the automotive industry took the initiative to work on a cybersecurity standard: ISO/SAE 21434 "Road vehicles—Cybersecurity Engineering." The standard requires a security risk assessment that includes the identification of assets and the determination of potential damages resulting from the security breach. The first draft of this standard is scheduled to be released in early 2020. One major component of this standard is the determination of the security risk level of a vehicle and its components.

The ISO/SAE 21434 Joint Working Group (JWG) is divided into four part groups: PG1: Risk Management, PG2: Product Development, PG3: Operation, Maintenance and other Processes, and PG4: Process Overview and Interdependencies [12].

Keen Security Lab researchers uncovered the vulnerability of Tesla's touch screen infotainment system and used that as a gateway to manipulate the driver's seat motor, the windshield wipers, the turn indicators, and the sunroof from a distance of 12 miles while the car was in motion [13].

3.2 Automotive Vehicle Security Attacks

Petit, Feiri, and Kargl [14] described an abstract model of attack surfaces on the vehicular communication domain. The attack model considers the sensor data in various stages: acquisition, processing, storing, and transmission. The generic attack model appears to be adaptable to any communication protocol. This seminal work has been extended by Monteuuis et al. [15] with the notion of a secured automotive perception consisting of two main components: objects and data stages. Various attack surfaces on vehicles ranging from the OBD port to the infotainment system were examined by Koscher et al. [16]. One such surprising revelation is the ease of embedding CAN messages into an audio file and transforming the infotainment system as a gateway for attack vectors.

Secure measures have been introduced to mitigate the vulnerabilities of the CAN protocol. An intrusion detection system based on the clock skew of the Electronic Control Unit (ECU) as a fingerprint to develop a reference behavior of legitimate devices was proposed by Cho and Shin [17]. Wang et al. [18] proposed a method wherein a CAN packet is augmented with an 8-byte message authentication code. In [19], the design, implementation, and evaluation of a hardware security module for a modern automotive vehicle is presented. Lokman et al. conducted a systematic review of Intrusion Detection Systems (IDS) for automotive CAN bus system based on detection approaches, deployment strategies, attacking techniques, and technical challenges [20]. The study categorized anomaly-based IDS into four methods, namely, frequency-based, machine learning-based, statistical-based, and hybrid-based. Kumar et al. focused on jamming signal-centric security issues for Internet of Vehicles (IoV) [21]. They proposed a machine learning-based protocol that focuses on jamming vehicle's discriminated signal detection and filtration for revealing precise location of jamming effected vehicles.

3.3 Industry and Government Initiatives

The U.S. Government Accountability Office (GAO) Report on Vehicle Cybersecurity [22] contains, among others, the key security vulnerabilities in modern vehicles, the key practices and technologies to mitigate vehicle cybersecurity vulnerabilities, the challenges facing stakeholders, and the Department of Transportation's (DOT) efforts in addressing the issues in vehicle cybersecurity.

The EVITA (E-Safety Vehicle Intrusion Protected Applications) project [6], which was co-funded by the European Commission and whose primary objectives are to design, verify, and prototype an architecture for automotive onboard networks to protect security-relevant components against tampering sensitive data against compromise when transferred inside a vehicle.

The Society of Automotive Engineers (SAE) Cybersecurity Guidebook for Cyber-Physical Vehicle Systems [23] provides and describes a cybersecurity process framework from which an organization can develop processes to design and build cybersecurity in vehicular systems. The process framework covers the entire product lifecycle, including postproduction aspects with respect to service, incident monitoring, incident response, etc.

The National Highway Traffic Safety Administration (NHTSA) Automotive Security Best Practices for Modern Vehicles [24] presents the results and analysis of a review of best practices and observations in the field of cybersecurity involving electronic control systems across a variety of industry segments where the safety of life is concerned.

The Intel® Automotive Security Research Workshops report [25] presents the findings that resulted from two automotive security workshops. The participants were afforded to perform hands-on work on a Linux®-based in-vehicle infotainment

(IVI) simulation platform to identify threats and vulnerabilities and to provide potential mitigation strategies.

4 Automotive Vehicle Security Metrics

A very well-known cliché states that "what cannot be measured cannot be improved." This is the motivation behind this research. To better develop security metrics, organizations must differentiate between measurement and metric [26]. Measurement represents raw data of a point in time while metric comes from the analysis of aggregate data overtime (e.g [27].). A good metric should measure the relevant data that satisfy the needs of decision makers and should be quantitatively measurable, accurate, validated on a solid base, inexpensive to execute, able to be verified independently, repeatable, and scalable to a larger scale [28]. By adapting security risk regression that is successful in predicting attacks from simple security threats, Schechter [29] concludes that security strength is a key indicator of security risks for more complex security threats in information systems. In congruence, Manadhata and Wing propose the attackability of a system as an indicator of security strength [30]. Their security metric is based on the notion of attack surface by comparing attackability of systems along three abstract dimensions: method, data, and channel. The attackability of a system is a cost-benefit ratio between efforts of gaining access and potential impacts of security failure among the three dimensions [31].

There exists notable works on automotive vehicle security metrics. In [32], a set of security metrics for the software system in a connected vehicle is proposed. The set of metrics provides a quantitative indicator of the security vulnerability of the following risks on the system software: ECU coupling, communication, complexity, input and output data, and past security issues. The ECU coupling metric is based on the connectivity of the ECUs. Simply put, the risk is proportional to the extent of the connectivity of the ECUs. This proposed metric failed to take into account the fact that most vehicle networks are using the bus topology for interconnection. The communication risk metric is based on the number of communication technologies that are enabled on-board the vehicle. These are further normalized by the level of risk assigned to each of those technologies. The issue with this metric is that the assignment of risk level is quite arbitrary. The metric on input and output data risk takes into account the number of input data, the fixed and fluctuating properties of the input data, and the sensitivity level of output data. The authors argue that fluctuating input data and sensitive output data are more significant and should be given more emphasis in the calculation of security vulnerability. This metric failed to account the level of security testing that was applied to the vehicle's embedded system before deployment. Finally, the metric on security history utilizes the number of past attacks that occurred on the vehicle. This metric appears to assume the recurrence of an attack and that the vulnerability was never fixed. With system patches actively being carried out during vehicle recalls, this assumption is rather weak.

Use cases of automotive security threats are described in [33]. The use cases include, among others, brake disconnect, horn activation, engine halt air bag, portable device injection, key fob cloning, cellular attack, and malware download. The threat matrix on each of these use cases includes attributes such as exploitable vulnerability, difficulty of implementation, resources needed, attack scenario, and outcome.

A Bayesian Network (BN) for connected and autonomous vehicle cyber-risk classification was developed by Sheehan et al. [34]. The BN model uses the Common Vulnerability Scoring System (CVSS) software vulnerability risk-scoring framework for input parameters specifically on the Global Positioning System (GPS) jamming and spoofing.

In the following section, we present a collection of vehicle security metrics similar to those in an earlier work on critical infrastructure and industrial controls systems security [31, 35].

4.1 Common Vulnerability Scoring System

CVSS is an open framework for communicating the characteristics and severity of software vulnerabilities. It consists of three metric groups: Base, Temporal, and Environmental [36]. The Base group characterizes the static intrinsic qualities of vulnerability; the Temporal group represents the vulnerability as it evolves over time; and the Environmental group depicts the characteristics of the vulnerability that are endemic to the user's environment. The third group of metrics lends itself perfectly with that of an automotive vehicle system.

The Base group consists of two metrics: exploitability and impact. The Temporal group consists of the following metrics: Exploit Code Maturity, Remediation Level, and Report Confidence. The Environmental metrics include the following: Security Requirements and Modified Base.

To illustrate, we examined two published vulnerabilities: the Tesla Model S firmware vulnerability (CVE-2016-9337) [37] and the Mobile Devices OBD-II dongles firmware vulnerability (CVE-2015-2908) [38].

The Tesla Model S firmware vulnerability applies to versions below version 7.1 with web browser functionality enabled. The vehicle with this firmware is susceptible to commands that may allow an attacker to execute a Command Injection attack on the CAN bus [37]. It has the following CVSS v3.1 Base vector:

$$AV : N/AC : H/PR : N/UI : R/S : U/C : N/I : H/A : H$$

This translates to a Network for the Attack Vector (AV), High for Attack Complexity (AC), None for Privileges Required (PR), Required for User Interaction (UI), Unchanged for Scope, None for Confidentiality (C), High for Integrity (I), and

High for Availability. The CVSS score for this Base vector is 6.8. This CVSS Base Score is calculated based on a table of metric values and the following formulae found in CVSS v3.1 Specification Document [36]:

$$\text{BaseScore} = \begin{cases} 0 & \text{if Impact} \leq 0 \\ (\text{Min}\left[(\text{Impact} + \text{Exploitability})\text{ , }10\right]) & \text{if Scope is Unchanged} \\ \left(\text{Min}\left[\left(1.08 * \left(\text{Impact} + \text{Exploitability}\right)\text{ , }10\right]\right) & \text{if Scope is Changed} \end{cases}$$

Where

$$\text{Impact} = \begin{cases} 6.42 * \text{ISS} & \text{if Scope is Unchanged} \\ 7.52 * (\text{ISS} - 0.029) - 3.25 * (\text{ISS} - 0.02)^{15} & \text{if Scope is Changed} \end{cases}$$

$$\text{Exploitability} = 8.22 * \text{Attack Vector} * \text{Attack Complexity} * \text{Privileges Required} * \text{User Interaction}$$

$$\text{ISS} = 1 - \left[(1 - \text{Confidentiality}) * (1 - \text{Integrity}) * (1 - \text{Availability})\right]$$

Extending this to include the Temporal and Environmental metrics, we derive the following CVSS v3.1 vector:

$$AV : N/AC : H/PR : N/UI : R/S : U/C : N/I : H/A : H/E : X/RL : O/RC : C/CR : X/IR : X/AR : X/MAV : N/MAC : H/MPR : N/MUI : R/MS : U/MC : N/MI : H/MA : H$$

An explanation of the Temporal and Environmental metric notation is in order. The three metrics under the Temporal Score indicates E:X for undefined Exploitability, RL:O for Official fix on remediation, and RC:C for a confirmed Report Confidence. The eight metrics under the Environmental Score include MAV:N for a Modified Attack Vector on the Network, MAC:H for High on Modified Attack Complexity, MPR:N for none for Modified Privileges Required, MUI:R for required Modified User Interaction, MS:U for unchanged Modified Scope, MC:N for none on Modified Confidentiality, MI:H for high impact on Modified Integrity, and MA:H for high impact on Modified Availability. The overall CVSS score for the vector is 6.5. The Common Vulnerability Scoring System Calculator result is depicted in Fig. 1.

For the second illustration, we use the Mobile Devices OBD-II dongles firmware vulnerability (CVE-2015-2908). This disputed vulnerability does not validate firmware updates which enables the execution of arbitrary code remotely [38]. It has the following CVSS v3.1 Base vector:

$$AV : A/AC : H/PR : N/UI : R/S : U/C : H/I : N/A : H$$

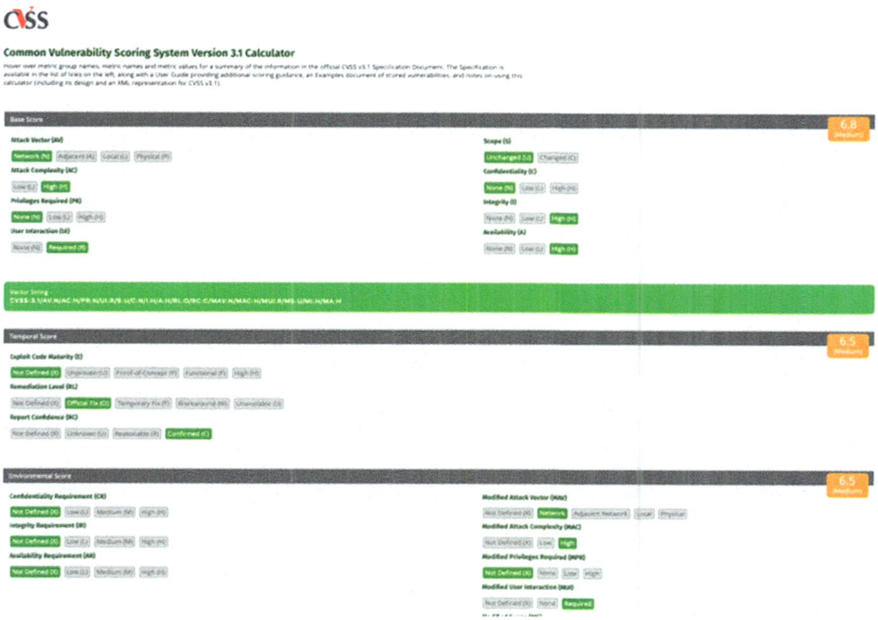

Fig. 1 CVSS calculator results

In this case, the Attack Vector (AV) is categorized as Adjacent Network (A) indicating that the vulnerable component is limited to the same physical network and cannot propagate beyond the layer 3 boundary of the network. The overall score for the Base metrics is 6.4.

4.2 Common Methodology for IT Security Evaluation (CEM) [39]

The CEM is a companion document to the Common Criteria for Information Technology Security Evaluation (CC) [40]. It defines the minimum actions to be taken by an evaluator conducting a CC evaluation utilizing the criteria and evidence as stated in the CC.

In this chapter, we specifically examine the attack potential on an automotive vehicle. The following factors need to be considered when performing an analysis of an attack potential:

- Elapsed time: Time taken by an attacker to identify a potential vulnerability, to develop an attack method, and to sustain effort required to execute the attack. Value ranges from 1 day to more than 6 months.

Table 1 Attack Potential calculation

Asset	Time	Expertise	Knowledge	Opportunity	Equipment	Total
False data from ECU	10	6	3	1	4	24
Blocking bus	4	3	3	0	4	14
Malicious software	10	6	3	1	4	24
Unauthorized access	1	3	3	1	2	10
Masquerading	4	3	5	5	5	22
Data tampering	1	2	4	1	2	10

- Specialist expertise: Describes the level of sophistication of the attacker. Levels include laymen, proficient personnel, expert and multiple experts.
- Knowledge of the target: Refers to the familiarity of the attacker on the target. Levels include public knowledge availability, restricted information, sensitive information, and critical information.
- Window of opportunity: This refers to the duration of time in which the vulnerability is exploitable. Window of opportunity includes unlimited, easy, moderate, difficult, and none.
- IT hardware/software or other equipment. This refers to the availability and the level of complexity of equipment/software needed to identify or exploit a vulnerability. Classes of equipment/software include standard, specialized, highly specialized, and multi-specialized.

Levels in each factor are assigned corresponding numeric values and illustrated in [39]. Bauer and Schartner demonstrate sample calculations of attack potential [10] on generic threat assets in an automotive vehicle. An excerpt of those calculations is shown on the first three rows of Table 1. The table is augmented by our own analysis of threats that are prevalent on connected automotive vehicles. Those last three rows represent unauthorized access, identity masquerading, and unauthorized data tampering.

An unauthorized access may originate locally, such as a cloning of key fob, or remotely through an internetwork communication channel. Time factor could be between 1 day and 1 week (1); expertise factor requires at least at the proficient level; knowledge of the vehicle assets will be most likely at the restricted level; the window of opportunity is unlimited; and the attack may not need specialized equipment.

The time to accomplish identity masquerading in connected vehicles may take a bit longer compared to unauthorized access; expertise factor requires at least at the proficient level; knowledge of the vehicle assets will be most likely at the sensitive level; the window of opportunity is very limited; and the attack may need some specialized equipment.

Data tampering can be accomplished by the widespread Man-In-The-Middle (MITM) attack tools. The time to accomplish such attack can take place very quickly; expertise factor requires at least at the semi-proficient level; knowledge of the vehicle assets will be most likely at the familiarity level; the window of opportunity is somehow large; and the attack may not need specialized equipment.

5 Conclusion and Future Research Directions

The recent developments in connected vehicle technology ushered newly found vulnerabilities in automotive vehicle systems. These vulnerabilities underscore the need to look closely at the state of automotive vehicle security. In conjunction with this effort, it is paramount that we investigate the metrics with which security can be measured. As a major component of continuous improvement, quantitative and qualitative measures must be devised to be able to make a full appreciation.

This chapter presents a comprehensive review of communication technologies found in modern vehicles. It covers the threats, vulnerabilities and attacks that are prevalent in modern automotive vehicles and the transportation infrastructure system. Further, widely recognized security metrics are adapted to automotive vehicle security. Sample metric calculations are illustrated to belabor the significance of the adaptations.

With the preceding discussions in mind, we offer the following future research directions:

- The development of a unified automotive vehicle security metrics framework that incorporates both the CVSS framework and the Common Criteria for Information Security Evaluation.
- The utilization of machine-learning techniques to predict the status of automotive vehicle security based on known vulnerability attributes. An ongoing research by the author in this area of applied ML appears to reveal promising results.

Acknowledgments This work is partially supported by the Florida Center for Cybersecurity, under grant # 3901-1009-00-A (2019 Collaborative SEED Program) and the National Security Agency under Grant Number H98230-19-1-0333. The U.S Government is authorized to reproduce and distribute reprints notwithstanding any copyright notation herein.

References

1. A. Karahasanovic, *Automotive Cyber Security* (Chalmers University of Technology University of Gothenburg, Gotehnburg, Sweden, 2016)
2. T. Micro, A Vulnerability in Modern Automotive Standards and How We Exploited It. (July 2017). [Online]. Available: https://documents.trendmicro.com/assets/A-Vulnerability-In-Modern-Automotive-Standards-and-How-We-Exploited-It.pdf. Accessed Nov 2018
3. SAE International, CAN Specification 2.0: Protocol and Implementations. (01 August 1998). [Online]. Available: https://www.sae.org/publications/technical-papers/content/921603/. Accessed 13 Oct 2019

4. CSS Electronics, A Simple Intro to LIN bus. (2019). [Online]. Available: https://www.csselectronics.com/screen/page/lin-bus-protocol-intro-basics/language/en. Accessed Oct 2019
5. National Instruments, FlexRay Automotive Communication Bus Overview. (28 May 2019). [Online]. Available: https://www.ni.com/en-us/innovations/white-papers/06/flexray-automotive-communication-bus-overview.html. Accessed 13 Oct 2019
6. EVITA Project, EVITA E-Safety Vehicle Intrusion Protected Applications. (01 December 2011). [Online]. Available: https://www.evita-project.org/. Accessed 13 Nov 2018
7. PRESERVE, About the Project. (June 2015). [Online]. Available: https://preserve-project.eu/about. Accessed 12 Oct 2019
8. SeVeCom, Security on the Road. (2008). [Online]. Available: https://www.sevecom.eu/. Accessed 13 Oct 2019
9. Society of Automotive Engineers (SAE), Cybersecurity Guidebook for Cyber-Physical Vehicle Systems J3061. (12 January 2012). [Online]. Available: https://www.sae.org/standards/content/j3061/. Accessed 13 Oct 2019
10. S. Bauer, P. Schartner, Reducing risk potential by evaluating specialized countermeasures for electronic control units, in *17th Escar Europe Conference 2019*, (Stuttgart, Germany, 2019)
11. D. Leopold, Relevance of ISO 21434 for the Automotive Development Process, ITEMIS. (20 December 2019). [Online]. Available: https://blogs.itemis.com/en/relevance-of-iso-21434-for-the-automotive-development-process. Accessed 12 Feb 2020
12. C. Schmittner, G. Griessnig, Z. Ma, Status of the Development of ISO/SAE 21434, in *Proceedings of the 25th European Conference, EuroSPI 2018*, (Bilbao, Spain, 2018)
13. D. Pauli, Hackers Hijack Tesla Model S from Afar, While the Cars are Moving. (16 September 2016). [Online]. Available: https://www.theregister.co.uk/2016/09/20/tesla_model_s_hijacked_remotely/. Accessed Oct 2019
14. J. Petit, M. Feiri, F. Kargl, Revisiting attacker model for smart vehicles, in *2014 IEEE 6th International Symposium on Wireless Vehicular Communications, WiVec 2014 Proceedings*, (2014)
15. J.-P. Monteuuis, J. Petit, J. Zhang, H. Labiod, S. Mafrica, A. Servel, Attacker model for connected and automated vehicles, in *ACM Computer Science in Cars Symposium (CSCS'18)*, (Berlin, Germany, 2018)
16. K. Koscher, A. Czeskis, F. Roesner, S. Patel, T. Kohno, S. Checkoway, D. McCoy, B. Kantor, D. Anderson, H. Shacham, S. Savage, Experimental security analysis of a modern automobile, in *2010 IEEE Symposium on Security and Privacy*, (Berkeley/Oakland, CA, 2010)
17. K.-T. Cho, K.G. Shin, Fingerprinting electronic control units for vehicle intrusion detection, in *Proceedings of the 25th USENIX Security Symposium (USENIX Security 16)*, (2016)
18. Q. Wang, S. Sawhney, VeCure: A practical security framework to protect the CAN bus of vehicles, in *International Conference on the Internet of Things (IOT)*, (Cambridge, MA, 2014)
19. M. Wolf, T. Gendrullis, Design, implementation, and evaluation of a vehicular hardware security module, in *14th International Conference on Information Security and Cryptology*, (Seoul, South Korea, 2011)
20. S. Lokman, T. Othman, M. Abu-Bakar, Intrusion detection system for automotive Controller Area Network (CAN) bus system: A review. EURASIP J. Wirel. Commun. Netw. **184** (2019)
21. S. Kumar, K. Singh, S. Kumar, O. Kaiwartya, Y. Cao, H. Zhao, Delimitated anti jammer scheme for internet of vehicle: Machine learning based security approach. IEEE Access **7**, 113311–113323 (2019)
22. Government Accountability Office (GAO), United States, Vehicle Cybersecurity: DOT and Industry Have Efforts Under Way, but DOT Needs to Define Its Role in Responding to a Real-world Attack. GAO Report 16-350. (2016). [Online]. Available: https://www.gao.gov/assets/680/676064.pdf. Accessed 14 Nov 2018
23. Society of American Engineers (SAE), Cybersecurity Guidebook for Cyber-Physical Vehicle Systems J3061. (17 January 2012). [Online]. Available: https://www.sae.org/standards/content/j3061/. Accessed 13 Nov 2018

24. C. McCarty, K. Harnett, A. Carter, *A Summary of Cybersecurity Best Practices* (National Highway Traffic Safety Administration, Washington, DC, 2014)
25. Intel Corporation, Intel Automotive Security Research Workshops. (2016). [Online]. Available: https://www.intel.com/content/www/us/en/automotive/automotive-security-research-workshops-summary.html?wapkw=automotive+security. Accessed 13 Nov 2018
26. S. Payne, A Guide to Security Metrics, SANS Institute. (19 June 2006). [Online]. Available: http://www.sans.org/readingroom/papers/5/55.pdf
27. K. Kark, P. Stamp, J. Penn, S. Bernhardt, A. Dill, Defining An Effective Security Metrics Program. (16 May 2007). [Online]. Available: https://www.forrester.com/report/Defining+An+Effective+Security+Metrics+Program/-/E-RES42354#. Accessed Feb 2020
28. S. Saydjari, Is risk a good security metric? in *Proceedings of the 2nd ACM Workshop on Quality of Protection*, (2006)
29. S. Schechter, Toward econometric models of security risk from remote attack. IEEE Secur. Priv., 40–44 (2005)
30. P. Manadhata, J. Wing, *An Attack Surface Metric–CMU-CS-05-155* (Carnegie Mellon University, Pittsburgh, PA, 2005)
31. G.A. Francia, Baseline operational security metrics for industrial control systems, in *International Conference on Security and Management*, (Las Vegas, NV, 2016)
32. L. Moukahal, M. Zulkernine, Security vulnerability metrics for connected vehicles in *2019 IEEE 19th International Conference on Software Quality, Reliability and Security Companion (QRS-C)*, (Sofia, Bulgaria, 2019)
33. C. McCarthy, K. Harnett, A. Carter, Characterization of Potential Security Threats in Modern Automobiles: A Composite Modeling Approach. (October 2014). [Online]. Available: https://rosap.ntl.bts.gov/view/dot/12119. Accessed 25 Feb 2020
34. B. Sheehan, F. Murphy, M. Mullins, C. Ryan, Connected and autonomous vehicles: A cyber-risk classification framework. Transp. Res. A **124**, 523–536 (2019)
35. G.A. Francia, X.P. Francia, Critical Infrastructure Protection and Security Benchmarks, in *Encyclopedia of Information Science and Technology*, 3rd edn., (IGI Global, Hershey, PA, 2014), pp. 4267–4278
36. Forum of Incident Response and Security Teams (FIRST), Common Vulnerability Scoring System version 3.1: Specification Document. (June 2019). [Online]. Available: https://www.first.org/cvss/specification-document. Accessed 13 Feb 2020
37. National Institute of Standards and Technology, CVE-2016-9337 Details (14 March 2017). [Online]. Available: https://nvd.nist.gov/vuln/detail/CVE-2016-9337. Accessed 13 Feb 2020
38. Common Vulnerabilities and Exposure. CVE-2015-2908. (3 April 2015). [Online]. Available: https://cve.mitre.org/cgi-bin/cvename.cgi?name=CVE-2015-2908. Accessed 13 Feb 2020
39. Common Criteria Portal, Common Methodology for Information Technology Security Evaluation. (July 2017). [Online]. Available: https://www.commoncriteriaportal.org/files/ccfiles/CEMV3.1R5.pdf. Accessed 24 Feb 2020
40. Common Criteria Portal, Common Criteria for Information Technology Security Evalaution. (April 2017). [Online]. Available: https://www.commoncriteriaportal.org/files/ccfiles/CCPART1V3.1R5.pdf. Accessed 24 Feb 2020

Requirements for IoT Forensic Models: A Review

Nawaf Almolhis, Abdullah Mujawib Alashjaee, and Michael Haney

1 Introduction

The increased use of IoT technologies by both individuals and enterprises for various applications will continue to spread. This has brought about the need for IoT forensics for investigations on security incidents in the IoT ecosystem. Consequently, this growth of usage has resulted in an increase of research in IoT forensics, and several IoT forensic methods and tools are proposed in the literature.

There are several issues and challenges investigators are encountering when conducting IoT forensics [1]. Many report that conventional digital forensic tools and methods cannot effectively be used in investigating IoT security incidents. In this light, new IoT forensic tools and process models that can guide procedures of investigators conducting IoT forensics are urgently required.

In this paper, the authors review the most prominent technical research publications to date in IoT forensics. As such, some IoT forensic challenges are identified. Subsequently, from these challenges, we present a set of requirements

N. Almolhis (✉)
Computer Science Department, University of Idaho, Moscow, ID, USA

Computer Science Department, Jazan University, Jazan, Saudi Arabia
e-mail: almo3113@vandals.uidaho.edu

A. M. Alashjaee
Computer Science Department, Jazan University, Jazan, Saudi Arabia

Computer Science Department, Northern Borders University, Arar, Saudi Arabia
e-mail: alas0145@vandals.uidaho.edu

M. Haney
Computer Science Department, University of Idaho, Moscow, ID, USA
e-mail: mhaney@uidaho.edu

K. Daimi et al. (eds.), *Advances in Security, Networks, and Internet of Things*,
Transactions on Computational Science and Computational Intelligence,
https://doi.org/10.1007/978-3-030-71017-0_25

that an IoT forensic process model should address. Furthermore, the IoT forensic process models identified in the literature have been evaluated based on the deduced requirements to shed light on gaps that remain.

This paper is organized as follows: Sect. 2 presents the related work, Sect. 3 presents challenges and issues about IoT forensics, Sect. 4 enumerates the identified IoT forensic model requirements, Sect. 5 maps the requirements to the existing models, and finally Sect. 6 concludes the paper.

2 Related Work

The "Internet of Things" (IoT) is a novel implementation of Internet-based technologies that integrates several collaborative and communication technologies that exist as objects in the real world, as opposed to traditional computers that exist with a monitor and keyboard for interaction with humans. IoT technologies have taken root in all sorts of lives of individuals and society at large, from small-scale consumer items such as dishwashers and refrigerators to larger moving objects such as automobiles, to even large-scale industrial systems that make up parts of manufacturing or energy plants. A significant characteristic of these IoT devices is their ability to collect data from many types of sensors. Much of the collected data may be considered personal to the devices' owners. IoT architectures include tracking and identification technologies integrated through actuator and sensor networks to allow communications between distributed intelligent objects including wearable smart devices that may collect information about locations without the consent of the user [2, 3].

Best practices and guiding principles of digital forensics and any sub-disciplines such as IoT forensics require not only the tools used for conducting the investigation but also the process of conducting the investigation to be reviewed, calibrated, verified, and approved such that they are reproducible independently. This helps ensure that digital evidence collected in a forensically sound way will be admissible by a court of law. Hence, to standardize and capture the process of conducting digital investigations, process models are considered vital to expedite the investigation and address issues investigators are encountering, especially with new technologies [4, 5]. It is based on the widely accepted models that different tools necessary for investigations of different technologies are developed.

There are some IoT-specific digital forensic models proposed in the literature. In [6], researchers have discussed the issues in digital forensics brought about by the use of IoT systems and proposed an IoT forensic deployment model that incorporates probable solutions in each of its phases *(M1)*. Researchers focused on the postmortem investigation, and they did not design and implement the model. Although the issue of user data privacy has been raised, researchers did not propose any probable solution in this regard.

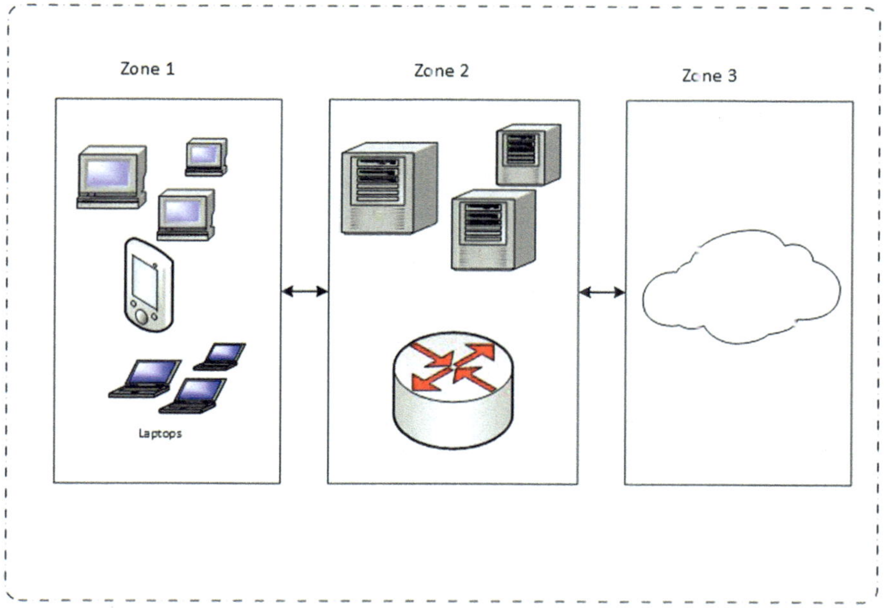

Fig. 1 Digital evidence acquisition model for IoT forensics [7]

The model proposed in [7] divides the IoT network into three zones including internal, middleware, and external networks as depicted in Fig. 1. Researchers adopted the triaging principle to their model in conjunction with the zones. Researchers claim that their model would be better suited for internal incident responders *(M2)*. The model does not cover the IoT devices and application aspect of the IoT forensic investigation. It rather works on the network layer of the IoT ecosystem. Besides, the model does not take into consideration the user privacy issues and, hence, does not give measures to protect user identity in the live data collected for analysis.

The model proposed in [8] is relatively more comprehensive. The model would increase the level of trust between different IoT consumers. In doing so, researchers proposed the employment of blockchain technology with a lightweight fragmented ledger in the IoT infrastructure as can be seen in Fig. 2 *(M3)*.

In [9], researchers proposed a blockchain-based IoT forensic model that preserves identity privacy throughout the lifecycle of the evidence. Researchers have given a working definition to IoT forensics. Figure 3 shows the sequential diagram for the evidence collection proposed. As the figure provides, the main roles are played by the attacker, the digital "witnesses" and custodians of data evidence, and the law enforcement agency. Collection steps depend on the case of whether there is sufficient evidence collected, if the victim has additional evidence or not. An issue

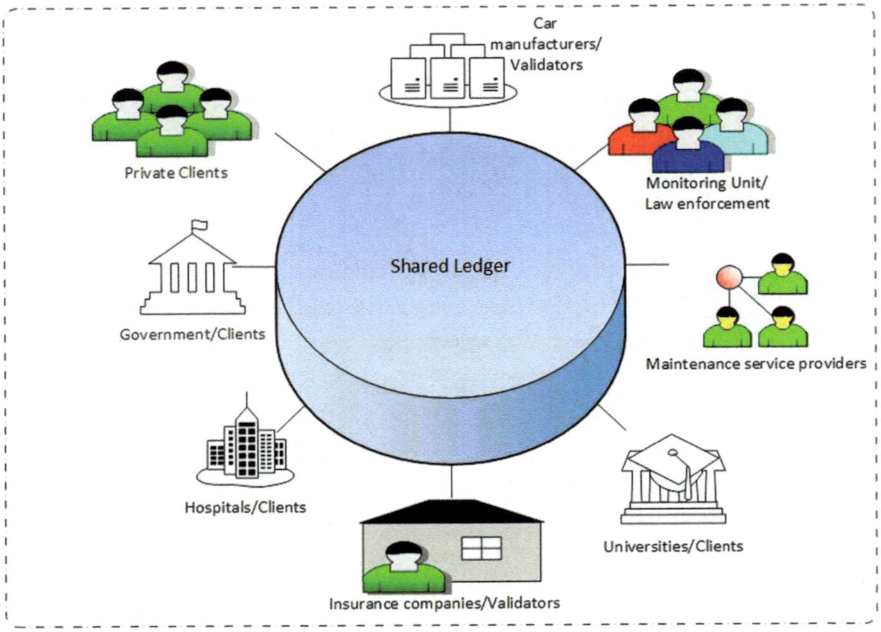

Fig. 2 Lightweight blockchain [8]

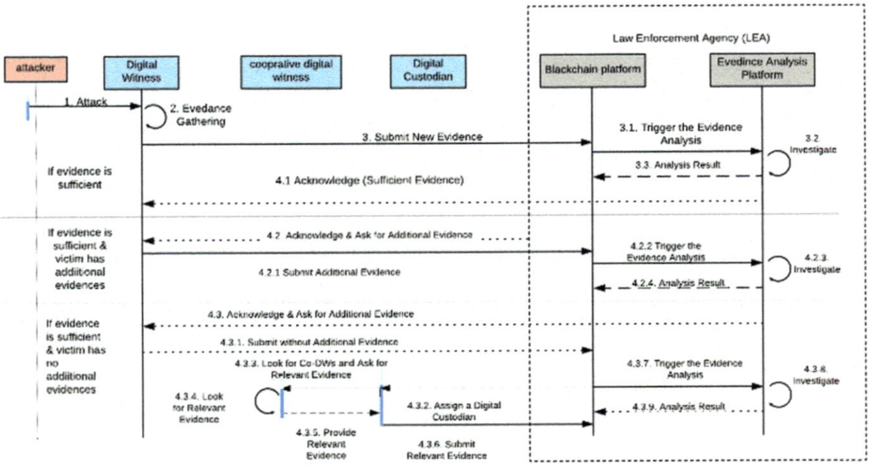

Fig. 3 Sequential diagram of evidence collection [9]

in this model is that it only focuses on the evidence collection after other important phases of the IoT investigation *(M4)*.

Researchers defined in [10] a conceptual model that uses a secure logging scheme that stores evidence in a centralized repository as shown in Fig. 4. In this model, an

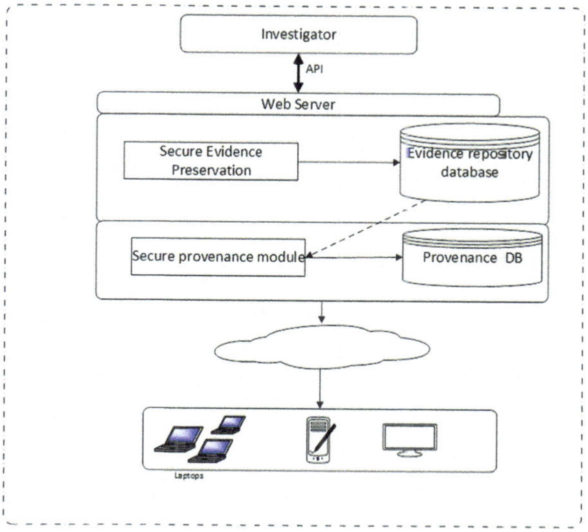

Fig. 4 A conceptual model for IoT forensics [10]

investigator interacts with a web-based system via an API. The system contains an evidence repository and an independent provenance database. The system automates the collection of data from mobile devices via the cloud infrastructure of the IoT system. Most of the work of the researchers focused on the preparation of the proactive part of the investigation *(M5)*. However, in the case of live forensics, most postmortem processes proposed in the model fall short.

In [11], the researchers introduce a fog-based IoT forensic framework in which they try to accommodate challenges associated with IoT forensics The framework uses fog computing to recover forensic evidence from the IoT ecosystem. The framework is proposed as a monitoring tool that identifies malfunctioning devices, collects associated evidence, and analyzes it to determine suspicious activities *(M6)*.

In [12], researchers proposed a privacy-aware model that stimulates the cooperation of different consumers in digital forensics. The model promotes a collection of evidence from surrounding IoT devices to fully describe the context of a crime scene. However, some researchers suggested that this solution does not mitigate the IoT forensic privacy issue and, hence, is not suitable for IoT devices [9] *(M7)*. One of the main issues associated with this model is that it directly applies an existing conventional digital forensic model to accommodate challenges faced by IoT forensic practitioners.

In [13], the researchers have based their model on three phase: forensic readiness, forensic initialization, and forensic investigation. The researchers did not raise the issue of privacy that most of the IoT forensic communities have pointed out as one of the main challenges IoT technology investigators are suffering from. Likewise, the live forensic measures reported in most papers as an integral part of the investigation in the IoT ecosystem are not discussed as the main part of the model *(M8)*. The researchers have raised relatively more general steps which complicate its suitability in IoT forensics. Nevertheless, the three phases proposed in the model are real parts of most investigations.

Most of the solutions in the literature focus on the postmortem or aftermath investigations rather than a readiness or proactive one. For instance, if a security incident did not disrupt the whole system, consumers cannot determine the status of a compromised IoT device. It is recommended to have some readiness measures such as monitoring mechanisms that can premeditatedly collect data that can be used as forensic evidence [11]. Similarly, extending conventional digital forensic tools and models to the IoT ecosystem is considered inefficient. Hence, a dedicated holistic model that takes together the readiness, live, and postmortem measures is needed for IoT forensics [14]. To that end, it is obvious that researchers are using either research products or their own experience to formulate the models. It is also notable that the models are indifferently accommodating several aspects of IoT forensics based on the need or experience of the developer. So to create some sort of uniformity, there must be some essential requirements the IoT forensic model should fulfill to be considered suitable in the IoT ecosystem.

3 Issues and Challenges in IoT Forensics

IoT is growing rapidly and being deployed in a wide range of applications starting from smart grids to healthcare and intelligent transport systems. The sensitive nature of private data that is collected and transmitted by IoT devices has attracted perpetrators [15]. IoT devices are highly constrained in terms of memory and processing to be secured properly, and their inability to support conventional security measures make them a low-hanging fruit for exploitation [16].

For instance, the Mirai attack in 2016, which is the prime example of such a notorious attack against the IoT ecosystem, propagated malware automatically that first infects IoT devices and then uses them as a launch platform for DDoS attacks [17, 18].

IoT security attacks may target and exploit vulnerabilities in any of the components of the IoT ecosystem, i.e., the sensor, the edge/fog, or the cloud layer. Digital forensics is needed to hold perpetrators accountable [19]. However, digital forensic practitioners are facing challenges in investigating IoT devices [20]. These challenges are described in the following paragraphs.

Scale-Up Heterogeneous Devices Different types of IoT technologies, operating systems, and network protocols from different vendors make it impossible to have a standardized approach in device-level forensics. For instance, device heterogeneity in the IoT ecosystem may call for specialized evidence collection tools [12]. For communication purposes, different wireless protocols may be associated with different IoT devices. Collecting evidence from Zigbee enabled devices which need a smart hub to connect to the Internet, and collecting evidence from Wi-Fi-enabled devices that may directly connect to a gateway router may not be the same. As a result, this makes it difficult to identify the source of the evidence in the IoT ecosystem.

Data Management in Devices Heterogeneous nature of devices regarding the methods in which data is distributed, aggregated, and processed poses challenges in forensic investigations. Such highly complex data collected from different sources in IoT may hinder the performance of the smooth analysis needed to make the required decisions [21].

User Privacy Collecting evidence only from suspected devices in a way that preserves the privacy of other innocent users is posing challenges to investigators. This is challenging not only at evidence acquisition but also when analyzing and correlating the collected evidence which may contain personal information. Privacy-preserving measures have to be taken throughout IoT forensic procedures [12, 21–23]. In the literature, it is obvious that IoT consumers are lacking the knowledge and understanding of their rights in their data. Most of the research and model developments with privacy preservation of consumer data are focused more on postmortem or passive preservation. Hence, the development of a holistic model that provides consumers the means of protecting their private data way before deployment IoT systems is necessitated.

Volatile Data Typically, through some type of wireless network, i.e., Wi-Fi, Bluetooth, or Zigbee. The volatile network traffic sent to and the corresponding data from the cloud by the IoT devices poses challenges to investigators.

Volume of Data The volume of data involved in IoT technologies is causing another challenge to IoT forensics, especially in situations where considerable time has elapsed after the incident has occurred. For example, the volume of IoT data captured by sensors and smart devices from networks and the cloud complicates the identification of relevant data.

Dependency on Cloud Service Provider Collection of evidence residing at the cloud data center frequently relies on the service provider. This would create unnecessary trust to the provider who may manipulate evidential data for preserving the reputation of his cloud services [16, 24].

Digital forensic models are used as a guide by practitioners as well as digital forensic tool developers. Hence, an IoT forensic model must get solution to the challenges reported in this section. Some of the challenges are directly related to the characteristics of the IoT ecosystem, while some are indirectly related such as those arising from other technologies that IoT devices usually use to accomplish daily activities including cloud computing, fog computing, and edge computing.

4 IoT Forensic Model Requirements

In an effort to overcome these challenges, researchers have formulated certain IoT forensic requirements that they think would provide solutions. In [1, 25], researchers presented essential requirement analysis approaches that can help digital forensic readiness (DFR) processes to be successfully implemented in an IoT environment. These include extraction of digital evidence, parsing forensic logs, digital preservation, creation of hash values, evidence storage, log analysis and characterization, and readiness report. Although digital forensic readiness is considered important in the IoT forensics, researchers did not take into consideration privacy measures that have to be taken at the readiness stage.

Requirements proposed for the digital forensic investigation of the IoT include integrity, non-repudiation, relieve single point of trust, persistence of forensic analysis, lightweights, and privacy [8, 26, 27]. Similarly, some key requirements are proposed in [21] including managing IoT data volume, mitigation of privacy risks, guidelines for the IoT deployment approaches, and dealing with system identification and human behaviors.

Some privacy-oriented IoT forensic requirements are proposed in [28]. Researchers translated privacy principles in ISO 29001 into requirements. These include consent and choice where an IoT consumer should give consent on the collection of his data. The purpose legitimacy and specification requirement promotes that the consumer has to be informed about the reason for the data collection. Collection limitation requirement refers to the collection of the data strictly relevant for the case in hand.

The data minimization requirement answers to the large volume of data that can be collected from the IoT ecosystem and hence promotes the reduction of the data to its minimum volume possible. Based on the use, retention and disclosure limitation requirement, data collected must not be used for a purpose other than the one originally specified. The accuracy and quality requirements ask an investigator that the data collection process must be done in a trusted and admissible way (Table 1).

In the openness, transparency, and notice requirement, a consumer must be informed on the procedure, policies and practices of the forensic analysis that his data is going to be subjected to. Similarly, according to the individual participation and access requirement proposed by the researchers, a consumer must have access to his data throughout the process of investigation.

The accountability requirement promotes that the investigator has to follow the privacy policies set forth for the collection and analysis of the evidence. And the information security control requirement protects collected personal data from unauthorized access, loss, and modification.

And finally, the compliance requirement incorporates the implementation of an auditing mechanism to ensure that the whole process of investigation complies with the investigative and privacy principles.

Table 1 Identified requirements

IID	Requirements
RQ1	Digital forensic readiness (DFR) processes include extraction of digital evidence, parsing forensic logs, digital preservation, creation of hash values, evidence storage, log analysis and characterization and readiness report
RQ2	The volume of IoT data captured by sensors and smart devices from networks and the cloud complicates the identification of relevant data
RQ3	The parties should be held responsible for their actions by providing proof of integrity
RRQ4	The system should have minimum overhead on endpoints since it includes multiple parties that may have different capabilities and resources
RRQ5	Consent and choice where an IoT consumer should give consent on the collection of his data
RRQ6	The purpose legitimacy and specification requirement promotes that the consumer has to be informed about the reason for the data collection
RRQ7	Collection limitation requirement refers to the collection of the data strictly relevant for the case in hand
RRQ8	Data minimization requirement answers to the large volume of data that can be collected from the IoT ecosystem and hence promotes the reduction of the data to its minimum volume possible
RRQ9	Based on the use, retention, and disclosure limitation requirement, data collected must not be used for a purpose other than the one originally specified
RRQ 10	The accuracy and quality requirements ask an investigator that the data collection process must be done in a trusted and admissible way
RRQ 11	In the openness, transparency, and notice requirement, a consumer must be informed on the procedure, policies, and practices of the forensic analysis that his data is going to be subjected to
RRQ 12	According to the individual participation and access requirement proposed by the researchers, a consumer must have access to his data throughout the process of investigation
RRQ 13	The accountability requirement promotes that the investigator has to follow the privacy policies set forth for the collection and analysis of the evidence
RRQ 14	The information security control requirement protects collected personal data from unauthorized access, loss, and modification
RRQ 15	The compliance requirement incorporates the implementation of an auditing mechanism to ensure that the whole process of investigation complies with the investigative and privacy principles

5 Mapping the Requirements to Existing Models

The summary of the IoT forensic models evaluated in this section is provided in Table 2. In this table, the sign ✓ shows that the requirement is supported by the model, and if the requirement is neither supported nor discussed by the IoT forensic model, then no sign is given.

To evaluate the forensic readiness (RQ1) in an IoT forensic model, the model is checked for its adoption of a pre-incident preparation approach to IoT forensic investigations. The model is also checked whether it provides a means of evaluating

Table 2 Mapping requirements against the model

Requirements	IoT forensic model							
	M1	M2	M3	M4	M5	M6	M7	M8
RQ1							✓	✓
RQ2		✓	✓					
RQ3							✓	
RQ4			✓	✓				
RQ5			✓	✓			✓	
RQ6							✓	
RQ7	✓						✓	
RQ8								
RQ9								
RQ10							✓	
RQ11							✓	
RQ12							✓	
RQ13							✓	
RQ14							✓	
RQ15	✓	✓	✓	✓	✓	✓	✓	✓

an IoT technology for its readiness for forensics and, specifically, whether technology takes a privacy-enhancing mechanism in order not to expose consumer private data in cases of investigation.

To evaluate the source of data (RQ2) generated by the IoT ecosystem, the model is assessed whether it implements data source triaging methods that would allow an investigator to identify the source of evidence for the case. Besides, the model is checked for its adoption of some non-repudiation measures where individuals involved in a case cannot deny the integrity of the collected data (RQ3).

Taking into consideration the resource-constrained nature of the IoT technologies (RQ4), the automated IoT forensic tool should not be creating a burden to the smart devices involved in the case. Hence, a model is checked for its employment measures that relieve the computational and storage capacity of the IoT devices.

Considering the privacy of the users of the IoT technologies, a model is assessed for its implementations of means that uphold the consent of the user when collecting evidence from their devices (RQ5). Besides, the model is checked whether it allows the investigator to inform the user for the reason of data collection (RQ6). The model is also assessed for implementing means of collecting the only case-specific data in order not to breach the privacy of other users not involved in the case (RQ7).

The large amount of data produced by the IoT ecosystem is one of the challenges investigators are facing when investigating the IoT ecosystem. The model is checked for its use of data reduction methods to ease the process of forensic analysis (RQ8). A model must also restrict data collected for a case to be used only in that specific case (RQ9). Principles of digital evidence collection have to be taken into consideration in a model (RQ10). A model must provide means that clarify to the owner about the policies and procedures that govern his data once collected for investigation (RQ11).

To evaluate for (RQ12), the model is assessed whether it has processes that allow the consumer to have a say on this collection at any time in the process of investigation. A model is also assessed for its implementation measures that can hold the investigator into account in case he/she perpetrates the privacy of the owner of the data (RQ13). Likewise, a model is evaluated whether it employs means of protecting collected data access from unauthorized individuals (RQ14). Finally, the model is checked for its inclusion of steps that ensure the compliance of the investigator to the process and procedure set for in compliance to the principles of the investigation and privacy (RQ15).

6 Conclusion

This paper highlights the state of the digital forensic process models specific to an IoT environment proposed in the literature. Given the increased impact of IoT forensics on conventional digital forensic process models, this paper identifies the requirements that an IoT forensic process model should cover to be used by IoT consuming enterprises. The requirements were collected from the literature with the IoT forensic issues and challenges faced by the digital forensic community. Subsequently, by evaluating the cloud forensic process models in the literature against the aforementioned requirements, gaps that need to be covered in standardizing cloud forensic have been identified and presented here.

References

1. U. Karabiyik, K. Akkaya, Digital forensics for IoT and WSNs, in *Mission-Oriented Sensor Networks and Systems: Art and Science*, (Springer, 2019), pp. 171–207
2. X. Caron et al., The Internet of Things (IoT) and its impact on individual privacy: An Australian perspective. Computer Law & Security Review **32**(1), 4–15 (2016)
3. A.N. Moussa, N.B. Ithnin, O.A. Miaikil, Conceptual forensic readiness framework for infrastructure as a service consumers, in *2014 IEEE Conference on Systems, Process and Control (ICSPC 2014)*, (IEEE, 2014)
4. M.D. Kohn, M.M. Eloff, J.H. Eloff, Integrated digital forensic process model. Comput. Secur. **38**, 103–115 (2013)
5. X. Du, N.-A. Le-Khac, M. Scanlon, Evaluation of digital forensic process models with respect to digital forensics as a service. arXiv preprint, arXiv:1708.01730 (2017)
6. R. Hegarty, D.J. Lamb, A. Attwood, Digital evidence challenges in the internet of things, in *INC*, (2014)
7. M. Harbawi, A. Varol, The role of digital forensics in combating cybercrimes, in *2016 4th International Symposium on Digital Forensic and Security (ISDFS)*, (IEEE, 2016)
8. M. Cebe et al., Block4forensic: An integrated lightweight blockchain framework for forensics applications of connected vehicles. IEEE Commun. Mag. **56**(10), 50–57 (2018)

9. D.-P. Le et al., BIFF: A blockchain-based IoT forensics framework with identity privacy, in *TENCON 2018–2018 IEEE Region 10 Conference*, (IEEE, 2018)
10. S. Zawoad, R. Hasan, Faiot: Towards building a forensics aware eco system for the internet of things, in *2015 IEEE International Conference on Services Computing*, (IEEE, 2015)
11. A. Nieto, R. Rios, J. Lopez, IoT-forensics meets privacy: Towards cooperative digital investigations. Sensors **18**(2), 492 (2018)
12. L. Sadineni, E. Pilli, R.B. Battula, A holistic forensic model for the internet of things, in *IFIP International Conference on Digital Forensics*, (Springer, 2019)
13. E. Al-Masri, Y. Bai, J. Li, A fog-based digital forensics investigation framework for IoT systems, in *2018 IEEE International Conference on Smart Cloud (SmartCloud)*, (IEEE, 2018)
14. J. Hou, L. Qu, W. Shi, A survey on internet of things security from data perspectives. Comput. Netw. **148**, 295–306 (2019)
15. M. Conti et al., *Internet of Things Security and Forensics: Challenges and Opportunities* (Elsevier, 2018)
16. M.G. Devi, M.J. Nene, Security breach and forensics in intelligent systems, in *Information and Communication Technology for Intelligent Systems*, (Springer, 2019), pp. 349–360
17. B. Herzberg, D. Bekerman, I. Zeifman, *Breaking Down Mirai: An IoT DDoS Botnet Analysis* (Incapsula Blog, Bots and DDoS, Security, 2016)
18. N. Koroniotis, N. Moustafa, E. Sitnikova, Forensics and deep learning mechanisms for botnets in internet of things: A survey of challenges and solutions. IEEE Access **7**, 61764–61785 (2019)
19. N.K. Bharadwaj, U. Singh, Acquisition and analysis of forensic artifacts from raspberry Pi an Internet of Things prototype platform, in *Recent Findings in Intelligent Computing Techniques*, (Springer, 2019), pp. 311–322
20. L. Babun et al., IoTDots: A Digital Forensics Framework for Smart Environments. arXiv preprint **arXiv:1809**, 00745 (2018)
21. I. Yaqoob et al., Internet of things forensics: Recent advances, taxonomy, requirements, and open challenges. Future Generation Computer Systems **92**, 265–275 (2019)
22. E. Oriwoh et al., Internet of things forensics: Challenges and approaches, in *9th IEEE International Conference on Collaborative Computing: Networking, Applications and Worksharing*, (IEEE, 2013)
23. Y.-N. Liu et al., Privacy-preserving raw data collection without a trusted authority for IoT. Comput. Netw. **148**, 340–348 (2019)
24. A.N. Moussa, N. Ithnin, A. Zainal, CFaaS: Bilaterally agreed evidence collection. J. Cloud Comput. **7**(1), 1 (2018)
25. V.R. Kebande, N.M. Karie, H.S. Venter, Functional requirements for adding digital forensic readiness as a security component in IoT. Environments (2018)
26. J.H. Ryu et al., A blockchain-based decentralized efficient investigation framework for IoT digital forensics. J. Supercomput., 1–16 (2019)
27. A.N. Moussa et al., A consumer-oriented cloud forensic process model, in *2019 IEEE 10th Control and System Graduate Research Colloquium (ICSGRC)*, (IEEE, 2019)
28. A. Nieto, R. Rios, J. Lopez, A methodology for privacy-aware IoT-forensics, in *2017 IEEE Trustcom/BigDataSE/ICESS*, (IEEE, 2017)

Mobile Malware Forensic Review: Issues and Challenges

Abdullah Mujawib Alashjaee, Nawaf Almolhis, and Michael Haney

1 Introduction

With the dramatic increase of mobile device usage, users need to protect sensitive data and information in their mobile devices more than ever before. The increased practice of storing confidential information in smartphones has increased the motive for hackers to gain access to this data. The widespread adoption of mobile technologies not only has attracted legitimate users but also has become the main target for malicious code writers. However, to protect their data against malware, mobile device users have adopted conventional approaches such as antivirus, despite not being efficient [1]. Once a security incident has occurred, digital forensic experts may be called upon to investigate. Subsequently, first responders are expected to use a proven step-by-step methodology to investigate the incident and then provide evidence accordingly [2]. Investigating mobile devices is defined by the National Institute of Technology as "the science of recovering digital evidence from

A. M. Alashjaee (✉)
Computer Science Department, University of Idaho, Moscow, ID, USA

Computer Science Department, Northern Borders University, Arar, Saudi Arabia
e-mail: alas0145@vandals.uidaho.edu

N. Almolhis
Computer Science Department, University of Idaho, Moscow, ID, USA

Computer Science Department, Northern Borders University, Arar, Saudi Arabia

Computer Science Department, Jazan University, Jazan, Saudi Arabia
e-mail: almo3113@vandals.uidaho.edu

M. Haney
Computer Science Department, University of Idaho, Moscow, ID, USA
e-mail: mhaney@uidaho.edu

K. Daimi et al. (eds.), *Advances in Security, Networks, and Internet of Things*,
Transactions on Computational Science and Computational Intelligence,
https://doi.org/10.1007/978-3-030-71017-0_26

a mobile device under forensically sound conditions using accepted methods" [3]. Nevertheless, it is often difficult for responders to properly identify mobile malware that is using obfuscation techniques to hide from malware detection software [4]. Hence, for a thorough and correct analysis, it is necessary to understand the unique characteristics of mobile malware forensics [5].

Mobile malware forensics can be understood as a process to analyze malicious software affecting mobile devices (e.g., smartphones, tablets, etc.) using reliable information that can be applied to the investigation of an incident. To make such malware forensics more effective and efficient, the investigator needs a thorough understanding of the mobile malware types, including both the event and the agents involved in the malware attack. In most cases, forensic experts use a controlled environment employing malware detection and analysis tools. During mobile malware forensic analysis, investigators need to understand the behavior of the malware being used for the incident and the underlying characteristics of the operating system and hardware that are unique to mobile devices.

The rest of the paper is organized as follows: In Sect. 2, the impact of malware on the mobile industry is reviewed. Section III discusses mobile malware threats and exposure, Section IV studies categories of mobile malware, and Sect. 5 introduces current mobile malware forensic best practices. In Sect. 6, the authors present the efficiency and effectiveness of the existing tools, Sect. 7 studies mobile malware detection techniques, and Sect. 8 concludes the paper.

2 Impact of Malware on the Mobile Industry

Malware is referred to as any software with the intention of mischievous activities. It disrupts normal functioning, bypasses access controls, displays unwanted marketing, gathers sensitive information about the user, or attempts to get control of proprietary or public systems without user's consent [6]. Malware includes all malicious software such as viruses, worms, Trojans, ransomware, rootkits, botnets, etc.

Mobile malware is the kind of malicious software primely prepared to target mobile devices such as smartphones and tablets. Such malware includes any kind of software that disrupts the functionality of mobile devices without the knowledge of the user [7]. Mobile malware is becoming sophisticated in a way that presents a serious threat to mobile devices due to malicious activities such as stealing sensitive data, exfiltrating premium messages, making calls, etc. It is reported that mobile malware has experienced an increase of more than 1800% in 2016 [8]. Because of the bring-your-own-device strategy employed by many organizations, businesses are experiencing huge numbers of mobile malware attacks. Kaspersky in a recent release stated that in 2019 mobile users encountering malware have more than tripled to 1.7 million globally [9]. According to McAfee Labs, they detected massive mobile malware attacks of 16 million events in the third quarter of 2017 alone [6]. Android malware alone has increased by 400% in 2019 [6].

What makes mobile malware so prevalent is that it is continuously updated with new features and techniques that take advantage of new distribution methods. Obfuscation methods, repackaging, and stealth techniques have also exacerbated the need for up-to-date analysis and detection techniques [10]. The majority of mobile malware uses legitimate mobile apps to bypass barriers [11, 12].

3 Mobile Malware Threat and Exposure Analysis

Mobile malware is often mistakenly considered to be just like any other computer malware. Mobile malware is somehow different from computer malware; in the same way, computer applications are different from mobile apps. In mobile malware, each specimen of malware has its specific security threats and characteristics which specifically target mobile devices and their capabilities, such as the presence of motion sensors, location sensors, cameras, and microphones, as well as the types of personal user data expected to be present, such as personal photos and videos. In this light, to conduct forensics of mobile malware, one must consider the kind of threats to mobile devices and their corresponding features [13].

Mobile malware attacks are most often either hardware- or software-based. Hardware-based attacks are more devastating because it generally leaves no room for repair but requires the user to replace their device [14]. This happens when attackers are trying to get super user privileges by inserting specific commands or firmware into the mobile device. With hardware-based malware attacks, attackers can easily achieve their goal of sabotage or espionage. Usually, this kind of malware attack results in permanent alteration of the device. On the other hand, software-based mobile malware attacks have effects that are not that much different from those of computer malware attacks [15]. This kind of attack makes use of existing mobile applications. With such attacks, benign Android apps are often changed into malicious ones and subsequently uploaded to the market for download.

4 Categorizing Mobile Malware

Mobile malware can be of several classes based on their malicious goals and behaviors. However, mobile malware uses two primary distribution strategies: self-propagation and social engineering. The self-propagation of "worms" results in automatically installing malware payloads into mobile devices. With the social engineering approach, the attacker would deceive users to allure them to manually install mobile apps [16]. The most popular types of mobile malware include but is not limited to banking malware, mobile ransomware, mobile spyware, MMS and SMS Trojan malware, and mobile adware [17]. Figure 1 shows mobile malware evolutions as of 2018.

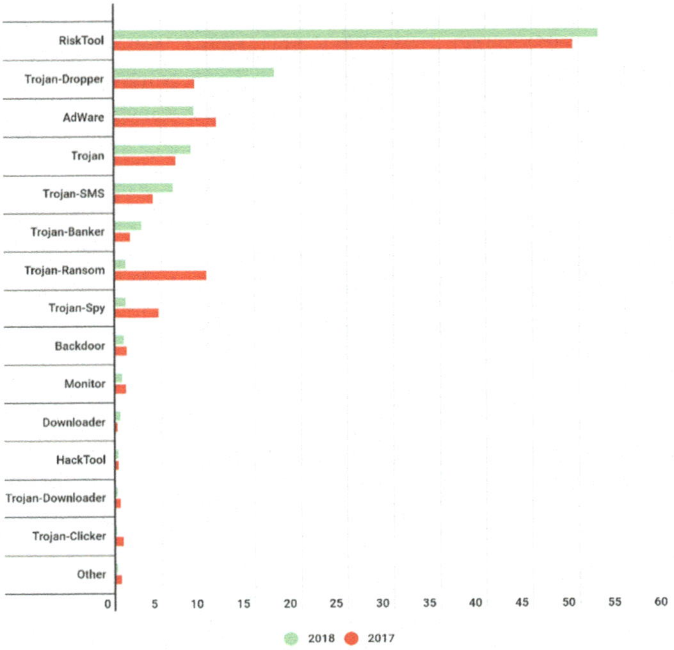

Fig. 1 Types of mobile malware [18]

5 Mobile Malware Forensics

Investigating security incidents involving mobile devices is the practice of mobile forensics. In the domain of digital forensics, the tools and techniques used for conducting an investigation, the procedures followed, and the competence of the investigator are important and specific to the sub-discipline. Compared to general computer forensics, mobile forensics is more complicated. That is, due to mobile devices being constantly active and updating information in their memory, the most important data is much more volatile, and the devices generally offer less access to the investigator [19]. Usually, mobile devices store sensitive data about the owners in their memory. And that is why mobile devices are of special interest to forensics. The recent shift of communication toward mobile devices and the use of different messaging applications with a wide variety of features have exacerbated issues and challenges pertaining to the mobile forensic investigations [20]. Hence, this environment has grasped the focus of many forensic institutions and individuals researching new tools and techniques for this sub-discipline.

Several researches have begun to define the process of conducting mobile forensics. For example, the United States Secret Service (USSS) and the National Institute of Justice have proposed two mobile forensic guidelines. However, both guidelines do not provide enough details on how to forensically approach smartphones, because they do not address smartphone storage capabilities and different applications of them [19, 21]. Mobile forensics like any other digital forensic model have proactive, active, and postmortem investigative mechanisms [12]. For instance, the proactive mobile forensics may include some monitoring measures that detect potential threats and issue alerts based on pre-assigned thresholds [22, 23]. Mobile intrusion detection systems can be considered in this part of the mobile forensic process. And it is in this phase of the mobile forensic process that mobile malware detection and analysis systems may fit in.

Malware analysis reveals detailed behavior of malware. Apart from using it for digital investigations, the extracted characteristics of malware can later be used to detect future strains of similar malware. In other words, understanding the functional goal and behavior of malware is a necessary step toward conducting an admissible malware forensic investigation. But it usually requires experience or a careful analysis of the targeted malicious activities. Hence, malware analysis can be regarded as a twofold process: Firstly, it is for investigating for legal purposes and, secondly, for the development of malware detection systems. In this paper, the research focuses on the analysis of malware to expose its creator or for the reconstruction of malicious events.

Mobile malware is controlled by a group of perpetrators who may include criminals and ethical practitioners with an organized plan that targets crimes involving mobile devices. Therefore, mobile malware forensics is the process of detecting, identifying, collecting, and analyzing evidence about malware involving mobile devices [20, 24]. Mobile malware forensics involves four aspects: identification of suspicious programs, defeating the anti-forensic code, extraction of malicious code from malware, and deduction of its malicious functions [6, 25].

6 Efficiency and Effectiveness of Mobile Malware Analysis

There is a plethora of research in mobile malware analysis systems. Efficiency and effectiveness are the major attributes of mobile malware analysis. To identify the admissibility of the evidence, the effectiveness of the process is considered. Likewise, efficiency in mobile forensics is related to the tools that are used for the evidence collection. The two attributes are different in terms of their functionality and the roles they play independently in mobile forensics [26].

Mobile malware developers are trying to employ different techniques to deter and hide from malware analysis tools so that they remain unseen. Together with the massive increase of malware released each year, there is a need for employing more appropriate tools and techniques [27]. Naturally, mobile malware analysis can be conducted manually; however, these are time-consuming and less accurate as opposed to automatic techniques [28]. In smartphones where different applications can communicate with each other, evidence can only be tracked by correlating between different sources [29]. Appropriate tools are required to execute mobile malware forensics to obtain usable forensic information for further investigation. A substantial amount of research has addressed this [15]. For instance, conventional digital forensic tools such as EnCase and Forensic Toolkit (FTK) cannot be directly applied in the domain malware forensics let alone mobile malware which happens to be more complicated. Hence, developing solutions that can analyze mobile malware for forensic purposes specific to certain mobile operating systems may result in a far more efficient methodology. Some early developments of this kind of solution can be evidenced in the literature [25]. Particularly, such tools have to be developed with more efficiency and effectiveness [6, 30].

7 Mobile Malware Detection Techniques

Malware detection systems are used for detecting malicious programs in the system. Malware detection and analysis approaches can be of static and dynamic techniques or both. For example, there are some static mobile malware detection tools in the literature [31–34]. Such tools examine the code of programs without running any executable application. They usually check the functionality of a program by means of signature comparison. But they are considered less effective compared to dynamic analysis tools that analyze behavior. Such static analysis is generally ineffective against packed or heavily obfuscated malware.

On the other hand, dynamic malware detection conducts investigations by running the whole program in a virtualized environment [6, 35]. Dynamic mobile malware detection performs checks to API or system calls, and instruction traces look for system modification and check for variability in system memory [27]. In [36], the researchers used the method of modifying system calls to capture traces of malware based on its behavior. The issue associated with the dynamic mobile malware analysis is that it creates overhead to the system and affects the performance of the mobile devices. That is because it requires resources such as memory and computing power to accomplish its job. Additionally, due to the burden of setting up a virtual environment that accommodates different mobile operating systems and dependent libraries, dynamic analysis is hard to automate [34].

There are mobile malware detection and analysis tools that use an amalgamation of static and dynamic techniques. Such tools are referred to as hybrid malware analysis tools, and they overlay the parameters of static and dynamic analysis [37]. There are fewer tools that employ a hybrid analysis method in the literature compared to static and dynamic analysis. The detection accuracy of malware is higher in the hybrid analysis as compared to adopting a static or dynamic method. However, the drawback of the hybrid analysis is that it is slower during malware investigation. Apart from the static and dynamic analysis, some researchers have employed data mining or machine learning mechanism for increased efficiency of mobile malware analysis. Table 1 compares and summarizes the advantages and limitations identified in each of the three mobile malware analysis types.

Commercial anti-malware programs usually consist of a scanner and a signature database. For example, scanner matches file on the mobile and matches them against

Table 1 Comparison between mobile malware detection techniques

Factors	Static analysis	Dynamic analysis	Hybrid analysis
Time required	Needs less time	Needs more time	Needs more time
Input	Binary file, scripting, language file, etc.	Memory snapshots, runtime API data	Data obtained from both static and dynamic
Code obfuscation	Less effective to analyze packed malware	Effective to analyze packed malware	More effective to analyze packed malware
Resource conception	Needs less resources	Needs more resources	Needs more resources
Effectiveness and efficiency	Less compared to dynamic analysis	Better than static analysis	Better than both static and dynamic analyses
Target code execution	Conducts analysis without running malware code	Conducts analysis on executable malware code	Conducts analysis with or without running the malware code
Advantages	Requires less time and low cost	Provides deeper analysis and higher detection rate with unknown malware detection	Extracts features of both static and analyses and provides more accurate analysis
Limitations	Limited to signature database and can only detect known malware	More time and resource consumption	High cost

the available signatures. Subsequently, an alert is generated informing the user when a match is found. Despite being extremely exact, it does not work against obscure dangers because it requires consistent signature updates. In cases of new malware, its signature has to be identified and added to the database [38]. Norton Mobile Security Lite and TrendMicro Mobile Security Personal Edition are two examples. In [39], the two tools have been subjected to a set of mobile malware software. The result showed that only 79.6% of the threats were detected the two commercial tools.

8 Conclusion

This paper reviews the recent literature related to mobile malware forensics. As a result, several malware forensic issues and challenges that need immediate attention of the research community are identified. To better understand the domain, the study was done in the context of issues and challenges associated with mobile malware detection and analysis. It was noted that there is a lack of widespread competence among investigators of mobile malware. To this end, mobile malware forensic processes intended for investigations involving malware forensics in mobile devices are discussed. The importance of efficient and effective mechanisms including robust hybrid analysis environments was found to be much needed for malware forensics.

References

1. Q. Zhou et al., A novel approach for mobile malware classification and detection in Android systems. Multimed. Tools Appl. **78**(3), 3529–3552 (2019)
2. Z. Grimmett, J. Staggs, S. Shenoi, Categorizing mobile device malware based on system side-effects, in *IFIP International Conference on Digital Forensics*, (Springer, 2017)
3. R.P. Ayers, S. Brothers, W. Jansen, in *Guidelines on Mobile Device Forensics* (2014)
4. C. Lim, K. Ramli, Y.S. Kotualubun, Mal-Flux: Rendering hidden code of packed binary executable. Digit. Investig. **28**, 83–95 (2019)
5. X. Zhang, T.T. Yuen, K.-K.R. Choo, Experiential learning in digital forensics, in *Digital Forensic Education*, (Springer, 2020), pp. 1–9
6. A. Qamar, A. Karim, V. Chang, Mobile malware attacks: Review, taxonomy & future directions. Futur. Gener. Comput. Syst. **97**, 887–909 (2019)
7. M. Alazab, Profiling and classifying the behavior of malicious codes. J. Syst. Softw. **100**, 91–102 (2015)
8. M. Alazab et al., Intelligent mobile malware detection using permission requests and API calls. Futur. Gener. Comput. Syst. (2020)
9. E.W. Burroughs, *Pocket Sized Threats: Discussing Malware Attacks on Android Smartphones* (Utica College, 2019)
10. J. Milosevic, M. Malek, A. Ferrante, Time, accuracy and power consumption tradeoff in mobile malware detection systems. Comput. Secur. **82**, 314–328 (2019)
11. N.K. Gyamfi, E. Owusu, Survey of mobile malware analysis, detection techniques and tool, in *2018 IEEE 9th Annual Information Technology, Electronics and Mobile Communication Conference (IEMCON)*, (IEEE, 2018)
12. A.N. Moussa et al., A consumer-oriented cloud forensic process model, in *2019 IEEE 10th Control and System Graduate Research Colloquium (ICSGRC)*, (IEEE, 2019)
13. D. Geneiatakis et al., Towards a mobile malware detection framework with the support of machine learning, in *International ISCIS Security Workshop*, (Springer, 2018)
14. P. Yan, Z. Yan, A survey on dynamic mobile malware detection. Softw. Qual. J. **26**(3), 891–919 (2018)
15. G. Suarez-Tangil et al., Evolution, detection and analysis of malware for smart devices. IEEE Commun. Surv. Tutor. **16**(2), 961–987 (2013)
16. V. Kouliaridis et al., A survey on mobile malware detection techniques. IEICE Trans. Inf. Syst. **103**(2), 204–211 (2020)
17. M. Fan et al., Graph embedding based familial analysis of android malware using unsupervised learning, in *2019 IEEE/ACM 41st International Conference on Software Engineering (ICSE)*, (IEEE, 2019)
18. K.L. products, Mobile malware evolution 2018, in *Kaspersky Official Web Page*, (2019)
19. R. Ahmed, R.V. Dharaskar, Mobile forensics: An introduction from Indian law enforcement perspective, in *International Conference on Information Systems, Technology and Management*, (Springer, 2009)
20. A.N. Moussa, N. Ithnin, A. Zainal, CFaaS: Bilaterally agreed evidence collection. J. Cloud Comp. **7**(1), 1 (2018)
21. K. Alissa et al., A comparative study of WhatsApp forensics tools. SN Appl. Sci. **1**(11), 1320 (2019)
22. V.R. Kebande, N.M. Karie, S. Omeleze, A mobile forensic readiness model aimed at minimizing cyber bullying. Int. J. Comp. Appl. **140**(1) (2016)
23. A.N. Moussa, N.B. Ithnin, O.A. Miaikil, Conceptual forensic readiness framework for infrastructure as a service consumers, in *2014 IEEE Conference on Systems, Process and Control (ICSPC 2014)*, (IEEE, 2014)
24. M. Spreitzenbarth, Tools and processes for forensic analyses of smartphones and mobile malware, in *SPRING-SIDAR Graduierten-Workshop über Reaktive Sicherheit*, (Bochum, Deutschland, 2011), p. 2011

25. J. Li, D. Gu, Y. Luo, Android malware forensics: Reconstruction of malicious events, in *2012 32nd International Conference on Distributed Computing Systems Workshops*, (IEEE, 2012)
26. M. Kim et al., A study on behavior-based mobile malware analysis system against evasion techniques, in *2016 International Conference on Information Networking (ICOIN)*, (IEEE, 2016)
27. K. Barmpatsalou et al., Mobile forensic data analysis: Suspicious pattern detection in mobile evidence. IEEE Access **6**, 59705–59727 (2018)
28. G. Suarez-Tangil et al., ALTERDROID: Differential fault analysis of obfuscated smartphone malware. IEEE Trans. Mob. Comput. **15**(4), 789–802 (2015)
29. L. Caviglione et al., Seeing the unseen: Revealing mobile malware hidden communications via energy consumption and artificial intelligence. IEEE Trans. Inf. Forensics Sec. **11**(4), 799–810 (2015)
30. J. Alhassan et al., Comparative evaluation of mobile forensic tools, in *International Conference on Information Theoretic Security*, (Springer, 2018)
31. Y. Shao et al., Kratos: Discovering inconsistent security policy enforcement in the android framework, in *NDSS*, (2016)
32. S.Y. Yerima, S. Khan, Longitudinal performance analysis of machine learning based android malware detectors, in *2019 International Conference on Cyber Security and Protection of Digital Services (Cyber Security)*, (IEEE, 2019)
33. A. Kumar, K. Kuppusamy, G. Aghila, FAMOUS: Forensic Analysis of MObile devices Using Scoring of application permissions. Futur. Gener. Comput. Syst. **83**, 158–172 (2018)
34. X. Lin et al., Automated forensic analysis of mobile applications on android devices. Digit. Investig. **26**, S59–S66 (2018)
35. H. Ruan et al., Analyzing android application in real-time at kernel level, in *2017 26th International Conference on Computer Communication and Networks (ICCCN)*, (IEEE, 2017)
36. X. Su, M. Chuah, G. Tan, Smartphone dual defense protection framework: Detecting malicious applications in android markets, in *2012 8th International Conference on Mobile Ad-hoc and Sensor Networks (MSN)*, (IEEE, 2012)
37. B. Team, *SandDroid: An APK Analysis Sandbox* (Xian Jiaotong University, 2014)
38. H. Alimardani, M. Nazeh, A taxonomy on recent mobile malware: Features, analysis methods, and detection techniques, in *Proceedings of the 2018 International Conference on E-Business and Mobile Commerce*, (2018)
39. Y. Zhou, X. Jiang, Dissecting android malware: Characterization and evolution, in *2012 IEEE Symposium on Security and Privacy*, (IEEE, 2012)

The Organizational Cybersecurity Success Factors: An Exhaustive Literature Review

Nancy Poehlmann, Kevin Matthe Caramancion, Irem Tatar, Yueqi Li, Mehdi Barati, and Terry Merz

1 Introduction

The research questions at the foundation of this paper are as follows: Where are the gaps in cybersecurity research? Where should future research focus its efforts? To that end, this paper investigates the secondary literature written about whether technology design has any effect on cybersecurity posture. The authors focused on five factors: (a) the technology itself; (b) the cybersecurity procedures formulated by the management of an organization; (c) the organizational structure and its effect on cybersecurity; (d) the laws that affect cybersecurity on national, state, and local levels; and (e) the human factors that affect correct implementation of cybersecurity procedures.

Be it through the Internet of Things, daily interactions, or business operations, technology is rapidly changing the way people think and behave. Broadly speaking, technology is the application of scientific knowledge in practice, including tools and machines, to assist human beings in solving real-world problems. In the context of cybersecurity, technology is a widely used tool in cyberattack prevention, deterrence, and detection.

Proper cybersecurity technology can prevent most attacks, quickly detect vulnerabilities, mitigate cybersecurity risks, and assure the security of strategic business initiatives (such as digital transformation). Today, everyone, from top management to staff employees, has to utilize technology to complete regular tasks. As tools used

N. Poehlmann (✉) · K. M. Caramancion · I. Tatar · Y. Li · M. Barati · T. Merz
Department of Information Science, University at Albany, State University of New York, Albany, NY, USA
e-mail: npoehlmann@albany.edu; kcaramancion@albany.edu; itatar@albany.edu; yli69@albany.edu; mbarati@albany.edu; tmerz@albany.edu

K. Daimi et al. (eds.), *Advances in Security, Networks, and Internet of Things*, Transactions on Computational Science and Computational Intelligence,
https://doi.org/10.1007/978-3-030-71017-0_27

by organizational users, especially those created for security purposes, technology has direct influence on cybersecurity posture in the organization.

A common, yet crucial, organizational factor affecting the success of cybersecurity posture design is how to handle threats when they occur. Based on existing best practices, these approaches should be documented in corporate manuals and registers. This paper will explore the elements contributing to successful development of these risk procedure manuals, including a discussion of the legal environment and human competencies.

1.1 Significance and Pragmatic Implications

In the study of technology design effects on cybersecurity posture, one must consider different factors independently as well as the interactions among the factors. Organizational factors play a key role in any technology design, by shaping and enforcing the behavior of the human players in the organization, as well as mediating the differences in management procedures and legal environments. The security behavior of the employees in an organization is affected by many factors including personal characteristics and background, educational level, and the industry's cultural norms, among others. However, procedures implemented in the organization cause variations in employees' behavior and in a company's technology choices. The literature shows this interdependence between organizational procedures and management procedures, as the authors explore below.

Also, a discussion of cybersecurity measures, training, and protocols must include a consideration of the laws written to ensure compliance from private and public sector companies and their employees and that such compliance will fall under the protection of the law. Recent literature on cybersecurity law has pointed out a lack of definition in federal legislation and confusion among the various federal agencies tasked with enforcing compliance with existing legislation. Further, this lack of coordination among federal agencies is exacerbated by a lack of harmonization among federal and state legislations. Without coordination and harmonization, especially of training and protocols, there is an unnecessary and dangerous duplication of effort, as outlined below.

Finally, human's ability to identify risks, prioritize mitigation actions, and make the right decision, both before an attack occurs and after a successful attack, is a critical component in cybersecurity. Conversely, human reactions are also the weakest link in cybersecurity. Whether maliciously or unintentionally, a user may easily cause a well-defended system to be bypassed by threat actors. Companies and agencies can use technology design methods, such as improving situational awareness, usability, effectiveness of visualization methods, and security compliance, to reduce the likelihood of cyber incidents due to human error.

2 Literature Overview

There are three aspects of technology that can affect cybersecurity posture in the organization: application of advanced technology, improving the usability of security technology, and design of technology.

First, the application of advanced technology in creating new cybersecurity technology or improving existing technology can directly influence the effectiveness of cybersecurity systems. Prevalent examples are the use of biometric technology and artificial intelligence. Detection, prevention, and automation are three technological attributes that organizations are creating to improve security technologies. In prevention-based strategies, organizations try to enhance their systems to stop attack; in detection-based strategies, security teams work proactively to identify and remediate threats that have undermined the organization's defenses. Automation levels, both attack and defense sides, reduce the number of threats and prevent new and previously unknown attacks more quickly. However, few studies have examined the singular or collective effects of these three elements on organizational cybersecurity posture.

Second, applying security usability principles to the risk management process and security solutions improves the overall usability of security technology. These principles require users to learn and understand security actions and conclusions; it also demands a mental and physical load of such security actions and whether the conclusion is tolerable. Whether the usability level of security technology affects overall cybersecurity posture remains to be answered.

Third, design of technology should also be considered because when and how to incorporate cybersecurity characteristics into technology design processes affect the way such technology contributes to cybersecurity usefulness and effectiveness. Although cybersecurity feature installation has been illustrated in I&C systems, there is no framework indicating how cybersecurity features and risk assessment should be incorporated specifically for security technologies. Furthermore, whether variances in technology design processes influence organizational cybersecurity posture should be examined in future studies.

The existing literature surrounding management procedures in building a firmly built and potent cybersecurity design focuses on risk management, including planning, response, and actors. Understandably, most benchmarked procedures start with a data-driven calculation of perceived risks that an organization may meet in the future. An important note, however, in these risk calculations is not all possible categories of threat may be planned and accounted for; these calculations will later be known as the "unknown unknowns." The usual design of risk mitigation procedure manuals follows the creation of a plan that will be later referenced on when a certain perceived risk occurs. However, this response manual is not final in nature, because it is continuously improved, based on the collective experience of the organization.

The next and extremely important phase is the response itself, which should be based on the balance between the organizational resources and the attempt to minimize the impact of risks as much as possible. The final phase in this iterative process uses the wisdom and knowledge collected from the experience of threat management. In the best scenario, this knowledge should increase the quality of an organization in the form of manuals, policies, and even culture development. A consistent recommendation of studies surrounding the success of posture design is the proper dissemination of these knowledge tools to everyone in an organization regardless of its structure.

Finally, all of the phases in managing these procedures rely on human involvement and their respective competencies as the entities performing all these procedures. Prevailing literature suggests the appointment of a leader—typically in the form of a chief information officer (CIO)—in this undertaking. An external influencing force, in the form of laws and regulations governing the organization in business process improvements and reengineering, also dictates the prospects of success of risk management and should work in harmonious conjunction with the policies and regulations of the organization.

Designing a robust and self-sustaining cybersecurity compliance program within an organization requires a focus on key aspects of compliance: processes, people, documents, and systems [1]. Cyber risk management must be bottom-up so that IT security operational staff actively participate in program design and management to ensure the alignment of practices and policies [2]. However, changing the policies and norms cannot be entirely successful if the engineers do not actively pursue the consideration of privacy and security in designing the technological products. Incorporation of security and privacy concerns in the design of technological products has proven to be challenging as engineers see themselves as responsible for designing secure technologies, but do not find any pleasure in doing so [3] found three factors that impede the design of secure systems, including weak organizational norms (normative beliefs), low engineering control (time, autonomy, knowledge), and limited perceived responsibility. The current literature lacks case studies of the successful and unsuccessful cybersecurity initiatives in organizations. Future studies should study the extent to which each of the factors (organizational norms, engineering control, and perceived responsibility) can predict the success of cybersecurity policies in organizations.

Much of the literature has focused on individual methods and strategies to improve the existing cybersecurity posture. However, to the best of our knowledge, there is no comprehensive conceptual model for designing, implementing, and revising the cybersecurity-related organizational procedures. Through the literature review, we found three common themes in studies on organizational procedures for cybersecurity: First, researchers have proposed strategies for strengthening cybersecurity measures in organizations and reducing the costs of cyberattacks. These studies include integration of cybersecurity in the corporate culture, creating incidence response teams and business continuation management. Second, researchers are concerned with designing a unified security platform to integrate all security tools. These studies make suggestions for organizations to create platforms that unify all existing cybersecurity solutions in the organization, from firewall systems to security information and event management (SIEM) systems. Third, researchers focus mainly on the importance of considering insider threats in the design of cybersecurity-related organizational procedures. These studies endeavor to make connections among processes, people, and documents in organizations.

All public and private sector organizations must comply with the legal environment in which they are situated. The literature on cybersecurity law in the past 3 years focuses on several difficulties: (a) no definition of what cybersecurity is, (b) no coordination or harmonization of the various pieces of legislation, (c) multiple federal agencies being responsible for oversight of the legislation, (d) the resultant difficulty of assigning responsibility for creating and executing training for frontline users, (e) the oversight and evaluation of training and protocols, (f) the coordination of public and private sector companies with federal and state agencies, and (g) the rate of human error.

The secondary legal literature points to two major concerns for the future: small and midsized businesses will be the majority of cyberattack victims, and human error continues to be the largest factor in cyberattack success. Current cybersecurity legislation does not address either of these concerns. Small businesses cannot afford to hire security specialists; therefore, federal or state legislation needs to create some structure of shared information, training, and protocols, as well as some economic incentives for small companies [4] includes recommendations for a better approach: shift the burden from end users to an automated system, for which the federal government should pay, share metrics, promote transparency, and look at threats to wireless capabilities.

Although the Congress has enacted several pieces of legislation concerning cybersecurity in the past 3 years, there is a lack of definition about what is being protected, who is protecting it, and what those entities are doing in order to protect it. Training and implementation of cybersecurity training and protocols are spread among several federal agencies; then 50 state legislatures amplify the confusion by adding their own protocols. Finally, the rate of human error in cybersecurity protocols is stunning.

3 Discussion and Analysis

3.1 Technology

Cybersecurity often involves technology, people, and process [5], and these elements constitute the people, process, and technology (PPT) framework. This framework describes the methods using these elements to protect networks, devices, programs, and data from attack, damage, or unauthorized access. The PPT framework, often known as the Golden Triangle, dates to Leavitt's diamond model [6], which considered four elements: people, structure, tasks, and technology. The term "technology," as discussed in both models, refers to what people use in the context of cybersecurity. To illustrate, "technology" in Leavitt's model is anything that helps people do innovative work more effectively—especially in the age of artificial intelligence.

3.1.1 Current Security Technology

Due to the current increased threat level, the security market is looking for more robust solutions. In recent years, companies have implemented artificial intelligence (AI) to enhance the capabilities of security technologies. Where conventional security systems are often slow and insufficient, AI techniques can improve overall security performance. Socially responsible use of AI will be essential to mitigate further related risks and concerns [7]. AI solutions are already being implemented to increase detection and automation, and the adoption rate for AI-enabled security innovation will only continue to grow [8].

Detection, prevention, and automation are the main attributes of security technologies discussed in the literature. Prevention-based technology deploys security solutions such as firewalls, antiviruses, and patches to identified vulnerabilities, which can dramatically reduce the likelihood of a successful attack. Intrusion prevention systems (IPS), which focus on the prevention side of the security technology, have also become popular.

With the rapid growth of malware and the explosion of software vulnerabilities, detection should be used to search actively and quickly for signs and indications of cyberattacks. Detection frameworks such as distributed intrusion detection system (DIDS) for supervisory control and data acquisition (SCADA) industrial control systems have been developed by [9]. DIDS is generated from the traditional intrusion detection system (IDS), which uses audit trails and network packets to detect intrusions. Compared to IDS, DIDS is able to detect attack patterns across an entire corporate network, with geographic locations separating segments by time zones or even continents; DIDS also allows early detection of an Internet worm making its way through a corporate network. Detection and prevention are combined in intrusion detection and prevention systems (IDPS), which identify possible incidents, log information about them, attempt to stop them, and report them to security administrators [10].

Modern cyberattacks have become highly automated. It is therefore crucial to fight machine with machine—by incorporating automation into cybersecurity efforts—to defend successfully against automated attacks. Automated security configuration requires reliably detecting potential known issues and remediating them through a reliable process that does not require a busy employee's oversight. For example, in current cloud systems, security requirements must be manually translated into controls. This process is labor-intensive, tedious, and error-prone [11]. Therefore, automation of the configurations of cloud systems according to the client's security requirements is crucial, allowing security processes to avoid the errors and inefficiencies of the manual approach.

Biometric technology offers data privacy and cybersecurity. Its use of physical characteristics and traits for the identification of individuals reduces cases of fraud and theft [12]. presented a biometric recognition system based on fingerprint recognition, which can be embedded in any system involving access control, e-commerce, online banking, or computer login to enhance security. The technology of biometric voice recognition and identification also lends itself well to a variety of uses and applications, including security access control for cell phones (to eliminate cell phone fraud), ATM manufacturers (to eliminate pin number fraud), and automobile manufacturers (to dramatically reduce theft and carjacking) [13].

3.1.2 Security Usability

Because usability is the weakest link in the security chain of many prominent applications, security usability principles should be considered when designing and engineering IT security solutions [14]. In [14], the authors present two sets of principles of security usability: security action usability principles, employed when users are required to produce information and security tokens or to trigger some security-relevant mechanism, and security conclusion usability principles, applied when users observe and assess security-relevant evidence in order to assess the security state of systems. These principles can be incorporated in risk assessment processes by considering violations of the security usability principles as measures of security usability vulnerabilities. The study proposes two approaches to specify suitable controls and mitigate the risk assessed: the sustaining approach consists of keeping the security building blocks more or less unchanged while improving the interface and changing the way users interact with the system; the disruptive approach consists of replacing existing security building blocks with totally new ones that have a better potential for being implemented in a user-friendly way. In addition, a metric is needed to assess the degree to which a particular security element is usable.

3.1.3 Design Process

Beyond improving the usability of security technology and applying advanced technology in the cybersecurity field, companies need to focus on the technology design process and its effect on cybersecurity posture. Focusing on authentication protocols, [15] discusses key design principles to help ensure the correctness and effectiveness of standards for security protocols, including explicit principals' names in the message, unique encoding, explicit trust assumptions, use of times-tamps, protocol boundaries, release of secrets, and explicit security parameters [16]. proposed a general life cycle process of instrumentation and control (I&C) systems in nuclear power plants, involving a system design (SD) phase, a component design (CD)/equipment supply (ES) phase, and finally an operation and maintenance phase. Within the life cycle, cybersecurity features should be incorporated in the SD phase with cybersecurity risk assessment being performed afterward; cybersecurity characteristics should be reassessed in the hardware and software design during the CD/ES phase to incorporate cybersecurity design features in the target systems.

3.2 Management Procedures

Management procedures in an organization's cybersecurity domain of cybersecurity involve an unending process of identifying, assessing, and responding to risks. Organizational management of these processes often begins with an assessment of the likelihood and potential impact of a calculated "risky" event [17]. This phase is followed by the determination of the best approach (avoid, transfer, accept, or mitigate) to deal with the perceived threats and risks. Even in ideal conditions, however, not all risks can be eliminated, nor is any organization equipped with unending financial resources, including the human assets to combat these risks. The biggest challenge in management procedures is mapping uncertainties to organizational objectives in a way that makes the most effective and efficient use of limited resources to fulfill the ultimate goal—minimization of threat impacts through a strong risk management plan [18].

An ideal cybersecurity management procedure establishes clear communications and situational awareness about threats and calculated risks to all employees in an institution. The resulting decisions based on these factors create a well-informed, carefully considered plan, made in the context of organizational objectives, such as opportunities to support the organization's mission or seek business rewards [19]. However, the concentration of responsibilities surrounding this task should involve significant contributions by everyone in the organization. After all, the development, implementation, and maintenance of a cybersecurity management program for an organization are no small undertaking [20].

Selection of a chief information officer (CIO), oftentimes referred to as chief digital information officer, fulfills an essential role in the management of existing policies touching the intersection of data and technology, functioning as a bridge between the business assets, customer demands, and technology used in the organization [21]. However, the emphasis on these roles rests on the specialization of the domains of both businesses, technologies, and even government [22] for open data ecosystems. In addition, although the CIO directs only part of the organization's workforce, decisions about technology and security affect the entire organization. A CIO is typically required to have strong organizational skills [23]. Ultimately, it is the CIO's final decision which risk response will be implemented. Some strategies include threat avoidance, typically a combination of reducing the negative consequence or probability of the threat, transferring all or part of the threat to another party, or even retaining some or all of the potential or actual consequences of a particular threat.

Although challenging, the overall value that organizations achieve through development and implementation of such management procedures include reduced instances and impacts of actual cyber offenses. Kevin Caramancion in his paper explained that the procedures are the extrinsic forces that compel all the humans in an organization to following the program [24]. A carefully documented cybersecurity management program provides a structure for organizations with the means to reduce the prospective impact on the institution and its workers, due to its planned, predefined approach to identify and respond to cybersecurity incidents [25]. On a hierarchical level, a company's CIO needs to ensure that any localized plan is in full compliance with Information Security Law at the local, national, and international levels [2].

3.3 *Organizational Structure*

3.3.1 Strategies for Strengthening the Cybersecurity in Organizations and Reducing the Costs of Cyberattacks

[2] suggest some strategies for solidification of cybersecurity in organizations including integration of cybersecurity into corporate culture, protection of corporate assets such as intellectual property and personal information through cybersecurity policies, and involvement of every level of the enterprise in program design and management. A successful cybersecurity program should require that policies be integrated into everyday business tasks and implemented cohesively, collaboratively, and transparently [26]. Conducting employee training, making connections between IT security and legal departments, and using the National Institute of Standards and Technology (NIST) Cybersecurity Framework to validate current policies are important practical steps toward strengthening the organization's cyber defenses [2].

However, with increasing sophistication of organizations' interconnected technologies, cyberattacks are inevitable [26]. proposes some processes to reduce the costs of a security breach including business continuation management, providing support by incident response teams and extensive encryption, and employee awareness and training. Business continuation management refers to company plans created for recovery from potential threats. When companies incorporate cybersecurity in these plans, they can achieve reduced costs due to avoiding cyberattacks. Extensive encryption after the breach makes it harder for the hackers to access the data, and the incident response team (including managers, analysts, and researchers) analyzes the access points to detect and contain the attack. Finally, training employees about potential vulnerabilities empower them to help the organization recover from the threat and be prepared for the future.

3.3.2 Designing a Unified Security Platform to Integrate all Security Tools

Organizations use various security solutions such as antiviruses, firewalls, intrusion detection systems/intrusion prevention systems (IDS/IPS), and SIEMs. Unifying these solutions can increase the effectiveness of the cybersecurity systems as well as the ability to deal with the security incidents within organizations [27]. Moreover, a national-level cybersecurity framework overseeing cybersecurity strategies and integrating collaborative mechanisms can help ensure early detection of likely threats and enable implementation of security solutions involving vast numbers of stakeholders, including private entities, government entities, and citizens [28].

3.3.3 Consideration of Insider Threat in the Design of Cybersecurity Systems

One of the easiest ways malware enters organizations' control systems is through links between these systems and the Internet, in other words, through employees' emails and laptops [29]. Employees' careless or malicious unauthorized access to the organizational information system poses a serious threat to the organizations' digital assets [30]. Technologies designed to improve collaboration, productivity, and innovation in organizations have contributed to employee-related breaches or insider attacks mainly because few employees are sufficiently trained in security issues; also, organizations do not have effective data protection strategies [31].

Although cybersecurity awareness training programs are crucial for organizations, [31] suggest that they should be enhanced with continuous assessments to prevent further employee-related breaches. Also, employees' unauthorized access has been found to be dependent not only on personal characteristics, but also on the environment at the time of access [30]. Analysis and quantification of human errors can identify the extent of systems' vulnerability and associated risks which can illuminate future protocols [31]. Furthermore, management should build a dynamic risk profile for employees, which considers the characteristics of the unauthorized access in context (e.g., when and where), as well as the static attributes of employees [30].

3.4 Legal Environment

Several federal statutes on cybersecurity have been enacted in the past 3 years. The resultant secondary literature discusses the positive and negative aspects of these statutes, as well as suggestions for a path forward in cybersecurity law. The difficulties with the legislation include (a) the lack of definitions for cybersecurity, an umbrella term for concerns from protection of personal information to denial of service attacks; (b) the multiple federal and state legislatures writing cybersecurity legislation, with overlapping areas of application; (c) the multiple federal and state agencies involved in applying acts and statutes; and (d) the private sector interests that do not always align with governmental priorities.

[32] discusses the need for a definition of cybersecurity law, stating "The statute, Cybersecurity Act of 2015, fails to provide a concrete definition that sets forth the scope and goals of cybersecurity law" [32, p. 987] [32]. further posits that the

"patchwork of U.S. statutes and regulations that constitute cybersecurity law is an uncoordinated mish-mash of requirements that mostly were conceived long before there were cyber-threats" [32, p. 988].

The US legal system creates another difficulty with cybersecurity law, because both the federal and state legislative bodies have the right to concurrently pass cybersecurity legislation. For example, [21, p. 2] refers to 42 states passing more than 240 bills related to cybersecurity concerns in 2017 alone [32, p. 1027]. outlines the multiple federal agencies involved in cybersecurity: the Department of Transportation oversees connected vehicles, the Food and Drug Administration medical devices, the Federal Energy Regulatory Commission the national electrical grid, and the Department of Health and Human Services health data security, to name only a few [32]. goes on to point out "it is difficult to imagine how they can work with any degree of precision to achieve common economic and national security goals" [32, pp. 1027–1028]. In addition to the broad or undefined cybersecurity terms such as privacy concerns, data breaches, or required training, cybersecurity law is difficult to assess and improve.

As proof of the difficulties of a coordinated, precise response, a Senate Permanent Subcommittee on Investigations, House Committee on Homeland Security, and Governmental Affairs issued a report on federal cybersecurity of 2019, citing a disturbing list of failures to enact security protocols:

Over the past decade, inspector generals (IGs) for all eight agencies reviewed by the subcommittee found each agency failed to timely remediate cyber vulnerabilities and apply security patches. For example, the HUD and State IGs identified the failure to patch security vulnerabilities seven of the last ten annual audits. HHS and education cybersecurity audits highlighted failures to apply security patches 8 out of 10 years. For the last 9 years, the USDA failed to timely apply patches. Both DHS and DOT failed to properly apply security patches for the last 10 consecutive years [21, p. 4].

A further problem noted in the subcommittee's study is a reliance on legacy systems: "All eight agencies examined by the Subcommittee relied on legacy systems. For example, the DHS IG noted the use of unsupported operating systems for at least the last four years, including Windows XP and Windows 2003."

The federal cybersecurity report does not break down these failures by cause. It would be fruitful to know how many of these issues are based on money-saving tactics versus human error versus workers not understanding the importance of protocols. This list indicates the depth of human error that [21] also mentions in the multiplicity of state and federal laws. H.B. 2371, a law enacted in Illinois, amended the Data Security on State Computers Act to require certain state employees to annually undergo training by the Department of Innovation and Technology concerning cybersecurity. However, [21] states that 50% of data breaches are due to human error, so that humans are an equal problem to any hardware or software systems.

The [33] has an interesting response to the problem of human error: "In short, we seek strategies that remove the major responsibilities and costs of cybersecurity from the end users of technology, in favor of higher level, international, public/private that inure [sic] to the common good" [35. p. 3]. The authors of the report suggest the US government leads the security effort in coordination with industry, acknowledging the shift in thinking from earlier security efforts that place the onus on the company and its employees. The authors offer lessons learned from security breaches and distributed denial of services in the past 4 years: there are billions of Internet of Things devices with little or no security, such that we are dependent on third party infrastructures for our data security; critical infrastructure is connected to the Internet, creating ready targets for cyberattacks; state-sponsored industrial espionage accounted for 90% of 600 data breaches in 2016; and our control systems were designed to withstand environmental conditions and accuracy standards "incompatible with consumer grade hardware and software and in conflict with common network protocols" [33, p. 5].

[32] suggests that policymakers should consider economic incentives for companies, due to the disparity between small and large corporations. Small and midsized businesses constitute the majority of cyberattack victims, which is not surprising, given smaller companies do not have the technology staff to devote time to maintaining cybersecurity protocols [32]. states "accordingly, government resources that help small businesses prepare for cyberattacks could be a worthwhile investment" [32, p. 1028] but also refers to the US Department of the Treasury recommendation in 2013 against consideration of such incentives, concluding they "would come at the expense of foregone revenue for the government or reallocation of existing fiscal obligations" [32, p. 1028].

Neither [32] nor the [33] find fault with the federal legislation as far as it goes, but both offer many improvements for the future [32]. allows that the Cybersecurity Act of 2017 is a step in the right direction for addressing cybersecurity in the United States, in that "the statute explicitly recognizes that companies and the federal government share an interest in securing information, systems, and networks, and are positioned to work toward a common goal of societal security" [32, p. 1025].

However, as outlined above, he posits a definition of exactly what we are securing and why, as well as who is doing what is necessary before moving forward. The [33] contains a statement from Steven R. Chabinsky, who maintains that while the NIST Cybersecurity Framework is well-crafted, the complex risk environment in which it must be applied makes implementing it "enormously difficult and costly" [33, p. 6]. His suggestions for the future include shifting the burden of cybersecurity away from the end user, as well as echoing [32]'s calls for developing and sharing metrics and promoting legal harmonization.

3.5 *Human Factors*

3.5.1 Human-Centric Technology Design

While the threat environment and cybersecurity practice have been developing over the last decades, humans have been considered the weakest link of cybersecurity

[34–36]. The importance of the involvement of humans in cybersecurity has two sides, either having sufficient experience to bridge the gaps where cybersecurity algorithms and systems lack or being the weakest link by intentional or unintentional actions that lead to a security breach.

In order to minimize the cyber incidents caused by human error, technology design needs to be human-centric. As the direct or indirect end users of the designed technology, humans have an active or passive role in security, no matter to whom the technology belongs (e.g., government, private sector, or individuals themselves). Several factors lessen the chance of human error, including situational awareness, usability, effective visualization methods, and security compliance with government and industry standards [34].

Situational awareness is an essential factor affecting cybersecurity posture. It includes having the capability to assess the available information, to evaluate the choices, and to provide conclusions when needed [34]. For cybersecurity analysts, situational awareness provides the ability to formulate the big picture from small irregularities in network traffic. Knowing the emerging threat environment offers trained end users the opportunity to indicate any anomalies in a timely manner and to be prepared to take quick and appropriate actions [37, 38]. The Department of Homeland Security initiatives to establish international partnerships for sharing information about emerging cyber threats enhance situational awareness [39]. Such information should be easily accessible to users, enabling them to consult relevant information rapidly and thereby reduce the impact of disruptions [34]. Providing situational awareness can also engage end users to take practical cybersecurity approaches when required [40].

Usability is a measure of effectiveness, efficiency, and satisfaction for a product addressing its specified goals [41]. Usability is another important factor affecting cybersecurity posture. Cybersecurity technology should be designed to provide usability by assessing the target user performing the specific tasks required by the end product and by consulting users' opinions [34]. In other words, the design should progress by building personas, after examining the target users, their goals, knowledge, behaviors, and activities [42].

Visualization, an integral element of human-machine interaction, is also an essential human factor regarding cybersecurity actions. Users need to be able to reach accurate information rapidly. Data visualization is also vital for improving users' situational awareness. Effective visualization method design should provide insights by separating beneficial information from arbitrary noise [34]. To make the visualizations usable, the designers should consider the expectations of the targeted users by distinguishing the focus areas of users in different roles within an organization [42]. Audio feedback for users—in the form of alarms and alerts—should support visualizations to improve the prioritized feedback mechanism [34]. Visual interface models should be evaluated for usefulness [38] and for an ability to adapt to emerging user expectations [42].

Another aspect of usability is its trade-off with security. Maximizing ease of use tends to increase the attack surface. Cybersecurity requirements should be in balance with usability. Developers should follow the guidelines established by government regulations or industry standards. While compliance improves cybersecurity posture, unfriendly security requirements tend to encourage users to create work-arounds that cause even more security problems [34]. Research by [43] indicates that software developers and policymakers' design decisions are not in alliance with user needs.

Technology design should consider any approach to reduce possible human errors while users are operating the product or software [34]. This includes, but is not limited to, an appropriate adjustment of interface elements, access and authentication procedures such as password policies, and security configurations of network components while ensuring users properly follow security procedures including maintaining security patches and auditing network logs [44].

Human-centric design that considers the aforementioned factors not only helps reduce human errors causing cyber incidents but also provides the foundation for well-informed decision-making, especially with improved situational awareness and visualization.

3.5.2 Cross-Cultural Technology Design

The extent to which users take precautionary actions against cyber risks is conditional upon how they perceive the value of information security relative to other important personal goals [45]. This study illustrates that users in most cybersecurity contexts are faced with trade-offs between information security and other important attributes that they desire to maximize. The authors found the value of information security was sensitive to usage context. For example, social media invoked greater security premiums in terms of productivity than email and web surfing.

One of the important aspects of usage context is culture, which plays an important role in the design field. Cross-cultural design will be a key design evaluation point in the future [46]. Designers of information visualization and user interfaces must take culture into account in the design of metaphors, mental models, navigation, interaction, and appearance, to better ensure usability, usefulness, and appeal [47]. While cross-cultural factors become important issues for product design in the global economy, the intersection of technology design and culture becomes a key issue in a lot of in-depth studies, discussed in the following part.

[48] gives specific information on the sensitivity of usage context, from the aspect of culture. In [48]'s study, some technologies have similar uses across cultures, like emails, pagers, instant messaging, and blogging. Other technologies like mobile text messaging, however, can be dramatically different in diverse cultural contexts in terms of public vs. private, formal vs. casual, orality vs. literacy, and social vs. technical. Nowadays, cross-cultural design has become standard practice and a daily test in many IT companies [48]. points to the phenomenon of mobile text messaging use as one of the many demanding challenges that cross-cultural technology design has faced in this increasingly globalized world with a rising participatory culture. A large number of today's IT products are consumer-oriented. Individual users are no longer passive users but active designers who shape, redesign, and localize an

available technology to fit into their own contexts. That is, a user would employ, and are able to reshape, a technology according to his or her own lifestyle.

Building upon the existing body of research in website design and anthropologists' cultural dimensions, [49] discussed cultural markers including interface design elements and cultural dimensions that are appropriately used for cultural-centered website design and localization. The research re-evaluated some websites studied earlier, indicating seven important cultural dimensions which play a role in designing websites for cross-cultural audiences. Those seven dimensions are (1) experience of technology, (2) context, (3) international trade and communication, (4) gender roles, (5) uncertainty avoidance, (6) human nature orientation, and (7) power distance. To establish absolute criteria for what is important and which cultural markers are applicable for cultural-centered website design, five levels of suitable cross-cultural markers for designing cultural-centered websites and localization have been proposed in the study: (1) context-dependent cultural markers (e-culture), (2) settled cultural markers, (3) broad cultural markers, (4) variable cultural markers, and (5) vista cultural markers. The findings suggested a grouping of the cultural markers into these five levels can be used for designing cultural-centered websites.

4 Conclusion and Recommendations for Future Research

In conclusion, the secondary literature in the five major topics studied in this paper points out the successes in cybersecurity technology and suggests further improvements as well as where further research is needed. Technology is one of the main tools to improve cybersecurity culture in the organization Serving as part of the cybersecurity supply chain, security technology can be improved by applying advanced technology, improving usability, and ameliorating the design process. All these three aspects are to be considered during the technology design process, with the exception that usability should also be tested while using the technology. Future studies should look into these three aspects while designing, implementing, and maintaining security technology and examine how each of them contribute to the improvement of cybersecurity posture in organizations, singularly or collectively.

Management procedure as an organizational factor plays a vital role in building the design posture of cybersecurity policies. This factor instantiates to risk mitigation strategies put in place to limit the impact of threats. The competencies of human actors in creation and implementation of these procedures are central themes to its success. Future studies involving management procedures and risk management can highlight the impact of laws and regulations on corporate policies, should there be a mismatch, so as to display its influence empirically on the success or failure of cybersecurity design posture. Researchers should also consider case studies exhibiting the individual prominence, and impact of management procedures in the success of posture design should be considered.

Many papers on organizational structure reviewed in this study are focused on methods and strategies to improve the existing cybersecurity posture, to employ a unified security platform to integrate all security tools, and to consider insider threats in the design of cybersecurity-related organizational procedures. The lack of strong quantitative analyses that compare the existing organizational procedures based on the cybersecurity outcomes is evident in the literature. Future studies should focus on running experiments in real organizational settings to evaluate the effectiveness of suggested interventions. To do so, there must be a clear set of measures for evaluating the cybersecurity posture in organizations. These measures should differ for different ecosystems. For example, cybersecurity outcome measures for government agencies are different from those for business corporations. These differences must be considered in evaluative studies and intervention design.

There are several steps that would improve the confused state of cybersecurity legislation. First, as [32] suggests, there needs to be a definition of cybersecurity. Once the various aspects have been defined, oversight of what portion of the training and protocols should be assigned to which federal agencies will be clearer. Once the federal oversight is clearly defined, the state legislatures can define the oversight of training and protocols for additional pieces of legislation that answer a state's particular needs. States will need to coordinate the public and private sector companies and agencies, all of whom are crucial in the implementation of cybersecurity laws. These laws, with one exception, address the rank-and-file user only as a focus of training and an agent of error. The [33], the sole exception, recommends a way to avoid the high rate of human error by not asking the human frontline user to learn and execute the cybersecurity protocols; rather, it recommends having a fully automated system in place to execute the protocols.

While several methods to reduce human error exist in the literature, the problem has yet to be solved entirely. Researchers need to conduct more studies to provide an assessment of human-machine interface design concepts, to improve usability based on the standards, and also to evaluate how the new design affects users' performance. For automated systems, where software handles a significant proportion of data analysis tasks, human-machine interaction gains more importance, and improvements of visualization methods require attention. Technology design should also provide the means to update users with the emerging threat environment and recent global trends on cyberattacks in order to strengthen the situational awareness.

Cybersecurity decision-making is one of the areas needing improvement. Since behavioral and nonbehavioral (rational) decision-makers tend to have different biases, they have diverse strategies to allocate the investment budget for cybersecurity activities where nonbehavioral decision-makers prefer to focus on the critical components of defense strategy [50]. Further research is needed to focus on this aspect of human behavior in order to implement the technology design more efficiently. Additionally, design studies are needed to improve the error tolerance of the decision support technology in order to reduce the possibility of a user making an error, as well as improving the ability to make suggestions on possible resolutions of the issues caused by the errors [34].

Further research is also needed on amelioration of human error, whether via more training or different training or removing humans from the protocols entirely. In the latter case, research studies need to determine the acceptable amount of error in a fully automated protocol, compared to that of untrained and trained human subjects.

Keeping technical features up to date does not ensure cybersecurity, given that the user is the weakest link of cybersecurity. Therefore, human-centric technology designers should create human-centric designs in addition to training personnel against cyber incidents. Global organizational cyber risk communication initiatives should promote situational awareness and deliver it appropriately to users. Procedures and interfaces should be intuitive, and their usability should be optimized by balancing the ease of use with the security level. Research on these areas of human-centric technology design is still needed.

Acknowledgments The authors will forever be in debt to Mr. Christian Poehlmann, Nancy Poehlmann's other half, for his voluntary proofreading of this paper. Furthermore, without the guidance of Dr. Terry Merz, this paper would not instantiate from a premature yet promising idea to a well-crafted manuscript. Finally, the exceptional leadership of Mr. Kevin Matthe Caramancion as principal investigator made working on this paper highly motivating.

References

1. J. Farquharson, Cyber security programs: Design, implementation & controls, and metrics & measurements. Power Eng. **122**(6), 6 (2018)
2. K. Peretti, J. Wool, K. Todt, R. Cressey, *Five Steps to Strengthening Cyber-Defenses* (CIO Insight, 2015), p. 1
3. S. Spiekermann, J. Korunovska, M. Langheinrich, Inside the organization: Why privacy and security engineering is a challenge for engineers. Proc. IEEE **107**(3), 600–615 (2019). https://doi.org/10.1109/JPROC.2018.2866769
4. Cyebersecurity Threat Landscape, *Commitee: Senate Homeland Security and Governmental Affairs* (CQ Congressional Testimony, 2017)
5. V. Loy, K. Mattar, T. Ye, B. Perera, J. Sng, M. Leong, *Reclaiming Cybersecurity: The Global State of Information Security Survey. Technical Report* (PwC, 2016), pp. 1–8
6. H.J. Leavitt, Applied organization change in industry: Structural, technical and human approaches. PsycEXTRA Dataset. (1964). https://doi.org/10.1037/e509852009-001
7. N. Wirkuttis, H. Klein, Artificial intelligence in cybersecurity. Cyber Intell. Sec. J. **1**(1), 103–119 (2017)
8. J. Kang, S. Westskytte, Diffusion of Cybersecurity Technology – Next Generation, Powered by Artificial Intelligence. Independent Thesis Advanced Level. (2018). Retrieved from http://www.diva-portal.org/smash/get/diva2:1295355/FULLTEXT01.pdf
9. Cruz et al., A cybersecurity detection framework for supervisory control and data acquisition systems. IEEE Trans. Ind. Inf. **12**(6), 2236–2246 (2016)
10. K.A. Scarfone, P.M. Mell, *Guide to Intrusion Detection and Prevention Systems (IDPS)* (2007). https://doi.org/10.6028/nist.sp.800-94
11. Tunc et al., Cloud scurity automation framework, in *2017 IEEE 2nd International Workshops on Foundations and Applications of Self* Systems (FAS*W)*, (Tucson, AZ, 2017), pp. 307–312
12. J. Kour, M. Hanmandlu, A.Q. Ansari, Biometrics in cyber security. Def. Sci. J. **66**(6), 600–604 (2016)

13. R.A. Rashid, N.H. Mahalin, M.A. Sarijari, A.A.A. Aziz, Security system using biometric technology: Design and implementation of Voice Recognition System (VRS), in *2008 International Conference on Computer and Communication Engineering*, (2008). https://doi.org/10.1109/iccce.2008.4580735
14. A. Jøsang, B. Alfayyadh, T. Grandison, M. Alzomai, J. Mcnamara, Security usability principles for vulnerability analysis and risk assessment, in *Twenty-Third Annual Computer Security Applications Conference (ACSAC 2007)*, (2007). https://doi.org/10.1109/acsac.2007.14
15. H. Khurana, R. Bobba, T. Yardley, P. Agarwal, E. Heine, Design principles for power grid cyber-infrastructure authentication protocols, in *2010 43rd Hawaii International Conference on System Sciences*, (2010). https://doi.org/10.1109/hicss.2010.136
16. J. Song, C. Lee, D. Lee, A cyber security risk assessment for the design of I&C systems in nuclear power plants. J. Nucl. Eng. Technol. **44**(8), 919–028 (2012). https://doi.org/10.5516/NET.04.2011.065
17. D. Hubbard, *The Failure of Risk Management: Why It's Broken and How to Fix It* (Wiley, 2009), p. 46
18. ISO/IEC Guide 73:2009, Risk management — Vocabulary. International Organization for Standardization (2009)
19. V.C. Wong, *Cybersecurity, Risk Management, and How Boards Can Effectively Fulfill Their Monitoring Role*, vol 15 (UC Davis Bus. LJ, 2014), p. 201
20. M. Goodyear, H. Goerdel, S. Portillo, L. Williams, Cybersecurity management in the states: The emerging role of chief information security officers. SSRN **2187412** (2010)
21. A. Hütter, R. Riedl, Chief information officer role effectiveness: Literature review and implications for research and practice, in *Chief Information Officer Role Effectiveness*, (Springer, Cham, 2017), pp. 1–30
22. K.M.M. Caramancion, Modelling Maternal Health Data in the Philippines Using Machine Learning (Doctoral Dissertation, De La Salle University—Manila), (2017)
23. L. Portela, R. Carvalho, J. Varajão, L. Magalhães, A review of chief information officer'main skills, in *World Summit on Knowledge Society*, (Springer, Berlin, Heidelberg, 2010), pp. 387–392
24. K.M. Caramancion, An exploration of disinformation as a cybersecurity threat, in *2020 3rd International Conference on Information and Computer Technologies (ICICT)*, (IEEE, 2020), pp. 440–444
25. CISCO Systems, Cybersecurity Management Program. (2017).Accessed via https://www.cisco.com/c/dam/en/us/products/collateral/security/cybersecurity-management-programs.pdf
26. R. Krishan, Corporate solutions to minimize expenses from cyber security attacks in the United States. J. Internet Law **21**(11), 16–19 (2018)
27. C. Islam, M.A. Babar, S. Nepal, A multi-vocal review of security orchestration. ACM Comput. Surv. **52**(2), 1–45 (2019). https://doi.org/10.1145/3305268
28. I. Atoum, A. Otoom, A.A. Ali, A holistic cyber security implementation framework. Inf. Manag. Comput. Secur. **22**(3), 251–264 (2014). https://doi.org/10.1108/IMCS-02-2013-0014
29. D. Ashenden, The human shield. TCE: The. Chem. Eng. **896**, 22–25 (2016)
30. J. Wang, Z. Shan, M. Gupta, H.R. Rao, A longitudinal study of unauthorized access attempts on information systems: The role of opportunity contexts. MIS Q. **43**(2), 601–622 (2019). https://doi.org/10.25300/MISQ/2019/14751
31. M. Evans, L.A. Maglaras, Y. He, H. Janicke, Human behaviour as an aspect of cybersecurity assurance. Sec. Commun. Networks **9**(17), 4667–4679 (2016). https://doi.org/10.1002/sec.1657
32. J. Kosseff, Defining cybersecurity law. Iowa Law Rev. **103**(3), 985–1031 (2018)
33. Cybersecurity Threat Landscape, *Commitee: Senate Homeland Security and Governmental Affairs* (CQ Congressional Testimony, 2017)
34. M.W. Boyce, K.M. Duma, L.J. Hettinger, T.B. Malone, D.P. Wilson, J. Lockett-Reynolds, Human performance in cybersecurity: A research agenda, in *Proceedings of the Human Factors and Ergonomics Society annual Meeting*, vol. 55, No. 1, (SAGE Publications, Sage, CA\Los Angeles, CA, 2011, September), pp. 1115–1119

35. G. Gross. Human error causes most security breaches. InfoWorld. (2003). https://www.infoworld.com/article/2680263/human-error-causes-most-security-breaches.html
36. M. Narendra, Human error remains the main cause of data breaches. PrivSec Report. (2019). https://gdpr.report/news/2019/06/20/human-error-remains-the-cause-of-data-breaches/
37. D.M. Best, S. Bohn, D. Love, A. Wynne, W.A. Pike, Real-time visualization of network behaviors for situational awareness, in *Proceedings of the Seventh International Symposium on Visualization for Cyber Security*, (2010, September), pp. 79–90
38. R.S. Gutzwiller, S. Fugate, B.D. Sawyer, P.A. Hancock, The human factors of cyber network defense, in *Proceedings of the Human Factors and Ergonomics Society Annual Meeting*, vol. 59, No. 1, (SAGE publications, Sage CA\Los Angeles, CA, 2015, September), pp. 322–326
39. T. Takahashi, H. Fujiwara, Y. Kadobayashi, Building ontology of cybersecurity operational information, in *Proceedings of the Sixth Annual Workshop on Cyber Security and Information Intelligence Research*, (2010, April), pp. 1–4
40. A. Cadzow, Are we designing cybersecurity to protect people from malicious actors? in *International Conference on Human Systems Engineering and Design: Future Trends and Applications*, (Springer, Cham, 2018), pp. 1038–1043
41. National Institute of Standards and Technology, *Digital Identity Guidelines (NIST SP 800-63-3)* (U.S. Department of Commerce, 2017). https://doi.org/10.6028/NIST.SP.800-63-3
42. S. McKenna, D. Staheli, M. Meyer, Unlocking user-centered design methods for building cyber security visualizations, in *2015 IEEE Symposium on Visualization for Cyber Security (VizSec)*, (IEEE, 2015), pp. 1–8
43. S. Parkin, A. Van Moorsel, P. Inglesant, M.A. Sasse, A stealth approach to usable security: Helping IT security managers to identify workable security solutions, in *Proceedings of the 2010 New Security Paradigms Workshop*, (2010), pp. 33–50
44. Government Accountability Office, *Cybersecurity for Critical Infrastructure Protection (GAO-04-321)* (United States General Accounting Office, 2004). https://www.gao.gov/new.items/d04321.pdf
45. K.D. Nguyen, H. Rosoff, R.S. John, Valuing information security from a phishing attack. J. Cybersec. **3**(3), 159–171 (2017). https://doi.org/10.1093/cybsec/tyx006
46. R. Lin, M.X. Sun, Y.P. Chang, Y.C. Chan, Y.C. Hsieh, Y.C. Huang, Designing "culture" into modern product: A case study of cultural product design, in *International Conference on Usability and Internationalization: Usability and Internationalization*, (2007), pp. 146–153
47. A. Marcus, Cross-cultural user-experience design, in *SIGGRAPH Asia 2013 Courses on - SA 13*, (2013). https://doi.org/10.1145/2542266.2542274
48. H. Sun, *Cross-Cultural Technology Design: Creating Culture-Sensitive Technology for Local Users* (Oxford University Press, Oxford, 2012)
49. A. Mushtaha, O.D. Troyer, Cross-culture and website design: Cultural movements and settled cultural variables. Lecture notes in computer science internationalization design and global. Development, 69–78 (2009). https://doi.org/10.1007/978-3-642-02767-3_8
50. M. Abdallah, D. Woods, P. Naghizadeh, I. Khalil, T. Cason, S. Sundaram, S. Bagchi, BASCPS: How does behavioral decision making impact the security of cyber-physical systems? arXiv preprint, arXiv:2004.01958 (2020)

A Hybrid Recommender System for Cybersecurity Based on a Rating Approach

Carlos Ayala, Kevin Jimenez, Edison Loza-Aguirre, and Roberto O. Andrade

1 Introduction

The cybersecurity analysts, according to Randall Fietzsche [1], are the professionals in charge of analysing the risk and threats that may compromise an organization. Then, they should plan and execute security measures with the aim of protecting the organizational networks and computer systems [2]. In other words, their job is to help the organization to understand what is happening and where it should go in terms of computer security [2].

Cybersecurity analysts' work involves dealing with risks, vulnerabilities and threats on a daily basis, leading them to search for a frame of reference to prioritize the most critical incidents and attacks in order to get the best actions to counter them. However, there are three factors that can affect their decisions [2]: (1) time, because cybersecurity analysts must resolve attacks as soon as possible; (2), 'manual processes and methodologies', because most of the process to identify and respond to attacks are manual; and (3) the 'subjectivity' of their decisions, because the analyst usually depends on his good judgement and experience at the time to make decisions and perform the tasks to solve a security incident. These three factors can affect the performance of any cybersecurity analyst regardless of the environment in which they work.

If we consider that, traditional practices tend to produce large numbers of alerts, which should be examined and verified with all the information available, be it structured or not [3]; it makes an analyst feel overwhelmed when trying to discriminate which product, content or service meets the correct need to optimally

---

C. Ayala · K. Jimenez · E. Loza-Aguirre (✉) · R. O. Andrade
Facultad de Ingeniería en Sistemas, Escuela Politécnica Nacional, Quito, Ecuador
e-mail: carlos.ayala01@epn.edu.ec; kevin.jimenez@epn.edu.ec; edison.loza@epn.edu.ec; roberto.andrade@epn.edu.ec

K. Daimi et al. (eds.), *Advances in Security, Networks, and Internet of Things*, Transactions on Computational Science and Computational Intelligence,
https://doi.org/10.1007/978-3-030-71017-0_28

solve a problem [4]. It is in this context that recommendation systems can be used to help them in their search for solutions. A recommendation system is a tool that produces personalized recommendations as output and can guide users to choose interesting or useful products in line with their needs [5]. In our case, we refer as a 'product' to all information pieces about a cybersecurity incident. The development of such a system in cybersecurity context would alleviate the tasks and problems that a cybersecurity analyst can present.

In this research, we propose a recommendation system that seeks to prioritize vulnerabilities, threats and risks and the possible solutions to them. The aim is to improve the analyst's response time to different incidents. The system will provide the best responses for each classified rating-based security incident with the help of experts in the area.

This article is organized as follows: Sect. 2 presents the theoretical background about security incident response processes and recommendation systems. Section 3 shows the process followed for implementing the recommendation system. Section 4 presents the developed system. Section 5 offers the results obtained from the evaluation with experts. Section 6 shows a brief discussion of the results obtained. Finally, Sect. 7 highlights the conclusions.

2 Theoretical Background

2.1 Cybersecurity Incident

The objective of digital attacks is to be able to access, modify or delete information. This is particularly challenging today because there are more devices than people and attackers are becoming more innovative [6]. Cybersecurity refers to the way to protect information from any digital attack. It focuses on providing defensive methods to detect and capture any intruder who wants to compromise any information system [7].

A cybersecurity incident is an unwanted or unexpected event or set of events, which negatively impacts the processes and operations of organizations. Its impacts include disabling the use of information or the elimination or modification of data by corrupting information systems through malware infections, phishing, etc. [8].

2.2 Recommendation Systems

Recommendation systems, also known as recommender systems or simply recommender, are software tools that provide, as suggestions, a subset of elements belonging to a universe of alternatives that are considered the most appropriate for a user. Thus, a recommendation system is a decision-making support [9, 10].

One of the fundamental pillars of recommendation systems is the large amount of information they can handle to deliver a valid recommendation. This becomes such systems as a good alternative to deal with information overload problems [11]. Most recommendation systems focus on the past behaviour of users as the recommendations are generated by the similarity of searches of the users of other similar users or by the rating that the user has given, in the past, to an item or an option. Indeed, to provide recommendations to a user, the system could consider the knowledge or experience of the same user. If the knowledge or experience of a user is not available, a valid recommendation can be provided in a category that the user has selected [12].

2.3 *Types of Recommendation Systems*

There are different types of recommendation systems, each of which has its own approach about how to provide recommendations. Accordingly, recommenders can be classified into:

1. Recommenders based on collaborative filtering. These systems focus on the items that received a rating from users [12]. This type of system is the most used since it helps to joint users with similar interests. This type of recommendation system does not need too much information about items, because the user is the one who really provides the information considered for the recommendation.
2. Content-based recommenders. These systems do not use the evaluation that a user provides about a product, but they use other parameters such as the information of the product itself or the user's profile [12]. This type of system is used within scenarios where there are a lot of new products for which good information about the product and its characteristics is available.
3. Knowledge-based recommenders. In order to deliver a recommendation, these systems take all the available explicit knowledge about a product and a user, past queries of the users and information about what the expected result should look like [13]. The user can control the recommendations provided by different filters.
4. Recommenders based on demographic information. To make a recommendation, these systems consider users' characteristics such as their gender, age, education, etc. [14].
5. Keyword-based recommenders. The operation of these systems is founded on measuring preferences based on keywords. For this, an analysis of texts written by users is used to generate recommendations. In this type of recommendation system, users are classified as previous and active. From previous users, a set of keywords is extracted from their reviews or comments and stored within a database. Thus, when an active user provides a new keyword and its weighted importance, the similarity of the keywords of previous users' texts with the new keyword is calculated [14]

Table 1 Advantages and disadvantages by recommendation system type [12, 14, 15]

Advantages	Disadvantages
Recommenders based on collaborative filtering	
No information about the products is needed Classifiers can be used to provide recommendations	Cold start problem High cost to find the best neighbour 'Black sheep' problem Data scarcity problem Scalability Quality
Content-based recommenders	
Provides recommendations as soon as it has information about product	Causality problem
Knowledge-based recommenders	
Acceptable quality and cost for finding the best neighbour Improves causality problems It facilitates cold start	Association rules between products and knowledge bases Complexity grows as the number of products grows
Recommenders based on demographic information	
Best quality recommendation for the user	Cold start problem
Keyword-based recommenders	
Can handle comments and text reviews Can be integrated with social networks It can incorporate multi-criterion rating Good precision	Difficulty for calculating similarities Keyword classification problem Weight calculation problem
Hybrid recommenders	
Improved precision Improved performance It can overcome the problems of other recommendation systems	Complex systems Expensive systems to implement

6. Hybrid recommenders. These systems combine two or more types of the recommenders listed above. The aim of these systems is to provide better performance and improve recommendations [13].

Each of the recommendation systems mentioned has advantages and disadvantages, which are listed in Table 1. For our research, it was essential that the recommendation system selected allows users' involvement as much as possible through ratings and that also takes into account the characteristics of items (anomalies and vulnerabilities that would compromise computer security). For this reason, we used a hybrid system that combines both a recommender based on a collaborative filter and a recommender based on knowledge. The selected types and strategies are detailed below.

2.4 Recommenders Based on Collaborative Filtering

These systems are useful in environments where there is few content or knowledge associated with the elements to recommend. The recommendation is made to users who have relevant interests and preferences by calculating similarities between their profiles and behaviours [9]. Users create a group, which is called a neighbourhood, where a user gets recommendations for items that have been rated, or not, by other users in the same neighbourhood [9]. For its operation, it is necessary that another user has read the same recommendation, which allows to group them as similar users (Fig. 1).

The response provided by a recommendation system based on a collaborative filter may be one of the two types: prediction and recommendation. Prediction is a rating of an item that would be given by a user, while a recommendation is the elements that the user likes or would like the most [9] (Fig. 2).

To implement this recommender in our project, the neighbour-based strategy was used to make recommendations based on ratings from similar users [16].

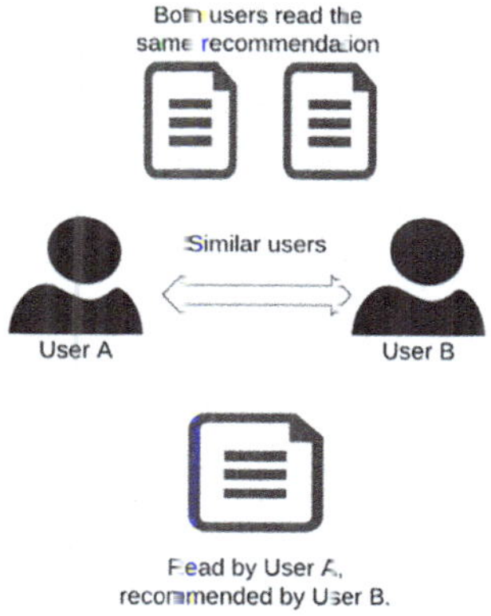

Fig. 1 Collaborative filtering working model

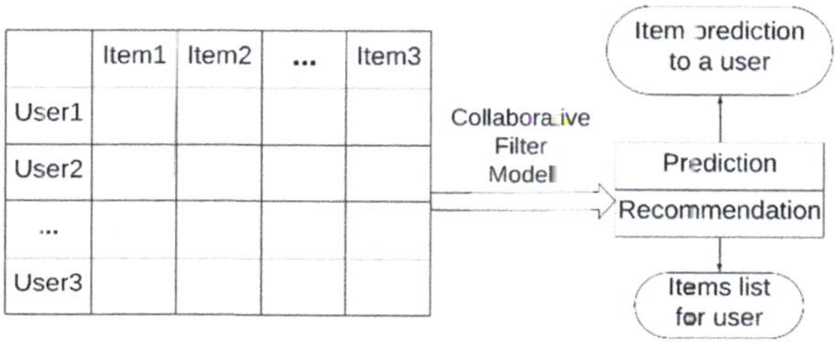

Fig. 2 Utility matrix

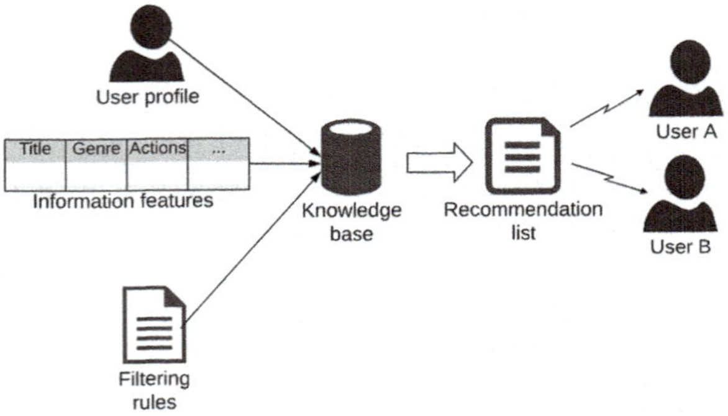

Fig. 3 Knowledge-based system operating model

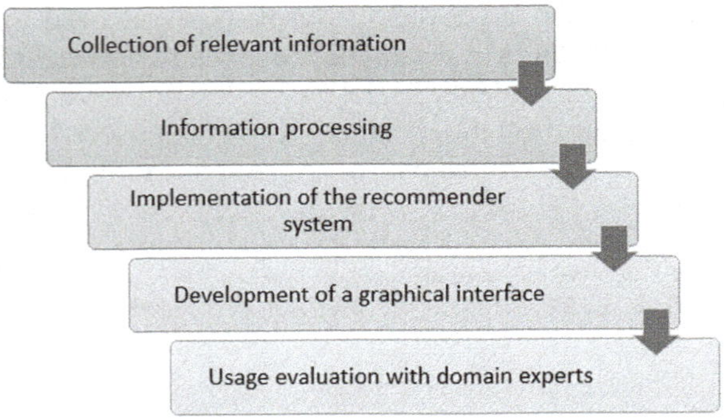

Fig. 4 Development process for the recommendation system

2.5 Knowledge-Based Recommender

This recommender offers the possibility of exploiting the knowledge of a specific domain and thus provides expert recommendations to solve a problem [13]. For its operation, it is necessary to create a database or knowledge base of preferred elements of the different users of the system (Fig. 3). In our case, the knowledge base was populated based on the knowledge of the domain defined in static information from security databases and the contribution of experts' knowledge [13] as detailed in the following section.

3 Research Methodology

For the implementation of the recommendation system, the steps outlined in Fig. 4 were followed.

1. Collection of relevant information: First, we collected information from the websites of Symantec; OWASP; NIST; the University of Trento, Italy; and the CSIRT of the Escuela Politécnica Nacional of Quito, Ecuador. The rating of each anomaly proposed by the University of Trento, Italy, was used to prepare

the recommendations. We used web scraping techniques to scan the content of the listed web pages to obtain the information with which the recommendation system will work.

2. Information processing: After a process of verification and cleaning of the data obtained, the information was classified according to its level of criticality. For it, the NumPy and Pandas libraries of Python were used. The name of the anomaly, a short description and its possible mitigation were identified. This information will be one input, in addition to the user's rating, to prepare recommendations.
3. Implementation of the recommender system: Once the information was verified, cleaned and classified, the recommendation system was implemented based on both collaborative filter and knowledge methods. The recommender was designed so that once an attack to be mitigated is chosen the system provides the recommendation with the best rating and the five best alternative recommendations to the solution presented.

   The system was developed applying an incremental iterative development model which consists of delivering functional prototypes. Python version 3 in an Anaconda environment was used at the core of the system. It was also used for the web scraping tool which gathers data from different sources. For the visualization of the data, the workflow and the results, Jupyter Notebook and JupyterLab were used. GitHub repositories were used to control versioning.
4. Development of a graphical interface: To facilitate the use and to measure the effectiveness of the system, a graphical interface was developed. It allowed final users interact with the recommendation system and easily visualize the anomalies and recommendations suggested. The graphical interface was developed with Python. The Tkinter graphical library was selected because of its clear, easy-to-code syntax and available documentation.
5. Evaluation: The recommendation system was evaluated by a group of experts in cybersecurity and data analysis who carried out their evaluation using the technology acceptance model (TAM) framework [17]. Thus, the evaluation was based on the two main dimensions of TAM: (1) the perceived usefulness defined as the subjective probability of a person that, by using a certain system, would improve their performance at work and (2) the perceived ease of use that refers to the degree to which a person believes that using a certain system will be effortless.

4 System Description

4.1 Logical Architecture

Our system is structured in two parts: a knowledge base and an anomaly ranking base. The knowledge base uses a flat file, where the information collected from the official security pages of Symantec; OWASP; NIST; the University of Trento, Italy; and the CSIRT of the Escuela Politécnica Nacional are stored. This information

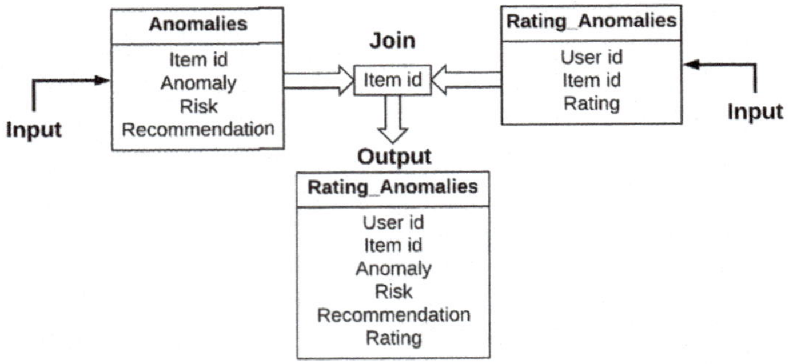

Fig. 5 Data structure

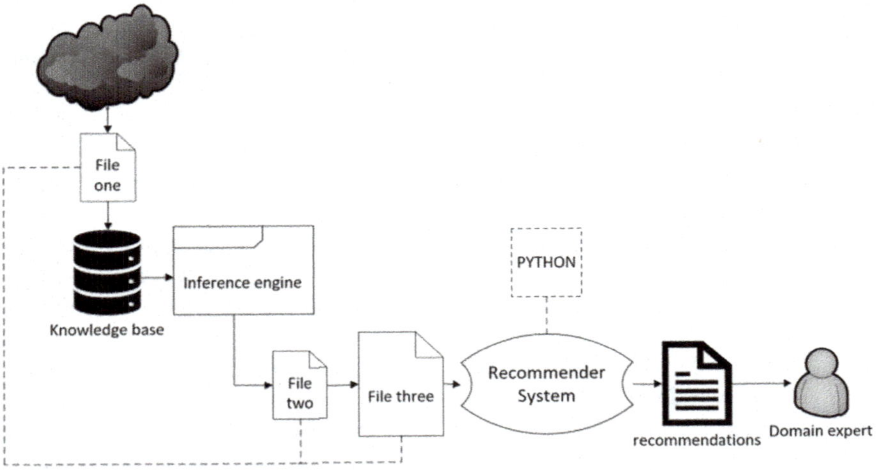

Fig. 6 Logical architecture of the system

consists of the attack identifier, the attack name, the criticality of the anomaly and the recommendation.

For the ratings, a second flat file is used. It stores the identifier of the user who rate the recommendation, the identifier of the attack and the rate (from 1 to 5). The joint of both files provides the necessary data for the operation of the recommender (Fig. 5).

For its operation, the system follows the following procedure (Fig. 6):

1. The knowledge base (file one) feeds the inference engine, which contains the rules that will be used to classify the collected information.
2. Once the anomalies have been ranked (file two), we proceed to join the knowledge base and the classified anomalies (file one and file two). The recommendation system base (file three) contains the anomaly, the user whoever rated it, its rating and its according recommendation.

3. For the operation, an end user is also considered an expert in the domain. However, as experts, they can contribute by rating a recommendation, adding their own recommendations, or modify and remove existing recommendations. This will add feedback into the system, which will lead to improve future recommendations or adapt them to particular environments.

4.2 *Physical Architecture of the Recommendation System*

The recommendation system is a local software that does not need any kind of installer and whose information is stored within the client computer that is running it. For collect external information, the system connects to the Internet to download and store data about anomalies in the database. Once these data are stored locally, no more Internet connection is required for its operation.

4.3 *User Interface*

System start screen: On this screen, the analyst will be able to enter the name of the anomaly for which he wishes to receive a recommendation. From the content entered, the analyst will be able to carry out a search in a proposed list of the different anomalies registered (Fig. 7). The name of the anomaly is autocompleted as it is typed. If the anomaly is not registered, the system will display an informative message.

Anomaly result screen: On this screen, the analyst will be able to see the five anomalies closest (similar) to the one required. The recommendations to these anomalies could help mitigate the actual security issue (Fig. 8).

Recommendation screen: On this screen, the analysts will be able to see the description and the recommendations that would serve him to mitigate the anomaly selected on the result screen (Fig. 9).

Fig. 7 System anomaly list

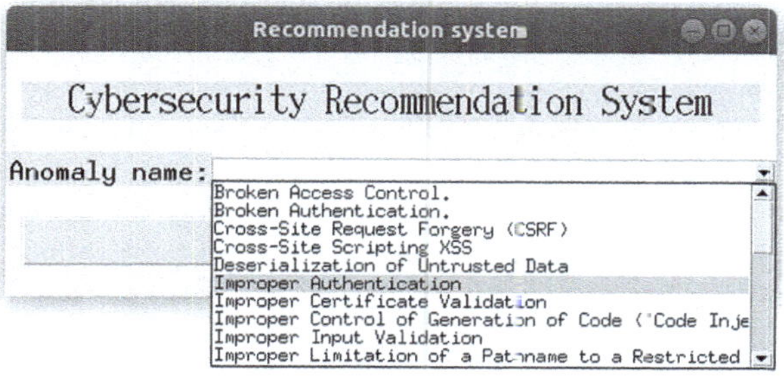

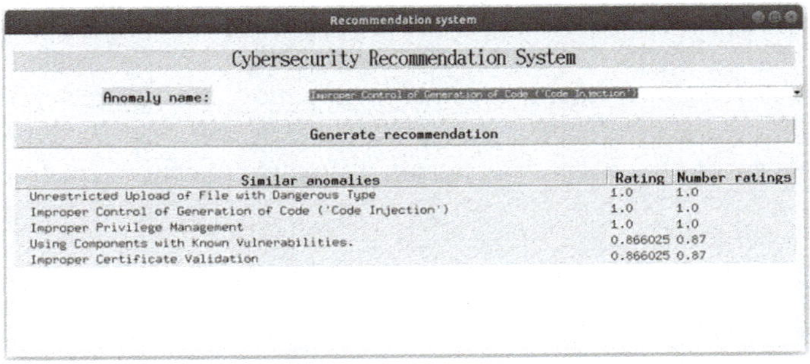

Fig. 8 Anomaly result screen

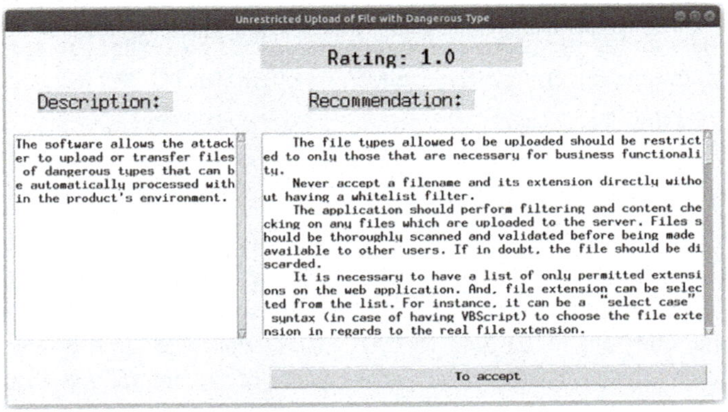

Fig. 9 Description and recommendation for the anomaly

5 Evaluation

The recommendation system was evaluated by a group of experts in cybersecurity and data analysis. These experts were selected with the aim of cover a wide range of criteria in the use and usefulness of the recommendation system [18]. Thus, six experts from the academic and professional fields and who belong to public and private sectors were considered [18]. For the evaluation, we used the criteria established by the TAM model [17] with focus on the perceived ease of use and the perceived usefulness of the system. Also, some questions to understand the work of the experts were asked.

The results found helped to understand the behaviour of each expert at the moment to provide a response to the anomalies that may arise in their daily work. This allowed to explore the way the recommender system can be useful to them. Thus, each of the experts reacts in different ways to an anomaly:

- By reporting the anomaly to the service desk
- By solving the anomaly immediately and looking for help from partners depending on the severity of the issue and budget limits
- By performing a process according to the recommendation of a CSIRT
- By carrying out a vulnerability analysis, in which the presence of the vulnerability is confirmed and is scaled to the respective system administrator
- Or by implementing a patch

Almost all the experts confirmed that in their daily activities, they frequently face to anomalies and vulnerabilities that can compromise computer security. Among these, the experts identified the following as the most recurrent: phishing, virus, malware, improperly configured services, SQL injection attacks, misleading window threats, cross-site scripting attacks and obsolete applications' vulnerabilities. Given these anomalies and depending on its severity, the experts estimated that they can provide a response in a delay between 1 hour and 1 day. The extra time it takes to each expert to write a report is not considered.

Concerning our recommender, four of the six experts affirmed that the system satisfy their needs to provide a quick response to anomalies and vulnerabilities. The other two gave a positive evaluation, pointing out improvements that can be introduced such as implement the recommendation system as a web-based solution or include the source of each recommendation.

All six experts believe that, thanks to the system, they were able to improve their response time to anomalies and vulnerabilities that could compromise the computer security of their organizations.

6 Discussion

From our preliminary analysis, it was determined that the type of recommendation system that best suits the needs of both the user and the elements (the anomalies and vulnerabilities that compromise computer security) is a hybrid recommender based on (1) a collaborative filter that allows the generation of a knowledge base, which is created jointly by the experts' judgements and information of security websites and (2) a knowledge-based recommender for dealing with the scarcity of the content or data that would be found in a recommender based only on a collaborative filter. This approach allowed us to provide recommendations qualified by expert judgement and, in this way, take advantage of both the knowledge and the experience of the computer security experts.

The expert judgement allowed to identify the worst vulnerabilities and anomalies within different entities. However, within each organization, the information about their worst vulnerabilities is confidential because the security of the organization could be very compromised if this information falls into wrong hands. The fear of exposing this information constituted an important barrier that was mitigated by using initial information from the official pages of different entities (Symantec;

OWASP; NIST; the University of Trento, Italy; and the CSIRT of the Escuela Politécnica Nacional). As each of these sites provides different datasets, some tools were used to identify and get the information necessary to generate the knowledge base.

Despite the fact that some of the initial recommendations regarding anomalies and vulnerabilities did not adequately adjust to the reality of the organizations of evaluators, it is possible to affirm that the purpose of the system was fulfilled. Indeed, based on the results obtained, the recommendation system helped the experts to reduce the time they spent to solve a cybersecurity issue, and it limited unnecessary manual processes and reduced the subjectivity of the cybersecurity analyst. This is because having a tool that consolidates information about anomalies or vulnerabilities with their respective recommendations makes it easier not only to provide and respond to problems with a shorter response time but also to prepare reports.

7 Conclusions

Recommendations based on collaborative filters help to generate a knowledge base with the help of expert judgement. In this way, the experts can provide adequate recommendations about cybersecurity anomalies and vulnerabilities and how to counter them. However, due to its main drawbacks with cold start issues (data scarcity and quality), the collaborative filter recommender was joined with a knowledge-based recommender to resolve the above-mentioned drawback. A recommendation system based on the content was discarded because they do not consider the assessment that a user can provide.

The datasets collected from universities and cybersecurity entities (i.e., Symantec, OWASP and NIST) allowed the elaboration of qualifications, the identification of the worst vulnerabilities and the best recommendations to counter them. However, it is data that could not always be used for every organization. In addition, the data present different nomenclature for anomalies and vulnerabilities, which can confuse end users. Thus, it is then necessary to include mechanisms that allow each organization to add its own information into the system or edit the existing one.

Finally, we conclude that the implementation of our recommendation system helped to improve the response time of cybersecurity analysts. It facilitated the execution of manual processes and reduced the subjectivity of the analyst. This was achieved, thanks to the fact that the users can access to expert judgement, getting truthful information that facilitates a better decision to respond to a cyberattack.

Acknowledgement The authors thank the support of the Ecuadorian Corporation for the Development of Research and the Academy (RED CEDIA) for funding this work under the Project Grant GT-II-2017.

References

1. Western Governors University, What does a cyber security analyst do? (13 Agosto 2018). [En línea]. Available at: https://www.wgu.edu/blog/what-does-cybersecurity-analyst-do1808.html. Last access: 2 Apr 2019
2. A. Oltramari, N. Ben-Asher, L. Cranor, L. Bauer, N. Christin, General requirements of a hybrid-modeling framework for cyber security, in *2014 IEEE Military Communications Conference*, 20 Nov 2014
3. I. Herwono, F.A. El-Moussa, A system for detecting targeted cyber-attacks using attack patterns, in *Information Systems Security and Privacy*, vol. 867, (Springer International Publishing AG, Cham, 2018), pp. 20–34
4. X. Wang, C. Wang, Recommendation system of e-commerce based on improved collaborative filtering algorithm, in *2017 8th IEEE International Conference on Software Engineering and Service Science (ICSESS)*, Beijing, 2017, pp. 332–335, https://doi.org/10.1109/ICSESS.2017.8342926
5. P.K. Singh, P.K. Dutta Pramanik, P. Choudhury, Collaborative filtering in recommender systems: Technicalities, challenges, applications, and research trends, in *New Age Analytics: Transforming the Internet Through Machine Learning, IoT, and Trust Modeling*, (Apple Academic Press, Toronto, 2020)
6. CISCO (2020). [Online]. Available: https://www.cisco.com/c/en/us/products/security/what-is-cybersecurity.html
7. N. Diakun-Thibault, Defining cybersecurity. Technol. Innov. Manag. Rev., 13–21 (Oct 2014)
8. Universidad Veracruz (2020). [Online]. Available: https://www.uv.mx/csirt/que-es-un-incidente-de-ciberseguridad. Last access: Jan 2020
9. F. Isinkaye, Y. Folajimi, B. Ojokoh, Recommendation systems: Principles, methods and evaluation. Egypt. Inform. J. **XVI**(3), 261–273 (2015)
10. A. Ferferning, L. Boratto, M. Stettinger, M. Tkalčič, *Group Recommender Systems: An Introduction*, vol I (Springer International Publishing, Cham, 2018)
11. J. Shu, X. Shen, H. Liu, B. Yi, Z. Zhang, A content-based recommendation algorithm for learning resources. Multimed. Syst. **24**, 163–173 (2017)
12. M.C. Martínez, Sistemas de Recomendación basados en técnicas de predicción de enlaces para jueces en línea, Madrid, 2017
13. H. Casanova, E. Ramos, H. Nuñez, Sistema Basado en Conocimiento para Recomendación de Información Turística Venezolana, in *III Simposio Científico y Tecnológico en Computación – SCTC 2014*, May 2014
14. N. Vaidya, A.R. Khachane, Recommender systems-the need of the ecommerce ERA, in *The 2017 International Conference on Computing Methodologies and Communication (ICCMC)*, July 2017
15. P. Pérez, Recomendaciones en tiempo real mediante filtrado colaborativo incremental y real-time Big Data, Madrid, 2016
16. A. Ruiz Iniesta, Estrategias de recomendación basadas en conocimiento para la localización personalizada de recursos en repositorios educativos, 2014
17. V. Mezhuyev, M. Al-Emran, M.A. Ismail, L. Benedicenti, D.A.P. Chandran, The acceptance of search-based software engineering techniques: An empirical evaluation using the technology acceptance model. IEEE Access **7**, 101073–101085 (2019). https://doi.org/10.1109/ACCESS.2019.2917913
18. E.F. Loza-Aguirre, A.F. Buitrago Hurtado, Qualitative assessment of user acceptance within Action Design Research and Action Research: Two case studies. Lat. Am. J. Comput. **1**(1), 4–14 (2014)

Secure Stor: A Novel Hybrid Secure Edge Server Architecture and CDN to Enhance the Security and Response Time for Edge Devices

Mais Nijim, Raghava Reddy Marella, Muhammad Aurangzeb, and Moustafa Nasralla

1 Introduction

Data is stored in more places in a different format in a large quantity than ever before. From autonomous cars, oil rigs, and factories, there's a need for real-time data processing. One solution for all such problems is edge computing. Nowadays, the public cloud, such as the CDN [1], is widely used due to its ability to compute and store data on a large scale. The limitations on the size of the Internet pipelines and the speed of light create some problems while moving the stored data from the cloud. With edge computing [2], data is acted at the source not sent to the cloud and creates real-time insights. This creates an environment that acts as a public cloud. In recent years, the rapid advances in smart mobile apps and the Internet of Things (IoT) [3] have significantly enabled edge computing [4, 5] to advance. In edge computing, it has given great support for IoT devices to perform difficult tasks in the best way. Although its bad implementations result in avoiding large threats happening in the factor of security, platforms, and applications. Edge computing is simply described as the processing and transmission of data from devices distributed worldwide. The rapid development of devices associated with networking technologies, IoT, and their applications keeps driving the edge computing frameworks. The fast growth in networking technologies, such as wireless

M. Nijim (✉) · R. R. Marella · M. Aurangzeb
Department of Electrical Engineering and Computer Science, Texas A&M University-Kingsville, Kingsville, TX, USA
e-mail: mais.nijim@tamuk.edu; Raghava_reddy.marella@students.tamuk.edu; Muhammad.aurangzeb@tamuk.edu

M. Nasralla
Department of Communications and Networks, Prince Sultan University, Riyadh, Saudi Arabia
e-mail: mnasralla@psu.edu.sa

K. Daimi et al. (eds.), *Advances in Security, Networks, and Internet of Things*, Transactions on Computational Science and Computational Intelligence,
https://doi.org/10.1007/978-3-030-71017-0_29

networking technology, accelerates development or enables real-time applications such as artificial intelligence, video processing and analytics, and self-driving vehicles. Edge computing operates on instant data, which is real-time data produced by a sensor or generated by the user in real time. In contrast, cloud computing is a technology that operates on big data.

Edge computing provides location-conscious, sufficient bandwidth, private, real-time, and low-cost infrastructure to keep up with emerging smart city applications [4]. These advantages over the cloud have contributed to the growth of edge computing. Edge computing's market size is increasing every day. The accomplishment of the IoT and the lush cloud services helped to establish a new generation of edge computing, where data processing takes place at the network edge, instead of processing the data entirely in the cloud. It could address issues such as cost of bandwidth, latency, limited battery life, privacy, and security. Moving computing activities to the cloud network has become an efficient method to handle data, as the cloud has more processing capacity than the computers at the edge network.

Moreover, while processing data speeds have risen exponentially, there has been no massive increase in the bandwidth of the networks that carry data to the cloud. While the system is becoming the bottleneck of a cloud with edge devices producing more data, for example, an autonomous vehicle camera can capture a large amount of raw data to be processed by the system in real time to generate better driving results without latency. If there is no edge server, the data needs to be sent to the cloud for processing, and the response time will be longer where the efficiency of the response is affected. Autonomous vehicles [5] are one area that would further strain network reliability and bandwidth. Processing data at the edge is more productive and more efficient; it produces shorter responses that will lower pressure on the network. The interconnectivity, which has increased tremendously, is resulting in providing access for more applications with improved edge computing. A new IoT and specific business use cases of the industry are used, and along with it, the edge computing-based infrastructure is approved to be one of the best storages and servers, which has immense growth in the future.

There are several factors [5] that determine cloud computing is not adequate to be used alone. The first important factor is latency. Edge computing can provide latency in milliseconds by sending data to edge server, and high response time can be obtained, whereas in the cloud computing model, applications need to send data to the data center and then attain a response, which will significantly increase the latency period of the system. The second factor is the high throughput: throughput is available to the user from the edge, which is served via locally generated or cached content. In our proposed framework, data will be cashed in an array of solid-state disks. The third factor is data reduction when running edge-based computing applications, for example, data analytics, operators, and application providers may significantly reduce the amount of data to be submitted upstream, which is often used to achieve cost savings. The fourth factor is data availability, where nowadays there are more and more cloud-based Internet services; the use of such services has become an essential part of our everyday lives. The fifth important factor is security; cloud service providers can also secure their networks against the attacks from customer premise equipment or user equipment using edge security. Lastly, the sixth factor is isolation. There is a concern with high-speed connections in the number of places that are not always linked to the Internet. During times of disrupted or missing connectivity, an edge cloud is still able to provide services that provide disaster resilience.

Hybrid storage systems [6] provide a reduced cost for data centers without scarifying the request response times. In our hybrid storage systems, we will use two levels in the centralized location, which consists of the solid-state disks in the upper level and the hard disk drives in the lower level. The most frequently used data will be placed on the upper level. If the request is not found in the upper level, the lower level will be checked for data retrieval. The data requests will be first checked at the higher level, which is the solid-state disks; if the data is not found, it will be checked from the lower level of the hard disk drives for prefetching and retrieval.

Security services for the above architecture are important to protect and secure the residing data from talented intruders [7]. Besides, many edge server requests require to be finished before their absolute deadline [8]. This paper proposed a hybrid storage framework that integrates several storage devices, i.e., CDN, edge server that is connected with the CDN, and an array of solid-state disks. In the central location, we will use a hybrid storage system architecture that consists of a combination of an array of solid-state disks and an array of hard disk drives. The proposed framework uses a security control protocols that can be adjusted to encounter security changes, necessities, and workload environments, therefore providing a high level of security for all edge device requests. A prominent feature of the proposed framework is the use of the solid-state disk partitioning mechanism that helps in incrementing the level of security of the proposed framework while maintaining the desired deadline.

The rest of the paper is organized as follows: Sect. 2 describes the architecture of the proposed system and the proposed solid-state disk partitioning mechanism. Section 3 describes the video on demand, a real-time application that can successfully use the proposed framework. Section 4 offers the processing algorithm for the Secure Stor framework. Section 5 analyzes the experimental results. Finally, Sect. 6 concludes the paper with future directions.

2 Architecture and Data Management

In this section, we describe the Secure Stor architecture and the set of its supporting features, which include the data partitioning, the security management, and the data placement. Figure 1 illuminates the hybrid architecture for the Secure Stor. The CDN is responsible for storing metadata and stream video. Hard disk drives (HDDs) and an array of solid-state disks (SSDs), parallel system server, and data handler request transmitter and receiver, which are directly connected to the edge devices. The number of hard disk drives and the solid-state disks is independent of each other. The array of the solid-state disks is the nonvolatile cache that will save the highly

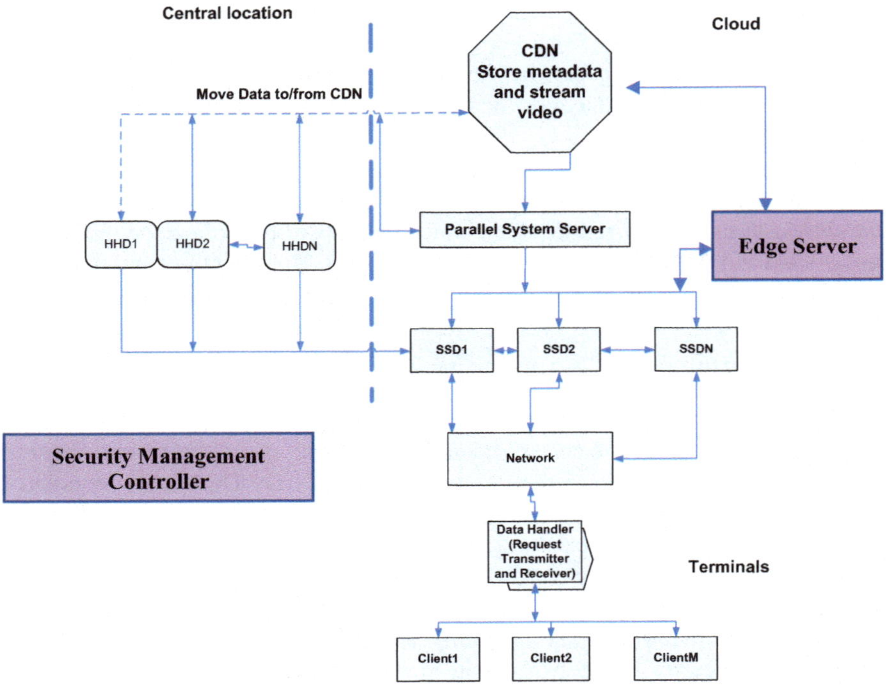

Fig. 1 The general architecture of Secure Stor

request data, and it is used to boost up the performance. If the requested content is not available at the solid-state disks, it will be requested from the hard disk drives, and if there is an error or a failure in the hard disk drives, the data will be requested from the central location server. The security management controller is the heart of Secure Stor architecture. It comprises of a security middleware services, a solid-state disk partitioning system, a real-time response time estimator, and a security controller. The security middleware services are responsible for assigning a level of security for each user requests based on the request response time. The security middleware services are extremely adaptable in the sense that it allows new security services such as new encryption algorithms to be added or to replace old security services with new ones.

2.1 *Security Management Controller*

The performance of the whole architecture can be significantly enhanced by employing an array of solid-state disks. Solid-state disks are vital because they can judiciously store the frequently used data for future access. Notably, the solid-state drives can help to shorten the response time for data requests if they are not available in the edge server, making it possible to increase the level of security for the whole system while meeting the users' deadlines. Note that we use the least recently used algorithm for moving data from solid-state disks to hard disk drives. To maximize the security of the requests without violating the deadline of the applications, the solid-state disks will be vigorously segregated among the hard disk drives. The total size of the solid-state disks will be divided into several equal-size partitions. Each hard disk drive will be assigned one or more partitions based on the hard disk drive workload. Each solid-state disk partition will be managed separately using the LRU replacement policy.

We model the collection of security services required by an edge device request Er_i as:

$$S_i = \left(S_i^1, S_i^2, S_i^3, \ldots, S_i^j\right)$$

where S_i^j is the required level of security range for each edge device request.

Let PA_i be the degree of parallelism of Er_i, then the security allowance of request Er_i can be calculated by the following equation:

$$S\left(\mathrm{Er}_i, \mathrm{PA}_d\right) = \sum\nolimits_{j=1}^{\mathrm{PA}_m} \mathrm{SecLevels}\left(\mathrm{ER}_i, \mathrm{PA}_d\right), \mathrm{PA}_{m \leq} K, \mathrm{PA}_d \leq \mathrm{PA} \tag{1}$$

where PA is the total size of the solid-state disks, PA_d is the dth hard disk drive partition size, and K is the number of hard disk drives available in the cloud system. It's clear that the parallelism degree PA cannot exceed in any way the total number of the hard disk drives in the whole system. Our main goal is maximizing the level of security of all requests, which can be formalized by the following nonlinear equation:

$$\mathrm{Maximize} \sum_{d=1}^{k} S\left(\mathrm{ER}_d, \mathrm{PA}_d\right) \tag{2}$$

3 Real-Time Application Example

Video on demand is considered to be a real-time application of the Secure Stor proposed architecture. In the video on demand, customers can do trick play mode faster when using the nDVR (network digital video recorder), by putting the I-frames that will allow the customer to fast forward (FF) or rewind appropriately. The I-frames will be stored per program on the CDN for all programs recorded from customers on that particular CDN. We need to collect information about the customers who are watching content on a specific CDN using data mining techniques. The data mining algorithm is out of the scope of this paper. For example, we could collect information on how many customers recorded on the nDVR for the same program and then push the trick play or trick mode I-frames to the CDN. The whole file could be, or just part of it, based on the watching behavior on that CDN across all customers. Suppose the I-frames are placed on the CDN or on closer storage such as the edge server rather than storing it on the centralized storage. This step will increase the request response time, which will result in better response time on the consumer side by avoiding the access to the centralized storage unit. The centralized storage unit consists of an array of low-cost hard disk drivers rather than high-cost storage drivers such as solid-state disks.

These files could be personalized per customer and per their usage of the nDVR content or their general usage of watching content. For example, a customer is watching a program from their nDVR, which is stored in centralized storage and will be passed through the CDN to allow them to watch. The customer halfway through the program decided to rewind 15 minutes in the program. The customer presses on their TV or set-top box remote control or from another device (slide the rewind on their other devices like the computer, tablets, etc.). We have information about either the customer's behavior when watching content or what they do when pressing rewind, in which they could be pressing rewind just 5 minutes, or rewind few seconds, or rewind 10 minutes at a time. We can capture all the watching behavior information and store that in the back office per customer. Figure 2 shows the algorithm for pushing the trick play files to the edge server of the CDN for easier access and better response time and performance.

4 The Proposed Algorithm of Secure Stor

Figure 3 describes the processing algorithm of Secure Stor.

5 Experimental Results

To evaluate the performance of the Secure Stor framework, we implemented a simulation toolkit, which consists of a set of core components. The core component includes a parallel data transfer between HDDs and SSDs and a parallel data transfer between the edge server, CDN, and solid-state disks. We use nine confidentiality service algorithms that are implemented within the simulation toolkit. We use eight SSD disks with a capacity of 1 GB and eight hard disk drives with a capacity of 200 GB. We compare Secure Stor with another framework that does not incorporate the hybrid storage system in the central location and also does not require solid-state disk partitioning. In our simulation experiments, we made use of the following two metrics to demonstrate the effectiveness of the Secure Stor: The first metric is the

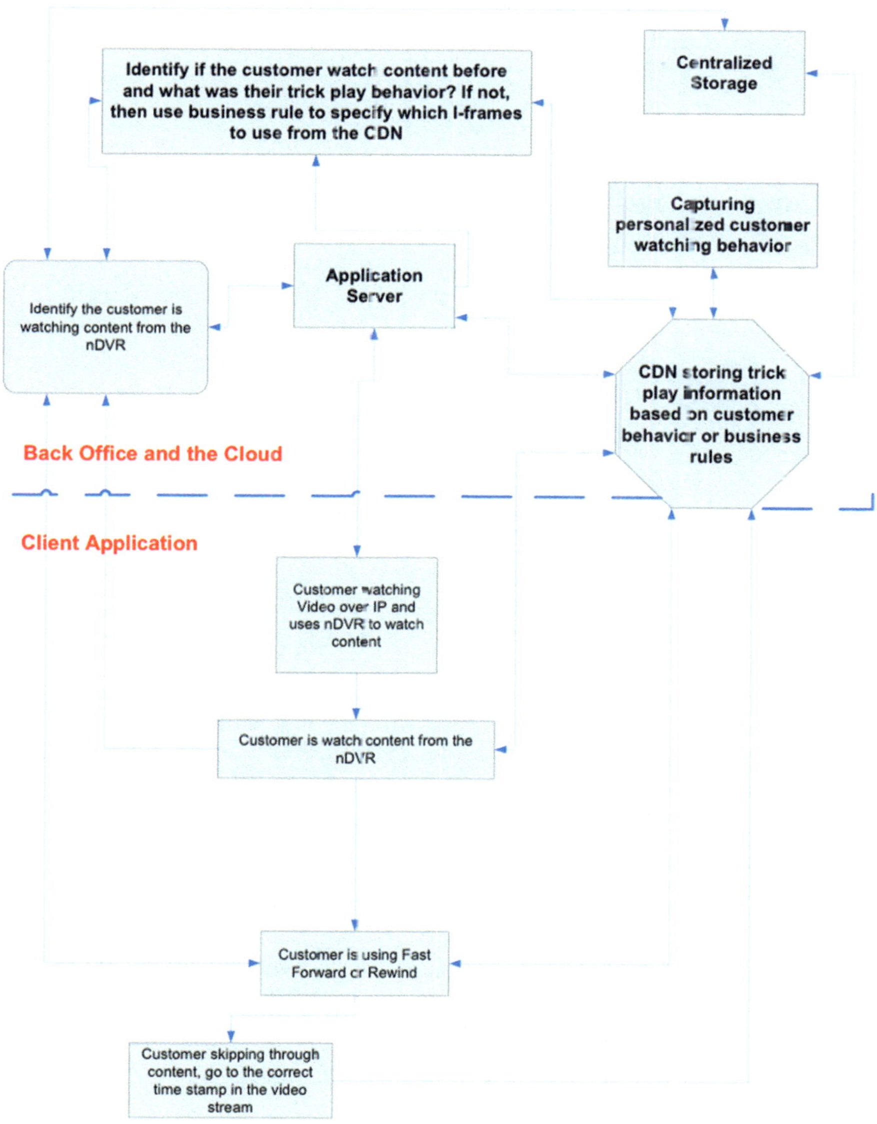

Fig. 2 The block diagram of the video on demand using the proposed framework

success ratio. It is the fraction of total arrived edge server requests that are finished before their absolute deadline. The second metric is the average level of security, which is the ratio of the sum of the level of security of all requests to the number of requests.

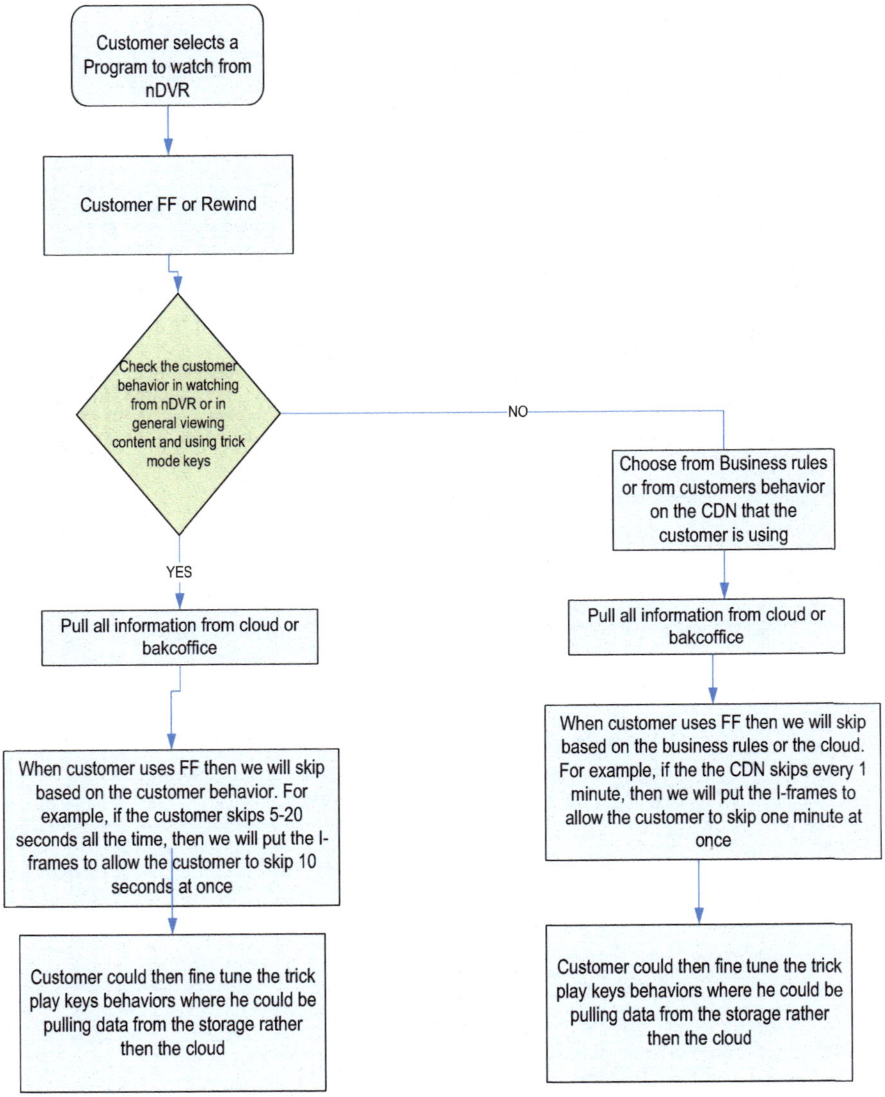

Fig. 2 (contindea)

5.1 *Impact of Arrival Edge Device Request and Level of Security*

In this experiment, we study the impact of edge device request arrival rate when we vary the solid-state disk size and the disk bandwidth. To accomplish this goal, the arrival rate was increased from 0.1 to 0.5 NO./SEC.; we set the edge device request

```
Input: ER : A newly arrived edge device request
       D: Absolute deadline of the request
       Sl : The lower bound of the edge device level of security
       Sm : The upper bound of the edge device level of security
       R: waiting que for the edge device request
1.   Insert the requests into R based on their absolute deadline
2.   Partition the solid state disks among the hard disk drives based on the hard disk drives workload
3.   For each request do
4.       Initiate the lower bound of the level of security for each request to 0.1 (0.1 means use a weak
encryption algorithm)
5.       While (Di < desired response time) do
6.           If the level of security < 1 (1 is the strongest encryption algorithm)
7.               Increase the level of security by 0.1 ( go to the next stronger encryption algorithm)
8.       End while
9.       If Di > desired response time
10.          Then decrease the level of security by 0.1 (go to the weaker encryption algorithm)
11.  Deliver the requests
```

Fig. 3 The processing algorithm of Secure Stor

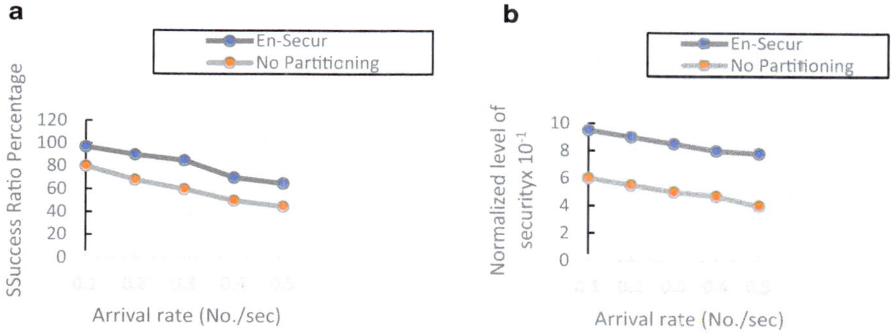

Fig. 4 (**a**) Impact of arrival rate on satisfied ratio, (**b**) impact of arrival rate on security level

data size to 1 GB and the disk bandwidth to 100 MB. Figure 3 reveals that Secure Stor outperforms other algorithms significantly. Also, Fig. 4 shows that Secure Stor outperforms the different algorithm that does not use the component proposed in the Secure Stor framework. Secure Stor delivers a higher level of security than other algorithms.

6 Conclusion

This paper presented a novel hybrid framework that integrates CDN with edge server in the cloud and a hybrid storage architecture that consists of hard disk drives and solid-state disks in the centralized storage. Consequently, security has become very crucial for cloud and edge computing. In this paper, we proposed a solid-state disk partitioning algorithm to increase the quality of security for the whole system. Our simulation results revealed that when comparing the performance and the security of Secure Stor with another architecture that does not use solid-state partitioning technique, Secure Stor significantly increases the level of security and the performance of the whole system by an averages of 85%.

References

1. L. Ling, M. Xiaozhen, H. Yulan, CDN cloud: A novel scheme for combining CDN and cloud computing, in *IEEE International Conference on Measurement, Information and Control*, 2013
2. M. Satyanarayanan, W. Shi, Overview of edge computing, in *IEEE Course*, 2018
3. T. Voigt, C. Rohner, What is the Internet of Things: An introduction, in *IEEE Course*, 2017
4. W. Shi, J. Cao, Q. Zhang, T. Li, L. Xu, Edge computing: Vision and challenges. IEEE Internet Things J. **3**(5), 637–646 (2016)
5. A. Koike, Y. Sueda, Content delivery for autonomous driving cars in conjunction with car navigation system, in *20th Asia Pacific Network Operations and Management Symposium (APNOMS)*, 2019
6. M. Nijim, Z. Zong, X. Qin, Y. Nijim, Multilayer prefetching for hybrid storage systems: Algorithms, models, and evaluations, in *IEEE International Conference on Parallel Processing Workshops*, 2010
7. J. Zhang, B. Chen, Y. Zhao, Z. Cheng, F. Hu, Data security and privacy-preserving in edge computing paradigm: Survey and open issues, in *IEEE Access*, 2018
8. Y. Xing, H. Seferoglu, Predictive edge computing with hard deadlines, in *IEEE International Symposium on Local and Metropolitan Area Networks*, 2018

Leveraging Security Management with Low-Level System Monitoring and Visualization

Karlen Avogian, Basel Sababa, Ioanna Dionysiou, and Harald Gjermundrød

1 Introduction

The challenge of identifying quickly security attacks is nontrivial and becomes even more complex when considering small enterprises with a single server or individuals running their personal server at home. In the first case, it is unlikely that there is a dedicated system security administrator managing the server resource and all server management is undertaken by the most technically oriented employee in the company. In the second case, users have at least a minimum technical skill set to configure and maintain their own home servers. This user group usually is excluded from published surveys of security incidents as it is intractable to reach neither every single user administrating his/her own personal server nor administrators of small enterprises. On the other hand, this user group is an attractive target for hackers who do not seek financial gain but instead they are interested in compromising machines to be used either as stepping stones for an attack or to expand an existing botnet.

This paper discusses `SMAD` [1, 7, 8], a configurable and extensible System Monitoring and Anomaly Detection framework that monitors kernel and system resources data (e.g. system calls, network connections, process info) based on user-defined configurations that initiate non-intrusive actions when alerts are triggered or anomaly behavior is detected. `SMAD` is envisioned to be used by users with some technical skills to track their local Linux server health, where the server is dedicated to run a few services (web, email, file, to just name a few). The *stable* nature of dedicated servers running a few services makes it manageable to

K. Avogian · B. Sababa · I. Dionysiou (✉) · H. Gjermundrød
Department of Computer Science, School of Sciences and Engineering, University of Nicosia, Nicosia, Cyprus
e-mail: avogian.k@live.unic.ac.cy; sababa.b@live.unic.ac.cy; dionysiou.i@unic.ac.cy; gjermundrod.h@unic.ac.cy

K. Daimi et al. (eds.), *Advances in Security, Networks, and Internet of Things*, Transactions on Computational Science and Computational Intelligence, https://doi.org/10.1007/978-3-030-71017-0_30

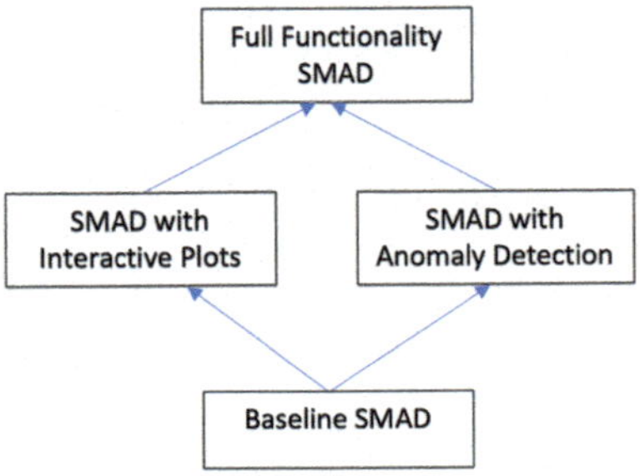

Fig. 1 SMAD

detect unexpected behavior (e.g. login attempts). Once it is configured properly, its behavior is predictable and abnormal activity can be easily detected.

Monitoring raw kernel and system resources data is a tedious and error-prone task, if done manually. The baseline `SMAD` [7, 8] (see Fig. 1) was developed using `Sysdig` [11] and eases this overwhelming task by allowing users launching several system monitors via `SMAD` in a user-intuitive manner, alerting him/her on any unexpected behavior based on user-defined metrics, including CPU usage, directories visited, system errors, commands executed, HTTP requests, IP addresses connected to the system, and files opened. The latest version of SMAD [1], which is the focus of the paper, integrates `Falco` [9] that leverages the system monitoring of the baseline system with an anomaly detection engine and also allows for interactive plots that enhance the visualization aspect of `SMAD`.

The paper contributions are twofold, as listed below:

1. Propose a configurable, open-source, user-friendly front-end for `Sysdig` and `Falco`, wrapping the gory low-level details surrounding system calls and scheduling anomaly detection tasks in a simple, yet effective way.
2. Evaluate the effectiveness of the aforementioned system.

The rest of the paper is organized as follows: Sect. 2 describes the two basic underlying technologies wrapped in `SMAD`, namely `Sysdig` and `Falco`. Section 3 presents the architectural details of the `SMAD` framework. An evaluation of `SMAD` is given in Sects. 4 and 5 concludes.

2 SMAD Underlying Technologies

The first underlying technology that `SMAD` is built on is `Sysdig`. `Sysdig` is an open-source command line utility for capturing all system calls residing within the Linux kernel. It could be perceived as the *Wireshark* for the end system. Every time an installed application performs a privileged operation (e.g. open/read file, open network port, read/write to any device), it invokes a system call that executes the operation on behalf of the user's process. Capturing all invoked system calls could be viewed as passive sniffing of all the operations performed within the server. A subset of the Linux monitoring and debugging tasks that are bundled by `Sysdig` is {`strace, tcpdump, netstat, htop, iftop, lsof`}. Additional features of `Sysdig` include provision of chisels (lightweight Lua scripts) for processing captured system events, provision of simple filtering of output, support of system and application tracing, and support of Linux server attack analysis features for ethical hackers. Last, but not least, state-of-the-art container visibility [5] is also supported.

The large number of the executed system calls would quickly overwhelm the system, with respect to both processing time and storage. As a countermeasure, `Sysdig` supports the configuration of filters in order to capture specific system calls or a subset of them. In this way, the filtered events are those of interest to the user. However, on the downside, it is nontrivial to formulate the appropriate filters tailored to the deployment, usage, and threat landscape of the specific server. There are commercial solutions that provide intuitive user interfaces for monitoring large deployments of servers (as well as container deployment) using `Sysdig`. An example is the Sysdig Monitor Dashboard [10] developed by Sysdig, a commercial product targeted for enterprises deploying applications in cloud infrastructures. Similarly, IBM offers a front-end to `Sysdig` [4] as part of its BlueMadator product [2].

The second underlying technology is `Falco`, a behavioral activity monitor designed to detect anomalous activity in applications making Linux system calls. Using powerful system call capture technology originally built by `Sysdig Falco` continuously monitors and detects application, host, and network activity, all in one place, from one source of data, with one set of rules [9]. Alerts can be triggered by the use of specific system calls, their arguments, and by properties of the calling process. Examples of such alerts are given below:

- A server process spawns a child process of an unexpected type
- A sensitive file, like /etc./shadow, is unexpectedly read
- A non-device file is written to /dev
- A standard system binary makes an outbound network connection

3 SMAD System Architecture

Figure 2 depicts the interactions among the `SMAD` baseline components. The `User Interface` redirects all user commands to the appropriate modules and at the same time it conveys output from those modules to the user in an intuitive user-friendly manner. A `User` can configure, launch, and manage monitors, with real-time data visualization support, via the `Monitoring Sensor`. In the event of an alert triggering, all events for a predefined time duration are optionally stored in `Captured Files` for post-mortem analysis, which allows the construction of attack profiles via event tracing. A user can also utilize the `Anomaly Detection Sensor` module by either configuring new anomaly rules or loading/modify-

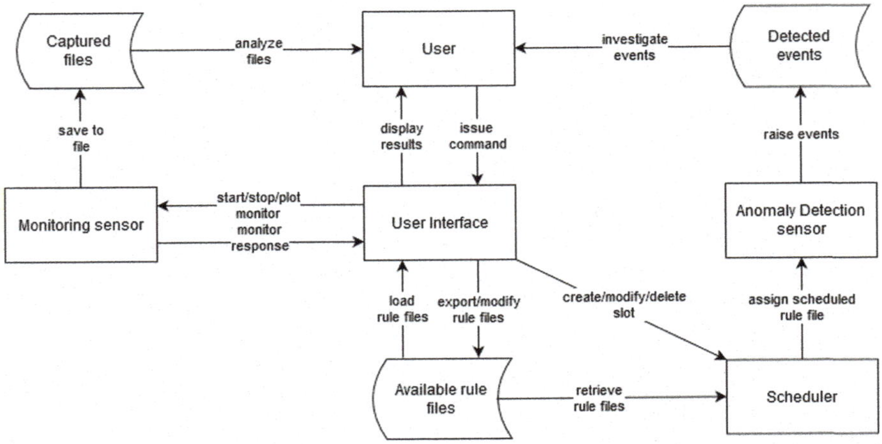

Fig. 2 SMAD system overview

ing existing ones stored in the `Available Rules Files` repository. The `Scheduler` component is responsible for assigning a scheduling slot for each anomaly rules file. More details on the baseline modules are given next.

3.1 SMAD Module: Anomaly Detection Sensor

The `Anomaly Detection Sensor` module is primarily responsible for configuring and managing user-defined anomaly rule files, including scheduling their running slot(s). All rule-related tasks are initiated via the `User Interface` that allows for an intuitive and error-free configuration and update of rule files. Once a scheduling slot is assigned to a rules file, the anomaly detection engine loads it at the appropriate time and starts its execution. `Falco` is the underlying technology integrated in `SMAD` that acts as the anomaly detection engine. In the event that anomalous behavior is detected, a notification is sent to the user. The processing tasks of this module are illustrated in Fig. 3.

3.2 SMAD Module: Monitoring Sensor

The `Monitoring Sensor` component is essentially a wrapper for `Sysdig`, executing `Sysdig` commands using `bash`, with the appropriate arguments that are automatically generated based on the user's preferences and selections set using the `User Interface`. `SMAD` supports a notification mechanism via alerts. An alert is assigned to a particular monitor and its configuration profile includes the metrics to

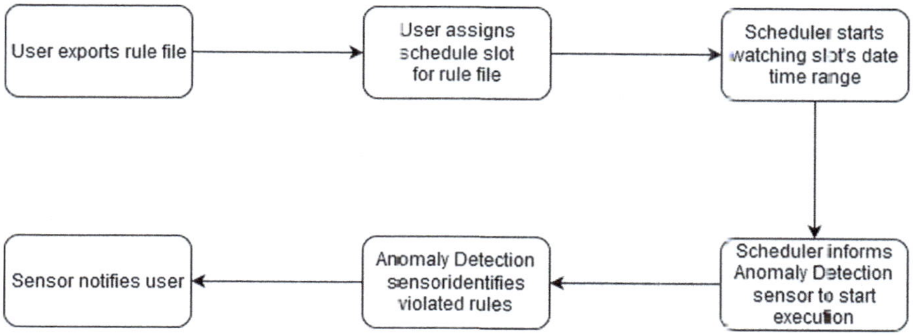

Fig. 3 Anomaly detection sensor module

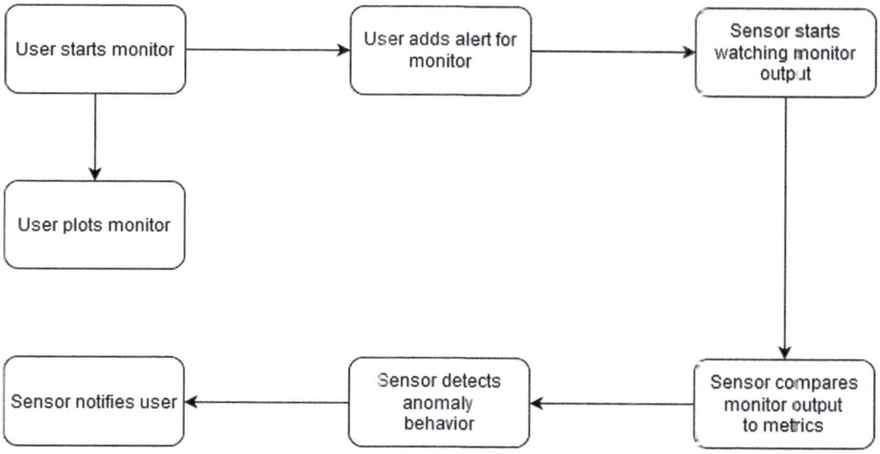

Fig. 4 Monitoring sensor module

be monitored, the user notification method, and whether or not event capturing upon triggering is required. A list of running monitors is maintained by the `Monitoring Sensor`, allowing the user at any given time to plot live monitor data graphs. In the case that a specific metric for an alert is violated, its corresponding alert profile is consulted in order to respond according to the user specifications. Figure 4 illustrates the overall processing of the `Monitoring Sensor` module.

3.3 *SMAD Module: User Interface*

The `User Interface` module was developed using PyQt5, a Python-binding of the cross-platform GUI toolkit Qt [6]. PyQtGraph [3] was used to create interactive graphs. The `User Interface` consists of three primary pages,

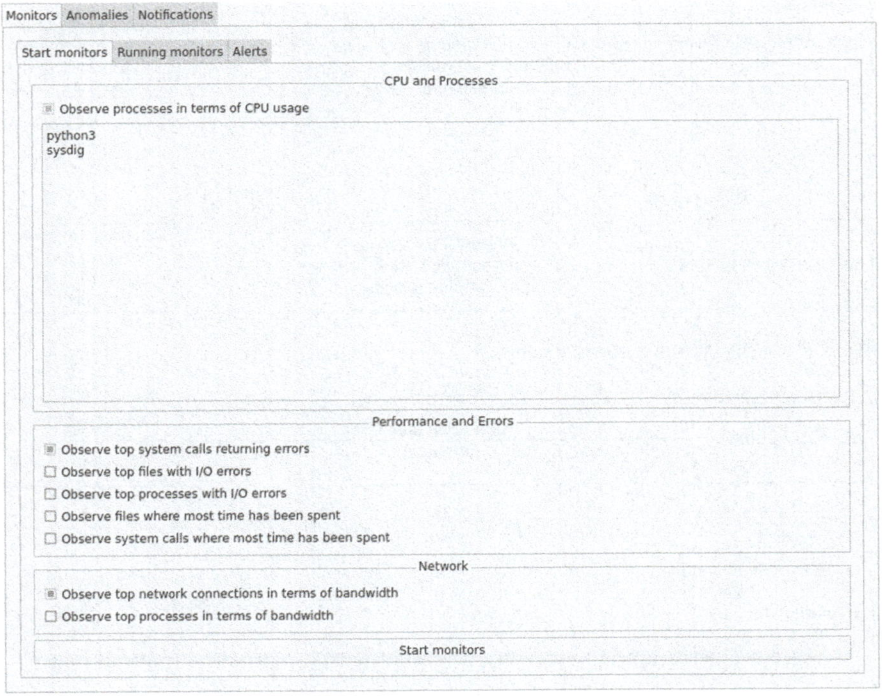

Fig. 5 Start monitors tab

namely, `Monitors`, `Anomalies`, and `Notifications`, as shown in Fig. 5. The underlying functionality of each page is described in detail next.

Monitors Page `Monitors` is the primary `SMAD` page responsible for managing monitors. `SMAD` supports three monitor categories (CPU and Processes, Performance and Errors, Network), as shown in Fig. 5, to monitor CPU usage, system calls triggered from system I/O, system errors and network usage. Analytically, a user could configure a monitor to track the:

- CPU usage of specific processes
- system calls that are returning the highest number of errors
- files with the most I/O errors
- processes with the most I/O errors
- files where the most time has been spent on
- system calls that consumed the most time to complete execution
- network connections that utilize the most bandwidth
- processes that consume the most bandwidth

Configuring and starting the execution of the monitors is done in a simplistic manner, requiring minimum user input. As a matter of fact, only the *CPU and*

Processes monitor category requires the user to explicitly specify the processes that their CPU usage will be monitored. The monitor configuration shown in Fig. 5 initializes four monitors; the first two monitor the CPU usage of the processes *python3, sysdig* respectively, the third one observes the top system calls returning errors, and the fourth monitor keeps track of the top network connections in terms of bandwidth utilization. The running monitors at any given time could be listed, terminated, or have their data plotted in real time. In all interactive plots the X-axis represents time whereas the Y-axis values depend on the monitor type, and specifically:

- Percentage, in the case of monitors observing CPU usage of user-specified processes
- Number of errors, in the case of monitors observing errors
- Time, in the case of monitors observing time
- Bandwidth, in the case of monitors observing network usage

In Fig. 6, live data is plotted for the monitor *networks_top_processes_bandwidth* responsible for observing bandwidth usage by network processes.

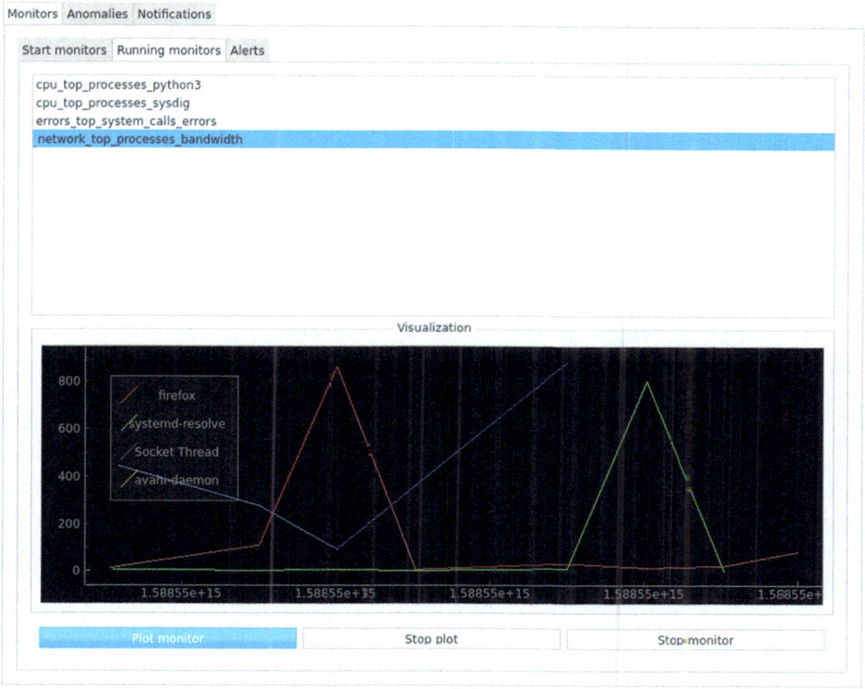

Fig. 6 Running monitors tab

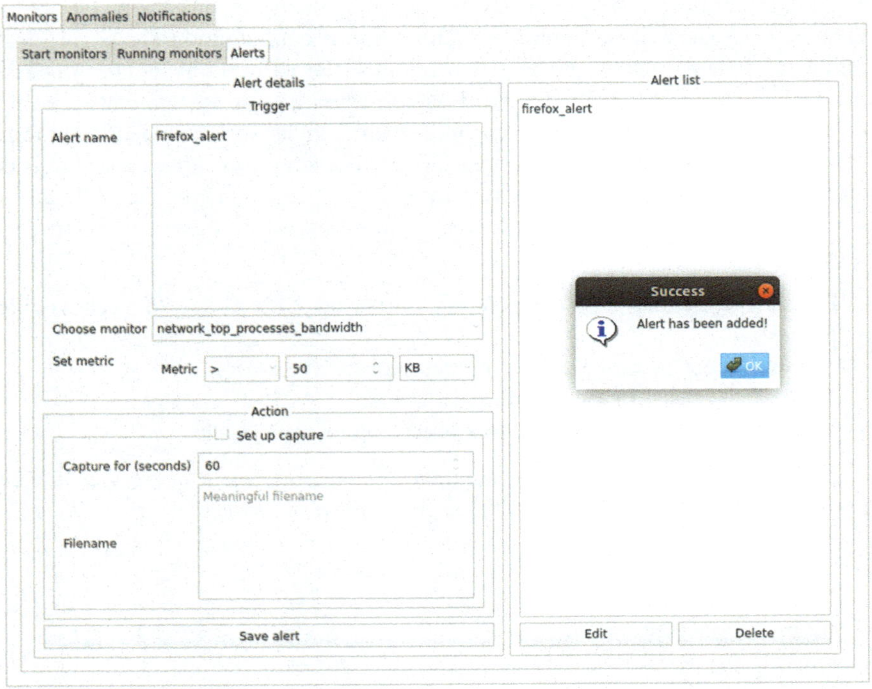

Fig. 7 Alerts tab

Once the monitor set is determined, alerts could be configured. Multiple alerts could be attached to a monitor, yielding in a one-to-many monitor-alert relationship. The alert profile is essentially a set of parameters attached to a specific monitor that has been already configured. The alert parameters are the monitor it is assigned to and the metric to be monitored. A metric violation yields an automatic notification posted on the `Notifications` page as well as initiating the capturing of events, if capturing is enabled for this alert. Figure 7 shows the configuration of the alert attached to the *networks_top_processes_bandwidth* monitor. The metric type is bandwidth and an alert is triggered for any process that transfers inwards or outwards data of more than 50 KB. Since the *Set up capture* checkbox is selected, event capturing will be initiated right after the metric violation for the selected duration of 60 s and stored in a user-specified file for further analysis.

Anomalies Page The `Anomalies` section leverages `SMAD` with anomaly detection capabilities, allowing a user to configure customized anomaly detection rules for the system and/or applications, store them in rule configuration files (with *.smadconf* extension), and create an execution schedule for each rule file.

`SMAD` supports multiple rule configuration files, run at different times. The motivation behind this underlying ability to load different rule configurations at different times is derived from the deployment scenarios envisioned for SMAD. Since the target SMAD user pool consists either small enterprise system administrators or personal system administrator hobbyists, it is expected that the usage patterns of their servers activity are predictable and stable. As an example, consider a small business **X** providing a specialized service, hosted on server **Y**, to local businesses during normal operational hours (09:00–17:00) on weekdays. Employees of company **X** only logon server **Y** during normal operational hours and *sudo* users never logon remotely. These are usage patterns that could be utilized to create different sets of rule configurations: one rule configuration running during normal operational hours, another one running off-business hours, etc.

The configuration of a rule file is shown in Fig. 8. In this example, the rule configuration file consists of two rules: detect the execution of the *sudo* command (potentially attempt to escalate process privileges) and keep track of the directories visited by the user *kosnet*. It is readily apparent that a user-centric design approach was employed to present the user with a simplistic, yet straightforward way to create and export a new rule file and also load and modify existing ones. The set of rules in a rule file is populated from the SMAD available rule options, where a user could set a rule to detect:

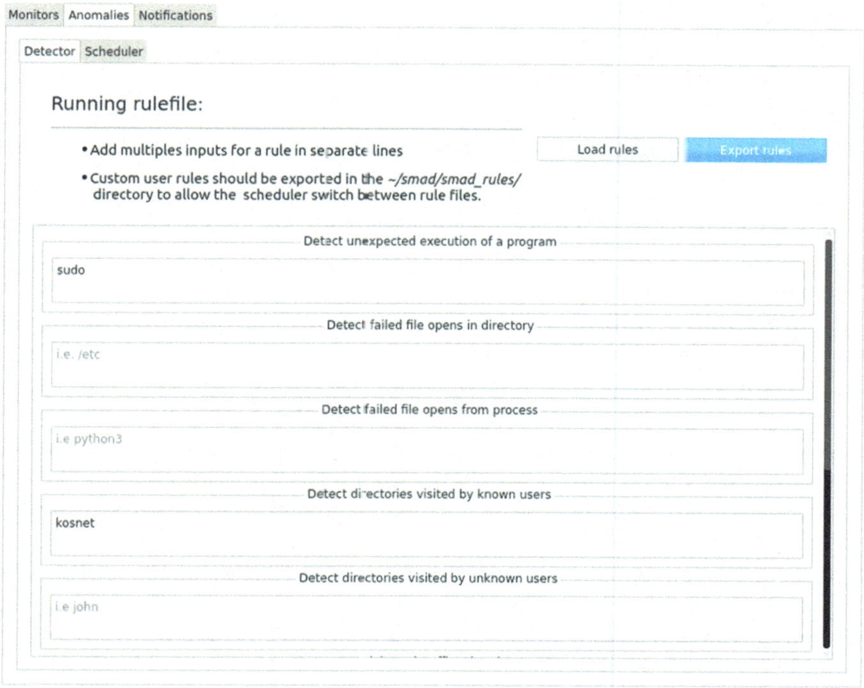

Fig. 8 Rules configuration

1. unexpected usage of one or more programs
2. failed file open operations in specified directories
3. failed file open operations from a specific process
4. directories visited by known users
5. directories visited by unknown users
6. incoming network traffic from specified IPs
7. outgoing network traffic from specified IPs
8. incoming or outgoing traffic from publicly known malicious IPs
9. unusual traffic on a MySQL server
10. unusual traffic on a HTTP server
11. unusual traffic on a Kafka server
12. unusual traffic on a MongoDB server

The available rule set empowers users with a simple mechanism to create powerful anomaly detection shields for their servers, test network connections, and also monitor activity from authorized system users in an attempt to detect at an early stage an ongoing insider attack.

All the available rule configuration files are located in the *smad_rules* directory and are in a dormant phase until they get activated. The activation is initiated by generating a schedule slot or even multiple schedule slots for a rule file. `SMAD` supports two scheduling modes:

- One-time execution mode, which yields in the generation of a single schedule slot that executes the rule file at the specified date and time range
- Repetitive execution mode, which yields in the generation of a single schedule slot that executes the rule file on a weekly basis at the specified day and time range (same for all weeks)

It is at the discretion of the user to delete active schedule slots or remove inactive ones.

Figure 9 is a snapshot of an instance of the `SMAD` scheduler. There are two execution schedule slots for rule file *test3.smadconf*. The first one is a repetitive execution schedule slot, where the schedule slot has a starting date Monday 09:15AM 4/5/2020 and ending date Tuesday 12:15AM 5/5/2020. That means, during the specified time interval and on a weekly basis the rule file *test3.smadconf* is loaded in the anomaly detection engine and its execution starts. The second one is also a repetitive execution schedule slot but as it is set to inactive it will not be considered by the scheduler module. In case the repetitive checkbox is unchecked, the execution mode is set to one-time execution schedule slot.

Once a scheduling slot is assigned to a rules file, the next step is to load its rules on `Falco` at the appropriate time and start their execution. The Python code segment in Listing 1 describes the initialization of a `Falco` thread.

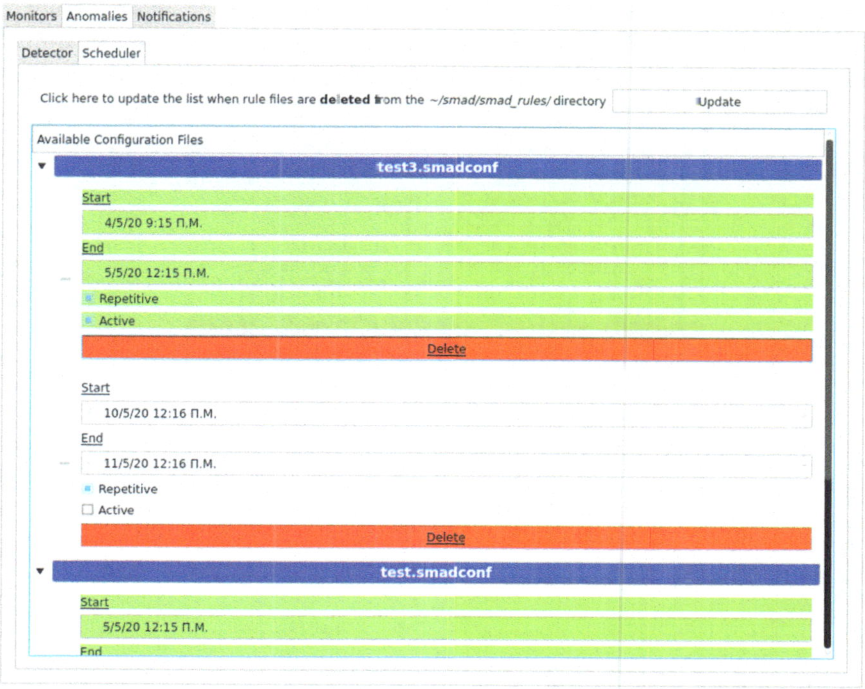

Fig. 9 Scheduler modes of execution

```
def run(self):
    # Create configuration file
    with open(self.user_generated_file, 'w+') as f:
        f.write(self.rules)
    # Start Falco
    process = Popen(shlex.split(self.command))
    while True:
        if self.stopped():
            process.terminate()
            os.wait()
            break
```

Listing 1 Falco thread initialization

It is the job of the scheduler thread to examine the scheduled rules and decide whether a running `Falco` instance should be stopped or a new one should be started. In the later case, the scheduler informs `SMAD` to load the file and a new thread gets spawned. The newly spawned `Falco` thread creates a new file *user_generated_file* that contains the rules in a `Falco`-required format and starts its execution until it is requested to stop.

Notifications Page The `Notifications` page is a tabular representation of either notifications yielded from metric violations of user-defined alerts or notification messages released by the anomaly detection engine. A color scheme is deployed to indicate the severity level of the notification event, as shown below:

User-defined Alert Notification
Anomaly Detector Warning Notification
Anomaly Detector Error Notification
Anomaly Detector Notice Notification
Anomaly Detector Emergency Notification
Anomaly Detector Informational Notification

An entry in the notification table consists of the date and time of the alert triggering, the alert responsible for the anomaly event, the name of the capture file if capturing was enabled for that alert, and other information relevant to the event. Figure 10 lists notifications derived from the anomaly detection engine, with varied severity levels.

4 SMAD Evaluation

Three different evaluation tests were conducted to assess the `SMAD` system: stress testing, vulnerability assessment, and intrusion detection testing.

4.1 Stress Testing

The stress testing was performed to assess the overhead of `SMAD` on the machine it runs. SMAD was tested under heavy workload with multiple monitors running,

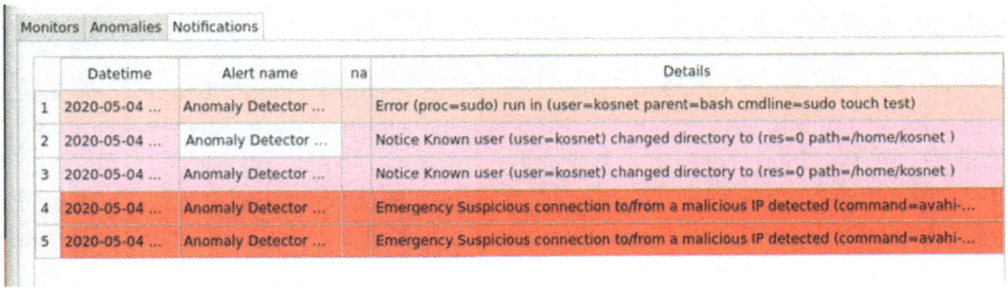
Monitors | Anomalies | Notifications

	Datetime	Alert name	na	Details
1	2020-05-04 ...	Anomaly Detector ...		Error (proc=sudo) run in (user=kosnet parent=bash cmdline=sudo touch test)
2	2020-05-04 ...	Anomaly Detector ...		Notice Known user (user=kosnet) changed directory to (res=0 path=/home/kosnet)
3	2020-05-04 ...	Anomaly Detector ...		Notice Known user (user=kosnet) changed directory to (res=0 path=/home/kosnet)
4	2020-05-04 ...	Anomaly Detector ...		Emergency Suspicious connection to/from a malicious IP detected (command=avahi-...
5	2020-05-04 ...	Anomaly Detector ...		Emergency Suspicious connection to/from a malicious IP detected (command=avahi-...

Fig. 10 Anomaly detector notifications

Table 1 SMAD stress testing

#monitors	Visualization enabled	Anomaly detector running	CPU usage (%)
3	No	No	12
3	Yes	No	14
3	No	Yes	13
3	Yes	Yes	15
5	No	No	25
5	Yes	No	27
5	No	Yes	26
5	Yes	Yes	28
10	No	No	59
10	Yes	No	61
10	No	Yes	60
10	Yes	Yes	62
20	No	No	100
20	Yes	No	100
20	No	Yes	100
20	Yes	Yes	100

each with multiple configured alerts. The tests were executed for all four SMAD configurations, as shown in Fig. 1. The testing environment was an Ubuntu 18.04 Linux distribution operating under a virtual environment. The computer running the virtual machine had a x64-based processor Intel Corei5-9300H with 2 40 GHz frequency and 8 GB RAM. However, the virtual machine was restricted to only 2GB RAM and 4 out of 8 cores.

Table 1 presents the results of the various tests conducted in relation to the CPU usage. Based on the findings, it is recommended not to exceed the execution of 10 monitors at any given time, with full functionality. Running 20 monitors with limited or full functionality will lead to 100% CPU consumption. Given the usage cases for SMAD, it may be safe to assume that users would be willing to use a relative large percentage of their computing resources for monitoring.

4.2 *Vulnerability Testing*

The goal of the vulnerability assessment was to uncover ways that SMAD could be exploited. It was discovered that SMAD was vulnerable to command injection, giving a malicious user access to the system. In particular, a number of text fields used for monitor configuration are subject to this attack. However, a user must be permitted to enter data into these fields otherwise the capabilities of SMAD would be limited. For example, a user would not be able to specify a specific process, user, or IP address to monitor.

The countermeasure against command injection is the filtering of special characters from the user input, whose presence could indicate a command injection. If those characters are present, the input is discarded, as shown in the regular expression listed in Listing 2.

```
if re.search('[&|;#$]', line):
    return False
```

Listing 2 Protection against command injection

4.3 *Intrusion Testing*

SMAD was used in a red–blue teams exercise. The objective of the red team was to penetrate the machine running SMAD, whereas the blue team was instructed to configure the system appropriately to detect abnormal behavior. The blue team's configuration setup included the following rules and monitors:

- Rule to identify usage of sudo command
- Rule to identify a known user behavior
- Rule to identify traffic from a specified malicious IP
- Rule to identify unusual traffic to a MongoDB server
- Monitor to observe files where most time has been spent

After analyzing the notifications produced from SMAD and the captured files, the blue team was able to identify several suspicious events:

- A known non-privileged user tried to use the sudo command multiple times in the */etc* directory
- An attempt was made to use the */etc/passwd* file that contains the hashes of user passwords
- One of the IPs in the malicious IP list tried to use the MongoDB server by accessing it on unusual ports.

The identification of the events was straightforward, which demonstrates that SMAD users who are aware of the typical behavior of their systems should be able to utilize the system extensively without any difficulties. More information on this testing could be found in [1].

5 Conclusion

This paper presented SMAD, a novel framework that monitors kernel and system resources data (e.g. system calls, network connections, process info) based on user-defined configurations that initiate non-intrusive actions when alerts are triggered or anomaly behavior is detected. SMAD architecture uses as its underlying foundation technologies `Sysdig` and `Falco`. The user-centric `SMAD` environment allows the specifications of monitors, alerts, and anomaly detection rules to be done in a free-of-errors manner. `SMAD` is envisioned to be used by security-enthusiastic users with some technical skills to track their local Linux server health. Unlike existing `Sysdig` commercial tools, the proposed system is open source in its entirety, welcoming new contributions to the existing source repository.

References

1. K. Avogian, Leveraging the visualization and analysis features of smad (system monitoring and anomaly detection) application (2020). Final Year Project Report, Department of Computer Science, University of Nicosia
2. BlueMatador: Alert automation for your cloud infrastructure. https://www.bluematador.com. Last accessed 30 January 2020
3. M. Fitzpatrick, Create Simple GUI Applications with Python and PyQT. Leanpub (2019)
4. IBM , IBM cloud monitoring with sysdig. https://www.ibm.com/cloud/sysdig. Last accessed 30 January 2020
5. R. McKendrick, *Monitoring Docker* (Packt, Birmingham, 2015)
6. Qt, Qt open source widget toolkit for GUI and cross-platform applications. https://www.qt.io. Last accessed 30 January 2020
7. B. Sababa, System monitoring and anomaly detection application (2020). Final Year Project Report, Department of Computer Science, University of Nicosia
8. B. Sababa, K. Avogian, I. Dionysiou, H. Gjermundrod, SMAD: a configurable and extensible low-level system monitoring and anomaly detection framework. In: Daimi K., Francia III G. (eds) *Innovations in Cybersecurity Education*. Springer, Cham. https://doi.org/10.1007/978-3-030-50244-7_2
9. Sysdig , Sysdig falco. https://sysdig.com/opensource/falco/. Last accessed 30 January 2020
10. Sysdig : Sysdig monitor dashboards. https://sysdig.com/products/monitor/dashboarding/. Last accessed 30 January 2020
11. Sysdig : Sysdig open source. https://github.com/draios/sysdig. Last accessed 30 January 2020

Lightweight Network Steganography for Distributed Electronic Warfare System Communications

Tim Lei, Jeremy Straub, and Benjamin Bernard

1 Introduction

Cyberwarfare is a modern form of warfare which transforms battles from occurring on the physical ground to battles across the virtual grounds of computer networks. It is the next generation of warfare [1], and it allows battles to be fought prospectively without actual human bloodshed. In physical warfare, humans with weaponry physically battle on a battlefield. In cyberwarfare, on the other hand, humans utilize computers as cyberweapons to remotely and virtually battle in a virtual battleground. Even though the physical damage inflicted can be reduced, significant damage can be inflicted upon both data and real-world assets. These damages can be extremely destructive to societal infrastructures. As cyberwar becomes more prominent, information security requires growing attention.

Cyberwarfare has been an ongoing, to various extents, between global powers, including among the United States and Russia. Cyberwarfare can be a war of information. Whichever force has access to more information and can exploit this information against an opponent or can defend its information from being accessed by opponent forces gains an advantage in cyberwarfare. It is essential to be on the side with an advantage in cyberwarfare as it provides opportunities to act

T. Lei
Department of Computer Science, San Francisco State University, San Francisco, CA, USA

J. Straub (✉)
Institute for Cyber Security Education and Research, North Dakota State University, Fargo, ND, USA
e-mail: jeremy.straub@ndsu.edu

B. Bernard
Department of Computer Science, North Dakota State University, Fargo, ND, USA
e-mail: ben.bernard@ndsu.edu

K. Daimi et al. (eds.), *Advances in Security, Networks, and Internet of Things*, Transactions on Computational Science and Computational Intelligence,
https://doi.org/10.1007/978-3-030-71017-0_31

advantageously, either offensively or defensively. Once one side has an advantage over the other, the trend is very likely to continue. In many cases, the side with disadvantage can only react to the actions from the advantaged side to neutralize possible damages.

Modern cyberwarfare, which utilizes attack systems, is growing rapidly. The development of artificial intelligence is growing at a swift rate. New technologies incorporate the power of artificial intelligence into their products to boost their performance. The application of artificial intelligence is expanding to more and more fields, and there is no doubt that cyberwarfare attack systems are part of this trend [2]. Attack systems that use artificial intelligence use these autonomous capabilities to greatly improve the functionality of the attack system.

Distributed AI systems require methods of communications. In some cases, there is significant benefit to this transmission being covert. Passing information through the internet is simple; however, sending or receiving information through the internet without being discovered by opponent forces is not. The Distributed Electronic Warfare System (DEWS) requires communications between the central blackboard and local blackboards to update them with the knowledge that has been gathered. This information is transmitted through the network, which raises security concerns. The information transmitted between central and local blackboards is sensitive, and in particular, a way to securely transmit information through the network without being detected is required. There is currently no implementation of an information security method for this purpose in the DEWS, which makes the system insecure and vulnerable to attacks. A network steganography method, based on the StegBlocks TCP method [3], can be applied to the DEWS to ensure the covert transmission of information within the DEWS.

The StegBlocks TCP method is a form of steganography which involves hiding data inside multimedia data or network data. Network steganography uses a network as cover media to hide data and transmit data without being detected. This paper presents the implementation of a network steganography StegBlocks TCP-derived method for the DEWS to secure the information transmitted between existing hosts and the assimilated machines. The client-server implementation (CSI) is used to improve the security of communication between machines. The implementation presented in this paper is compared to another implementation [4] of the same method on the metrics of usability, versatility, and applicability of the method.

2 Background

This section reviews prior work, in several areas, that provides a foundation for the current work. First, prior work on the Blackboard Architecture is reviewed. Next, prior work on a Distributed Electronic Warfare System, based on the Blackboard Architecture, is presented. Then, steganography is discussed. Finally, prior work on network steganography and the StegBlocks TCP method is covered.

2.1 Blackboard Architecture

The Blackboard Architecture [5] takes the concept of an expert system and transforms it into a task-solving architecture with three main components: the blackboard, knowledge sources, and the control. Similar to the expert system, which passively infers a possible solution, the Blackboard Architecture works on a task or a goal to generate solutions or partial solutions. Enhancements have been made on the Blackboard Architecture concept proposed by Hayes-Roth [6]. The Blackboard Architecture adds a layer of control to exploit the capabilities of AI systems and to adapt to the changing environment which may include newly generated tasks or partially known solutions. A distributed Blackboard Architecture implements the Blackboard Architecture with a host/central blackboard which generates a hierarchical tree of local blackboards, and the central blackboard distributes tasks to the local blackboards to solve. After a partial or the whole solution is accomplished, the solution is sent back to the central blackboard knowledge base, where all the information and tasks are stored.

2.2 Distributed Electronic Warfare System

The DEWS was developed based on the aforementioned distributed Blackboard Architecture, as shown in Fig. 1. The DEWS [8] has been proposed to be used in cyberwarfare against other forces to control opponents' complex networks and computing systems. The proposed system implemented artificial intelligence to gather information about targets, to make the decision on suitable methods to exploit to attack these targets, to launch payloads, and to propagate deeper into the adversary's connected network of computers. The DEWS is based on a central blackboard and local blackboards hosted on both command stations and assimilated machines. Each local blackboard contains a portion of the central blackboard's knowledge, based on the local machine's capabilities, location, and logistics of data transfer. The communications between the central blackboard and local blackboards are crucial, and the information sent between them needs to be secured, covert, and traceless. At present, the DEWS has not incorporated information security or covertness in the communications between the blackboards.

2.3 Steganography

Steganography is defined as "the art or practice of concealing a message, image, or file within another message, image, or file" [9]. It has been used with a variety of forms of media. Network steganography is gaining its importance from its powerful capability to transmit secret data through a network. There are four main attributes

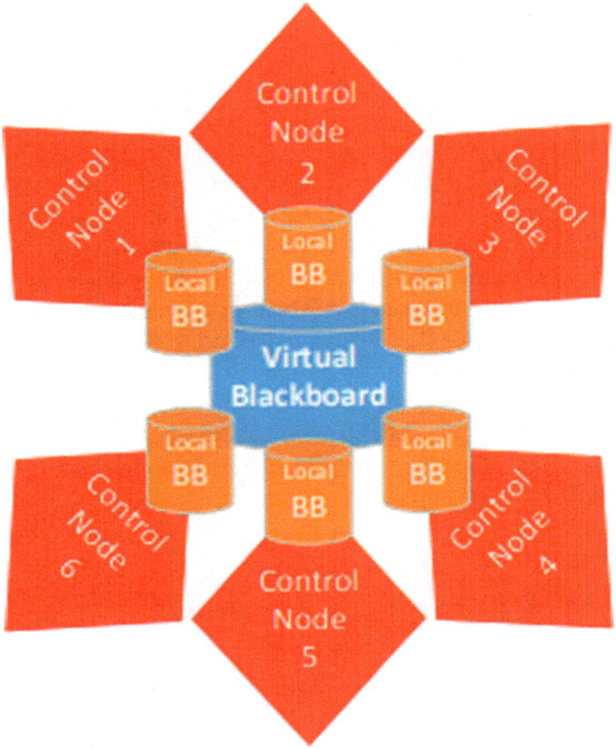

Fig. 1 Blackboard-based electronic warfare system [7]

of network steganography communications [3]. Each is now briefly discussed. Bandwidth is the amount of data that can be handled at once. Undetectability is the extent to which the hidden message is undetectable and untraceable. Robustness characterizes the integrity of the hidden message after the carrier media is altered or damaged. Finally, cost refers to the level of distortion on the message carrier media caused by the steganography method.

There are many forms of steganography. These include text, image, voice/video, and network steganography methods. Steganography in text hides a secret message within a block of text using an encoding scheme. Image steganography hides a secret message in an image file by modifying certain bits within the image. Voice/video steganography is similar to image steganography and uses similar techniques to hide a secret message within the bits of the voice or video datagrams. Network steganography hides secret messages in network traffic.

2.4 Network Steganography

Most steganography studies were conducted on text, image, and voice/video steganography as opposed to network steganography. However, a number of methods have been developed for network steganography. These include StegBlocks [3], PadSteg [10], HICCUPS [11], RSTEG [12], WiPad [13], and ReLACK [14]. These methods are based on different layers of the Open Systems Interconnection Reference Model [15].

2.5 StegBlocks TCP Method

In the StegBlocks TCP method [3], which the CSI method proposed herein is based on, TCP connections are established between two machines. The two select a steganographic key and blocks are identified. Each block has a value which is based on "the last x bits of the number of TCP segments" in it [3]. If this block does not have the desired value for the message that will be transmitted with it, it must be changed to have this value.

3 Implementation

The DEWS uses network traffic to transfer data information; however, sending data through the network without disguise can be easily tracked and intercepted by cybersecurity professionals or detected by programmes searching for intrusions or anomalies. Network steganography helps to obfuscate the transmission of data that systems receive or send to, from, or between assimilated machines through the network. In the DEWS, agreements exist between the nodes with local blackboards that comprise the virtual central blackboard. Only the machines with applicable agreements can readily discover the hidden data that is transferred.

The work presented herein uses the conceptual model of the StegBlocks TCP method, with some changes. These changes, in particular, are designed to minimize the computing overhead of the protocol to facilitate its use on computationally limited devices, such as the Internet of Things devices.

Under the method used herein (the CSI), like with the StegBlocks TCP method, connections are established between two nodes. Among these TCP connections, two gatekeeper connections are selected, and the rest of the connections are treated as connections to transport payload data. The functionality of the connections depends on the ports they are connected to. The first port, port A, of the sender's connection is selected as the first gatekeeper. The last port, port Z, of the sender's connection is selected as the second gatekeeper. The rest of the ports, ports B to Y, of the sender's

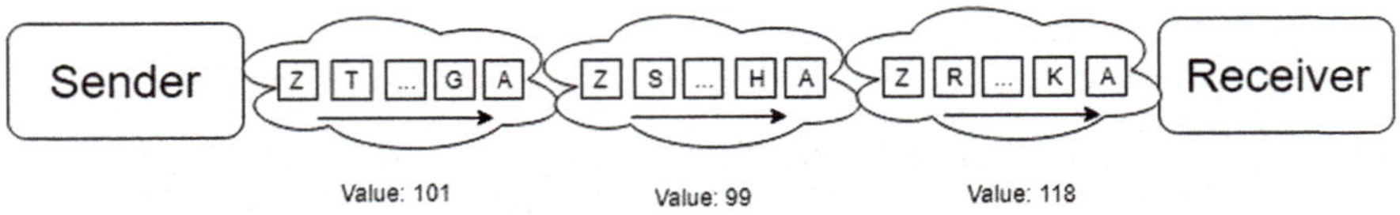

Fig. 2 Example of sending the word ‘cat’

Table 1 Encoding table of three different digits of character values

Character	ASCII decimal value coding	New value coding	Packets needed for new coding	Total packets needed
TAB	9	‘009’	$0+0+9=9$	$9+6=15$
K	75	‘075’	$0+7+5=12$	$12+6=18$
d	100	‘100’	$1+0+0=1$	$1+6=7$

connection are selected for transmission, based on the value of the data. The ports are selected randomly.

Using CSI to send the word ‘cat’ with ASCII encoding, the values for the characters are 99, 97, and 116. The total number of packets required to send the word is the sum of its ASCII values plus two packets for the gatekeepers, which is 318 packets in total. Figure 2 demonstrates this for sending the word ‘cat’ with this method.

3.1 Encoding Scheme

When using the ASCII value encoding to send characters, the number of packets required to represent a character can be as high as 127. It is, thus, not always efficient to send a character based on the character’s ASCII decimal value. To optimize efficiency, the method of sending a character is modified from sending a number of packets based on its ASCII value to a number based on the digits of the ASCII value, as shown in Table 1. For example, the character ‘A’ has an ASCII value of 65. By applying the new method, the value becomes ‘065’, and the number of packets required to represent the character is $0 + 6 + 5 = 11$. By including two packets for the two gatekeeper ports to send each digit across, the number of packets required is $11 + 6$ (2 packets for gatekeepers * 3 digits) $= 17$. By converting to this new method, the largest number of packets required is $18 + 6 = 24$ for the character ‘c’ with an ASCII value of 99. This can be compared to the unmodified method, where the largest number of packets used is 129 packets. Given this, the modified version can be up to five times as efficient as the old method.

3.2 Scenario

A simulation of multiple clients communicating with a server was implemented to simulate data transfer between nodes in the DEWS using the CSI method. A central server was implemented to simulate the blackboard command station, and clients were implemented to simulate the assimilated and other geographically diverse nodes controlled by the DEWS. The functions of the server node are to listen for incoming packets, decode the carried message from the number of packets sent to the server, and store the message into a text file for future use. The functions of the clients are to capture packets, store them in a pool of packets, modify the headers of the packets, and send the modified packets through a predefined set of ports using a predefined encoding scheme. The server acknowledges the encoding scheme used by the clients. Both the server and clients use the Scapy tool in their implementation. The data in the text file that was used in this scenario is 10kB.

3.3 Client-Side Implementation

The client side of the implementation has three clients which use the same interface. Three clients connect to different sets of ten open server ports. These ports were preselected and assigned to each client. Two of the open ports were used as gatekeeper ports; one was used to tell the server when the end of information transmission was reached. The rest of the ports were used as transport ports. The encoding scheme applied to the message was converting each character to its ASCII value, converting this value to its three-digit integer string, and filling empty digits with zeros. Additionally, the software modifies the source IP address to the IP address of the client and the destination IP address to the server's IP address, in the packet's IP header. It also modified the destination port in the packet's TCP header and sent the packet. This process of modifying and sending packets was repeated until the number of packets required for each digit was sent. For testing, each client opens a text file containing a message to be transferred. It then reads a character, calculates the number of packets required to send the character, reads in the requisite number of packets from the pool of packets, and then sends each character using the aforementioned encoding scheme.

3.4 Server-Side Implementation

The server software is equipped with a sniffer to listen on its open ports assigned to the three clients and receive packets from the three clients. There are three counters, one for each set of ports, to count how many packets are sent through transport ports of the three clients. The count is converted to a string representation of the ASCII value of the character being sent. Once the string has all three digits, the string is converted to an integer value and converted to an ASCII character. The decoded character is then written into the respective resulting text file for the particular client. Once all three clients finish their activities, the server stops listening on the ports and is shutdown.

4 Data Collection

To evaluate the performance of this approach, a script was written to run both the server and the clients. Network packets were pre-captured through the Ethernet port and imported into the programme for the use of the CSI method to transfer hidden messages. The implementation of the method in [3] (client-server method) was also evaluated. This method (from [3]) uses a single server to send data to a single client rather than having a server send data to multiple clients. In both scenarios, the same text file is used.

The average elapsed time for transmission for three clients sending to a server (client-server method) is 384.59 seconds with 111,738 packets required to send the text file. The average elapsed time for the client-server method was 282.68 seconds with 86,176 packets required to send the text file. It is notable that the client-server programme skips characters which are not in its lossy collection, which only includes lower case letters and the space character. On the other hand, the CSI can transfer all ASCII characters. The byte rate for the client-server programme, considering that there are three clients, is 77.57 bytes per second. The byte rate for client-server programme is 25.86 bytes per second. Test run results are shown in Figs. 3 and 4.

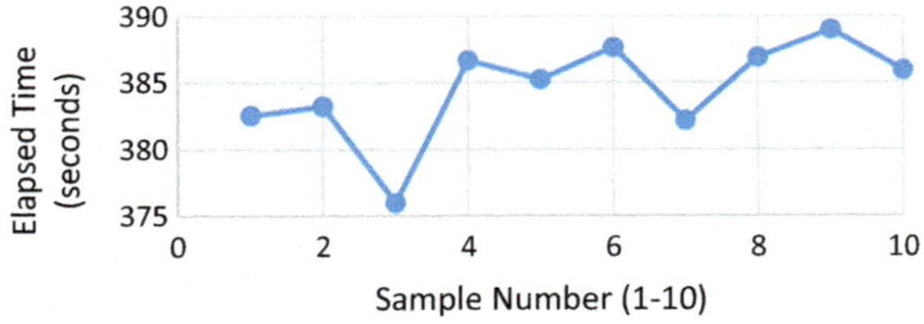

Fig. 3 Sending sample text ten times from three clients to a server

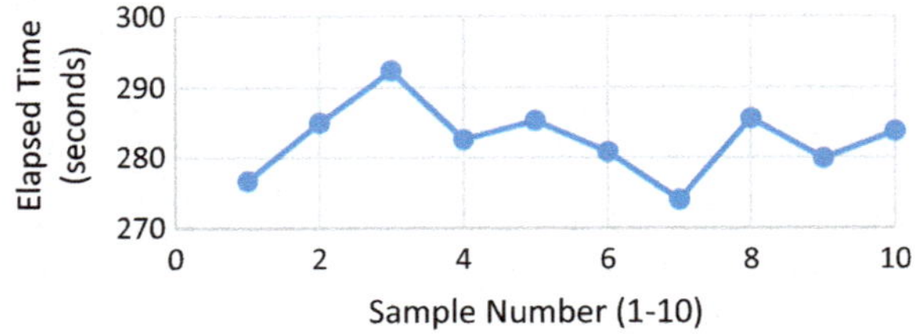

Fig. 4 Sending sample text ten times from a client to a server

5 Assessment

This section assesses the CSI (which is based on the StegBlocks method), in terms of multiple criteria. The criteria considered include usability, versatility, applicability, and undetectability. Each is now discussed.

5.1 Usability

The CSI, based on the StegBlocks method, was implemented with multiple clients to allow its use for DEWS communications between hosts. The DEWS is a highly interconnected network where multiple instances of multiple-to-one client-server structures are required. Thus, a single client to a single-server communication paradigm would not be appropriate for this application. The multiple-to-one client-server structure is needed for the system to operate with its full capabilities. Once the attack system breaches an adversary's machine and takes control of the machine, the CSI begins to operate by gathering the required IP address and the ports of the machine. It also collects the information for use by the attack system for the purpose of further propagation of the infection of machines or leaving behind backdoors for future access.

5.2 Versatility

The encoding scheme used in the CSI can transmit all ASCII characters, while the implementation in [3] can only transmit lowercase letters and the space character. The CSI made it possible for uppercase letters, numbers, punctuation, and special characters to be transferred. The more limited implementation of the StegBlocks method would not be useful for some applications if the only data which can be transported is lowercase letters and the space character. For example, if a webpage link needed to be transmitted, the link would contain numbers and the special character '/' in addition to letters.

5.3 Applicability

The implementation of the client-server structure of the StegBlocks method and its derivatives can be applied to other future attack systems, in addition to DEWS, which utilize the network to communicate. Many attack systems may need the capability for transmitting secret data from multiple assimilated machines back to a central point of command. In such a case, the implementation of a single client to a single server would require excessive management on numerous individual machines.

5.4 Undetectability

The level of undetectability is the core requirement for steganographic methods. The CSI did not use the Vernam cipher as suggested in [3] for perfect undetectability. The client-server implementation, instead, focuses on the transfer of overt text to support covert communication between multiple hosts. Perfect undetectability is not provided by the CSI because there is the possibility of detection using statistical analysis of the number of packets sent. Analysis may reveal the order of digits and the payload. The CSI does not include a mechanism which alters the packets' payload; thus, the analysis of the contents of packets has no effect and does not increase the probability of triggering detection.

6 Conclusions and Future Work

In this paper, a modified lightweight implementation of the StegBlocks method was presented and implemented for evaluation for possible incorporation into the DEWS. Comparing the two, the client-server implementation did not outperform the client-server implementation. The CSI has a multiple-to-one client-server structure and is able to transfer any ASCII characters, while the client-server implementation has a one-to-one client-server structure and is only able to transfer lowercase letters and the space character.

While the implementation shows promise for future use, a number of enhancements are needed as future work. First, the programme should be optimized to run in the background. The implementation is more useful for the DEWS if it can run covertly. The CSI did not implement a mechanism where the programme can be run in the background without being detected. This improvement could hide the activity of the programme and delete all traces of it after the programme finishes running.

Another prospective improvement is to randomize port selection. The proposed implementation uses pre-assigned ports for both the gatekeeper and transmission ports. This impairs the level of undetectability if statistical analysis is run on the system. The process of randomly choosing ports and repeating the selection process frequently can greatly reduce the chance of being detected using statistical analysis. It is planned that these improvements could be implemented as future work.

Acknowledgements This research was supported by the US National Science Foundation (Award # 1757659). Some facilities and equipment were provided by the NDSU Institute for Cyber Security Education and Research and the NDSU Department of Computer Science.

References

1. T. Franz, The cyber warfare professional: realizations for developing the next generation. Air Sp Power J **25**(2), 87–99 (2011)
2. J. Straub, Artificial intelligence is the weapon of the next Cold War. *The Conversation*, 29-Jan-2018
3. W. Fraczek, K. Szczypiorski, Perfect undetectability of network steganography. Secur. Commun. Netw. **April**, 2998–3010 (2014)
4. P. Bak, J. Bieniasz, M. Krzeminski, K. Szczypiorski, Application of perfectly undetectable network steganography method for malware hidden communication. *2018 4th Int. Conf. Front. Signal Process. ICFSP 2018*, pp. 34–38, 2018
5. H.P. Nii, Blackboard systems: Part I. AI Mag. **7**(3), 38–53 (1986)
6. B. Hayes-Roth, A blackboard architecture for control. Artif. Intell. **26**(3), 251–321 (1985)
7. J. Straub, Blackboard-based electronic warfare system. *In Proceedings of the ACM Conference on Computer and Communications Security*, 2015, vol. 2015-Oct
8. I. Burton, J. Straub, Autonomous distributed electronic warfare system of systems. *2019 14th Annu. Conf. Syst. Syst. Eng.*, pp. 363–368, 2019
9. "Definition of Steganography," *Merriam-Webster Dictionary*, 2020. [Online]. Available: https://www.merriam-webster.com/dictionary/steganography. Accessed: 08-Jun-2020
10. B. Jankowski, W. Mazurczyk, K. Szczypiorski, PadSteg: introducing inter-protocol steganography. Telecommun. Syst. **52**(2), 1101–1111 (2013)
11. J.P. Black, et al., Steganography in TCP/IP networks. Outline. *Proc. – 2010 2nd Int. Conf. Multimed. Inf. Netw. Secur. MINES 2010*, vol. 4, no. 3, pp. 225–229, 2014
12. W. Mazurczyk, M. Smolarczyk, K. Szczypiorski, Retransmission steganography and its detection. Soft. Comput. **15**(3), 505–515 (2011)
13. K. Szczypiorski, W. Mazurczyk, Steganography in IEEE 802.11 OFDM symbols. *Secur. Commun. Networks*, no. March 2011, pp. 118–129, 2014
14. M. Hamdaqa, L. Tahvildari, RELACK: a reliable VoIP steganography approach. *Proc. – 2011 5th Int. Conf. Secur. Softw. Integr. Reliab. Improv. SSIRI* 2011, pp. 189–197, 2011
15. J. Lubacz, W. Mazurczyk, K. Szczypiorski, Principles and overview of network steganography. IEEE Commun. Mag. **52**(5), 225–229 (2014)

Security of DBMSs

Suhair Amer

1 Security of ADABAS

ADABAS is a software that is compatible with many operating systems such as Linux, Unix, and Windows and has been updated continuously throughout the years. ADABAS, while extremely useful, has some security issues. One security risk occurs from direct calls to the database from an unauthorized third-generation language program, such as COBOL, FORTRAN, or Assembler. An unauthorized user gains full access to the data and can modify the stored data even if the database has limited access to users with need-to-know-and-modify privileges because the Natural Security definitions don't apply to these cases [1]. To defend against such risk, the Natural SAF-compliant security system must be configured placing authentication checks that are checked against the ACL list.

Hacking of non-encrypted data stored in the database is another security risk where a hacker can gain access to the data and read sensitive data stored. Therefore, data should be encrypted at hardware level by using, for example, a security code entered by the user. Another way to protect the data is to mask sensitive fields such as social security numbers and credit card numbers when copying data from one environment to another so that it isn't stored in plain text exposing this information to developers, testers, and other privileged users. Improper sending of data from one destination to another is a security risk. This can be between an application and a database or between programs. To secure the transmissions of data, public networks need to be secured and the user verified. A secure connection request will be sent to

S. Amer (✉)
Department of Computer Science, Southeast Missouri State University, Cape Girardeau, MO, USA
e-mail: samer@semo.edu

K. Daimi et al. (eds.), *Advances in Security, Networks, and Internet of Things*, Transactions on Computational Science and Computational Intelligence,
https://doi.org/10.1007/978-3-030-71017-0_32

the database where an encryption certificate is read and the encryption component is setup. The data is sent and decrypted on the database or vice versa [1].

In addition, ADABAS security can be achieved using an event analytic function that is compatible with the Apama Streaming Analytics platform. Apama that utilizes security technologies, market trading technologies, in-memory data management, and many other tools to monitor live data and provide reports on possible threats [2]. Event analytics has its own functions including alerts and the ability to reduce time between detection and reaction to a security breach.

ADABAS can also utilize ADABAS security options which include password protection and maintaining data in encrypted form while it is in the database. The passwords provide security for the file level, data field level, and data value level. Every data field potentially can have up to 15 different levels of read and update security assigned to it. The password checks if the user has permission to do execute the requested operation after the correct password is inputted. Access is then granted if the user enters a password with equal or greater permission level. ADABAS also uses cipher codes which utilize simple numeric codes assigned through ADACMP. The point of it is to encrypt the data so that it is unreadable until it is displayed with the ADABAS program or utility. If the data is accessed in some other way than the intended method, it is encrypted and cannot be read. NATURAL can also be sued with its ability to restrict the use of the entire application or just specific sections or operations. These updates and configurations are stored in the NATURAL environment of the ADABAS files. To restrict certain aspects of the database, NATURAL defines users, libraries, and files [3, 4].

2 Security of Adaptive Server Enterprise

Sybase Incorporated developed Adaptive Server Enterprise (ASE) and is now owned and operated through System Applications and Products in Data Processing (SAP). AES is a high-performance relational database management system that can be on site or can be incorporated into a cloud infrastructure. The purpose of ASE is to manage large volumes of data and thousands of users while protecting the data and systems from cybersecurity threats. It utilizes technologies for data replication, encryption, and security that support confidentiality, integrity, and availability. One of the features for handling security is creating alerts and reporting by extending an SQL database with a workload analyzer to increase reporting capabilities. This also supports query execution while following the desired specific requirements configured. The customer can create a standard for SSL implementations by utilizing a common crypto library. Implementing important or sensitive commands is also a security function that guards certain commands from being executed even if the main database API was breached. Granular audits reduce the number of records logged. In addition to that, ASE offers a compatibility function that enables common dialect of SQL scripts across the different SAP database platforms [5]. The SQL Server means that ASE is prone to SQL injection attack [6].

In-memory databases (IMDBs) allow storing integration functionality which increases the speed because they are much faster than the alternative, disk-optimized databases that take longer to access memory due to more CPU instructions and longer algorithms. IMDBs also allow and facilitate virtualizing data and scaling which are vital to companies that process big data and many users that require access at the same time. Many telecommunication networks utilize IMDBs because the response time is important. One flaw with IMDBs is a result from utilizing RAM which is volatile causing data to be lost or deleted with power loss or failure. In such a situation, the system needs to reload all the data it previously had. One way around this is to use nonvolatile random-access memory (NVRAM) technology, allowing the data to be maintained in case of power loss. ASE can be used in private data centers or in cloud environments. Due to this feature, configurations of hardware and software are critical to security. Even a single misconfiguration on an outward pointing device can compromise the entire database. This is a cloud service security vulnerability, and the only way to secure such a vulnerability, in addition to proper configurations and best security practices, is to host the cloud environment on a secure intranet [7].

Another security issue that affects this database is a potential denial of service attack, which renders the database inoperable. As stated on Packet Storm, "an attacker can trigger a condition in which the process ceases to run. This condition can be intentionally provoked by an attacker to cause a denial of service," which can be exploited remotely through the network [8]. This affects many versions of the Adaptive Server Enterprise database systems such as server versions 15.7 and 16.0 and SQL anywhere personal servers.

Buffer overflows are another type of attack and can be used to gain access to different memory addresses or to overwrite preexisting data in lower memory locations. CVE Details states that "SAP Adaptive Server Enterprise (ASE) 15.7 before SP122 or SP63, 15.5 before ESD#5.4, and 15.0.3 before ESD#4.4 does not properly restrict access, which allows remote authenticated database users to overwrite the master encryption key or trigger a buffer overflow [9]." This is a problem because the master encryption key is used to encrypt the entire database and to decrypt the database as well since it is a symmetric AES key [10].

Another security issue is caused by the ability to log in with a default log-in that is installed on the ASE database from the factory called probe that is used for the two-phase commit probe process. This uses a challenge and response mechanism to access Adaptive Server. The implementation flaw of the challenge and response mechanism allows anyone to access the server as "probe" log-in and elevate their privileges to database administrator, in doing so compromising the entire database confidentiality and integrity [11].

3 Security of Advantage Database Server

Advantage Database Server (ADS) is a relational database management system developed by SAP. Advantage Database Server can extend a company's existing applications. It is a full-featured, easily embedded, relational database management system which can be implemented into an existing system easily and is a DBMS. The first version was released in August 1993 called Advantage xBase Server by Extended Systems, while the current version is Advantage Database Server 12.0 and is now owned by SAP. One of ADS features is being able to be managed from any device or location even while the database is in use. This allows adjusting security settings and configurations while allowing employs to continuous access. It also allows multiple different types of platforms and development languages like SQL or ISAM providing more flexibility and possible functionality. Another functionality is allowing scalability from peer-to-peer to client-server environments with only source code provided. One can also install and manage without a database administrator. This has both positive and negative effects on security. It might make management more difficult and forces the security experts to rely on logs and reports developed by the program. Also, when an intrusion is detected, it makes it difficult to fix the issues. On the other hand, with no database administrator to exploit, the damages are lessened because if a database administrator account is exploited, the database is entirely owned by the attacker [12].

A security issue is dynamic-link library (DLL) injections. These injections are used to invoke malicious DLL inside a process to escalate privileges in the database or to perform a portable executable injection which causes a remapping of the memory writing the malicious code into the executable similar to viruses being imbedded into applications by using assembly language jump statements. Since this database can be a client to server database for web applications as well, this attack could infect one of the applications stored on the database and be accessed by many users who interact with the database every day. There aren't a lot of ways to prevent DLL injections except to analyze apps processes and use third-party scanners [13].

Another security issue is heap overflows. The Advantage Database Server is prone to heap overflows when a user opens a crafted script file (.SQL) with a long query inside. This issue causes a function pointer to be overwritten allowing the execution of arbitrary code. This means that the heap or dynamically allocated memory in the server overflows into the stack memory where local variables and function calls are stored overwriting the function pointer stored in this location [14].

4 Security of Datacom

Datacom is a company focused on information technology services. It made a partnership with AirDefence Inc. in 2003 to help increase the security of their local area networks (LANs). A local area network is a small network that usually made up of the network in a home or a business. The LAN may consist of Wi-Fi signals and Ethernet connections. Often in businesses, the LAN is guarded by a firewall that protects it from many outside threats. However, any client connected to the LAN directly can bypass the firewall which can cause vulnerabilities in that LAN network. These malicious users usually have access to direct connections with other computers which often might include the company's main server. These computers are usually protected by a firewall over the Internet but not on a local area network. Datacom offers a managed service called AirPatrol that help protect businesses from these vulnerability issues. Datacom's AirPatrol service offers many features such as 802.11 wireless LANs with 24×7 monitoring to identify rogue wireless LANs, detecting intruders and attacks, enforcing network security policies, and deflecting intruders from the network and monitor the health of the wireless LAN. These security services aim to monitor the network and report suspicious activity. Their system works by using remote access point sensors and a server appliance. A remote sensor is placed at each access point in the network and monitors the wireless LAN and transmits the data collected to the main server appliance [15].

Datacom also has a wide variety of database features designed for ease of access and security. Datacom's database services have a unique sign in process. Once a user is logged into any of Datacom's applications, the user can connect to their database without entering any password if their account has been authenticated. The user's client stores all tables that have been accessed in the past, and if the user tries to view or edit a table that has been stored on that list, no server-side authentication is required. This allows a faster response time. In addition, a unique feature is the data dictionary and data query security. Their data dictionary can store all the functions that a program connected to which allows programs to essentially be stored on the database until they need to be loaded and executed. Storing program functions in the data dictionary helps to prevent them from being manipulated by any code injection attempts. Datacom's data query feature allows table authorizations for external applications to be created and allows table permissions to be set for a certain application to restrict that application access to the database [16].

5 Security of FileMaker

FileMaker is a cross-form relational database developed by Claris International. There are many embedded security features such as account authentication, privilege sets established via access control, data encryption, and server monitoring administration. Authentication permits users to authenticate their account using Active Directory. Access control assigns privileges to users, which then determines the data the individual may access. Encryption conceals data from unwanted eyes. Administration services allow the administrator to invoke redundancy and supervise user activity. FileMaker is a platform that employs a unified security model where the security established for a solution is in effect across all clients. In the product documentation, the security guards have two versions: FileMaker Pro and FileMaker

Server. Security settings established in Pro are strictly applicable to the information and schema definition(s), whereas security configurations established in server apply to all solutions hosted by the server [17].

Unfortunately, FileMaker has many known vulnerabilities. There have been nine vulnerabilities detected since the year 2000. Two of these nine had a high severity that registered a score of 7.5. The CVE numbers are CVE-2000-0386 and CVE-2000-0123, respectively. CVE-2000-0386 is an issue residing in FileMaker Pro, which allowed remote attackers to send anonymous or forged email by means of the web companion. The later exploit, CVE-2000-0123, allowed those with malintent to manipulate the shopping cart application. Consequently, remote users are equipped to modify sensitive purchase information residing in hidden form fields [9].

FileMaker 13 and lower are vulnerable to a Secure Sockets Layer (SSL) vulnerability known as FREAK. SSL encrypts its information after a handshake by using a certain encryption suite. This allows for SSL to be compatible with many common encryption methods including DES, AES, and RSA. SSL encryption in the 1990s worked by limiting the RSA encryption key length of any SSL implementation which is known as the export suite. FREAK works by using clients that have support enabled for the older export SSL encryption methods by using a man in the middle attack which tricks the client into using the older export suite SSL rules with only 40-bit or 56-bit encryption. This means that the encryption key can be broken by modern computers in only a few hours. Once the encryption key is broken, the encrypted contents can be read. Users of FileMaker need to make sure they use FileMaker 14 or higher to prevent a FREAK attack from occurring [18].

6 Security of IDMS

IDMS stands for Integrated Database Management System. It was created in the 1960s by B.F. Goodrich Chemical Company to manage their computer mainframe systems. It went through several different company ownerships and currently is owned by Computer Associates known as CA IDMS and offered as an online cloud database service [19].

The CA IDMS uses a centralized security approach to protect resources when an external security system is not used, when user exits are not enforced, and when an external security system only has partial protection. This approach allows CA IDMS security to be implemented with a variety of other applications and provides many features including client/server environments, a useful command facility tool, ANSI-compliant security syntax, and availability to user-written applications as well as front-end software. There are three levels of security privileges within the CA IDMS that function in a hierarchical format, which are application security, CA IBMS centralized security, and host access security. The centralized security approach allows easy integration with the various external servers that its clients may have [20].

The CA IDMS uses a table for storing security information executed and checked at runtime called the Security Resource Type Table (SRTT). It stores each resource type to be secured, the system that enforces the security, and the information external security systems need to operate. Each resource type is categorized as internal or external. Certain security specifications can be applied for each of the two resource types. Internal security specifications are listed by specifying certain permissions to each user, while the external security specifications are listed by mapping routes of resources to external resources for use of security checks. The SRTT enhances security by listing many of the security protocols and definitions all in one place for easy maintenance [20].

Many security improvements were made to the CA IDMS software after the US Department of Homeland Security started using the software to manage critical citizenship, immigration, and noncitizen resident systems. The main improvement was to make managing and updating security principles significantly easier. A better API was developed that had support for Java and. NET development environments. A new web-based interface called SiteMinder was also implemented and incorporated their centralized management and security approach [20].

7 Security of Informix Dynamic Server 2000

Informix is an object relational database management system that was first developed by the Informix Corporation in 1994. At the time of release, it was able to pass performance benchmarks due to the newly implemented Dynamic Scalable Architecture (DSA). Most of the DBMS core engine had to be rewrote to support both horizontal and vertical parallelisms based on multi-threaded cores that would synergize with symmetric multiprocessing systems. The result of adding the horizontal and vertical parallelism made the Informix Dynamic Server 2000 the top of the market for scalability decision support systems (DSS) and online transaction processing (OLTP). Later, Informix was bought by IBM and renamed it IBM Informix Dynamic Server. The Informix Dynamic Server (IDS) provides all aspects of data security, which includes authentication before access; access control after authentication; confidentiality after access; supporting the Pluggable Authentication Module (PAM) mechanism for authentication, network, and data encryption mechanisms; role separation; label-based access control; and discretionary access control [21].

The IDS can also provide data compression, data transformation, and data encryption mechanisms for offline data to ensure the security of data that is out of the control of IDS. These measures prevent unauthorized access of the database leading to a breach of data integrity that is stored in the IDS. There are audit features also implemented in the IDS, which allow review of all user activities in the event of any malicious activity. IDS is based on client-server architecture, but also supports server-server communication between servers on the same machine or over the network. It also provides various options for preventing unauthorized access such as supporting Pluggable Authentication Module (PAM), supporting password and data encryption using CSM modules, and limiting denial-of-service attacks. Another feature is separation for database administration. From a security perspective, configuration and monitoring are the major activities. Different roles are assigned to different user preventing one user from having more privilege than necessary. The audit administrator later analyzes the audit log for unauthorized or suspicious database activity and allows an administrator to become aware of a breach or an attempted breach of the database so that they can respond accordingly. They can counter an attack and go into damage control and find if a user is abusing their privileges. These logs can then be used as evidence in any future legal matters regarding the incident [21].

8 Security of Ingres

The Ingres DMBS is a large, open-source SQL relational DBMS mostly intended to support large company or large government applications. Even though Ingres is fully open source, the Actian Corporation controls the development of Ingres and provides worldwide support. Ingres was developed in the mid-1970s, and the community rewrote Ingres several times to a finished product that was finally completed in 1985. Examples of the security features implemented are data-at-rest encryption, PAM, audit logs, privilege separation, and roles. Ingres was developed mostly for UNIX and today POSIX and can run on Windows and Linux systems and has some security measures specifically created for Linux systems. Ingres uses RSA and AES both 128- and 256-bit encryption. In some environments, Ingres uses the ingvalidpw program to validate user passwords that may originate from any local application or from a remote application coming through Ingres Net or the Data Access Server. Ingvalidpw is used depending on the requirements of the platform where the password is validated. For example, it is used to validate shadow passwords on UNIX or to enforce C2 security in some UNIX environments [22].

Security alarms are another unique feature of Ingres DBMS. It allows an administrator to look at specific events that were recorded in the security audit log for individual tables and databases. They can then place triggers on important databases and tables to detect when users attempt to perform access operations that are not normally expected. If a security alarm is triggered, it means the security tests for a certain query were passed [22].

9 Security of InterBase

InterBase is a relational database management system marketed by Embarcadero Technologies. It can run on macOS and iOS. InterBase full installation only takes up about 40 megabytes of disk space. It is fully open-source and still in use today.

It is equipped with basic security features such as maintaining usernames and passwords in a security database. The security database allows clients to connect to an InterBase database on a server if the username and password supplied by the client match a valid username and password combination in the InterBase security database on the server. The encryption can be either very good or very poor. This is because, by default, InterBase allows only the use of weak encryption (DES) if the Strong Encryption License has not been activated. If strong encryption has been activated, the user can choose between DES and AES. The AES is a 256-bit encryption, which is uncrackable to all but quantum computers [23]. DES is considered not secure because in early 1998, the Electronic Frontier Foundation built a DES-cracking machine that found a DES key in few days, and it costs $200,000 to design and build it. Today, it is easier to defeat DEC and very affordable. Therefore, when using InterBase if the Strong Encryption License wasn't activated, an attacker could crack the encryption for data very easily [24].

InterBase security is based on the concept of the user. Authorized users are stored in a security database, which is called ISC4.GDB. Each server has its own security database, which means that a user definition is bound to the server where it is stored. The same user may exist on several servers, but it must be created on all servers where it is required. The security database also stores an encrypted password for each user. Users are handled at the core of the security of database. That means that gaining access to one account on a server doesn't give access to all the servers because it is separated out. There are two levels of security in InterBase. In the first level, users are validated at the time of connection against InterBase's security database. The second level of security is implemented at the database level which involves privileges and permissions. Each user's privilege is a list of operations that the user can perform for a certain view or table which is a good security feature because it uses an implicit deny [25].

However, according to cvedetails.com, InterBase had a major vulnerability in 2001. InterBase had a built-in backdoor which was built into the system and left all InterBase databases vulnerable giving the attacker full access to the system with very little effort to launch the attack [9].

10 Security of InterSystems Caché

InterSystems Caché is a database management system that specializes in transaction processing. It is a very high-performance database management system and can process a large number of queries while maintaining stability and supporting a large number of active users. It uses a "bottom-up" approach for modeling complex real-world information and does need to use extra libraries. InterSystems Caché's security infrastructure is strong and consistent and meets all the standards for security certification. It also places minimal burden on the performance and operations of the system which minimizes the chance that the application will crash.

It can be accommodated into existing security architecture without many problems and provides infrastructure needed to write security policies and enforce them [26].

InterSystems Caché's security is based on authentication, authorization, encryption, and auditing. With authentication, a user's identity is established which is necessary because authorization is worthless without authentication. Without authenticating a user, they can gain unauthorized permissions. InterSystems Caché uses several authentication mechanisms that are available to the developer, such as Kerberos. If the organization using Caché is not storing sensitive data, authentication can be completely disabled [26].

Caché's authorization manages what permissions a user has and what data they can or cannot control. Each user can have one or more roles, each of which authorizes the user to perform activities with resources. Caché also supports role-assignment mechanisms that are used for assigning the lowest permission available by default making sure that new users are not accidentally given more roles than they are supposed to have [26].

InterSystems Caché's auditing provides a way for a developer to see a log of all actions performed on the database. This is used to resolve a problem or by forensics after an attack. Another purpose of this feature is to act as a deterrent for attackers as their actions are being logged with their information. InterSystems Caché also allows the developer to define extra log events, and viewing the audit log also causes a log to be made, because sensitive information can be stored in a log [26].

Finally, InterSystems Caché provides encryption for all block-level databases, encryption key management, and data element encryption for applications. It uses the Advanced Encryption Standard. Encryption or decryption occurs when writing or reading data on the disk, respectively [26].

11 Security of Microsoft SQL Server

Microsoft SQL Server is a relational database management system (RDBMS) that supports a variety of transaction processing, business intelligence, and analytic applications in a corporate IT environment. Microsoft SQL Server, along with Oracle Database and IBM's DB2, is one of the three market-leading database technologies. "Like other DBMS software, Microsoft SQL Server is built on top of SQL, a standardized programming language used by database administrators (DBAs) and other IT professionals to manage databases and query the data they contain. SQL Server is associated with Transact-SQL (T-SQL), Microsoft's SQL implementation that adds a proprietary set of programming extensions to standard languages" [27].

SQL Server is a Microsoft database management system that is built on top of the SQL language. SQL Server implements a defense in-depth strategy for security and has overlapping layers of security. Microsoft considers this as the best way for developers to effectively secure their applications and to counter threats. However, the security features are not built into every application written with SQL Server.

Different applications require different levels of security, and developers need to understand these features and to incorporate them. It implements security of entities by using authentication and authorization. Authentication supports two modes. Windows authentication mode is the default because the SQL Server security model is integrated within Windows. Mixed authentication mode uses Windows authentication along with authentication parameters set in SQL Server such as usernames and passwords. The SQL Server authorization method is based on the principle of least privilege where new users are given the least amount of privilege by default. System admin's user account has very tight restrictions on what it has privileged over. In addition, SQL Server supports encryption with certificates, along with symmetric and asymmetric key encryption. Although not the most secure method of encryption, symmetric key encryption is the fastest and is best used for large quantities of data. SQL Server supports symmetric key encryption in DES, Triple DES, DESX, and AES. Finally, SQL Server encrypts the encryption keys using certificates or symmetric key encryption [28].

12 Security of MS-Access

MS-Access 2010 is used to create web applications that will usually work in conjunction with SQL Server. MS-Access is not a very secure DBMS, but it possesses some security features. It supports the encryption standard used by Microsoft and a third-party software can also be used. This third-party encryption is the option that most choose when securing their applications. MS-Access also features the Trust Center which is where developers manage the security settings of MS-Access and determine trusted locations. MS-Access also has macro actions that run when a database is disabled. These macro actions can be embedded into existing forms and reports and allow developers to detect errors. MS-Access also allows developers to disable key features that prevent untrusted sources from viewing databases. In addition, action queries, VBAs, macros, and certain expressions are disabled by default because they might pose security risks. This would happen when a trusted source is viewing the database and these features are left on. This poses a security risk if a malicious source manages to become trusted [29].

References

1. B. Johnson, [Online], Adabas-Natural Best Practices for Multi-level Security. Available: https://techcommunity.softwareag.com/techniques-blog/-/blogs/adabas-natural-best-practices-for-multi-level-security. Retrieved: February 10, 2020
2. Anonymous [Online], Germany: Bringing the Latest Big Data Security Technology to the Mainframe, MENA Report, 2015. Available: https://library.semo.edu:2443/login?url=https://library.semo.edu:4836/docview/1662311993?accountid=38003. Retrieved: February 10, 2020

3. Anonymous [Online], ADABAS DBMS. Available: http://support.sas.com/documentation/cdl/en/acadbas/59521/HTML/default/viewer.htm#a000606259.htm. Retrieved: February 10, 2020
4. Anonymous [Online], ADABAS and NATURAL Security Options. Available: http://support.sas.com/documentation/cdl/en/acadbas/59521/HTML/default/viewer.htm#a000382555.htm. Retrieved: February 20, 2020
5. Anonymous [Online], Relational Database Server: Sybase: SAP ASE. (n.d.). Available: https://www.sap.com/products/sybase-ase.html#security-compliance. Retrieved February 20, 2020
6. M. Rouse [Online], What is Adaptive Server Enterprise (ASE)? Available: https://whatis.techtarget.com/definition/Adaptive-Server-Enterprise-ASE. Retrieved February 10, 2020
7. Anonymous [Online], Adaptive Server Enterprise 16.1. Available: http://infocenter.sybase.com/help/index.jsp?topic=/com.sybase.infocenter.help.ase.16.0/doc/html/title.html. Retrieved February 25, 2020
8. V. Vardanyan [Online], SAP ASE ODATA Server 16 Denial of Service. Available: https://packetstormsecurity.com/files/140610/SAP-ASE-ODATA-Server-16-Denial-Of-Service.html. Retrieved February 20, 2020
9. Anonymous [Online], CVE Details. (2016, November 3). Vulnerability Details: CVE-2016-7402. Available: https://www.cvedetails.com/cve/CVE-2016-7402/. Retrieved: February 25, 2020
10. P. Dobler, [Online], SAP Sybase ASE – Keeping Private Data Private with Data Encryption. Available: https://www.doblerconsulting.com/db-tech-trends/sap-sybase-ase-keeping-private-data-private-data-encryption/. Retrieved: February 25, 2020
11. Anonymous [Online], TrustWave SpiderLabs. TrustWave. Available: https://www.trustwave.com/en-us/resources/security-resources/security-advisories/?fid=18846. Retrieved: February 25, 2020
12. D.E. Denning, S.G. Akl, M. Heckman, T.F. Lunt, M. Morgenstern, P.G. Neumann, R.R. Schell [Online], Views for multilevel database security. IEEE Transactions on Software Engineering, 13(2), 129–140. Available: https://doi.org/10.1109/TSE.1987.232889. Retrieved: February 25, 2020
13. Anonymous [Online], MITRE Corporation. Process Injection. Available: https://attack.mitre.org/techniques/T1055/
14. Anonymous [Online], Exploit Database. Sybase Advantage Data Architect – '.SQL' Format Heap Overflow. Available: https://www.exploit-db.com/exploits/15378. Retrieved: February 5, 2020
15. Anonymous [Online]. Beach, NCS DATACOM AND AIRDEFENSE LAUNCH WLAN SECURITY SERVICE. LAN Product News, 15(3) . Available: https://library.semo.edu:2443/login?url=https://library.semo.edu:4836/docview/204364487?accountid=38003. Retrieved: February 5, 2020
16. Anonymous [Online], Security Overview. Available: https://techdocs.broadcom.com/content/broadcom/techdocs/us/en/ca-mainframe-software/database-management/ca-datacom/15-1/administrating/security-overview.html. Retrieved: February 5, 2020
17. Anonymous [Online].FileMaker Platform Security – Overview. Available: https://support.filemaker.com/s/article/FileMaker-Platform-Security-Overview-1503693058473?language=en_US. Retrieved: February 16, 2020
18. M. Woodfield [Online], FREAK attack: what you need to know. Available: https://www.digicert.com/blog/freak-attack-need-know/#targetText=Currently known as 'FREAK', can then easily be decrypted. Retrieved: March 5, 2020
19. C. Hoelscher [Online], IDMS History. Available: http://www.manmrk.net/tutorials/database/IDMS/IDMSHistory.htm. (1999). Retrieved: February 25, 2020
20. CA IDMS, Centralized Security Overview. Available: https://techdocs.broadcom.com/content/broadcom/techdocs/us/en/ca-mainframe-software/database-management/ca-idms/19-0/administrating/administrating-security-for-idms/ca-idms-centralized-security-overview.html. Retrieved: February 5, 2020

21. W.-J. Chen, H. Kirstein, R. Pachipala, V.S. Dantale [Online], Security and Compliance Solutions for IBM Informix Dynamic Server [First Edition]. Available: http://www.redbooks.ibm.com/redbooks/pdfs/sg247556.pdf. Retrieved: February 10, 2020
22. Anonymous. Actian Corporation, Ingres 10.2 Security Guide. (2016). doi: ING-102-SG-04
23. Anonymous [Online].InterBase 2017 Operations Guide. Available: http://docs.embarcadero.com/products/interbase/IB2017/OpGuide.pdf. Retrieved: February 5, 2020
24. S. Harris [Online], DES is not Secure. Available: https://www.freeswan.org/freeswan_trees/freeswan-1.5/doc/DES.html. Retrieved: February 8, 2020
25. M. Kemper, B. Bandy, [Online], SQL roles: users and security in InterBase. (n.d.). Available: https://www.ibphoenix.com/resources/documents/general/doc_59. Retrieved: February 20, 2020
26. Anonymous [Online], About Caché Security. Available: https://cedocs.intersystems.com/latest/csp/docbook/DocBook.UI.Page.cls?KEY=GCAS_intro. Retrieved: February 8, 2020
27. Anonymous [Online], Microsoft SQL server. In F. Botto, Dictionary of e-business (2nd ed.). Hoboken, NJ: Wiley. Available:https://library.semo.edu:2443/login?url=https://search.credoreference.com/content/entry/dictebusiness/microsoft_sql_server/0?institutionId=1804. Retrieved: February 11, 2020
28. Anonymous [Online], Overview of SQL Server Security. Available: https://docs.microsoft.com/en-us/dotnet/framework/data/adonet/sql/overview-of-sql-serversecurity. Retrieved: February 11, 2020
29. Anonymous [Online], Introduction to Access 2010 security. Available: https://support.office.com/en-us/article/introduction-to-access-2010-security-cae5d764-0318-4622-955f-68d9f186d6ca. Retrieved: February 10, 2020

Static Analysis for Software Reliability and Security

Hongjun Choi, Dayoung Kang, and Jin-Young Choi

1 Introduction

As computer systems become complex and integrated, software reliability and security are becoming increasingly important [12]. We often see software disasters in news or articles. This software disaster is caused by software bugs. Bugs can be divided into reliability bugs and security bugs. Examples of incidents caused by reliability bugs are Y2K [16], and incidents caused by security bugs are often found in CVE [5], such as Heartbleed [8]. There are already books and papers dealing with software reliability and security together [1, 2, 10, 11, 18, 19]. However, there are insufficient prior studies that look at weaknesses in code from two perspectives. Weaknesses [6] mean bugs, errors, and the root cause of software vulnerabilities; we must diagnose and minimize them at every stage of the software development process.

Typically, programmers assume that input comes in as specified when writing code. If you analyze the program code written with the assumption using a static analysis tool, the corresponding weakness alert will occur. However, it is ambiguous for the tool used to determine whether the vulnerability is due to reliability or security issues. So we have to follow the code flow from that line of weakness to see the perspective.

In this paper, we analyze the weakness of integer overflow [6] that can occur in the code of binary search from two aspects. Also, it shows the need for secure coding by analyzing the results of static analysis. When the reliability tester and the security tester use the static analysis tool, the range of input values from the two perspectives is different, so the task of checking for the presence of false positives

H. Choi · D. Kang · J.-Y. Choi (✉)
The Graduate School of Information Security, Korea University, Seoul, Republic of Korea
e-mail: hjcho1@korea.ac.kr; dayokiki@korea.ac.kr; narnia@korea.ac.kr

K. Daimi et al. (eds.), *Advances in Security, Networks, and Internet of Things*,
Transactions on Computational Science and Computational Intelligence,
https://doi.org/10.1007/978-3-030-71017-0_33

[4] is performed. This work takes a lot of time and effort. Instead of dividing the perspective and checking for false positives, we should combine the two aspects. When vulnerability occurs, we must correct the code by applying secure coding. And we claim that all vulnerability must be eliminated by input validation or secure coding, not by checking whether false positive or true positive.

2 Background

Before analyzing with two perspectives with an example of binary search code, we introduce software reliability, security, and static analysis.

2.1 *Software Reliability*

Software reliability [17] Informally, the reliability of a system is the probability, over a given period of time, that the system will correctly deliver services as expected by the user. An order can be considered reliable if its behavior matches that defined in the specification for the correct input. During program development, programmers perform tasks called debugging to detect logical errors or abnormal computations in the system. Programmers usually enter a value between the minimum and the maximum amount of the variable if there is no defined range. Programmers are debugging if the actual output value differs from the expected output value. This process can be seen as an act to ensure software reliability.

Safety-critical systems [13] are those systems whose failure could result in loss of life, significant property damage, or damage to the environment. It is related to software reliability in that lack of software reliability can result in human casualties.

2.2 *Software Security*

Software security [14, 17] Informally, the security of a system is a judgment of how likely it is that the system can resist accidental or deliberate intrusions. An outsider uses unauthorized input values to cause attacks such as data modification and deletion, backdoor installation, and elevation of privilege. All members, designers, developers, testers, should consider security to create secure software from the requirements phase until the software is retired. It is also repeatedly necessary to perform a security development lifecycle because code changes every time update or function is added. And software security issues arise not only in specific security functions but also in the entire implemented system. Therefore, programmers should consider software security and software reliability together.

2.3 *Static Analysis*

Static analysis [7] analyzes code without program execution. The static analysis tool automatically checks for weaknesses such as code grammar, coding rules, and execution errors. Static analysis cannot diagnose all alarms due to different each tool-specific rules and patterns(false negative). Also, not all alarms are true positive (false positive). Although these shortcomings exist, static analysis tools are widely used because they detect defects in the early stages of development through static analysis and have advantages in development efficiency and cost reduction. In particular, static analysis tool can effectively diagnose for weaknesses such as overflow, division by zero, and out of bounds array access.

3 Binary Search Code

This section will show the integer overflow weakness from two perspectives in the following binary search codes. We may write other sample codes, but we use a common code of binary search that can be understood easily.

```
/* Code of Binary Search */
#include<stdio.h>
#define INT_MIN -2147483648
#define INT_MAX 2147483647

int main() {
    int number = 0; // Array size
    int index = 0; // Array index
    int search = 0; // Element to find
    int first = 0; // The position of the first element
    int middle = 0; // The position of the middle element
    int last = 0; // The position of the last element
    int *array; // Dynamic memory allocation

    printf("Enter the number of element : ");
    scanf_s("%d", &number);

    array = (int *)malloc(sizeof(int)*number);
    last = number - 1;

    for (index = 0; index < number; index++) {
        printf("Enter element in array : ");
        scanf_s("%d", &array[index]);
    }

    printf("Select the element to find : ");
    scanf_s("%d", &search);

    while (first <= last) {
        middle = (first + last) / 2;
```

```
        if (array[middle] < search)
            first = middle + 1;
        else if (array[middle] == search) {
            printf("%d found at position %d.\n", search, middle+1);
            break;
        }
        else
            last = middle - 1;
    }

    if (first > last)
        printf("Not found %d!\n", search);

    return 0;
}
```

It is an example code of a binary search. Binary search is an algorithm that finds a specific value in an array sorted in ascending order. Briefly, the program user enters the size of the array. After that, the user puts values into the array as much as the size of the array. Then, the user specifies the value to find. The first variable has the first index of the array, and last has the last index of the array. As the code implemented with the binary search algorithm operates, first and last values are modified according to each condition. The middle value is used to search for the specific value in the array.

3.1 Software Reliability Perspective

The (1) below is integer overflow weakness in a binary search code.

$$\mathbf{middle} = \frac{(\mathbf{first} + \mathbf{last})}{\mathbf{2}} \tag{1}$$

In the algorithm books and papers [3, 9, 15], the middle value is frequently used as (1). It cannot say that it is wrong at the result of the calculation. The first value and the last value are defined in the specification. But, it is a problem in terms of reliability because an integer overflow occurs due to the corresponding values. Assume that the requirement is specified as follows:

Requirement 1: Reliability Perspective

- A normal program user enters a number between 1 and 100 as the size of the array.
- The elements of the array are natural numbers.

In this case, the first and last are natural numbers because they are the index of the array, and the range is as follows:

$$\mathbf{0 \le first, last < 100} \quad (2)$$

Even if each maximum value is added and divided by 2, the middle's maximum value does not exceed 99. Therefore, it does not exceed INT_MAX. However, suppose that the requirement has been modified as follows:

Requirement 2: Reliability Perspective

- A normal program user enters a number between 1 and INT_MAX as the size of the array.
- The elements of the array are natural numbers.

$$\mathbf{0 \le first, last < INT\_MAX} \quad (3)$$

Since the first value and the last value can have INT_MAX, an integer overflow occurs when two values are added. It is defined in the specification, but this may cause bugs or weaknesses. Even if both values are not the maximum value, an integer overflow occurs if the result added by both values exceeds the INT_MAX.

3.2 *Software Security Perspective*

In terms of software security, it must be a situation where input values that are not defined in the requirement or specification are entered. Assume the specification is specified as follows:

Requirement 1: Security Perspective

- A normal program user enters a number between 1 and 100 as the size of the array.
- The elements of the array are natural numbers.

A malicious user may enter a value other than the required input range. In this case, the scope of the first and last values is as (4).

$$\mathbf{100 \le first, last < INT\_MAX} \quad (4)$$

If first and last come in values between 1 and 100, as explained in Requirement 1: Reliability, there is no integer overflow. However, in terms of security, an integer overflow occurs because other positive integers can be entered. What is unusual is that the result of Requirement 2: Reliability and Requirement 1: Security perspectives are the same as an integer overflow. It means that vulnerabilities occur in the absence of reliability or security.

4 Static Analysis and Secure Coding

The static analysis results on the code of binary search using the Commercial Static Analysis A tool are as shown in Fig. 1. Four vulnerabilities have occurred, including the integer overflow, (first+last)/2, addressed in this paper. The overflow checker diagnosed the operation. If the code's bifurcation is semantically well-separated, it will be easy to see whether it is a reliability or security issue by looking at the alarms in the static analysis tool. If not, reliability and security testers can take a lot of time and effort to determine whether they are false positive in terms of each. To solve this, the programmer must do secure coding. However, there should be no problem with prerequisites. The proposed secure coding includes input value validation and modifying the sequence of operations.

If it is defined as Requirement 1: Reliability, integer overflow does not occur when the following input validation is added:

$$\textbf{if}(\textbf{number} \geq \textbf{1} \ \&\& \ \textbf{number} \leq \textbf{100}) \tag{5}$$

Figure 2 shows the result of static analysis adding (5) to the original code. (5) removed the overflow from the addition operation of (first+last)/2 and subtraction operation of last=n-1.

Modifying the operations sequence is the same as the computational result in (1) and can remove weaknesses from a defensive perspective. Although (1) used in the above code is much more intuitive and readable, it is a dangerous computational

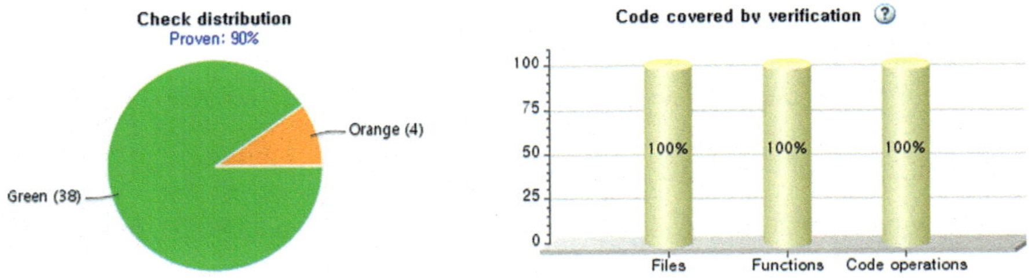

Fig. 1 The result of static analysis for code of binary search

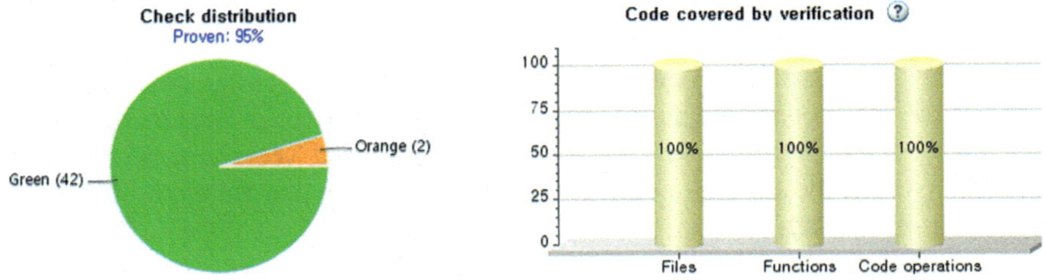

Fig. 2 The result of static analysis for code including input validation

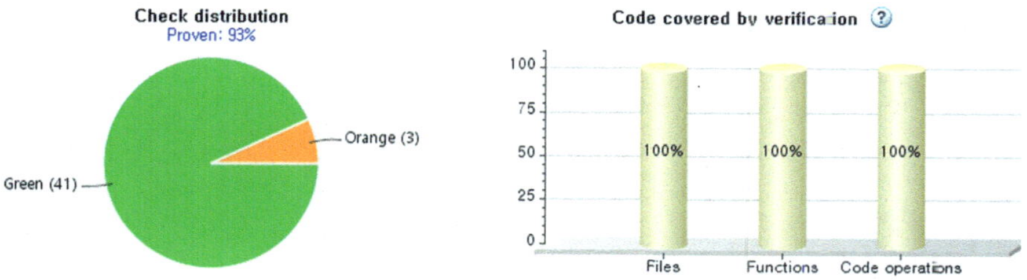

Fig. 3 The result of static analysis for code modifying the operation sequence

method because it has been shown earlier that integer overflow can occur in terms of reliability and security. (1) shall be replaced by

$$\mathbf{middle} = \mathbf{first} + \frac{\mathbf{last} - \mathbf{first}}{\mathbf{2}} \tag{6}$$

Figure 3 shows the results of a static analysis after changing (1)–(6). The difference from Fig. 2 is that the overflow from the last=n-1 subtraction operation was not eliminated.

The best secure coding for integer overflow is input validation, but we recommend using both methods. It was confirmed that the integer overflow did not occur when the static analysis tool was executed after the secure coding in the two ways presented above.

5 Conclusion

In this paper, we analyzed that the integer overflow could be occurred in terms of reliability and security with the binary search code by using the static analysis tool. We confirmed that the integer overflow was solved after modifying the original code by applying secure coding. If the programmer does not correctly implement the parts defined in the requirements, it is not easy to anticipate a reliability problem or a security issue by looking at the static analysis tool alert.

Therefore, before using the static analysis tool, we should confirm that the code has secure coding applied. If not, when a vulnerability occurs, we must correct the weakness code by using secure coding. Since the weakness caused by reliability issues becomes a software security problem in the future, an adversary attack may occur, so we should consider both perspectives when developing the software.

Future studies will show a more explicit relationship between reliability and security. We also compare the execution time between an original code and the code applying security and reliability.

Acknowledgments This research was supported by Next-Generation Information Computing Development Program through the National Research Foundation of Korea (NRF) funded by the Ministry of Science and ICT (No. 2017M3C4A7083676).

References

1. H. Adkins, B. Beyer, P. Blankinship, P. Lewandowski, A. Oprea, A. Stubblefield, *Building Secure and Reliable Systems: Best Practices for Designing, Implementing, and Maintaining Systems*, 1st edn. (O'Reilly Media, Newton, 2020)
2. E. Burtescu, Reliability and security-convergence or divergence. Inf. Econ. **14**(4), 68–77 (2010)
3. A.R. Chadha, R. Misal, T. Mokashi, Modified binary search algorithm. Int. J. Appl. Inf. Syst. **7**(2), 37–40 (2014)
4. B. Chess, G. McGraw, Static analysis for security. IEEE Secur. Priv. **2**(6), 76–79 (2004)
5. Common vulnerabilities and exposures. https://cve.mitre.org/. Accessed 7 July 2020
6. Common weakness enumeration. https://cwe.mitre.org/. Accessed 8 June 2020
7. L.N.Q. Do, J. Wright, K. Ali, Why do software developers use static analysis tools? a user-centered study of developer needs and motivations. IEEE Trans. Softw. Eng. (2020). https://doi.org/10.1109/TSE.2020.3004525
8. Z. Durumeric, F. Li, J. Kasten, J. Amann, J. Beekman, M. Payer, N. Weaver, D. Adrian, V. Paxson, M. Bailey et al., The matter of heartbleed, in *Proceedings of the 2014 Conference on Internet Measurement Conference* (2014), pp. 475–488
9. G.T. Heineman, G. Pollice, S. Selkow, *Algorithms in a Nutshell: A Practical Guide* (O'Reilly Media, Inc., Newton, 2016)
10. J. Henkel, V. Narayanan, S. Parameswaran, R. Ragel, Security and dependability of embedded systems: a computer architects' perspective, in *2009 22nd International Conference on VLSI Design* (IEEE, New York, 2009), pp. 30–32
11. M. Howard, S. Lipner, *The Security Development Lifecycle*, vol. 8 (Microsoft Press, Redmond, 2006)
12. R.K. Iyer, Z. Kalbarczyk, K. Pattabiraman, W. Healey, W.H. Wen-mei, P. Klemperer, R. Farivar, Toward application-aware security and reliability. IEEE Secur. Priv. **5**(1), 57–62 (2007)
13. J.C. Knight, Safety critical systems: challenges and directions, in *Proceedings of the 24th International Conference on Software Engineering* (2002), pp. 547–550
14. G. McGraw, Software security. IEEE Secur. Priv. **2**(2), 80–83 (2004)
15. A. Oommen, C. Pal, *Binary Search Algorithm*, vol. 1 (Codility Limited, London, 2014), pp. 1–4
16. H. Pham, *Software Reliability* (Springer Science & Business Media, Berlin, 2000)
17. I. Sommerville, *Software Engineering*, 10th edn. (Addison-Wesley, Boston, 2015)
18. J. Viega, G. McGraw, *Building Secure Software: How to Avoid Security Problems the Right Way (Paperback)*. Addison-Wesley Professional Computing Series (Addison-Wesley Professional, Boston, 2011)
19. J. Whittaker, Reliability vs. security (2007). https://www.microsoft.com/security/blog/2007/12/07/reliability-vs-security/. Accessed 6 July 2020

Part V
Wireless Networks, Novel Technologies and Applications

A Tool for the Analysis of MANET Routing Protocols Based on Abstract State Machines

Alessandro Bianchi, Emanuele Covino, Giovanni Pani, and Sebastiano Pizzutilo

1 Introduction

Mobile Ad-hoc NETwork (MANET) is a technology used to establish and to perform wireless communication among nomadic hosts in absence of physical infrastructure [1]. Hosts in a MANET are intended as autonomous agents: during their lifetime, they can enter or leave the network, and they can continuously change their relative position; thus, the network lacks of a predefined topology. Each host is able to communicate with hosts inside its radio range only; outside this area, communication is possible only by means of cooperation between intermediate hosts. They can act as initiator, intermediate, and destination of a communication. Concepts and protocols have been developed a few decades ago; however, this research area is receiving special attention in the last few years, in the context of smart mobile computing, cloud computing, Cyber Physical Systems, and Internet of Things [2, 3].

This technology is often used in a number of sensitive applications, such as rescue operations in case of disasters, data tracking of environmental conditions, health-care, intelligent transportation, environmental emergency management, and energy saving; possible malfunctioning, improper behavior, or not adequate performance could result in severe damage to people, environment, or other systems. This raises several problems about the analysis of performance, synchronization,

This work is dedicated to the memory of Alessandro Bianchi, who passed away in the summer of 2019.

A. Bianchi · E. Covino (✉) · G. Pani · S. Pizzutilo
Dipartimento di Informatica, Universitá di Bari, Bari, Italy
e-mail: emanuele.covino@uniba.it; giovanni.pani@uniba.it; sebastiano.pizzutilo@uniba.it

K. Daimi et al. (eds.), *Advances in Security, Networks, and Internet of Things*, Transactions on Computational Science and Computational Intelligence,
https://doi.org/10.1007/978-3-030-71017-0_34

and concurrency of the network. Moreover, the request of computing services characterized by high quality levels, broad and continuous availability, and interoperability over heterogeneous platforms increases the complexity of the systems' architecture. A sound analysis of their behavior and their properties is of particular interest.

Therefore, it is important to be able to verify qualities like responsiveness, robustness, correctness, and performance, starting from the early stages of the development. To do this, most studies are executed with the support of simulators. For example, they have been used to compare the performances of some network's protocols [4]; to evaluate a topology control approach [5]; to study congestion adaptive routing [6]. Simulators are suitable to evaluate performance and to compare different solutions, but they do not provide a formal model of the MANETs. They implement the network at a low abstraction level, and they consider only a limited and predictable range of scenarios. They cannot support specification at higher level, so they do not adequately help the study of typical problems of MANETs, and they are not able to prove correctness properties. Conversely, formal methods are satisfactory for reasoning about correctness properties, but they rarely are useful for studying performance properties [7]. To this end, it seems appropriate to provide a formal approach to the system architecture: recent examples are [8–10].

Generally speaking, correctness properties are formally proved, while the performance properties are investigated through simulations of the system. Some conceptual tools enable the coexistence of the two approaches for both verifying correctness and validating performance; for example, Petri Nets and Abstract State Machines.

In this paper, we introduce MOTION (MOdeling and simulaTIng mObile adhoc Networks), a Java application in which the behavior of MANETs is modeled by means of an Abstract State Machine representation, and then simulated with the simulation engine ASMETA. MOTION has the twofold ability to formally prove system's properties as well as to simulate the system behavior.

2 Related Works

As we stated before, the analysis and the evaluation of MANET's properties can be done by means of simulators (focusing on defining performance's metrics in standard or unusual scenarios), or by means of formal models of the system (studying its behavior and computational properties from a more abstract point of view). For example, [3, 4, 11, 12], and [13] compare some routing protocols performances; [6, 14], and [15] study congestion adaptive routing; [16] and [17] discuss the issues in managing synchronization among components involved in simulation; [5] and [18] evaluate a topology control approach. Nevertheless, some authors show that the results obtained using simulators can be inaccurate or unreliable [14, 15, 19], and [20].

The previous approaches are all based on simulators: they measure performance, but they cannot model MANETs, formally. They implement the network at a low level, while a higher abstraction level of specification is needed in order to study problems such as concurrency, synchronization, deadlock. On the other hand, some examples of the application of formal methods to the analysis of MANETs (based on process calculi) have been proposed, such as ω-calculus [17], CMN (Calculus of Mobile Ad-Hoc Networks) [20], and AWN (Algebra for Wireless Networks) [15]. They capture some essential characteristics of nodes, such as mobility or packet's broadcasting and unicasting.

Another class of formal methods used for studying MANETs is represented by state-based models, such as Finite State Machines [21] and Petri nets [18]. In particular, Petri nets have been employed to study modeling and verification of routing protocols [22], evaluation of protocol performance [23], and application to vehicular networks [24]. With respect to process calculi, state-based models provide a more suitable way of representing algorithms. Moreover, they are typically equipped with tools, such as CPN Tools [25], that allow to simulate the algorithms, directly. However, state-based models lack of expressiveness: basically, they provide only a single level of abstraction, and cannot support refinements to executable code.

One of the most popular routing protocols for MANETs is the Ad-hoc On-demand Distance Vector (AODV), that is standardized as RFC (Request For Comments) by the IETF MANET working group [26]. Several variants of the protocol have been introduced in order to reduce communication failures due to topology changes. For example, Reverse-AODV (R-AODV, [27] and [16]) overcomes this problem by building all possible routes between initiator and destination: in case of failure of the primary route (typically the shortest one), communication is still provided by the alternative routes. More recently, variants have been proposed for coping with congestion issues [28, 29]. Finally, some improvements of AODV are related to security, such as the use of cryptography for securing data packets during their transmission (e.g., Secure-AODV [30]), and the adoption of the so-called *trust methods*, in which nodes are part of the communication if and only if they are considered trustworthy (e.g., Trusted-AODV, [29, 31]).

Our approach, using Abstract State Machines [32], is similar to [22], in which Colored Petri Nets are used to model the AODV routing protocol, and CPN model is used to simulate the MANET behavior. As an improvement our approach is more general purpose, meaning that the implementation of the routing protocol is only one of the several services that can be provided. Each intended service can be modeled in our layered framework, and implemented in a simulator. Thanks to the structured approach, services can be easily added, removed, and replaced by changing some transitions and nested Petri nets, as well as changing classes in software implementation.

The ASM approach also provides a way to describe algorithmics in a simple abstract pseudo-code, which can be translated into a high-level programming language source code [32]. Finally, from the implementation point of view, the capability of translating formal specifications into executable code, in order to carry out simulations of the models, is provided by tools like CoreASM [33] and ASMETA [34].

In this paper, we provide a detailed description and a platform-independent version of the MOTION environment; the initial interface of the application and the dialogue with AsmetaS are coded entirely in Java, in order to ensure compatibility with the main Operating Systems.

3 Background

In this section, we summarize concepts related to the routing protocols implemented within MOTION, to the Abstract State Machines formalism, and to the ASMETA framework.

3.1 MANET and Routing Protocols

Mobile Ad-hoc NETworks are wireless communication systems whose topology can change dynamically; each host of the network is an autonomous agent, and it can re-arrange itself without conforming to a fixed topology. During its lifetime, it can enter or leave the network, and it can continuously change its position; this means that routes connecting the hosts can rapidly change.

Several routing protocols have been proposed; among them, the *Ad-hoc On-demand Distance Vector* (AODV) is one of the most popular. Indeed, a large number of simulation studies are dealing with it, representing a reliable baseline for comparison to the results of simulations executed with MOTION. Moreover, we add two variants of AODV: *NACK-based Ad-hoc On-demand Distance Vector* (N-AODV, [35]) that improves the awareness that each host has about the network topology, and *Blackhole-free N-AODV* (BN-AODV, [36]) that detects the presence of malicious hosts leading to a blackhole attack.

Ad-Hoc On-Demand Distance Vector (AODV) This routing protocol has been defined in [26]: it is a reactive protocol that combines two mechanisms, namely the *route discovery* and the *route maintenance*, in order to store some knowledge about the routes into *routing tables*. The routing table associated with each node is a list of all the discovered (and still valid) routes towards other nodes in the network, together with other information. In particular, for the purposes of the present paper, an entry of the routing table of the host i concerning a node j includes: the *address* of j; the last known *sequence number* of j; the *hop count* field, expressing the distance between i and j; and the *next hop* field, identifying the next node in the route to reach j.

The sequence number is an increasing number maintained by each node, that express the freshness of the information about the respective node. When an *initiator* wants to start a communication session towards a *destination*, it first checks if a route is currently stored in its routing table. If so, the protocol ends and the communication starts. Otherwise, the initiator broadcasts a control packet called *route request* (RREQ) to all its neighbors.

An RREQ packet includes the initiator address and broadcast id, the destination address, the sequence number of the destination (i.e., the latest available information about destination), and the hop count, initially set to 0, and increased by each intermediate node. The pair <*initiator address*; *broadcast id*> identifies the packet, uniquely; this implies that duplications of RREQs already handled by nodes can be ignored.

When an intermediate node *n* receives an RREQ, it creates the routing table entry for the initiator, or updates it in the fields related to the sequence number and to the next hop. Then, the process is iterated: *n* checks if it knows a route to destination with corresponding sequence number greater than (or equal to) the one contained into the RREQ (this means that its knowledge about the route is more recent). If so, *n* unicasts a second control packet (the *route reply*—RREP) back to the initiator. Otherwise, *n* updates the hop count field and broadcasts once more the RREQ to all its neighbors.

The process successfully ends when a route to the destination is found. While the RREP travels towards initiator, routes are set up inside the routing tables of the traversed hosts, creating an entry for destination, when needed. Once the initiator receives back the RREP, the communication session can start. If the hosts' movements break a link (i.e., a logical link stored in a routing table is no more available), a route maintenance is executed in order to notify the error and to invalidate the corresponding routes: to this end the control packet *route error* (RERR) is used.

NACK-Based AODV (N-AODV) One of the main disadvantages of the AODV protocol is the poor knowledge that each host has about the network topology. In fact, each node *n* is aware of the existence of a node *m* only when *n* receives an RREQ, either originated by, or directed to *m*. In order to improve the network topology awareness of each host, the NACK-based AODV routing protocol has been proposed and modeled by means of a Distributed ASM in [35].

This protocol is a variant of AODV: it adds a *Not ACKnowledgment* (NACK) control packet in the route discovery phase. Whenever an RREQ originated by *n* and directed to *m* is received by the node *p* that does not know anything about *m*, *p* unicasts the NACK to *n*. The purpose of this control packet is to state the ignorance of *p* about *m*. In this way, *n* (as well as all the nodes in the path to it) receives fresh information about the existence and the relative position of *p*. Therefore, on receiving the NACK, all the nodes in the path to *p* add an entry in their respective routing tables, or update the pre-existing entry. N-AODV has been experimentally validated through simulations, showing its efficiency and effectiveness: the nodes in the network actually improve their knowledge about the other hosts and, in the long run, the number of RREQ decreases, with respect to the AODV protocol [37].

Black Hole-Free N-AODV (BN-AODV) All routing protocols assume the trustworthiness of each host; this implies that MANETS are very prone to the *black hole attack* [38]. In AODV and N-AODV a black hole node produces fakes RREPs, in which the sequence number is as great as possible, so that the initiator sends the message packets to the malicious node, and the latter can misuse or discard them. The black hole can be supported by one or more *colluders*, that confirm the trustworthiness of the fake RREP. The Black hole-free N-AODV protocol [36] allows the honest nodes to intercept the black holes and the colluders, thanks to two control packets: each intermediate node n receiving an RREP must verify the trustworthiness of the nodes in the path followed by the RREP; to do this, n produces a *challenge packet* (CHL) for the destination node, and only the latter can produce the correct *response packet* (RES). If n receives RES, it sends the RREP, otherwise the next node towards the destination is a possible black hole.

3.2 Abstract State Machines

An Abstract State Machine (ASM, [32]) M is a tuple (Σ, S, R, P_M). Σ is a *signature*, that is a finite collection of names of total functions; each function has -arity n, and the special value *undef* belongs to the range. Relations are expressed as particular functions that always evaluate to *true*, *false*, or *undef*.

S is a finite set of *abstract states*. The concept of abstract state extends the usual notion of state occurring in finite state machines: it is an algebra over the signature Σ, i.e., a non-empty set of objects of arbitrary complexity together with interpretations of the functions in Σ.

R is a finite set of *rules* of the form "**if** *condition* **then** *updates*," which transform the states of the machine. The concept of rule reflects the notion of transition occurring in traditional transition systems: *condition* is a first-order formula whose interpretation can be true or false; *updates* is a finite set of assignments of the form $f(t_1; t_2; \ldots; t_n) := t$, whose execution consists in changing in parallel the value of the specified functions to the indicated value.

P_M is a distinguished rule of -arity 0, called the *main rule* of the machine M, which represents the starting point of the computation.

Pairs of function names together with values for their arguments are called *locations*: they are the abstraction of the notion of memory unit. Since a state can be viewed as a function that maps locations to their values, the current configuration of locations, together with their values, determines the current state of the ASM.

In order to better understand the semantics of the states with respect to the computational behavior of the modeled system, it is worth remarking that each ASM state can be characterized by one or more predicates over the states. More precisely, a predicate H over an ASM state s is a first-order formula defined over the locations in s, such that $s \models H$. Each predicate allows us to focus on the subsets of locations that turn out to be interesting for verification purposes.

The execution of an ASM consists in iterating computational steps. A *computational step* in a given state consists in executing all the rules whose condition is true in that state. Since different updates could affect the same location, it is necessary to impose a consistency requirement: a set of updates is said to be *consistent* if it contains no pairs of updates referring to the same location. Therefore, if the updates are consistent, the result of a computational step is the transition of the machine from the current state to another. Otherwise, the computation does not produce a next state. A *run* is a (possibly infinite) sequence of steps: the computational step is iterated until no more rules are applicable.

The aforementioned notions refer to the so-called *basic* ASMs. However, there exist some generalizations, e.g., parallel ASMs and Distributed ASMs (DASMs) [39]. Parallel ASMs are basic ASMs enriched with the forall construct, to express the simultaneous execution of the same ASM (i.e., of rules satisfying a given condition) over a number of independent agents. A Distributed ASM is intended as a finite number of independent agents, each one executing its own underlying ASM: it is capable of capturing the formalization of multiple agents acting in a distributed environment. A run of a DASM is a partially ordered set of the runs of its ASMs: the underlying synchronization scheme reflects causal dependencies; determining which agents move comes before a move is a single computational step of an individual agent, and is only restricted by the consistency condition, which is mandatory. Roughly speaking, a global state corresponds to the union of the signatures of each ASM together with interpretations of their functions.

3.3 ASMETA

The ASM-based method consists in development phases, from requirements' specification to implementation, supporting developers in realizing complex systems. Some environments support this method, and among them we use the ASMETA (ASM mETAmodeling) framework [11, 40]. This framework is characterized by logical components that capture the requirements by constructing the so-called *ground models*, i.e., representations at high level of abstraction that can be graphically depicted. Starting from ground models, hierarchies of intermediate models can be built by stepwise refinements, leading to executable code: each refinement describes the same system at a finer granularity. The framework supports both verification, through formal proof, and validation, through simulation.

In order to implement MOTION, we considered three among these logical components. The basic component is the Abstract State Machines Metamodel (AsmM), i.e., the description of a language for ASMs, expressed as an abstract syntax that represents domains, functions, axioms, rules; the syntactic constructs occurring in the ASM's states; the syntactic elements enabling the transition rules, and so on. According to the rules of the abstract syntax, we then use the ASMETA Language (AsmetaL) and the ASMETA Simulator (AsmetaS): essentially, they are an interpreter that navigates through the ASM specifications and performs its computations.

4 MOTION

4.1 General Behavior

As we have stated before, MOTION (MOdeling and simulaTIng mObile ad-hoc Networks) is a Java application that allows to specify the simulation parameters, to execute the network described, and to collect the output data of the simulation. Its web pages, with the complete package, can be found at https://sourceforge.net/projects/motion-project/.

MOTION is developed within the ASMETA framework thanks to the abstract syntax defined in the AsmM metamodel; the behavior of the MANET is modeled using the AsmetaL language, and then the network is executed by the AsmetaS simulator. Since AsmetaS simulates instances of the model expressed by means of the AsmetaL, the information concerning each instance, (number of agents and their features, for instance), must be recorded into the AsmetaL file.

The executions of MOTION and ASMETA are interleaved: MOTION provides the user interface and captures the data inserted by the user, representing the parameters of the simulation. MOTION then includes these data into the AsmetaL file, and it runs AsmetaS. AsmetaS executes an ASM move, simulating the behavior of the network protocol over the current data, and it records the values of the locations in a log file, for each state. At the end of each move the control goes back to MOTION: it gets the information about the results of the ASM move, such as the relative position of the hosts, the sent/received packets, and the values of waiting time, and it records them into the AsmetaL file. Then, MOTION invokes AsmetaS for the next move. At the end of the simulation session, MOTION reads the final log file, parses it, and stores the collected results in a csv file.

4.2 Defining the Mobility Model

In a realistic scenario, the hosts of a MANET behave according to the rules expressed by a specific routing protocol, and they are characterized by a set of features. More precisely, each host is a computational agent, which plays two different roles. On one hand, it is a communicating agent, acting as an initiator, destination, or as an intermediate host of a communication. At the same time, it is a mobile agent, moving into the MANET space, and changing speed and direction. Moreover, due to the wireless nature of MANET, each host is associated with a radio range, which specifies the maximum distance the signal sent by a host can be received by another station. The movement of the hosts determines the current topology and, together with the amplitude of the radio range, it affects the current set of physical connections among hosts.

A realistic simulation should take into account all these features. However, simulating all aspects of a MANET can be cumbersome, and sometimes impossible; according to [41], the model of the systems to be simulated must be tailored depending on the goals of the simulation project. Therefore, the movement issues, as well as the amplitude of the radio range, are abstractly defined within the mobility model. In this sense, we assume that the whole network topology is expressed by the connections among hosts, implicitly, and for each host we consider only its current neighborhood. More precisely, in MOTION the network topology is logically expressed by a *connectivity matrix* C, such that $c_{ij} = 1$ if i and j are neighbors; 0 otherwise, for each pair of hosts i and j.

Within the ASM model, C is expressed by the predicate isLinked(a_1, a_2), which evaluates to *true* when the agent a_1 is linked to the agent a_2; to *false* otherwise. Changes of isLinked represent the transitions of each host from one set of neighbors to another.

Within MOTION, the mobility model is implemented into a Java class that, before executing any ASM move, updates the connectivity matrix. To this end, each c_{ij} is randomly set to 0 or 1, according to a parameter defined by the user (see below). The new values of the connectivity matrix are then set within the AsmetaL file, so that the ASM move can be executed, accordingly.

4.3 The Abstract State Machine-Based Models

The AODV routing protocol has been formally modeled through ASMs in [42]. More precisely, it is described as a set of agents, each one representing a host. The high-level machine in MOTION is:

MAIN RULE =
    forall a $\in$ Agents **do** AODVSPEC(a)

where

AODVSPEC(a) =
  forall dest $\in$ Agents **with** dest $\neq$ self **do**
    if *WaitingForRouteTo*(self, dest) **then**
      if *Timeout*(self, dest) $>$ 0 **then**
        Timeout(self, dest) := *Timeout*(self, dest)-1
      else
        WaitingForRouteTo(self, dest) := false
  if *WishToInitiate*(self) **then** PREPARECOMM
  if not *Empty*(Message) **then** ROUTER

If the host needs to start a communication (i.e., the function *WishToInitiate* evaluates to true), then the PREPARECOMM submachine is called. The function *WaitingForRouteTo* expresses that the discovery process previously started by the host is still running; in this case, if the waiting time for RREP is not expired (i.e., *Timeout*() > 0), the time-counter is decreased. Finally, if the host has received a message (either RREQ, RREP, or RERR), the ROUTER submachine is called.

ROUTER =
    ProcessRouteReq
    ProcessRouteRep
    ProcessRouteErr

where each submachine expresses the behavior of the host that depends on the type of the message received.

The ASM model for N-AODV is similar to [35]: the main difference concerns the ROUTER submachine that includes the call to the submachine PROCESS-NACK, in order to unicast the NACK packet, if needed. The BN-AODV model is more structured [36], because it has to describe the behavior of three different kinds of agents: honest hosts, black holes, and colluders. So, the main rule has the form:

MAIN RULE =
  forall a ∈ Blackhole **do** BLACKHOLESPEC(a)
  forall a ∈ Colluder **do** COLLUDERSPEC(a)
  forall a ∈ Honest **do** HONESTSPEC(a)

where the HONESTSPEC submachine describes the behavior of the honest nodes, and it is analogous to AODVSPEC. BLACKHOLESPEC and COLLUDERSPEC are the specifications for the non-honest nodes and the colluders, respectively. Moreover, the ROUTER submachine for the honest nodes includes a submachine for verifying the trustworthiness of the received RREPs.

Thanks to the formalization of the protocols, some correctness properties have been proved in the past, such as the starvation freeness for the AODV protocol, the properness of the packet (either NACK or RREP) received back by the initiator of any communication, when it is not isolated for N-AODV, and the capability to intercept all blackhole attacks for BN-AODV [35, 36].

4.4 *Specific Behavior of the Tool*

A simulation in MOTION is performed in a number of sessions, established by the user; during each session, the MANET includes a number of hosts defined by the user that depends on the specific evolution of the network (due to movements, some of them can be disconnected, in the sense that they cannot be reached by the other hosts). Moreover, during each session, each host is the initiator for a number of attempts, trying to establish a communication, each towards a destination different from the initiator itself: the user expresses the probability that each host acts as an

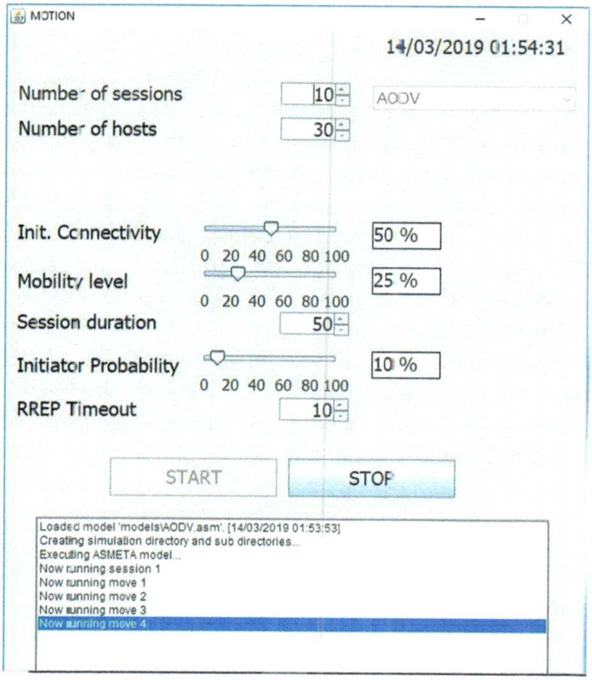

Fig. 1 MOTION user interface

initiator by setting the parameter *Initiator Probability* (in Fig. 1, the value of this parameter is 10%). For each communication attempt (in what follows, CA), both initiator and destination are randomly defined. Thanks to the intrinsic parallelism in the execution of the ASM's rules, more attempts can be simultaneously executed. A CA is considered successful if the initiator receives an RREP packet within the waiting time expressed by the parameter *RREP Timeout*; otherwise, the attempt is considered failed. Note that the time is measured as the number of times the main rule of the ASM is executed.

In MOTION, the hosts mobility is defined by the user by means of two parameters, namely *Initial Connectivity* and *Mobility level*. The former defines the initial topology of the MANET: it expresses the probability that each host is directly linked to any other host. During the simulation, the hosts mobility is expressed by the random redefinition of the values of the isLinked predicate. More precisely, for each pair of agents $< a_i, a_j >$, and for each move of the ASM, the value of isLinked(a_i, a_j) predicate is changed with a probability expressed by *Mobility level*.

When the BN-AODV routing protocol is simulated, the MOTION user interface includes the definition of the number of black holes and colluders, and two parameters establishing the increment of the fake sequence number produced by the black hole. Figure 1 shows the current state of the simulation in the panel under the two buttons START and STOP.

From the ASM perspective, there are two different machines, both called by the ASMETA's main rule. The first one is the OBSERVERPROGRAM: it is not part of the MANET, but it is used in order to manage the execution. It initializes the locations and data structures for all the agents, manages the mobility (setting the initial topology and resetting the connectivity matrix at each move), and updates the counter for the time expiration. The second machine, called by the main rule, is the model of the hosts' behavior. Currently, MOTION allows the users to study AODV, N-AODV, and BN-AODV, specified according to the ASMs presented in [35, 42], and [36], respectively. Note that, for all of them, the MANET is modeled by a Distributed ASM. In both AODV and N-AODV all the nodes behave in the same way, described by the respective DASM, so the machine specifying the protocol is called; at each move the machine randomly decides if the current host will initiate new communication attempts by invoking the R-PREPARECOMM submachine, then it acts as a router by processing the proper control packets (R-ROUTER submachine).

5 The Empirical Study

In order to evaluate the performance properties of the AODV and the N-AODV routing protocols, a number of simulations have been executed, and their results have been compared to the major findings discussed in literature. Since empirical studies about BN-AODV are not available, we have considered simulations of AODV and N-AODV, only. The first analysis compares some performance results measured by MOTION to those obtained with other simulators. The second one deepens into the relationships among some simulation parameters.

Each simulation is characterized by a specific number of hosts in the MANET: 10, 20, and 30 hosts, respectively. For each MANET population, three different cases for mobility have been considered, expressed by the values 25, 50, and 75% for the *Mobility level* parameter, respectively. Each one of the remaining parameters is set to the same value in the nine simulations: each simulation includes ten sessions, each of which lasting 50 ASM moves; the initial connectivity value is 50%; each host is an initiator of a CA with a probability of 10%; a CA is considered successful if the RREP is received by the initiator within 10 ASM moves.

During each simulation, the following metrics have been collected:

- M_1 the *rate of success* that is the ratio between successful CA's and the overall number of CA's;
- M_2 the *control overhead,* that is the total amount of control packets, produced for each CA (i.e., RREQs, RREPs, and REERs for both protocols);
- M_3 the RERR amount that is the total number of RERR packets produced as a result of a link breakage;
- M_4 the RREQ percentage that is the percentage of RREQs over the total of control packets.

Fig. 2 Rate of success

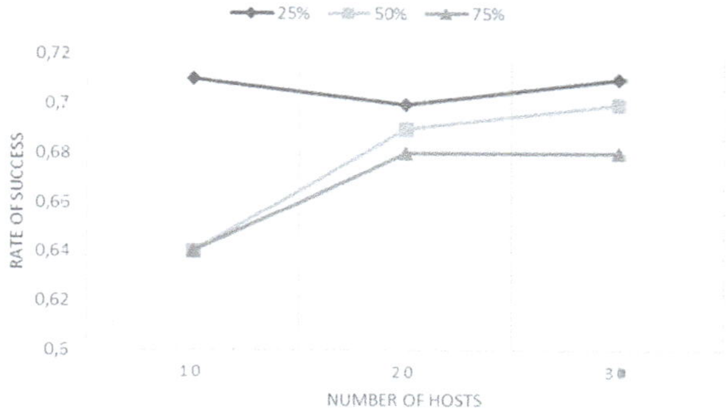

Fig. 3 Control overhead

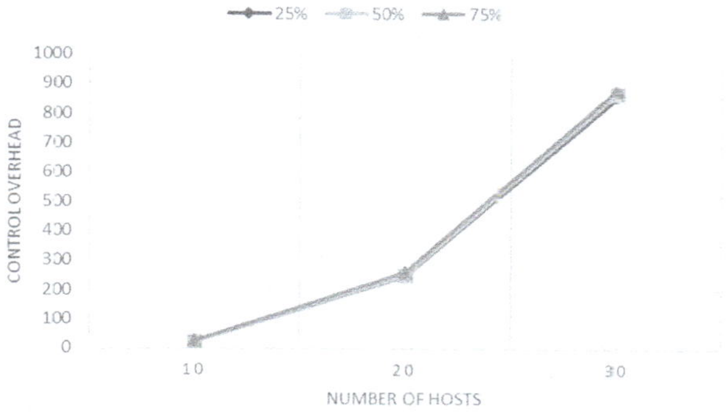

Fig. 4 Number of RERRs

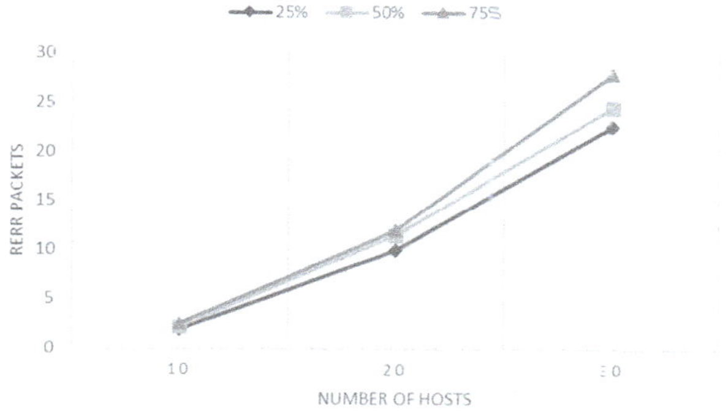

The results of the simulations are summarized in Figs. 2, 3, 4, and 5. Each data point represents an average of 10 simulation sessions with identical parameter setting, but with different random initialization of the connectivity matrix. The figures show the protocol's rate of success (Fig. 2), the control overhead (Fig. 3), the number of RERRs (Fig. 4), and the percentage of RREQs (Fig. 5) for each MANET population and for each mobility level.

We have performed the Kruskal–Wallis test to check the null hypothesis, i.e., to check if the median of control overhead and route errors are equal for the MANET populations under consideration. The null hypothesis has been tested either for groups with different mobility levels and fixed size of the MANET, or groups with

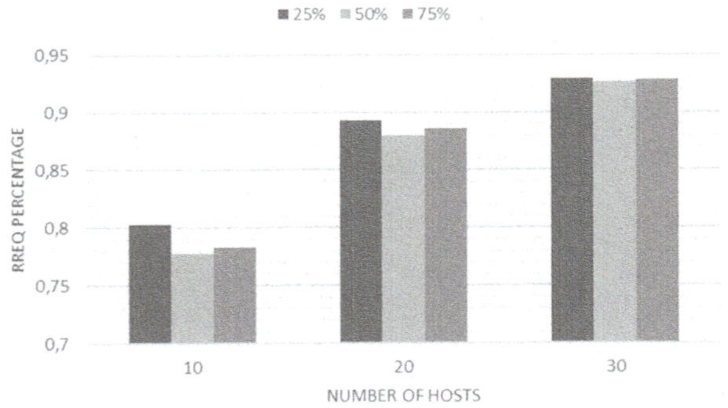

Fig. 5 Percentage of RREQs over the overall number of control packets

different MANET sizes and a fixed mobility level. We used this non-parametric test because we have more than two independent groups to be compared, and the normality assumption is violated. The same approach has not been adopted for the rate of success and for the RREQ percentage, because they are only expressed as percentages. We have found that there is no statistically significant difference (at the significance level 0.01) between the control overhead induced by networks with the same population (10, 20, or 30 hosts), varying the mobility level (25, 50, and 75%). Conversely, there is always a statistically significant difference (p-value < 0.0001) between the control overhead induced by networks with different populations and fixed mobility level. This confirms that the increasing of control overhead mainly depends on the increasing of the network size.

As for the spread of route errors along the network, we have found that there is no statistical difference (at the significance level 0.01) between networks with 10 or 20 hosts, with a variable mobility level; this difference is statistically significant in the case of 30 hosts. Since rejecting the null hypothesis does not indicate which of the groups differ, we refined the analysis performing a pairwise comparison by using the Mann-Whitney test. We observed that there is a statistical difference at the significance level 0.01 only between 25 and 75% of mobility level (p-value $= 0.0002$). Instead, there is always a statistically significant difference (p-value < 0.0001) between the route errors injected into the networks with different populations and fixed mobility level. These suggest that the increasing of RERRs largely depends on the network size too.

6 Conclusions and Future Work

In this paper, we have introduced MOTION, a Java environment for modeling MANETs and for simulating their behavior. We have used our tool to analyze the performances of three routing protocols, and to compare the results to those that can be found in the literature.

A sensible prosecution of this work could be the attempt to modeling a larger set of MANET behavior, in order to establish the usefulness of the tool, and to improve the user interface of our system, showing the dynamic evolution of the network, during the computations.

Acknowledgments We are very grateful to Gianluca Gennaro Bevilacqua and Marco Pinto for the help they provided in the development of MOTION.

References

1. D.P. Agrawal, Q.-A. Zeng, *Introduction to Wireless and Mobile Systems*, 4th edn. (Cengage Learning, Boston, 2016)
2. A.P. Pandian, J.I.-Z. Chen, Z.A. Baig, Sustainable mobile networks and its applications. Mobile Netw. Appl. **24**(2), 295–297 (2019)
3. A. Garcia-Santiago, J. Castaneda-Camacho, J.F. Guerrero-Castellanos, G. Mino-Aguilar, V.Y. Ponce-Hinestroza, Simulation platform for a VANET using the truetime toolbox: further result toward cyber-physical vehicle systems, in *IEEE 88th Vehicular Technology Conference (VTC-Fall)* (IEEE, Piscataway, 2018), pp. 1–5
4. S. Basagni, M. Mastrogiovanni, A. Panconesi, C. Petrioli, Localized protocols for ad hoc clustering and backbone formation: a performance comparison. IEEE Trans. Parallel Distrib. Syst. **17**(4), 292–306 (2006). https://doi.org/10.1109/TPDS.2006.52
5. J. Wu, F. Dai, Mobility-sensitive topology control in mobile ad hoc networks. IEEE Trans. Parallel Distrib. Syst. **17**(6), 522–535 (2006). https://doi.org/10.1109/TPDS.2006.73
6. D.A. Tran, H. Raghavendra, Congestion adaptive routing in mobile ad-hoc networks. IEEE Trans. Parallel Distrib. Syst. **17**(11), 1294–1305 (2006). https://doi.org/10.1109/TPDS.2006.151
7. R. Calinescu, C. Ghezzi, M. Kwiatkowska, R. Mirandola, Self-adaptive software needs quantitative verification at runtime. Commun. ACM **55**(9), 69–77 (2012)
8. E. Cavalcante, J. Quilbeuf, L.-M. Traonouez, F. Oquendo, T. Batista, A. Legay, Statistical model checking of dynamic software architectures, in *European Conference on Software Architecture* (Springer, Berlin, 2016), pp.185–200
9. S. Mesli-Kesraoui, D. Kesraoui, F. Oquendo, A. Bignon, A. Toguyeni, P. Berruet, Formal verification of software-intensive systems architectures described with piping and instrumentation diagrams, in *European Conference on Software Architecture* (Springer, Berlin, 2016), pp. 210–226
10. P. Arcaini, R. Mirandola, E. Riccobene, P. Scandurra, A DSL for MAPE patterns representation in self-adapting systems, in *European Conference on Software Architecture* (Springer, Berlin, 2018), pp. 3–19
11. P. Arcaini, A. Gargantini, E. Riccobene, P. Scandurra, A model-driven process for engineering a toolset for a formal method. Softw. Pract. Exp. **41**(2), 155–166 (2011)
12. J. Broch, D.A. Maltz, D.B. Johnson, Y.-C. Hu, J.G. Jetcheva, A performance comparison of multi-hop wireless ad hoc network routing protocols, in *Proceedings of the Fourth Annual ACM/IEEE International Conference on Mobile Computing and Networking MobiCom*, vol. 98 (1998), pp. 85–97
13. S.R. Das, R. Castaneda, J. Yan, R. Sengupta, Comparative performance evaluation of routing protocols for mobile, ad-hoc networks, in *Proceedings of 7th International Conference on Computer Communications and Networks (Cat. No. 98EX226)* (IEEE, Piscataway, 1998), pp. 153–161
14. S. Kurkowski, T. Camp, M. Colagrosso, Manet simulation studies: the incredibles. ACM SIGMOBILE Mobile Comput. Commun. Rev. **9**(4), 50–61 (2005)

15. A. Fehnker, R. van Glabbeek, P. Höfner, A. McIver, M. Portmann, W.L. Tan: a process algebra for wireless mesh networks, in *European Symposium on Programming* (Springer, Berlin, 2012), pp. 295–315
16. L. Bononi, G. D'Angelo, L. Donatiello, HLA-based adaptive distributed simulation of wireless mobile systems, in *Proceedings of the Seventeenth Workshop on Parallel and Distributed Simulation* (IEEE Computer Society, Washington, 2003), p. 40
17. A. Singh, C. Ramakrishnan, S.A. Smolka, A process calculus for mobile ad-hoc networks. Sci. Comput. Program. **75**(6), 440–469 (2010)
18. A. Bianchi, S. Pizzutilo, Studying MANET through a Petri net-based model, in *2010 2nd International Conference on Evolving Internet* (IEEE, Piscataway, 2010), pp. 220–225
19. D. Cavin, Y. Sasson, A. Schiper, On the accuracy of manet simulators, in *Proceedings of the Second ACM International Workshop on Principles of Mobile Computing* (ACM, New York, 2002), pp. 38–43
20. M. Merro, An observational theory for mobile ad hoc networks. Inf. Comput. **207**(2), 194–208 (2009)
21. G. Delzanno, A. Sangnier, G. Zavattaro, Parameterized verification of ad hoc networks, in *International Conference on Concurrency Theory* (Springer, Berlin, 2010), pp. 313–327
22. C. Xiong, T. Murata, J. Leigh, An approach for verifying routing protocols in mobile ad hoc networks using Petri nets, in *Proceedings of the IEEE 6th Circuits and Systems Symposium on Emerging Technologies: Frontiers of Mobile and Wireless Communication*, vol. 2 (IEEE, Piscataway, 2004), pp. 537–540
23. F. Erbas, K. Kyamakya, K. Jobmann, Modelling and performance analysis of a novel position-based reliable unicast and multicast routing method using coloured Petri nets, in *2003 IEEE 58th Vehicular Technology Conference. VTC 2003-Fall*, vol. 5 (IEEE, Piscataway, 2003), pp. 3099–3104
24. M.H. Jahanian, F. Amin, A.H. Jahangir, Analysis of Tesla protocol in vehicular ad hoc networks using timed colored Petri nets, in *2015 6th International Conference on Information and Communication Systems (ICICS)* (IEEE, Piscataway, 2015), pp. 222–227
25. K. Jensen, L.M. Kristensen, L. Wells, Coloured Petri nets and CPN tools for modelling and validation of concurrent systems. Int. J. Softw. Tools Technol. Transf. **9**(3–4), 213–254 (2007)
26. C.E. Perkins, E.M. Belding-Royer, S.R. Das, Ad hoc on-demand distance vector (AODV) routing. RFC 3561 (2003), pp. 1–37. https://doi.org/10.17487/RFC3561
27. C. Kim, E. Talipov, B. Ahn, A reverse AODV routing protocol in ad hoc mobile networks, in *International Conference on Embedded and Ubiquitous Computing* (Springer, Berlin, 2006), pp. 522–531
28. N. Kaur, R. Singhai, Analysis of traffic impact on proposed congestion control scheme in AODV. Wireless Personal Communications (2019), pp. 1–24
29. N. Das, S.K. Bisoy, S. Tanty, Performance analysis of TCP variants using routing protocols of manet in grid topology, in *Cognitive Informatics and Soft Computing* (Springer, Berlin, 2019), pp. 239–245
30. M.G. Zapata, Secure ad hoc on-demand distance vector routing. ACM SIGMOBILE Mobile Comput. Commun. Rev. **6**(3), 106–107 (2002)
31. X. Li, M.R. Lyu, J. Liu, A trust model based routing protocol for secure ad hoc networks, in *2004 IEEE Aerospace Conference Proceedings*, vol. 2 (IEEE, Piscataway, 2004), pp. 1286–1295
32. E. Börger, R. Stärk, *Abstract State Machines: A Method for High-Level System Design and Analysis* (Springer, Berlin, 2003)
33. R. Farahbod, V. Gervasi, U. Glässer, Coreasm: an extensible ASM execution engine. Fundam. Inform. **77**(1–2), 71–103 (2007). http://content.iospress.com/articles/fundamenta-informaticae/fi77-1-2-04
34. A. Gargantini, E. Riccobene, P. Scandurra, Model-driven language engineering: the ASMETA case study, in *Proceedings of the Third International Conference on Software Engineering Advances, ICSEA 2008, October 26–31 Sliema, Malta* (2008), pp. 373–378. https://doi.org/10.1109/ICSEA.2008.62

35. A. Bianchi, S. Pizzutilo, G. Vessio, Preliminary description of nack-based ad-hoc on-demand distance vector routing protocol for MANETS, in *2014 9th International Conference on Software Engineering and Applications (ICSOFT-EA)* (IEEE, Piscataway 2014), pp. 500–505
36. A. Bianchi, S. Pizzutilo, G. Vessio, Intercepting blackhole attacks in manets: an ASM-based model, in *International Conference on Software Engineering and Formal Methods* (Springer, Berlin, 2017), pp. 137–125
37. A. Bianchi, S. Pizzutilo, G. Vessio, Coreasm-based evaluation of the NAODV protocol for MANETS. J. Mobile Multimedia **12**(1–2), 31–51 (2016). http://www.rintonpress.com/xjmm12/jmm-12-12/031-051.pdf
38. F.-H. Tseng, L.-D. Chou, H.-C. Chao, A survey of black hole attacks in wireless mobile ad hoc networks. Human-centric Comput. Inf. Sci. **1**, 4 (2011)
39. U. Glässer, Y. Gurevich, M. Veanes, Abstract communication model for distributed systems. IEEE Trans. Softw. Eng. **30**(7), 458–472 (2004). https://doi.org/10.1109/TSE.2004.25
40. A. Gargantini, E. Riccobene, P. Scandurra, A metamodel-based language and a simulation engine for abstract state machines. J. UCS **14**(12), 1949–1983 (2008). https://doi.org/10.3217/jucs-014-12-1949
41. A. Boukerche, L. Bononi, Simulation and modelling of wireless, mobile and ad hoc networks, in *Mobile Ad Hoc Networking*, ed. by S. Basagni, M. Conti, S. Giordano, I. Stojmenovic (IEEE Press, New York, 2004), pp. 373–410
42. E. Börger, A. Raschke, *Modeling Companion for Software Practitioners* (Springer, Berlin, 2018). https://doi.org/10.1007/978-3-662-56641-1

A New Real-Time Geolocation Tracking Tool Enhanced with Signal Filtering

Erkan Meral, Mehmet Serdar Guzel, Mehrube Mehrubeoglu, and Omer Sevinc

1 Introduction

Estimation of the exact object positions located on Earth is a crucial problem, and the GPS is usually used to solve positioning operations. A unique system based on GPS technology and Arduino board was previously introduced [1]. The flowchart of the previously proposed system is illustrated in Fig. 1. In this study, an improved system is presented that processes data obtained from the GPS sensor integrated into an electronic circuit which estimates real-time locations and displays them on a map via a website. However, the accuracy of the position data obtained from GPS includes error. This error can easily increase if the GPS signals are obtained in a noisy environment. Noisy environments may cause distortions, such as those occurring in signal lines, and must be eliminated to increase the overall reliability of the data. One of the most efficient ways of noise removal is to apply proper signal filtering techniques. Averaging and Kalman [2] filters are two popular and suitable filters for signal enhancement. The goal of implementing filters is to prevent spikes or large changes in the signal from suddenly appearing with respect to the previous

E. Meral
Gazi University, Ankara, Turkey
e-mail: emeral@gazi.edu.tr

M. S. Guzel
Ankara University, Computer Engineering, Ankara, Turkey

M. Mehrubeoglu
Texas A&M University-Corpus Christi, Corpus Christi, TX, USA
e-mail: ruby.mehrubeoglu@tamucc.edu

O. Sevinc (✉)
Ondokuz Mayis University-Vezirkopru, Samsun, Turkey
e-mail: osevinc@omu.edu.tr

K. Daimi et al. (eds.), *Advances in Security, Networks, and Internet of Things*, Transactions on Computational Science and Computational Intelligence,
https://doi.org/10.1007/978-3-030-71017-0_35

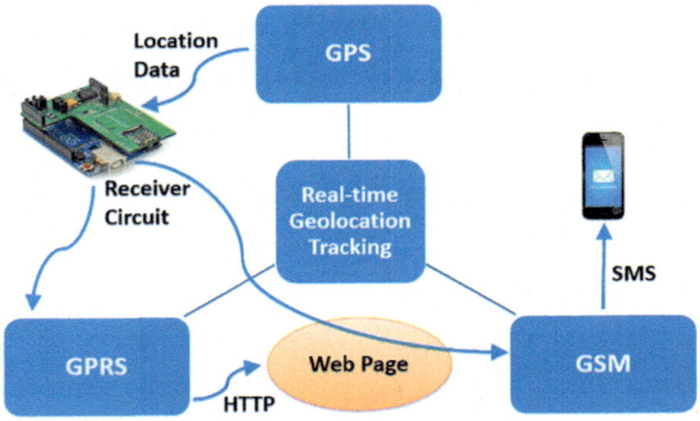

Fig. 1 Architecture of the used real-time geolocation tracking and geofencing system [1]

data values obtained from the system, thus preventing such changes from affecting the system at a large scale.

In this study, a simplified version of an adaptive Kalman filter is applied to the data to prevent sudden changes in the positions obtained by the proposed device [3]. The main reason that lies behind this post-processing operation, as mentioned, is to prevent these undesirable fluctuations in the system to allow more accurate and practical resultant position data.

Kalman filters have been used for GPS signals in previous studies [4]. Zheng et al. applied extended Kalman filter (EKF) to track high dynamic GPS signals [5]. The authors' results showed that EKF improved dynamic acceleration as well as tracking accuracy in simulations. Zhou et al. incorporated Kalman filter and backpropagation neural networks (BPNN) to GPS/INS data to improve UAV navigation. Kalman filtered data was fed into BPNN to compensate for missing GPS/INS integrated data to allow UAV position information from Kalman filter estimations [6]. A similar study by Kanghui and Chaoyang incorporated fuzzy strong tracking EKF to integrated GPS data to improve the accuracy of position in UAV navigation in simulations [7]. Kalman filters have been used in other applications for signal smoothing [8, 9].

In addition to the Kalman filter, the averaging filter has been tested in this study. The averaging filter is normally used to eliminate sharp transitions in single and multidimensional data, such as images. As such, the averaging filter is mainly designed based on the arithmetic mean of the data and, as a general filter, can be employed for different problems [10]. In this study, an averaging filter is applied to data from multiple satellites to improve GPS positions obtained from the tool (see Fig. 2).

Overall, this study incorporates two different filtering approaches into the developed tool separately to enhance the position data and increase the overall accuracy of the device. The results of the two filtering operations are presented and compared in this paper.

Sections 2 and 3 summarize the details of the proposed Kalman and average filtering techniques, respectively. Section 4 covers experimental results and analysis. Finally, conclusions are presented in Sect. 5.

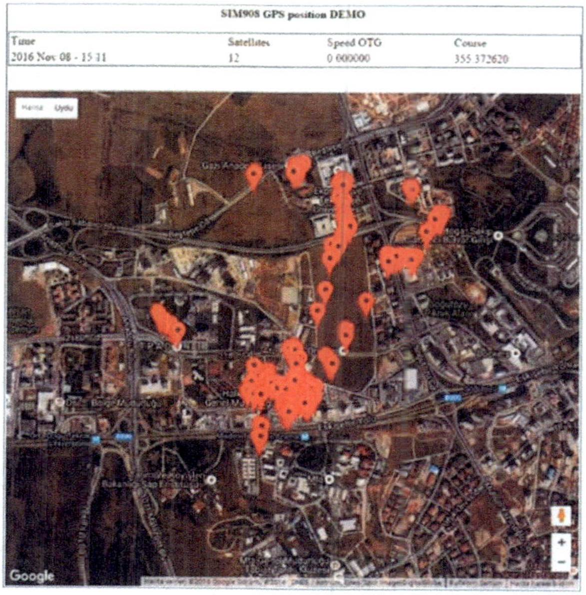

Fig. 2 GPS data obtained from Google Maps

2 Kalman Filter for GPS Data Enhancement

The Kalman filter is a reliable and efficient method for noise removal in signal and image processing applications. The position data obtained by the proposed device consists of the latitude and longitude data. Accordingly, errors occurring in the location data are the result of errors occurring in the values of these data. Hence, the Kalman filter is applied separately on the latitude and longitude values. The new latitude and longitude values, enhanced by the Kalman filtering process [2], are reassembled to determine the new location. Kalman filtering produces far more accurate results than the previous and original measurements. The Kalman filter offers complex equations for different problems and systems. When the filter is applied, the appropriate equations for each problem is considered to reduce the overall complexity of the system. Essentially, the matrices that are not needed in equations are omitted. In this study, Eq. (1) is obtained by subtracting the unused condition matrices from the main formulas when the Kalman filter is applied to the signal processing problem:

$$X_k = K_k Z_k + (1 - K_k)\, X_{k-1}. \tag{1}$$

In the equation above, k is used as a subindex denoting the operating states of the system. The purpose of the formula is to compute each of the predicted values, X_k, of the signal. Z_k represents each of the original values of the signal obtained from the receiver continuously which encompasses a certain amount of error. Similarly,

X_{k-1} denotes the estimated value of the signal of the previous state, whereas K_k is called the Kalman gain. Kalman gain is the only unknown value in the equation; for each case, K_k is recalculated with Eq. (2) based on the values of the previous error covariance, P_k, and the standard deviation of the measurement, R:

$$K_k = P_{k-1}/\left(P_{k-1} + R\right). \tag{2}$$

In cases where the Kalman gain is taken as a constant value of 0.5, the equation will behave as if it is an average filter [9]. By recalculating the Kalman gain in each step, the optimum average value can be calculated, taking advantage of the capabilities of the Kalman filter [11]. In order to calculate the previously mentioned values preferred in this algorithm, some initial values and parameters need to be determined. The standard deviation of measurement, R, which is to be used for recalculating the Kalman gain at each step, is set to 1 in practice. X_0, the first estimated value of X_k at time $k = 0$, is set as the first position data received from the device. Kalman filter is then applied to latitude and longitude values separately, each denoted by X_k, and first data point received from the device for each represented as X_0. The first value to be used for the error covariance, P_k, at time $k = 0$, is set to 4 ($P_0 = 4$). This value can be set to any reasonable nonzero value. Setting P_0 to zero means that there is no noise in the environment. The P_k value for each case is then recalculated using the Kalman gain, K_k, and the previous error covariance value, P_{k-1}, using Eq. (3):

$$P_k = (1 - K_k)\, P_{k-1}, \tag{3}$$

where P_k is the error covariance matrix at time k. Essentially, the Kalman filter is executed with the algorithm in Table 1.

The steps of applying the Kalman filter to an example set of latitude values obtained from the receiver device are illustrated in Table 2, at times $k = 0$ and $k = 1$.

Table 1 Algorithm representing GPS enhancement method based on 1D Kalman filter

```
While y > 0
Zk ← Receiver_Values[k]
Xk′ ← Xk
Pk′ ← Pk
Kk ← Pk′ / (Pk′ + R)
Xk ← Kk * Zk + (1 – Kk) * Xk
Pk ← (1 – Kk) * Pk′
Kalman_Values[k] ← Xk
end (while)
```

y, number of receiver data points to be filtered;
Pk′, P_{k-1}; Xk′, X_{k-1}; Kk, K_k; Xk, X_k; Pk, P_k; Zk, Z_k

Table 2 Example scenario with Kalman filter applied to latitude values

	Initial value	Final value
y	30	0
R	1	1
K	0	1
Z_k	39.953250	39.953200
X_k	39.953250	39.953250
P_k	4	0.8
X_{k-1}	39.953250	39.953250
P_{k-1}	4	0.8
K_k	$K_k = P_{k-1}/(P_{k-1} + R)$	
	$K_0 = 4/(4 + 1) = 0.8$	$K_1 = 0.8/(0.8 + 1) = 0.44$
X_k (new)	$X_k = K_k Z_k + (1 - K_k) X_{k-1}$	
	$X_0 = 0.8*39.953250 +$ $(1-0.8)*39.953250$ $= 39.953250$	$X_1 = 0.44*39.953200 +$ $(1-0.44)*39.953250$ $= 39.953228$
P_k (new)	$P_k = (1 - K_k) P_{k-1}$	
	$P_0 = (1-0.8)*4$ $= 0.8$	$P_1 = (1-0.44)*0.8$ $= 0.448$

3 Averaging Filter for GPS Data Enhancement

The average filter is usually applied on images to eliminate sharp transitions from the picture. The value of each pixel that creates the picture is changed with the arithmetic average of its pixel value with its neighbors. Thanks to this filter, sharp passes on the picture are transformed into soft passes. The average filter is applied in many areas where noisy signals exist to reduce the effects of signal noise. In this study, the average filter is applied to the latitude and longitude data, taken from the GPS receiver, separately. Thirty latitude and longitude data representing 30 location data received from the device are stored in the data arrays. After the averaging filter is applied, the values in the array are moved to another array. The first value of the array handled from the device is the same with the array of average filter. The second value of the averaging filter array is the average of the first and second values of the array of the device. The third value is calculated by calculating the average of the first through third values of the array and so on. At the end, a cumulative average value is calculated for 30 values of latitude and longitude. Equation (4) summarizes the averaging process:

$$X_N = \frac{1}{N} \sum_{k=0}^{N} X_k, \tag{4}$$

where N represents the average number of data points which corresponds to the row number (Data ID + 1) in the data array, X_N represents the cumulative average value, and k and X_k are as defined before.

4 Results and Analysis

Within the scope of this study, several experiments have been carried out to calculate the position error for the data from the receiver. It has been determined that the accuracy of the position data given by the receiver during operation varies based on weather conditions, such as clear and cloudy weather. For this reason, experiments were carried out separately for clear and cloudy weather conditions. For a comprehensive outdoor experiment, the position data were taken 30 times in a clear day. Table 3 illustrates the tested scenario with position data received by the tool and position error in meters with respect to the reference position obtained from Google's location API shown on top of Tables 3 and 4. Table 4 represents a similar experimental scenario on a cloudy day under overcast weather conditions.

The Record ID, latitude, and longitude columns shown in Tables 3 and 4 reflect the data obtained from the receiver tool. The satellite column indicates how many GPS satellites are communicated with when the receiver received the data. Error columns indicate the distance error of the position data obtained by the receiver, Kalman filtered data, and average filtered data with respect to the reference location. To determine the location, the receiver must exchange signals with at least three GPS satellites. However, it is noted that the number of satellites in the tables is three or greater. Considering that the sequence is sorted according to increasing time, it is observed that the number of satellites communicating with the receiver increases as time progresses.

Figure 3 illustrates 30 data locations obtained from the receiver tool and shown on Google Maps in a clear weather day. The location illustrated with the green color is the actual location including the receiver location and used as a reference, whereas the locations shown in red are the positions taken during the experiments. It is apparent from Table 3 and Fig. 3 that as the number of satellites incorporated into the system increases, the error gradually decreases. Table 4 shows similar results from an experiment in a cloudy day. Figure 4 shows the variation of position error calculated using the original data obtained from the receiver on the clear day, as well as position error using Kalman and averaging filtered data. Figure 5 represents the error graphs for the receiver data (original, Kalman filtered, average filtered) from the cloudy and overcast day.

As can be seen in Fig. 4, the error in the last value recorded on the receiver in clear weather record measures 9.40 meters for the original data. When the values are passed through the Kalman filter, the error in the same register is measured as 3.64 meters. When evaluated based on the last recorded value, it can be seen that the Kalman filter provides an improvement by 5.76 meters. The averaging filter shows an error value of 4.18 meters, with a 5.22-meter improvement, when the last record (30th position value) is considered.

When comparing earlier values, it is clear that both the Kalman and averaging filters reduce the original receiver data error significantly well before the 30th data point, just after the third data record. The two filters demonstrate notable improvement in position error compared to the original receiver position data, and

Table 3 GPS position values from the receiver in clear weather

Data ID	Latitude	Longitude	Satellites	Error: original data (meters)	Error: after Kalman filter (meters)	Error: after average filter (meters)
0	39.953250	32.796365	3	80.22	80.22	80.22
1	39.953200	32.796365	3	74.90	77.90	77.59
2	39.951920	32.796900	3	74.50	30.90	26.94
3	39.951901	32.796570	4	74.20	11.30	9.29
4	39.951961	32.796604	4	67.30	12.76	15.61
5	39,952,102	32.796604	5	54.70	17.86	21.01
6	39.952711	32.797189	5	48.02	11.34	14.41
7	39.952711	32.797189	5	48.02	9.15	11.85
8	39.952711	32.797189	6	48.02	10.40	12.50
9	39.952201	32.796580	6	40.90	11.46	13.65
10	39.952231	32.796580	7	37.70	12.86	14.99
11	39.952287	32.796580	7	31.60	13.83	15.82
12	39.952303	32.796589	7	29.70	14.71	16.59
13	39.952771	32.796851	8	28.20	12.60	14.24
14	39.952753	32.796824	8	25.20	11.00	12.42
15	39.952732	32.796801	9	22.20	9.79	11.11
16	39.952703	32.796797	9	22.20	8.97	10.21
17	39.952741	32.796598	10	20.30	7.56	8.62
18	39.952726	32.796588	10	18.90	6.40	7.30
19	39.952709	32.796561	11	18.10	5.28	6.14
20	39.952678	32.796525	11	17.00	4.23	5.06
21	39.952624	32.796534	12	12.50	3.47	4.24
22	39.952460	32.796701	12	12.17	3.77	4.45
23	39.952460	32.796701	12	12.17	3.97	4.75
24	39.952460	32.796701	12	12.17	4.25	4.98
25	39.952460	32.796701	13	12.17	4.48	5.29
26	39.952670	32.796678	13	11.80	4.10	4.75
27	39.952501	32.796761	13	11.20	4.31	5.02
28	39.952651	32.796653	14	9.62	3.90	4.52
29	39.952648	32.796641	14	9.40	3.64	4.18

Reference position: latitude, 39.9525646; longitude, 32.7966589

for initial values, averaging filter performs better than the Kalman filter in these experiments; however, at this initial stage, the error is still too large to be tolerable.

Similar results are observed for the position data collected on the cloudy and overcast weather. The receiver data on a cloudy day resulted in significant position errors over the 30 location records in the register. The Kalman filter reduced the position error from 19.50 meters down to 14.09 meters (5.41-meter improvement), whereas the averaging filter reduced the error down to 14.63 meters (4.87-meter improvement). Similar to the results from the clear day experiments, the error results for cloudy and overcast day experiments were significantly reduced by both Kalman

Table 4 GPS position values from the receiver in cloudy and overcast weather.

Data ID	Latitude	Longitude	Satellites	Error: original data (meters)	Error: after Kalman filter (meters)	Error: after average filter (meters)
0	39.953307	32.797505	3	109.61	109.61	109.61
1	39.953260	32.797503	3	100.70	105.62	105.14
2	39.952842	32.797071	4	87.70	46.74	41.60
3	39.952621	32.796843	4	87.70	16.89	12.29
4	39.952712	32.796701	5	76.30	16.77	13.17
5	39.952768	32.796612	5	70.30	22.96	21.30
6	39.952803	32.796549	5	66.30	28.11	27.41
7	39.952704	32.796546	5	65.40	18.24	18.10
8	39.952628	32.796531	6	66.20	12.98	14.10
9	39.952569	32.796521	6	62.40	11.76	13.88
10	39.952601	32.796498	6	52.60	14.30	15.84
11	39.952631	32.796477	7	57.70	17.17	18.25
12	39.952607	32.796507	7	34.30	13.17	14.99
13	39.952572	32.796508	7	50.50	12.88	14.33
14	39.952544	32.796511	8	49.10	12.81	14.54
15	39.952519	32.796525	9	48.10	12.48	14.34
16	39.952539	32.796511	9	46.80	12.92	14.47
17	39.952557	32.796598	10	46.80	13.74	15.04
18	39.952573	32.796487	11	46.80	14.68	15.85
19	39.952580	32.796478	11	33.90	15.51	16.54
20	39.952586	32.796489	11	15.70	14.67	15.58
21	39.952589	32.796481	12	30.90	15.40	16.29
22	39.952592	32.796474	12	30.90	16.05	16.84
23	39.952594	32.796467	13	30.90	16.68	17.47
24	39.952598	32.796482	13	21.80	15.52	16.20
25	39.952602	32.796489	13	21.80	15.06	15.70
26	39.952607	32.796495	13	20.70	14.74	15.28
27	39.952609	32.796494	14	19.90	14.89	15.43
28	39.952610	32.796499	14	19.50	14.53	14.97
29	39.952611	32.796505	14	19.50	14.09	14.63

Reference position: latitude, 39.9525646, longitude, 32.7966589

and averaging filters after the third data record. Kalman filter performed better in this experiment compared to the averaging filter in reducing position error in the 30th data record.

Table 5 summarizes the position error among the original tool and Kalman and averaging filtered data from clear and overcast weather conditions for the 30th and fourth position records. Both filtering operations demonstrate significant improvement in position error, particularly in clear weather. In the case of Kalman filter applied to the GPS tool's original data, the error is reduced by over 61% in

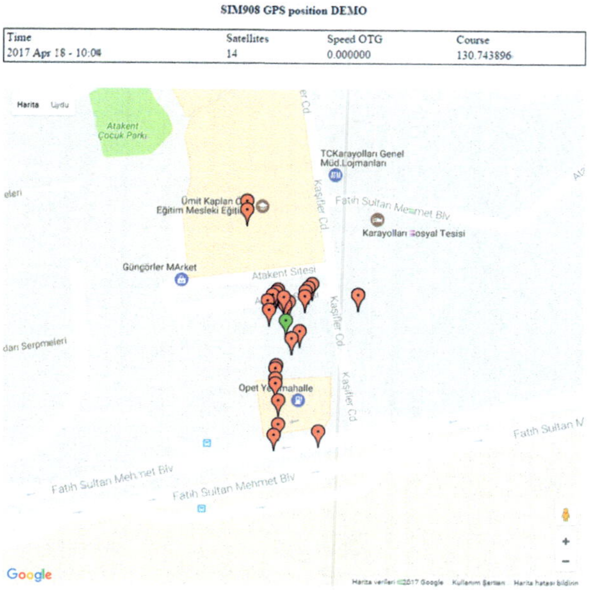

Fig. 3 30 data locations obtained from the receiver tool and shown on Google Maps in clear weather. The green location pointer is used as a reference to calculate position error

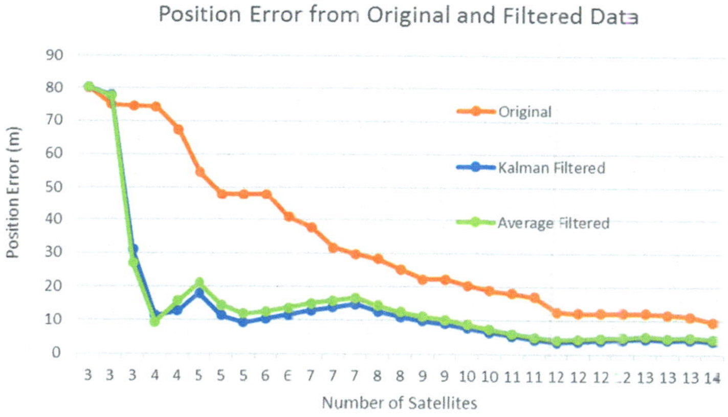

Fig. 4 Error graph of the original receiver values and the Kalman and averaging filter results in clear weather

clear weather and over 27% in cloudy overcast weather by 30th data record. The averaging filter improved the results for the same data by over 55% in clear weather and 24% in cloudy weather. In the case of the fourth data record, Kalman filter reduced position error by over 84% in clear weather and 80% in cloudy weather. Averaging filter improved results by over 87% and 85%. It is important to note that in cloudy weather, the original signal data included much higher error to start with as expected due to signal interference. Even then, the filtering operations show a notable improvement in error compared to the original unprocessed data from the

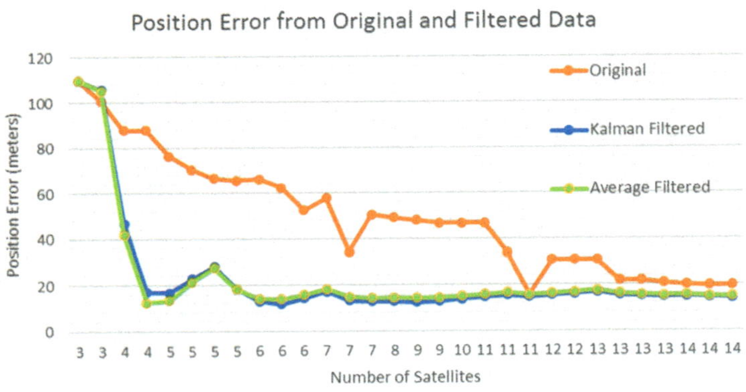

Fig. 5 Error graph of the original receiver values and the Kalman and averaging filter results in cloudy weather

Table 5 Error comparison between the tool and filters applied in different environmental conditions

Error	Original receiver (meters)	Kalman filter results (meters)	Kalman filter improvement rate (%)	Averaging filter results (meters)	Averaging filter improvement rate (%)
30th data record (record ID 29)					
Position error in clear weather	9.40	3.64	61.28	4.18	55.53
Position error in cloudy weather	19.50	14.09	27.74	14.63	24.97
4th data record (record ID 3)					
Position error in clear weather	74.20	11.30	84.77	9.29	87.48
Position error in cloudy weather	87.70	16.89	80.74	12.29	85.99

tool. Another important point is that although the error improvement rates for the fourth data record is significantly higher than error improvement rates for the 30th data record, the position error in meters may not yet be at tolerable levels, requiring data from more satellites for increased accuracy further down.

5 Conclusions

This paper introduces efficient post-processing techniques to enhance GPS data obtained from a variety of satellites. Position data is obtained from a previously designed device [1]. The device is able to connect different satellites simultaneously and obtain location data using latitude and longitude values. However, due to signal noise, significant position errors may occur in recorded location data, deviating from the exact location. In order to reduce this undesirable position error, the corresponding noise must be reduced into a tolerable level. Accordingly, Kalman and averaging filters have been implemented and results compared to enhance the performance of the tool's overall position estimation. The results show that both filtering approaches improved the overall performance of the system successfully. However, Kalman filter generally showed higher performance over the averaging filter in both clear and cloudy weather when the last record is considered. Both filters demonstrated lower performance in data from cloudy weather in absolute position error values, as expected, since signal interference resulted in higher positional error in the original receiver data. Overall, both filtering operations improved the position accuracy in both weather conditions improving the performance of the tool.

Acknowledgments This study is adapted from the master thesis of the first author [12]. An early preprint version of this paper can be found in the following link: https://arxiv.org/ftp/arxiv/papers/1803/1803.08325.pdf

References

1. E. Meral, M. Güzel, Real-time geolocation tracking and geofencing using gprs+gps technologies with sim908 shield over Arduino. *Communications Faculty of Sciences University of Ankara Series*, A2-A3 58 (2), 14–27 (2016)
2. Y. Geng, J. Wang, Adaptive estimation of multiple fading factors in Kalman filter for navigation applications. GPS Solut. **12**(4), 273–279 (2008)
3. H. Zhou, H. Huang, H. Zhao, X. Zhao, X. Yin, Adaptive unscented Kalman filter for target tracking in the presence of nonlinear systems involving model mismatches. Remote Sens. MDPI **9**(657), 1–20 (2017)
4. Z. Özdemir, M.S. Güzel, Comparison of Kalman filters and LSTM networks for error reduction problem. *2019 3rd Int'l Symp. Multidisciplinary Studies and Innovation Technologies (2019 ISMSIT)* (2019)
5. C. Zheng, W. Li, Application of extended Kalman filter for tracking high dynamic GPS signal. *Proc. IEEE Int. Conf. Signal and Image Processing*, pp. 503–507 (2016)
6. Y. Zhou, J. Wan, Z. Li, Z. Song, GPS/INS integrated navigation with BP neural network and Kalman filter. *IEEE Proc. 2017 IEEE Int.l Conf. on Robotics and Biomimetics*, Dec. 5–8, 2017
7. H. Kanghui, D. Chaoyan, A fuzzy strong tracking extended Kalman filter for UAV navigation considering interruption of GPS signal. *Proc. IEEE Int.l Conf. Power, Intelligent Computing and Systems (ICPICS)*, pp. 254–259 (2019)
8. M.A. Reche, S. Kanarachos, M.E. Fitzpatrick, Optimized tire force estimation using extended Kalman filter and fruit fly optimization. *43rd Annual Conf. IEEE Industrial Electronics Society (IECON 2017)*, pp. 4074–4079 (2017). https://doi.org/10.1109/IECON.2017.8216698
9. V.A. Filimonov, V.V. Shavrin, V.I. Tislenkoz, A.P. Kravets, V.Y. Lebedev, V.N. Shkolniy, Coordinate and time-frequency support of a spacecraft flight by means of autonomic navigation using sigma-point Kalman filter algorithm. J. Siberian Univ. Math. Phys. **8**(4), 385–393 (2015)
10. J. Alvarez-Ramirez, E. Rodriguez, C.J. Echeverría, Detrending fluctuation analysis based on moving average filtering, in *Physica A: Statistical Mechanics and its Applications*, (Division de Ciencias Basicas e Ingenieria, Mexico, 2005), pp. 199–219
11. J. Gomez-Gil, R.R. Gonzalez, S.A. Garcia, F.J. Gil, A Kalman filter implementation for precision improvement in low-cost GPS positioning of tractors. Sensors, 15307–15323 (2013)
12. E. Meral, Real Time Geolocation Tool, M.S. Thesis, Ankara University, Turkey (2017)

A Self-adaptivity Indoor Ranging Algorithm Based on Channel State Information with Weight Gray Prediction Model

Jingjing Wang and Joon Goo Park

1 Introduction

There are two types of wireless positioning systems according to the positioning environment: indoor wireless positioning systems and outdoor wireless positioning systems. The Global Positioning System (GPS) [1, 2] is a wireless positioning technology. For indoor positioning, the navigation satellite signal is difficult to cover, and the wireless base station signal multipath and attenuation characteristics are complicated, resulting in low positioning accuracy and large jitter. Nowadays, with the expansion of cities and the increase in people's needs, indoor wireless positioning technology has become the focus of people's attention. However, it is much more difficult to perform accurate target positioning indoors than outdoors because the indoor environment is complex, such as the random movement of object locations, multipath scattering, electromagnetic interference, and so on.

Due to the complexity of the indoor environment, the signal strength indicators (RSSI) [3] will be affected by factors such as wall reflections and interference from obstacles, which will cause the RSSI value received by the node under test to fluctuate impact greatly. To improve the accuracy of indoor positioning, it is necessary to propose a measurement index that is stable over time and is less affected by multipath effects. Currently, in the widely used Orthogonal Frequency Division Multiplexing (OFDM) [4] system, data is modulated into multiple subcarriers of different frequencies to propagate simultaneously. For each subcarrier, we will get an estimated channel value, which is Channel State Information (CSI) [5]. Reference [6] compared CSI and RSSI.

J. Wang · J. G. Park (✉)
Kyungpook National University, Electronics Engineering, Daegu, South Korea
e-mail: wjj0219@knu.ac.kr; jgpark@knu.ac.kr

K. Daimi et al. (eds.), *Advances in Security, Networks, and Internet of Things*, Transactions on Computational Science and Computational Intelligence,
https://doi.org/10.1007/978-3-030-71017-0_36

The latest research [7] leverages the CSI to build a propagation model and a fingerprinting system at the receiver. FIFS [8] uses the weighted average CSI amplitude values on multiple antennas to improve the performance of indoor fingerprint recognition methods. Due to the influence of multiple paths, the weighted average CSI amplitude value obtained does not reflect the location fingerprint information well. DeepFi [9] learns a large amount of CSI data from three antennas for indoor positioning based on deep networks. The information obtained on the third antenna is defective, which will cause the trained model to not accurately reflect the fingerprint information. At the same time, the learning process requires a lot of CSI data. The gray theory holds that the behavior of the system is still hazy, and the data is complex, but after all, it is orderly and has overall functions. The generation of gray numbers is to find the rules from the clutter. Luo et al. [10] uses the gray prediction method in a wireless sensor network and employs a wireless LAN medium (Zigbee/802.15.4). The gray prediction is used to predict the tendency of RSSI. However, this time-varying and vulnerable RSSI value create undesirable localization errors.

According to the characteristics of signal attenuation in an indoor environment, this paper uses the method of least square curve fitting to obtain the nonlinear relationship between CSI and distance. By measuring the CSI value of the position information of different reference points, the current distance information is estimated by using the gray prediction model combined with the signal attenuation model.

In summary, the contributions of this article are as follows:

- The proposed algorithm can obtain accurate ranging information in indoor environment areas where Wi-Fi signals cannot be obtained. It can further obtain high precision position information.
- The gray prediction model is used to predict the CSI value, which greatly reduces the amount of repeated sampling and calculation of the CSI value. Because the time for building an indoor ranging model is reduced, it meets the real-time requirements of indoor positioning algorithms.
- The higher accuracy and robustness of the proposed localization method under a representative indoor environment are shown by experiments.

The rest of this paper is organized as follows. In Sect. 2, we introduce preliminaries, which is divided into the CSI ranging model and a gray prediction GM (1,1) model, respectively. We next explain the proposed indoor positioning model based on CSI measurements with gray prediction GM (1,1) model in Sect. 3. The implementation of the enhanced indoor positioning model and experimental evaluations are presented in Sect. 4. Finally, conclusions and future research are presented in Sect. 5.

2 Preliminaries

2.1 The Indoor Ranging Model of Channel State Information

CSI contains rich, fine-grained channel state information. In 802.11a/g/n network, the OFDM modulation technology is used for data transmission. In the OFDM system, the received signal after multipath channel can be represented by formula (1):

$$Y = HX + N \tag{1}$$

where Y and X represent the received signal vector and the transmitted signal vector respectively, H and N represent the channel matrix and the additive Gaussian white noise respectively. A group of CSI can be obtained from each received packet by using intel5300, a wireless network card compatible with ieee802.11a/g/n.

$$H = [H_1, H_2, H_3, \ldots, H_k, \ldots, H_n] \tag{2}$$

where H_k describes the CSI of the k-th subcarrier. By modifying the firmware, the common Wi-Fi device can obtain CFR samples on 30 OFDM subcarriers, so $n = 30$. Each group of CSI represents the amplitude and phase of an OFDM subcarrier:

$$H_k = \|H_k\| \, e^{j\angle H_k} \tag{3}$$

where $\|H_k\|$ and $\angle H_k$ represent the amplitude and phase of k-th subcarriers respectively. Intel 5300 wireless network card works in the high throughput mode (HT mode) of 20 MHz.

In MIMO systems with p transmit antennas and q receive antennas, CSI is a matrix of pxq dimension, which can be expressed as follows:

$$H(f_x) = \begin{bmatrix} h_{11} & h_{22} & \cdots & h_{1q} \\ h_{21} & h_{22} & \cdots & h_{2q} \\ \vdots & \vdots & \ddots & \vdots \\ h_{p1} & h_{p2} & \cdots & h_{pq} \end{bmatrix} \tag{4}$$

where H_{pq} exists in the form of a complex number, which represents the amplitude and phase of the subcarrier of the antenna stream. Wu et al. [7] proposed a fine-grained indoor localization based on CSI data. FILA weights the filtered CSI and normalizes the power to the center frequency in the band:

$$CSI_{eff} = \frac{1}{K} \sum_k \frac{f_k}{f_c} \times ||A||_k \tag{5}$$

where CSI_{eff} is the effective CSI for distance estimation, K is the number of subcarriers, f_c is the calculated center frequency, and $||A||_k$ is the amplitude of the filtered CSI on the kth subcarrier. The propagation distance between the transmitting end and the receiving end can be represented by effective channel state information.

$$d = \frac{1}{4\pi}\left[\left(\frac{c}{f_0 \times |CSI_{eff}|}\right)^2 \sigma\right]\frac{1}{n} \tag{6}$$

where d is the distance between the transmitter and receiver in indoor environments, c is the radio velocity, f_0 is the central frequency of CSI, n is the path loss attenuation factor, and σ is the environmental factor.

2.2 Gray Prediction GM (1,1) Model and Residual Test

Gray system theory is a method to study the problem of uncertainty with little data and poor information. To predict the future development of things, this model can effectively improve the smoothness of the modeled data series.

The gray prediction is based on the gray model [11]. Among the many gray models, the GM (1,1) model is the most commonly used.

Equipped with the original data column $x^{(0)} = (x^{(0)}(1), x^{(0)}(2), \ldots, x^{(0)}(n))$, n is the number of data.

1. The original data is accumulated in order to weaken the fluctuation and randomness of the random sequence, and a new data sequence is obtained:

$$x^{(1)} = (x^{(1)}(1), x^{(1)}(2), \ldots, x^{(1)}(n)) \tag{7}$$

where,$x^{(1)}(t) = \sum_{k=1}^{t} x^{(0)}(k)$, $t = 1, 2, \ldots, n$ or $x^{(1)}(t+1) = \sum_{k=1}^{t+1} x^{(0)}(k)$, $t = 1, 2, \ldots, n$

2. Establish a first-order linear differential equation for, $x^{(1)}(t)$:

$$\frac{dx^{(1)}}{dt} + ax^{(1)} = u \tag{8}$$

where a is called the development gray number; μ is the endogenous control gray number, and the matrix formed by a and μ is $\hat{a} = \binom{a}{u}$

3. Mean the accumulated generated data to generate B and the constant term vector Y_n,

$$B = \begin{bmatrix} \frac{1}{2}(x^{(1)}(1) + x^{(1)}(2)) \\ \frac{1}{2}(x^{(1)}(2) + x^{(1)}(3)) \\ \cdots \\ \frac{1}{1}(x^{(1)}(n-1) + x^{(1)}(n)) \end{bmatrix} \tag{9}$$

$$Y_n = (x^{(0)}(2), x^{(0)}(3), \ldots, x^{(0)}(n))^T \tag{10}$$

4. Use the least square method to solve the gray parameter $\hat{a}$, then

$$\hat{a} = \begin{pmatrix} a \\ u \end{pmatrix} = (B^T B)^{-1} Y_n \tag{11}$$

5. Bring the gray parameter $\hat{a}$ into Eq. (8) and solve it, get

$$\hat{x}^{(1)}(t+1)) = \left(x^{(0)}(1) - \frac{u}{a}\right) e^{-\alpha} + \frac{u}{a} \tag{12}$$

It is worth noting that, because the gray parameter $\hat{a}$ is the approximate value obtained by the method of least squares, $\hat{x}^{(1)}(t+1))$ is an approximate expression, to distinguish it from $x^{(1)}(t+1))$, it is recorded as $\hat{x}^{(1)}(t+1))$

6. Discrete $\hat{x}^{(1)}$ and $\hat{x}^{(1)}(t)$ and make a difference to restore the $\hat{x}^{(1)}(t+1))$ sequence as

$$\hat{x}^{(0)}(t+1)) = \hat{x}^{(0)}(t+1)) - \hat{x}^{(1)}(t)) \tag{13}$$

7. Test the established gray model. Calculate residual $e^{(0)}(t)$ and relative error $q(x)$ between $\hat{x}^{(0)}$ and $\hat{x}^{(0)}(t)$

$$e^{(0)}(t) = \hat{x}^{(0)} - \hat{x}^{(0)}(t) \tag{14}$$

$$q(x) = \frac{e^{(0)}(t)}{x^{(0)}(t)} \tag{15}$$

3 Methodology

3.1 Indoor Positioning System Architecture

The Wi-Fi-based positioning method uses a nonlinear relationship between CSI value and distance to perform positioning. The method is divided into two phases: Predicting the CSI phase and the positioning phase. The block diagram of the positioning system of the algorithm proposed in this paper is shown in Fig. 1. When the mobile terminal is positioned online, the position of the mobile terminal can be estimated according to the obtained ranging model.

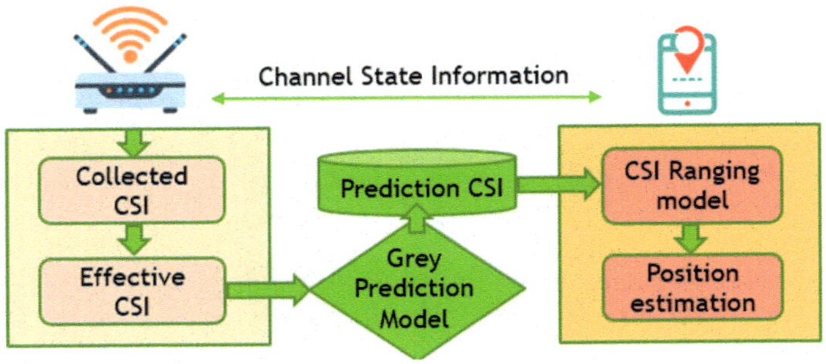

Fig. 1 The architecture of indoor positioning system

3.2 CSI Ranging Algorithm Based on the Weight Gray Prediction Model

At the same time, the gray theory establishes the data model, not the original data model. Therefore, the data of gray prediction is the inverse processing result of the predicted value obtained by the GM (1,1) model. We equipped with the effective CSI data column:

$$\begin{aligned} x^{(0)}(1) &= CSI_{eff}^{i} \\ x^{(0)}(2) &= CSI_{eff}^{i+1} \\ x^{(0)}(3) &= CSI_{eff}^{i+2} \\ &\cdots \\ x^{(0)}(n) &= CSI_{eff}^{i+n-1} \end{aligned} \tag{16}$$

where, n is the number of the effective CSI data. Next, we can use Eq. (7) to generate data $x^{(1)}$. According to the process of the gray prediction model, we can obtain the predicted CSI value, that is,

$$CSI_{pred}^{i} = \hat{x}^{(0)}(n) \tag{17}$$

The key indicator for evaluating positioning technology is the accuracy of the estimated position coordinates. To illustrate the high accuracy of the proposed algorithm, this paper uses the mean square error (MSE) to compare the performance of the signal attenuation model obtained with the gray prediction model and the signal attenuation model obtained with the effective CSI value. We put the CSI measured from the mobile client into the positioning system and then get the predicted CSI value.

The weight gray prediction method used the following equation to obtain

$$CSI_{pred(weight)} = w_1 * CSI^i_{pred} + w_2 * CSI^{i-1}_{eff} \tag{18}$$

where, CSI^i_{pred} represents the ith predicted CSI in the gray prediction system. CSI^{i-1}_{eff} represents $(i-1)^{\text{th}}$ effective CSI value. w_1 and w_2 represent the weights of the effective CSI value and the predicted CSI value. The other is SPKF. In the experiments, we let $w_1 = 0.5$ and $w_2 = 0.5$, and the performance of MSE was the best.

We compared the proposed positioning method of the weighted gray prediction CSI value with the positioning method based on the effective CSI value. MSE_{eff} and MSE_{GP} are used to evaluate the accuracy of the CSI values in the proposed positioning system and the existing positioning system, respectively:

$$MSE_{eff} = \sum_{i=1}^{N} \sqrt{(CSI^i_{eff} - CSI^i_{mean})^2} \tag{19}$$

$$MSE_{WGP} = \sum_{i=1}^{N} \sqrt{(CSI^i_{pred(weight)} - CSI^i_{mean})^2} \tag{20}$$

where, N is the number of the input data. CSI_{mean} is an average of subcarriers from different antennas.

4 Experiments and Performance Evaluation

4.1 Experimental Scenarios

In experiments, we use ipTIME N3004 as a node and place it at a height of 0.2 m on the ground. The reference node place on 0.2 m high to reduce the impact of ground reflection on CSI value. An Intel 5300 wireless network card built-in notebook computer DELL Inspiron n4010 is used as a mobile node and is located 0.5 m away from a fixed node for measurement start. Considered reference points are 0.5 m apart sequentially to each other.

4.2 Performance Evaluation

To verify the proposed adaptive ranging model, experiments were performed using the aforementioned method. Before performing the verification experiment, first use the above experimental method to measure the CSI value of 0 m to 6 m in the

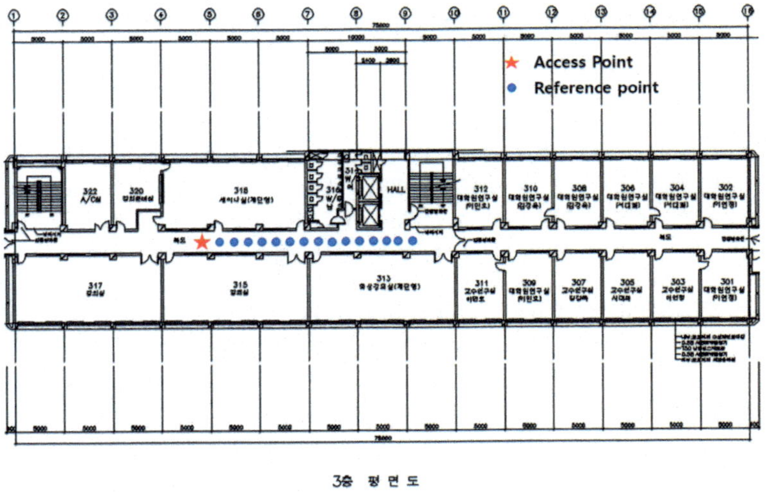

Fig. 2 The 7th floor of Kyungpook National University IT-1 building

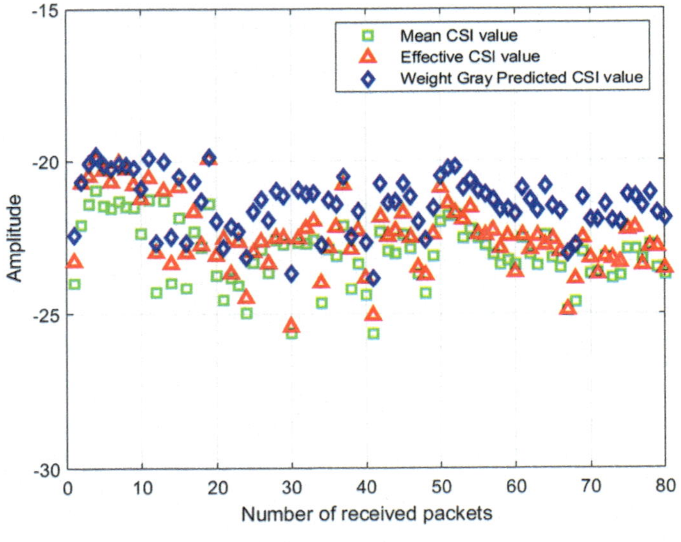

Fig. 3 Comparison amplitude of mean CSI value, gray prediction CSI value, and effective CSI value

experimental scene, and continuously collect 80 sets of data at each measurement point. The transmitting node AP (red pentagram in Fig. 2) is fixed, and the receiving node (blue dot in Fig. 2) is sequentially placed at a distance of 0.5 m to measure the CSI value, thereby obtaining a signal intensity distribution in a two-dimensional space. The parameter settings and measurement methods of each node device are the same as above. Figure 3 shows the amplitude of mean CSI value, gray prediction

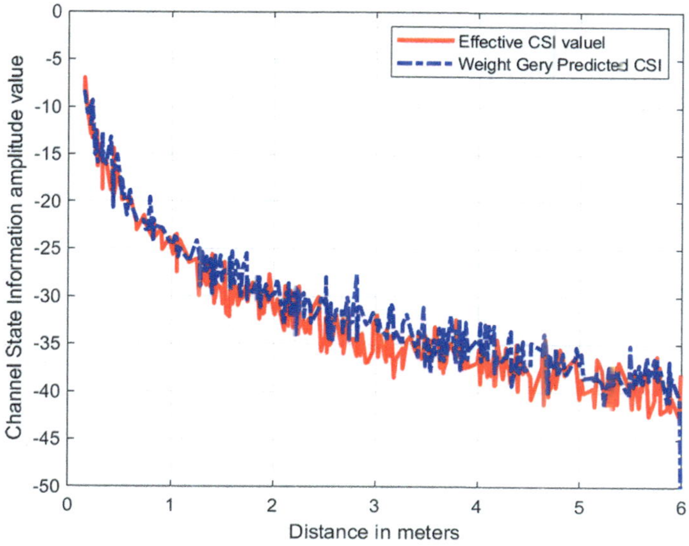

Fig. 4 Comparing the range model of gray prediction CSI value with the range model of effective CSI value

Table 1 Comparison of the distance errors of the proposed method and existing method

Ranging model	Average error	Error range
The effective CSI (the existing method)	1.21 m	Maximum:4.372 m Minimum:0.394 m
Gray prediction CSI (the proposed method)	1.07 m	Maximum:3.769 m Minimum:0.261 m

CSI value, and effective CSI value in 0.5 m. We take the effective CSI value and the gray prediction CSI value as the effective value, and take the average CSI as the measurement value. The signal environmental loss coefficient is estimated through the measured distance and CSI, and the relationship between the signal strength and distance is determined according to Eq.(6). Then the performance of the two ranging models is compared. A comparison of ranging models is shown in Fig. 4.

This paper analyzes the positioning performance of the system by comparing the average error of the ranging model algorithm based on the CSI value obtained by the gray prediction model and the ranging algorithm based on the CSI effective value. In Table 1, the distance error accuracy obtained by the two methods is compared. The experimental results show that the average distance error of the model algorithm proposed in this paper can reach a smaller value of 1.07 m in the two positioning systems. Compared with the ranging system based on the effective CSI value, its average error is reduced by 11.6%. It can be seen that the proposed ranging model algorithm has higher accuracy and can improve the positioning performance of the system.

5 Conclusions and Future Research

Based on collecting CSI of the physical layer, this paper analyzes various indoor positioning defects and proposes a new CSI-based adaptive modified model positioning algorithm. In the actual indoor environment, this paper obtains a more accurate signal attenuation model by using the CSI prediction value obtained by the gray prediction model. At the same time, the gray prediction has the advantages of requiring fewer sample data, simple principle, convenient operation, high short-term prediction accuracy, and testability, so it can be widely used. Experimental results show that the obtained model has improved positioning accuracy and positioning time. However, the gray prediction model also has certain limitations, and the prediction accuracy is sometimes unsatisfactory. In future research work, we plan to further improve the signal attenuation model obtained through the gray prediction model. Based on the current basis, a new prediction model is established by improving the smoothness of the original data sequence, which overcomes the shortcomings in the traditional gray prediction model.

Acknowledgments This study is supported by the BK21 Plus project funded by the Ministry of Education, Korea (21A20131600011). This work is supported by the Smart City R&D project of the Korea Agency for Infrastructure Technology Advancement (KAIA)grant funded by the Ministry of Land, Infrastructure and Transport (Grant 18NSPS-B149843-01).

References

1. M.J. Rycroft, Understanding GPS. Principles and applications. J. Atmos. Solar-Terrestrial Phys. **59**, 598–599 (1997)
2. A. El-rabbany, *Introduction to GPS: The Global Position System* (Artech House, Boston, MA, 2006)
3. J. Xu, W. Liu, F. Lang, Y. Zhang, C. Wang, Distance measurement model based on RSSI in WSN. Wirel. Sens. Netw. **2**(8), 606–611 (2010)
4. J. Armstrong, OFDM for optical communications. J. Lightw. Technol. **27**(3), 189–204 (2009)
5. D. Halperin, W. Hu, A. Sheth, D. Wetherall, Tool release: gathering 802.11n traces with channel state information. Comput. Commun. Rev. **41**, 53 (2011)
6. Z. Yang, Z. Zhou, Y. Liu, From RSSI to CSI: indoor localization via channel response. ACM Comput. Surv. (2013). Article No: 25. https://doi.org/10.1145/2543581.2543592
7. K. Wu, J. Xiao, Y. Yi, M. Gao, L.M. Ni, FILA: fine-grained indoor localization, in *Proceedings – IEEE INFOCOM*, 2012
8. J. Xiao, K. Wu, Y. Yi, L.M. Ni, FIFS: fine-grained indoor fingerprinting system, in *2012 21st International Conference on Computer Communications and Networks, ICCCN 2012 - Proceedings*, 2012
9. X. Wang, L. Gao, S. Mao, S. Pandey, DeepFi: deep learning for indoor fingerprinting using channel state information, in *2015 IEEE Wireless Communications and Networking Conference, WCNC 2015*, 2015
10. R.C. Luo, O. Chen, S.H. Pan, Mobile user localization in wireless sensor network using grey prediction method, in *IECON Proceedings (Industrial Electronics Conference)*, 2005
11. Z. He, Y. Shen, Q. Wang, Boundary extension for HilbertHuang transform inspired by gray prediction model. Signal Process. **92**, 685–697 (2012)

Autonomous Vehicle Security Model

Noha Hazzazi, Kevin Daimi, and Hanady Issa

1 Introduction

Autonomous vehicles augmented the traditional vehicle features with the autonomy taste. They are envisaged to gather massive data from various sources and replace human drivers. These built-up data will be huge and will attract further research settings for numerous fields including technological data science, safety, and security. Autonomous vehicles are anticipated to play a significant role in the future of transportation systems, as they entail a potential for more safety, expanded productivity, elevated accessibility, improved road efficiency, and encouraging effect on the environment. Autonomous vehicles (AVs) are broadly expected to lessen road congestion through greater throughput, enhance road safety by getting rid of human error, and free drivers from the burden of driving [1]. Autonomy as defined by the National Highway Traffic Safety Administration varies based on the approach the control functions are conducted by the vehicle. The full autonomous vehicle obtains information from the surrounding environment via various sensors.

N. Hazzazi (✉)
Department of Electrical Engineering and Computer Science, Howard University, Washington, DC, USA
e-mail: noha.hazzazi@howard.edu

K. Daimi
Electrical and Computer Engineering, and Computer Science, University of Detroit Mercy, Detroit, MI, USA
e-mail: daimikj@udmercy.edu

H. Issa
Electronics and Communications Department, Arab Academy for Science, Technology and Maritime Transport, Alexandria, Egypt
e-mail: hanady.issa@aast.edu

K. Daimi et al. (eds.), *Advances in Security, Networks, and Internet of Things*, Transactions on Computational Science and Computational Intelligence,
https://doi.org/10.1007/978-3-030-71017-0_37

analyzes the received signals, and carries out applicable path of movement [2]. With these high-level control functions, the vehicle becomes more reliant on communication networks both within the vehicle and within exterior environment [3]. This inevitable reliance on communication networks will widely open the doorways for even more sophisticated security attacks.

There are five levels of autonomous vehicle representing the diverse degree of automation that the vehicle is capable of. The transition from lower level to higher level signifies an increase in automation. The levels continue to elevate until the last level (level 5) is reached. These levels include driver assistant (level 1), partial automation (level 2), conditional automation (level 3), high automation (level 4), and full automation (level 5). At level 1, the driver is in charge of monitoring the outer environment and making decisions to control the vehicle movement. Controlling the steering and acceleration tasks with limited driving conditions is taken care of in level 2. Level 3 ensures the vehicle is fully responsible for monitoring the environment and performing the safety-critical functions. At level 4, the vehicle is fully responsible for controlling driving. The difference between levels 3 and 4 is characterized by the needed interference of the driver in case of failure. Finally, with level 5, the vehicle performs all environmental analysis and planning techniques to reach destination without any driver interference [4–9].

The approach that autonomous vehicles follow relies on three systems: perception, planning, and control. The perception system is responsible for sensing the environment and finding out the location of the autonomous vehicle. Lane line identification, obstacle detection, and road sign analysis are captured through cameras, Light Detection and Ranging (LIDAR) device, and radar. The planning system receives data from perception system, analyzes it, and makes the appropriate decision (plans) for movement. The data received by planning system stems from the perception system's output data, feedback from the control system, and inter-vehicle communication data. Finally, the control system implements the decision taken by the planning system through a large number of electronic control units (ECUs). For proper performance, each system needs three components: sensors, processors, and communication technologies [10–13].

There are two types of communications in autonomous vehicles, inter-vehicle and intra-vehicle communications. Intra-vehicle communications, represented by buses, are responsible for data transfer between the autonomous vehicle's components. Inter-vehicle communications deal with transferring of data between the vehicle and the external environment including other vehicles, infrastructures, and smart road signs. This makes the autonomous vehicle more vulnerable to various security attacks that are classified based on the type of the attacker, motivation for the attack, type of the attack, and target for the attack [14].

Thing and Wu [14] discussed a number of attacks on autonomous vehicles including side-channel attacks, code modification, code injection, packet sniffing, packet fuzzing, in-vehicle spoofing, and jamming. Consequently, powerful and efficient security countermeasures should be followed to protect autonomous vehicles. To avoid such security attacks, security requirements should have been considered before the definite design of these vehicles [15]. Message encryption techniques by themselves do not enforce data integrity and confidentiality [16]. To help with the security efforts, a number of tools to enhance the security of data transmission for both inter-vehicle and intra-vehicle communications were introduced [17]. For this purpose, Schlatow et al. [18] advocated relying on trust management and control. Furthermore, the use of polices to restrict access to these resources was proposed in [19]. Further attempts to deepen the security of the autonomous vehicles involved counting on tamper proof microkernels, proxies, and network stacks to enhance the security of vehicle networks [20]. Attaching more sensors to the autonomous vehicle helps to observe performance with respect to integrity and availability [21].

The aim of this paper is to suggest an approach to secure autonomous vehicles based on cryptography. The communication between the systems, perception and planning, planning and control, and planning and other vehicles and road infrastructure will be protected using various types of encryptions. To assist this approach, an Authentication Center is proposed. The remainder of the paper is organized as follows: Section II introduces an overview of the functions of each of the three systems of autonomous vehicles: perception, planning, and control. Section III discusses the security architecture. The various cryptographic techniques applicable to the parties discussed in Section III will be applied in Sect. IV. Finally, Section V concludes the paper.

2 Autonomous Vehicle Systems Operation Overview

As stated above, autonomous vehicles rely on three systems: perception, planning, and control [22]. These are briefly explained below. For details, the following references provide excellent discussion [23–36].

2.1 Perception

Perception includes environmental perception and localization. Environmental perception furnishes the vehicle with critical details regarding the driving environment, which include the unoccupied driving areas and neighboring obstacles' locations and velocities, in addition to predicting their potential changing states. This kind of perception depends on a number of sensors. These sensors include LIDAR and vision sensors (cameras). LIDAR transmits millions of light pulses per second in a well-designed pattern. It is the core element for object detection for many of the existing autonomous vehicles. The vision system typically includes road detection and on-road object detection. The road detection comprises lane line marking detection and detection of road surface. Lane line marking detection identifies lane line markings on the road and assesses the vehicle position and orientation pertaining to the detected lines. This can fulfill the vehicle position feedback to vehicle control system. Traffic road conditions are associated with many uncertainties including cars and tree shadows, deviation of lighting conditions, unclear lane markings, directional arrows, warning text, and zebra crossings. Road surface detection notifies the autonomous vehicle about free space locations where it is safe to drive. This is the prerequisite for any online path planning and control operations.

Localization is the process of concluding the position and orientation of the vehicle and determining its own motion. It is one of the key capabilities that makes autonomous driving possible. However, it is often hard and impractical to specify the exact position and orientation of the vehicle. Hence, localization is often articulated as a position and orientation estimation challenge. A Global Positioning System (GPS) is used for localization but entails reliable signals from external satellites.

2.2 Planning

Planning permits autonomous vehicles to manage a broad range of urban driving scenarios. The mission planner (or route planner) takes into consideration high-level tasks, such as mission of pickup/drop-off charges, and what roads should be followed to accomplish these tasks. The behavioral planner (or decision-maker) takes care of unprepared decisions to appropriately interact with other parties and abide by rule restrictions. These result in decisions regarding various objectives including lane change, overtaking, or proceeding through an intersection. The motion planner (or local planning) produces proper paths and sets of actions to achieve the above objectives with the most distinctive objective of reaching a destination region while avoiding obstacle collision.

2.3 Control

The execution proficiency of an autonomous system (referred to as motion control) is the process of converting plans into actions. The foremost purpose is to implement the planned goals through the necessary inputs to the hardware level that will cause the desired motions. Controllers map the interaction in the real world via forces and energy, while the cognitive navigation and planning of an autonomous system are normally concerned with velocity and position of the vehicle as related to its environment. Measurements within the control system are deployed to determine how intelligent the system is behaving. Therefore, the controller's reaction can result in discarding disturbances and adjusting the dynamics of the system to the anticipated state. The actions of the control system instruct various ECUs in the vehicle to carry out their functions. In other words, the various plans will be translated by the control system to instruction that the ECUs will follow.

3 Autonomous Vehicle Security Architecture

The security architecture composes of the autonomous vehicle's internal parties represented by sensors, perception unit, planning unit, and control unit and the external units vehicle-to-vehicle (V2V), vehicle-to-infrastructure (V2I), and authentication authority. This is illustrated in Fig. 1.

The sensor unit includes LIDAR, ultrasonic radar, GPS, inertial measurement unit (IMU), and camera. These sense the environment and send various messages to the perception unit, such as unoccupied driving areas, neighboring obstacles' locations, velocities, potential changing states, lane line markings, vehicle position and orientation, and free space locations. These messages must be encrypted before being sent to the perception unit. Instead of repeating the protocol for each sensor, M_S will refer to all these messages. Because the majority of these messages are relatively large, symmetric encryption is used to speed up the process. All the messages will be received and validated by the perception unit.

The perception unit forwards M_P, which includes the critical messages of the sensors after validating them, to the planning unit. The planning unit will make decisions on high-level tasks, such as pickup/drop-off, roads to follow, lane change, overtaking, or proceeding through an intersection. These decisions will be denoted by M_{PL}. These decisions are transmitted to the control unit who will transform them into control actions. For this security model, the planning unit will be responsible for creating and exchanging keys with both the perception and the planning units.

Communication with external units includes other vehicles on the road (denoted by V2V) and the road infrastructure (represented by V2I). V2V permits vehicles to communicate wirelessly to exchange various information, such as speed, location, direction of travel, braking, and loss of stability. This contributes to avoiding crashes and reduces traffic congestion. The broadcasting of information should not reveal the identity of the vehicle. V2I facilitates bidirectional transfer of information between vehicles and road infrastructure, such as lane marking, road signs, and

Fig. 1 Security architecture

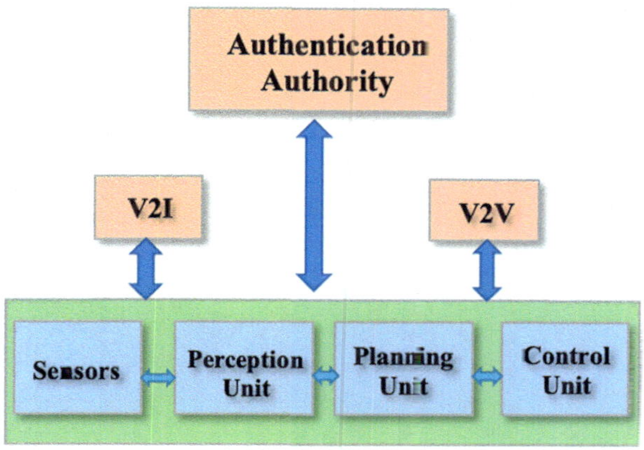

traffic lights. The exchanged information includes weather, road conditions, traffic conditions, traffic jams, sharp curves, and accidents. This paper will use M_V to denote messages from other vehicles and M_I for messages from the infrastructure. As these are external units, a trusted authentication authority is needed for key exchange between the planning unit and both the V2V and V2I.

4 Autonomous Vehicle Security Protocol

The security protocol will be divided into initialization, communication between sensors (S) and perception (P), communication between P and planning (PL), communication between PL and control (C), communication between PL and infrastructure (I), and communication between PL and other vehicles (V). The parties involved and the notation used are presented in Tables 1 and 2.

4.1 Initialization

To start with, all the parties of the autonomous vehicle will have their public key and private keys installed. In addition, S has PU_P and P has PU_S stored, P and PL own each other's public keys, and PL and S will follow this approach. In a similar manner, each pair of parties involved in the communication will have their IDs and the ID of the other party stored.

With regard to the external communication, PL, V, and I share public keys and their IDs with A. A will create pseudo-IDs to replace the original IDs of PL and V. This guarantees privacy of vehicles and infrastructure. Here, PL refers to the vehicle in question and V to other vehicles (other vehicles' PL).

The perception unit will create the symmetric key, K_{SP}, shared with sensors, and the planning unit will create the keys, K_{PPL}, and K_{PLC} shared with PL and C, respectively. The same process will be followed when keys need to change.

The keys K_{PLI} and K_{PLV} that are essential for communications between PL and I and PL and V, respectively, will be created by the authentication authority. It is up

Table 1 Communication parties

Party	Meaning
S	Sensors
P	Perception unit
PL	Planning unit
C	Control unit
I	Infrastructure (V2I)
V	Other vehicles (V2V)
A	Authentication authority

Table 2 Notation used in protocol

Party	Meaning
K_{SP}	Shared symmetric key (S and P)
K_{PPL}	Shared symmetric key (P and PL)
K_{PLC}	Shared symmetric key (PL and C)
K_{PLI}	Shared symmetric key (PL and I)
K_{PLV}	Shared symmetric key (PL and V)
PU_S, PR_S	Public/private key of S
PU_P, PR_P	Public/private key of P
PU_{PL}, PR_{PL}	Public/private key of PL
PU_C, PR_C	Public/private key of C
PU_I, PR_I	Public/private key of I
PU_V, PR_V	Public/private key of V
PU_A, PR_A	Public/private key of A
ID_S	Sensor ID
ID_P	Perception unit ID
ID_{PL}	Planning unit ID
ID_C	Control unit ID
ID_I	Infrastructure ID
ID_V	Other vehicle ID
ID_A	Authentication authority ID
ID_{PLS}	Pseudo-ID of PL
T_X	Time stamp by party X
SIG_X	Message signed by party X
M_S	Messages sent by sensors to P
MP	Messages sent by P to PL
M_{PL}	Messages sent by PL to C
M_X	Message sent by party X
||	Concatenation

to this authority to decide when to change keys. It informs the three parties PL, I, and V when the new keys are available to request them provided the vehicles are within the responsibility of the infrastructure. The infrastructure location is fixed and can always get a new key. However, autonomous vehicles may leave the area or the section that the road infrastructure is responsible for. If that happens, there will be no need to update the key.

4.2 S and P Communication

Sensors sense their environment and forward these messages to P. First P creates the shared key, K_{SP}; finds its hash code, $H(K_{SP})$; and signs with its private key, PR_P, using a digital signature algorithm. As this is a security model, no algorithm will be specified. Hence, SIG_P $(H(K_{SP}))$ is used here. The shared key, its signature, ID_S, and ID_P are encrypted with the public key of S and then forwarded:

$$P \rightarrow S : E\ [PU_S, ID_S \| ID_P \| K_{SP} \| SIG_P (H (K_{SP})) \| T_P]$$

After decrypting the message with PR_S, S is assured the message is coming from P through the two IDs and the message is currently based on T_P. S then verifies the signature using the algorithm they agreed on, finds the hash code of the shared key, and compares it with the received code. At this point, S is ready to send what it senses in the environment to P. To do that, S signs M_S and attaches it to the message together with both IDs and a time stamp T_S and encrypts them all with K_{SP}:

$$S \rightarrow P : E\ [K_{SP}, ID_S \| ID_P \| M_S \| H (M_S) \| T_S]$$

P will decrypt this message with K_{SP}, assuring it came from one of the sensors (based on ID_S and ID_P) and that the message is not a replay (based on T_S). Furthermore, P will verify the signature. If the signature is valid, the message M_S will be accepted.

4.3 P and PL Communication

P will process these signals (messages) from various sensors and combine them before sending them to PL. As noted above, PL creates the symmetric keys for both P and C. Because the approach is similar to the communication between S and P, only the symbolic notation is stated here:

$$PL \rightarrow P : E\ [PU_P, ID_P \| ID_{PL} \| K_{PPL} \| SIG_{PL} (H (K_{PPL})) \| T_{PL}]$$

$$P \rightarrow PL : E\ [K_{PPL}, ID_P \| ID_{PL} \| M_P \| H (M_P) \| T_P]$$

4.4 PL and C Communication

PL will collect all these inputs from P, in addition to further input from V and I as explained below, and creates plans (action paths) that will be sent to C. The control unit, C, translates these into actions and asks respective ECUs to execute these actions to control vehicle movement and maneuvering. Once again, the description is similar to the one above:

$$PL \rightarrow C : E\ [PU_C, ID_C \| ID_{PL} \| K_{PLC} \| SIG_{PL} (H (K_{PLC})) \| T_{PL}]$$

$$PL \rightarrow CL : E\ [K_{PLC}, ID_C \| ID_{PL} \| M_{PL} \| H (M_{PL}) \| T_{PL}]$$

4.5 PL and I Communication

Unlike other vehicles, the infrastructure, *I*, has a fixed location. When the autonomous vehicle is within the range of *I*, A receives the ID_{PL} (referring to vehicle in question) and ID_I encrypted with PU_A. A creates a shared key between PL and *I*, K_{PLI}, and a pseudo-ID for PL, ID_{PLS}, to preserve the privacy of the vehicle. At this point, A signs the shared key and concatenates it with the pseudo-ID, the shared key, and a time stamp T_A.

$$PL \rightarrow A : E\ (PU_A,\ ID_{PL} \| ID_I \| T_{PL})$$

$$A \rightarrow PL : E\ [PU_{PL},\ ID_{PLS} \| ID_I \| K_{PLI} \| SIG_A\ (H\ (K_{PLI})) \| T_A]$$

After applying the required decryptions and verifications, PL will have both the pseudo-ID (ID_{PLS}) and the shared key K_{PLI}. *I* does not need a pseudo-ID because it is fixed and known to all vehicles within its authority (range). Hence, *I* receives the key as follows:

$$I \rightarrow A : E\ (PU_A,\ ID_{PL} \| ID_I \| T_I)$$

$$A \rightarrow I : E\ [PU_{PL},\ ID_{PLS} \| ID_I \| K_{PLI} \| SIG_A\ (H\ (K_{PLI})) \| T_A]$$

Here, *I* got the ID_{PLS} and the shared key, K_{PLI}. PL and *I* can share information about weather, road conditions, traffic conditions, traffic jams, sharp curves, and accidents. M_{IPL} denotes messages sent by *I* to PL, and M_{PLI} denotes messages sent by Pl to *I*:

$$I \rightarrow PL : E\ [K_{PLI},\ ID_I \| ID_{PLS} \| M_{IPL} \| H\ (M_{IPL}) \| T_I]$$

$$PL \rightarrow I : E\ [K_{PLI},\ ID_I \| ID_{PLS} \| M_{PLI} \| H\ (M_{PLI}) \| T_{PL}]$$

4.6 PL and V Communication

In this communication, a PL of one autonomous vehicle will interface with a PL of another vehicle. Vehicles broadcast messages. This implies all vehicles nearby and the infrastructure, I, will receive the broadcasted message. However, to ensure the messages do not represent an attack, the infrastructure will be resending the broadcasted message to all vehicles. Here, it is assumed the message is sent from vehicle to infrastructures and then back to all vehicles in the range. In other words, the broadcasted message by another vehicle will be discarded by all nearby vehicles until it is received from I. Here, I will retransmit the broadcasted message. The last two communications of Section (E) will be reversed below:

$$PL \rightarrow I : E\ [K_{PLI},\ ID_I \| ID_{PLS} \|\ M_{PLI} \| H(M_{PLI}) \|\ T_{PL}]$$

$$I \rightarrow PL : E\ [K_{PLI},\ ID_I \| ID_{PLS} \|\ M_{IPL} \| H(M_{IPL}) \|\ T_I]$$

5 Conclusion

A fully autonomous vehicle is a vehicle capable of finding its way through roadways and accounting for traffic control devices without any human driver interfering with any of the vehicle's control systems. The high and ongoing sophistication of such vehicles is exceedingly demanding a very robust and very effective protection against security attacks. A simple error or inaccurate decision caused by an attack on autonomous vehicle will result human loss and could be disastrous when multiple vehicles are involved. In an effort to contribute to protecting autonomous vehicles against various security attacks, this paper presented an autonomous vehicle security architecture. Both inter-vehicular and intra-vehicular communications are included in this architecture. Symmetric key cryptography is used to protect various exchanged messages between the internal units of the vehicle and the external units (other vehicles and infrastructure). Public key cryptography was used for key exchange and signature purposes only. For future work, further external units, such as vehicle cloud, can be included in the architecture. The cloud will serve as a channel of accumulating and delivering information that help vehicles to avoid accidents, acquire updated maps on the best route to reach a destination, and receive various traffic details. As the above approach represents a model, algorithms for encryption, hashing, and digital signature are not specified.

References

1. S.D. Pendleton, H. Andersen, X. Du, X. Shen, M. Meghjani, Y.H. Eng, D. Rus, M.H. Ang Jr., Perception, planning, control, and coordination for autonomous vehicles. Mach. Des. **5**(6), 1–54 (2017)
2. J. Zhao, Q.C.B. Liang, The key technology toward the self driving car. Int J Intell Unmanned Syst **6**, 2–20 (2018)
3. J. Wang, J. Liu, N. Kato, Networking and communications in autonomous driving, a survey. *In IEEE Communication Surveys and Tutorials* (2018)
4. I. Harner, The 5 autonomous driving levels explained, https://www.iotforall.com/5-autonomous-driving-levels-explained/, October 2017, (Retrieved: May, 2019)
5. M. Burgess, When does a car become truly autonomous? Levels of self-driving technology explained, 2017, https://www.wired.co.uk/article/autonomous-car-levels-sae-ranking, (Retrieved: May, 2019)
6. M. Burgess, " We Went Off-Road In Jaguar Land Rover's Autonomous Car," 2016, https://www.wired.co.uk/article/self-driving-autonomous-land-rover-jaguar-technology, [Retrieved: May, 2019]

7. National Highway Traffic Safety Administration, Preliminary statement of policy concerning autonomous vehicle, https://www.nhtsa.gov/.../rulemaking/pdf/Automated_Vehicles_Policy.pdf, (Retrieved: May, 2019)
8. S. Lin, Y. Zhang, C. Hsu, M. Skach, E. Haque, L. Tang, J. Mars, The architectural implications of autonomous driving: constraints and acceleration. *In Proc. the Twenty-Third International Conference on Architectural Support for Programming Languages and Operating Systems (ASPLOS'18)*, Williamsburg, VA, USA, 2018, pp. 751–766
9. K. Hyatt, C. Paukert, Self driving cars: a level-by-level explainer of autonomous vehicle, https://www.cnet.com/roadshow/news/self-driving-car-guide-autonomous-explanation/, (Retrieved: May, 2019)
10. M. Mody, J. Jones, K. Chitnis, R. Sagar, G. Shurtz, Y. Dutt, M. Koul, M. G. Biju, A. Dubey, Understanding vehicle E/E architecture topologies for automated driving: system partitioning and tradeoff parameters. *In Proc. the Autonomous Vehicles and Machine Symposium*, 2018, pp. 358(1)–358(5)
11. J.R. Van Brummelen, M. O'Brien, D. Gruyer, H. Najjaran, Autonomous vehicle perception system: the technology of today and tomorrow. Transp. Res. C **89**, 384–406 (2018)
12. R. Blake, M. Shiffrar, Perception of human motion. Annu. Rev. Psychol. **58**, 47–73 (2007)
13. A.M. Wyglinski, X. Huang, T. Padir, L. Lai, T.R. Elsenbarth, K. Venkatasubramanian, Security of autonomous systems employing embedded computing and sensors. IEEE Micro. **33**(1), 80–86 (2013)
14. V.L.L. Thing, J. Wu, Autonomous vehicle security: a taxonomy of attacks and defences. *In Proc. the IEEE International Conference on Internet of Things (iThings), IEEE Green Computing and Communications (GreenCom), IEEE Cyber, Physical and Social Computing (CPSCom), IEEE SmartData (SmartData)*, Chengdu, China pp. 164–170, 2016
15. E. Yagdereli, C. Gemci, A.Z. Aktas, A study on cyber-security of autonomous and unmanned vehicles. Proc. J. Defense Model. Simulat. Appl. Methodol. Technol. **12**, 369–381 (2015)
16. J. Yoshida, EE Times, "CAN Bus Can Be Encrypted, Says Trillium," [online]. Available: http://www.eetimes.com/document.asp?docid=1328081, Oct. 2015. Retrieved: May 2019
17. A. Lima, F. Rocha, M. Volp, P. Esteves-Verissimo, Towards safe and secure autonomous and cooperative vehicle ecosystems. *In Proc. the 2nd ACM Workshop on Cyber-Physical Systems Security and Privacy CPS-SPC'16*, Vienna, Austria, Oct. 2016, pp. 59–70
18. J. Schlatow, M. Moestl, R. Ernst, An extensible autonomous reconfiguration framework for complex component-based embedded systems. *In Proc. 12th International Conference on Autonomic Computing (ICAC)*, Grenoble, France, July 2015, pp. 239–242
19. V. Prevelakis, M. Hammad, A policy-based communications architecture for vehicles. *In Proc. International Conference on Information Systems Security and Privacy*, Angers, France, 2015, pp. 155–162
20. M. Hamad, J. Schlatow, V. Prevelakis, R. Ernst, A communication framework for distributed access control in microkernel-based systems. *In Proc. the 12th Annual Workshop on Operating Systems Platforms for Embedded Real-Time Applications (OSPERT16)*, Toulouse, France, July 2016, pp. 11–16
21. E. Villani, N. Fathollahnejad, R. Pathan, R. Barbosa, J. Karlsson, Reliability analysis of consensus in cooperative transport systems. *In Proc. 32nd International Conference on Computer Safety, Reliability and Security, SAFECOMP 2013 – Workshop ASCoMS (Architecting Safety in Collaborative Mobile Systems)*, Toulouse, France, Sept. 2013, pp. 1–8
22. J. Wang, J. Liu, N. Kato, Networking and communications in autonomous vehicles driving: a survey. IEEE Commun. Surv. Tutor. **21**(2), 1243–1274 (2019)
23. A. Asvadi, C. Premebida, P. Peixoto, U. Nunes, 3D Lidar-based static and moving obstacle detection in driving environments: an approach based on voxels and multi-region ground Planes. Robot. Auton. Syst. **83**, 299–311 (2016)
24. A.B. Hillel, R. Lerner, D. Levi, G. Raz, Recent progress in road and lane detection: a survey. Mach. Vis. Appl. **25**, 727–745 (2014)

25. S. Sivaraman, M.M. Trivedi, Looking at vehicles on the road: a survey of vision-based vehicle detection, tracking, and behavior analysis. IEEE Trans. Intell. Transp. Syst. **14**, 1773–1795 (2013)
26. A. Mukhtar, L. Xia, T.B. Tang, Vehicle detection techniques for collision avoidance systems: a review. IEEE Trans. Intell. Transp. Syst. **16**, 2318–2338 (2015)
27. P. Dollar, C. Wojek, B. Schiele, P. Perona, Pedestrian detection: an evaluation of the state of the art. IEEE Trans. Pattern Anal. Mach. Intell. **34**, 743–761 (2012)
28. A. Kelly, *MobileRobotics* (Cambridge University Press, New York, 2013)
29. B. Qin, Z.J. Chong, T. Bandyopadhyay, M.H. Ang, Metric mapping and topo-metric graph learning of urban road network. *In Proc. the 2013 IEEE Conference on Robotics, Automation and Mechatronics (RAM)*, Manila, Philippines, 2013, pp. 119–123
30. W. Liu, S.W. Kim, M.H. Ang, Probabilistic road context inference for autonomous vehicles. In *Proc. the 2015 IEEE International Conference on Robotics and Automation (ICRA)*, Seattle, WA, USA, 2015, pp. 1640–1647
31. C.R. Baker, J.M. Dolan, Traffic interaction in the urban challenge: putting boss on its best behavior. *In Proc. the 2008 IEEE/RSJ International Conference on Intelligent Robots and Systems*, Nice, France, 2008, pp. 1752–1758
32. J.C. Latombe, *Robot Motion Planning* (Springer Science & Business Media, Berlin, 2012)
33. M. Salazar, A. Alessandretti, A.P. Aguiar, C. N. Jones, An energy efficient trajectory tracking control of car-like vehicles. *In Proc. the 2015 54th IEEE Conference on Decision and Control (CDC)*, Osaka, Japan, 2015, pp. 3675–3680
34. T. Faulwasser, R. Findeisen, Nonlinear model predictive control for constrained output path following. IEEE Trans. Automat. Contr. **61**, 1026–1039 (2016)
35. T. Kunz, M. Stilman, *Time-Optimal Trajectory Generation for Path Following with Bounded Acceleration and Velocity* (MIT Press, Cambridge, MA, 2013)
36. M.W. Park, S.W. Lee, W.Y. Han, Development of lateral control system for autonomous vehicle based on adaptive pure pursuit algorithm. *In Proc. the International Conference on Control, Automation and Systems*, Seoul, Korea, 2014, pp. 1443–1447

Wi-Fi Direct Issues and Challenges

Rabiah Alnashwan and Hala Mokhtar

1 Introduction

The widespread use of smartphones with advanced communication technologies opens new opportunities for users, providers, developers, and manufacturers. Billions of devices around the world form a huge connected network, where each device is connected to other devices by wired or wireless links. Formally, wireless connection leads the current technological revolution, where most devices rely on the access point (AP)/base station-based paradigm. It is important, however, to mention that the maintenance of a constant connection with a wireless infrastructure base is expensive and not always possible. Weather, natural and human-made disasters, or even electricity problems can damage or simply deactivate a base station. Critical applications, such as military and safety applications, should therefore not be fully dependent on infrastructure-based wireless networks.

Infrastructureless wireless ad hoc networks provide a cheap solution to the need for more independent reliable networks. Bluetooth [1], Wi-Fi Direct [2], and Zigbee [3] are the most well-known technologies that are used for ad hoc wireless connectivity. By exploiting wireless ad hoc technology, users will be able to establish an offline communication channel.

Recently, the Wi-Fi Alliance released an ad hoc wireless technology called Wi-Fi Direct [2] that is superior to other existing ad hoc technologies in terms of the coverage area, throughput, and power consumption. Wi-Fi Direct supports users' direct connection without requiring a physical central AP. A Wi-Fi Direct group consists of multiple group members (GMs) and one group owner (GO) that acts as

R. Alnashwan · H. Mokhtar (✉)
Department of Information Technology, College of Computer and Information Sciences, King Saud University, Riyadh, Saudi Arabia
e-mail: hmokhtar@ksu.edu.sa

K. Daimi et al. (eds.), *Advances in Security, Networks, and Internet of Things*, Transactions on Computational Science and Computational Intelligence,
https://doi.org/10.1007/978-3-030-71017-0_38

a soft AP to provide basic service set (BSS) functionalities and services. The GO is responsible for the participation, advertising, and communications among GMs. One of the main problems from which Wi-Fi Direct suffers is that communication between peers in a Wi-Fi Direct network occurs within a single group only and different groups in Wi-Fi Direct cannot communicate with each other even if they are in the coverage range. This limits the packet/message propagation in the network. An additional problem is its complete dependence on GOs. The GO is the centric of the group, and therefore no group can survive after its disconnection. After the disappearance of the GO, the whole group is disassembled, and all connections between GMs are broken.

In this work, we provide an in-depth analysis of the intergroup communication problem and possible group formation optimization solutions to improve Wi-Fi Direct. To identify their advantages and disadvantages, we also investigate and compare existing research methods proposed to solve Wi-Fi Direct issues.

The remainder of the paper is structured as follows: Section II provides an overview of Wi-Fi Direct technology and describes its main functionalities: the device discovery procedure, service discovery procedure, and peer-to-peer (P2P) group formation. In Sect. III, we focus on the main problems of Wi-Fi Direct technology. An in-depth analysis of the previous research work on Wi-Fi Direct is presented in Sect. IV. Finally, the conclusions drawn are provided in Sect. V.

2 Wi-Fi Direct

Wi-Fi Direct, also known as Wi-Fi P2P, is a standard released by the Wi-Fi Alliance in October 2010 [2]. Wi-Fi Direct is aimed to allow device-to-device (D2D) direct communication without a central AP. Basically, Wi-Fi Direct inherits all the enhanced Wi-Fi infrastructure mechanisms, such as quality of service (QoS), power-saving, and security [4]. Because of the inheritance relationship, Wi-Fi Direct technology permits devices to maintain a connection with other Wi-Fi Direct devices and an infrastructure network simultaneously. As compared to other ad hoc technologies, such as Bluetooth, Wi-Fi Direct achieves the highest data transfer rate by using many energy-saving mechanisms. With 250 Mbit/s and 100 m of coverage, Wi-Fi Direct is more attractive especially for exchanging text and voice messages, location information, and photos [2].

In general, a Wi-Fi Direct group contains multiple GMs that are connected to one GO forming 1:n topology. Each client in a group may be either a Wi-Fi Direct client or a legacy client (LC), that is, a client who is Wi-Fi certified but not P2P compliant (i.e., does not support Wi-Fi Direct). A Wi-Fi Direct certified device can concurrently connect to both Wi-Fi Direct devices and legacy devices over a Wi-Fi Direct network. The network establishment consists of three main phases: device discovery, service discovery, and group formation.

2.1 *Device Discovery*

Device discovery is the first phase of the group establishment process. It is responsible for quickly finding and determining the P2P device to which another device will attempt to make a connection. To be discoverable, unconnected peers should be in the listening state, where the listening channel should be chosen at the beginning of the discovery phase and retained until the connection is completed. Basically, the P2P device discovery phase consists of two major stages: scan and find [5]. In the scanning stage, devices scan all supported channels to gather information about the surrounding devices or groups to find the most suitable device to which to connect. The finding stage is used to provide a common channel to enable communication between simultaneously searching P2P devices. When a P2P device wants to connect to a group, it sends a provision discovery request to the GO to indicate its willingness to join the group. When a searching P2P device discovers a peer that is already connected to a group, it sends the device discoverability frame to the GO indicating the target device ID so that the peer is available to interchange the discovery information or to begin a group formation. This case does not, however, occur when the searching device is an LC, because it can discover only GOs [2]. Meanwhile, the P2P GO is either waiting for other devices, whether legacy or P2P, to discover it or searching for other desirable devices or services in other channels.

2.2 *Service Discovery*

Service discovery is an optional procedure that is usually performed after the device discovery phase and prior to the group formation phase. Its purpose is to determine the compatibility of services offered by a discovered P2P device through exchanging generic advertisement service (GAS) frames [2]. Its main objective is to find a list of all the services together with their information that are offered by a P2P device and determine whether the services offered by that device have been updated. For example, a printer can advertise itself as a printing service to which devices interested in printing can connect.

2.3 *Group Formation*

In the group formation phase, a P2P device decides to be a part of a group by creating either its own group of which it becomes the GO or a group with other devices that have already been discovered. In the second case, a group formation procedure is used to set the GO through comparing the exchanged credentials of the candidate devices for the P2P group and determining the group formation type. Basically, there are three types of P2P group formation: standard, persistent, and autonomous.

In the standard formation, the devices first discover each other and then negotiate over which device will act as the P2P GO. During the discovery process, the devices listen to the social channels, namely, 1, 6, or 11 in the 2.4 GHz band. Then, after finding each other, they start the GO negotiation which is performed using a three-way handshake, namely, GO negotiation request/response/confirmation. After roles are distributed to the potential GMs, they perform the processes of establishing a secure communication using Wi-Fi Protected Setup (WPS), followed by Dynamic Host Configuration Protocol (DHCP) IP assignment [6]. Persistent formation permits the storage of all credentials exchanged in the initial formation. This feature replaces GO negotiation with the invitation procedure (two-way handshake) and reduces the WPS provisioning process for faster group reinitiation. In autonomous group formation, the device creates a group and assigns itself as the GO [6].

2.4 Power-Saving Schemes

An important aspect of Wi-Fi Direct technology is its energy efficiency. In a Wi-Fi Direct network, most connected devices are battery-constrained. In Wi-Fi Direct, two power-saving mechanisms are therefore defined: the Opportunistic Power Save protocol and the Notice of Absence (NoA) protocol. In the first mechanism, the legacy power-saving protocol is assumed. The GO specifies the minimum amount of time for which it will stay awake after the reception of a beacon through advertising a time window (CTWindow) in all beacons and responses. After advertising the CTWindow, the GO can enter the sleep mode if and only if all GMs are in the saving mode. In the second mechanism, NoA, the GO has greater control of deciding the time at which it enters the sleeping mode and its duration. It allows the GO to announce absence periods which prevents the GMs, regardless of their state, from accessing the channel [6].

3 Wi-Fi Direct Network Limitations

Interconnection between different Wi-Fi Direct groups is one of the main challenges facing its large-scale deployment. The Wi-Fi Direct standard [2] defines only intragroup communications where the GO acts as an AP. The standard does not specify any rules or protocols for multigroup communication; the network is thus essentially based on single-hop communication which connects peers of a single group and ignores other devices or other existing groups even if within the detection range. The standard does not, however, preclude a Wi-Fi Direct device from becoming a member of more than one group; thus, intergroup communication is possible, but the standard does not describe in what manner. In a Wi-Fi Direct network, the GO acts as a DHCP server to provide IP addresses to all connected GMs. All GOs have the same IP address (192.168.49.1), whereas GMs are assigned random IP addresses in the range (192.168.49.2–254). The communication between two Wi-Fi Direct GOs cannot be established because the packet is dropped if the sender and receiver acquire the same IP address. Additionally, no GM can communicate with other GMs, because they rely solely on the GOs to forward their packets, and the GO cannot propagate anything to either unidentified members or other GOs. The question is, therefore, how different Wi-Fi Direct groups can be bridged if the GO cannot communicate with other GOs and/or members outside its own group.

Additional important challenges facing ad hoc Wi-Fi Direct are the optimization of a single group to extend the lifetime of the network and the reduction of power consumption to maximize the battery lifetime of the different nodes. The existence of a Wi-Fi Direct group relies entirely on the GO. If the GO leaves the group, it is destroyed, and the group formation process must be repeated to establish a new group. To extend the lifetime of the group, the following questions need to be addressed: How is it possible to reduce the energy consumption of GOs? What is the optimum group size? What recovery schemes can be used when the GO has left?

4 Research Work

Recently, there has been increasing interest in searching for efficient solutions for ad hoc connectivity and Wi-Fi Direct as the most suitable ad hoc technology currently available. A survey and analysis of various research studies aimed at improving Wi-Fi Direct are presented in this section. We first discuss the research studies in which the problem of connectivity between two or more Wi-Fi Direct groups was addressed and then describe the research studies aimed at improving the general performance of a single Wi-Fi Direct group.

4.1 Wi-Fi Direct Intergroup Communications

Several research studies addressed the group connection problem and their authors proposed various solutions to extend the range of Wi-Fi Direct and allow multigroup communication.

A first solution for providing intergroup communication is the time-sharing mechanism, i.e., disconnection from one group and connection to another when there are data to be transmitted. Sunil et al. [7] addressed this problem and proposed a solution for exchanging data between two groups based on the time-sharing mechanism. The main idea is that when there are data to be forwarded to an external group, the sender disconnects from the group temporarily and connects to the second group as a member. The authors described four operation modes to differentiate between the required processes: register, update, transfer, and scatter. The scatter mode is used to achieve multigroup communication over Wi-Fi Direct.

It is activated when the source peer wants to send a message to a peer in another group. After checking that the recipient is not in the local group, the source peer sends the intended message to the GO with the mode set to scatter. Then the GO performs an additional verification operation to ensure the nonexistence of the recipient in its group. After validation, the source peer temporarily disconnects itself from the current group. It then requests a connection to the second group. When the connection has been established, the message is sent to the new GO. Finally, the peer disconnects from the new group and reconnects to the original group by using the saved MAC address of the GO. Unfortunately, the scatter mode could exhaust the power of the network if the entered IP address is not correct. In this case, the data are transmitted across all groups in the network until it is eventually realized that no device with such an IP exists. Moreover, the operations of disconnecting from a group, joining a new group, and then reconnecting to the original group waste a considerable amount of time and resources for every data transfer between the two groups.

Another solution based on time-sharing was proposed by Liu et al. [8]. They proposed a simple method of developing a mobile ad hoc network using Wi-Fi Direct on Android devices. Although this method does not use any assistive technologies or sophisticated techniques, it successfully manages to disseminate data between devices outside their range. Basically, in the proposed method, all devices in the network must act as GOs. When a device decides to transmit a message, it first removes its GO status and connects to the target device as a GM-Wi-Fi P2P client. When the connection is completed and the data transmitted, the device disconnects from the current group and removes its client status and then returns as a GO of its own group to prepare for the subsequent transmission cycle. In this technique, devices are allowed to transmit data to devices outside their range in a multi-hop manner with the assistance of routing tables as shown in Fig. 1. Although this method is simple, it does not take advantage of Wi-Fi Direct's broadcasting nature or power-saving mechanisms. Moreover, the connection and disconnection of the different GOs waste many resources.

ADDR	GTW	HOPS
B	A	0
C	B	1
D	B	2

ADDR	GTW	HOPS
A	B	0
C	B	0
D	C	1

ADDR	GTW	HOPS
A	B	1
B	C	0
D	C	0

ADDR	GTW	HOPS
A	C	2
B	C	1
C	D	0

Fig. 1 Wi-Fi P2P MANET routing table in a multi-hop [3]

Felice et al. [9] also proposed a time-sharing solution for the assistance of delay-tolerant networks (DTN). The solution depends mainly on the role of group relay (GR), which is responsible for transmitting data among different groups in the network. The P2P GR device switches among different groups on a temporal basis, following a round-robin scheduler. The data propagation method in this solution is based on offloading and buffering data between the GO and GR. When a GR device has joined a group, it sends an Arrival message to notify its presence. It then waits for the GO to transmit all buffered data and offloads the data previously received by other groups. To limit the amount of data transmitted by GR, a sequence number is attached to the Alert message to avoid the transmission of redundant messages to the same GO and limit the number of hops over which the message is transmitted. The scheme, however, still suffers from the performance delay and the resource wastage incurred by the switching between the different groups.

A different solution for intergroup communication proposed in [10] is based on noticing that a Wi-Fi Direct peer is able to connect as an LC in a traditional Wi-Fi network while connected to a Wi-Fi Direct group. The main idea is therefore to use simultaneous connections of Wi-Fi Direct and conventional legacy Wi-Fi. Duan et al. considered a scenario where ad hoc connectivity is needed for public safety in emergency situations when both APs and cellular networks are unreachable. The authors proposed a logical topology that exploits Wi-Fi Direct's features and overcomes its limitations on Android devices. According to the proposed topology, peers in a Wi-Fi Direct group operate also as LCs, as shown in Fig. 2, through leveraging the use of UDP connections with IP broadcast packets. The authors designed a tunneling mechanism that enables multigroup communication in Wi-Fi Direct using application-layer addressing, which enables peers to have two virtual network interfaces. The idea behind the design is to allow the GO of one group to behave as an LC in another group, thus acting as a gateway between the two groups. The design allows each GO to have two network interfaces with two different IP addresses, one for the conventional Wi-Fi and the other for P2P Wi-Fi Direct. The proposed approach is mainly a bus topology that connects all groups unidirectionally. This scheme allows GMs to disseminate data between groups but does not allow direct dissemination between GOs. The following example provides clarification. As shown in Fig. 2 when GO2 wants to send a message to GO1 or GO3, the source address of that message is 192.168.49.1 which coincides with that of other GOs. However, when Client 2 wishes to disseminate data to other groups, the message reaches GO3 successfully, because the source address of that message is 192.168.49.59, whereas GO2 does not receive the message because it is not part of the Wi-Fi Direct group of Client 2.

A similar solution utilizing the idea of simultaneous connection was presented in [11]. In this paper, the authors proposed an efficient multigroup formation and communication (EMC) protocol for Wi-Fi Direct. EMC utilizes the battery specifications of the devices to qualify potential GOs. In addition the EMC approach extends the protocol to support Wi-Fi Direct to allow dynamic group creation and multigroup communication by providing two main features: initial data exchange support and intragroup and intergroup communication support. The authors utilized

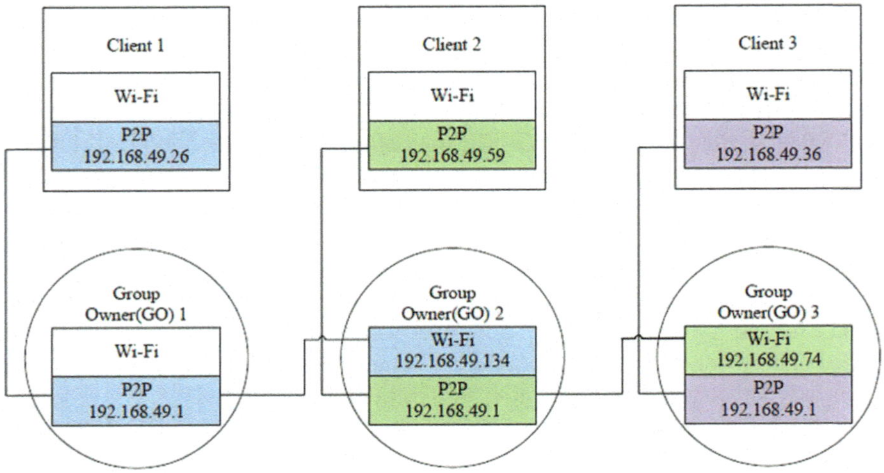

Fig. 2 Multigroup physical topology with two interfaces [5]

the alert dissemination using service discovery (ADS) protocol [12] to permit the initial data exchange with the help of Wi-Fi Direct service discovery. The ADS protocol takes advantage of the service discovery announcements and requests in Wi-Fi Direct to propagate data between devices without a prior connection. The authors utilized the ADS protocol to exchange specific types of information, such as battery information and the software AP (SAP) credentials (SSID and Key). The SAP credentials are used to allow devices to connect to Wi-Fi Direct groups as LCs. Intergroup communications are achieved by allowing GMs (proxy members) to be connected to two groups simultaneously: as a regular GM in one group and as an LC in the other group. The EMC protocol consists of several stages. The first and second stages address the choice of GOs and the creation of the group which depend mainly on the shared battery information. The next stage describes the process of selecting which group to join, where each (non-GO) device chooses the desired group and shares all collected SAP records with the GO. After a successful group creation/joining, the GO starts selecting GMs for each known SAP; the assignment is sent to the selected GM which in turn acts as a GM for the assigned group and connects to it as an LC. This is similar to the scheme proposed by Duan et al. in [10] but makes a GM instead of GOs connect to other GOs as an LC. The final stage in EMC is about tearing and restarting the entire protocol to allow the energy load to be balanced and the devices to be synchronized. The EMC protocol cannot be applied on regular devices since the author implemented some changes in the source code; the protocol is thus limited to rooted Android devices and is not expected to operate successfully on most regular users' devices. Additionally, the tearing process raises the overhead of GMs and consequently requires additional time and energy.

An additional solution based on the use of simultaneous Wi-Fi legacy/Wi-Fi Direct topologies was proposed in [13]. This paper provides an in-depth analysis

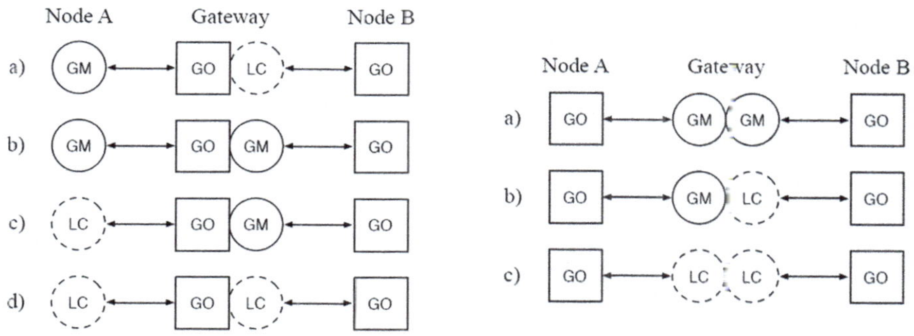

Fig. 3 Multigroup communication scenarios [6]

for the simultaneous solution where several scenarios were considered as shown in Fig. 3 (gateway can be a GO in one group and GM/LC for the other group or a GM/LC in both groups). It also compares the performance of the network using a time-sharing solution and a simultaneous solution. Several interesting results were obtained. For a node to act as a GM/LC gateway, a UDP multicast socket should be used, and it is not possible to forward data using unicast TCP or UDP. This is mainly because the Android platform prioritizes a Wi-Fi link over a Wi-Fi Direct link, and therefore no data can be routed with a unicast socket over the Wi-Fi Direct link. A multicast socket allows the particular interface to be used by the socket for receiving and transmitting data packets to be specified. Multicast UDP is not, however, reliable and cannot fully utilize the total available bandwidth. Consequently, the authors proposed a hybrid solution based on a combination of the two solutions: the time-sharing mechanism and the simultaneous approach. The solution uses unicast sockets for transmitting over Wi-Fi Direct and Wi-Fi legacy links and multicast for a control channel. The control channel triggers a gateway node configuration to change, if such is required. Because the Android platform prioritizes a Wi-Fi link over a Wi-Fi Direct link, the LC gateway can receive and transmit data, whereas the Wi-Fi Direct gateway cannot. The gateway configuration should thus always be LC/GM to allow transmission between groups. If the gateway is an LC in the sending group, data are transmitted immediately; the gateway then disconnects from the first group and connects to the second group and uses a TCP connection to forward the data. If, however, the gateway is set as GM/LC, then it is triggered to change its configuration. The new solution achieves a better performance but still suffers from a waste of resources and delay during connection/disconnection to the different groups.

The final solution we present is a hybrid scheme that uses Wi-Fi Direct and Bluetooth to achieve multi-hop communication [14]. The scheme was proposed mainly for connecting heterogeneous devices that support devices having one wireless technology only (either Bluetooth or Wi-Fi Direct) using a bridging device that supports both technologies. The proposed solution, BWMesh, is a framework that runs on top of the Android OS to integrate two typical D2D wireless technologies,

Bluetooth and WI-FI Direct, to alleviate the limitation of single-hop networking. Basically, the framework's logical architecture contains three layers, the networking (technology), middle (function), and typical application layer. Although the solution is not specifically for supporting intergroup communication in Wi-Fi Direct, the main idea can be used for such scenarios to connect, for example, two GOs using a Bluetooth link. However, further investigations of the feasibility of the idea are required.

4.1.1 Discussion

Many researchers have designed and implemented techniques to overcome the intergroup communication issue in Wi-Fi Direct. The solutions can be classified into two main categories: time-sharing and simultaneous connection. The time-sharing solution is a concept that is based on peers detaching from and attaching to different groups for the purpose of exchanging data among groups. This TCP-based mechanism requires a gateway node to switch between two (or more) groups. Because there is no built-in switching functionality, the switching is based on disconnecting from the current group, scanning for active nodes, and then connecting to a new group. The required time for a successful switching operation between groups depends on the type of the gateway, i.e., whether it is a GO, GM, or LC. If the gateway is a GO, the entire process of group formation is repeated every time the gateway node disconnects. Although the time-sharing solution solves the intergroup communication issue, its use has several drawbacks. In addition to the delay and energy dissipation and resource wastage, the solution does not capitalize on Wi-Fi Direct's power-saving mechanism if the gateway is an LC in both groups because of the complete dependence on the WiFiManager (standard Wi-Fi). This means that if the gateway is an LC in both groups, the attaching, detaching, and transmitting operations purely depend on the Wi-Fi legacy mode. Furthermore if the gateway is a GO, the detaching step leads to the destruction of the entire group, which means that the group reconfiguration takes more time than when a gateway is a regular GM. In addition, in a case where a regular GM acts as a gateway, there is a high probability of failure in the operation of joining the original group, as many research studies have shown that many cases occur where peers are unable to reconnect to their original group after joining other groups.

Simultaneous connection is a better solution that can achieve a superior network performance in terms of delay and energy consumption. In this solution, a gateway node uses two different wireless technologies, Wi-Fi Direct and Wi-Fi, with two different interfaces and two different IP addresses. However, real implementation on the Android platform shows that it is not possible to use UDP or TCP unicast and that the only possible connection is based on UDP multicast. Multicast UDP cannot, however, fully utilize the total available bandwidth and is not reliable where packets can be lost and/or duplicated easily.

4.2 Single-Group Optimization

An additional important challenge facing ad hoc Wi-Fi Direct is finding means of extending the lifetime of the network and reducing power consumption to maximize the lifetime of the batteries of the various nodes. As described previously, a Wi-Fi Direct group is destroyed when the GO leaves the network or if its batteries are exhausted. It is therefore important to select the GO efficiently and reduce its load as much as possible to extend the lifetime of the group. Some scholars addressed these issues and proposed different solutions.

The problem of efficiently assigning a GO automatically without user interaction was discussed in [15], and three possible schemes for achieving self-organizing, self-healing Wi-Fi Direct groups were proposed. The first scheme, backup-based group formation, is a scheme in which the GO is initially elected following the standard approach of Wi-Fi Direct, and then the GO elects a backup GO after the group formation is complete. In the case of a GO disconnection, the backup node declares itself as the new GO and reforms the group and chooses a new backup GO. This scheme provides a higher level of robustness and reliability in the case of GO failure and disconnection. The solution is not, however, suitable for highly dynamic networks. The second scheme, ID-based group formation, depends on the GO being chosen based on the device ID where the node with the smallest ID wins the GO election. Because this scheme requires that all peers exchange their IDs, it may not be practical for large networks because of the long peer discovery times. The third scheme, random device group formation, includes a stochastic component for the selection of the GO. This is a random selection scheme based on random timers and therefore may result in the selection of more than one GO. Thus, a step-back algorithm is used to ensure that there is only one GO at any given time. According to this algorithm, if a GO detects another GO during the peer invitation time, it will disconnect and restart the Scheme. A performance evaluation showed that the three schemes are completely automatic with self-adaptation to changes in network topology, such as node mobility or GO failure. It also showed that the random GO selection scheme is the most suitable solution in terms of group formation/reformation time.

A similar method was proposed in [16, 17] to enhance the method of GO selection and to reduce the time required for the group formation process. In [16], the authors proposed method accelerates the group formation and eliminates the three-way handshake standard negotiation phase for selecting GOs by exchanging intent values during the device discovery phase. Through utilizing the scan/find phase and sending the intent values as the information element in the probe request/probe response frames, the node with the highest intent value is selected. Moreover, each node can save a list of neighbor nodes with their intent values to simplify group reformation when a GO leaves. In this case, the node with the second highest intent value acts as a backup for the GO and can start group formation autonomously. The authors' results show that the proposed method reduces the time required for group formation and group recovery.

Chaki et al. [18] proposed a new mechanism, seamless group reformation (SGR), that helps maintain semicontinuous connectivity and alleviates overdependence of a group on its GO. The SGR mechanism is coupled with a technique, the Dormant Backend Links (DBL), to reduce the total group disruption time. The proposed solution consists of four processes: identifying emergency GO's (EGOs) intent and credentials, calculating the EGO's metric, updating and sharing the EGO list, and creating DBLs and activating them. The process of identifying emergency GOs is performed in the GO Negotiation Request/Response frame, where all participating devices may add a 1-bit flag to their existing GO intent, called Emergency GO Intent. Emergency GO Intent is set to 1 if the device wants to act as a GO in the future (EGO). The EGO shares his/her EGO intent value and credentials with the actual GO after joining the group. Then the GO calculates the shared information based on a specific EGO metric and updates the EGO list in decreasing order. The EGO metric is based on a predefined scale which may contain the following data: residual battery power, CPU speed, primary memory size, transmission range, closeness quotient, and sojourn time. The GO shares the EGO list with all associated clients periodically or when it is updated. Upon receipt of the EGO information from the GO, the DBL is created. The creation of a DBL is the creation of a virtual persistent session history that is achieved by replacing the old configuration with a new configuration; this new configuration represents the EGO itself as the GO. All the previously mentioned processes occur while the GO is still extant. However, when the GO disappears, the DBLs are activated. The activation is triggered when the GMs, previously configured as a virtual persistent group, reconfigure themselves with the new configuration. Immediately after the GO's disconnection, the EGO creates an autonomous group (autonomous persistent group) with the credentials previously assigned by the previous GO. This enables all members to connect to the group created by the EGO in a persistent manner.

Santos et al. [19] proposed a framework that combines P2P and DTNs for disaster management. This framework allows users to create applications that could benefit from the capabilities of the combined technologies. In the framework, the DTN component is responsible for handling situations where the data connection is intermittent by using a store-and-forward mechanism. However, the Wi-Fi Direct component is responsible for data propagation between the different peers. As a part of the framework, the authors suggested several modifications to improve the performance of Wi-Fi Direct. For selecting the most appropriate GO, they proposed using the battery percentage as a criterion. Instead of using the intent value as a criterion, the system elects the node with the highest battery percentage as a GO. Such a method was also adopted in the study in [9]. Furthermore, to increase the robustness of Wi-Fi Direct and decrease the recovery time, an EGO acts as a GO backup if the original GO leaves. To connect the different groups, a time-sharing mechanism is used where a proxy peer in each group is used to transmit data from one group to another by detaching itself from its current group and attaching to a new group, as a client then returns to its original group.

Minimizing energy consumption in Wi-Fi Direct is a vital aspect for battery-powered devices especially in emergency situations, which have been addressed in

[20, 21]. Usman et al. proposed a scheme that is aimed at optimizing the energy consumption and throughput of multi-hop Wi-Fi Direct networks. To achieve this, the scheme analyzes the effect of two parameters, group size and the transmit power of different nodes (GO, GM, or LC), on the throughput and energy consumption. Using simulations and based on standard group formation, the study showed that when the group is larger than a threshold (optimum size), energy consumption increases and throughput decreases. The authors suggested tuning the transmission power of the devices which would consequently affect the transmission range and group size. The effect of tuning on throughput and energy dissipation was analyzed. Again, it was shown that reducing the transmit range results in a decrease in the energy consumption and also the total throughput. Simulation results also showed that by reducing the group size, multi-hop communication increases, and thus gateway nodes between the different groups become a bottleneck hindering high throughput. Consequently, to achieve a better performance, a medium-sized group should be used, and the transmit range should be selected to balance energy-saving and throughput.

4.2.1 Discussion

Energy dissipation and lifetime are critical issues for Wi-Fi Direct networks, especially in emergencies. Several researchers proposed ideas and solutions that can be merged to improve the performance of the network. Most solutions can be classified into the following categories:

- Solutions to optimize GO selection: Whereas most studies were based on random selection, some authors proposed using the battery level as a criterion. Other criteria that may be used are the processing capability, the number of neighbors, communication with fixed networks, etc. In addition, autonomic solutions to change the GO to save its energy would need further investigations.
- Solutions to accelerate the group formation/reformation process: Most researchers suggested the use of a backup GO that would follow the autonomous group formation if the previous GO leaves. However, the criterion for choosing the backup GO has not been investigated. An additional solution that requires investigation is the use of persistent group reformation to reduce the time and resources required.
- Solutions to control the parameters of the single group in terms of size and transmission range: These solutions need to be extended to cover multigroup architecture and the possible use of gateways to reduce the load on the GO, thus extending the group lifetime.

5 Conclusion

In this paper, we presented an overview of Wi-Fi Direct technology and an in-depth analysis of its main problems and the various existing solutions. Two main areas that need enhancement were identified, namely, intergroup communication and the group formation process. For the intergroup communication problem, solutions were classified into two categories: time-sharing and simultaneous connection. The time-sharing solution suffers from delay, energy dissipation, and resource waste, in addition to insufficient utilization of the power-saving mechanism provided by Wi-Fi Direct. The simultaneous connection solution shows a better network performance than the time-sharing solution in terms of delay and energy consumption; however, this solution is more complex, introduces packet loss, or is not applicable in all cases. Thus, more research work is required to improve existing solutions.

For the group formation process, research studies have focused on improving the network performance by optimizing the GO selection, accelerating the group formation process, or controlling the size and transmission range of the single group. However, no single solution considers all three aspects. Thus, a comprehensive solution that considers the three aspects with further improvements and enhancements is needed.

Acknowledgments This research project was supported by a grant from the "Research Center of the Female Scientific & Medical Colleges," Deanship of Scientific Research, King Saud University. The authors thank the Deanship of Scientific Research and RSSU at King Saud University for their technical support.

References

1. "Markets | Bluetooth Technology Website." [Online]. Available: https://www.bluetooth.com/markets. Accessed: 02-Mar-2018
2. "Wi-Fi Peer-to-Peer (P2P) Technical Specification Version 1.7." Wi-Fi Alliance
3. "White Papers | Zigbee Alliance."
4. C. Funai, C. Tapparello, H. Ba, B. Karaoglu, W. Heinzelman, Extending volunteer computing through mobile ad hoc networking, in *2014 IEEE Global Communications Conference*, (2014), pp. 32–38
5. M. Conti, F. Delmastro, G. Minutiello, R. Paris, Experimenting opportunistic networks with WiFi Direct. *In 2013 IFIP Wireless Days (WD)*, 2013, pp. 1–6
6. D. Camps-Mur, A. Garcia-Saavedra, P. Serrano, Device-to-device communications with Wi-fi direct: overview and experimentation. IEEE Wirel. Commun. **20**(3), 96–104 (2013)
7. S. Sunil, A. Mukhopadhyay, C. Gujjar, Multi-group message communication on android smartphones via WiFi direct. *In 2017 International Conference on Advances in Computing, Communications and Informatics (ICACCI)*, 2017, pp. 1994–1999
8. K. Liu, W. Shen, B. Yin, X. Cao, L.X. Cai, Y. Cheng, Development of Mobile ad-hoc networks over Wi-Fi direct with off-the-shelf android phones, in *2016 IEEE International Conference on Communications (ICC)*, (2016), pp. 1–6
9. M. Di Felice, L. Bedogni, L. Bononi, The emergency direct mobile app: safety message dissemination over a multi-group network of smartphones using Wi-Fi direct. *In Proceedings*

of the 14th ACM International Symposium on Mobility Management and Wireless Access, New York, 2016, pp. 99–106
10. Y. Duan et al., Wi-fi direct multi-group data dissemination for public safety, in *WTC 2014; World Telecommunications Congress 2014*, (2014), pp. 1–6
11. A.A. Shahin, M. Younis, Efficient multi-group formation and communication protocol for Wi-Fi Direct. *In 2015 IEEE 40th Conference on Local Computer Networks (LCN)*, 2015, pp. 233–236
12. A.A. Shahin, M. Younis, Alert dissemination protocol using service discovery in Wi-Fi direct. *In 2015 IEEE International Conference on Communications (ICC)*, 2015, pp. 7018–7023
13. C. Funai, C. Tapparello, W. Heinzelman, Enabling multi-hop ad hoc networks through WiFi direct multi-group networking, in *2017 International Conference on Computing, Networking and Communications (ICNC)*, (2017), pp. 491–497
14. Y. Wang, J. Tang, Q. Jin, J. Ma, BWMesh: a multi-hop connectivity framework on android for proximity service, in *2015 IEEE 12th Intl Conf on Ubiquitous Intelligence and Computing and 2015 IEEE 12th Intl Conf on Autonomic and Trusted Computing and 2015 IEEE 15th Intl Conf on Scalable Computing and Communications and Its Associated Workshops (UIC-ATC-ScalCom)*, (2015), pp. 278–283
15. U. Demir, C. Tapparello, W. Heinzelman, Maintaining connectivity in ad hoc networks through WiFi direct. *In 2017 IEEE 14th International Conference on Mobile Ad Hoc and Sensor Systems (MASS)*, 2017, pp. 308–312
16. W. Cherif, M.A. Khan, F. Filali, S. Sharafeddine, Z. Dawy, P2P group formation enhancement for opportunistic networks with Wi-Fi direct. *In 2017 IEEE Wireless Communications and Networking Conference (WCNC)*, 2017, pp. 1–6
17. K. Jahed, O. Farhat, G. Al-Jurdi, S. Sharafeddine, Optimized group owner selection in WiFi direct networks, in *2016 24th International Conference on Software, Telecommunications and Computer Networks (SoftCOM)*, (2016), pp. 1–5
18. P. Chaki, M. Yasuda, N. Fujita, Seamless group reformation in WiFi Peer to Peer network using dormant backend links. *In 2015 12th Annual IEEE Consumer Communications and Networking Conference (CCNC)*, 2015, pp. 773–778
19. J. Santos, T. Danah, M. Villanoy, C. Festin, J. Doctor, R. Ocampo, Rapid mobile development with ARC: application framework for robust communications for disaster risk reduction and management – semantic scholar. *Presented at the Third International Workshop on Emergency Networks for Public Protection and Disaster Relief*, 2016
20. M. Usman, M.R. Asghar, I.S. Ansari, F. Granelli, K. Qaraqe, Towards energy efficient multi-hop D2D networks using WiFi direct, in *GLOBECOM 2017–2017 IEEE Global Communications Conference*, (2017), pp. 1–7
21. A. Laha, X. Cao, W. Shen, X. Tian, Y. Cheng, An energy efficient routing protocol for device-to-device based multihop smartphone networks, in *2015 IEEE International Conference on Communications (ICC)*, (2015), pp. 5448–5453

RFID Assisted Vehicle Navigation Based on VANETs

Yang Lu and Miao Wang

1 Introduction

As urbanization deepens, the structure of the urban area is becoming more complicated with the increased number of buildings, tunnels, etc. Therefore, navigation becomes more important with complex road conditions. Currently, vehicle navigation is mostly based on global positioning system (GPS) that is a network for positioning and path planning based on 30 satellites orbiting the Earth [1]. However, buildings in urban canyons block the GPS satellite signals [2], which let positioning satellites failed or with unignorable errors. In order to eliminate the inaccuracy or failure problems of GPS, one possible solution is to integrate inertial navigation systems (INS) or dead reckoning (DR) to help a GPS when it is failed. However, this method has significant disadvantages. That is, INS and DR are self-contained navigation techniques in which measurements are provided by accelerometers and gyroscopes. They integrate the measured angular rate and linear acceleration to obtain position change, velocity change, and attitude change. Hence, small errors in the measurement of acceleration and angular velocity will gradually become larger errors as the number of integrations increases [3]. In other words, the positioning error of INS and DR increases as the working time increases. Therefore, the position must be periodically corrected by input from GPS. In the case of GPS failure, the error of INS and DR cannot be corrected and becomes larger, and the vehicle will be positioned incorrectly making drives in the wrong path. Thus, it is still challenging to navigate when GPS is failed.

To improve positioning accuracy, we first choose VANETs that can share local real-time information to assist GPS. Vehicular ad hoc network (VANET) is a

Y. Lu (✉) · M. Wang
Department of Electrical and Computer Engineering, Miami University, Oxford, OH, USA
e-mail: I.luy33@miamioh.edu; II.wangm64@miamioh.edu

K. Daimi et al. (eds.), *Advances in Security, Networks, and Internet of Things*,
Transactions on Computational Science and Computational Intelligence,
https://doi.org/10.1007/978-3-030-71017-0_39

wireless network that consists of groups of moving or stationary vehicles and roadside unit (RSU). VANETs support vehicle-to-vehicle (V2V) and vehicle-to-road-side-unit (V2R) communications, which can collect and send real-time traffic information to each other [4]. Through these communications, vehicles can transfer their local information to other vehicles and RSUs. The server uses the data uploaded by RSU for path planning. VANET allows us to independently plan the path or use server to compare local data and GPS data to improve GPS accuracy. However, the vehicle through VANETs cannot obtain the surrounding information. It is still challenging to obtain the building's accurate coordinates.

Thanks to the rapid development of wireless location technologies, radio frequency identification (RFID) has attracted full attention and becomes a possible solution for sharing the positioning information in a mobile vehicle network. RFID system uses electromagnetic waves for data transmission, which operates in various frequency bands and provides the corresponding radio ranges. The RFID system is composed of an RFID reader and an RFID tag. Tags store some buildings and street location information. The vehicle can read the location information stored in the RFID tag through an RFID reader. Since the RFID system cannot transfer information to other vehicles, VANETs can help to deliver information to the server for further operations, e.g., path planning. Hence, by combining RFID data and real-time vehicular information from VANETs, we can make positioning more accurate and perform better path planning.

In this chapter, to solve independent positioning and path planning for mobile vehicles, we propose an RFID-enhanced VANET system to achieve more accurate and reliable positioning for path planning. The contributions of this paper are threefold. First, we propose a real-time traffic navigation architecture based on the RFID and VANETs, which can achieve independent communication. Second, a path planning algorithm is proposed based on the proposed navigation architecture. Thirdly, the proposed navigation algorithm's performance is demonstrated based on a simulator VISSIM that is the trace generator to record the vehicle information in a real-time manner.

The rest of this chapter is organized as follows. Section 2 provides related works on navigation methods. Section 3 discusses the system model. Problem formulation is presented in Sects. 4, and 5 shows the simulation results. At last, Sect. 6 concludes this chapter.

2 Related Works

GPS positioning failure is due to multiple reasons. At least four satellite pseudoranges are taken at the same time [5]. In the urban area, there are many high buildings, tunnels, and viaducts that will make satellite positioning less than four. This will cause huge errors or failure of GPS, and let the vehicle have an error destination or get lost, which may become a big problem in the urban transportation.

For example, Google map will fail in GPS-denied environment, which makes the user miss the intersection and take more time to return to the right path. However, by the help of RFID system, these conditions can be effectively improved.

There are many works that have studied RFID assisted vehicle positioning in VANETs. A RFID assisted localization system is put forward in [6], which uses DGPS principle. Differential GPS was used first for a maritime navigation system [7]. DGPS takes advantage of scientific observations that the distance from the satellite is so far that the GPS receivers near the Earth experience the same signal propagation delay, resulting in the same GPS error. In the DGPS system, the reference point is installed near the shoreline, such as a lighthouse, and calculates the GPS error that is the difference between the GPS value received from its own GPS receiver and the exact position it was measured at the time of installation. RFID assisted GPS system (RF-GPS) uses reference vehicles on a road to improve position accuracy. RF-GPS is different from the traditional DGPS system is that RF-GPS uses the GPS vehicle becomes a moving reference point. When a vehicle passes a stationary RFID tag, the vehicle obtains Abs coordinates and calculates the GPS error value using its own GPS coordinates. Then, it broadcasts GPS error to adjacent vehicles to help them to correct their GPS coordinates.

However, this theory may not work in the GPS-denied environment. In [8], the authors propose to use GPS–INS–RFID system to locate the train in GPS-denied environment. When the train runs in the tunnels, mountains, and remote areas, GPS receiver will lose signal and the coordinate is not available. INS (inertial navigation system) is a kind of autonomous navigation system, and it does not rely on outside information [9]. Providing the location information and speed to the INS, the INS updates the current position coordinates and speed through the calculation of information. But the positioning error increases with time, so it can only be used for a short time. GPS-denied environment will lead to long term use INS to calculate the coordinates that will cause coordinate deviation. RFID system can solve this problem. When GPS loses signal, the train can read the RFID tag to get location coordinates and replace the GPS coordinates to correct the error of INS. By this method, the train can be precise in positioning. He achieved the positioning in the GPS-denied environment, because the train has tracks, the author does not need to consider the path planning problem.

To transfer location information and real-time traffic status information, V2R and V2V communications can make real-time message delivery much quicker and more efficient in short distance transmissions. RSUs in VANETs can collect data and update to server to process. At the same time, RSUs can also transfer the processed data to other RSUs and vehicles.

Therefore, in this chapter, using RFID to assist navigation based on VANET is proposed for positioning and shortest time path planning in the urban area. Using information collection, transfer, and vehicle positioning with VANET and RFID, the additional cost will be reduced. In addition, the shortest path planning algorithm will discuss.

3 System Model

The architecture of the RFID-enhanced VANET system is firstly proposed in the urban area, as shown in Fig. 1. It consists of GPS system, RSUs, server, vehicles, and RFID systems.

GPS is referred as the global positioning system. It consists of satellites and GPS receiver. Satellites constantly send out signals. Once a GPS receiver listens for these signals and calculates its distance from four or more GPS satellites, it can figure out the location.

RSUs (roadside units) are the part of VANETs, and it uses IEEE 802.11 radio to exchange data with vehicle. The RSUs are deployed at each road intersection and collect real-time traffic information from vehicle via V2R communication in VANETs. The RSUs store two types of information. One is the map information, including road name, building name, and their coordinates. When the GPS fails, the vehicle can request coordinates from nearby RSUs to determine real-time location. The other one is the real-time data collected for vehicles that is provided by RFID. The real-time information contains traffic density and the number of vehicles.

Server is the calculation module. It can exchange information with RSUs. Vehicles arrive at new section and request information from RSUs. RSUs upload the map information and real-time information to the server. Server calculates the traveling time estimation, chooses the optimal path from vehicle to destination by shortest time path planning algorithm, and transfers this information to vehicle by RSUs.

Vehicle can provide real-time information via V2V and V2R communications. Vehicle can transfer information to other vehicles via V2V communication, and vehicle can transfer information with RSUs via V2R communication.

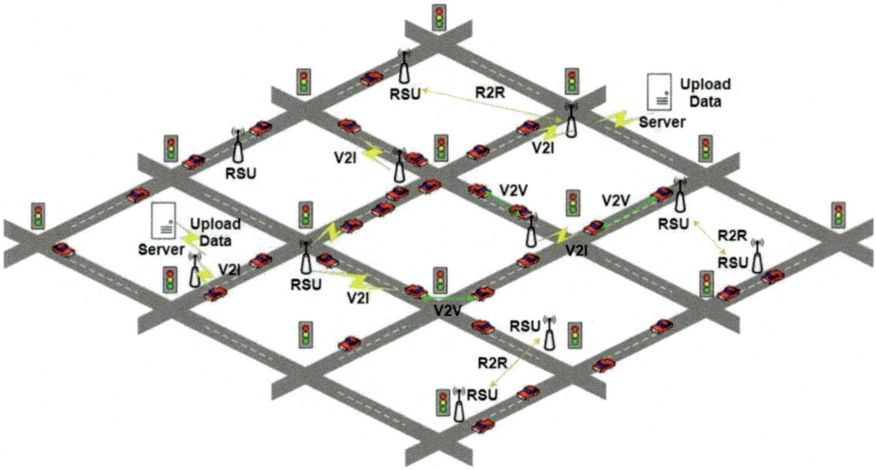

Fig. 1 RFID-enhanced VANET system model

RFID system is composed of the RFID tag and RFID reader. RFID tag is a short radio range passive tag fixed on the roadside unit or the road surface, such as road sign and speed sign. The RFID tag can provide accurate position coordinates, the lane's current traffic direction, and the street name that can help the vehicle identify itself its own locations. A vehicle can obtain its current position when passing by these tags. RFID reader is installed at the center of the vehicle front bumper. When the vehicle travels into the range of the RFID tag, RFID reader can read and collect the information that is saved in the RFID tags. The RFID system transfers the information to RSUs and other vehicles via VANET. After calculation, the vehicle can be used to plan the path.

The model works as follows which shown in Fig. 2: we consider the vehicle driving in the urban area which is GPS-denied/GPS error. The target vehicle's GPS signal is missing and cannot be located itself and its intended destination. At this time, the vehicle uses RFID reader to read the tags around it. The vehicle transmits the information to the RSUs via V2R. RSU uploads this information and the map information that is stored in the RSUs to the server. After server analysis and calculation, this processed information is sent back to the vehicle via RSUs. The vehicle uses this information to determine its own location and do path planning. At the same time, other vehicles can also upload the information that is read by their RFID reader, and this information can be used by every vehicle. When the vehicle arrives near the destination, it can use the information uploaded by other vehicles to determine the exact location of the destination. In addition, we use VANETs to get real-time traffic information in order to choose the shortest path when doing path planning Fig. 3.

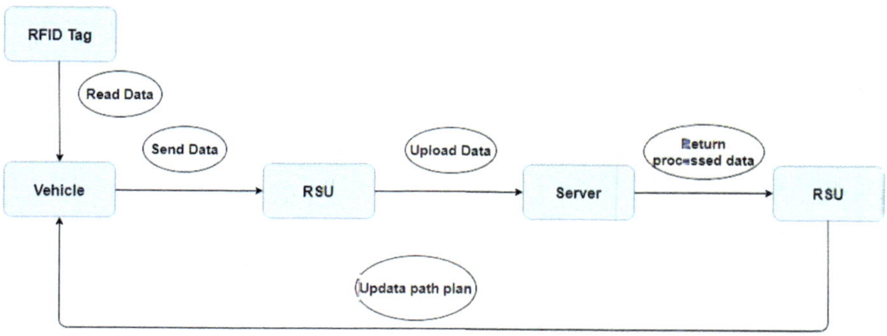

Fig. 2 Path planning procedure in RFID-enhanced VANET system

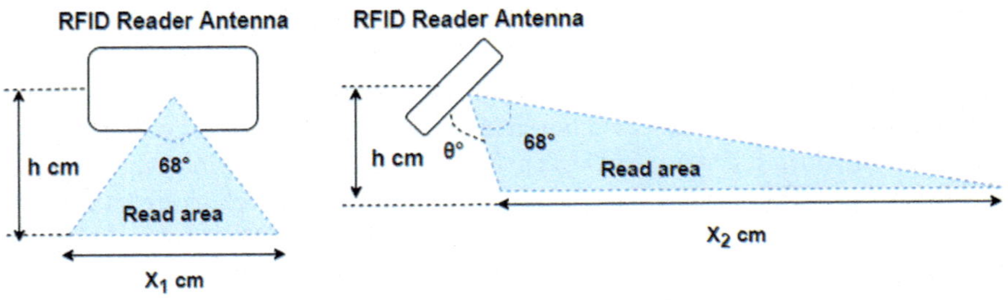

Fig. 3 RFID reader angle placement

4 Problem Formulation

4.1 Placement of RFID System

In this architecture, we use passive RFID system. It uses ultrahigh frequency (UHF, 860 MHz–2.45 GHz) because of lower tag price (only a few cents) and long radio range (up to 10m) [6]. An RFID reader and an RFID tag encounter during a very short time which vehicle moves at a very fast speed. RFID reader's read length follows below:

$$\frac{\bar{V}\times R_{data}}{R_{tag}} < L_{read} < \min\{D_{tag}, L_{V\,\min}\}. \quad (1)$$

L_{read} is the RFID's read length, and R_{data} and R_{tag} are the RFID tag's data size and transmission rate. V is the road speed limit. RFID reader's width follows below:

$$W_{read} < W_{lane}. \quad (2)$$

The L_{read} (x_1) and W_{read} (x_2) of the read area are calculated by equation:

$$x_1 = 2 \times h \times \tan 34^\circ \quad (3)$$

$$x_2 = \frac{h}{\tan(56^\circ + \theta^\circ)} + \frac{h}{\tan(56^\circ - \theta^\circ)}, \quad (4)$$

where the h and θ are the height and pitch angle of the reader. If h = 37.5 cm and $\theta = 45$, then we can compute $x_1 = 58.58$ cm and $x_2 = 185.63$ cm. Using these values, we can compute the maximum time in which a tag can stay in the moving RFID read area. The "Theoretical values" are computed by Eqs. (3) and (4), and the experimental values result from our test.

Data transmission rate is another important parameter. According to explanation, the tag has 256 kbps data rate, and it takes 0.22 ms to transmit 64 bit data, but in actual situations, it cannot achieve. To improve the read rate, we use duo RFID antennas. The duo antennas can increase the read width. Thus, the RFID read rate is also increased, and a vehicle can read tags even though it deviates from the center of a lane.

For tag placement, we want to make sure that the vehicle should not read two tags at the same time, so the distance between two tags should be lower bounded by the RFID reader's read area that is shown as below:

$$D_{\text{tag}}^2 > W_{read}^2 + L_{read}^2. \tag{5}$$

But the navigation system requires a vehicle to read RFID tags by RFID reader once in every distance D_{ANS}. To ensure seamless navigation in the GPS failure environment, we assume the following scenario: when the vehicle twice tries to read the D_{tag1} and its distance is larger than D_{tag2}. The vehicle will not read tag 2, and the vehicle obtains the tag 3's data. The distance between two success reads is D = 2 × D_{tag2}. We propose the distance between tag 1 and tag 2, D = D_{tag1} + 2 × L_{read}. Therefore, the seamless navigation should follow D <D_{ANS}, and the requirement is shown as

$$D_{tag} < D_{ANS} - 2 \times L_{read}. \tag{6}$$

According to Eq. (5) and Eq. (6), we can get

$$D_{ANS} - 2 \times L_{read} > D_{tag} > \sqrt{W_{read}^2 + L_{read}^2}. \tag{7}$$

4.2 RFID-enhanced VANET Bellman–Ford Algorithm

The Bellman–Ford algorithm is an algorithm that computes shortest paths from a single source vertex to all of the other vertices in a weighted digraph [10].

The traffic graph can be described as G (V, E, W). V is the set of nodes, and points on the streets such as starting points, destination points, and intersection points are chosen as nodes. E is the set of edges that are the segments between two nodes. W is the set of edge weight functions. For an edge e_{ij} = (V_i, V_j), its weight function W_{ij} means the traveling time from node V_i to node V_j. W consists of traffic flow and the length of road that are collected by our system.

For the traditional Bellman–Ford algorithm, the primary operation is "relax." Relax is to estimate the shortest path of the source node S to node V for each node V. The algorithm records its precursor node pre [V] for each non-source node V during the relaxation process. We defined distance as "dis," if the shortest path between S and U follows dis [U] + W [U, V] < dis [V], the dis [V] update. The relax will produce a "shortest path tree." Because each non-source node V, its source node to the shortest path has been calculated in this process.

All shortest paths starting from vs would be calculated when Bellman–Ford algorithm is terminated, which indicates a lot of unnecessary calculations. Since we only focus on the shortest path from V_s to V_e, there are many unnecessary calculations. To avoid this, we need to optimize Bellman–Ford algorithm.

After each round of "relax," there are some nodes that have already found the shortest path and these nodes are no longer affected by next "relax." We should exclude these nodes and only relax the edges whose shortest path has changed. Here we can use a queue to store these nodes. Select the node u of the head of queue. If the shortest distance from the source node to the node v becomes shorter by the edge u → v, then the node v is placed at the end of the queue.

In our algorithm, "W" (weight) consists of traffic flow and emergency situations. We consider q is the number of electric vehicles passing a reference point per unit of time (veh/h). The inverse of flow is headway (h), which is the time that elapses between the ith electric vehicle passing a reference point in space and the (i + 1)th electric vehicle. In emergency situations, h approaches infinity.

$$q = kv \tag{8}$$

$$q = \frac{1}{h}. \tag{9}$$

v is the speed of electric vehicle, and k is the density that is defined as the number of electric vehicles per unit length of the roadway. The flow (q) passing a fixed point (x_1) during an interval (T) is equal to the inverse of the average headway of the n electric vehicles.
v is the speed of vehicle, and k is density that is defined as the number of vehicles per unit length of the roadway. The flow (q) passing a fixed point (x_1) during an interval (T) is equal to the inverse of the average headway of the n vehicles.

$$q(T, x_1) = \frac{n}{T} = \frac{1}{\bar{h}(x_1)}. \tag{10}$$

In a time–space diagram, the flow may be calculated in the road R, where td is the total distance traveled in R.

$$q(R) = \frac{n}{T} = \frac{mdx}{Tdx} = \frac{td}{|R|}. \tag{11}$$

Our path planning algorithm calculates the shortest time path through W. W represents the length of the transit time, and a high W means that the road passes slowly, which in turn means fast. If there is a traffic jam such as traffic accident or traffic congestion, W will equal to infinity and abandon this road.$\sum_i W = \sum_i q$, and our algorithm is to find the lowest $\sum_i W$ to find the shortest time path.

4.3 *VANET (Vehicular Ad Hoc Network)*

In this architecture, VANET is mainly used to exchange real-time information. When the vehicle reads the RFID tag, it transfers tag information to the RSU via V2R communication. Also, vehicle can transfer some dynamic traffic information, such as the traffic accident and the traffic congestion. RSU's information sharing mechanism can immediately share this dynamic real-time traffic information. RSU uploads information to the server to calculate the shortest path and transmits it back to vehicle. Meanwhile, vehicle can transfer latest information through V2V communication. What is more, since we put the map in RSU, when the GPS is denied, vehicle can use the map directly for shortest time path planning.

5 Simulation

In this section, we consider a realistic area as shown in Fig. 4, which is the region around the campus of Miami university. To keep our simulation data highly realistic, a highly realistic microscopic vehicle traffic simulator, VISSIM [11], is used to generate vehicle parameters. However, we cannot change or control the paths of vehicle by the algorithm in VISSIM, and we use C++ to run the algorithm to prove the performance of our system.

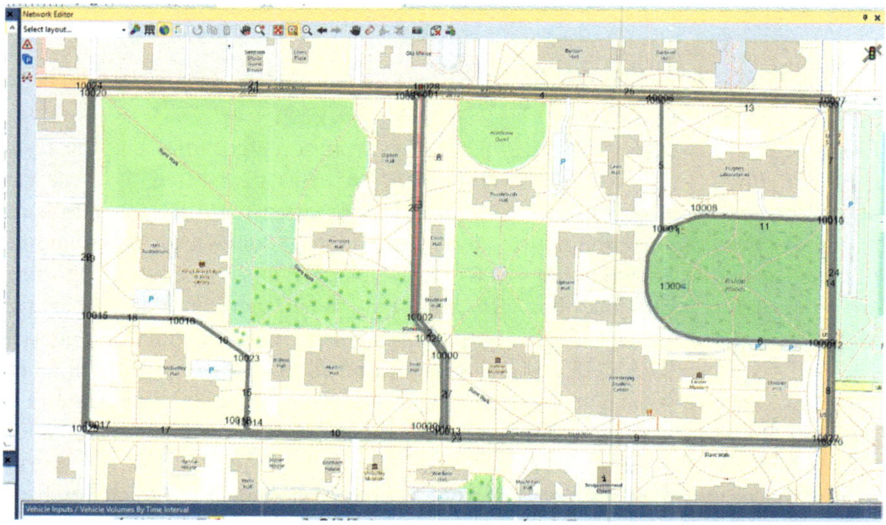

Fig. 4 Simulated map

5.1 Simulation Setup

Traffic Generation in VISSIM: To simulate real traffic flow in Miami university region, we set the vehicles coming in and going out at the intersection are random in VISSIM, which is a simulator as the realistic traffic generator. At the warm-up time duration in VISSIM, vehicles are set to enter the region from the initial booking, following a Poisson process at a rate 2000 vehicle/hour/entry. After the duration of the first 200s, stop the vehicles entering to reach a medium density scenario. In the VISSIM, vehicle information is recorded every 0.2s (e.g., traffic flow and speed, etc.). All the vehicles' speed follows actual road speed limit, not higher than 25 mph/hour. In addition, we can set the reduce speed area at any time during the simulation in VISSIM, which can be present as the different kinds of accidents/jams in the scenarios.

Navigation Evaluation in C++ algorithm: To evaluate the accuracy of the system and the performance of the path planning algorithm in C++ in Miami University region, every intersection is set as the node. RSU will be deployed at every intersection, and every road will place tags as required. The traffic flow of the road between every two nodes is taken as weight. To prove the impact of the accident on path planning, the weight of the accident road is infinite. To prove the accuracy of the system, the destination is set in a slightly blurred place, such as east or west sides of a building that will affect the path planning.

In order to prove the system can navigate in GPS-denied environment, we set up a situation where only RFID system is used for navigation.

5.2 Experimental Results

First, we have tested performance of the two proposed algorithms and the result shown in Fig. 5. It shows the travelling time of paths discovered by these two algorithms, and we compare the travelling time by choosing different number of nodes. Here, we choose the intersection as the node. It shows that as the number of nodes increases, the advantages of RFID-enhanced VANET navigation algorithm are more and more obvious than GPS algorithm. When there are only one or two nodes, their performance is basically the same. As the number of nodes increases, the RFID-enhanced VANET navigation algorithm shortens the traveling time and saves up to nearly 16%. There are two main reasons for this situation. The first is our algorithm excludes the nodes that are no longer affected by the next relaxation. The second is due to the information transmission by VANETs. We can get the newest information, which means that we can find the sum of the road weights is lowest for path planning.

To compare the accuracy, the starting point we choose for each experiment is the same and the destination we choose from near to far. Because the more accuracy the positioning, the shorter the distance traveled. The result is shown in Fig. 6. Except

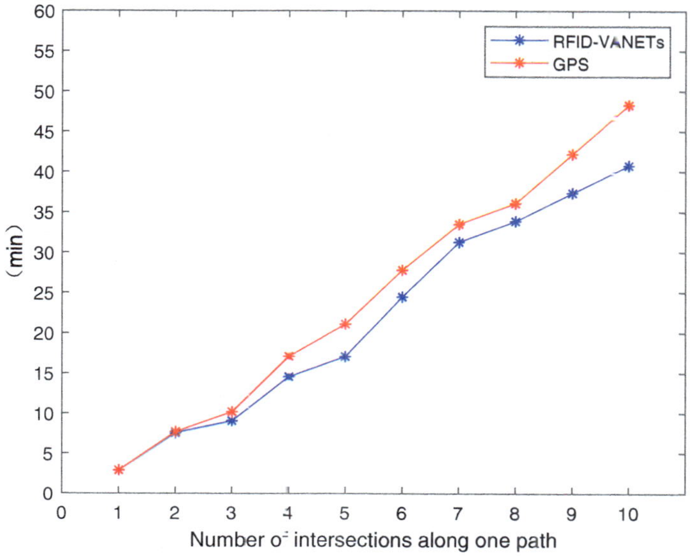

Fig. 5 Travelling time for different destinations

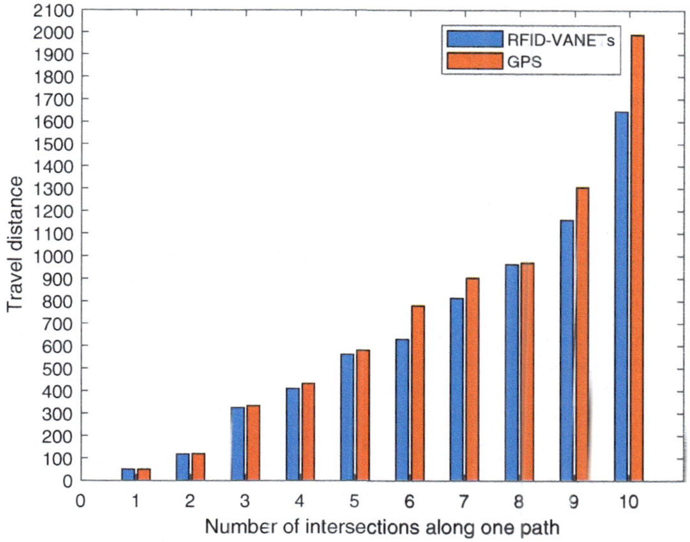

Fig. 6 Travelling distance for different destinations

for a few close destinations at the beginning, the remaining travelling distance for RFID-VANETs is shorter than GPS. The reason for this is that the deviation of the destination coordinates leads to different path planning. Vehicle via RSU corrected the coordinates and obtained a new path plan. Also, we can collect real-time information through V2V and V2R communications, and vehicle gets real-time traffic conditions faster than only GPS navigation, which let the shortest time path algorithm choose the exact shortest path. In addition, since the vehicle can receive the information delivered by RSUs at every intersection, it can update the weight of the road section in real time through the traffic flow. This information allows the shortest time path algorithm to quickly update the path selection to ensure that the time to reach the destination is the shortest.

6 Conclusions

In this chapter, we have proposed an architecture involving RFID and VANETs to achieve an accurate path planning for vehicles in urban areas. Specifically, RFID system is utilized for recording the accurate positioning information that is further delivered to server via VANETs. In addition, through VANETs, the real-time vehicular information can be collected for estimating traffic conditions for path planning. The simulations have been conducted to demonstrate that the proposed system based on RFID and VANETs can achieve better performance than conventional GPS system. In our future work, we are planning to do the implementation of our proposed navigation system in reality.

References

1. G. Wang, X. Xu, Y. Yao, J. Tong, A novel BPNN-based method to overcome the GPS outages for INS/GPS system. IEEE Access **7**, 82134–82143 (2019)
2. A. Abosekeen, A. Noureldin, M.J. Korenberg, Improving the RISS/GNSS land-vehicles integrated navigation system using magnetic azimuth updates. IEEE Trans. Intelligent Transp. Syst. (2019)
3. S.K. Shukla, J.-P. Talpin, *Synthesis of Embedded Software: Frameworks and Methodologies for Correctness by Construction* (Springer Science & Business Media, 2010)
4. M. Wang, H. Shan, R. Lu, R. Zhang, X. Shen, F. Bai, Real-time path planning based on hybrid-VANET-enhanced transportation system. IEEE Trans. Vehicular Technol. **64**(5), 1664–1678 (2014)
5. D. Špoljar, N. Črnjarić-Žic, K. Lenac, V. Perinović, Characterisation of multipath-caused commercial-grade GPS positioning error in intelligent transport systems (ITS), in *2019 International Symposium ELMAR* (IEEE, 2019), pp. 27–30
6. E.-K. Lee, S.Y. Oh, M. Gerla, RFID assisted vehicle positioning in VANETs. Pervasive Mobile Comput. **8**(2), 167–179 (2012)
7. N. Drawil, Improving the VANET vehicles' localization accuracy using GPS receiver in multipath environments. Master's thesis, University of Waterloo, 2007

8. Z. Wei, S. Ma, Z. Hua, H. Jia, Z. Zhao, Train integrated positioning method based on GPS/INS/RFID, in *2016 35th Chinese Control Conference (CCC)* (IEEE, 2016), pp. 5858–5862
9. K.R. Britting, Inertial navigation systems analysis (1971)
10. Y. Sun, X. Yu, R. Bie, H. Song, Discovering time-dependent shortest path on traffic graph for drivers towards green driving. J. Network Comput. Appl. **83**, 204–212 (2017)
11. G. Gomes, A. May, R. Horowitz, Congested freeway microsimulation model using VISSIM. Transp. Res. Record **1876**(1), 71–81 (2004)

Regular Plans with Differentiated Services Using Cuckoo Algorithm

John Tsiligaridis

1 Introduction

The most familiar aspect of mobile computing technology is the hand phone, and it is based on the communication between clients and large-scale distributed databases. A broadcast schedule program has to address the queries minimizing the expected delay that is the average waiting time, until the clients receive the requested items. The objective of any broadcasting plan is to reduce the expected delay. From the three basic data broadcasting design methods, the flat, the skewed, and the regular [1–5], our interest focuses on the last one. The flat design with the bigger size of data is transmitted by the server with the higher expected delay in large cycles. In this context, the use of a higher number of channels can reduce the waiting time. The skewed design sends the hot data to the fast channels and the cold data to slow channels. The regular design offers equal spacing considering the popularity of data, channel availability, and energy conservation. In [6], a set of algorithms has been developed for RBP discoveries. Depending on channel availability, data can be sent by a single or a set of channels. A number of channels can send a group of data providing also equal spacing of repeated instances of items.

In [7], a proposed Multichannel Square Root Rule (MSRR) for variable-length data with skewed access probabilities on variable-bandwidth channels has been presented. Data is partitioned and allocated to different channels according to MSRR, and different scheduling strategies are adopted. Data scheduling algorithms play significant role for the real-time on-demand data broadcast systems considering the bandwidth-limited wireless environment. To this end, two profit-based scheduling algorithms, [8], the profit versus cost (PVC) and single-slot

J. Tsiligaridis (✉)
Math and Computer Science Department, Heritage University, Toppenish, WA, USA
e-mail: tsiligaridis_j@heritage.edu

K. Daimi et al. (eds.), *Advances in Security, Networks, and Internet of Things*, Transactions on Computational Science and Computational Intelligence,
https://doi.org/10.1007/978-3-030-71017-0_40

allocation (SSA), are proposed. The PVC is for single-channel and multichannel scheduling, respectively, both of which utilize our new concepts "profit" of pending items and "opportunity cost" of pending requests. The SSA allocates some items, selected from scheduled requests to available channels, to a time slot. A tree-based adaptive broadcasting (TAB) algorithm for data dissemination to improve data access efficiency is developed in [9]. TAB first constructs a broadcast tree to determine the broadcast frequency of each data, and it splits the broadcast tree into some broadcast wood to generate the broadcast program. In [10], an adaptive algorithm is presented that exploits the trade-offs between blocking of requests reduced interference and guaranteed resources for individual data transmissions in each cell.

The server needs to discover regular plans to address the users' mobile queries. The problem lies in the computation of the size of data of different services so that an RBP or RBPs can be feasible. An RBP is considered feasible if it follows certain criteria, as it will be developed next, and it uses the smallest number of channels. For this purpose, a framework with a set of service queues, along with the main queue, is considered. In addition, solutions for high- and low-capacity channels are given based on the grouping length.

A set of theorems provide criteria that ensure the creation of an RBP with differentiated services. Based on CS algorithm's operation, there are two approaches for differentiated services that are developed. The CSDS can discover RBPs with the differentiated ratios. The ECSDS can provide an RBP with a desired AWT ratio for two high-priority services along with a minimum AWT for the remaining ones.

The server works with a set of different message (service) queues and has to define the size of data of each set to be processed in order to create an RBP. Many types of data sets make the problem of RBP creation more complicated. Considering the average waiting times (AWT), the RBPs can solve the broadcasting problem. For each service type of the various data sets, there is a queue, and the scheduler extracts packets in round-robin way and sends them to the main queue. The size of the data of each service type and the number of times data sending can be repeated into a sequence constitute the fundamental points of the construction of any RBP. Frequently requested data (hot data) have service priority over rarely requested ones (cold data). The basic purpose of this work is to prepare RBPs by using the metaheuristic Cuckoo search (CS) algorithm [11, 12] considering also the desired AWT.

Carrier aggregation (CA) [13] was introduced to meet the increasing demands in terms of throughput and bandwidth and to ensure *the* quality of service (QoS) for different classes of bearers in LTE networks. However, such solution is still inefficient before implementing a good resources' management scheme. Several scheduling mechanisms have been proposed in the literature, to guarantee the QoS of different classes of bearers in LTE-A and 5G networks. In [13], a new approach of uplink scheduling resources has been developed. It aims to ensure service fairness of different traffic classes that allocates bearers over LTE-A and 5G networks. The [14] proposes an RPL compliant solution, MI-RPL, to provide differentiated services exploiting the multi-instance support of RPL for industrial low-power and lossy networks (LLN).

The design of a robust dynamic congestion control system called modified sliding mode controller (MSMC) in differentiated service communication network has been developed [15] providing high utilization and less delay, while the network fulfills with the demands of each type of traffic flow. The differentiated service network outline has been expected, and the control approach was framed for three types of services such as premium service and ordinary service as well as best effort facility as a new idea [15]. For the differentiated services, a new mechanism allows the extra bandwidth of a class to redistribute of the unused bandwidth to the backlogged flows according to their required rates and the backlog status [16]. The virtual networks are not served equally, but they are classified into different service levels. This is achieved through the DiffServ approach [16]. A variant of the accumulating priority queue, called the delayed APQ, is developed [17], which involves a period of delay for the low-priority class of customers before they can accumulate priority. The waiting time distribution for the lower class of customers when an APQ is determined and the impact of the initial delay upon that distribution are assessed [17]. The one-server case was considered where service times are selected from a common general distribution and the multi-server case where all service times are exponentially distributed with a common mean [17].

The paper is organized as follows: In Sections 2 and 3, model description and mathematical analysis are provided. In Sections 4, 5, and 6, the CS, CSDS, and ECSDS are developed, respectively. Finally, simulation results are provided in Sect. 7.

2 Model Description

This section contains the definitions for relations and the criteria for the broadcast plan. Mainly three or four sets (for pedagogic reasons) are considered $S_i(i = 1,2,3)$ with their sizes S_{is} so that $S_{3s} \geq S_{2s} \geq S_{1s}$. The relations are examined iteratively from the last level of hierarchy S_3 in order to provide an RBP. In the RBP, no empty slot is included. It is considered that at least one item from the last (cold) set, S_3, will be sent and from the other two sets at least two.

Any relation includes subrelations for each set, s_sub $_i$ $(i = 1,2,3)$ or *i* subrelation. A set of relations is created including their subrelations considering definite number of items from each set. There are also examples developed with more than three sets. It is considered that $a|b$ (a divides b) only when b mod $a = 0$ (f.e. 14 mod 2 = 0).

Some definitions of concepts relevant to this work are given below:

Definition 1 The *size of a relation (s_rel)* is the number of items that belong to the relation, and it is equal to the sum of the sizes of the three subrelations. The *number of relations (n_rel)* is the number of relations that include items of $S1$ and $S2$ only once.

Example 1 The relation A = (a, b, c, d, f) has the following three subrelations:

(a) $s\_sub_i = 1$ with a, (b) $s\_sub_2 = 3$ with (b,c,d), and (c) $s\_sub_3 = 1$ with f. The s_rel = 5.

This work is based on equal size relations.

Definition 2 *i subrelation$_j$* means the subrelation that comes from the *j* relation, and *i* stands for the number of subrelations in the *j* relation. *Vector of subrelations (v_s)* of a relation is given by the integer numbers of the subrelations that compose a relation.

Example 2 Consider the sets $S_1 = \{1\}$, $S_2 = \{2,3,4\}$, and $S3 = \{5,6,7,8,9,10\}$ and the relations (1, 2, 3, 5, 6, 7), (1, 3, 4, 8, 9, 10) with $s\_sub_1 = 1$, $s\_sub_2 = 2$, and $s\_sub_3 = 3$,(v_s = [1,2.3]) and then 2-subrelation1 = (2, 3) and 2-subrelation2 = (3, 4). The last two subrelations ((2, 3), (3, 4)) come from $S2 = \{2, 3, 4\}$ having 3 as a repeated item. This design cannot provide a relation that can lead to an RBP, due to the repetition of 3.

Definition 3 A BP is *full* for *S*1, *S*2, and *S*3 if it provides at least two repetitions (without duplicates) of items of *S*1 and *S*2 and it does not include empty slots. A BP is regular if it is full and provides equal spacing property [1].

Considering now four sets, the SD4 is the set of divisors of the size of the last set (S_{4s}).

Example 3 If $S_{4s} = 200$, the $SD_4 = \{10, 20, 40, 50,100\}$.

Definition 4 *Group Length(gl)* is any number of SD4, and pv*i*: pv*i*|gl holds for all pv*i*. Moreover, gl SD4. The symbol *d*4 represents any divisor of *S*4s (*d*4 SD4) with gl as the final value of *d*4 that makes the RBP feasible. For an RBP, the gl term is used.

Definition 5 *Partition value (pv)* is the number that comes after the definition of s_sum_i and $pv_i = S_{is}/s\_sum_{i.}$ It is a common divisor of *Sis* ($i = 1,.., k$) and gl for a given size of s_sumi. Hence, pv*i* | *Sis* and pv*i* | gl. Each set must have its own pv. The pv*i* vector is pv*iv* = [pv1,pv2,..pvn], and n is the number of services.

Definition 6 *Item multiplicity (it_mu)* is the number of times that all the items of a set can be repeated in an integrated relation.

Definition 7 *Integrated relations* is a set of repeated relations for each set when pv*i* | gl. (the criterion of homogenous grouping). The number of integrated relations for a set of data is the number of channels (n_ch) that can be used for an RBP broadcasting. The integrated relation dimension vector (idv) is simply the it_mu*i* €I that appears in an integrated relation given gl.

Definition 8 *Homogenous grouping (hg)* is the kind of grouping where all the data of *S*1 and *S*2 remain in the group without having any kind of partition or empty slots.

Definition 9 The number of channels (n_ch): n_ch = S_k / gl (where S_k is the last set). The n_ch is equal to the number of relations.

Considering the hg, the gl can be defined as the number of relations (n_rel) that can provide homogenous grouping. The gl is a divisor of S_{ks} (1,.., *k*).

Example 4 If $S_{3s} = 40$, gl = 20, considering that s_sum3 = 8, then pv3 = 5 (=40/8). Hence, pv3| *S*3s and pv_3 |gl.

The vector of the ratio values of differentiated services is dif[*a*1,*a*2,..*an*], and n is the number of services. Second, the definitions of the criteria are as follows:

The criterion of homogenous grouping (chg): when pv_i | gl.
The criterion of multiplicity constraint (cmc): This happens if it_mu $_{i+1}$ < it_mu_i ($i = 1$,.., $n-1$) (loose cmc) and it_mu_i *I*. (tight cmc).
The criterion of PV (cpv): when pv*i* < pv*j* (for *i* < *j*) (loose cpy) and pv*i* *I* (tight cpv).

Example 5 Consider the sets *S*1 = {1}, *S*2 = {2,3,4}, and *S*3 = {5,6,7,8,9,10} with their sizes *S*1s = 1, *S*2s = 3, and *S*3s = 6, respectively. Consider 6 as a divisor of *S*3s (gl = 6). If s_sum1 = 1, s_sum2 = 1, and s_sum3 = 1, then pv1 = 1/1 = 1, pv2 = 3/1 = 3, and pv3 = 6/1 = 6. The cpv is valid.

The it_mu*i* = 6/1 = 6, it_mu2 = 6/3 = 2, it_mu3 = 6/6 = 1. The cmc is valid.
So an integrated relation considering one channel (n_ch = 1) and cir are valid ((n_ch | S_{3s} = 1|6) with this design is valid. This integrated relation can be denoted as vector: intv[1,1,1].

Example 6 For *S*1s = 10, *S*2s = 20, *S*3s = 40, and *S*5s = 120, the v_s could be [5,5,8,1] the pv*i* = [2,4,5], and for given gl = 10, it is not possible to have integrated relations since the it_mu*i* is not possible to exist (because: 2|10, 4 not| 10, 5|10).

But if gl = 20, then there is an integrated relation since there is the it_mu*i* vector [10 = 20/2, 5 = 20/4, 4 = 20/5).
In the next example, the average waiting time (AWT) is introduced.

Example 7 Let us consider *S*1s = 10, *S*2s = 20, *S*3s = 40, and *S*5s = 120. Taking s_sub1 = 5, s_sub2 = 5, s_sub3 = 5, and s_sub4 = 1 with s_sum = 16, the AWT1 = 32(=4 + 5 + 5 + 1 + 5 + 5 + 5 + 1 + 1) for a single-channel service. In addition, the pv1 = 10/5 = 2, pv2 = 20/5 = 4, pv3 = 40/5 = 8, and pv4 = 8/1 = 8. It is considered that the gl = 8 (because 120|8 = 15). This means that only eight items will be sent in this integrated relation. For this relation, the pv*i* criterion is valid. The it_mu1 = 8/2 = 4, it_mu2 = 8/4 = 2, it_mu3 = 8/8 = 1, and it_mu4 = 8/8 = 1. The cmc criterion is valid. Since both criteria (cpv, cmc) are valid, an RBP can be constructed. Considering this design, all the S_4 data will be sent with 15 integrated relations.

3 Mathematical Analysis

A set of theorems dealing with the criteria (cmc and pv$i$) has been developed. These theorems are fundamentals for the characteristics of an RBP. To this end, they are presented in this work. The following theorem is developed for an RBP construction, and it examines the feasibility of an RBP.

Theorem 1 Consider the four S_{is} so that $S_{4s} \geq S_{3s} \geq S_{2s} \geq S_{1s}$. With the factorization of the S_{4s}, a set of divisors ($D4$) are found. After the selection of a divisor, d, from $D4$ set, so that pv$i$ (=S_{is} / s_sum$_i$) I, $\text{pv}_1 \leq \text{pv}_2 \leq \text{pv}_3$ with pv$i$ I (the tight cpv criterion), and it_mu$_i$ < it_mu$_{i+1}$ with it_mu$_i$ (=D_4/pv$i$) I (the tight cmc criterion), then there is an RBP based on the d. The validity of cmc can be obtained as well only by the condition of pv$i$ | d. (or gl) (chg criterion). Both criteria, without the constraint of the integer values, are considered as loose criteria.

Proof Since pv$i$ (=S_{is} / s_sum$_i$) I (=m_i) and it_mu$_i$ (=D_4/pv$i$) I (=k_i), then it means that the total items of Si will be included in an RBP, ki times. The m_i values show the number of the relations that could contain all the items of Si. The pv$i$ criterion with the multiplicity constraint criterion (with integer value) can guarantee the existence of an RBP. The condition of pv$i$ | $D4$ means that D_4/pv$i$ I.

From the above, the pv$i$ criterion must be accompanied from chg criterion (or the same from cmc with integer vales) for an RBP construction.

Example 6 provides an explanation of this theorem since the two criteria (cpv, cmc) are valid and an RBP can be constructed.

The next example is referred to as the RBP with the use of SD4 and a minimum number of channels (thr_ch) that are available in the server for serving the RBP.

Example 8 Let us consider S_{1s} = 10, S_{2s} = 20, S_{3s} = 40, S_{4s} = 120, thr_ch = 4, SD4 = {10, 20, 30, 40, 60}, and s_sum4 = 1. Considering (a) for s_sum1 = 2, s_sum2 = 2, s_sum3 = 2, s_sum4 = 1 (or [2, 2, 2,1]), and $d4$ = 20. The pv1 = 5(10/2), pv2 = 10 (20/2), and pv3 = 20 (40/2). Also pv1 | $d4$ (=8 = 20/5), pv2|$d4$ (=2 = 20/10), pv3|$d4$ (=1 = 20/20), and $\text{pv}_1 \leq \text{pv}_2 \leq \text{pv}_3$. The pv and multiplicity criterion are valid. n_ch = 120/20 = 6 > thr_ch. So [2, 2, 2] with $d4$ = 20 cannot provide the desired RBP. (b) for s_sum1 = 2, s_sum2 = 2, s_sum3 = 2, s_sum4 = 1 (or [2, 2, 2,1]), and $d4$ = 40(=2*20). The pv1 = 5(10/2), pv2 = 10 (20/2), and pv3 = 20 (40/2). Also pv1 | $d4$ (=8 = 40/5), pv2|$d4$ (=4 = 40/10), pv3|$d4$ (=2 = 40/20), and pv1 ≤ pv2 ≤ pv3. The pv and multiplicity criterion is valid. n_ch = 120/40 = 3 < thr_ch. So [2, 2, 2] with $d4$ = 40 can provide the desired RBP.

Theorem 2 For multiple-channel allocation with sets of different multiplicity (such as $S1$, $S2$, $S3$) in an RBP, if pv$i$|gl, the validity of multiplicity constraint (it_mu$i$ + 1 < it_mu$i$ (i = 1,.., k−1) can be achieved from the pv criterion (pv$i$ < pv$i$ + 1, i < k, k = #sets). Similarly, the pv criterion can guarantee the multiplicity constraint criterion (the loose one).

Proof If

$$\mathrm{pv}i < \mathrm{pv}i + 1, \tag{1}$$

then

$$\mathrm{it\_mu}i > \mathrm{it\_mu}i + 1. \tag{2}$$

From (1), = > 1/ pv$i$ > 1/ pv$i$ + 1 = > gl/ pv$i$ > gl/ pv$i$ + 1. If (gl/pv$i$) I, = > it_mu$i$ > it_mu$i$ + 1. Following the reverse order, we can go from (2) to (1). Therefore, it is not necessary to examine the multiplicity criterion and the pv criterion can provide the multiplicity.

Example 9 Let us consider the four sets as in the previous example, having $d = 20$ (divisor of S4s), gl (group length) = d, and n_ch = 120/20 = 6. For the v_s = [5,5,8,1], the pv1 = 10/5 = 2, pv2 = 20/5 = 4, and pv3 = 40/8 = 5. The pv criterion is valid. Also pv$i$|20, pv2|20, and pv3|20, the chg is valid and an RBP can be created. But if d = 10 (another divisor of S4s), then the cpv is valid (pv$i$ | 10), but the cmc is not valid (2| 10 but 4 ~ | 10), and there is not an RBP.

Theorem 3 If pv_i ($i < k$, k = #sets) are analogous to a_i, the AWT_i are also analogous to the a_i and

$$\mathrm{pv}_1/\mathrm{AWT}_1 = \mathrm{pv}_2/\mathrm{AWT}_2 = \cdots = \mathrm{pv}_{k-1}/\mathrm{AWT}_{k-1} \tag{3}$$

Proof Let us consider $n = 4$ (the number of sets) and pv1/a1 = pv2/a2 = pv3/a3.

Finding AWT1 (if pv1 = 2) AWT1 = s_sum*pv1 = ((s_sum$_1$-1 + s_sum$_2$ + s_sum$_3$ + s_sum$_4$) + (s_sum$_1$ + s_sum$_2$+ s_sum$_3$ + s_sum$_4$) + 1)). In analogous way, AWT$_2$ = s_sum*pv$_2$ and AWT$_3$ = s_sum*pv$_3$. For $n = k$, the hypothesis is

$$\mathrm{pv}_1/a_1 = \mathrm{pv}_2/a_2 = \mathrm{pv}_3/a_3 = .. = \mathrm{pv}_{k-1}/a_{k-1} \tag{4}$$

and AWT $_{k-1}$ = s_sum*pv $_{k-1}$. The equivalence is

$$\mathrm{pv}_1/\mathrm{AWT}_1 = \mathrm{pv}_2/\mathrm{AWT}_2 = \mathrm{pv}_3/\mathrm{AWT}_3 = \cdots = \mathrm{pv}_{k-1}/\mathrm{AWT}_{k-1} \tag{5}$$

Moreover, dividing the ratios (4) and (5)

$$\mathrm{AWT}_1/a_1 = \mathrm{AWT}_2/a_2 = \cdots = \mathrm{AWT}_{k-1}/a_{k-1}. \tag{6}$$

.

From Theorem 3, it is evident that from $\mathrm{pv}_i/\mathrm{AWT}_i = \mathrm{pv}_j/\mathrm{AWT}_j$, there is $\mathrm{pv}_i/\mathrm{pv}_j = \mathrm{AWT}_i / \mathrm{AWT}_j$. Hence, the pv$i$ ratio is analogous to AWT_i ratio. This is significant to provide services with predefined AWT ratio.

Generally, the

$$\text{AWTi} = \text{s\_sum}^* \, \text{pv}i. \quad (7)$$

Example 10 Let us consider again the set of services: *S*1s = 10, *S*2s = 20, *S*3s = 40, and *S*5s = 120 (RBP1). Taking s_sub1 = 5, s_sub2 = 5, s_sub3 = 5, and s_sub4 = 1 with s_sum = 16, the AWT1 = 32(=4 + 5 + 5 + 1 + 5 + 5 + 5 + 1 + 1) and AWT3 = 128 (=5 + 16*7 + 10 + 1). All the data of *S*4 will be sent with 15 integrated relations. The ratio of AWT3/AWT1 = 128/32 = 4 . This is also the same with PV3/PV1 = 8/2 = 4 (Theorem 3).

With the new parameters, s_sub1 = 5, s_sub2 = 5, s_sub3 = 8, s_sub4 = 1, and s_sum = 19, the AWT1 = 38 another plan (RBP2) can be created. In addition, pv1 = 10/5 = 2, pv2 = 20/5 = 4, pv3 = 40/8 = 5, and pv4 = 20/1 = 20. Therefore, the pv*i* criterion is valid. Also, it_mu1 = 120/2 = 60, it_mu2 = 120/4 = 30, it_mu3 = 40/8 = 5, and it_mu4 = 20/1 = 20. Thus, the cmc criterion is valid.

Since both criteria (cpv, cmc) are valid, an RBP can be constructed. All the data of *S*4 will be sent with six (120/20) integrated relations. The ratio of AWT3/AWT1 = 90/38 = 2.5. This is also the same with PV3 /PV1 = 5/2 = 2.5 (Theorem 3).

From the two RBPs, it is evident that when s_sum3 (from 5 to 8) in the RBP2 increases, for the same s_sum1 (=5), the ratio AWT3/AWT1 decreases.

4 Cuckoo Search

Cuckoo search (CS) algorithm is also a nature-inspired algorithm, based on brood reproductive strategy of cuckoo birds to increase their population. Cuckoo search algorithm is a nature-inspired algorithm based on reproduction of cuckoo birds [14]. The aim is to use Lévy flights which are walks whose directions are random, and their step lengths are derived from the Lévy distribution. Compared to normal random walks, Lévy flights are more efficient in exploring large-scale search areas. That is mainly due to the fact that Lévy flight variance increases much faster than that of the normal random walk. Lévy flights can reduce the number of optimization algorithms' iterations by about four orders compared to a normal random walk [12].

There are times when the cuckoos discover that the eggs in their nests do not belong to them; in those cases, either the foreign eggs are thrown out of the nests or the whole nests are abandoned.

The solution is represented by one egg in a nest, and a cuckoo egg represents a new solution. The aim is to use the new and potentially better solutions (cuckoos) to replace worse solutions that might be in the nests.

5 CSDS

The CSDS is based on Theorem 1 considering the tight cpv and cmc criterion for the RBP creation. Theorem 3 provide the opportunity to have delay differentiation services with the ratios of $\mathrm{AWT}_1/a_1 = \mathrm{AWT}_2/a_2 = \mathrm{AWT}_3/a_3$ (Theorem 3, (6)). The initial population is randomly generated. The fitness function evaluates each individual of the CS. The step size for the Lévy flight is set to the $\mathrm{upperbound}_i$ /100. The $\mathrm{upperbound}_i$ is the size of each queue set. The number of available messages in the service queues represents the size of population n. The next generations are created using the GA operators. The fitness function consists of two parts, $f1$ and $f2$. The $f1$ (for the RBP existence) tests if the sum (pv$i$ /d4 + it_mu$i$) I. The $f2$ tests the existence of delay ratio for differentiation services (AWTi /$ai = c$, c N) as (6), given the ai. Both $f1$ and $f2$ are computed for each individual.

The divisors of the last set S_{4s} are extracted, and the sets of s_sumi are created according to Theorem1. The analogous values (ai) to the AWTi ratios are given. The AWTi for all sets of services are computed along with the two fitness functions. Individuals who do not follow fitness functions are eliminated.

The CSDS is a static algorithm since it discovers an RBP along with the differentiated ratios.

The pseudocode for CSDS is as follows:

```
CSDS:  input:  initial population of n host nests (the size of
data in the service queues) and the desired values of ai for
AWTi ratios
output:  discover an RBP  and an accepted AWTi ratio analogous
to ai values
Generate initial population of n host nests xi
while {(t<MaxGenerations) and (! termin.condit.)
        get a cuckoo randomly via Lévy flights
        //in order to find the next sizes of the sets
          evaluate its fitness Fi
        //if the numbers follows f1,f2
           f1 = sum (pvi /d4 + it_mui )   I  and
           f2 = AWTi /ai = c, c  N (so that:AWTi/a1 =AWT2/a2
                =AWT3/a3, as  (6))
           randomly choose nest among n available nests
          (for example j) if(Fi > Fj)
                {replace j by the new solution;}
         fraction pd of worse nests are abandoned
         and new nests are being built;
         //search around pd (n*pd) for best nests (A)
         // and replace if a better solution exists(A)
         for the rest of n keep the best solutions or
             nests with quality solutions;
         rank the solutions and find the current best
}//end while
//Post process and visualize results
```

6 ECSDS

The extended version of CSDS (ECSDS), which is a static solution and an extension of CSDS, can directly discover the analogous ratio for all the rest of the services using the predefined ratio of two services as pilot after searching for the appropriate integrated relation vector.

The ECSDS can provide the AWTi ratios having as a pilot ratio value (r) of AWT2/AWT1 $= c, c\ I$ considering $a1 = 1$. From the initial population (individuals) and the fitness function $f1$ and after discovering the RBP with the r ratio of two services, the ratio values ai are found with the use of Theorem 3. The RBPs provided by the CSDS are examined so that the minimum AWT3 will be discovered. From (7), we can have various values of AWT3 dependent on the proposed RBP. The RBP with the smallest AWT3 value is the ideal case in order to include the lowest waiting time for the low-priority mobile services.

A queuing system can be used to provide differentiated services (Diffserv) for ECSDS. The appropriate RBP with the desired AWTi ratio will be discovered using ECSDS (as in Scenario 2). This selection can also provide the opportunity for a minimum AWT of the lower-priority service. The differentiated service code point (DSCP) can be used into an IP header to market it according to the class of traffic it belongs in. The scheduler serves the data of the RBP ensuring the equal spacing of services. By having more channels, a smaller AWT can be achieved for the low-priority service. The higher-priority tag is used in order to provide the shorter AWT among the other services.

7 Simulation

Simulation focuses on various scenarios potentially on the discovery of RBPs and differentiated services using CSDS and ECSDS. The items are separated into four categories using the Zipf distributions. The scenarios are as follows:

Scenario 1 Let us consider $S1$s $= 200$, $S2$s $= 400$, $S3$s $= 800$, and $S5$s $= 2400$ (RBP1). Using the CSDS, one choice could be s_sub1 = 100, s_sub2 = 100, s_sub3 = 100, and s_sub4 = 1 with s_sum = 301 considering also their pv$i$ (pv$i$ = 2, pv2 = 4, pv3 = 8). After getting different values for differentiated services $a1 = 1$, $a2 = 2$, and $a3 = 4$ (or dif[1,2,4]), from Theorem 3, Equation (6) was found valid (AWTi/ai =2). Figure 1 depicts this scenario.

Scenario 2 Let us consider the sets $S1$s $= 100$, $S2$s $= 200$, $S3$s $= 400$, and $S5$s $= 1200$ with AWT2/AWT1 = 5. With CSDS, one choice for RBP1 could be s_sub1 = 50, s_sub2 = 20, s_sub3 = 10, and s_sub4 = 1 with s_sum = 81 considering also their pv$i$ (pv$i$ = 2, pv2 = 10, pv3 = 40). For the ECSDS, the AWT1 (2*81 = 162), AWT2 (10*81 = 810), and the ratio AWT2/AWT1 = 5, or AWT1/1 = AWT2/5. The AWT3 is 3240 (81*40), and from the ratio, 162/1 = 3240 /$a3$ => $a3 = 20$.

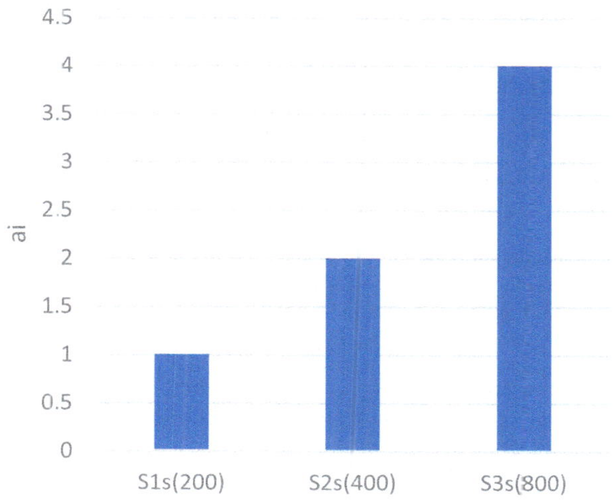

Fig. 1 CSDS for RBP and a_i values

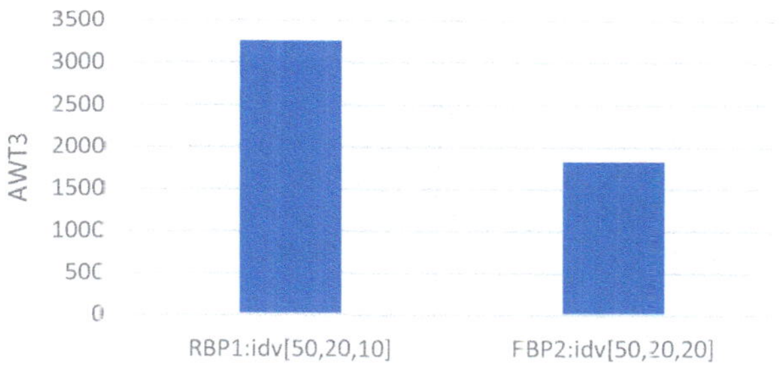

Fig. 2 ECSDS for RBP discovery

Considering another choice, RBP2 with s_sub1 = 50, s_sub2 = 20, s_sub3 = 20, and s_sub4 = 1 with s_sum = 91 considering also their pv$i$ (pv$i$ = 2, pv2 = 10, pv3 = 20). AWT1 (2*91 = 182), AWT2 (10*91 = 910), and the ratio AWT2/AWT1 = 5, or AWT1/1 = AWT2 /5. AWT3 is 1820 (91*20), and from the ratio, 182/1 = 1820 /$a3$ => $a3$ = 10.

From the above, the RBP2 is preferable since it has the minimum AWT3 (Fig. 2).

8 Conclusions

Two algorithms for discovering RBPs with delay differentiated services are presented. Theorems that provide criteria and useful results are produced. Basically, the CSDS can discover RBPs with differentiated services using the ratio values as parameters. The ECSDS discovers from a set of RBPs, the one that has minimum AWT3 so that the low-priority services have lowest waiting time.

Servers with their clusters will enhance their self-efficacy, the sustainability of self-monitoring of these new operations, and their abilities of providing better quality of service in an automated way considering the delay ratio.

References

1. S. Acharia, M. Franklin, S. Zdonik, R. Alonso, Broadcast disks: Data Management for asymmetric communications environments, in *Proceeding of International Conference on Management of Data (SIGMOD)*, (ACM, San Jose, CA, May 1995), pp. 199–210
2. W. Wee, S. Navathe, E. Omiecinski, C. Jermaine, Efficient data allocation over multiple channels of broadcast servers. IEEE Trans. Comput **51**(10), 1231–1236 (2002)
3. E. Ardizzoni, A. Bertossi, M. Pinotti, S. Ramaprasad, R. Rizzi, M. Shashanka, Optimal skewed data allocation on multiple channels with flat per channel. IEEE Trans. Comput **54**(5), 558–572 (2005)
4. A. Bertossi, M. Pinotti, S. Ramaprasad, R. Rizzi, M. Shashanka, Optimal multi-channel data allocation with flat broadcast per channel, in *Proceedings of 28th International Parallel and Distributed Processing Symposium (IPDPS)*, (IEEE, Phoenix, April 2004), pp. 18–27
5. N. Vaidya, S. Hameed, Scheduling data broadcast in asymmetric communication environment. ACM/Baltzer Wirel. Netw **5**(3), 171–182 (1999)
6. J. Tsiligaridis, Static and dynamic algorithm for regular data broadcasting plans in wireless mobile environment. Int. J. Adv. Comput. Sci **2**(2), 42–48 (2012)
7. P. Yu, W. Sun, Y. Qim, A. Zhang, B. Shi, A data partition based near optimal scheduling algorithm for wireless multi-channel data broadcast, in *Proceeding of the 13thInternational Conference on Database Systems for Advanced Applications(DASFAA)*, (New Delhi, India, 2008), pp. 188–203
8. J. Lv, V. Lee, M. Li, E. Chen, Profit-based scheduling and channel allocation for multi-item requests in real time on-demand data broadcast systems. Data Knowl. Eng, Elsevier **73**, 23–42 (2012)
9. G. Horng, C. Wang, C. Chou, Adaptive broadcasting mechanism for bandwidth allocation in mobile services. Sci. World J. Hindawi, Publishing Corporation **2014**, Article ID: 735457, 14 (2014)
10. H. Klessig, G. Fettweis, Adaptive admission control in interference-coupled wireless data networks: A planning and optimization tool set, in *Proceeding of International Conference on Communications (ICC)*, (IEEE, Sydney, Australia, 2014), pp. 2375–2380
11. X. Yangm, S. Deb, Engineering optimization by Cuckoo search. Int. J. Math. Model. Numer. Optimiz **1**(4), 330–343. ISSN 2040-3607 (2010)
12. X. Yang, *Nature-Inspired Optimization Algorithms*, 1st edn. (Elsevier, London, 2014)
13. C. Tata, N. Fellag, M. Kadoch, New courteous algorithm for uplink scheduling in LTE-advanced and 5G networks. J. Comput. Netw. Commun, Hindawi **2020**, Article ID 4189789, 15 (2020)
14. M. Monowar, M. Basheri, On providing differentiated service exploiting multi-instance RPL for industrial low-power and lossy networks. J. Wireless Commun. Mobile Comput **2020**, Article ID 1748647, 12 (2020)
15. S. Khodhair, M. Naghmash, R. Abduljabbar, A. Alrawi, Minimum delay congestion control in differentiated service communication networks. Open Electr. Electron. Eng. J., Bentham **12**, 42–51 (2018)
16. J. Shi, S. Chung, A traffic-aware quality-of-service control mechanism for software-defined networking-based virtualized networks. Int. J. Distrib. Sensor Netw. **13**(3), 13p (2017)
17. M. Mojalal, D. Stanford, R. Canon, The lower-class waiting time distribution in the delayed accumulating priority queue. J. Inf. Syst. Operation. Res **58**(1), 60–86 (2020)

Using Multimodal Biometrics to Secure Vehicles

Kevin Daimi, Noha Hazzazi, and Mustafa Saed

1 Introduction

Modern vehicles encompass many electronic control units (ECUs) and a number of buses to facilitate the communications between the ECUs. They are responsible for many functions in the vehicle. In addition, a number of electric vehicles are on the roads that add more complexity to vehicles. The current advances in autonomous vehicle has increased the sophistication of vehicle design and networking. This increased sophistication demands enormous efforts to secure vehicles and their networks. Such networks are more susceptible to diverse attacks, such as wormhole attacks, denial of service (DoS) attacks, and black hole attacks [1]. To withstand the enrichments in safe vehicle technologies, it is vital to develop a robust vehicle network security system, which detects security vulnerabilities, threats, and attacks facing vehicle network [2]. The ECUs can also be exposed to security attacks that could be disastrous and can cause casualties. Hence, there is a significant need to guard the ECU infrastructure [3]. The success of the autonomous vehicle is

K. Daimi
Electrical and Computer Engineering, and Computer Science,
University of Detroit Mercy, Detroit, MI, USA
e-mail: daimikj@udmercy.edu

N. Hazzazi (✉)
Department of Electrical Engineering and Computer Science, Howard University,
Washington, DC, USA
e-mail: noha.hazzazi@howard.edu

M. Saed
HATCI Electronic Systems Development, Hyundai-Kia America Technical Center, Superior
Township, MI, USA
e-mail: msaed@hatci.com

K. Daimi et al. (eds.), *Advances in Security, Networks, and Internet of Things*,
Transactions on Computational Science and Computational Intelligence,
https://doi.org/10.1007/978-3-030-71017-0_41

fundamentally reliant on three technologies: embedded processors, sensors, and the communication systems [4]. These increased levels of communications of the autonomous vehicle make it more vulnerable to security attacks including spoofing, sender-/receiver-related errors, segmented network-related errors, and communication corruption [5, 6]. To avoid security attacks based on these vulnerabilities, security requirements should be enforced prior to the actual design of these vehicles [7].

Biometrics play a decisive role in authentication and identification of individuals for many devices and applications. This is simply because these biometrics are unique to individuals. However, a single biometrics is possibly prone to errors. For this purpose, some applications and devices are moving toward multimodal biometrics. These multimodal biometrics can substantially contribute to vehicle security.

Nagamma, Lakshmaiah, and Narmada [8] suggested using fingerprints to prevent unauthorized entry to vehicles. Their work demanded the scanner to be placed at vehicle's door locking system and the use of GSM module to forward messages to the vehicle owner's mobile device. Their setting required the use of Raspberry Pi 3 processor to control the whole process. Gill and Sachin [9] concentrated on using fingerprint for the ignition system in case drivers forget their keys. Their proposed work suggested the use of two modules that consist of an LCD crystal for displays of values and fingerprint sensor which receives input from the driver side. An anti-theft vehicle security system that allows access to the vehicle only if the individual's fingerprint matches the stored template in the system was introduced by Kaushik, Veralkar, Parab, and Nadkarny [10]. The check for the biometric match was implemented with MATLAB and the result was displayed using LCD. In case of an illegal access to the vehicle, the vehicle's fuel tank will be locked via a relay circuit. As a result, unauthorized drivers will no longer be able to refuel the gas tank when empty. Further work on using fingerprints for vehicles could be found in [11–15].

In addition to using fingerprints to protect vehicles, face images were also suggested. Jaikumar and Jaiganesh [16] presented a system that allows capturing the face of the intruder to trace the vehicle. They relied on an embedded car security system consisting of a face detection subsystem (FDS), a Global Positioning System (GPS) module, a GSM (Global System for Mobile Communications) module, and a control unit. The face detection technique can cause an alarm signal to force a call to the police. In an effort to minimize vehicle theft, a method involving a GSM controlled by the Renesas microcontroller is utilized to determine the location of the vehicle [17]. The vehicle ignition will only fire when the person accessing the vehicle is authorized with three-phase security system. An SMS alert is sent to the vehicle owner if an unauthorized person trying to access the vehicle fails the three-phase security system. Haar-like features were employed to recognize the face, and adaptive boosting classifier was used to combine all weak classifiers into strong classifier for determining whether the captured image from the video matches the template (face). More attempts on employing face biometric are introduced in [18, 19].

Iris scan is one of the most accurate biometrics. Sreekala, Jose, Joseph, and Joseph [20] adopted iris scan to allow vehicle door unlocking. If the iris template does not match the incoming iris sample, their system allowed using passwords. If both methods fail, the door will remain locked. Punnoose and Kumar [21] implemented an Iris Recognition System (IRS) for ensuring security and safety of the vehicle's owner. Iris image acquisition and preprocessing were carried out using MATLAB. The edges of iris and pupil were detected using the Canny edge detection technique. If successful, the ignition circuit will be turned on. Otherwise, an alert signal will be sent to the owner. Their work represents a good attempt on using single biometrics to authenticate drivers.

An interesting in-vehicle driver recognition technique was demonstrated in [22]. They proposed a method and system that enable the identification of the driver through information extracted from electrocardiographic (ECG) signals collected between both hands of an individual. Their method used a custom sensing device that allowed data acquisition through nonintrusive techniques. The hardware was specifically designed to be mounted on the steering wheel. They compared different ECG representations using simple classification method to verify the identity of the driver.

Few attempts on deploying multimodal biometrics are available. To start a vehicle, Lupu and Lupu [23] suggested inserting a fingerprint sensor at the vehicle door and steering wheel, a camera for iris scan on the mirror, and a microphone for voice recognition. If a person is recognized, they can take control of the car. Otherwise, the police are informed using a complex GPRS system. Two existing approaches of combining speech, face, and additional biometric modalities in automotive applications are studied in [24]. Their goal was to achieve improvements of the feasible comfort and security. They also used additional soft biometric modalities to offset failures of the biometric sensors to guarantee business continuity through enriched availability. Khan, Khan, Zhang, and Zhang [25] highlighted the advantages and disadvantages of both unimodal and multimodal biometric verification processes.

In all of the above literature, no attempt has been made to safeguard both the vehicle entry and vehicle start together. Furthermore, none of the papers above attempted to secure the stored biometric templates and incoming samples. In this paper, the multimodal biometric iris scan, face scan, and voiceprint are used to safeguard vehicle start, and the biometric fingerprint and face scan are applied to vehicle entry. The stored biometrics are protected with cryptography as well as the communications between various components. Section 2 provides an overview of the available biometrics. In Sect. 3, the vehicle owner registration is dealt with. The vehicle security initialization, drivers' multimodal biometrics, and vehicle entry multimodal biometrics are introduced in Sects. 4, 5, and 6 respectively. Section 7 handles securing communications. Overriding biometrics with passwords and PINs is explained in Sect. 8. Finally, the paper is concluded in Sect. 9.

2 Biometric Overview

In biometrics world, the sensed input sample will be digitized and subject to some processing to extract a set of features that will be stored as a set of numbers. This set of numbers represents unique biometric characteristic referred to as an individual's template. Biometrics can produce false-negative and false-positive results. The false rejection rate (FRR) represents the ratio of false rejections to valid acceptances. Dividing the total number of false acceptances by valid acceptances results in false acceptance rate (FAR). Sebastin [26] indicated that the FARs for face, fingerprint, iris, and voice are 1%, 2%, 0.01%, and 6%, respectively, and the FRRs are 10%, 2%, 0.99%, and 10%, respectively. In this section, various biometrics will be briefly introduced [27–30]:

(a) Fingerprints: Fingerprints are observable patterns on the fingers and thumbs of individuals. Each individual has unique fingerprint patterns.
(b) Face Scans: Face scans rely on geometric patterns of faces to recognize individuals.
(c) Retina Scans: Retina scans center at the pattern of blood vessels at the back of the eye. They are the most accurate biometrics but involve some privacy issues of revealing some medical conditions. Individuals need to be about 3 inches away from the scanner to function properly.
(d) Iris Scans: Iris scans pay attention to the colored area around the pupil of the eye. They are considered the second most accurate biometrics. Scans can be executed from a distance of up to 40 feet.
(e) Palm Scans: Palm scans concentrate on the vein patterns in the palm of the human hand.
(f) Hand Geometry: Hand geometry focuses on the width and length of the palm and fingers of the individuals. These are rarely used.
(g) Voice Pattern Recognition: This biometrics relies on the voiceprint (characteristics of a person's speaking voice). Various factors can impact the voice of individuals.
(h) Signature Scans: Individuals have different unique style of writing including signatures. This is not easy to adopt due to the fact that slight variations in the individual's signature can cause matching failures.
(i) Facial Features: These concentrate on details of the face, such as the eyes, eyebrows, nose, lips, and ears.
(j) DNA: As a biometric, DNA segments are analyzed for possible matching.

3 Vehicle Owner Registration

The owner of the vehicle should register with the registration server (Vehicle Registration Authority) through secure connection. This is achieved via an online session. First the owner is authenticated by the server. Later, the owner should create a new account containing name, address, telephone, and other information. Later, a username, password, and PIN must be created. Having done that, the owner of the vehicle creates a profile, which includes security setting. During the security setting, the owner can specify the types of biometrics to be used and how many biometric matching out of the total number of biometrics used are acceptable to allow the operation (vehicle entry, vehicle start) to succeed. Furthermore, the owner indicates how to override the required biometrics when they fail. The options are using the key and using password and PIN. For security purposes, password and PIN are selected.

4 Vehicle Security Initialization

Various initializations should take place prior to applying vehicle entry and vehicle start biometrics. These include preinstalled cryptographic keys, passwords, and PINs. The connections between various units that will be referred to in this section are provided in Fig. 1. The components of this figure will be secured in Sect. 7. Note that password, PIN, and username of the owner are not stored in the vehicle. Only the owner's ID is stored.

The Vehicle Registration Authority (VRA) shares a symmetric key, (K_{RT}), with the telematics (TCU) and another symmetric key, (K_{RO}), with the owner registration (OR) unit. Future keys will be provided by the VRA. In addition, VRA will verify the username, password, and PIN of the vehicle's owner for TCU and send the types of biometrics to be used, allowable number of biometric matches for both

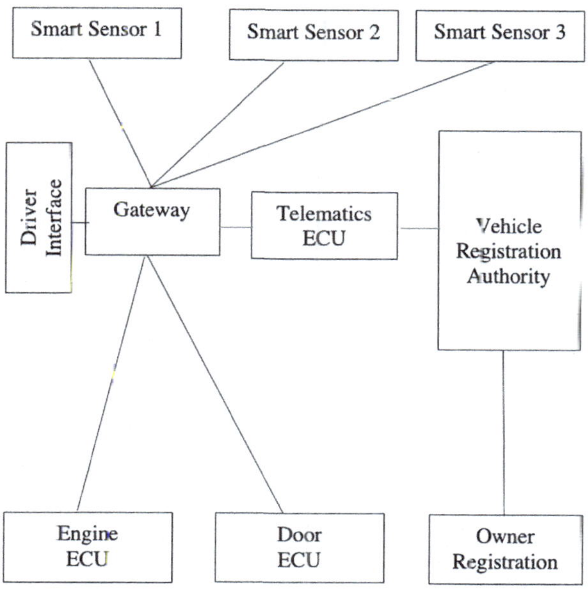

Fig. 1 Biometric security architecture

vehicle entry and start, and confirmation on whether to use passwords to override the biometrics or not and ID of the vehicle owner to the TCU.

The gateway (G) and TCU share a symmetric key, (K_{GT}). Updated keys will be created by G. In addition, the TCU receives the username, password, and PIN of the owner from G to be verified by the VRA. TCU sends types of biometrics needed, how many biometric template matches are acceptable to grant starting the vehicle or unlocking the door, the override choice, and ID_O to G. For the vehicle start, the valid number is one, two, or three matches and one or two for the vehicle entry.

The engine control unit (ECU) and the door control unit (DCU) will share their public keys, PU_{ECU} and PU_{DCU}, with G. G shares its public key PU_G with both ECU and DCU. The private keys of ECU, DCU, and G are preinstalled too. They also share the secret value (S_i). The keys are updated.

The smart sensors (SS_i) share symmetric keys (K_{SSGi}) with G. For vehicle entry, SS_1 is related to the fingerprint biometric and SS_2 to the face biometric. For the vehicle start, the smart sensors SS_1, SS_2, and SS_3 represent iris, face, and voice biometric sensors, respectively.

The driver interface (DI) unit shares a symmetric key (K_{DIG}) with G. G will provide DI with all future key updates. DI will allow the owner to register other drivers with G.

Furthermore, various components have shared MAC and HASH keys preinstalled as explained in Sect. 7.

5 Drivers' Multimodal Biometrics

In this section, the biometric templates are created, secured through cryptographic operations, and stored. Later, possible drivers' biometric samples are compared with the stored templates. If they match, the vehicle will start. This is illustrated in Fig. 2. Steps (D) and (E) are demonstrated in Fig. 3.

A. Biometric Processing and Feature Extraction

The first step in the biometric system discussed in this paper is acquiring templates of the iris, face, and voice through their corresponding sensors. The scans are preprocessed. The outcome of biometric processing will be extracted features or templates of the iris, face, and voice. The message, T, represents the template in Fig. 2. T will be secured with cryptography.

B. Applying Cryptography

The first step is to add (concatenate) the secret value, S_1, S_2, and S_3, to the templates of the sensors, iris sensor (SS_1), face sensor (SS_2), or voice sensor (SS_3), to obtain the concatenation $T||S_i$ where T is the template. As stated above, S_i is created by G and shared with the smart sensors. The message authentication code (MAC) is then calculated resulting in MAC ($T||S_i$). Driver ID, T, and MAC($T||S$) are encrypted using symmetric encryption.

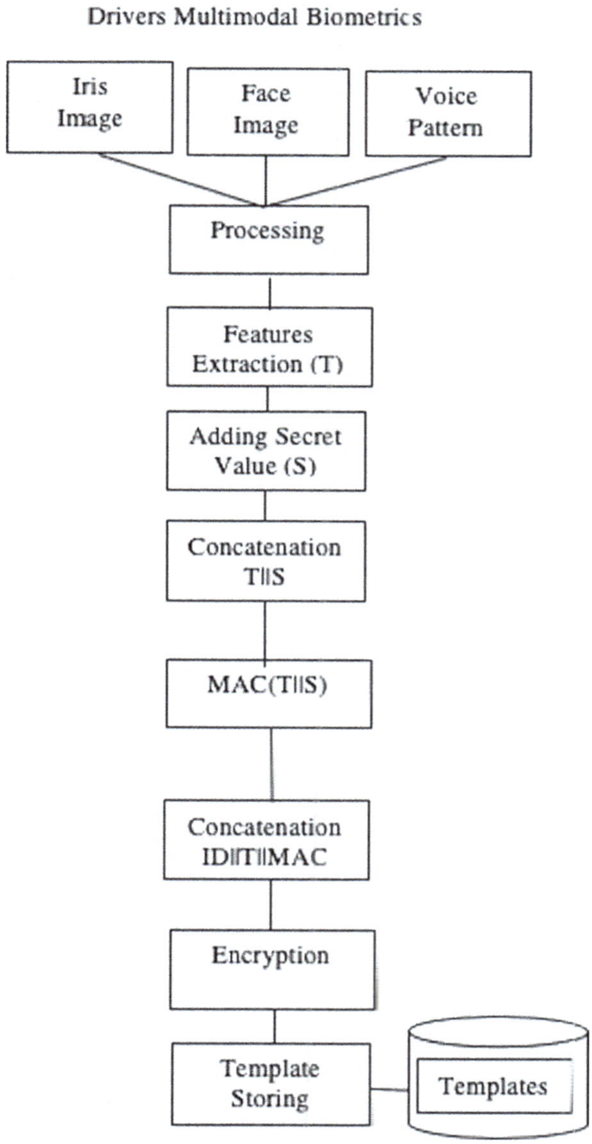

Fig. 2 Creating and storing templates

C. Storing Templates

The encrypted message of subsection (B) is forwarded to G to be stored in the database of biometrics. G also already stored the number of acceptable number of biometric template matches that were received from TCU.

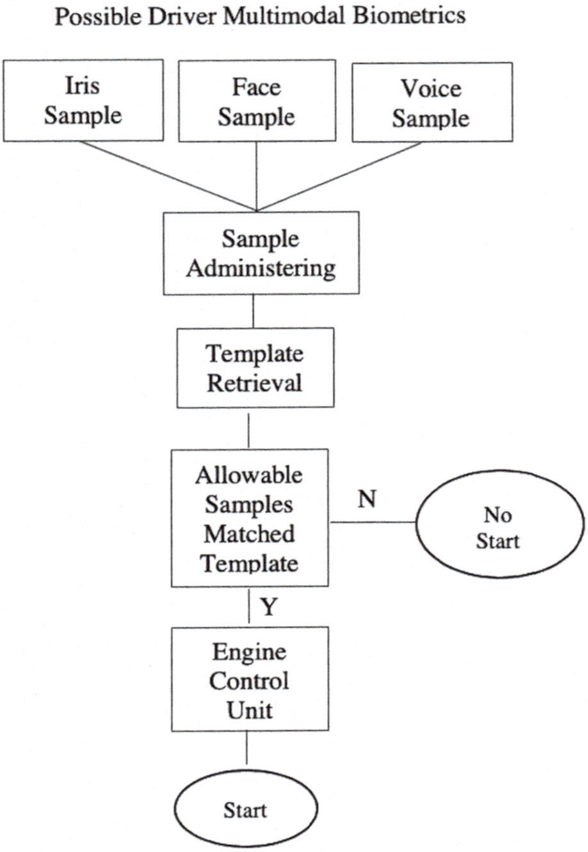

Fig. 3 Verifying biometrics of claimed driver

D. Sample Administration

Any claimed driver will go through step A to get the sample (actual biometrics before applying cryptography) for each of the biometrics, iris, face, and voice. Then, step (B) will be applied to the samples of each of the three biometrics. These samples are forwarded to G but not stored in the database.

E. Sample Template Matching

The templates are retrieved from the database indexed by the ID. These templates are compared with the entered samples following the stored number of matches. If a match is successful, G will send a message to the ECU to start the vehicle. Otherwise, the vehicle will not turn on.

6 Vehicle Entrance Multimodal Biometrics

For vehicle entrance, SS_1 represents fingerprint smart sensor, and SS_2 stands for face smart sensor. The steps of Fig. 2 apply to vehicle entrance using two sensors. The number of allowable matches is one or two.

Similar approach to the driver multimodal biometrics is used, and the verification of the person aiming at unlocking the vehicle door is similar too with one exception as illustrated in Fig. 4. If the number of matches is satisfied, G sends a message to DCU to unlock the door. Otherwise, the door will continue to be locked.

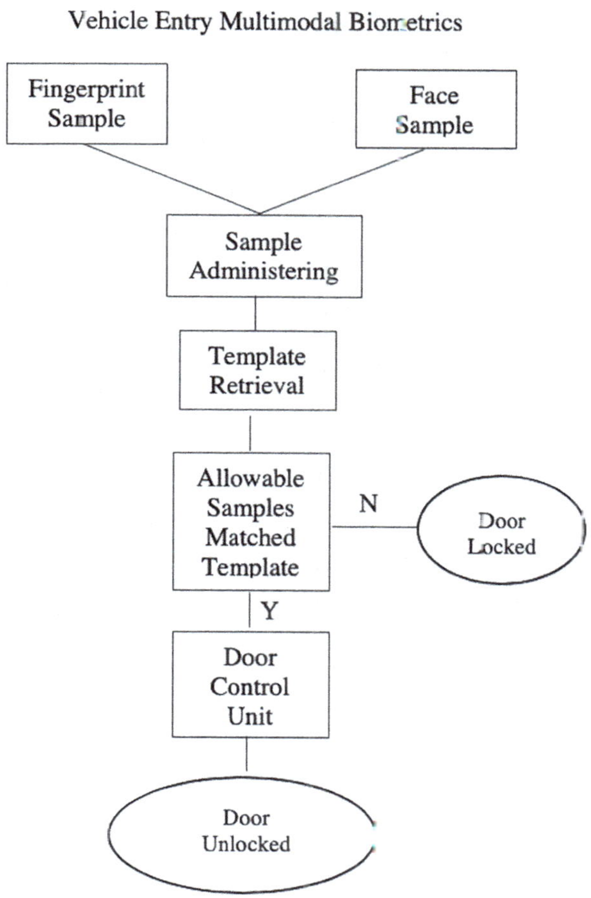

Fig. 4 Verifying biometrics for vehicle entry

7 Securing Communications

In this section, the units of Fig. 1 will be referred to. Both symmetric and asymmetric cryptographies are relied on to secure the communications. Public key cryptography is only employed to secure the communication between G and ECU and DCU, respectively, due to the fact that the exchanged messages are short. The parties involved and the notations used are presented in Tables 1 and 2.

A. Owner Registration Authority Communication

The vehicle owner communicates with the Vehicle Registration Authority at the manufacturer's site to create an account and profile. The owner also specifies the biometrics to be used and the number of allowable biometric matches to allow the vehicle start or vehicle entry. In addition, biometrics override with password choice is specified. This communication is secured as below:

OR concatenates the owner name and Vehicle Identification Number (VIN) to get the message M_1. This is encrypted with K_{RO} to get X1. The MAC of M_1, MAC (K_{ROM}, M_1), is calculated and concatenated with X_1. The resulting message is sent to VRA:

$$M_1 = \text{Name} \,||\, \text{VIN}$$
$$X_1 = E\ (K_{RO}, M_1)$$

$$OR \rightarrow VRA : X1 \,||\, MAC\ (K_{ROM}, M_1)$$

VRA decrypts X_1, verifies the name and VIN, and ensures the MAC of M_1 is equal to the received MAC. It then starts the registration process. A new account including the profile of the owner is created by the VRA. The vehicle owner creates username (USR), password (PWD), and PIN. The VRA concatenates these with the owner ID (ID_{OR}) that it generates to get M_2 and then encrypts M_2 with K_{RO}. The MAC for the concatenation of USR, PWD, PIN, and ID_{OR} is obtained and concatenated with the previous encrypted result and sent to O:

Table 1 Communication parties

Party	Meaning
VRA	Vehicle Registration Authority
TCU	Telematics
OR	Owner registration unit
G	Gateway
ECU	Engine control unit
DCU	Door control unit
SS_i, i = 1–3	Smart sensors 1–3
DI	Driver interface unit

Table 2 Notation used in protocols

Party	Meaning
K_{RT}	VRA/TCU symmetric key
K_{RO}	VRA/OR symmetric key
K_{GT}	G/TCU symmetric key
K_{SSGi}	G/SSi symmetric key
K_{DIG}	G/DI symmetric key
PU_{ECU}, PR_{ECU}	Public/private key of ECU
PU_{DCU}, PR_{DCU}	Public/private key of DCU
PU_G, PR_G	Public/private key of G
S_i, i = 1–3	Secret value
T	Template
SMP	Sample
ID_D	Driver ID
ID_{VRA}	VRA ID
ID_{TCU}	TCU ID
ID_{OR}	OR ID
ID_G	G ID
ID_{ECU}	ECU ID
ID_{DCU}	DCU ID
$IDSS_i$	SS_i ID, i = 1–3
PIN	Personal identification number
USR	Username
PWD	Password
K_{RTM}	VRA/TCU MAC key
K_{ROM}	VRA/OR MAC key
K_{GTM}	G/TCU MAC key
K_{SSGiM}	G/SSi MAC key
K_{DIGM}	G/DI MAC key
H(M)	Hash of message M

$$M_2 = USR\,||PWD||\,PIN\,||ID_{OR}$$
$$VRA \rightarrow OR : E\,(K_{RO}, M_2)\,||MAC\,(K_{ROM}, M_2)$$

OR decrypts the received message M_2 with K_{RO}, calculates the MAC of M_2, and ensures both MACs are equal. If successful, the owner has now its USR, PWD, PIN, and ID_O confirmed.

At this point, the owner updates the security settings of the account. These include specifying the type of biometrics to be applied and how many biometrics out of three for vehicle start and out of two for vehicle entry need to match to allow the operation. In addition, the owner will specify if a password would be used to override biometrics in case of any malfunction or just a preference. For this purpose, the OR sends the VRA the types of biometrics for vehicle start (BS = {iris, face, voice}); the types of biometrics for vehicle entry (BE = {fingerprint, face}); two numbers, N_1 (1–3 for vehicle start) and N_2 (1–2 for vehicle entry); a confirmation,

C, that is equal to Y or N for password override; and its ID encrypted with K_{RO}. This is concatenated with the MAC of the message M_3 as below:

$$M_3 = BS \,||\, BE \,||\, N_1 \,||\, N_2 \,||\, C \,||\, ID_{OR}$$
$$OR \rightarrow VRA : E\,(K_{RO}, M_3) \,||\, MAC\,(K_{ROM}, M_3)$$

VRA decrypts the M_3, verifies ID_{OR}, determines the MAC for M_3, and compares with the MAC at hand. At this point, VRA saves the selections BS, BE, N_1, N_2, and C. This concludes the registration.

B. Telematic Registration Authority Communication

The TCU facilitates the interface of the vehicle with the outside world (outside the vehicle). For the current biometric system of this paper, its role is to get the owner's security settings to G. Since VRA does not allow the USR, PWD, and PIN to be stored in the vehicle for security purposes, TCU will forward them to the VRA for verification whenever G requires that.

The TCU of the vehicle sends a message (MSG_1) to VRA indicating a request for verification of USR, PWD, and PIN of the owner. This message is acknowledged by VRA. When an acknowledgment is received by TCU, it forwards USR||PWD||PIN||ID_{OR} to VRA. This is demonstrated below:

TCU sends VRA the message M_4 encrypted with K_{RT} and concatenated with MAC (K_{RTM}, M_4):

$$M_4 = ID_{VRA} \,||\, ID_{TCU} \,||\, MSG_1$$
$$TCU \rightarrow VRA : E\,(K_{RT}, M_4) \,||\, MAC\,(K_{RTM}, M_4)$$

VRA decrypts E (K_{RT}, M_4) to get M_4, verifies both IDs, checks the request message (MSG_1), calculates the MAC of M_4, and compares it with the MAC in the forwarded message. If everything is fine, it replies using both IDs that were received plus an acknowledgment (ACK1). The MAC of this generated message is also attached:

$$M_5 = ID_{TCU} \,||\, ID_{VRA} \,||\, ACK1$$
$$VRA \rightarrow TCU : E\,(K_{RT}, M_5) \,||\, MAC\,(K_{RTM}, M_5)$$

Now, TCU sends the owner's credentials to VRA as represented below for verification:

$$M_6 = ID_{VRA} \,||\, ID_{TCU} \,||\, USR \,||\, PWD \,||\, PIN \,||\, ID_{OR}$$
$$TCU \rightarrow VRA : E\,(K_{RT}, M_6) \,||\, MAC\,(K_{RTM}, M_6)$$

VRA decrypts M6 and verifies the IDs are valid and the MACs match. Here, VRA will send a message indicating the details of the owners are verified:

$$M_V = ID_{TCU} \,||\, ID_{VRA} \,||\, VERF$$
$$VRA \rightarrow TCU : E\ (K_{RT}, M_V)\ ||\ MAC\ (K_{RTM}, M_V)$$

After carrying the necessary decryption and checking, TCU is ascertained the owner is authenticated.

The other secure communication between VRA and TCU includes receiving the security setting, BS || BE || N_1 || N_2 || C, of the owner from VRA:

$$M_S = ID_{TCU} \,||\, ID_{VRA} \,||\, BS \,||\, BE \,||\, N_1 \,||\, N_2 \,||\, C \,||\, ID_{OR}$$
$$VRA \rightarrow TCU : E\ (K_{RT}, M_S)\ ||\ MAC\ (K_{RTM}, M_S)$$

Once this message is verified, TCU will extract the security settings and forward them to G.

C. Telematic Gateway Communication

The TCU should send the vehicle owner's security settings to G. G needs two things from TCU: the security setting and the verification of USR, PWD, and PIN of the owner. This communication is secured as in subsection (B). For this reason, the description will be omitted. The protocol is stated as follows

$$M_7 = ID_{TCU} \,||\, ID_G \,||\, MSG_2$$
$$G \rightarrow TCU : E\ (K_{GT}, M_7)\ ||\ MAC\ (K_{GTM}, M_7)$$

$$M_8 = ID_G \,||\, ID_{TCU} \,||\, ACK2$$
$$TCU \rightarrow G : E\ (K_{GT}, M_8)\ ||\ MAC\ (K_{GTM}, M_8)$$
$$M_9 = ID_G \,||\, ID_{TCU} \,||\, ID_{OR} \,||\, BS \,||\, BE \,||\, N_1 \,||\, N_2 \,||\, C$$
$$TCU \rightarrow G : E\ (K_{GT}, M_9)\ ||\ MAC\ (K_{GTM}, M_9)$$

$$M_{9'} = ID_G \,||\, ID_{TCU} \,||\, ID_{OR} \,||\, USR \,||\, PWD \,||\, PIN$$
$$G \rightarrow TCU : E\ (K_{GT}, M_{9'})\ ||\ MAC\ (K_{GTM}, M_{9'})$$

TCU will perform the required decryption, MAC matching and ID checking, to extract USR||PWD||PIN and forward them to VRA for verification as explained in Sect. 7-B.

D. Driver Interface Gateway Communication

The DI allows the vehicle owner to create an ID (ID_O) for themselves that could be changed anytime the owner wishes instead of using the original ID (ID_{OR}) that was created during the registration of the vehicle with the VRA. DI also permits the owner to create two local (in vehicle) passwords, PWD_{OE} and PWD_{OS}, for vehicle entry and vehicle start, respectively.

To grant the owner access to G, DI sends the message, M_{10}, encrypted with K_{DIG} and then attaches the MAC of M_{10}:

$$M_{10} = ID_G \,||\, ID_{OR} \,||\, USR \,||\, PWD \,||\, PIN$$
$$DI \rightarrow G : E\ (K_{DIG}, M_{10})\ \|\ MAC\ (K_{DIGM}, M_{10})$$

After the necessary decryption and certification of the terms in M_{10} are performed by G, it forwards USR||PWD||PIN to TCU for verification by VRA as above. Once verified, G returns the same message it received to DI, but with ID_{OR} leading (referred to as $M_{10'}$) as a confirmation to start the actual communication:

$$G \rightarrow DI : E\ (K_{DIG}, M_{10'})\ \|\ MAC\ (K_{DIGM}, M_{10'})$$

Here, DI takes care of decrypting, checking contents of message are valid, and matching the MACs. It will then create and forward ID_O, PWD_{OE}, and PWD_{OS} to G to save as follows:

$$M_{11} = ID_G \,||\, ID_{OR} \,||\, ID_O \,||\, PWD_{OE} \,||\, PWD_{OS}$$
$$DI \rightarrow G : E\ (K_{DIG}, M_{11})\ \|\ MAC\ (K_{DIGM}, M_{11})$$

Once G verifies everything, DI will use ID_O for all future communications with G. ID_O will act as a pseudo-ID, and the owner can change it occasionally. Having done that, the owner will go ahead to add drivers. For each driver, an ID (ID_D), username (USR_D), password (PWD_D), and PIN (PIN_D) will be generated and sent to G. After that, the vehicle entry password PWD_{DE} and vehicle start password (PWD_{DS}) for the driver are created and sent to G. The process will be repeated for each driver. This is illustrated below:

$$M_{12} = ID_G \,||\, ID_O \,||\, ID_D \,||\, PWD_D \,||\, PWD_D \,\|\, PIN_D$$
$$DI \rightarrow G : E\ (K_{DIG}, M_{12})\ \|\ MAC\ (K_{DIGM}, M_{12})$$

$$M_{12} = ID_G \,||\, ID_O \,||\, ID_D \,||\, PWD_{DE} \,||\, PWD_{DS}$$
$$DI \rightarrow G : E\ (K_{DIG}, M_{11})\ \|\ MAC\ (K_{DIGM}, M_{11})$$

By now, the owner and all the drivers are granted their vehicle entry and start passwords. In addition, all drivers are registered with G and allowed to use the vehicle once their biometrics match the stored biometrics.

E. Smart Sensor Gateway Communications

As mentioned earlier, vehicle entry (unlocking door) relies on two biometrics and vehicle start on three biometrics. Due to the fact that all smart sensors function in a similar fashion as illustrated in Fig. 2 above, only the security of one sensor's communication with G is pursued. For this purpose, *i* will be replaced with 1. The template (T) refers to the extracted features of the iris image, face image, fingerprint, or voice pattern depending on the smart sensor used.

Initially, T is concatenated with the secret value S_1. The MAC for $T||S_1$ is found. The ID of the smart sensor (ID_{SS1}) ID_G, ID_O (or if another driver is involved), and the MAC are concatenated with $T||S_1$. The resulting expression is encrypted and sent to G. Note that ID_G is added to assure G:

$$M_{13} = ID_G \,||\, ID_{SS1} \,||\, ID_O \,||\, T \,||\, S_1 \| MAC\left(K_{SSG1M}, T \parallel S_1\right)$$
$$SS1 \rightarrow G : E\ (K_{SSG1}, M_{13})$$

Upon receiving this message, G will decrypt the message, verify ID_G and ID_{SS1}, calculate the MAC for T || S_1, and ensure that it matches the incoming MAC. If successful, ID_{SS1}, ID_O, and E (K_{SSG1}, T || S_1) are stored in the template database.

The sample to be matched with template will go through the same process but without being stored. In addition, T will be replaced with SMP:

$$M_{14} = ID_G \,||\, ID_{SS1} \,||\, ID_O \,||\, SMP \,||\, S_1 \| MAC\left(K_{SSG1M}, SMP \parallel S_1\right)$$
$$SS1 \rightarrow G : E\ (K_{SSG1}, M_{14})$$

Upon catching this message, G decrypts it with K_{SSG1}; verifies ID_G, ID_{SS1}, and S_1; obtains the MAC for SMP || S_1; and ensures the two MACs are equal. Then, G retrieves E (K_{SSG1}, T || S_1) using the index ID_{SS1} and ID_O, decrypts it with K_{SSG1}, ensures the same secret value S_1 is used, and then verifies T matches SMP. This approach above is repeated for each smart sensor.

F. Engine/Door Gateway Communication

When N_1 matches for vehicle start are guaranteed, the ECU should receive a start signal (ST) from G. To secure this communication, the hash of ST, H(ST), is found and signed by PR_G and then concatenated with ST. The result is further secured by encrypting it with the public key of the receiver PU_{ECU}.

$$M_{15} = ID_G \,||\, ID_{ECU} \,||\, ST \,\|\, E\ [PR_G, H\,(ST)]$$
$$G \rightarrow ECU : E\ (PU_{ECU}, M_{15})$$

ECU decrypts the arriving message with PR_{ECU}, verifies ID_G and ID_{ECU}, decrypts E [PR_G, H(ST) with PU_G to get H(ST), finds the hash of the received ST, and compares the two hashes. At this point, ECU will take care of starting the vehicle.

When N_2 matches for the vehicle entry are ensured, the message UNLOCK (ULK) needs to be received by DCU to carry out unlocking the vehicle door. The approach is similar to the vehicle start:

$$M_{16} = ID_G \,||\, ID_{DCU}|| \; ULK \,\|\, E\; [PR_G, H\,(ULK)]$$
$$G \rightarrow DCU : E\; (PU_{DCU}, M_{16})$$

8 Overriding with Passwords and Pins

If the vehicle owner selected Y (yes) for the confirmation, C, passwords (PWD_{OE}, PWD_{DE}, PWD_{DS}, PWD_{OS}) for the vehicle entry and vehicle start can be used to override biometrics. It is up to the owner/driver to apply passwords to both entry and start or only one of them. The driver will just key in the password at the door or at the driver interface inside the vehicle. Passwords can be used as a preference or in case any biometric sensor malfunctions.

9 Conclusion

With all the ongoing sophistication in vehicle industry, vehicle security is becoming more complicated and critical. Biometric identification and authentication are gaining more popularity in many applications. Using single biometrics should not be the preferred choice for vehicle security based on the FAR and FRR values presented above for the biometrics used in this paper. Hence, multimodal biometrics are essential to protect vehicles. Furthermore, leaving the biometric templates and samples unprotected creates a critical vulnerability. Stored plaintext templates can be attacked at their storage locations and modified. The same applies to the received plaintext samples. The consequences would be either others (nonlegal drivers) can control the vehicle or legal drivers will be forced to use their passwords. Furthermore, if these passwords are not protected, attackers can get hold of them and owners/drivers will face further problems. Therefore, biometric samples and templates (and passwords) should be encrypted and signed and have their integrity verified.

References

1. P. Tyagi, D. Dembla, Performance analysis and implementation of proposed mechanism for detection and prevention of security attacks in routing protocols of Vehicular Ad-Hoc Network (VANET). Egyptian Inf. J **18**, 133–139 (2017)
2. S. Rizvi, J. Willet, D. Perino, S. Marasco, C. Condo, A threat to vehicular cyber security and the urgency for correction, in *Proc. Complex Adaptive Systems (CAS 2017)*, (Chicago, Illinois, USA, 2017), pp. 100–105

3. M. Kang, J. Kang, Intrusion detection system using deep neural network for in-vehicle network security. Plos One J, 1–17 (2016). https://doi.org/10.1371/journal.pone.0155781 (Accessed: May 25, 2021).
4. A.M. Wyglinski, X. Huang, T. Padir, L. Lai, T.R. Elsenbarth, K. Venkatasbramanian, Security of autonomous systems employing embedded computing and sensors, in *Proc*, (IEEE Computer Society, 2013), pp. 80–86
5. V.L.L. Thing, J. Wu, Autonomous vehicle security: A taxonomy of attacks and defences, in *Proc. The 2016 IEEE International Conference on Internet of Things (iThings); IEEE Green Computing and Communications (GreenCom); IEEE Cyber, Physical and Social Computing (CPSCom), IEEE SmartData (SmartData)*, (2016), pp. 534–539
6. M. Gerla, P. Reiher, Securing the future autonomous vehicles: A cyber-physical systems approach, in *Securing Cyber-Physical Systems*, ed. by K. P. Al-Sakib, (CRC Press, London, 2015), pp. 197–217
7. E. Yagdereli, C. Gemci, A.Z. Aktas, A study on cyber-security of autonomous and unmanned vehicles. J. Defense Model. Simul. Appl. Method. Technol **12**, 369–381 (2015)
8. N.N. Nagamma, M.V. Lakshmaiah, T. Narmada, Raspberry Pi based biometric authentication vehicle door locking system, in *Proc. The 2017 IEEE International Conference on Power, Control, Signals and Instrumentation Engineering (ICPCSI)*, (Chennai, India, 2017), pp. 2348–2351
9. K.R. Gill, J. Sachin, Vehicle ignition using fingerprint sensor. Int. J. Innov. Res. Sci. Technol **2**(12), 357–363 (2016)
10. N. Kaushik, M. Veralkar, P. Parab, K. Nadkarny, Anti-theft vehicle security system. Int. J. Sci. Res. Dev **1**(12), 2845–2848 (2014)
11. N. Kiruthiga, L. Latha, S. Thangasamy, Real time biometrics based vehicle security system with GPS and GSM technology. Comput. Sci **47**, 471–479 (2015)
12. N. Kiruthiga, L. Latha, A study of biometric approach for vehicle security system using fingerprint recognition. Int. J. Adv. Res. Trends Eng. Technol **1**(2), 10–16 (2014)
13. C.S. Kumar, A.S.K. Reddy, J.R. Praveen, Biometric authentication based vehicular safety system using arm processor. Int. J. Eng. Sci. Adv. Technol **4**(5), 410–413 (2014)
14. N. Manikandan, K. Manikandan, K.E. Vishnu, R.T. Raja, K. Kanthaboopathi, T. Senthilnathan, Biometric vehicle security system and pollution monitoring. Int. Res. J. Eng. Technol **5**(1), 1126–1131 (2018)
15. R.M. Vithlani, S. Shingala, H.N. Pandya, Biometric automobile ignition locking system. Int. J. Electron. Commun. Eng. Technol **7**(5), 28–37 (2016)
16. K. Jaikumar, B. Jaiganesh, An economical car security authentication system based on face recognition structure. Int. J. Technol. Enhance. Emerg. Eng. Res **2**(8), 28–31 (2014)
17. J. Rajeshwari, K. Karibasappa, M.T. Gopalakrishna, Three phase security system for vehicles using face recognition on distributed systems, in *Information Systems Design and Intelligent Applications, Advances in Intelligent Systems and Computing*, ed. by S. C. Satapathy et al., (Springer, New Delhi, 2016), pp. 563–571
18. V.B. Seshasayee, E. Manikandan, Automobile security system based on face recognition structure using GSM network. Adv. Electron. Electr. Eng **3**(6), 733–738 (2013)
19. A.P. Sreedevi, B. Sarath, S. Nair, Image processing based real time vehicle theft detection and prevention system, in *Proc. The 2011 International Conference on Process Automation, Control and Computing*, (Coimbatore, India, 2011), pp. 1–6
20. P. Sreekala, V. Jose, J. Joseph, S. Joseph, The human Iris structure and its application in security system of car, in *Proc. The 2012 IEEE International Conference on Engineering Education. Innovative Practices and Future Trends*, (Kottayan, India, 2012), pp. 1–5
21. S. Punnoose, J.S.J. Kumar, Iris recognition for security & safety of automobiles. Int. J. Innov. Sci. Eng. Technol **2**(4), 961–966 (2015)
22. H. Silva, A. Lourenco, A. Fred, In-vehicle driver recognition based on hands ECG signals, in *Proc. The 2012 ACM International Conference on Intelligent User Interfaces (IUI'12)*, (Lisbon, Portugal, 2017), pp. 25–28

23. C. Lupu, V. Lupu, Multimodal biometrics for access control in an intelligent car, in *Proc. The 2007 International Symposium on Computational Intelligence and Intelligent Informatics*, (Agadir, Morocco, 2007), pp. 261–267
24. M. Biermann, T. Hoppe, J. Dittmann, C. Vielhauer, Vehicle systems: Comfort & security enhancement of face/speech fusion with compensational biometric modalities, in *Proc. The 10th ACM Workshop on Multimedia and Security*, (Oxford, UK, 2008), pp. 185–194
25. M.B. Khan, M.K. Khan, J. Zhang, D. Zhang, Enhancing the security of intelligent transportation systems (ITS) using Iris/finger-based multimodal biometrics, in *Proc.The 2006 IEEE International Conference on Engineering of Intelligent Systems*, (Islamabad, Pakistan, 2006), pp. 1–6
26. S. Sebastin, Literature survey on automated person identification techniques. Int. J. Comput. Sci. Mobile Comput **2**(5), 232–237 (2013)
27. M. Chapple, J.M. Stewart, D. Gibson, *Certified Information Systems Security Professional* (SYBEX, Indiana, 2018)
28. S. Stallings, L. Brown, *Computer Security Principles and Practice* (Pearson, Hoboken, 2018)
29. C.P. Pfleeger, S.L. Pfleeger, J. Margulies, *Security in Computing* (Printice Hall, Upper Saddle River, 2015)
30. Biometrics Institute, "Biometrics Types," 2019, https://www.biometricsinstitute.org/what-is-biometrics/types-of-biometrics, Retrieved: November, 2019

Part VI
Internet Computing, Internet of Things, and Applications

Per-user Access Control Framework for Link Connectivity and Network Bandwidth

Shogo Kamata, Chunghan Lee, and Susumu Date

1 Introduction

IHS Technology has forecast that 75.4 billion devices will be connected to the Internet by 2025 [1]. This forecast includes a diversity of devices such as sensors, mobile phones, tablets, electric appliances, and even cars. In response to the rapid growth and diversification of the Internet of Things (IoT) devices, a variety of IoT applications are now being developed in the world. Examples of such applications include those for health management services, energy-saving devices in factories, and applications that alert drivers to the need for vehicle maintenance.

Inherently, these types of IoT applications require strict access control to the IoT devices used by such applications because devices in close proximity to users may bring about danger to users if they are illegally used by malicious users. For this reason, such IoT applications usually adopt an access control mechanism on the IoT devices. However, most IoT applications do not consider access control to network resources and thus allow sharing with other applications and users, even though IoT applications often use cryptographic technologies to ensure confidentiality and integrity of communication.

S. Kamata (✉)
Graduate School of Information Science and Technology, Osaka University, Osaka, Japan
e-mail: kamata.shogo@ais.cmc.osaka-u.ac.jp

C. Lee
Toyota Motor Corporation, Tokyo, Japan
e-mail: chunghan_lee@mail.toyota.co.jp

S. Date
Cybermedia Center, Osaka University, Osaka, Japan
e-mail: date@cmc.osaka-u.ac.jp

K. Daimi et al. (eds.), *Advances in Security, Networks, and Internet of Things*,
Transactions on Computational Science and Computational Intelligence,
https://doi.org/10.1007/978-3-030-71017-0_42

With today's IoT applications, users and a variety of resources including devices, data resources, and computing resources are connected on the Internet. Usually, such resources have different types of security requirements and policies, and the users have different attributes. For example, data resources such as CT and MR images can be accessed by users whose attribute is a medical doctor, while the data resources cannot be accessed by users who have the medical student attribute. Taking the necessity of strict access control to such resources into consideration, network resources such as links, network switches, and bandwidth that are accessed by IoT applications should be controlled and managed, so that only users who are permitted by network administrators can utilize such network resources within permitted authority.

In this chapter, we propose a per-user access control framework for link connectivity and network bandwidth as network resources. The access control framework aims to control not only the network connectivity to a variety of resources including IoT devices, but also network resources using the network programmability brought by software defined networking (SDN) [2]. The key idea behind this approach is that we overturn the assumption that IoT devices should always be connected on the Internet and the network cannot be controlled dynamically. To realize fine-grained access control, we have built an access control framework based on the role based access control (RBAC) concept [3].

This chapter is structured as follows. Section 2 explains the building block technologies for the access control framework. Section 3 presents the proposed per-user access control framework for link connectivity and network bandwidth. Section 4 investigates the feasibility and practicality of the proposed access control framework. Section 5 introduces related work. Section 6 concludes this chapter.

2 Building Block Technologies

2.1 Software Defined Networking (SDN)

SDN is a network architecture that enables the dynamic control of packet flows in a software programming fashion. This architecture separates the control plane, which determines how packets are handled, and the data plane, which performs the actual packet forwarding, into separate network devices.

OpenFlow [4] is a network flow control protocol and the de facto standard implementation of SDN. Figure 1 shows the SDN architecture with OpenFlow. The SDN architecture is composed of the SDN switch that provides data plane functionality and the SDN controller that provides control plane functionality. The SDN controller and SDN switches can communicate with each other using the OpenFlow message defined by the OpenFlow implementation. Each SDN switch has one or more flow tables, each of which is a list of rules (flow entries) describing how packets are forwarded. The SDN controller adds, updates, and deletes the flow entries and inquires about packet statistics to the SDN switches. The SDN switch

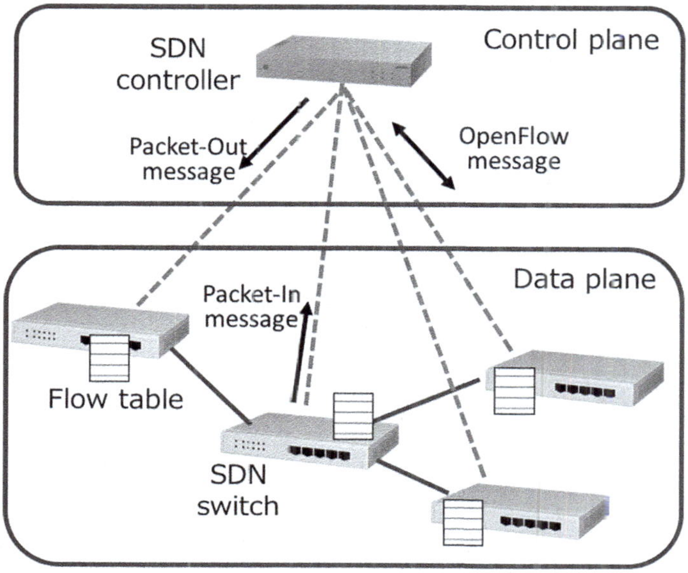

Fig. 1 SDN architecture with OpenFlow

(Table ID 0)		**Match field**			**...**	**Action**
MAC src	MAC dst	IP src	IP dst	TCP port		
	dd:ee:ff...					Forward from port 1
		10.1.2.3		443		Forward to SDN controller
			10.2.4.5			Drop
				80		Set IP src to 10.0.0.1
		10.3.5.6				Store in queue 1 Go to table 1

Fig. 2 Example of a flow table

sends the packet-in message, which is a type of OpenFlow message, if necessary. After that, the SDN controller analyzes this message and then determines how the packet should be processed. The SDN controller sends a packet with arbitrary content to an SDN switch using a packet-out message that is an OpenFlow message

Figure 2 shows an example of a flow table. The flow entry consists of a match condition field and an action condition field. The match condition is described to select packets for Layer 1 (physical port), Layer 2 (MAC address), Layer 3 (IP address), and Layer 4 (TCP/UDP port). The action condition describes how to process packets that match the match condition. Examples of the action condition include discarding packets, outputting to a port, storing in a queue, rewriting packet

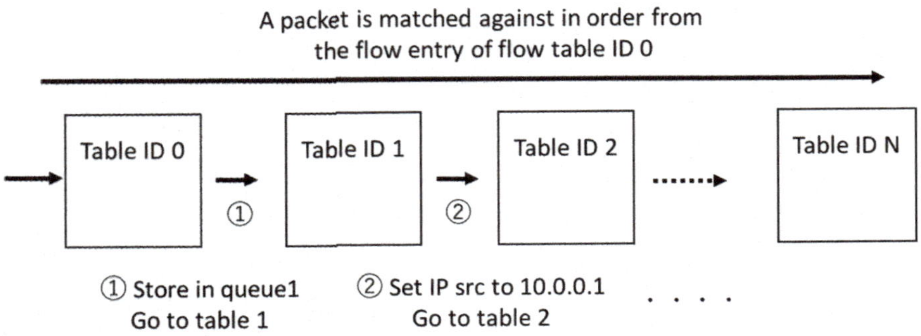

Fig. 3 Pipeline processing with multiple flow tables

Fig. 4 Format of the X. 509 certificate

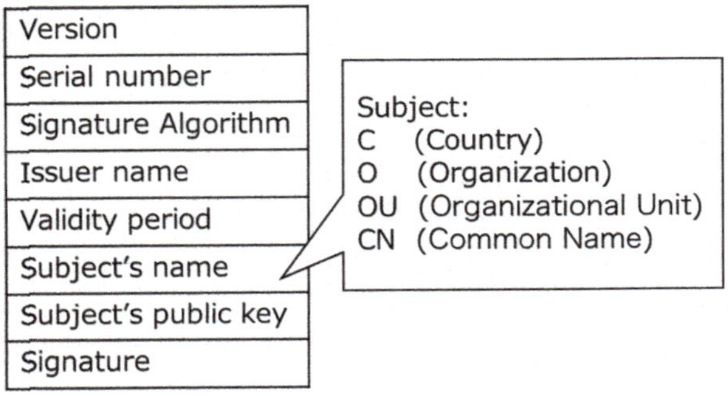

header fields, and forwarding packets to the SDN controller. In OpenFlow 1.3, an SDN switch can hold multiple flow tables. Each flow table is assigned an ID of 0, 1, 2… in that order. Figure 3 shows the pipeline processing with multiple flow tables. When a packet arrives at an SDN switch, the packet is investigated as to whether it matches with any match conditions of the flow table. The flow table is searched in ascending order. On this occasion, the action condition can be described to refer to a different table ID. For example, using this pipeline processing, it becomes possible to store the packet in a queue by the action condition in flow table 0 and then rewrite the source IP address by the action condition in flow table 1.

2.2 Public-Key Infrastructure Using X. 509 (PKIX)

PKI is an authentication infrastructure that leverages public key cryptography technology. The purpose of PKI is to guarantee the correspondence between a public key and the owner of that public key. For this purpose, X. 509 certificates [5] are used as the standard format for public key certificates.

Figure 4 shows the format of the X. 509 certificate. The X. 509 certificate has a list of fields. The signature algorithm field describes the algorithm used by the certificate issuer to sign certificates, such as RSA [6], DSA [7], and ECDSA [8]. The issuer name field describes the name of the authority that signed the certificate and issued it, i.e., the Certification Authority (CA). The subject's name field describes the information of the owner of the certificate. The owner's information includes C (Country), O (Organization), OU (Organizational Unit), and CN (Common Name). The subject's public key field contains the public key of the certificate owner. The signature value field contains the digital signature of the CA.

3 Proposed Architecture

3.1 *Architecture Overview*

To realize the access control to the network resources, we have applied the network programmability explained in Sect. 2.1. Figure 5 shows an overview of our proposed access control framework. This framework basically allows each user to dynamically obtain the network connectivity to resources on-demand and then utilize a network path composed of the links permitted for use within the permitted bandwidth.

The proposed framework functions based on the following key elements:

1. access control policy,
2. authentication and authorization function,
3. resource assignment function.

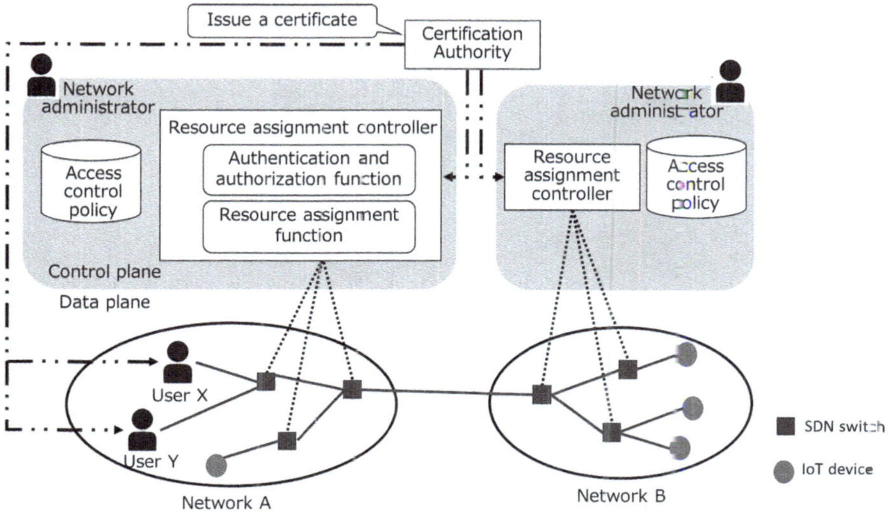

Fig. 5 Proposed framework

The access control policy links users to roles, each of which defines the set of permitted operations, based on the RBAC concept. Importantly, this access control policy is defined for a network domain by the network administrator and then applied on the network. In other words, if there are several network domains between a user device and a data resource to be accessed, the traffic between them has to satisfy the network policies by all of the network domains. More technically, the traffic can only use the network links permitted on each network domain within a permitted bandwidth.

In order to control links and bandwidth as network resources on a user-by-user basis, an authentication and authorization function is performed for each user. In addition, the network administrator for each network can make an access control policy. Based on this policy, users are allowed to perform the set of operations corresponding to the role.

The resource assignment controller in charge is responsible for assigning the network resources on a network domain as well as connecting and disconnecting the network. These functionalities are achieved through the use of network programmability brought by SDN.

As a result of the interaction of these three key elements, each user or user application is able to dynamically use a dedicated virtual network in an on-demand way. Figure 6 shows what the users' view of the virtual network established by the proposed framework is like. Under the proposed framework, each user is allowed to use only permitted links within a permitted bandwidth. In this case, since a link is assigned to each user, different connectivity can be provided even if the same link is available for user X and user Y.

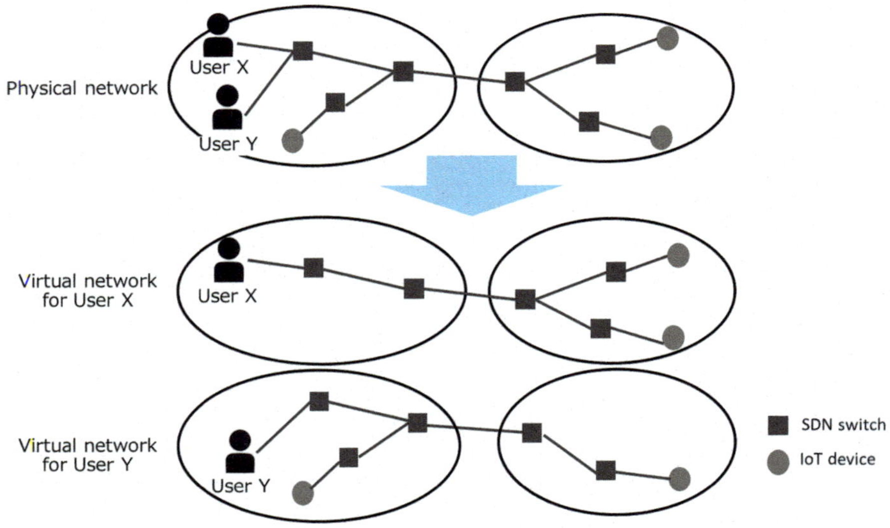

Fig. 6 User's view of the virtual network established

3.2 *Access Control Policy*

In this research, we assume the following simple access control to network resources. There are only two user categories: users who attempt to access to data resources and network administrators in charge who implement their own policy to the users' traffic on the network domain. Network resources are links and bandwidth.

Based on this simple assumption, we have defined an access control framework based on the RBAC concept. Figure 7 shows an example of an access control policy defined by a network administrator for a network domain. An access control policy is designed by a network administrator in each network. The RBAC concept consists of users, roles, and authorizations, and the authority to use a particular resource is assigned to a particular role. A user is assigned a particular role and then granted a set of authorizations associated with the role. In other words, the access control based on the RBAC concept gives the set of operation permissions through roles, rather than directly assigning them to users.

Based on the RBAC concept, we have designed the User-Role, Role-Authoriz ation ID, and Authorization list tables. The User-Role table contains users' names and their corresponding roles. The Role-Authorization ID table contains the roles and their corresponding authorization ID. The Authorization list table contains the authorization ID and their corresponding authorization list. The access control policy in Fig. 7 shows that User X is assigned to Role α on the network domain where this policy is applied and thus User X can use links 1, 2, 4, 5, and 7 within 400 Mbps.

User-Role table

User	Role
User X	Role α
User Y	Role β
User Z	Role β

Role-Authorization ID table

Role	Authorization ID
Role α	1,2,4,5,7,8,9
Role β	1,3,5,6,10

Authorization list table

Authorization ID	Authorization
1	Link 1
2	Link 2
3	Link 3
4	Link 4
5	Link 5
6	Link 6
7	Link 7
8	Link 8
9	400Mbps
10	100Mbps

Access control policy

Fig. 7 Example of access control policy

3.3 Authentication and Authorization Function

The proposed access control framework relies on PKI using the X. 509 certificate for authentication. In the proposed access control framework, it is assumed that each user and resource assignment controller have the X.509 certificates issued by a trusted CA and that the CA is trusted by each network domain. In other words, this authentication infrastructure is universal and common among all participating network domains. On the other hand, it is assumed that the authorization is independently performed by each network domain. As described, an access control policy is defined by each network administrator responsible for a network domain, meaning that each network administrator can assign a user to a dedicated role with a different authorization set.

The authentication function performs the mutual authentication between the resource assignment controller and the user as follows (Fig. 8 (1)–(7)). When a user attempts to use a network, the mutual authentication between the user and the resource assignment controller that is in charge of the network domain is performed. In detail, they exchange each other's X. 509 certificates via the SDN switch in which the packet from the user arrives through the use of the packet-in message and the packet-out message of the OpenFlow protocol.

After authentication, the authorization function performs (Fig. 8 (8)–(9)). For the authorization function, the resource assignment controller performs the following two actions. First, the resource assignment controller assigns network resources based on the access control policy. The details are described in Sect. 3.4. Second,

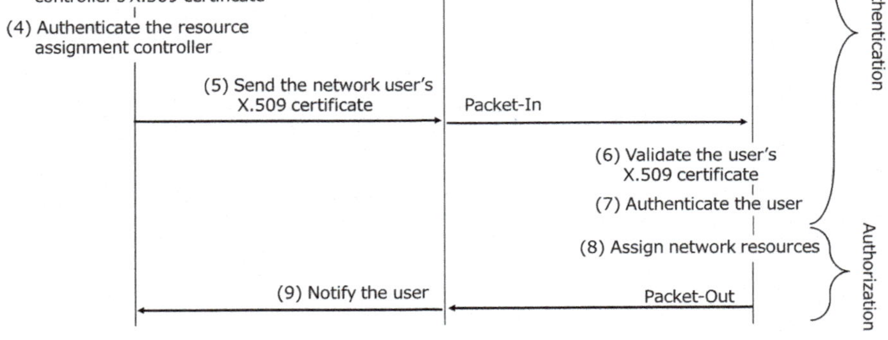

Fig. 8 Authentication and authorization procedure

the resource assignment controller notifies the user that the resource assignment has been completed through the use of the packet-out message. After the user receives the notification, the user will be able to use the network.

3.4 *Resource Assignment Function*

After finishing the authentication procedure, the user is assigned an appropriate role based on the resource management policy. After that, the actual assignment of network resources is implemented by the resource assignment controller. The resource assignment controller is basically an SDN controller responsible for the dynamic establishment of network paths between a user and IoT devices on behalf of the user. The resource assignment controller is deployed onto a network domain and performs the following functions: network path configuration and bandwidth configuration.

Network Path Configuration As described, the resource assignment controller assigns links and bandwidth on a network domain for the traffic between users and IoT devices. This function works as follows. Immediately after a user attempts to connect to an SDN switch and completes the authentication procedure described in Sect. 3.3, the resource assignment controller checks which physical links are available on the network domain based on the user's role in the network domain, by referring to the access control policy. After that, the resource assignment controller uses the Dijkstra method [9], which is the best-first search algorithm for solving the single starting point shortest path problem when the weights of the edges in graph theory are non-negative, to explore the network path between the user and the IoT device that the user wants to access. Then, the resource assignment controller inserts the packet forwarding rules onto the flow table on the SDN switches on the calculated network path. At this moment, the network connectivity between the user and the IoT device is achieved and as the result packets are reachable.

Figure 9 shows an example case of path assignment. In this example case, it is assumed that the network domain consists of two IoT devices and eight links. Also, it is assumed in this example case that the access control policy is the one shown in Fig. 7. Now suppose that User X wants to access IoT device A. Under this assumption, the resource assignment controller learns that User X should be assigned to role and that links 1, 2, 4, 5, 7, and 8 are available to users with role (blue link). The results of the path search using the Dijkstra method are, with the edge weight of each link being 1, the resource assignment controller grants the user the authorization to use only the links shown in red in the figure. Once the network path to be assigned to the user is finalized, the resource assignment controller inserts the flow entries corresponding to packet forwarding rules onto the flow table deployed on the SDN switch on the path. In this way, the resource assignment controller gives the users the minimum necessary authorization.

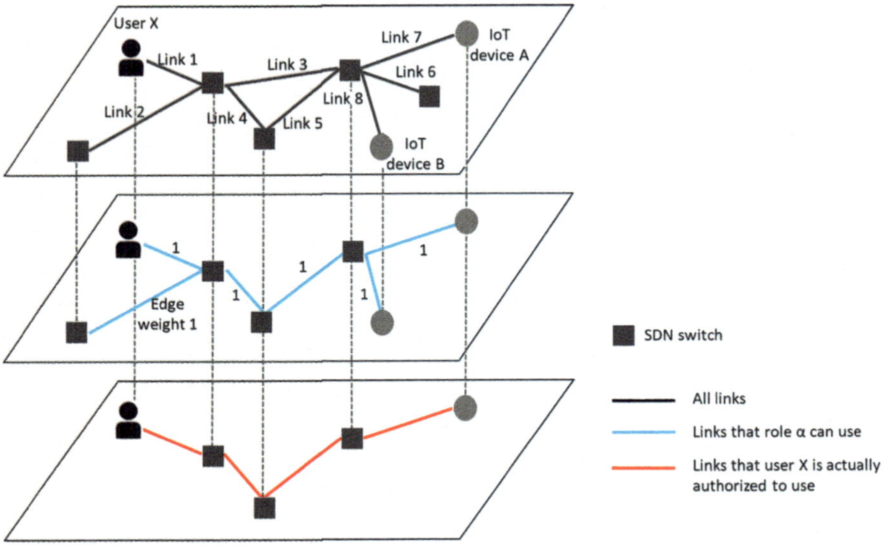

Fig. 9 Path assignment example

Bandwidth Configuration Figure 10 shows how the bandwidth configuration mechanism works. This mechanism basically relies on the throughput control utilizing packet queuing and scheduling. Figure 10a shows how packets are processed on a switch. The incoming packets are processed as follows. When packets arrive at the switch, the packets are queued based on the queuing, and then the packets are scheduled and taken out of the queue based on the scheduling. For queuing, if multiple queues are waiting ahead of time, packets can be classified based on any content in the L2-L4 header fields of packets such as MAC address, IP address, or port number, etc. In this chapter, we have used the source IP address and the destination IP address of the packets sent by the user for packet queuing to identify the user's packet. For scheduling, it is possible to guarantee or limit the amount of packet transfers by determining the amount of packets to be taken out of the queue. In the proposed method, we assume that the administrator sets queues that specify the maximum bandwidth at the time of packet output to the SDN switch.

The flow tables of (b) and (c) in Fig. 10 show examples of flow entries inserted by the resource assignment controller. Figure 10b shows a flow table (table ID 0) that is set up for queuing. Figure 10c shows a flow table (table ID 1) that is set up for scheduling. The action condition in table 0 executes queuing. The packets matching the match conditions are stored in the switch's queue and then set to refer to table 1. On the other hand, the action condition in table 1 executes scheduling. The packets stored in the queue are retrieved, and packets are forwarded from the specified physical port. In the proposed framework, we assume that the IP address of the user device and the IoT device is used as the match condition for the flow entry. Then, in order to enable the communication between the user and the IoT device,

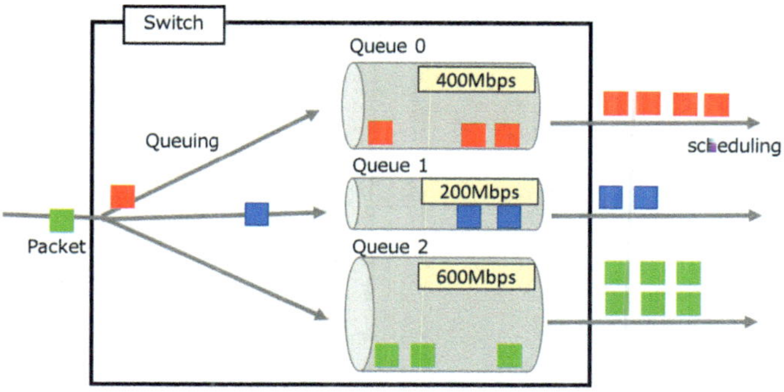

(a)

(Table ID 0) Match field: Destination IP address	Source IP address	...	Action
10.0.1.2	10.0.2.2		Store in queue 0, go to table 1
10.0.2.2	10.0.1.2		Store in queue 1, go to table 1

Queuing

(b)

(Table ID 1) Match field: Destination IP address	Source IP address	...	Action
10.0.1.2	10.0.2.2		Forward from port 1
10.0.2.2	10.0.1.2		Forward from port 1

Scheduling

(c)

Fig. 10 Bandwidth control mechanism. (**a**) Packet processing on a switch. (**b**) Flow table (Table ID 0). (**c**) Flow table (Table ID 1)

the resource assignment controller inserts a flow entry that not only allows packets to be sent from the user to the IoT device (forward direction), but also from the IoT device to the user (reverse direction).

4 Evaluation

4.1 Experimental Environment

In this section, we investigate the feasibility and practicality of the proposed access control framework through the simulation. For the simulation, we have built the evaluation environment assuming the LAN environment as shown in Fig. 11. The evaluation environment is assumed to be composed of three networks, each of which is administered by an independent network administrator. The network topology of each network is basically assumed to be tree-shaped. Open vSwitch (OVS) [10] and KVM [11], which makes the Linux kernel act as a hypervisor, were utilized to build this evaluation environment on a single machine. The specification of the machine used for this evaluation was shown in Table 1. In the machine, 10 IoT devices and 2 user devices were emulated as virtual machines on the guest OS. 11

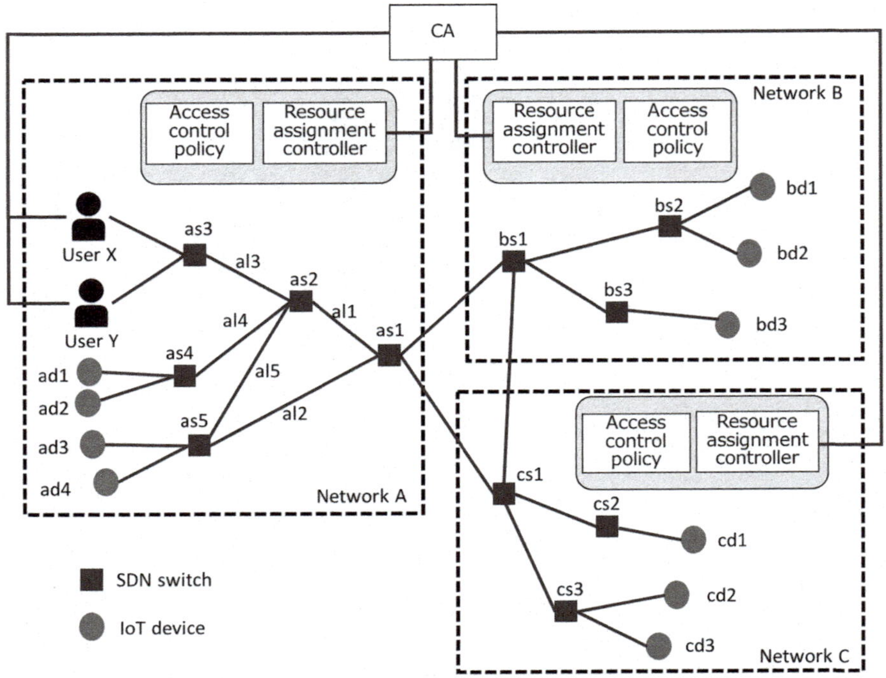

	SDN switch	IoT	Network C
Network A	as1, as2, as3, as4, as5	bs1, bs2, bs3	cs1, cs2, cs3
Network B	ad1, ad2, ad3, ad4	bd1, bd2, bd3	cd1, cd2, cd3
Network C	User X, User Y	---	---

Fig. 11 Evaluation environment

Table 1 Specification of the machine

CPU	Intel(R) Xeon(R) CPU E5-2430 v2 @2.50 GHz (6 cores total)
Memory	4 GB × 6
OS	CentOS 7.6.1810
Open vSwitch	2.0.0
Hypervisor	Kernel-based Virtual Machine 4.5.0

Table 2 Specification of the virtual machine

Memory	1 GB
OS	CentOS 7.5.1804

SDN switches were emulated with OVS on the host OS. The specification of the virtual machine is shown in Table 2. Each link was configured at 500 Mbps, using the performance limitation of the machine used in this evaluation. The round-trip time (RTT) between *as1* and *bs1* is set to various values to investigate the impact of the proposed framework due to RTT. The RTT between *as1* and *cs1* is fixed at 10 ms to investigate the impact of the proposed framework when the data size is changed.

In each network, to deploy the control plane at each network, a resource assignment controller is deployed with an access control policy on the host OS. To run multiple resource assignment controllers on the host OS, we execute each resource assignment controllers' processing as different programs. In the evaluation, Ryu 4.29 was used as the SDN controller for the resource assignment controller. Three access control policies with different table sets were configured in MySQL 5.7.24 and deployed on the host OS. The access control policies used in this evaluation were shown in Fig. 12. In the network configuration on each network, the resource assignment controller and the access control policy are linked on the control plane. SDN switches and IoT devices were placed on the data plane. The switches at the network were set up with two queues that specify the maximum bandwidth for packet output. The maximum output bandwidth of queue 0 was set at 400 Mbps, and the maximum output bandwidth of queue 1 was set at 100 Mbps.

4.2 Evaluation Method

In this evaluation, we now assume that User X and User Y attempt to collect the data from IoT devices. For this purpose, we measured the total time needed for User X and User Y to collect the data from IoT devices in each network. In this measurement, we investigated the following times. First, the authentication/authorization time refers to the time from when a user attempts to use the network until when the network becomes available. In other words, this authentication/authorization time is the amount of time the procedure takes, or (1)–(9) in Fig. 13. Second, the resource assignment time is the time from when the resource assignment controller authenticates the user until the time when it has completed inserting flow entries for resource assignment. This resource assignment time is (8) in Fig. 13. Third, the data collection time is the time from when the SDN controller completes the authentication and authorization of the user to the time when the data is collected. This time is (10) and (11) in Fig. 13. The total time refers to the sum of the authentication/authorization time and the data collection time.

We conducted three experiments based on the following three scenarios. In Experiment 1, we suppose that User X and User Y simultaneously attempt to collect

Network A

User-Role table

User	Role
User X	Role α
User Y	Role β

Role-Authorization ID table

Role	Authorization ID
Role α	1,3
Role β	2,4

Authorization list table

Authorization ID	Authorization
1	All links
2	Links except for link al5
3	400Mbps (Queue 0)
4	100Mbps (Queue 0)

Network B

User-Role table

User	Role
User X	Role γ
User Y	Role δ

Role-Authorization ID table

Role	Authorization ID
Role γ	1,2
Role δ	1,2

Authorization list table

Authorization ID	Authorization
1	All links
2	400Mbps (Queue 0)
3	100Mbps (Queue 1)

Network C

User-Role table

User	Role
User X	Role ε
User Y	Role ε

Role-Authorization ID table

Role	Authorization ID
Role ε	1,2

Authorization list table

Authorization ID	Authorization
1	All links
2	400Mbps (Queue 0)

Fig. 12 Access control policies used for evaluation

1 GB of data from *ad3* and *ad4* in network A, respectively. Under this scenario, we measured the authentication/authorization time, the resource assignment time, and the data collection time. Also, we measured the bandwidth used by each user with *iperf3*.

In Experiment 2, User X attempts to collect the data from *bd1* in network B. In this case, we fixed the data size at 100 MB and configured the RTT between switches *as1* and *bs1* from 10 ms to 100 ms. We measured the authentication/authorization time in each network, the resource assignment time, and the data collection time to investigate the impact of the proposed framework when the networks are geographically separated and the RTT is changed by network conditions, such as congestion and link utilization.

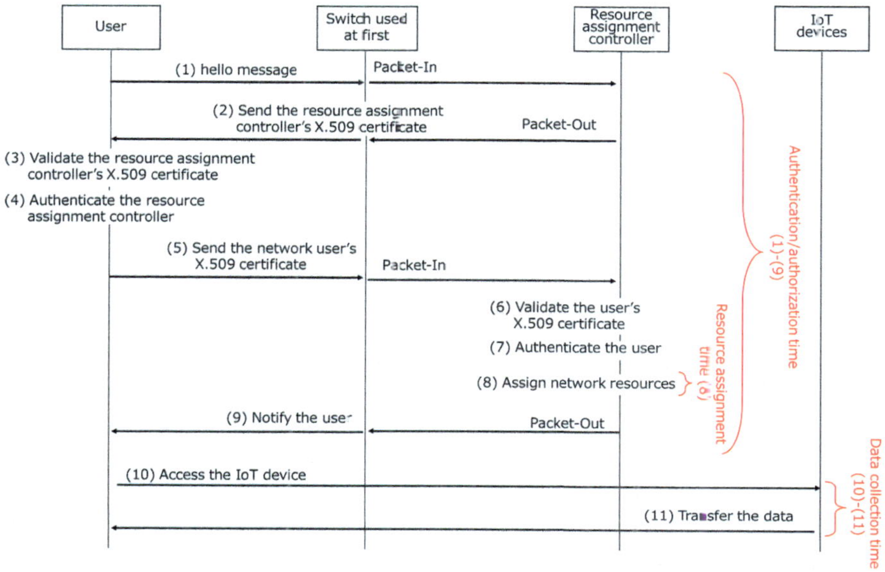

Fig. 13 Measurement of the proposed framework

In Experiment 3, User X attempts to collect the data from *cd1* in network C. In this case, we fixed the RTT at 10 ms and configured the data size from 100 MB to 1 GB. We measured the authentication/authorization time in each network, the resource assignment time, and the data collection time to investigate the impact of the data size on the proposed framework. Each experiment was measured 10 times each, and the average value was the result.

4.3 Result

Experiment 1 Before we measured the total time with the proposed framework, we measured the data collection time without the proposed framework. This measurement was conducted 10 times repeatedly. The average measurement time for User X was 22.8 s and for User Y was 86.2 s.

Figure 14 shows the result of the authentication/authorization time, the resource assignment time, and the data collection time for User X and User Y to collect data from *ad3* and *ad4*, respectively, with the proposed framework. The red shows the authentication/authorization time, the yellow shows the resource assignment time, and the gray shows the data collection time. The authentication/authorization time was 0.33 s in both User X and User Y cases. Then the resource assignment time was 0.04 s for both User X and User Y. The resource assignment time also remained unchanged in the subsequent experiments 2 and 3. The data collection time for User

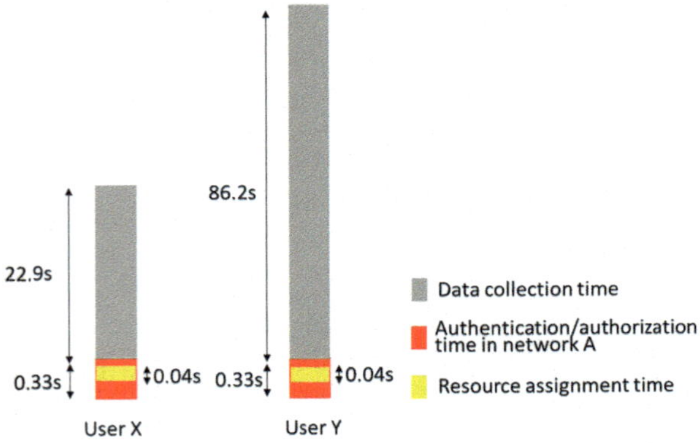

Fig. 14 Total time for Users X and Y

X was 22.9 s and for User Y 86.2 s. Therefore, the total time for User X was 23.2 s and for User Y 86.5 s. This means the impact of the proposed framework was small. In fact, the total time with the proposed framework increased by 1.7% in comparison with the one without it for User X and increased by 0.35% for User Y.

The data collection time was different because the privileges assigned to User X and User Y were different. In other words, the bandwidth each user can use was different, and as a result, the bandwidth resource each user can use was differentiated. As shown in Fig. 12, in network A, User X is assigned to Role α and thus can use up to 400 Mbps. On the other hand, User Y is assigned to Role β and thus can use up to 100 Mbps. In fact, we measured TCP throughput 10 times. On average, User X was using 339 Mbps and User Y 95.1 Mbps. Therefore, the resource assignment controller assigned the bandwidth according to the access control policy.

Experiment 2 Figure 15 shows the result of the authentication/authorization time and the data collection time for the RTT from 10 ms to 100 ms. The red shows the authentication/authorization time in network A, the blue shows the authentication/authorization time in network B, and the gray shows the data collection time. In the access control policy on network A, User X is assigned to Role α and thus can use up to 400 Mbps. In the access control policy on network B, User X is assigned to Role γ and thus can use up to 400 Mbps. The authentication/authorization time in network A for User X is approximately 0.32 or 0.33 s.

The authentication/authorization time in network B for User X is shown in Fig. 16. The longer the RTT between networks A and B, the longer the authentication/authorization time in network B. However, as shown in Fig. 15, the data collection time was also longer, so the percentage of the authentication/authorization time is smaller. In fact, when the RTT was set to 10 ms, the percentage of the authentication/authorization time to the total time was 19.6%, but when the RTT was set to 100 ms, the percentage was reduced to 9.3%.

Experiment 3 Figure 17 shows the result of the authentication/authorization time and the data collection time for the data size from 100 MB to 1 GB. The red shows the authentication/authorization time in network A, the blue shows

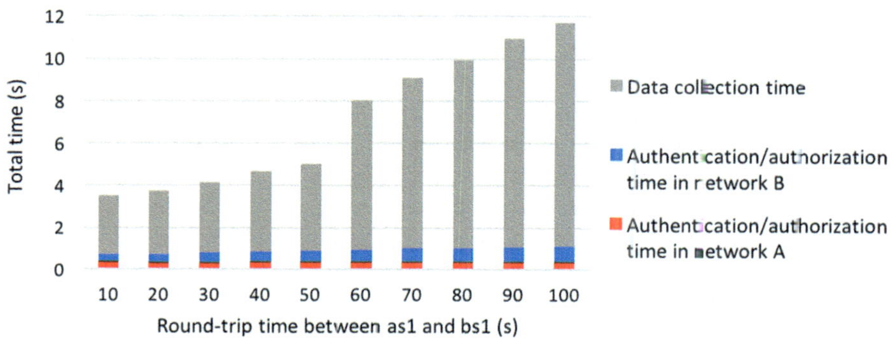

Fig. 15 Authentication/authorization time and data collection time for the RTT from 10 ms to 100 ms

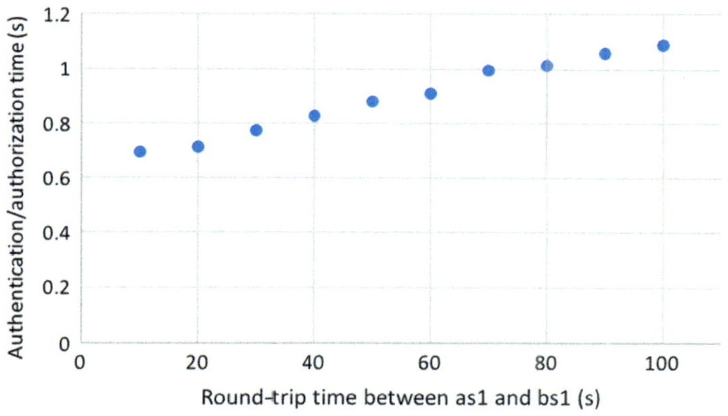

Fig. 16 Authentication/authorization time in network B

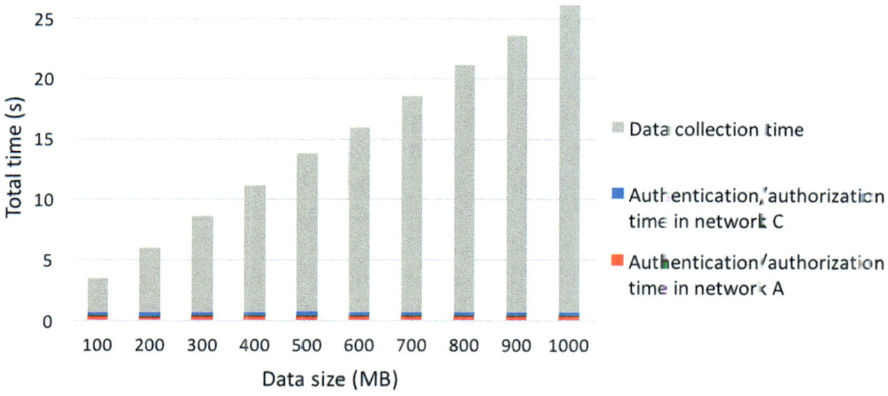

Fig. 17 Authentication/authorization time and data collection time for the data size from 100 MB to 1 GB

the authentication/authorization time in network C, and the gray shows the data collection time. In the access control policy on network A, User X is assigned to Role α and thus can use up to 400 Mbps. In the access control policy on network C, User X is assigned to Role ε and thus can use up to 400 Mbps. The authentication/authorization time in network A for Use X is approximately 0.32 or 0.33 s and in network C is approximately 0.36 or 0.37 s. As shown in Fig. 17, the larger the data size, the longer the data collection time and the smaller the percentage of the authentication/authorization time. In fact, when 100MB data was collected, the percentage of the authentication/authorization time to the total time was 20.0%, but when 1GB data was collected, the percentage was reduced to 2.7%.

From Experiment 1, we learned the authentication/authorization time and the resource assignment time were small compared to the data collection time. Furthermore, from Experiment 1, we found that the resource assignment controller assigns the bandwidth according to the access control policy. From Experiment 2, we found the authentication/authorization time increased as the RTT between the networks increased. However, the longer the RTT, the longer the data collection time, and the smaller the impact of the authentication/authorization time. In other words, the authentication/authorization time is not a bottleneck even though in the wide area computing environment, the geographical distance of the network is far. From Experiment 3, we found the larger the size of the data collected, the smaller the impact of the authentication/authorization time.

It is expected that the more networks are used by a user, the longer the authentication/authorization time. As an improvement measure, it is necessary to reduce the number of packets exchanged between the user and the SDN controller during authentication and authorization, and this is a challenge for the future.

5 Related Work

Recently, a variety of security technologies have been proposed and implemented to provide resources such as devices, data resources, and computing resources to any user. In the context of grid computing, a variety of authentication and authorization technologies have been proposed to realize coordinated resource sharing and problem solving in dynamic, multi-institutional virtual organizations. In [12], the concept of VO was defined and proposed. To embody the concept, a GSI (grid security infrastructure) [13] was used for the authentication infrastructure of grid computing. GSI uses X. 509 certificates for a user and a host authentication. The proposed framework in this chapter shares the idea of using the X. 509 certificate. Also, PRIMA [14] and VOMS [15] have been proposed for authentication and authorization to realize fine-grained access control to resources on the grid. The authentication and authorization technologies developed for grid computing do not treat network resources.

There are studies on isolating or controlling network resources on a user-by-user basis. FlowVisor [16] can slice a physical network into multiple virtual networks.

FlowVisor is a special SDN controller that acts as a proxy between the SDN controller and the SDN switch, enabling multiple network slices to be built. By dividing the physical network into multiple logical network slices, a user-specific virtual network can be realized. In [17], a role based campus network slicing was developed by using an authentication controller and FlowVisor. The slicing of the network was based on the VLAN tags attached to the packets. However, to slice a network using FlowVisor, the same number of SDN controllers that control the forwarding plane as the number of slices is required. In this way, the more roles that are managed, the more SDN controllers are needed.

A method to virtually integrate logically distributed networks is described in [18]. A distributed cloud environment based on SD-WAN was assumed, and an environment where distributed cloud resources can be integrated was proposed. Therefore, virtual dedicated networking (VDN) is generated quickly on-demand, enabling virtual networks to be built without degrading performance. However, VDN does not take into account the division of the network into different users or roles.

The science-DMZ [19] is also a technology for access control over a network. The science-DMZ is a network design pattern for efficient transfer of data used in scientific calculations. Moreover, access is controlled on the basis of ACL on routers and switches installed for the science-DMZ. However, changing access control policies dynamically is difficult.

6 Conclusion

We proposed an access control framework to network resources such as links and network bandwidth. We have applied the network programmability brought by SDN. This mechanism basically allows each user to dynamically connect to resources on-demand and then utilize a network path composed of the links permitted for use within a permitted bandwidth. The proposed mechanism functions are based on the following key elements: the access control policy based on the RBAC concept, the authentication and authorization function, and the resource assignment function. We mainly focused on the feasibility of the proposed architecture and the impact of authentication and authorization. In order to clarify the feasibility and the impact of the proposed framework, we designed three networks and deployed the networks on a single machine. Through the simulation, we evaluated the total collection time from the target IoT device to the network user with the authentication and the authorization. Although the total collection time with our method is slightly higher in comparison with that without the authentication and authorization, its impact is small and negligible.

In future work, we intend to reduce the authentication and authorization time when the number of networks used is large.

Acknowledgments This work was partially supported by Toyota Motor Corporation. Also, this work was partly supported by JSPS KAKENHI Grant Number JP17KT0083.

References

1. S. Lucero, *IoT Platforms: Enabling the Internet of Things*, https://cdn.ihs.com/www/pdf/enabling-IOT.pdf, Mar. 2016
2. G. Tank, A. Dixit, A. Vellanki, D. Annapurna, Software-defined Networking: The New Norm for Networks, Technical report, Open Networking Foundation, Apr. 2012
3. D. Ferraiolo, D.R. Kuhn, R. Chandramouli, *Role-based Access Control* (Artech House, 2003)
4. N. McKeown, T. Anderson, H. Balakrishnan, G. Parulkar, L. Peterson, J. Rexford, S. Shenker, J. Turner, OpenFlow: Enabling innovation in campus networks. ACM SIGCOMM Comput. Commun. Rev. **38**(2), 69–74 (2008)
5. D. Cooper, S. Santesson, S. Farrell, S. Boeyen, R. Housley, W.T. Polk, et al., Internet X. 509 Public Key Infrastructure Certificate and Certificate Revocation List (CRL) Profile. RFC vol. 5280, pp. 1–151 (2008)
6. D. Eastlake, RSA/SHA-1 SIGs and RSA KEYs in the Domain Name System (DNS), Technical report, RFC 3110, May 2001
7. D. Eastlake, DSA KEYs and SIGs in the Domain Name System (DNS), Technical report, RFC 2536, Mar. 1999
8. O. Sury, Use of the SHA-256 Algorithm with RSA, Digital Signature Algorithm (DSA), and Elliptic Curve DSA (ECDSA) in SSHFP Resource Records, Technical report, RFC 6594, Apr. 2012
9. E.W. Dijkstra, et al., A note on two problems in connexion with graphs. Numerische Mathematik **1**(1), 269–271 (1959)
10. B. Pfaff, J. Pettit, T. Koponen, E.J. Jackson, A. Zhou, J. Rajahalme, J. Gross, A. Wang, J. Stringer, P. Shelar, The design and implementation of open vswitch, in *Proceedings of the 12th Symposium on Networked Systems Design and Implementation (NSDI)*, pp. 117–130 (Apr. 2015)
11. A. Kivity, Y. Kamay, D. Laor, U. Lublin, A. Liguori, KVM: The Linux Virtual Machine Monitor, in *Proceedings of the 26th International Symposium on Ottawa Linux (OLS)*, pp. 225–230 (Jun. 2007)
12. I. Foster, C. Kesselman, S. Tuecke, The anatomy of the grid: Enabling scalable virtual organizations. Int. J. High Performance Comput. Appl. **15**(3), 200–222 (2001)
13. R. Butler, V. Welch, D. Engert, I. Foster, S. Tuecke, J. Volmer, C. Kesselman, A national-scale authentication infrastructure. Computer **33**(12), 60–66 (2000)
14. M. Lorch, D. Kafura, The PRIMA grid authorization system. J. Grid Comput. **2**(3), 279–298 (2004)
15. R. Alfieri, R. Cecchini, V. Ciaschini, L. dell'Agnello, A. Frohner, A. Gianoli, K. Lorentey, F. Spataro, VOMS, an authorization system for virtual organizations, in *European Across Grids Conference* (Springer, Feb. 2003), pp. 33–40
16. R. Sherwood, G. Gibb, K.K. Yap, G. Appenzeller, M. Casado, N. McKeown, G. Parulkar, FlowVisor: A network virtualization layer, OpenFlow Switch Consortium, Tech. Rep, vol. 1, p. 132, 2009
17. C.H. Chen, C. Chen, S.H. Lu, C.C. Tseng, Role-based campus network slicing, in *2016 IEEE 24th International Conference on Network Protocols (ICNP)* (IEEE, Nov. 2016), pp. 1–6
18. D. Kim, Y.H. Kim, K.H. Kim, J.M. Gil, Cloud-centric and logically isolated virtual network environment based on software-defined wide area network. Sustainability **9**(12), 2382 (2017)
19. E. Dart, L. Rotman, B. Tierney, M. Hester, J. Zurawski, The Science DMZ: A Network Design Pattern for Data-intensive Science," vol. 22(2), pp. 173–185 (2014)

Comparative Study of Hybrid Machine Learning Algorithms for Network Intrusion Detection

Amr Attia, Miad Faezipour, and Abdelshakour Abuzneid

1 Introduction

1.1 Background

Due to the increasing importance of intrusion detection in network security, devising a fast and reliable system that could detect the intrusion attacks effectively is very crucial. As for the state of the art, Network Intrusion Detection Systems (NIDS) are mainly based on two types of approaches: signature-based and anomaly-based detection [1]. In anomaly-based detection, machine learning techniques are widely used.

The performance of the machine learning algorithm is highly dependent on the dataset itself in terms of how large the samples are, as well as whether the classes are well balanced or not. On the other hand, it has less correlation with the pattern nature of the attack. Accordingly, it is highly recommended to acquire enough number of samples for the training phase along with reasonably balanced classes. To this end, applying feature extraction followed by Artificial Neural Network (ANN) as a classifier on a mega dataset could achieve very promising results with high accuracy close to 100% of detection if the training dataset is large enough and quite balanced. Dimensionality of the features (attributes) is also an important factor that could be reduced using Principal Component Analysis (PCA) or Linear Discriminant Analysis (LDA) to reduce the complexity of the system while maintaining high accuracies.

A. Attia · M. Faezipour (✉) · A. Abuzneid
Department of Computer Science & Engineering, University of Bridgeport, Bridgeport, CT, USA
e-mail: amrattia@my.bridgeport.edu; mfaezipo@bridgeport.edu; abuzneid@bridgeport.edu

K. Daimi et al. (eds.), *Advances in Security, Networks, and Internet of Things*,
Transactions on Computational Science and Computational Intelligence,
https://doi.org/10.1007/978-3-030-71017-0_43

1.2 Related Work

The domain of machine/deep learning for intrusion detection has become increasingly popular. In [2], the KitNET algorithm has been introduced using incremental statistics for feature extraction followed by a Neural Network Autoencoder as an unsupervised learning model. The model sets a threshold where malicious attacks are identified above the threshold [2]. The output of the model is a Root Mean Square anomaly score. The higher the score, the higher the probability of anomaly.

Supervised learning has also achieved very promising results in NIDS. The performance of the several classifier models, such as J48, ZeroR, Random Forest, AdaBoost, Multilayer Perceptron, and Logit Boost, has been evaluated in [3, 4]. The models have been evaluated with different number of extracted features.

Several deep and machine learning approaches for NIDS have been recently studied in the literature. Some approaches deal with imbalanced datasets [5, 6], while others also employ feature dimensionality reduction using PCAs [7], autoencoders [6], and spare autoencoders along with well-known classifiers such as Random Forest [8]. Most of the prior work in these domains use the CICIDS2017 dataset [9], while very few studies employed the Kitsune family dataset [2].

1.3 Contribution

Effective intrusion detection mainly relies on timely detection of malicious data. Reducing the complexity of the system is also required to ensure that the detection speed can comply with the packet speed. Recently, machine-learning techniques have been applied for NIDS and achieve promising results [10]. Combined or hybrid machine learning algorithms can overcome this problem and achieve high accuracy of detection in various domains with lower complexity [11]. To this end, this paper introduces the deployment of hybrid machine learning algorithms such as PCA with LDA, PCA with Random Forest, and LDA with Random Forest for efficient NIDS. In addition, we applied ANN, Naïve Bayes, LDA, and Random Forest as independent classifiers. The introduced algorithms are applied to the Kitsune dataset family [2]. Finally, a comprehensive comparison is provided between the designed hybrid algorithms and the independent algorithms. The experiments and simulations were performed on different percentages of the training set and a constant number of training samples to ensure the reliability of the results.

Initially, we started applying Principal Component Analysis (PCA) for feature dimensionality reduction followed by Linear Discriminant Analysis (LDA) as the classifier. However, LDA and PCA themselves are, by nature, primarily, dimensionality reduction algorithms. Applying LDA and PCA together could yield redundancy and increase the complexity without achieving better results [12].

Various machine learning algorithms such as PCA, LDA, Random Forest, Naïve Bayes, and ANN have been applied to the Mirai, OS Scan, and SSDP flood attacks of the Kistune family dataset [2]. We applied hybrid algorithms to study the effects for better accuracy and less complexity. High accuracy with low false negatives has been achieved in the independent classifiers where the training samples with balanced classes are increased. The most decent performances and best accuracy results were achieved when ANN, LDA, and LDA with Random Forest were applied in the SSDP dataset. We noted that in the cases that the classes are imbalanced, or the number of samples are not relatively large enough, the performance of the algorithms was negatively impacted.

2 Methodology and Procedure

2.1 Proposed Idea

In this paper, we present a comparative study of the performance and efficiency of various machine learning techniques for network intrusion detection. We introduce mix and matching well-known machine learning algorithms as hybrid approaches, via applying dimensionality reduction techniques along with classifiers. In the hybrid techniques, we apply PCA for dimensionality reduction along with LDA and Random Forest as classifiers. We also apply LDA for dimension reduction and Random Forest as a classifier. We additionally applied ANN and Naïve Bayes as independent classifiers. Moreover, we provided a comprehensive comparison among the introduced approaches in terms of accuracy and confusion matrix elements. The proposed hybrid use of machine learning algorithms for NIDS has not been studied earlier in the literature on the Kistune family dataset. The flowchart of our machine learning-based NIDS is shown in Fig. 1.

The reason behind using the selected machine learning, dimensionality reduction, and/or classifier techniques is apparent from the details presented hereafter.

1. *Principal Component Analysis* reduces the dimensions of the data by looking for the orthogonal vectors that best represent the data and also usually give the relationship that does not exist in the original features [13].

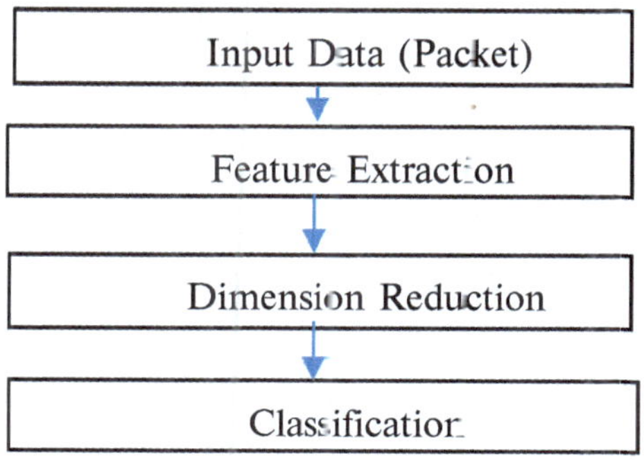

Fig. 1 Flowchart of the model applied for NIDS

2. *Linear Discriminant Analysis* is a technique for both data classification and dimensionality reduction. It looks for a direction in the space that has maximum discriminability.

   This supports class recognition tasks, where the within-class frequencies are unequal, and their performances have been examined on randomly generated test data [14, 15].
3. *Random Forest* is an ensemble method which consists of individual decision trees that select the attributes randomly at each node. Random Forest is more immune to error and outliers. The nodes are split using the best predictors selected randomly at each node. Random Forest gets its strength from its nonparametric nature, high classification accuracy, and its ability to determine variable importance [16, 17].
4. *Naïve Bayes* is a simple probabilistic classifier model with an independent assumption based on Bayes' theorem that the attribute probability does not affect each other. The Naïve Bayes results are usually accurate [18, 19]. However, our experiments show that it has lower performance compared to other algorithms.
5. *Artificial Neural Network* is used as a classifier with supervised training. Neural network includes a structure of input, hidden, and output layers and is a very powerful tool for pattern classification and logistic regression [20].

2.2 *Kitsune Dataset*

The dataset used in this paper has been introduced in [2] and is publicly available. It was collected using a real IP camera video surveillance network. Feature extraction was then applied using incremental statistics to maintain the current behavior of the data stream by decreasing the weight of the old instances [12]. Table 1 shows a characteristic summary of the Kitsune dataset used in this work.

3 Results and Discussion

Table 2 shows the performance comparison of seven different techniques applied on the three different datasets of the Kistune family. The best accuracy and lowest false negative (FN) came from ANN, LDA classifier, as well as LDA for dimension

Table 1 Characteristic summary of Kitsune family dataset

Attack type	Attack name	Dataset size	# of features	True negative percentage
Botnet malware	Mirai	764,136.00	115	84.08%: Imbalanced
Recon	OS_SCAN	1,697,850.00	115	3.87%: Heavily imbalanced
Denial of service	SSDP Flood	4,077,265.00	115	35.31%: Semi-balanced

Table 2 Summary of the results with validation rate = 0.2

Attack name	Algorithm	Accuracy	Confusion matrix		TPR	FNR
OS Scan	PCA (1) + LDA	96.0297%	TN = 313480	FP = 12935	0.9584	0.0416
			FN = 547	TP = 12608		
	LDA	96.0326%	TN = 313140	FP = 13275	0.9850	0.0150
			FN = 197	TP = 12953		
	PCA (1)+Random Forest	96.1275%	TN = 326396	FP = 19	0.0018	0.9982
			FN = 13131	TP = 24		
	Random Forest	96.1269%	TN = 326415	FP = 0	0.0002	0.9998
			FN = 13152	TP = 3		
	LDA + Random Forest	96.0842%	TN = 319820	FP = 6595	0.4905	0.5095
			FN = 6702	TP = 6453		
	ANN	96.1501%	TN = 326396	FP = 118	0.0077	0.9923
			FN = 12955	TP = 101		
	Naïve Bayesian	96.0948%	TN = 326238	FP = 276	0.0054	0.9946
			FN = 12985	TP = 71		
SSDP Flood	PCA (1) + LDA	99.7946%	TN = 526449	FP = 1095	0.9980	0.0020
			FN = 580	TP = 287329		
	LDA	99.9996%	TN = 527542	FP = 2	1.0000	0.0000
			FN = 1	TP = 287908		
	PCA (1) + Random Forest	99.7993%	TN = 526368	FP = 1176	0.9984	0.0016
			FN = 461	TP = 287448		
	Random Forest	99.9917%	TN = 527544	FP = 0	0.9998	0.0002
			FN = 68	TP = 287841		
	LDA + Random Forest	99.9998%	TN = 527543	FP = 1	1.0000	0.0000
			FN = 1	TP = 287908		
	ANN	100.0000%	TN = 527785	FP = 0	1.0000	–
			FN = 0	TP = 287668		
	Naïve Bayesian	35.4592%	TN = 1485	FP = 525300	1.0000	–
			FN = 0	TP = 287668		
Mirai	PCA (1) + LDA	85.5046%	TN = 6412	FP = 18020	0.9678	0.0322
			FN = 4133	TP = 124263		
	LDA	91.2333%	TN = 24074	FP = 358	0.8984	0.1016
			FN = 13040	TP = 115356		
	PCA (1) + Random Forest	89.7074%	TN = 22051	FP = 2381	0.8960	0.1040
			FN = 13349	TP = 1[illegible]5047		
	Random Forest	91.8307%	TN = 23714	FP = 7[illegible]8	0.9084	0.0915
			FN = 11767	TP = 116629		
	LDA + Random Forest	89.0275%	TN = 16004	FP = 8428	0.9350	0.0650
			FN = 8341	TP = 120055		
	ANN	94.3499%	TN = 22548	FP = 1769	0.9466	0.0534
			FN = 6866	TP = 121645		
	Naïve Bayesian	33.6849%	TN = 243	FP = 24074	0.9933	0.0067
			FN = 860	TP = 127651		

reduction, followed by the Random Forest classifier on SSDP, which is the largest and most balanced set. The hybrid approach of using LDA with Random Forest is considered the most efficient algorithm in terms of accuracy and complexity when applied to limited size data (e.g., 1% of the data) on the SSDP dataset. Our experiments show that PCA along with LDA speeds up the learning phase, increases the accuracy, and decreases the system complexity.

We repeat the experiments while maintaining the exact same size of training data (around 611,308 samples) for the three datasets. The results are pretty close as indicated in Table 3. However, LDA has better accuracy and lower FN. Moreover, LDA improves the performance of Random Forest as depicted in Table 3.

In order to deeply understand how the training data size affects the algorithm efficiency, we go to the extreme case and apply the models to only 1% of the training set of OS Scan and SSDP, as shown in Table 4. LDA also achieved the highest accuracy and the least FN. LDA improves the Random Forest classifier performance on OS Scan attack in terms of FN.

The results depicted in Fig. 2 indicate that changing the number of the PCA components equally affects the accuracy for both the Random Forest and LDA classifiers. With dimensionality reduction, a less complex and better system performance is achieved. Principal Component Analysis can interpret the relationship between the attributes that do not exists before and can, thus, contribute to better classification. As shown in Table 5, with very limited training size, the hybrid

Table 3 Performance results with constant training sizes for all datasets

Attack name	Algorithm	Accuracy	Confusion matrix		TPR	FNR
OS Scan test 64%	PCA (1) + LDA	96.0246%	TN = 1003102	FP = 41518	0.9600	0.0400
			FN = 1680	TP = 40324		
	LDA	96.0232%	TN = 1002010	FP = 42610	0.9856	0.0144
			FN = 603	TP = 41401		
	PCA (1) + Random Forest	96.1333%	TN = 1044454	FP = 166	0.0036	0.9964
			FN = 41851	TP = 153		
	Random Forest	96.2372%	TN = 1034110	FP = 10510	0.2768	0.7232
			FN = 30377	TP = 11627		
	LDA + Random Forest	96.0541%	TN = 1023125	FP = 21495	0.4910	0.5090
			FN = 21382	TP = 20622		
	ANN (hidden layer 10)	96.1220%	TN = 1044097	FP = 460	0.0092	0.9908
			FN = 41679	TP = 388		
	Naïve Bayesian	96.0781%	TN = 1043744	FP = 813	0.0063	0.9937
			FN = 41803	TP = 264		

(continued)

Table 3 (continued)

Attack name	Algorithm	Accuracy	Confusion matrix		TPR	FNR
SSDP Flood test 85%	PCA (1) + LDA	99.7969%	TN = 2237526	FP = 4479	0.9979	0.0021
			FN = 2560	TP = 1221110		
	LDA	99.9997%	TN = 2241995	FP = 10	1.0000	0.0000
			FN = 2	TP = 1223668		
	PCA (1) + Random Forest	99.8024%	TN = 2237155	FP = 4850	0.9984	0.0016
			FN = 1998	TP = 1221672		
	Random Forest	99.9929%	TN = 2242005	FP = 0	0.9998	0.0002
			FN = 246	TP = 1223406		
	LDA + Random Forest	99.9997%	TN = 2241995	FP = 10	1.0000	0.0000
			FN = 2	TP = 1223668		
	ANN (hidden layer 10)	99.9999%	TN = 2242605	FP = 0	1.0000	0.0000
			FN = 5	TP = 1223065		
	Naïve Bayesian	52.1733%	TN = 585089	FP = 1657516	1.0000	0.0000
			FN = 1	TP = 1223069		
Mirai test 20%	PCA (1) + LDA	85.5046%	TN = 6412	FP = 18020	0.9678	0.0322
			FN = 4133	TP = 124253		
	LDA	91.2333%	TN = 24074	FP = 358	0.8984	0.1016
			FN = 13040	TP = 115356		
	PCA (1) + Random Forest	89.7074%	TN = 22051	FP = 2381	0.8960	0.1040
			FN = 13349	TP = 115047		
	Random Forest	91.8307%	TN = 23714	FP = 718	0.9034	0.0916
			FN = 11767	TP = 116629		
	LDA + Random Forest	89.0275%	TN = 16004	FP = 8428	0.9350	0.0650
			FN = 8341	TP = 120055		
	ANN (hidden layer 10)	93.9893%	TN = 23501	FP = 816	0.9349	0.0651
			FN = 8370	TP = 120141		
	Naïve Bayesian	83.6849%	TN = 243	FP = 24074	0.9933	0.0067
			FN = 860	TP = 127651		

technique of PCA with 55 components followed by the LDA classifier achieves an accuracy of 99.9999%, FP = 0, True Positive Rate (TPR) of 0.999996, and False Negative Rate (FNR) of nearly zero. Moreover, the 80 PCA components with LDA achieves zero FN, TPR of one, and FNR of zero. The five-component PCA with LDA on 1% of OS Scan's total data size (very limited training size) and more

Table 4 Results when training size is 1% of the total data size

Attack name	Algorithm	Accuracy	Confusion matrix		TPR	FNR
SSDP Flood	PCA (1) + LDA	99.804%	TN = 2605637	FP = 5627	0.9984	0.0016
			FN = 2274	TP = 1422954		
	LDA	99.994%	TN = 2611014	FP = 250	1.000	0.000
			FN = 1	TP = 1425227		
	PCA (1) + Random Forest	99.798%	TN = 2606076	FP = 5188	0.9979	0.0021
			FN = 2950	TP = 1422278		
	Random Forest	99.990%	TN = 2611157	FP = 107	0.9998	0.0002
			FN = 307	TP = 1424921		
	LDA + Random Forest	99.994%	TN = 2611014	FP = 250	1.000	0.000
			FN = 1	TP = 1425227		
	ANN (10)	99.995%	TN = 2611178	FP = 210	1.000	0.000
			FN = 6	TP = 1425098		
	Naïve Bayesian	50.619%	TN = 618149	FP = 1993239	1.000	0.000
			FN = 1	TP = 1425103		
OS Scan	PCA (1) + LDA	96.035%	TN = 1551838	FP = 64036	0.9599	0.0401
			FN = 2605	TP = 62393		
	LDA	96.020%	TN = 1550024	FP = 65850	0.9839	0.0161
			FN = 1048	TP = 63950		
	PCA (1) + Random Forest	96.133%	TN = 1615874	FP = 0	0.000	1.000
			FN = 64998	TP = 0		
	Random Forest	96.146%	TN = 1598568	FP = 17306	0.2697	0.7303
			FN = 47470	TP = 17528		
	LDA + Random Forest	96.032%	TN = 1582113	FP = 33761	0.4934	0.5066
			FN = 32929	TP = 32069		
	ANN (10)	96.108%	TN = 1608799	FP = 7073	0.1024	0.8976
			FN = 58346	TP = 6654		
	Naïve Bayesian	96.045%	TN = 1614329	FP = 1543	0.0009	0.9991
			FN = 64941	TP = 59		

imbalanced classes achieves FN = 932, TPR = 0.98566, and FNR = 0.01434. Hybrid algorithms beat the ANN performance with 10 layers, which achieves FN = 58,346, TPR = 0.102369, and FNR = 0.89763. By introducing more complex structures of the ANN with hidden layers (50, 10, 10), the performance of FN = 971, TPR = 0.0985, and FNR = 0 0.01494 is achieved which is still lower than the hybrid algorithm. Accordingly, hybrid PCA-LDA achieves better results with much less complexity. Hybrid PCA-LDA is more immune for limited training sizes and imbalanced classes.

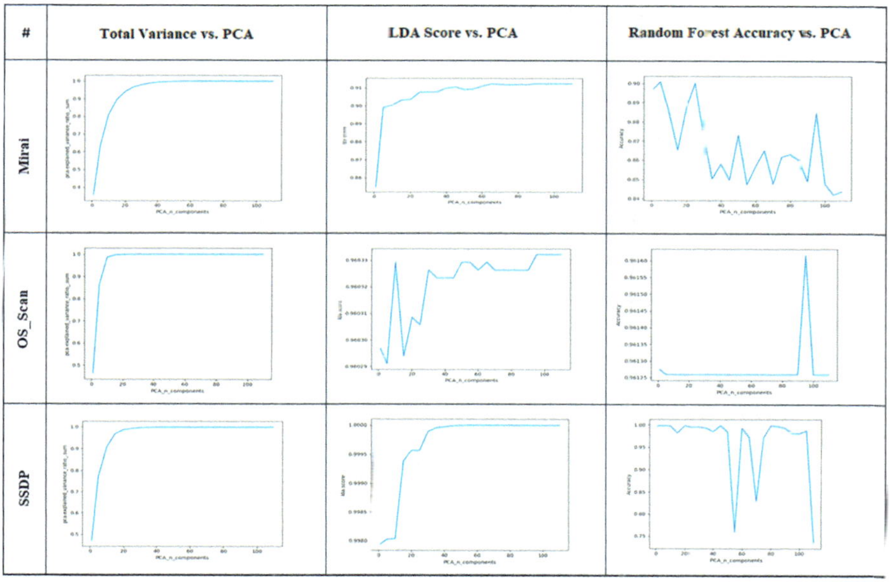

Fig. 2 Total variance, LDA score, and Random Forest accuracy vs. PCA components plots

Naïve Bayes achieves perfect results in terms of False Negative Rate (FNR) on SSDP, while the precision is only 35.341%. This is due to the fact that Naïve Bayes works with independent assumptions applied to features extracted via incremental statistics. Incremental statistics work on different directions that calculate the relation between the data stream such as mean, sigma, and covariance. Accordingly, how the features are extracted might confuse Naïve Bayes in terms of precision (high false positives), as shown in Tables 2 and 3.

LDA seeks to find the optimal projection vector by defining measures of separation between the projections, including the means of the two classes along with decreasing the variance in each class, and the covariance between the two classes. Such measures are applied to the attributes that are originally computed by the incremental statistics, where the standard deviation, mean, and covariance of the data stream are considered [21]. However, the performance of LDA starts to become influenced by the training size and the balance of classes. PCA improves this problem as shown in Table 5.

The ANN yielded great performance results in the accuracy and false-negative metrics. The structure applied here was quite simple: 10 neurons in the hidden layer and one neuron in the output layer for the classification applied to the whole dataset. This achieve 100% accuracy in the SSDP Flood dataset. However, for OS Scan, the accuracy is approximately 96.15%, with quite high number of false negatives around 13,000 which represents a real threat to the system. Accordingly, we apply more complex structures of hidden layer (50, 10, 10) on a more restricted dataset, only 1% of the total sample size (16,000), and achieved better FN results, as shown in Table 4.

Table 5 PCA with LDA algorithm and different ANN structures

Attack name	Algorithm	Accuracy	Confusion matrix		TPR	FNR
SSDP Flood (Test = 20%)	PCA (1 component) + LDA	99.795%	TN = 526449	FP = 1095	0.9980	0.0020
			FN = 580	TP = 287329		
	PCA (55 components) + LDA	100.000%	TN = 527544	FP = 0	1.000	0.000
			FN = 1	TP = 287908		
	LDA	99.996%	TN = 527542	FP = 2	1.000	0.000
			FN = 1	TP = 287908		
SSDP Flood (test = 99%)	PCA (1 component) + LDA	99.804%	TN = 2605637	FP = 5627	0.9984	0.0016
			FN = 2274	TP = 1422954		
	LDA	99.994%	TN = 2611014	FP = 250	1.000	0.000
			FN = 1	TP = 1425227		
	PCA (55 components)_LDA	100.00%	TN = 2610860	FP = 0	1.000	0.000
			FN = 5	TP = 1425223		
	PCA (80 components)_LDA	99.990%	TN = 2611257	FP = 404	1.000	–
			FN = 0	TP = 1425228		
OS Scan (test = 99%)	PCA (1 component)_LDA	96.035%	TN = 1551838	FP = 64036	0.9599	0.0401
			FN = 2605	TP = 62393		
	PCA (5 components)_LDA	96.029%	TN = 1550063	FP = 65811	0.9857	0.0143
			FN = 932	TP = 64066		
	LDA	96.020%	TN = 1550024	FP = 65850	0.9839	0.0161
			FN = 1048	TP = 63950		
	ANN (10)	96.108%	TN = 1608799	FP = 7073	0.1024	0.8976
			FN = 58346	TP = 6654		
	ANN (50,10,10)	96.027%	TN = 1550063	FP = 65809	0.9851	0.0149
			FN = 971	TP = 64029		

Although PCA does yield the best results when considering only one principal component of the extracted features, it decreases the dimension of the system significantly from 115 to only one dimension and maintains a very high accuracy, as depicted in Fig. 2. As seen from the figure, the accuracy changes according to the number of the principal components with LDA and Random Forest classifiers. The optimum number chosen for PCA is ranging from five to fifteen principal components. However, it jumps to 55 or more if the training size is limited, and the classes are imbalanced.

Moreover, the number of principal components does not relate to the classifier-type LDA or Random Forest and rather depends on the attack type and the dataset used. Choosing the optimized number of the principal components is really challenging. Therefore, we applied our experiments on 23 trials for 110 principal components to study their effects on the behavior and the accuracy of detection. Finally, PCA with the optimized number of principal components and LDA as the classifier on a restricted dataset with imbalanced classes achieve the best performance with low complexity.

We also illustrated the number of the total variations represented by the principal components to figure out its relation with the efficiency of the system. As a result, applying PCA is very powerful for dimension reduction. It could work together with other classifiers such as LDA and Random Forest to reduce the dimensionality and thus complexity and achieve a better representation of the data before applying the classifier.

Random Forest is a very strong classifier. However, its response while classifying OS Scan attack is not appreciably confirmed. The FN results of LDA on OS Scan is relatively high. LDA is applied to the features extracted from the incremental statistical analysis of the data stream and tries to maximize the mean between the two classes. Accordingly, even if the classes are not well balanced, LDA can overcome this challenge as LDA is compatible with the way features are extracted. PCA is also performed through subtracting the mean of each measurement and calculating the eigenvectors, which also match how the features are extracted [22].

Subsequently, it is very useful to apply statistics to extract the most discriminant features of the data, such as mean, standard deviation, and covariance, before employing the classifier. In this case, the efficiency of the classifier is improved for ANN and LDA in contrast to Random Forest and Naïve Bayes, as shown in Tables 2, 3, 4, and 5.

Naïve Bayes works by strong independent assumptions which may contradict the way features are extracted on the Kistune dataset. This leads to poor results though Naïve Bayes is a strong classifier. Moreover, this could also be the case for Random Forest as the features are chosen randomly at each node. Accordingly, not each tree perceives all the features. However, the features extracted here are highly correlated because they are extracted by the damped incremental statistics. The damped incremental statistics features along with the imbalanced classes might be the reason of lowered performance of the Random Forest classifier while predicting the decision.

It is also noticed that the applied LDA as a dimension reduction improves the efficiency of Random Forest, especially for FN, as shown in the confusion matrix of Tables 2, 3, and 4. Changing the PCA with LDA achieves FP = zero and FN = zero at 55 components and 80 components, respectively.

We also apply the algorithms to only 1% of the dataset and not the whole dataset as previous models. PCA with LDA achieves very good results. LDA also improves the performance of Random Forest and decreases the complexity. ANN achieves accuracy of 100% when applied to the whole dataset of SSDP. However, its performance slightly decreases when it is applied to only 1% of the dataset. Furthermore, the imbalanced classes negatively affect the performance. Surprisingly, hybrid PCA with LDA achieve the best results as depicted in Table 5 when applied to a restricted dataset and/or imbalanced classes. Hybrid PCA-LDA improves the results in terms of accuracy and lower FN.

4 Conclusion and Future Work

Applying machine learning techniques achieves decent results in the area of network security and intrusion detection, especially in the false-negative measures, which is very important to detect all potential attacks on the system. Though some algorithms are black boxes when it comes to how or why they perform in a certain manner, it is still important to choose algorithms that could work efficiently with one another. Hybrid algorithms that consider the nature of the data and how the features are extracted could reduce the model complexity while achieving with very high accuracies. This allows the system to perform better with low complexity even on a limited size data. Here, the hybrid PCA-LDA achieved the best results for different types of attacks with limited training set and imbalanced classes, as they are compatible with how the features are extracted. This is not the case for Random Forest and Naïve Bayes. If the algorithm used is not compatible with how the features are extracted, the dimension reduction approach could hinder the model performance.

As a future direction, both supervised and unsupervised learning could be combined together to achieve a more robust system. Unsupervised learning can create the labeled dataset by detecting the malicious behavior and setting the threshold for any unknown behavior or attack on the system. The labeled dataset can further be trained to create an efficient model for NIDS using hybrid algorithms. Applying a dimensionality reduction with strong classifiers is an area of interest for NIDS. Applying PCA with neural networks could further achieve an effective intrusion detection system with low complexity.

References

1. F. Anjum, D. Subhadrabandhu, S. Sarkar, Signature based intrusion detection for wireless ad-hoc networks: A comparative study of various routing protocols, in *2003 IEEE 58th Vehicular Technology Conference. VTC 2003-Fall (IEEE Cat. No. 03CH37484)*, vol. 3, (IEEE, 2003), pp. 2152–2156
2. Y. Mirsky, T. Doitshman, Y. Elovici, A. Shabtai, Kitsune: An ensemble of autoencoders for online network intrusion detection. arXiv preprint arXiv:1802.09089 (2018)
3. R. Abdulhammed, M. Faezipour, A. Abuzneid, A. Alessa, Effective features selection and machine learning classifiers for improved wireless intrusion detection, in *2018 International Symposium on Networks, Computers and Communications (ISNCC)*, (2018), pp. 1–6
4. R. Abdulhammed, M. Faezipour, A. Abuzneid, A. Alessa, Enhancing wireless intrusion detection using machine learning classification with reduced attribute sets, in *Proceedings of the IEEE International Wireless Communications and Mobile Computing Conference*, (2018), pp. 524–529
5. R. Abdulhammed, M. Faezipour, A. Abuzneid, A. AbuMallouh, Deep and machine learning approaches for anomaly-based intrusion detection of imbalanced network traffic. IEEE Sensors Lett. **3**(1), 1–4 (2019)
6. R. Abdulhammed, H. Musafer, A. Alessa, M. Faezipour, A. Abuzneid, Features dimensionality reduction approaches for machine learning based network intrusion detection. Electronics **8**(322), 1–27 (2019)

7. R. Abdulhammed, M. Faezipour, H. Musafer, A. Abuzneid, Efficient network intrusion detection using PCA-based dimensionality reduction of features, in *Proceedings of the IEEE International Symposium on Networks, Computers and Communications*, (Jun. 2019), pp. 1–6
8. H. Musafer, A. Abuzneid, M. Faezipour, A. Mahmood, An enhanced design of sparse autoencoder for latent features extraction based on trigonometric simplexes for network intrusion detection systems. Electronics **9**(259), 1–12 (2020)
9. I. Sharafaldin, A.H. Lashkari, A.A. Ghorbani, Toward generating a new intrusion detection dataset and intrusion traffic characterization, in *Proceedings of the 4th International Conference on Information Systems Security and Privacy, Madeira, Portugal*, (2018), pp. 108–116
10. P. Garcia-Teodoro, J. Diaz-Verdejo, G. Maciá-Fernández, E. Vázquez, Anomaly-based network intrusion detection: Techniques, systems and challenges. Comput. Secur. **28**(1–2), 18–28 (2009)
11. A. Sharma, K.K. Paliwal, G.C. Onwubolu, Class-dependent PCA, MDC and LDA: A combined classifier for pattern classification. Pattern Recogn. **39**(7), 1215–1229 (2006)
12. A.L. Buczak, E. Guven, A survey of data mining and machine learning methods for cyber security intrusion detection. IEEE Commun. Surv. Tutorials **18**(2), 1153–1176 (2015)
13. M. Abuzneid, A. Mahmood, Performance improvement for 2-D face recognition using multi-classifier and BPN, in *IEEE Long Island Systems, Applications and Technology Conference (LISAT), 2016*, (2016), pp. 1–7
14. S. Balakrishnama, A. Ganapathiraju, Linear discriminant analysis-a brief tutorial. Inst. Signal Inf. Process. **18**, 1–8 (1998)
15. S.J. Prince, J.H. Elder, Probabilistic linear discriminant analysis for inferences about identity, in *IEEE 11th International Conference on Computer Vision, 2007*, (2007), pp. 1–8
16. A. Liaw, M. Wiener, Classification and regression by randomForest. R News **2**(3), 18–22 (2002)
17. M. Belgiu, L. Drăguţ, Random forest in remote sensing: A review of applications and future directions. ISPRS J. Photogramm. Remote Sens. **114**, 24–31 (2016)
18. S. Mukherjee, N. Sharma, Intrusion detection using naive Bayes classifier with feature reduction. Procedia Technol. **4**, 119–128 (2012)
19. P. Tsangaratos, I. Ilia, Comparison of a logistic regression and Naïve Bayes classifier in landslide susceptibility assessments: The influence of models complexity and training dataset size. Catena **145**, 164–179 (2016)
20. Z. Zhang, J. Li, C. Manikopoulos, J. Jorgenson, J. Ucles, HIDE: A hierarchical network intrusion detection system using statistical preprocessing and neural network classification, in *Proc. IEEE Workshop on Information Assurance and Security*, (2001), pp. 85–90
21. E. Alexandre-Cortizo, M. Rosa-Zurera, F. Lopez-Ferreras, Application of fisher linear discriminant analysis to speech/music classification, in *IEEE EUROCON 2005-The International Conference on "Computer as a Tool", 2005*, vol. 2, (2005), pp. 1666–1669
22. J. Shlens, A tutorial on principal component analysis. arXiv preprint arXiv:1404.1100 (2014)

Unquantize: Overcoming Signal Quantization Effects in IoT Time Series Databases

Matthew Torin Gerdes, Kenny Gross, and Guang Chao Wang

1 Introduction

Quantized signals are prevalent in many Internet-of-Things (IoT) industries, including utilities, oil and gas, transportation, manufacturing, and even in business-critical enterprise computing data centers. Quantized signals originate from inexpensive analog-to-digital (A/D) chips with low-bit resolution, which are used to convert the analog transducer signals into digitized time series. Machine learning (ML) algorithms perform poorly with quantized signals, prohibiting its wide applications in many dense-sensor IoT industries. As the price of high-bit resolution analog-to-digital (A/D) converters has decreased, their adoption throughout sensor dense industries would seem to be inevitable; however, transition to 12-bit and 16-bit A/D chips is slow in new assets. Furthermore, even as high-resolution A/D chips become more widespread in future assets, older legacy assets will still need to be serviced, and data from this hardware will need to be analyzed for the remainder of their lifetimes.

For large-scale IoT prognostic applications, it is not possible to have humans look at all the signals to decide between quantized vs unquantized, and to quantify the number of quantization levels for each quantized signal. As an example, a modern oil refinery has 1M sensors recording time series signals 24 × 7 × 365. As a second example, in commercial aviation, one Airbus airplane has 75,000 sensors. Similarly, due to the proliferation of sensors inside enterprise IT servers and storage platforms in data center, a medium size enterprise or cloud data center today has 1M sensors, many of which are quantized at various levels of quantization. What is needed is a technique that can examine a large universe of sensor time series signals,

M. T. Gerdes · K. Gross · G. C. Wang (✉)
Oracle Physical Sciences Research Center, Oracle Corporation, San Diego, CA, USA
e-mail: matthew.gerdes@oracle.com; kenny.gross@oracle.com; guang.wang@oracle.com

K. Daimi et al. (eds.), *Advances in Security, Networks, and Internet of Things*,
Transactions on Computational Science and Computational Intelligence,
https://doi.org/10.1007/978-3-030-71017-0_44

automatically identify signals that are quantized, determine the number of levels of quantization, and "unquantize" the signals identified as quantized. It is also essential to unquantize the signals in a manner that is optimized per the number of levels of quantization, including for the lowest resolution, which yields only 2 quantization levels (and we've seen two-level quantization in many collections of real telemetry measurements). This paper explores a novel analytical framework that automates the discovery of quantized signals in "big data" databases that may contain thousands of sensors signals, identifies the exact quantization levels for all of those signals, and unquantizes the signals with a novel algorithm that is optimized to different quantization levels for the various individual signals.

The remainder of this paper is organized as follows. Section 2 introduces the implementation of our unquantize framework step-by-step. Sections 3.1 and 3.2 illustrate the impact of quantized signals and demonstrate the performance of our unquantize methodology, and Sect. 3.3 presents how our solutions address the challenges in ML prognostic caused by quantized signals. Section 4 provides the conclusions.

2 Methodology

This paper focuses on two aspects of unquatization. The first aspect is the accuracy of unquantized signals, when compared to the known high-resolution signal, relative to their quantized counterparts. The second aspect is the prognostic performance gains that unquantization imparts to the performance of ML pattern recognition and automated anomaly discovery. When quantized signals are included in training data sets, there is a much higher likelihood for false alarms and missed alarms. In most industries where ML prognostic surveillance is valuable, false alarms are very costly, resulting in taking revenue-generating assets out of service unnecessarily. Moreover, missed alarms can be catastrophic. We demonstrate in this paper that the introduction of an autonomous framework for unquantization of signals requires no hardware modifications and, when combined with Oracle's advanced ML pattern recognition, is helping to substantially increase component reliability margins and system availability goals while reducing (through improved root cause analysis) costly sources of "no trouble found" events that have become a significant warranty-cost issue for asset manufactures.

2.1 Overview of Unquantize Methodology

The telemetry time series signals are unquantized on a signal-by-signal basis. Figures 1 and 2 exemplify quantized sensor readings and demonstrate the drastic difference between the recorded signal and the genuine signal characteristics. The

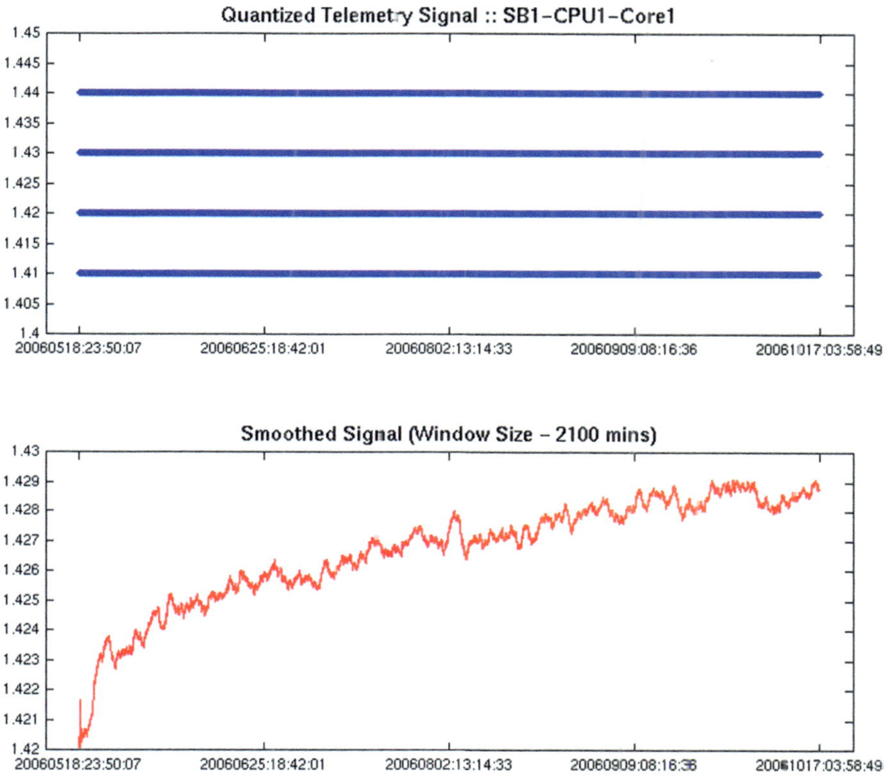

Fig. 1 Example of quantized telemetry signal reported from 8-bit A/D chips used in the servers (top) and the unquantized telemetry signal produced by our technique (bottom). Note that the four-line segments in the top plot only constitute one quantized signal, and each segment consists of discrete measurements. Because the signal sampling rate was high, the measurements on each of the four quantization levels were compressed, making appear like a continuous signal

first step of unquantizing is to identify the number of quantization levels in each signal. If the computed number of quantization levels is smaller than 20 or smaller than 5% of number of observations, the signal is deemed "quantized." Signals possessing a number of quantization levels greater than four are unquantized by Fourier decomposition and reconstruction. When there are between two and four quantization levels (e.g., -1 and 1 or -2, -1, 1, and 2), the signal is unquantized by computing the bin-switching frequency between higher and lower quantization levels in a sliding window. The scaled bin frequency to match the quantization levels serves as the unquantized signal. Figure 3 shows the flowchart of the unquantizing process.

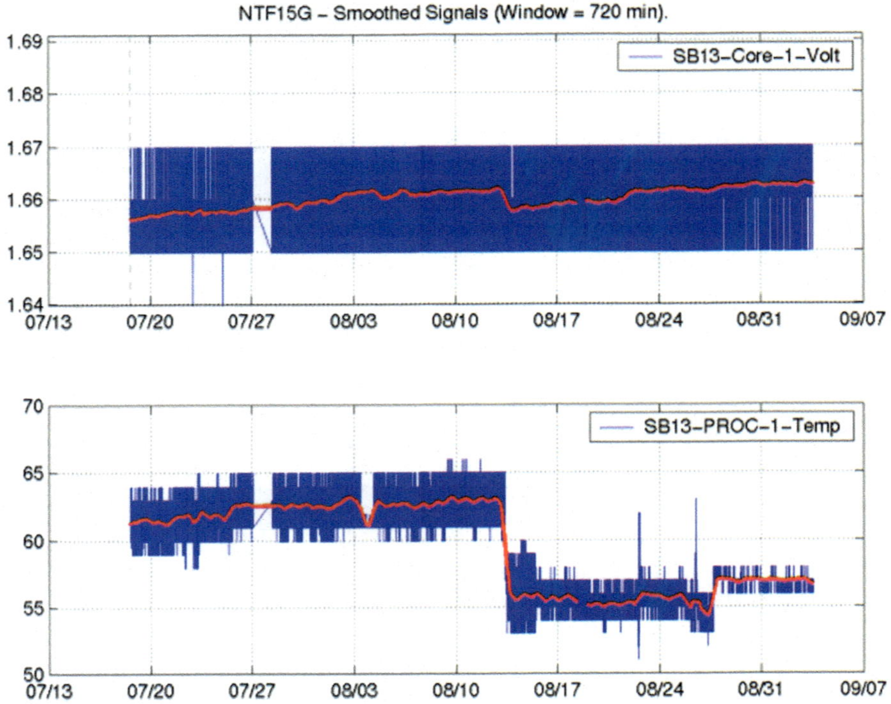

Fig. 2 The raw voltage (upper) and temperature (lower) signals reported from 8-bit A/D chips used in many electronic systems. The red signal shows the actual values of the monitored variables

2.2 Testing Signals

The time series signals used in the case study have been synthesized with a high-fidelity signal synthesis algorithm from real time series signatures across a variety of IoT industrial use cases. These signals are synthesized, not simulated, which match real IoT sensor signals in all statistical characteristics important to ML prognostics, including serial correlation content, cross correlation between/among signals, and stochastic content (variance, skewness, kurtosis), as real IoT sensor signals. For the large-scale database of synthesized signals used in this investigation, Oracle Labs' Telemetry Parameter Synthesis System (TPSS) has been employed [1–3].

Once the signals are synthesized, they are quantized by the mid-rise and mid-tread uniform quantizing method depending on their quantization levels (QL), as presented in Eq. (1). To emulate the low-resolution A/D chips more realistically, Δ is calculated using the minimum and maximum values from the original, noiseless signal. The quantized signals are then passed through the unquantize framework.

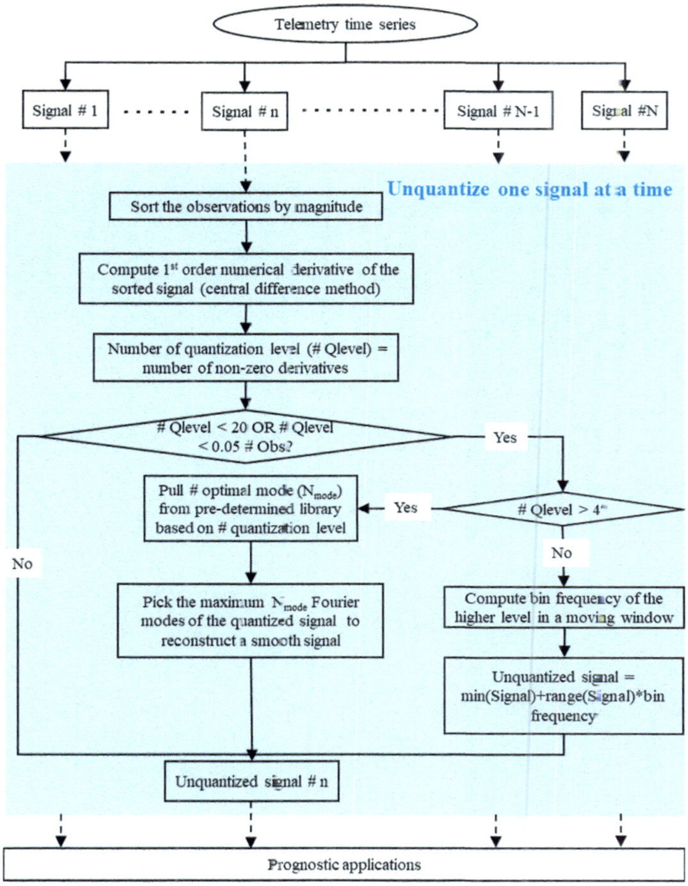

Fig. 3 Flowchart of framework for autonomous unquantization analysis that is utilized upstream of any ML algorithms

$$\begin{cases} \Delta \cdot \text{floor}\left(\frac{x}{\Delta} + 0.5\right), \text{even QL} \\ \Delta \cdot \left(\text{floor}\left(\frac{x}{\Delta}\right) + 0.5\right), \text{odd QL} \end{cases} \tag{1}$$

where $\Delta = \frac{\max(x) - \min(x)}{\text{QL} - 1}$.

2.3 *Determine QL in the Signals*

To determine the quantization level, the signal is sorted in ascending order, and then, a numerical central difference scheme is applied to find the first-order derivative of

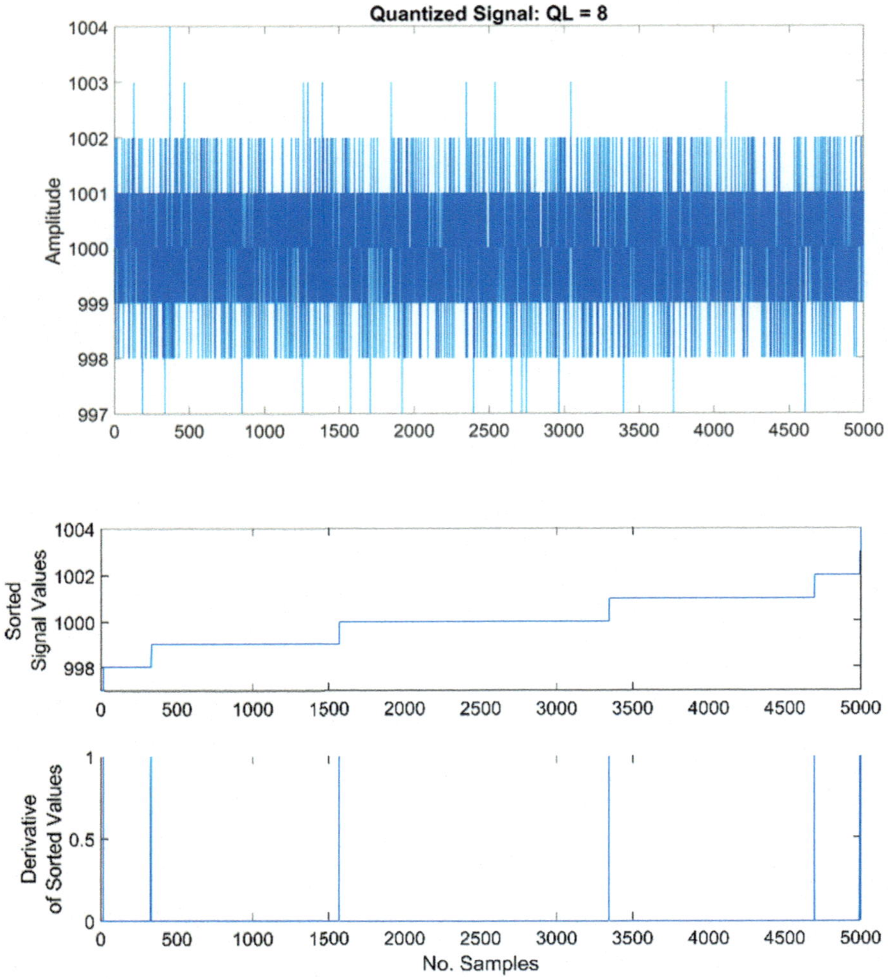

Fig. 4 Illustration of finding the number of quantization levels in a quantized signal. The upper plot showcases a typical quantized signal. The middle plot presents the sorted observations in ascending order, and the lower plot shows the first-order derivative of the sorted values. The number of quantization levels is found to be eight for this example

the sorted signals. The sum total of the nonzero values in the derivative of the sorted signal determines the QL of the signal. If QL is 20 or less or the QL is less than 5% of the number of observations, the signal is determined to be quantized. Figure 4 illustrates the process of determining the quantization level of a signal.

2.4 *Fourier Decomposition (for QL > 4)*

If QL is greater than four, Fourier decomposition is used to unquantize the signals. The quantized signal is converted into the frequency domain using a Fourier transform (FFT). In the frequency domain, the most prominent harmonic modes are extracted to generate a composite frequency signal. The number of Fourier modes (N largest modes) used is precomputed and stored in a mode library, where the number of modes is a function of the number of quantization levels. Then the new composite signal is converted back to the time domain through an inverse Fourier transform (iFFT) [6].

2.5 *Bin-Switching Frequency (for QL ≤ 4)*

If QL is two, the signal is processed with the bin-switching frequency algorithm. The algorithm passes a sliding window over the quantized signal, determines the frequency of the highest level in the window, and normalizes that frequency value by the length of the window. The normalized value becomes one data point in the unquantized signal (see Fig. 5 for a detailed illustration). Because the output of the bin-switching frequency algorithm is normalized, the range is between 0 and 1. This requires a rescaling of the signal.

If QL is three or four, additional upstream and post processing are required to use the bin-switching frequency algorithm. The upstream addition consists of splitting the quantized signal into multiple quantized signals where QL is two. For example, if the signal is simple and has levels that are equivalent to -1, 0, and 1, QL would be three and can be separated in two signals where the signals would have QL equivalent to two: -1 and 0 and 0 and 1. The new split signals are now processed with the bin switching in the same manner as the when QL is two. The postprocessing addition sums the scaled signals and then subtracts off the mean of quantized signal.

3 Evaluation and Discussions

3.1 *Negative Impact of Quantized Signals*

When low-resolution A/D converters record physical phenomena, the resulting signal is an abstraction. Many meaningful physical characteristics, such as periodicity, noise ratio, and number of modes, are lost due to this abstraction or quantization. Figure 6 illustrates the loss of information that can occur from quantization. When comparing the sinusoidal signal to its quantized version, it is very apparent that

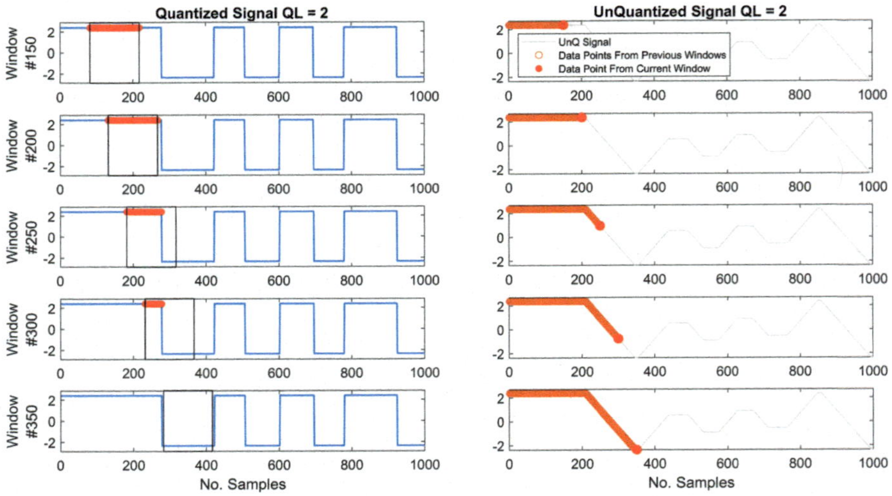

Fig. 5 Illustration of bin-switching frequency method. The left column contains a portion of the quantized signal (in blue) as the sliding window (in black) moves forward in time. The data points in red determine the corresponding red point in the right column. The right column illustrates the construction of the unquantized signal (in gray). The point in red indicates the unquantized sample determined by the current window (e.g., window 300 is equivalent to the 300th unquantized sample). The points in orange are the points determined by the previous windows. The windows are increasing by increments of 50 from top to bottom

many of the identifying patterns that characterize the time series are indecipherable, such as the number of modes and the underlying frequencies.

Unfortunately, the signal dynamics that quantization obscures are of importance for any ML algorithms to build accurate and meaningful models.

3.2 Comparisons of Signal Reproductions between Quantized and Unquantized Signals

Oracle innovators have developed an algorithm that analytically unquantizes signals in effect taking low-resolution input signals and turning them into high-accuracy output signals. These unquantized signals extract the dynamics of the ground truth much more closely and accurately than their quantized counterparts, making them much more conducive to prognostic ML modeling. Figure 7 quantitatively assesses and compares the deviations of the quantized and unquantized signal from the original signal. The continuous signal in Fig. 7(a) is quantized into three different quantization levels: 2, 3, and 4. In Fig. 7(b, c), the quantized and unquantized

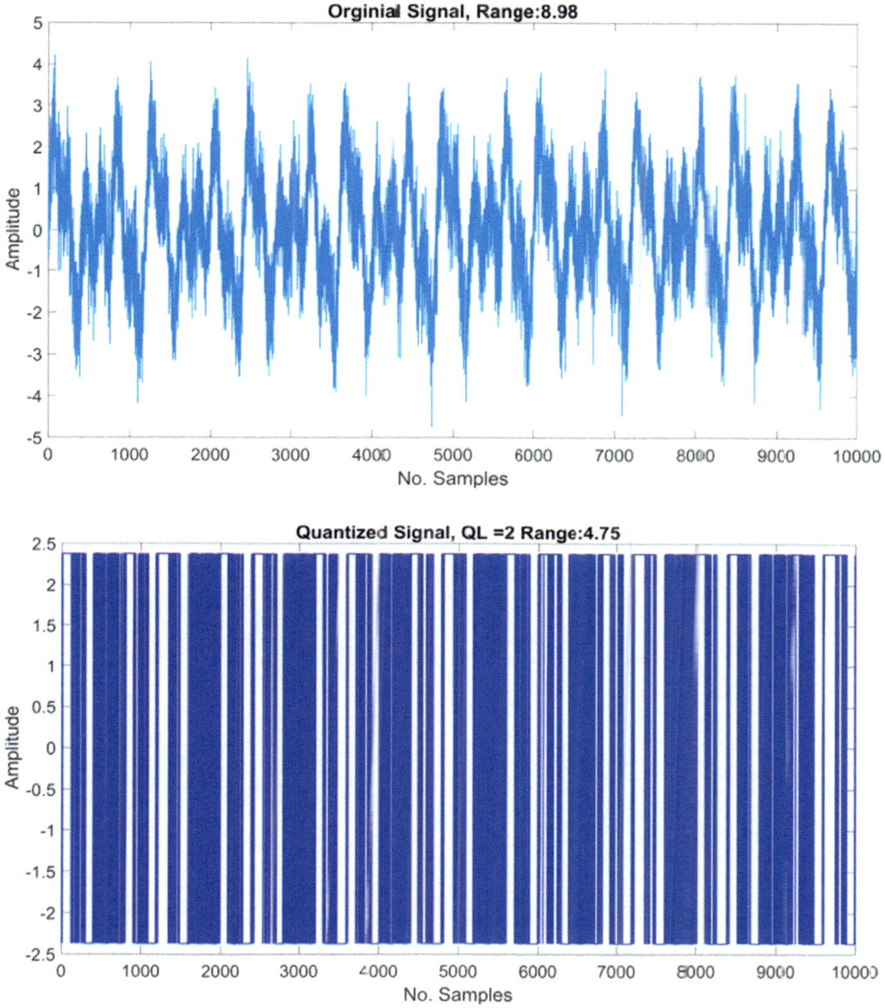

Fig. 6 Comparison between the intact telemetry signal (top) and the quantized version with QL = 2 (bottom). While the original signal was found to be a composition of three sin waves, the quantized signal only exhibits periodic oscillation between two points

signals overlaid on top of the corresponding original signals (left column) yield the deviations that are evaluated by RMSE metric (right column), respectively. As evidenced by the smaller RMSE values and more consistent residuals, the unquantized signals are much closer reproductions of the original signals.

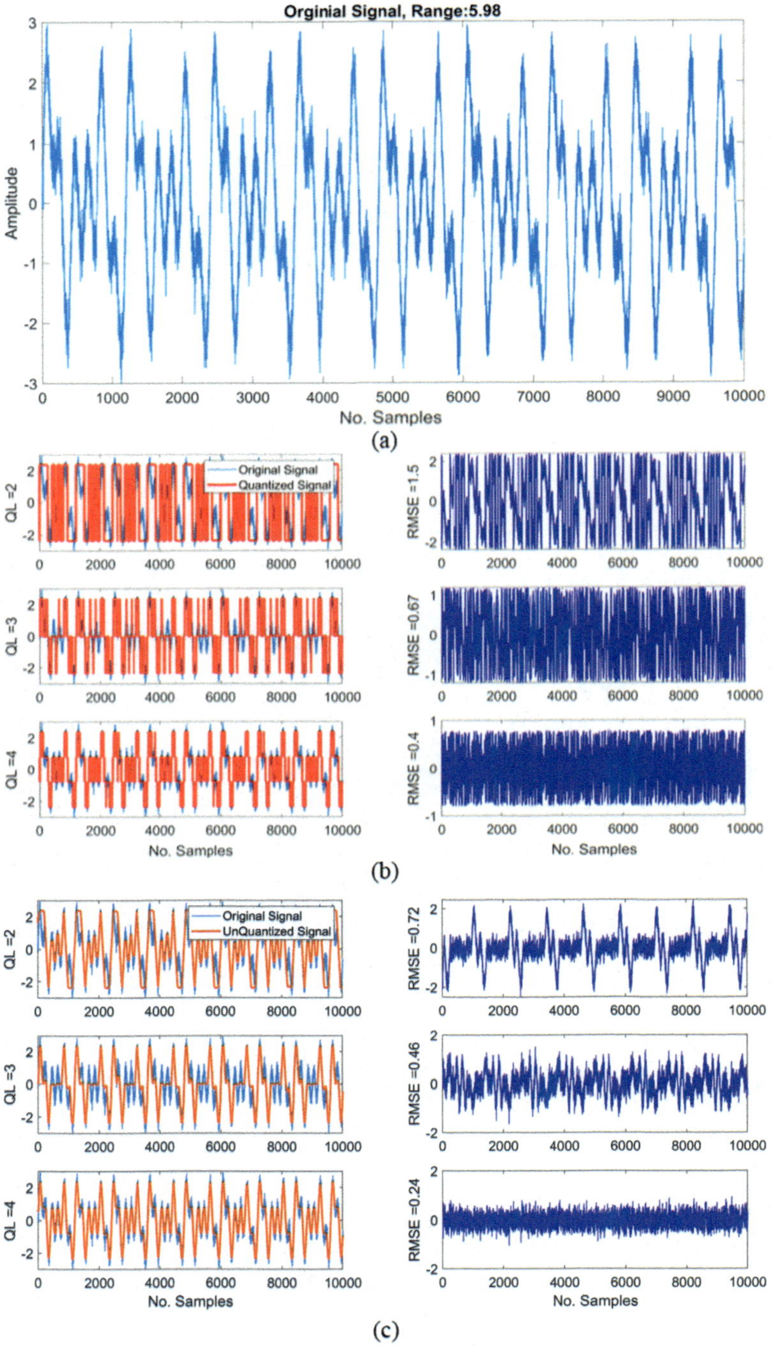

Fig. 7 (**a**) The original testing signal is quantized into three different levels, resembling the typical outcome of the low-bit A/D chips. (**b**) The quantized signals (red) in three different levels are compared with the original signal (blue) in the left column, and their respective deviations characterized by RMSE are presented on the right. (**c**) Same as (**b**) except the quantized signals have been unquantized according to their quantization levels

3.3 *Performance Gains with Unquantized Signals in ML Prognostics*

The case study presented herein demonstrates the unquantize framework upstream of the Oracle's preferred ML prognostic solution, which is the Multivariate State Estimation Technique (MSET; refer to [4–7] for more details). MSET provides high sensitivity for proactive warnings of incipient anomalies, and ultralow false-alarm and missed-alarm probabilities. The increased prognostic accuracy afforded by the unquantization of signals allows for better anomaly detection performance, which is evaluated and demonstrated in this section.

Figures 8, 9, and 10 demonstrate the unquantization of signal yields better false-alarm probability (FAP) in an anomaly detection example. A 50 sec long continuous signal was equally divided into two parts: the first part (Fig. 8) was used for building up an MSET model, which was then used to examine the second part as the surveillance data that had been quantized (QL = 4) deliberately (Fig. 9a). While zero false alarm is expected since the surveillance data has the same characteristic as in the training data, the fact that quantization causes loss of meaningful physical characteristics leads to significant deviations between the two data and subsequently yields false alarms (Fig. 9b). However, if we had the quantized signal undergo the unquantization process, the prior false alarms (red dots in Fig. 9b) were eliminated, as demonstrated in Fig. 10.

Figures 11, 12, 13, and 14 illustrate the prognostic performance gains of the unquantization technique with respect to lower missed-alarm probability (MAP) through another anomaly detection example. Figure 11 presents the testing signal

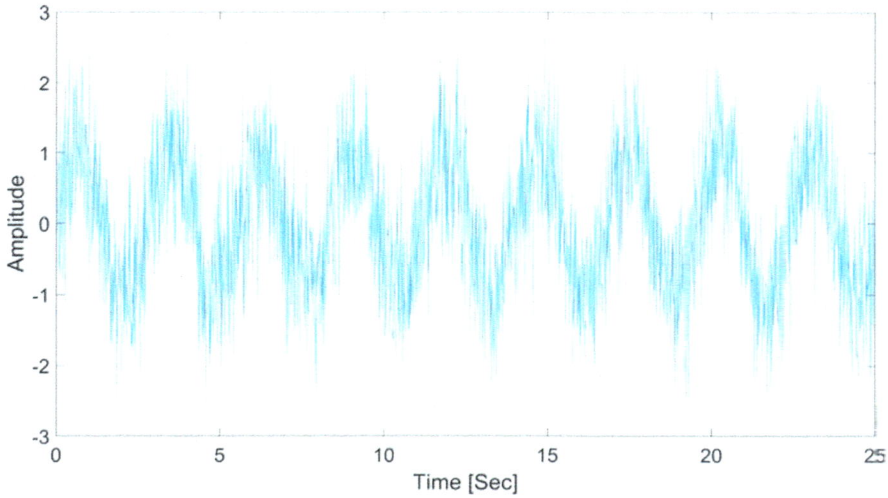

Fig. 8 The first half of a 50 sec long signal sampled at 100 Hz is used as the training data to create an MSET model

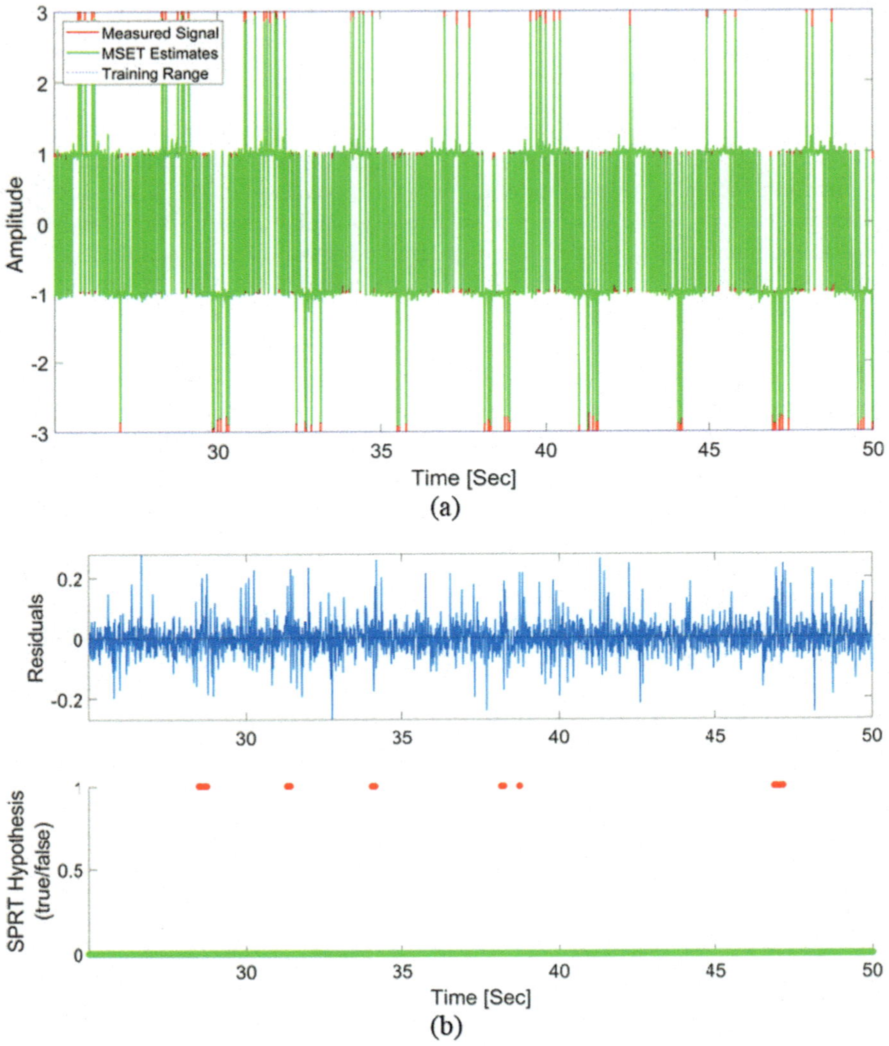

Fig. 9 The second half of the same continuous signal is quantized at QL = 4 before sending to the prior built MSET model as the surveillance data for anomaly detection (**a**). The residuals between the surveillance data (green) and the corresponding MSET estimates (red) and the subsequent anomaly detection results are illustrated in the top and bottom subplots in (**b**), respectively

with degradations starting at observation #3750. Similar to the previous example, the first half of the signal was used to train an MSET model, which was then used to find the degradations in the surveillance data (i.e. the second half). The deviations between the surveillance data and the corresponding MSET estimates become

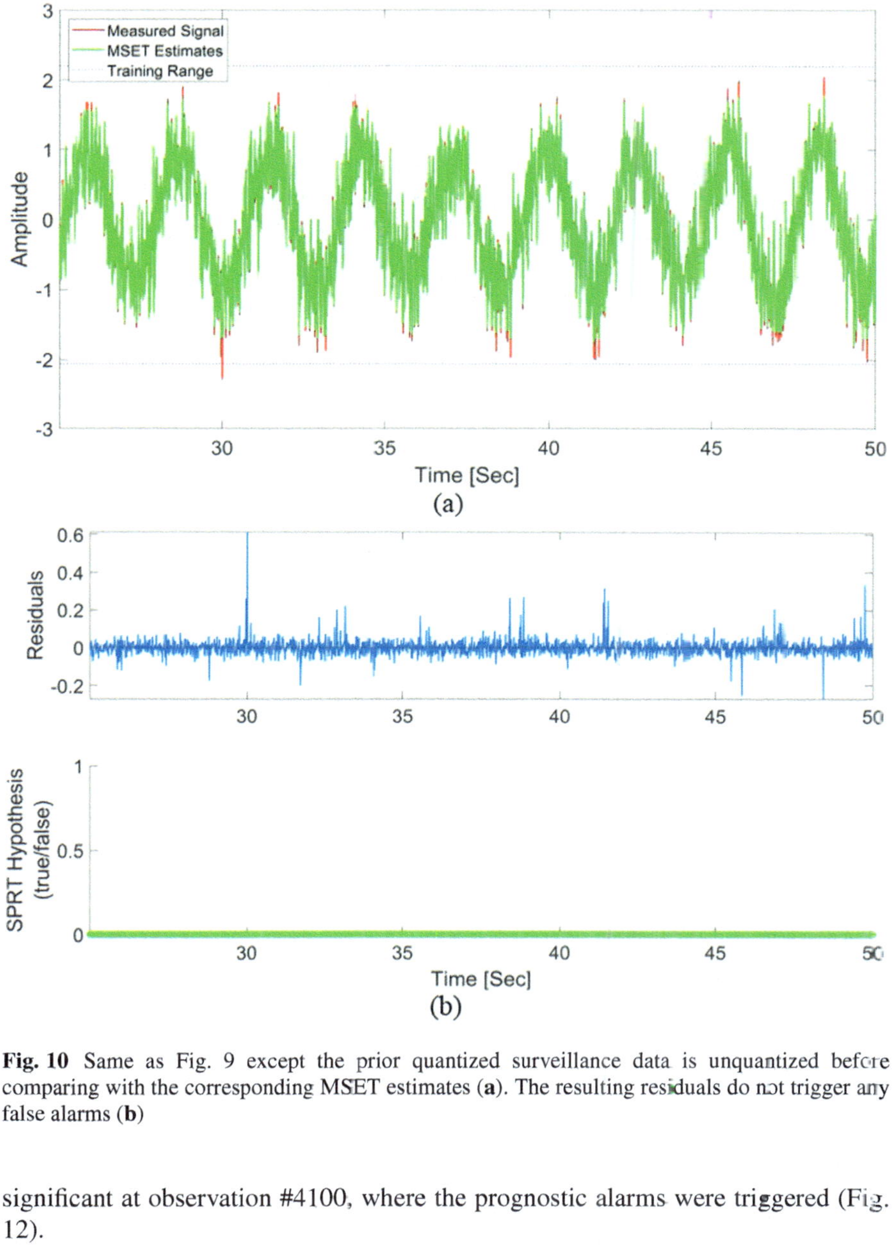

Fig. 10 Same as Fig. 9 except the prior quantized surveillance data is unquantized before comparing with the corresponding MSET estimates (**a**). The resulting residuals do not trigger any false alarms (**b**)

significant at observation #4100, where the prognostic alarms were triggered (Fig. 12).

On the other hand, the surveillance data was quantized (QL = 3), and the same anomaly detection process was repeated with poor results. Figure 13 illustrates how the quantized version of the degraded signal caused much fewer prognostic alarms, indicating significant missed alarms which can be costly in the safe critical industries. To proceed, we applied the unquantization technique to the quantized

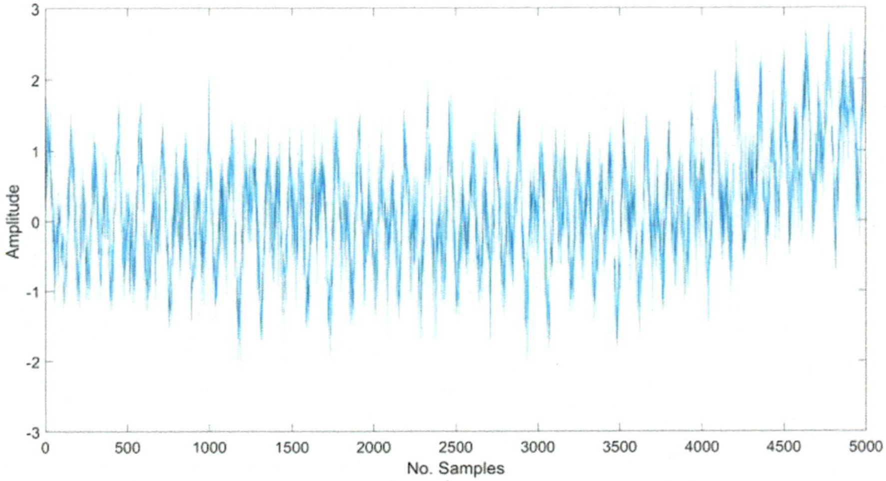

Fig. 11 A time series testing signal with a ramp inserted between #3750 and #5000 resembling degradations

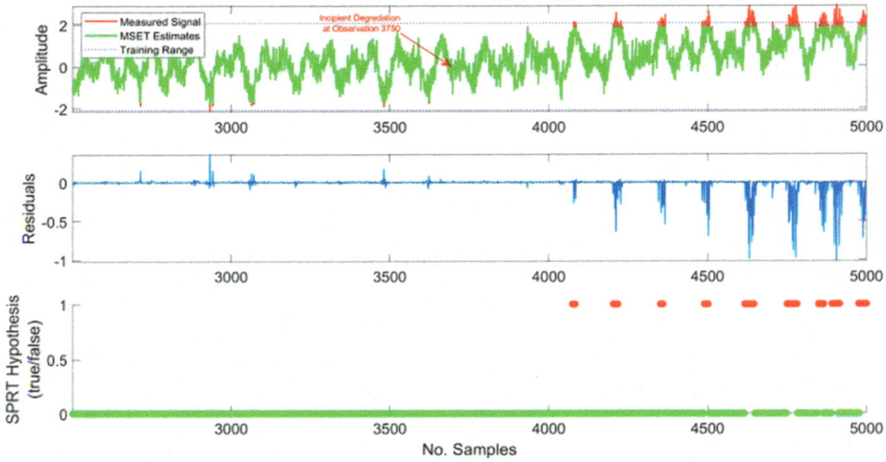

Fig. 12 An MSET-based anomaly detection monitors the second half of the signal where the degradation is located; the first half of the signal is the training data. The top plot compares the surveillance data (green) with the corresponding MSET estimates (red). The resulting residuals and the trigged alarms are presented in the middle and bottom plots, respectively

surveillance data and repeated the anomaly detection again. As illustrated in Fig. 14, the resulting alarms did not begin as early as in the original example in Fig. 12; however, when comparted to Fig. 13 the alarms appear to begin much earlier and reveal the severity of the degradation more accurately. The expansion of prognostic

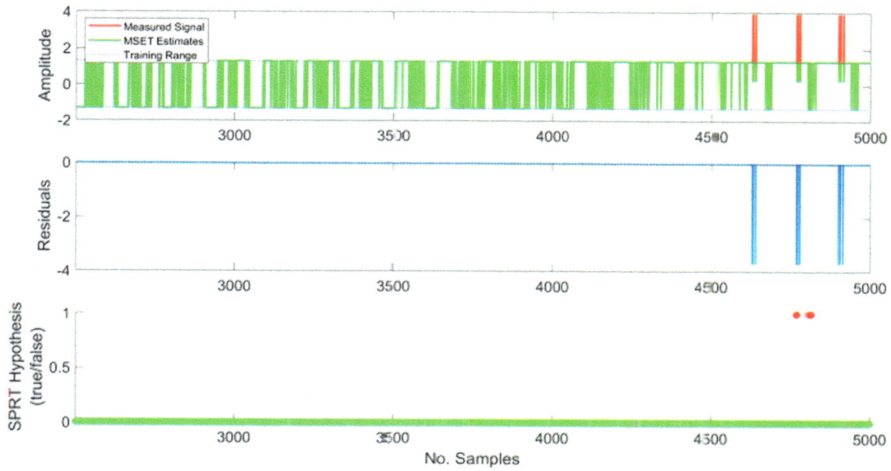

Fig. 13 Same as Fig. 12 except the surveillance data has been quantized at QL = 3 before the same anomaly detection is executed again. Substantial missed alarms are observed in reference to Fig. 12

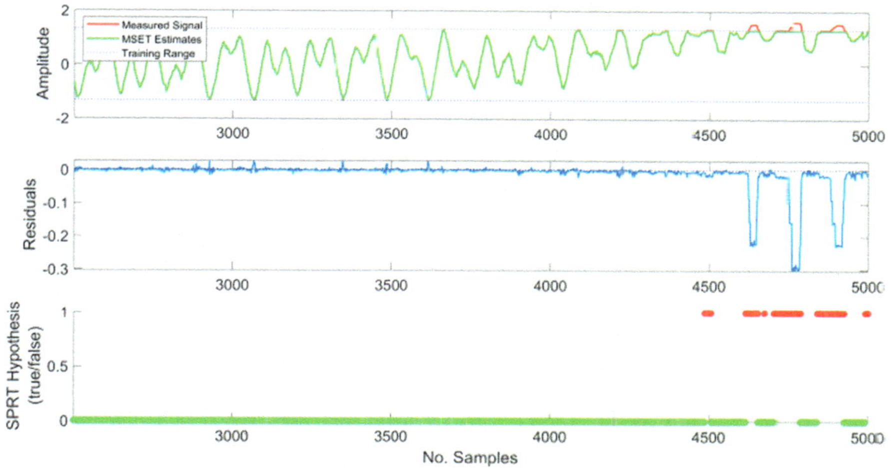

Fig. 14 Same as Fig. 13 except the quantized surveillance data has been unquantized before the same anomaly detection is executed once again. Most of the missed alarms in Fig. 13 are discovered when compared to Fig. 12

capacity was a direct result of the unquantization process, which was largely able to retrieve the meaningful physical characteristics from the quantized signal.

4 Conclusion

Addressing the negative impacts of quantized signals benefits the IoT sectors of utilities, oil and gas, manufacturing, transportation, and other sensor dense industries when it comes to ML prognostics. In this paper, we propose a novel technique that is able to convert the quantized signals that are typically unanalyzable to smooth signals that matches the original sensor output as closely as possible. With this technique, signals that previously required human surveillance or were monitored with simple thresholds can now be analyzed automatically with greater precision. Another major benefit is that this technique can improve all types of ML algorithms, such as Neural Nets or Support Vector Machine. Any ML algorithm intended for time series analysis will attain higher prognostic accuracy for discovering subtle anomalies in critical assets and processes and with much lower false-alarm and missed-alarm probabilities. While unquantization already presents majors strides in time series analysis, more research on this topic is currently being done at Oracle Lab.

References

1. G.C. Wang, K.C. Gross, Telemetry parameter synthesis system for enhanced tuning and validation of machine learning algorithmics, in *IEEE 2018 Intn'l Symposium on Internet of Things & Internet of Everything (CSCI-ISOT), Las Vegas, NV*, (2018)
2. R.C. Dhanekula, K.C. Gross, High fidelity telemetry signal synthesis for improved electronic prognostics, in *IEEE World Congress in Computer Science, Computer Engineering, and Applied Computing (WorldComp2012), Las Vegas, NV*, (2012)
3. K.C. Gross, E. Schuster, Spectral decomposition and reconstruction of telemetry signals from enterprise computing systems, in *CDES*, (2005), pp. 240–246
4. R.M. Singer, K.C. Gross, J.P. Herzog, R.W. King, S. Wegerich, Model-based nuclear power plant monitoring and fault detection: Theoretical foundations, in *Proc. 9th Intnl. Conf. On Intelligent Systems Applications to Power Systems, Seoul, Korea (July 6-10, 1997)*, (1997), pp. 60–65
5. K.C. Gross, R.M. Singer, S.W. Wegerich, J.P. Herzog, R. VanAlstine, F. Bockhorst, Application of a model-based fault detection system to nuclear plant signals, in *Proc. 9th Intnl. Conf. On Intelligent Systems Applications to Power Systems, Seoul, Korea (July 6-10, 1997)*, (1997), pp. 66–70
6. F. Zhang, S. Boring, J.W. Hines, J. Coble, K.C. Gross, Combination of unquantization technique and empirical modelling for industrial applications, in *2017 American Nuclear Society Intn'l Conf., Washington DC*, (2017)
7. K.C. Gross, G.C. Wang, AI decision support prognostics for IoT asset health monitoring, failure prediction, time to failure, in *IEEE 2019 Intn'l Symposium on Artificial Intelligence (CSCI-ISAI), Las Vegas, NV*, (2019)

Information Diffusion Models in Microblogging Networks Based on Hidden Markov Theory and Conditional Random Fields

Chunhui Deng, Siyu Tang, and Huifang Deng

1 Introduction

Social networks, which are affecting our daily lives and activities, are just like frame-by-frame of graphs composed of individual users and their relationships. In recent years, studies on dissemination of information in social networks, which involve analysis of network structure, information contents, and large-scale data processing, have aroused many scholars' interest in related fields. These types of studies have great merits in public opinion directing or rumor spreading controlling, economic efficiency enhancing, and other associated disciplinary research propelling.

Microblogging networks, a subset of social networks, have drawn researchers' large attention from traditional social networks. Microblog is dedicated to disseminating information and successfully distinguishes itself from traditional social networks with its own features. These features include fragmental contents, ease of use, rapid spread, emotional interaction and expression, and wide range of influence. Users can express themselves in microblog to realize the real-time information sharing anywhere through a message in many forms (including text, numbers and symbols, images, video, and audio) via a variety of ways (including Web pages, instant messaging, mobile phone, blog, and forums) under a strict limit of 140 characters. Generally, there are four basic functions in microblog without

C. Deng (✉)
School of Computer Engineering, Guangzhou College, South China University of Technology, Guangzhou, China
e-mail: dengch@gcu.edu.cn

S. Tang · H. Deng
School of Computer Science and Engineering, South China University of Technology, Guangzhou, China

K. Daimi et al. (eds.), *Advances in Security, Networks, and Internet of Things*, Transactions on Computational Science and Computational Intelligence,
https://doi.org/10.1007/978-3-030-71017-0_45

big changes on different microblogging platform. They are posting, commenting, reposting, and following. When a user "post" a message, his or her "followers" can either "repost" or "comment" on that message. Moreover, common practice of responding to a microblog message has evolved into a well-defined markup culture: RP stands for reposting, "//@" followed by a user identifier addresses the user, and "#" followed by a word represents a topic-tag. In addition, the repost mechanism empowers users to spread information of their choice beyond the reach of the original microblog message's followers.

Information diffusion is also known as information propagation, information spread, and information flow. Early studies on information diffusion were mainly concerned about the dissemination of innovation, epidemic, and product in real social networks by some sociologists, epidemiologists, and economists. In this context, some classic information diffusion models appeared, such as independent cascade model (ICM) [1], linear threshold model (LTM) [2], epidemic models [3], and game theory models [4]. Along with the development of social networks or the advent of microblogging networks, huge and rich large-scale online social network data as well as data with respect to information diffusion are readily available, which brings one new opportunity to study the information diffusion and makes it become a hot spot. However, modeling of the information diffusion in microblogging networks has proven to be very challenging. It is difficult to obtain all elements with respect to (1) information contents, (2) users, and (3) their relationships, such as emotions, links, tags, topics, user's interests, user's activeness, network structure, common interests or common reposts, spreading, and propagating through microblogging networks. Existing studies on information diffusion in microblogging networks have mainly focused on certain impact factors of the above three main elements [5, 6] and user's behavior [7, 8]. And most of the few studies on multi-information (a sequence of posted information pieces ordered by time) diffusion just simply assumed that multi-information are mutually exclusive [9, 10] instead of competitive and collaborative. In order to merge information contents, users, and their relationships in microblogging networks, we can use statistical modeling method in a generative or discriminative manner according to users' repost behavior to model dissemination of information in it. Thus, we propose an information diffusion model based on the hidden Markov theory (IDMBHMT) and a multi-information diffusion model based on conditional random fields (MIDMBCRF).

The rest of this paper is organized as follows. In Sect. 2, we described how to construct the IDMBHMT and MIDMBCRF models in detail. In Sect. 3, we presented microblogging networks' properties as well as how to take advantage of them to improve the performances of our models. In Sect. 4, we applied our models to predict users' reposting decisions and showed our experiment results. Finally, we concluded the paper in Sect. 5.

2 Proposed Models

Since the dissemination of information in microblogging networks is directed by users' behavior, study on information diffusion is in nature the study on users' behavior. In this paper, we focus on the users' reposting behavior.

Let $G = (U, E)$ represent a microblogging network where each node denotes a user $u \in U$, and each edge $(u, v) \in E$ denotes the following relationship between u and v, and let the user entities $x = \{x_1, \ldots, x_{|U|}\}$ such that each x_u corresponds to an $u \in U$, and the binary reposting decisions $y = \{y_1, \ldots, y_{|U|}\}$ represent the reposting decision of all users. The information diffusion problem is defined as predicting the reposting decisions for all users in U given a message m and the user entities x, and that is the task of finding the maximum posterior probability (MAP):

$$\widehat{y} = \arg\max_{y} P\,(y \mid m, x) \tag{1}$$

The graphical structure is depicted in Fig. 1. There are three characteristics associated with the above equation: (1) There are dependencies between y_i's. (2) There exists interaction between m and x that can aid predicting users' reposting decisions. (3) We can use the joint distribution $P(y, m, x)$ in a generative manner or the conditional distribution $P(y|m, x)$ in a discriminative manner to calculate the MAP. These characteristics motivate us to model the information diffusion problem using the hidden Markov theory (HMT) or the conditional random fields (CRF).

Fig. 1 The graphical structure of models

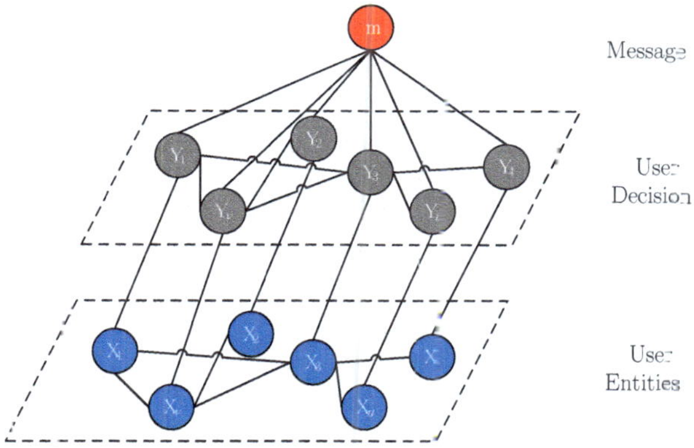

2.1 HMT-Based Information Diffusion Model

2.1.1 HMT-Based Mapping

HMT is composed in a quintuple form $\lambda = (M, N, \pi, A, B)$ such that M is the observation character set representing the user entities x in our model, N is the number of states representing the binary reposting decision values, and the remaining parameters (π, A, B) represent the initial probability distribution, the state transition probability distribution, and the.

observation probability distribution, respectively. The key issue in constructing this model is to determine the three parameters (π, A, B) via the parameter learning from dataset. In our model, we represent them in a matrix form as follows:

$$A = \left[a_{ij}\right]_{N\times N} \quad B = \left[b_j(k)\right]_{N\times M} \quad \pi = (\pi_i)$$

where a_{ij} denotes the transition probability from state S_i at time t to the state S_j at time $t + 1$, $b_j(k)$ denotes the observation probability of v_k under state S_i at time t, and π_i denotes the probability of state S_i at time $t = 0$. Each element of the above is calculated by synthesizing main factors that impact the reposting decision, including information contents, users' own features, and relationship between users.

2.1.2 Parameter Estimation

In this paper, we use the Baum-Welch algorithm [11] to estimate the three parameters (π, A, B). Table 1 gives the Baum-Welch algorithm in detail.

Table 1 Baum-Welch algorithm

Algorithm 1. Baum-Welch algorithm
Input: observation data $O = (O_1, O_2, \ldots, O_L)$
Output: HMT parameters $\lambda = (A, B, \pi)$
Steps: 1. Initialization: when $t = 0$, choose $a_{ij}^{(0)}, b_j(k)^{(0)}, \pi_i^{(0)}$ and get $\lambda^{(0)} = (A^{(0)}, B^{(0)}, \pi^{(0)})$. 2. Iteration: when $t = 1, 2, \ldots,$ $\pi_i^{(t+1)} = \gamma_1(i) \quad a_{ij} = \frac{\sum_{l=1}^{L-1} \xi_l(i,j)}{\sum_{l=1}^{L-1} \gamma_l(i)} \quad b_j(k)^{(t+1)} = \frac{\sum_{l=1, O_l=v_k}^{L} \gamma_l(j)}{\sum_{l=1}^{L} \gamma_l(j)}$ The values on the right sides are calculated according to $O = (O_1, O_2, \ldots, O_L)$ and $\lambda^{(t)} = (A^{(t)}, B^{(t)}, \pi^{(t)})$ using the EM (expectation-maximization) algorithm, and values of $\gamma_l(i)$ and $\xi_l(i,j)$ are calculated according to the forward-backward algorithm. 3. Termination: when the model parameters $\lambda^{(t+1)} = (A^{(t+1)}, B^{(t+1)}, \pi^{(t+1)})$ are obtained.

2.2 *CRF-Based Multi-information Diffusion Model*

2.2.1 CRF-Based Formulation

When we assume the decision of each user x_u follows the Markov property with respect to G, the probability in Eq. (1) can be modeled using CRF as

$$P\left(y|\,m, x\right) = \frac{1}{Z(m)} \prod_{C \in \{C_U, C_E\}} \Phi_C\left(m, x, y_C\right) \tag{2}$$

where $Z(m)$ is a normalization term to ensure the sum of probability equal to 1 for a given m. It is calculated by

$$Z(m) = \sum_y \prod_{C \in \{C_U, C_E\}} \Phi_C\left(m, x, y_C\right) \tag{3}$$

C denotes the set of *cliques*, which are grouped into *node cliques* C_U and *edge cliques* C_E. The set of Φ_C denotes the *potential functions* defined over cliques, having the form

$$\Phi_C = \exp\left\{\sum_k^{K(C)} \lambda_{ck} f_{ck}\left(m, y_C, x_C\right)\right\} \tag{4}$$

Where $f(\cdot)$ and λ denote the feature functions and their respective coefficients, and $K(\cdot)$ denotes the number of features of a type. The key issues to construct this model are the selection of feature functions $f(\cdot)$ and the estimation of their respective parameters λ.

2.2.2 Feature Functions Definition

Feature functions have a direct impact on the performance of model. They are a set of user-defined functions that well fit the real data and can be used to describe the distribution of random variables. The strength and positive and negative polarities of feature functions are represented through weights of data training. Moreover, the potential functions of cliques are in essence the linear combination of the corresponding feature functions.

In MIDMBCRF, we group features into three types: the first two defined for nodes and the last one defined for the edges.

Information features $f_I^u\left(m, y_u\right)$ incorporate features that impact on the message being reposted in its own right. They are independent of users, e.g., whether the message m contains a topic. There are 12 features in this type as shown in Table 2.

User features $f_U^u\left(m, x_u, y_u\right)$ integrate subjective views and objective relations of users that affect the message being reposted. These features depend on the user

Table 2 Information features

Feature	Description
Content similarity	Proportion of class contains message *m* in the entire classes composed of original messages and the reposted messages of the entire network
URL	Whether message *m* contains a URL; how frequently does the URL domain appear in global messages
Topic	Whether message *m* contains a topic; how frequently does the topic appear in global messages
@others	Whether message *m* contains @others
Author	Number of followers, friends, posted messages, and reposted messages; whether being labeled "V," i.e., verified

Table 3 User features

Feature	Description
Content similarity	Proportion of class contains message *m* in the six classes composed respectively of user original and global messages, user followers' original and global messages, and user friends' original and global messages
URL	How frequently does the URL domain appear in the user's messages
Topic	How frequently does the topic appear in the user's messages
User similarity	Similarity between user and author of message m; similarity between user and @others that appear in message *m*
Information interactions	The influence of the user historically browsed messages on the message *m*

Table 4 Relationship features

Feature	Description
Content similarity	Proportion of class contains message m in the classes composed of global messages of the user-pair
User similarity	Number of common messages among authors and user-pairs (connected by edge)

u, e.g., the similarity between the user u and the message's author. There are 11 features in this type as shown in Table 3.

Relationship features $f_R^{u,v}(m, x_u, x_v, y_u, y_v)$ include the perspectives whether the message is reposted simultaneously for two users u and v due to some characteristics matches of them. These features therefore depend on the interactions between u and v, e.g., the number of reposted message m among u, v, and the message's author. There are two features in this type as shown in Table 4.

There are many information pieces spreading in the microblogging networks simultaneously. When they are displayed in front of us, we can view them as a chronological sequence of information. And since we take into account the interaction between historical messages and the current message m, we refer the model as multi-information diffusion model rather than the information diffusion model.

We introduce three methods, namely, a Chinese short text classification method based on feature extension [12], a user similarity method based on attributions [13], and a method for the measurement of interactions between historical information and the current information [14], with respect to classification, user similarity, and the measurement of multi-information interactions in the three tables, respectively.

Based on these features' definitions, we can rewrite our CRF formulation as follows:

$$P\left(y|m,x\right)=\frac{1}{Z(m)}\exp\left\{\sum_{u}\left(\sum_{k}^{K(I)}\lambda_{Ik}f_{Ik}\left(m,y_u\right)+\sum_{k}^{K(U)}\lambda_{Uk}f_{Uk}\left(m,x_u,y_u\right)\right)\right.$$
$$\left.+\sum_{u,v}\left(\sum_{k}^{K(R)}\lambda_{Rk}f_{Rk}\left(m,x_u,x_v,y_u,y_v\right)\right)\right\} \tag{5}$$

2.2.3 Parameter Estimation

In this paper, we use the L-BFGS algorithm [15] to estimate the parameters $\eta=\{\lambda_k\}$ in Eq. (5). Table 5 shows the L-BFGS algorithm in detail.

Table 5 L-BFGS algorithm

Algorithm 2. L-BFGS algorithm
Input: Feature functions $f_1, f_2, \ldots, f_n$, empirical distribution $\tilde{P}\left(X,Y\right)$
Output: Optimal parameter values $\hat{\lambda}$ and optimal model $P_{\hat{\lambda}}\left(y|x\right)$
Steps: 1. Select initial point $\lambda^{(0)}$, take B_0 to be a positive definite symmetric matrix, and set $k=0$. 2. Calculate the gradient function $g_k=g(\lambda^{(k)})$. If $g_k=0$, then terminate; otherwise, go to 3. 3. Determine p_k by $B_k p_k=-g_k$. 4, One-dimensional search: Find w_k that optimizes the target function $f(\lambda^{(k)}+w_k p_k)=\min f(\lambda^{(k)}+w p_k)\quad (w\geq 0)$ 5. Set $\lambda^{(k+1)}=\lambda^{(k)}+w_k p_k$. 6. Calculate $g_{k+1}=g(\lambda^{(k+1)})$. If $g_k=0$, then terminate; otherwise, determine B_{k+1} by $B_{k+1}=B_k+\frac{y_k y_k^T}{y_k^T\delta_k}-\frac{B_k\delta_k\delta_k^T B_k}{\delta_k^T B_k\delta_k}$ where $y_k=g_{k+1}-g_k$, $\delta_k=\lambda^{(k+1)}-\lambda^{(k)}$. 7. Set $k=k+1$ and go to 3.

3 Network Structure of Models

The two models we proposed are based on the entire microblogging network. As we can see, an oversimplified structure is not capable enough of capturing the characteristics of the target network, while an overcomplex structure with excessive edge is very expensive to train and infer. In order to make our models more applicable, it is necessary to investigate the properties of the microblogging network.

3.1 Network Features

The microblogging network structure based on users following relationship directly affects the depth and breadth of information diffusion. The social network theory is derived from *Six Degrees of Separation* and the *150 Law*, and studies have shown that the microblogging network is a classical complex network with small-world, scale-free, high-clustering properties [16] and meets the power law distribution [17]. In other words, the Average Path Length (APL) between any arbitrary two users is small, while the Aggregation Coefficient in the network is very high. In addition, a small-world network observes the 80–20 rule: An individual in the network is primarily influenced by only a minor portion of his connections. Therefore, we may keep only a fraction of edges that reserves the essential clustering structures of the microblogging network such that we can construct fine-grained models to improve performance via the graph partitioning.

3.2 Graph Partitioning

The general graph partitioning problem is NP-complete. We use the open-source tool named METIS to partition the microblogging network in this paper.

The algorithms implemented in METIS are based on the multilevel recursive bisection, multilevel k-way, and multi-constraint partitioning schemes. The key features of METIS about graph partitioning do meet our needs: (1) METIS offers high-quality partitions, and experiments on a large number of graphs demonstrate that the partitions produced by METIS are consistently 10–50% better than those produced by spectral partitioning algorithms; and (2) it is extremely fast and experiments on a wide range of graphs have shown that METIS is one to two orders of magnitude faster than other widely used partitioning algorithms.

4 Experiments Results and Analysis

We apply our models to predict each users' reposting decision in Sina microblog, which is very popular and famous in China, that is calculating the maximum probability of the binary output sequence value (repost or not) y for all the user entities x in our dataset given a certain message m. This problem belongs to the inference problem of HMT and CRF and, in essence, is the inference problem of probabilistic graphical model. We can use different algorithms to solve this problem for different graphical structure. Since our microblogging network structure is complex, we adopt the junction tree algorithm in our experiments, which is applicable to any graph and ensure an accurate inference for any graph.

4.1 Dataset

We use the Sina API to obtain a dataset of 5148 users and 772,200 messages from Sina microblogging network. Using the breadth-first search (BFS) algorithm in a snowball sampling method, the dataset is collected by the end of December 2012 according to a reverse chronological order. This dataset consists of 86 user tags, 112 domains, and 348 topics.

We divide this dataset into three parts: 90% of it is used for models training, 5% for parameters optimization, and the rest 5% for testing. Using the first two parts of message data, we calculate parameters of the two models we proposed and the features for all 5418 users as mentioned in Sect. 3. And using the remaining 5% of message data, we conduct an analysis on the impact factors of models and compare the two models with other two reference models, respectively. All experiments are conducted on a Windows-based machine with Intel Core™ 2 Duo 2.93GHz and 4GB memory.

4.2 Metrics

The evaluation metrics used in our experiments are the precision, the recall, and F1 measure. They can be computed by the following equations:

$$P = U_{pr}/U_{all} \tag{6}$$

$$R = U_{rp}/U_{rall} \tag{7}$$

$$F = \frac{2PR}{P+R} \tag{8}$$

Where U_{pr} denotes numbers of users being predicted correctly, U_{all} denotes all users, U_{rp} denotes numbers of users being predicted for reposting correctly, and U_{rall} denotes numbers of users reposting the message m. Besides, F1 measure, which is the combination of precision and recall, is a metric that comprehensively reflect the performance of our models. The higher the metrics' values, the better the performances of our models. In fact, precision and recall are influenced with each other, and ideally, we want to have both high but normally a high-precision rate along with a low-recall rate.

4.3 Impact Factors Analysis

There are three main factors we considered in our experiments that influence the performance of the two models, IDMBHMT and MIDMBCRF, including the following: (1) The performance of models under different network sizes (120, 240, and 400 users) and the results are shown in Fig. 2. (2) The performance of models with graph partitioning using common network (800 users) but different subnetwork sizes (25, 48, 100, 200, and 400 users) and the results are shown in Fig. 3. (3) The performance of models with graph partitioning using common subnetwork (48 users) but different total network sizes (800, 2000, and 4000 users) and the results are shown in Fig. 4.

Before doing experiments, we need to determine some other factors associated with MIDMBCRF as well. Through continuously fine-tuning and optimizing, we finally obtained the weights settings as follows: (1) The support and confidence thresholds of FP-growth algorithm in the Chinese text classification method [12] are, respectively, set to 0.01 and 0.5, (2) coefficients setting for the user similarity algorithm [13] are shown in Table 6, and (3) the information browsed in the past is backdated to $k = 5$, and the categories of information is set as $T = 20$ in measuring the interactions between information pieces [14].

From the curve trend of Fig. 2, we can see that the values of precision, recall, and F1 for IDMBHMT and MIDMBCRF decline as the increasing network size, especially the recall value, which indicates the difficulty in predicting users' behavior with the complexity of network. Figure 3 tells us both models have a peak performance at 48 users, which imply that 48 may be close to the natural. Figure 4 illustrates that the graph partitioning does improve the performance of models. When the total network size is 800, values of the precision, recall, and F1 remain at 83.68%, 73.52%, and 78.27%, respectively, for IDMBHMT and 88.33%, 76.15%, and 81.79%, respectively, for MIDMBCRF, while in the unpartitioned case, the results drop to 45.64%, 17.38%, and 25.18%, respectively, for IDMBHMT and 49.82%, 21.14%, and 29.68%, respectively, for MIDMBCRF. MIDMBCRF contains a wealth of features that makes more factors affecting its performance. But we only focus on the feature of interactions between information pieces. From Figs. 2b, 3b, and 4b, we notice that the interactions between information pieces make very little contribution to the performance of MIDMBCRF.

Fig. 2 Precision, recall, and F1 scores of IDMBHMT (**a**) and MIDMBCRF (**b**) performance measurements under different settings. In **b**, "Y" means inclusion of the interactions between information pieces, while "N" means no inclusion

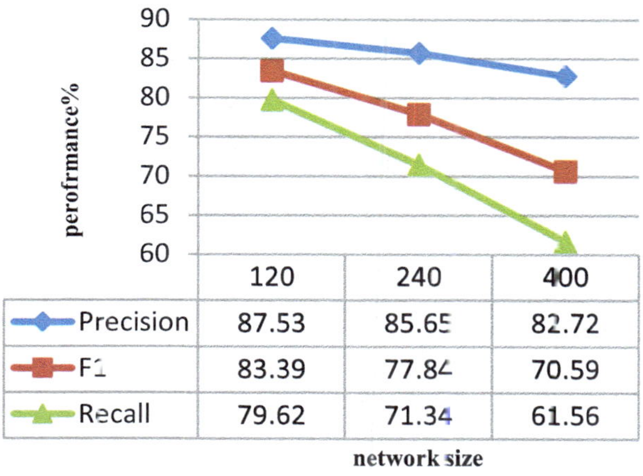

a IDMBHMT

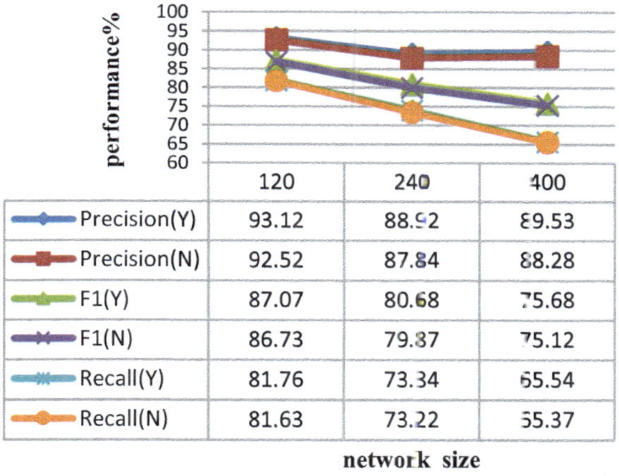

b MIDMBCRF

In reality, only a very few numbers of possible pairs of information interact with each other. Furthermore, many of the pairs of information do interact only a few numbers of times, whereas there are other information pairs that interact several thousands of times. To account for this, we use the relative change in probability to measure interactions impact:

$$P_{\text{relative}} = \frac{P\left(X|\left\{Y_k\right\}_{k=1}^{K}\right) - P\left(y|\,m, x\right)}{P\left(y|\,m, x\right)} \tag{9}$$

Fig. 3 Precision, recall, and F1 scores of IDMBHMT (**a**) and MIDMBCRF (**b**) performance measurements with graph partitioning using common network (800 users) but different subnetwork sizes (25, 48, 100, 200, and 400 users). In **b**, "Y" means inclusion of the interactions between information pieces, while "N" means no inclusion

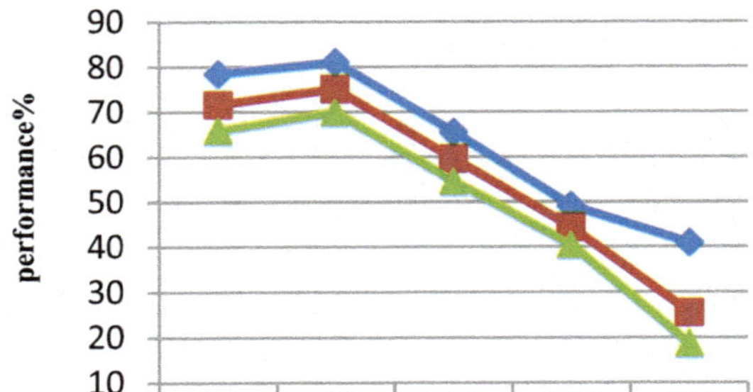

	25	48	100	200	400
Precision	78.64	81.24	65.66	49.28	40.98
F	71.66	75.14	59.61	44.41	25.45
Recall	65.81	69.89	54.58	40.42	18.46

Sub-newtork size

a IDMBHMT

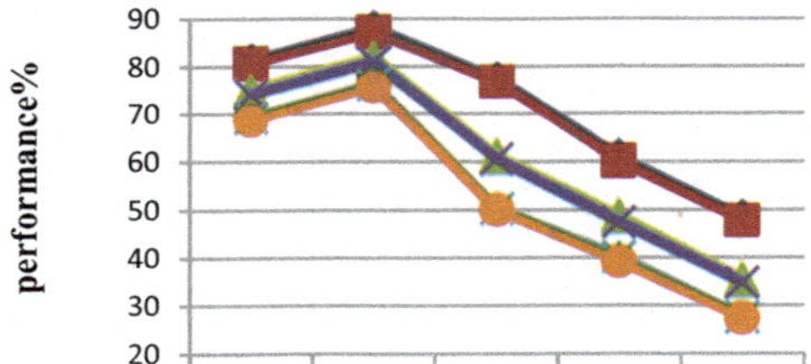

	25	48	100	200	400
Precision(Y)	81.56	88.37	77.65	61.24	48.68
Precision(N)	80.86	87.32	76.84	60.21	47.83
F(Y)	75.07	81.96	61.21	48.64	35.3
F(N)	74.36	81.19	60.67	47.51	34.77
Recall(Y)	69.53	76.41	50.51	40.34	27.69
Recall(N)	68.82	75.87	50.12	39.24	27.31

sub-network size

b MIDMBCRF

In the above equation, $P\left(X|\{Y_k\}_{k=1}^{K}\right)$ denotes the probability of interactions between different information pieces (details refer to [14]), and $P(y|m,x)$ refers to Eq. (5)). With this measure, the contribution of each piece of information pair to the distribution of interaction is proportional to the frequency they interact, as shown in Fig. 5. This plot tells us a very different story that the contribution of information interactions to MIDMBCRF varies sharply, and the curve shows a heavy tail that reaches 830% relative change in the reposting probability. In fact, the average

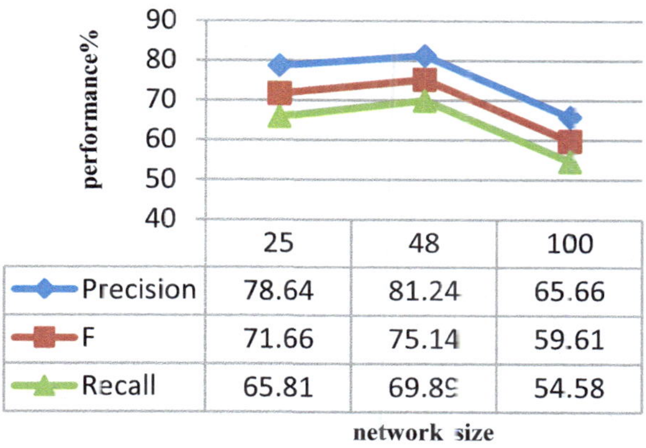

a IDMBHMT

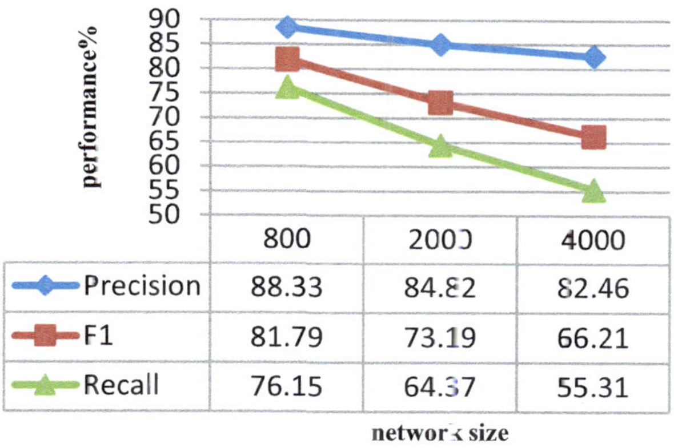

b MIDMBCRF

Fig. 4 Precision, recall, and F1 scores of IDMBHMT (**a**) and MIDMBCRF (**b**) performance measurements with graph partitioning using common subnetwork (48 users) but different total network sizes (800, 2000, and 4000 users)

absolute value of relative change is 43%, indicating that on average, 43% of the repost probability comes from interactions between different information pieces. Since there are so many complex factors impacting MIDMBCRF, the proportion of 43% is sufficient to highlight the importance of interactions.

4.4 *Comparisons and Analysis*

In order to better understand the performance of our models, we compare IDMBHMT and MIDMBCRF, respectively, with two reference models: RPMBLR [18] and MIDMBLT [19]. The results are shown in Fig. 6.

Table 6 Coefficients setting for the user similarity algorithm

Formula	Weights setting	Component weights setting
sim(*Bg*(*u*), *Bg*(*v*)) Background similarity	$\omega_1 = 0.12$	Position: $\omega_{11} = 0.28$ Tags: $\omega_{12} = 0.55$ Descriptions: $\omega_{13} = 0.17$
sim(*Text*(*u*), *Text*(*v*)) Text similarity	$\omega_2 = 0.06$	None
sim(Re*l*(*u*), Re*l*(*v*)) Social relationship similarity	$\omega_3 = 0.74$	Followees: $\omega_{31} = 0.18$ Followers: $\omega_{32} = 0.82$
Inter(*u*, *v*) Interaction similarity	$\omega_4 = 0.08$	Reposting: $\omega_{41} = 1$

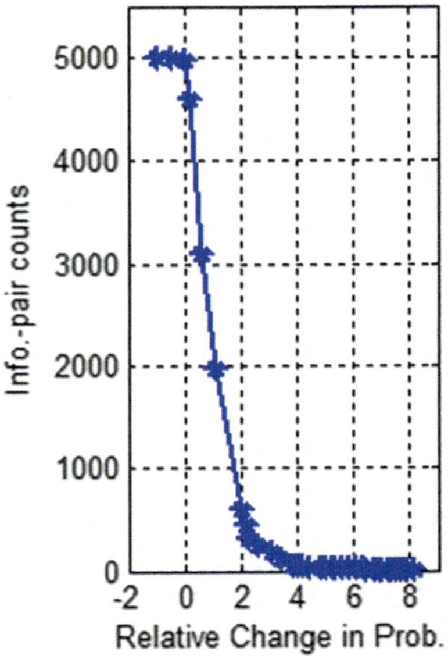

Fig. 5 The distribution of relative change in probability caused by interactions

In these experiments, for RPMBLR, we used the reposting probability equation of a user i based on the logistic regression as

$$P\left(y_i = 1 | x\right) = \frac{1}{1 + \exp^{-\vec{w}^T h_u(x;G)}} \tag{10}$$

Here, we only consider the contents influences and the network structure influences of all features $h_u(x; G)$ in the above equation without the time-decaying factors. And for MIDMBLT, we used the "activation" probability that node v is influenced by its neighbor node u:

$$p_T^i(v) = \alpha_i \sum\nolimits_{u \in A_T^i(v)} b_{u,v} \tag{11}$$

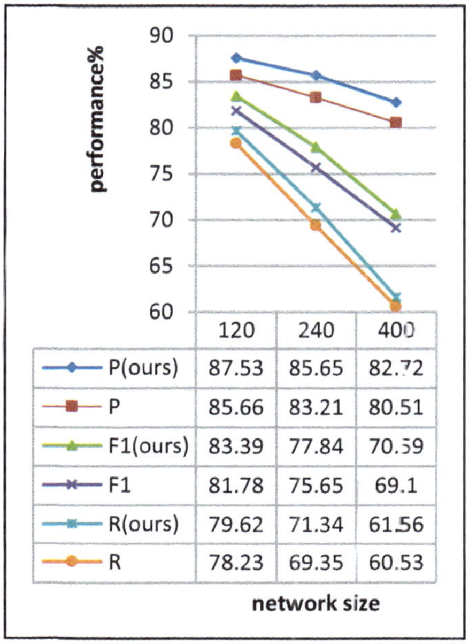

a IDMBHMT vs RPMBLR

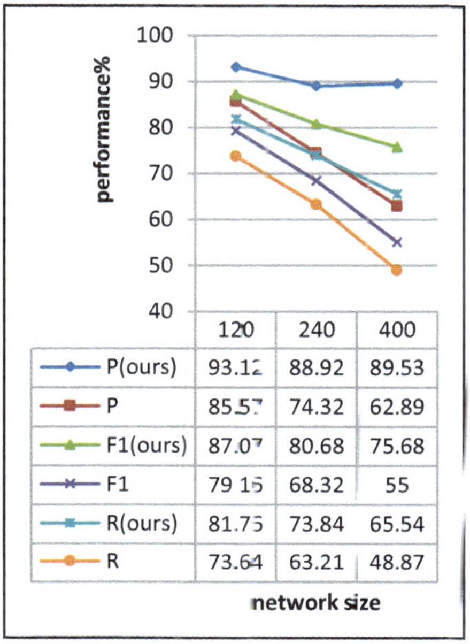

b MIDMBCRF vs MIDMBLT

Fig. 6 Precision, recall, and F1 scores of IDMBHMT and MIDMBCRF performance measurement, respectively, to the performance of RPMBLR (**a**) and MIDMBLT (**b**). P and R stand for precision and recall. In **a**, P(ours), F1(ours), and R(ours) are measurements for IDMBHMT, while the rest are for RPMBLR. In **b**, P(ours), F1(ours), and R(ours) are measurements for MIDMBCRF while the rest are for MIDMBLT

where α_i denotes the attractiveness of information i, and $b_{u,v}$ denotes the impact of node u on v. We use the proportion of the class that contains information i in the entire classes to represent α_i and set the specific threshold to 0.5.

By examining Fig. 6a, we learn that the performance of IDMBHMT without graph partitioning is higher than RPMBLR on average 2% under different network sizes (120, 240, and 400 users). The reason for the small differences in performance is that both of the two models considered the contents and network structure, and they are also based on a certain assumption of independence. The difference is that IDMBHMT takes user impact into account except for the contents and network structure, and parameter estimation on IDMBHMT is more direct, while parameter estimation process of RPMBLR is more indirect and cumbersome. Thus, the indirectly obtained parameter values make some negative impact on the performance of RPMBLR, and users' own features also make some contribution to the performance of IDMBHMT.

From the analysis of Fig. 6b, we can find that the performance of MIDMBCRF without graph partitioning is better than MIDMBLT on average 13% under the same settings in Fig. 6a. Moreover, the drop slope of the curves for MIDMBCRF is much gentler than the ones for MIDMBLT, meaning that degrading speed of the performance of MIDMBCRF is relatively slower with the increasing of network size, and its better performance will become more and more evident.

Based on the above analysis, when the network size is at 120, 240, and 400, the performances of the four models in a descending order are as follows: MIDMBCRF, IDMBHMT, RPMBLR, and MIDMBLT.

5 Conclusion

We used HMT and CRF to model and predict users reposting decision patterns in microblogging network through considering three types of main factors, i.e., information contents, users influence, and network structure. To improve the effectiveness and efficiency of models, we also investigated graph partitioning algorithm. The performances of the proposed models IDMBHMT and MIDMBCRF are evaluated by analyzing binary reposting decisions on 5148 sample users and 772,200 pieces of messages. The experimental results enable us to draw the conclusions as (1) interactions between information pieces make an important contribution to MIDMBCRF; (2) partitioning original microblogging network can significantly improve the prediction accuracy of models; and (3) both IDMBHMT and MIDMBCRF outperform the reference models of RPMBLR and MIDMBLT.

Acknowledgments This work is partially supported by Department of Education of Guangdong under Special Innovation Program (Natural Science) with Project No. 2015KTSCX183.

References

1. T.C. Schelling, *Micromotives and Macrobehavior* (Norton, New York, 1978)
2. J. Golderberg, B. Libai, E. Muller, Using complex systems analysis to advance marketing theory development. Acad. Mark. Sci. Rev. **2001**(9), 1–19 (2001). Available: http://www.amsreview.org/articles/goldenberg09-2001.pdf
3. F. Zhang, G. Si, P. Luo, A survey for rumor propagation models. Complex Syst. Complex. Sci. **6**(4), 1–11 (2009)
4. S. Sun, J. Wu, Z. Xuan, Knowledge diffusion on networks through the game strategy. MCDM **2009**, 282–289 (2009)
5. J. Yang, Scott counts, predicting the speed, scale, and range of information diffusion in twitter, in *Proceedings of the 4th International AAAI Conference on Weblogs and Social Media, ICWSM*, vol. 2010, (2010), pp. 355–358
6. L. Zheng, S. Li, A novel information diffusion model based on micro-blog network. Commun. Technol. **2**(45), 39–41 (2012)
7. M.G. de Bayser, A.P. Appel, C.N.D. Santos, et al., A simulation-based approach to analyze the information diffusion in micro-blogging online social network, in *Proceedings of the IEEE International on Simulation*, (2013), pp. 1685–1696
8. A. Cuilile, H. Hacid, A predictive model for the temporal dynamics of information diffusion in online social networks, in *Proceedings of the 21st International Conference Companion on World Wide Web*, (ACM Press, New York, 2012), pp. 1145–1152

9. B. Karrer, M. Newman, Competing epidemics on complex networks. Phys. Rev. E (2011)
10. S. Goyal, M. Kearns, Competitive contagion in networks. STOC (2012)
11. H. Li, *Statistical Learning Methods* (Tsinghua University Press, Beijing, 2012), pp. 181–184
12. X. Wang, X. Fan, J. Zhao, et al., J. Comput. Appl. **29**(3), 843–845 (2009)
13. Z. Xu, D. Li, T. Liu, et al., Measuring similarity between microblog users and its application. Chinese J. Comput. **37**(1), 207–218 (2014)
14. A. Seth, Myers and Jure Leskovec, clash of the contagions: Cooperation and competition in information diffusion. ICDM **2012** (2012)
15. H. Byrd, J. Nocedal, R.B. Schnabel, Representations of quasi-Newton matrices and their use in limited memory methods, in *Math Program*, (1994), pp. 129–156
16. S. Kang, C. Zhang, Z. Lin, et al., Complexity research of massively microblogging based on human behaviors, in *Proceeding of the 2nd International Workshop on Database Technology and Applications (DBTA)*, (Wuhan, China, 2010), pp. 1–4
17. P. Li, Z. Jiang, W. Li, H. Wang, Measurement and analysis of topology and information propagation on sina microblog, in *Proceedings of the IEEE International Conference on Intelligence and Security Informatics (ISI)*, (IEEE Computer Society, Washington, DC, 2011), pp. 396–401
18. J. Zhu, F. Xiong, D. Piao, et al., Statistically modeling the effectiveness of disaster information in social media, in *Proceeding of the IEEE Global Humanitarian Tecnology Conference*, (IEEE Computer Society, Washington, DC, 2011), pp. 431–436
19. L. Zheng, *The Research on Information Diffusion Modeling Over Social Networks*. Dissertation, (Shanghai Jiao Tong University, 2011)

ISLSTM: An Intelligent Scheduling Algorithm for Internet of Things

Fred Wu, Jonathan Musselwhite, Shaofei Lu, Raj Vijeshbhai Patel, Qinwen Zuo, and Sweya Reddy Dava

1 Introduction

Internet of Things (IoT) is a combination of embedded technologies including wired and wireless communications, sensor and actuator devices, and the physical objects connected to the Internet. Emerging technologies in recent years, and major enhancements to Internet protocols and computing systems, have made communication between different devices easier than ever before. According to various forecasts, around 25–50 billion devices are expected to be connected to the Internet by 2020 [1–3]. The purpose of IoT is to develop a smarter environment and a simplified lifestyle by saving time, energy, and money. Through this technology, expenses in different industries can be reduced. The enormous investments and many studies running on IoT have made IoT a growing trend in recent years. IoT consists of a set of connected devices that can transfer data among one another in order to optimize their performance; these actions occur automatically and without human awareness or input. IoT includes four main components: (1) sensors, (2) processing networks, (3) data analysis, and (4) system monitoring.

F. Wu (✉) · J. Musselwhite · R. V. Patel · S. R. Dava
Department of Mathematics and Computer and Science, West Virginia State University, Institute, WV, USA
e-mail: heng.wu@wvstateu.edu; jmusselwhite@wvstateu.edu; rpatel6@wvstateu.edu; sdava@wvstateu.edu

S. Lu (✉)
College of Electrical and Information Engineering, Hunan University, Changsha, China
e-mail: sflu@hnu.edu.cn

Q. Zuo
State Key Laboratory of NBC, Protection for Civilian, Beijing Institute of Chemical Defense, Beijing, China

K. Daimi et al. (eds.), *Advances in Security, Networks, and Internet of Things*, Transactions on Computational Science and Computational Intelligence,
https://doi.org/10.1007/978-3-030-71017-0_46

The development of the Internet is accompanied by the emergence and development of the IoT. The early stages of the Internet were characterized by WWW (World Wide Web) with linked static HTML (Hyper Text Markup Language) documents. This concept evolved to Web 2.0 which enabled user interaction through social networks, forums, blogs, e-learning platforms, CMS (Content Management Systems), etc. The next step in the evolution of the Internet is referred to as Web 3.0 or the Semantic Web. The main goal of Web 3.0 is to make Web content and services understandable by devices without human involvement. IoT takes Web 3.0 to a new level by enabling seamless connectivity anytime and anywhere by anyone and anything. It enables the creation of novel value-added services by dynamically assembling different types of capabilities (sensing, communication, data processing, actuation, etc.) [4, 5].

IoT can be considered a global network infrastructure composed of numerous connected devices that rely on sensors, communication, networking, and information processing technologies [6]. A foundational technology for IoT is RFID technology, which allows microchips to transmit identification information to a reader through wireless communication. By using RFID readers, people can identify, track, and monitor any objects attached with RFID tags automatically [7]. RFID has been widely used in logistics, pharmaceutical production, retailing, and supply chain management since the 1980s [8, 9]. Another foundational technology for IoT is wireless sensor networks (WSNs), which mainly use interconnected intelligent sensors to monitor and sense properties of their targets. Its applications include environmental monitoring, health-care monitoring, industrial monitoring, traffic monitoring, and so on [10, 11]. The advances in both RFID and WSN significantly contribute to the development of IoT. In addition, many other technologies and devices such as barcodes, smartphones, social networks, and cloud computing are being used to form an extensive network for supporting IoT [12–18].

As the IoT deploys a considerable number of sensors, the generated data is also extremely large. It is unacceptable for these data to be transmitted directly to cloud servers without any compression or processing. The massive data will consume immense network bandwidth and lead to a number of issues, such as transmission delay and packet loss. Thus, it is necessary for IoT gateways to perform data preprocessing and even aggregation before forwarding them to remote cloud servers. The challenge, then, is to control the traffic flow by optimally migrating data processing and aggregation tasks to reduce the bandwidth requirements of the end users while maintaining the quality of data.

There have been a number of research efforts devoted on this issue. For example, Abdelwahab et al. in [19] proposed an LTE-aware edge cloud architecture and an LTE-optimized memory replication protocol, called REPLISOM. The designed protocol can effectively schedule the memory replication operations. In this way, contentions among radio resources from devices accessing the resources simultaneously can be addressed. Sajjad et al. in [20] proposed a scheme to unify stream processing across the central and the near-the-edge data centers, which is called SpanEdge. With this scheme, the stream processing applications can be optimally deployed in a geo-distributed infrastructure so that bandwidth consumption and response latency can be significantly reduced.

In addition, Zhang et al. in [21] designed a mobile edge computing off-loading framework in cloud-enabled vehicular networks. In this study, a contract-based computation resource allocation scheme is designed. With this scheme, the utility of MEC service providers can be maximized and the off-loading requirements of the tasks can be satisfied, leading to the reduction of latency and transmission costs of computation off-loading. Nunna et al. in [22] proposed a real-time context-aware ad hoc collaboration system which combines the novel communication architectures for 5G with the principles of mobile edge computing. Thus, it can be used in geographically bound low latency use cases. Papageorgiou et al. in [23] proposed a stream processing framework extension, which considers topology-external interactions (interactions with databases, users, critical actuators, and more). With this solution, the latency requirements violations can be eliminated, and cloud-to-edge bandwidth consumption can be reduced [24].

Deep learning has been widely used in data post-processing of the IoT. Deep learning has recently been highly successful in machine learning across a variety of application domains, including computer vision, natural language processing, and big data analysis, among others [25, 26]. For example, deep learning methods have consistently outperformed traditional methods for object recognition and detection in the ISLVRC Computer Vision Competition since 2012 [27]. However, deep learning's high accuracy comes at the expense of high computational and memory requirements for both the training and inference phases of deep learning. Training a deep learning model is space and computationally expensive due to millions of parameters that need to be iteratively refined over multiple time periods. Inference is computationally expensive due to the potentially high dimensionality of the input data (e.g., a high-resolution image) and millions of computations that need to be performed on the input data. High accuracy and high resource consumption are defining characteristics of deep learning [28].

To meet the computational requirements of deep learning, a common approach is to leverage edge/cloud computing. To use edge/cloud resources, data must be moved from the data source location on the network edge (e.g., from smartphones and IoT sensors) to a centralized location in edge/cloud computing resources.

This paper is organized as follows. We first provide a brief background on IoT and deep learning (see Sect. 1). We then describe our scheduling algorithm for IoT where deep learning is used to predict the location and performance. In Sect. 3, we show the test data for IoT. Finally, we finish with discussion and future work (see Sect. 4).

2 Methodology

2.1 Architecture

The architectures of IoT are needed to represent, organize, and structure the IoT in a way that enables it to function effectively. In particular, the distributed, heterogeneous nature of the IoT requires the application of hardware, network, software, and process architectures capable of supporting these devices, their services, and the workflows they will affect [29]. The IoT should be capable of interconnecting billions or trillions of heterogeneous objects through the Internet, so there is a critical need for a flexible, layered architecture. The ever-increasing number of proposed architectures has not yet converged to a reference model [30, 31]. From the pool of proposed models, the basic model is a three-layer architecture, consisting of the Application, Network, and Perception layers. Figure 1 illustrates some common architectures. Among them is the five-layer model (not to be confused with the TCP/IP layers) which has been used in [32–34].

The intelligent scheduling algorithm locates at the Network layer in (a), the Coordination layer in (b), and the Service Management layers in (c) and (d).

In Fig. 2, the data collected by the sensors are preprocessed by an embedded program and will be send to either an edge computing or cloud computing resource. That choice is determined by ISLSTM (Intelligent Scheduler by Long-Short Term Memory). The intelligent scheduling algorithm chooses an available server to receive the task and data according to the location of the sensor and the status of the server. Different tasks have different requirements for the computing system. As the sensor position is changing or moving, the algorithm selects the appropriate server

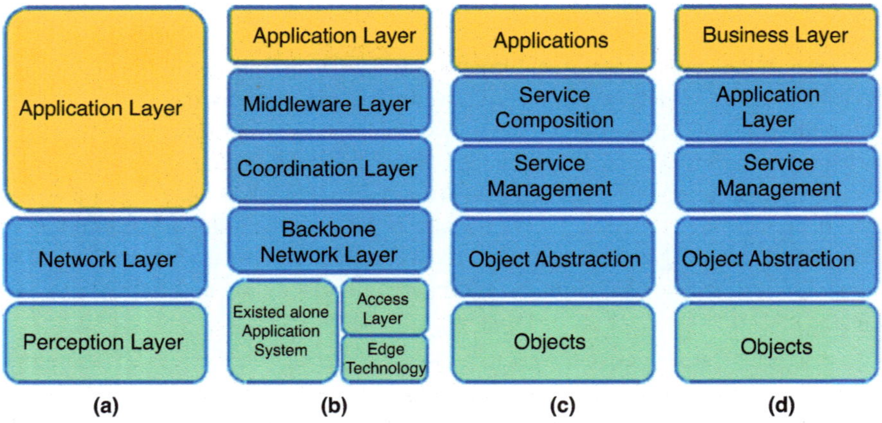

Fig. 1 The IoT architecture. (**a**) Three-layer. (**b**) Middle-ware based. (**c**) SOA based. (**d**) Five-layer

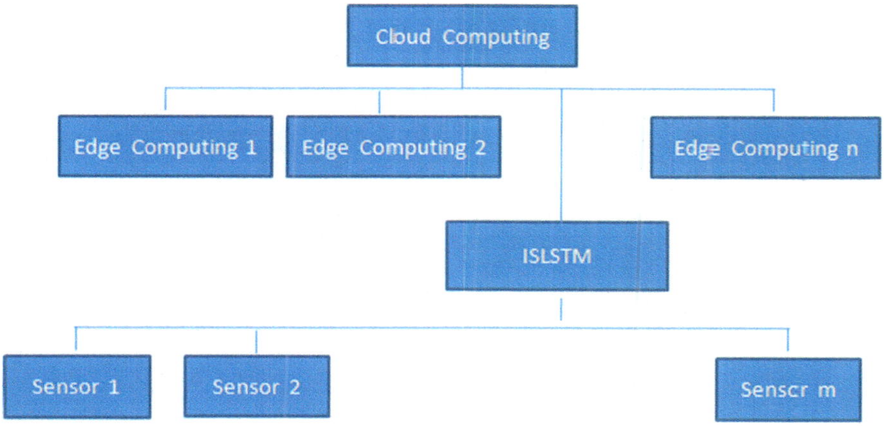

Fig. 2 The IoT architecture with ISLSTM

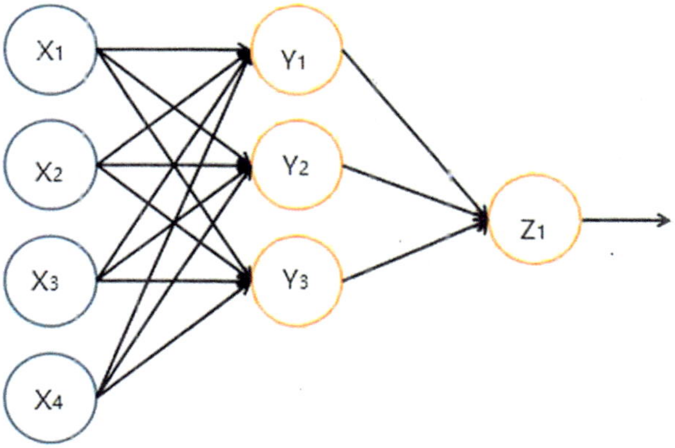

Fig. 3 The structure of neural network

in real time to accept data and tasks. This can minimize communication congestion and unfair assignment of tasks.

2.2 *Long-Short Term Memory*

As shown in Fig. 3, a neural network is the connection of many single neurons such that an output of a neuron can be an input of another neuron. Each single neuron has an activation function. The left layer of the neural network is called the input layer. It includes X1,X2,X3,X4. The right layer is the output layer, involving Z1. The other layer is a hidden layer, including Y1,Y2,Y3.

Recurrent neural network (RNN) is a typical class of neural network, as shown in the leftmost part of Fig. 4.

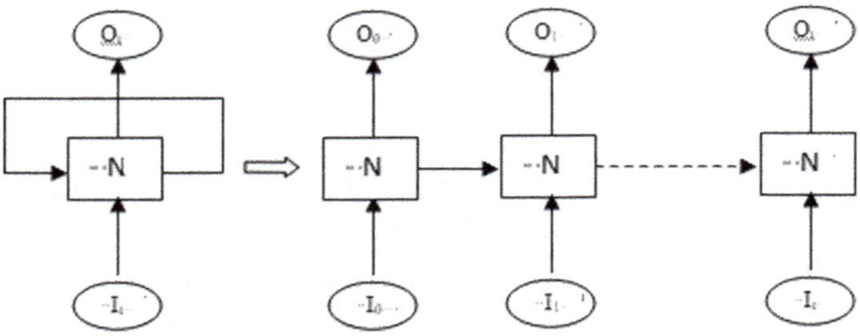

Fig. 4 The structure of recurrent neural network and its unfolding

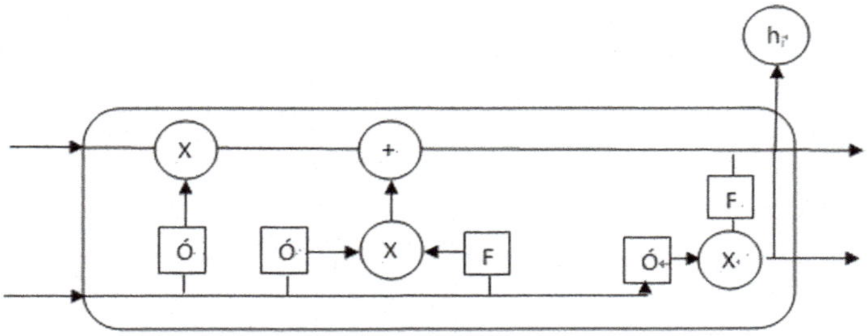

Fig. 5 The structure of an LSTM node

Like the leftmost of Fig. 4, RNN is a neutral network containing loops. N is a node of neural network. I represents input and O represents output. Loops allow information to be transmitted from the current step to the next step. RNN can be regarded as a multiple assignment of the same neural network, and each neural network module transmits the message to the next one. The right of Fig. 4 is the unfolding of the structure to the left of it. The chain feature of RNN reveals that RNN is essentially related to sequences and lists. RNN applications have been successful in speech recognition, language modeling, translation, and picture description, and this list is still growing. One of the key features of RNN is that it can be used to transmit the previous information to the current task.

LSTM overcomes this shortcoming. LSTM is a special type of RNN. LSTM solves the problem of long-term dependence of information. LSTM avoids long-term dependencies through deliberate design. Figure 5 shows the structure of a node of LSTM.

There is a forget gate in Fig. 5. The output of the forget gate is "1" or "0." "1" means full reserve, and "0" is abandon completely. The forget gate determines which data will be retained and which others will be abandoned. The upper horizontal line allows the input information to cross the neutral node without changing in Fig. 3. There are two gates (input and output gates) in LSTM. The middle gate is an input gate in Fig. 3, which determines what data will be saved in the neutral node. F means function modular and create a new candidate value vector. The right gate is

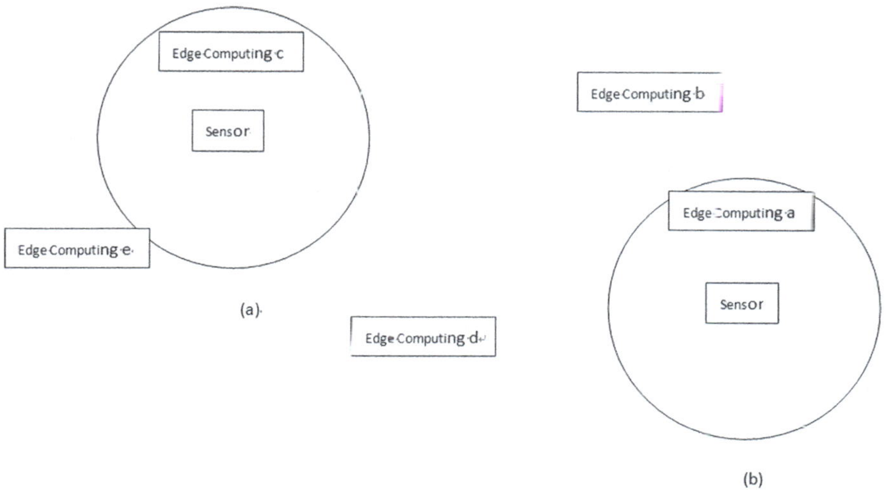

Fig. 6 The moving sensor and how to choose the server at the current time (**a**) and future time (**b**)

the output gate. The F module close to the output gate determines which information of the neutral node will be transmit to the output gate.

A node has three gates and a cell unit in Figs. 5 and 6. The gates use sigmoid as activation function. The tanh function is used to transfer from input to cell states. The following are to define a node [26].

For the gates, the functions are

$$i_t = g\left(W_{xi}x_t + W_{hi}h_{t-1} + b_i\right)$$

$$f_t = g\left(W_{xf}x_t + W_{hf}h_{t-1} + b_f\right)$$

$$O_t = g\left(W_{xo}x_t + W_{ho}h_{t-1} + b_o\right)$$

The transfer for input status is

$$c\_\text{in}_t = \tanh\left(W_{xc}x_t + W_{hc}h_{t-1} + b_{o\_\text{in}}\right)$$

The status is updated by

$$c_t = f_t c_{t-1} + i_t c\_\text{in}_t$$

$$h_t = o_t \tanh\left(c_t\right)$$

2.3 ISLSTM: The Intelligent Scheduling Algorithm

ISLSTM helps a sensor to determine a target server according to its location and the usage of nearby servers. This process begins with Algorithm 1.

Algorithm 1

```
Input: the location of sensor, the data size, task size
   Output: the IP address of Edging/Cloud Computing
   Beginning of Algorithm
   if counter less than the training size
      train the LSTM model 1
      get the usages of CPU and memory from servers
      train the LSTM model 2 and 3
      choose one server with lower usage of CPU and memory
      return
   endif
   predict the location of the sensor at the next time
   predict the future usage of memory and CPU of servers.
   calculate the weight according to the distance from servers to the sensor, the
usages of memory, and CPU.
   choose the lowest weight server and return its IP address
   End of Algorithm
```

The second algorithm (Algorithm 2) is to calculate the weight of servers.

Algorithm 2

```
Input: the location of a sensor
   Output: the weight of servers
   Beginning of Algorithm
      using the position of the sensor as the center, find the five nearest edge
computing/cloud computing servers
      get their CPU and memory usage
      get their IP address
      for each server
         weigh = α*the usage of memory + β* the usage of CPU + γ* distance/R
      return the lowest weigh and its IP address
      end for
   End of Algorithm
```

In Algorithm 2, the coefficients α, β, and γ. Usually, $\alpha = 0.2$, $\beta = 0.8$, and $\gamma = 1$, R is an input parameter. The first LSTM model predicts the location of the sensor after the task is completed; the second one predicts the CPU usage of a server, and the third predicts the memory usage of a server.

In Fig. 6, the sensor chooses the nearest edge computing resource – edge computing c – and uploads the data and tasks to it. The ISLSTM predicts edge computing a will be next optimal server in the future according to its predicted location and the memory and CPU usage of servers and will upload the new data to edge computing a and download the results from it.

Parameter selection is another hard problem. According to the 20/80 rule, 80% of system performance is determined by CPU usage and 20% by memory usage. They can be adjusted individually according to the actual situation. The choice of R is based on the distance the sensor moves in unit time. α, β, γ, and K will be adjusted according to different usage environments.

ISLSTM predicts the next server at the current moment, which is edge computing a. Edge computing c will migrate the task and data of the sensor to the new server at the next moment. This greatly reduces communication and computing bottlenecks. The edge computing a has predicted the next optimal server when the moving sensor reaches a new position.

3 Experiment and Test

In order to check the accuracy and feasibility of those algorithms, we have some dataset to test. For moving sensors, we use a public GPS trajectory dataset – UCI Machine Learning Reposition (https://archive.ics.uci.edu/ml/datasets/GPS+Trajectories). In the dataset, localization points of each trajectory are as follows: id: unique key to identify each point; latitude: latitude from where the point is; longitude: longitude from where the point is; track_id: identify the trajectory which the point belong; time: datetime when the point was collected (GMT-3). For the usages of memory and CPU, we also test on the public cloud platform.

Figure 7 is the comparison of training and prediction of longitude data, and Fig. 8 is for latitude data. We use 1000 locations to demonstrate the comparison. The training size is 667, and the prediction size is 333. The X-axis is the sequence number, and the Y-axis is the longitude or latitude value. The yellow part is the training data, and green is the prediction part. We chose the 800–810 data to show

Fig. 7 The comparison of training and prediction of longitude data

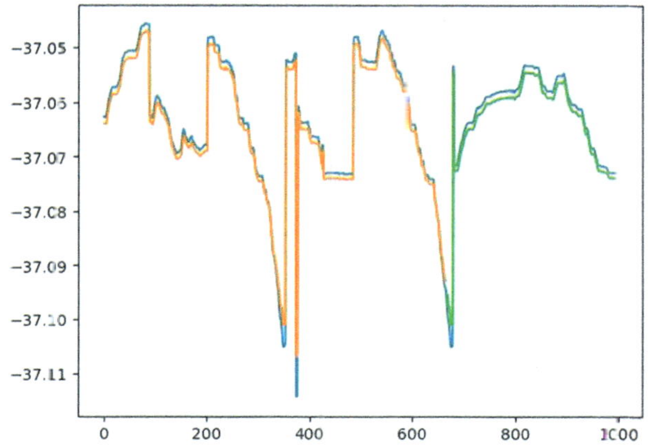

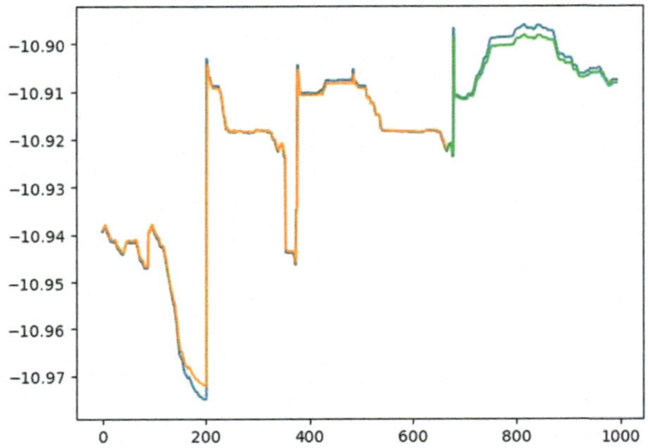

Fig. 8 The comparison of training and prediction of latitude data

Table 1 The comparison of real and predicted longitude values

Predicted values (*P*)	Real values (*R*)	(| *P*–*R* |/*R*)*100%
−37.0592	−37.05778	0.0038%
−37.059086	−37.057728	0.0036%
−37.059036	−37.05764	0.0038%
−37.05895	−37.05759	0.0037%
−37.0589	−37.0576	0.0035%
−37.058907	−37.057583	0.0036%
−37.05889	−37.05737	0.0041%
−37.05868	−37.05693	0.0047%
−37.05825	−37.05657	0.0045%
−37.057884	−37.05639	0.004%
−37.05771	−37.055973	0.0047%

Table 2 The comparison of real and predicted longitude values

Predicted values (*P*)	Real values (*R*)	(| *P*–*R* |/*R*)*100%
−10.899271	−10.897177	0.019%
−10.899141	−10.897079	0.018%
−10.899058	−10.89701	0.019%
−10.899	−10.897009	0.018%
−10.898999	−10.896991	0.018%
−10.898985	−10.896977	0.018%
−10.8989725	−10.896946	0.019%
−10.898945	−10.89683	0.019%
−10.898848	−10.896812	0.019%
−10.898833	−10.896709	0.019%
−10.8987465	−10.8966255	0.019%

in Tables 1 and 2. The left column is predicted data and followed the real data in the center column. The right column is the absolute value of the difference of the prediction value minus the real data, divided by the real data.

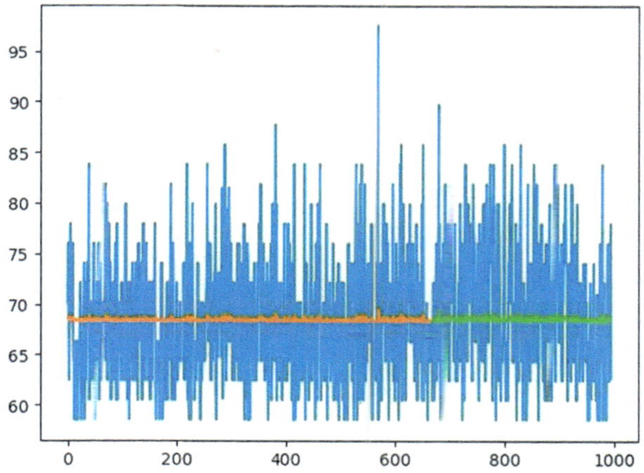

Fig. 9 The comparison of training and prediction of CPU usage

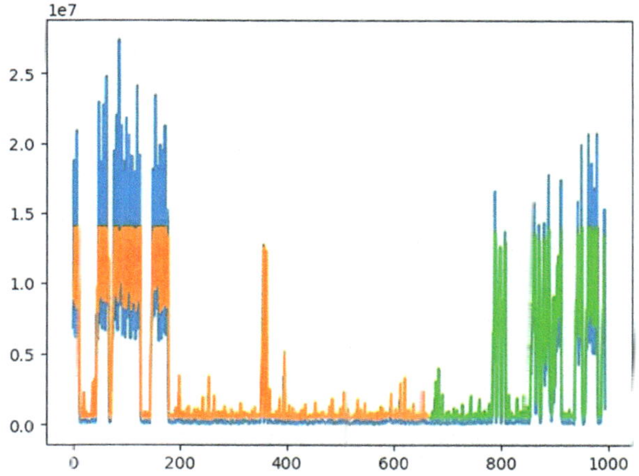

Fig. 10 The comparison of training and prediction of memory longitude values

Figure 9 is the comparison of training and prediction CPU usage, while Fig. 10 is for the memory usage. Yellow part is the training part, and green parts are the prediction ones.

The test sample is also 1000, the X-axis is the sequence number, and the Y-axis is the CPU usage in Fig. 9 and memory usage in Fig. 10. The training size is 667, and prediction size is 333.

In Tables 3 and 4 we use the first column to represent the predicted value, the second is real data, and the third is the absolute value of the difference of the prediction value minus the real data, divided by the real data.

Table 3 The comparison of real and predicted CPU usage values

Predicted values (*P*)	Real values (*R*)	(| *P*–*R* |/*R*)*100%
69.037346	70.22399	1.7%
68.52498	66.322655	3.3%
68.423676	60.470657	9.8%
68.29251	62.421326	9.4%
68.33343	66.322655	3%
68.423676	64.37199	6%
68.37716	60.470657	13%
68.29251	74.12532	7.8%
68.63715	74.12532	7.4%
68.63715	79.97732	14%
68.82548	70.22399	2%

Table 4 The comparison of real and predicted memory usage values

Predicted values (*P*)	Real values (*R*)	| *P*–*R* |/*R*
3421915.8	134217.6	24.8
503876.2	134217.6	2.75
503876.2	5055532.5	0.9
7035157.4	6621406.5	0.06
8628723.6	6845102.5	0.27
8835648.5	6487189.7	0.36
8502097.8	11632202.8	0.27
12113135.4	13734945.5	0.118
12957084.1	2863309.2	3.52
4385093.6	89478.4	48
436169.1	805305.6	10.39

4 Discussion and Conclusion

This paper proposed the intelligent scheduling algorithm for Internet of Things. The new algorithm borrows four LSTM model to predict the location of moving sensor, CPU, and memory usages, respectively, in next moment. ISLSTM chose a nearest and lightest workload server to process the task of the sensor.

The test data show the accuracy of position prediction can reach 99.99%, while which of performance is not as good as the location. For the CPU usage, close 90% of memory usage is very bad. Therefore, we do not need to use it. Fortunately, each server has a large amount of memory.

For the IoT where billions of devices are connected, the closest distance transmission can greatly alleviate network congestion; choosing a server with a lighter workload can also alleviate the server's computing bottleneck. This is a clear advantage of this algorithm proposed in here.

Next, we continue to improve the performance of the ISLSTM, replacing LSTM with reinforcement learning that can reduce training time and size. In addition, we will combine ISLSTM and process/task migration.

Acknowledgments This work is supported by the HPC and the Department of Mathematics and Computer Science of West Virginia State University. This work is also partially supported by the Industrial Internet Innovation and Development Project of China: Digital twin system for automobile welding and casting production lines and its application demonstration (TC9084DY).

References

1. L. Atzori, A. Iera, G. Morabito, The internet of things: A survey. Comput. Netw. **54**(15), 2787–2805 (2010)
2. C. Cecchinel, M. Jimenez, S. Mosser, M. Riveill, An architecture to support the collection of big data in the internet of things, in *2014 IEEE World Congress on Services*, IEEE, 2014, pp. 442–449
3. M.S. Mahdavinejad, M. Rezvan, M. Barekatain, P. Adibi, P. Barnaghi, A.P. Sheth, Machine learning for internet of things data analysis: A survey. Digit. Commun. Netw. **4**(3), 161–175 (2018)
4. A. Colakovic, M. Hadžialic, Internet of things (IoT): A review of enabling technologies, challenges, and open research issues. Comput. Netw. **144**, 17–39 (2008)
5. F. Paganelli, D. Parlanti, A DHT-based discovery service for the Internet of Things. J. Comput. Netw. Commun. (2012)
6. L. Tan, N. Wang, Future internet: The internet of things, in *Proceedings of the 3rd International Conference on Advanced Computer Theory and Engineering (ICACTE)*, Chengdu, China, 20–22 Aug 2010, pp. V5-376–V5-380
7. X. Jia, O. Feng, T. Fan, Q. Lei, RFID technology and its applications in internet of things (IoT), in *Proceedings o the 2nd IEEE International Conference on Consumer Electronics, Communications and Networks (CECNet)*, Yichang, China, 21–23 Apr 2012, pp. 1282–1285
8. C. Sun, Application of RFID technology for logistics on internet of things. AASRI Procedia **1**, 106–111 (2012)
9. E.W.T. Ngai, K.K. Moon, F.J. Riggins, C.Y. Yi, RFID research: An academic literature review (1995-2005) and future research directions. Int. J. Prod. Econ. **112**(2), 510–520 (2008)
10. S. Li, L. Xu, X. Wang, Compressed sensing signal and data acquisition in wireless sensor networks and internet of things. IEEE Trans. Industr. Informat. **9**(4), 2177–2186 (2013)
11. W. He, L. Xu, Integration of distributed enterprise applications: A survey. IEEE Trans. Industr. Informat. **10**(1), 35–42 (2014)
12. D. Uckelmann, M. Harrison, F. Michahelles, An architectural approach towards the future internet of things, in *Architecting the Internet of Things*, ed. by D. Uckelmann, M. Harrison, F. Michahelles, (Springer, New York, 2011), pp. 1–24
13. S. Li, L. Xu, X. Wang, J. Wang, Integration of hybrid wireless networks in cloud services oriented enterprise information systems. Enterp. Inf. Syst. **6**(2), 165–187 (2012)
14. L. Wang, L. Xu, Z. Bi, Y. Xu, Data filtering for RFID and WSN integration. IEEE Trans. Industr. Informat. **10**(1), 408–418 (2014)
15. L. Ren, L. Zhang, F. Tao, X. Zhang, Y. Luo, Y. Zhang, A methodology towards virtualization-based high performance simulation platform supporting multidisciplinary design of complex products. Enterp. Inf. Syst. **6**(3), 267–290 (2012)
16. F. Tao, Y. Laili, L. Xu, L. Zhang, FC-PACO-RM: A parallel method for service composition optimal-selection in cloud manufacturing system. IEEE Trans. Industr Informat. **9**(4), 2023–2033 (2013)
17. Q. Li, Z. Wang, W. Li, J. Li, C. Wang, R. Du, Applications integration in a hybrid cloud computing environment: Modelling and platform. Enterp. Inf. Syst. **7**(3), 237–271 (2013)
18. L. Da Xu, W. He, S. Li, Internet of things in industries: A survey. IEEE Trans. Industr. Inform **10**(4), 2233–2243 (2014)

19. S. Abdelwahab, B. Hamdaoui, M. Guizani, T. Znati, REPLISOM: Disciplined tiny memory replication for massive IoT devices in LTE edge cloud. IEEE Internet Things J. **3**(3), 327–338 (2016)
20. H.P. Sajjad, K. Danniswara, A. Al-Shishtawy, V. Vlassov, SpanEdge: Towards unifying stream processing over central and near the-edge data centers, in *Proceedings of the IEEE/ACM Symposium on Edge Computing (SEC)*, Oct 2016, pp. 168–178
21. K. Zhang, Y. Mao, S. Leng, A. Vinel, Y. Zhang, Delay constrained offloading for mobile edge computing in cloud-enabled vehicular networks, in *Proceedings of the 8th International Workshop on Resilient Networks Design Modeling (RNDM)*, Sept 2016, pp. 288–294
22. S. Nunna et al., Enabling real-time context-aware collaboration through 5G and mobile edge computing, in *Proceedings of the International Conference on Advances in Information Technology - New Generation (ITNG)*, Apr 2015, pp. 601–605
23. A. Papageorgiou, E. Poormohammady, B. Cheng, Edge-computing aware deployment of stream processing tasks based on topology-external information: Model, algorithms, and a storm-based prototype, in *Proceedings of the IEEE International Congress on Big Data (BigData Congress)*, Jan 2016, pp. 259–266
24. W. Yu, F. Liang, X. He, W.G. Hatcher, C. Lu, J. Lin, X. Yang, A survey on the edge computing for the internet of things. IEEE Access **6** (2018). https://doi.org/10.1109/ACCESS.2017.2778504
25. L.U. Shao-fei, Z. Qian, W.U. Heng, A new power load forecasting model (SIndRNN): Independently recurrent neural network based on Softmax Kernel Function, HPCC-2019
26. H. Wu, S. Lu et al., Temperature prediction based on long short term memory networks, CSCI'19, *In 2019 International Conference on Computational Science and Computational Intelligence (CSCI), IEEE,2019, pp. 312–317*
27. O. Russakovsky et al., ImageNet large scale visual recognition challenge. Int. J. Comput. Vis. **115**(3), 211–252 (2015)
28. J. Chen, X. Ran, Deep learning with edge computing: A review. Proc. IEEE **107**(8), 1655–1674 (2019)
29. A. Whitmore, A. Agarwal, L. Da Xu, The internet of things– A survey of topics and trends. Inf. Syst. Front. **17**, 261–274 (2015). https://doi.org/10.1007/s10796-014-9489-2
30. S. Krco, B. Pokric, F. Carrez, Designing IoT architecture(s): A European perspective, in *Proceedings of the IEEE WF-IoT*, 2014, pp. 79–84
31. A. Al-Fuqaha, M. Guizani, M. Mohammadi, M. Aledhari, M. Ayyash, Internet of things: A survey on enabling technologies, protocols, and applications. IEEE Commun. Surv. Tutor. **17**(4), 2347–2376 (2015)
32. R. Khan, S. U. Khan, R. Zaheer, S. Khan, Future Internet: The Internet of Things architecture, possible applications and key challenges, in *Proceedings of the 10th International Conference on Frontiers of Information Technology*, 2012, pp. 257–260
33. Z. Yang et al., Study and application on the architecture and key technologies for IOT, in *Proceedings of the ICMT*, 2011, pp. 747–751
34. M. Wu, T.J. Lu, F.Y. Ling, J. Sun, H.Y. Du, Research on the architecture of Internet of Things, in *Proceedings of the 3rd ICACTE*, 2010, pp. V5-484–V5-487

The Implementation of Application for Comparison and Output of Fine Dust and Public Database Using Fine Dust Sensor

YunJung Lim

1 Introduction

As the fine dust problem has become more serious recently, the need to know the concentration of fine dust in the surrounding areas where individuals are active has also increased. Interest is rising in portable fine dust measurement sensors, which allow individuals to measure the real-time concentration of fine dust themselves. The study developed and tested a system that can check the value of portable fine dust on smartphones and Web pages. The fine dust measuring sensor value is sent to Raspberry Pi via Wi-Fi and stored in the database. The stored sensor values can be found on smartphones and Web pages. It can also provide users with data accuracy by comparing public data with fine dust concentrations. The system developed in this study turned out to be useful in life. It is also expected to reduce the damage from fine dust pollution as it can quickly recognize and cope with changes in surrounding fine dust levels using portable fine dust measurement sensors.

Health damage caused by fine dust is becoming an important social problem not only in Korea but also in the world. Superfine dust, which is less than 2.5 mm in particle size, is highly likely to cause respiratory diseases, lung cancer, and circulatory machinery problems, as well as heavy metals that are harmful to human body such as lead, arsenic, and water silver. However, there are many difficulties in coming up with effective countermeasures as fine dust sources and routes are diverse, and scientific investigation of the mechanism has yet to be carried out properly. Fine dust exists in the air in a particle state, which is classified as ultrafine dust less than 2.5 μm (PM2.5)1 in diameter and less than 10 mm (PM10) in general fine dust. The smaller the diameter of fine dust, the higher the concentration

Y. Lim (✉)
Department of R&D, Seiwoong Meditech. Cooperation, Seoul, South Korea

K. Daimi et al. (eds.), *Advances in Security, Networks, and Internet of Things*, Transactions on Computational Science and Computational Intelligence,
https://doi.org/10.1007/978-3-030-71017-0_47

of element particles such as heavy metals, and the greater the adverse effects on the human body, such as penetration into wastepaper and easier movement from the windpipe to other institutions, the greater the need for ultrafine dust and countermeasures.

According to data released in 2013 by the Cheong-wa University in China and the Health Effects Institute in the United States, 3.2 million people die early each year due to fine dust, and 76 million people see health damage. In particular, about 14.9% of China's dead population, or 1.234 million, said fine dust was the cause. In the case of Japan, it has been making efforts to reduce fine dust by drawing up comprehensive measures including domestic and external measures against fine dust since 2013. However, it has limitations. In other words, although fine dust varies depending on the region, Japan's efforts alone are difficult to effectively respond to, with more than 60% of the sources of fine dust reportedly coming from China in the Kyushu region.

This paper aims to help health by providing data provided by public institutions and measured data from actual locations to users on the Web and smartphone in real time. In this study, the fine dust sensor is connected to the MCU board with built-in Wi-Fi as shown in Fig. 1, and the fine dust measurement is sent to the MCU board in real time. This value is parsed to PHP and stored in the Raspberry Pi database. The same value is stored on cloud servers (CloudMQTT) to facilitate data access. The stored values can be parsed into PHP for real-time verification with Web pages and smartphones, enabling comparison of the fine dust concentration in the current location with public data that provides information in a wide area. Users can contribute to health improvement by utilizing this data for outdoor activities, etc.

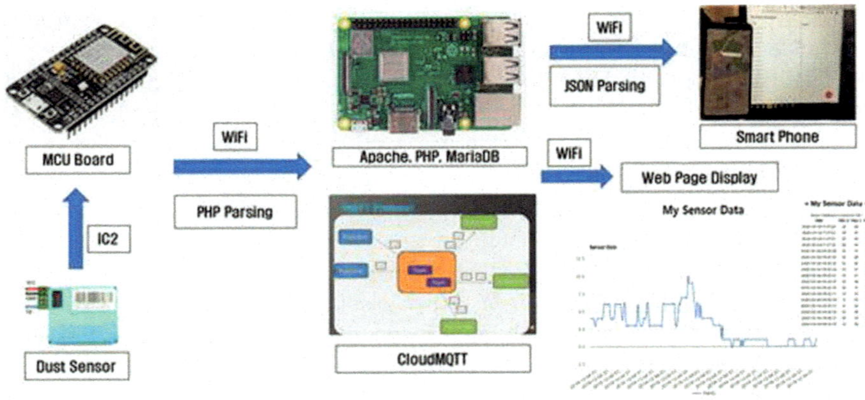

Fig. 1 System configuration

2 Theory

Although the government is making efforts to address fine dust such as implementing various fine dust reduction policies, it is difficult to fully solve fine dust problems due to secondary generation and effects of external factors as well as primary emission areas. As a result, modern people have become interested in fine dust responses and adaptations. such as cleaning up the air quality of indoor spaces and wearing yellow dust masks during outdoor activities. Also, the desire to know the concentration of fine dust indoors and outdoors has increased, increasing demand for portable fine dust sensors that can directly measure the concentration of fine dust in real time.

Figure 2 shows the source of fine dust contamination. A ubiquitous detailed air exchange light service system has been proposed to ensure that information on fine dust pollution, which changes rapidly in time and space, is available to users anytime, anywhere. However, this study also defined virtual sensors using detailed scale modeling because actual small sensors that can be used realistically are not present. According to the Korean Intellectual Property Office, patent applications for fine dust measurement technology in 2016 increased by about 11 times compared to 2013.

However, portable fine dust measurement sensors are produced and sold without a set certification standard, and the Ministry of Environment plans to seek ways to enhance the reliability of the measuring instruments, including setting standards for authentication, so that people can get access to accurate information about fine dust. In this study, the reliability of a portable fine dust measurement sensor is a common

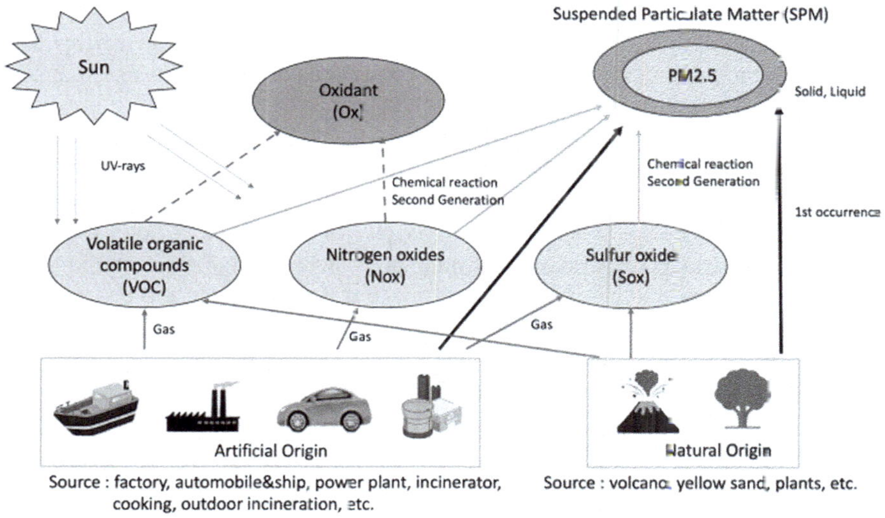

Fig. 2 Source of fine dust contamination

citizen's perspective. As a result, there was a problem to trust the absolute values of the concentrations of portable sensors as they were, but its reproducibility and linearity were found to be useful for practical use.

It is also expected that users will be able to quickly recognize and cope with the changing levels of fine dust around them by using handheld fine dust measuring sensors.

Also, the current small sensors are small, relatively inexpensive, and equipped with wireless communication functions, which can be a good alternative to overcome these limitations. Among the widely used methods of measuring fine dust are the Weight Act, the Be Thaw Measurement Act, and the Mine Ran Act, which is the Korea's certified method of measuring fine dust in the environment air, the Weight Act and the Beta Line Measurement Act.

The weight method is the most common method of refining the particle matter of the air through constant velocity sampling, and then the volume of air collected and the mass of the collected material are measured to obtain the particle concentration of the contaminant. Since this method has the advantage of measuring the weight of particulate matter directly and thus measuring the relatively accurate mass, it is recommended that the concentration of ultrafine dust be measured using the weight method for accurate observation.

However, the time required to measure is longer and more difficult to automate than other methods of measurement, and there are disadvantages of not being able to identify changes in concentration. Beta line measurement method identifies the degree of purity attenuated by the material collected by the energy generated by the beta decay of radioactive materials and converts the collected fine dust into the mass value to show the concentration. It can be measured at a relatively short time interval and is simple to measure, so it is currently being used to automatically measure fine dust in the atmosphere. The mining method measures the concentration of particles through scattered light so that scattered light can be collected into the hydroponics, and then the amount of light collected can be measured through electrical signals to measure the number and size of particles.

Based on the size and number of particles, the negative blood of the entire particle can be obtained, and the density can be corrected for this value to output the mass concentration. However, the method of measurement by light detection is sensitive to the effects of the density and humidity conditions of the dust, making it difficult to simulate the exact concentration. Despite these shortcomings, the mining method is widely used in portable fine dust meters due to its small volume and light volume. Japan is the only country that has type approval for the Mine Enclosures Act, and the US EPA (US EPA) also partially recognizes it. Through this study, we conducted a study on identifying real-time fine dust, which can contribute to understanding fine dust in real life and protecting the health of individuals in demand, which is already widely used in our daily lives.

3 Implementation

In this study, air quality is measured using fine dust sensor (PMS7003). The sensor is a laser-based fine dust measurement sensor that can measure fine dust PM10, ultrafine dust PM2.5, and ultrafine dust PM1.0 at the same time and has an air circulation FAN inside for uniform measurement.

Figure 3 shows the AirVisual data value and the MCU hardware system fine dust measurement result.

Measured values are transferred to the MCU board reliably because they are converted to digital data and output without noise effects. In this paper, sensor numbers 1 and 2 are connected to 5-volt Vin each, and pins 3 and 4 are connected to ground. RX 7 and TX 9 were then connected to MCU boards 4 (D2) and 5 (D1), respectively (Fig. 4).

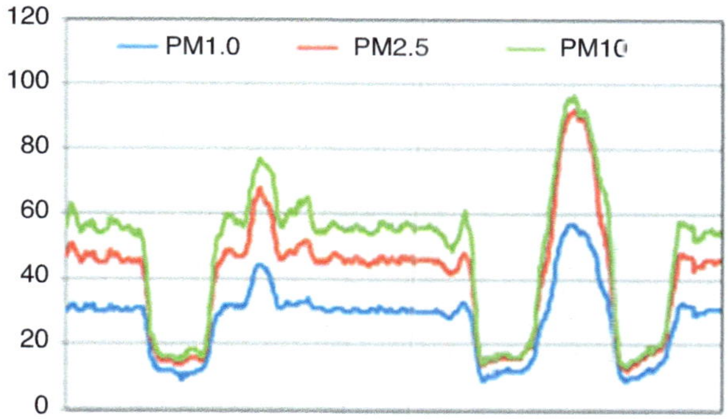

Fig. 3 Schematic diagram of this system

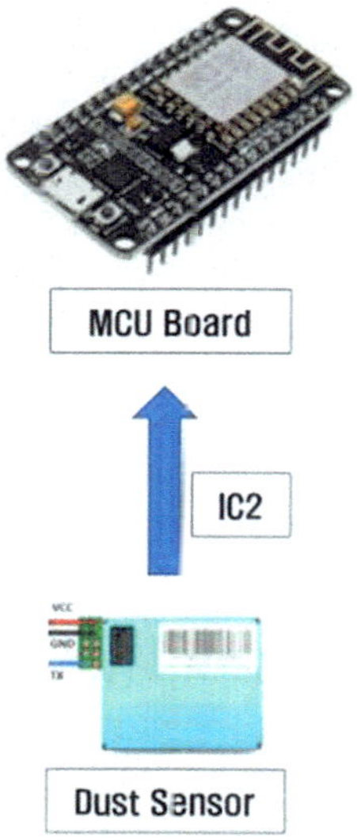

Fig. 4 MCU board and dust sensor

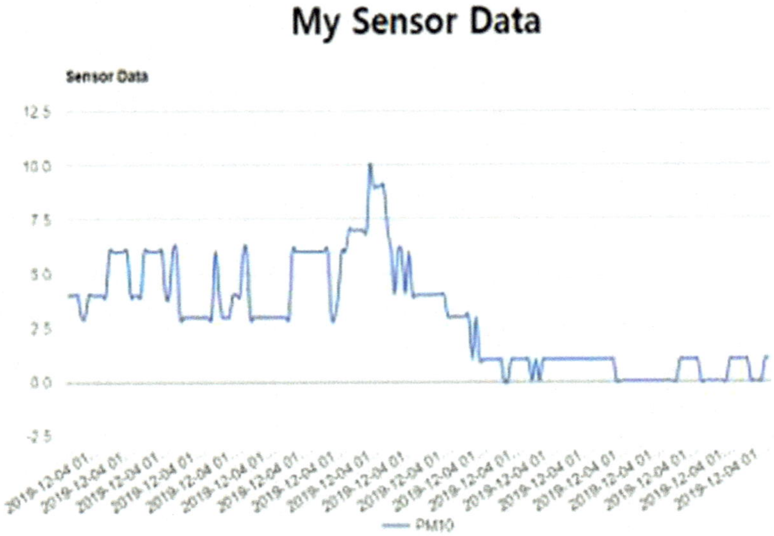

Fig. 5 Sensor data (graph)

The data measured here is transmitted over Wi-Fi to raspberry pies and stored in the database (mySQL). It is stored on a cloud server at the same time.

The values sent to the Raspberry Pi database (MySQL) are entered into the Web server in real time as PHP. Pars the data on the Web page entered into Python and saves it. Store the same value on cloud servers to make data more accessible. Figure 5 shows an example of a graph output of dust sensor values stored on a database. This data is the value measured by the sensor that is provided to the user via the Web page in real time.

Figure 6 shows an example of outputting a character data value from a dust sensor. The value of the dust sensor is measured in two seconds and output values of fine dust (PM10), ultrafine dust (PM2.5), and ultrafine dust (PM1.0) on the screen.

Next, it explains how to use Android studios to output sensor data from smartphones. Android version 9.0 was used.

Figure 7 shows a screenshot of the Android development environment used in this study.

The values stored on the server can be checked in real time on smartphone applications and Web pages, enabling comparison of the fine dust concentration in the current location with public data that provides information in a wide area.

Figure 8 shows the schematic diagram of smartphone app. The Naver Map API is JavaScript-type map flat, providing various classes and methods to implement map functions in Web services or applications, which are used to bring in Naver Map APIs and print maps on smartphone screens.

The values measured through public fine dust APIs and direct fine dust sensors are output above the current location. Naver map storage path, Naver map SDK,

Fig. 6 Sensor data (Text)

= My Sensor Data =

Sensor Database Connection OK !

TIME	PM1_0	PM2_5	PM10
2020-03-04 11:27:44	29	46	61
2020-03-04 11:27:42	29	45	55
2020-03-04 11:27:42	29	46	56
2020-03-04 11:27:40	28	44	54
2020-03-04 05:42:49	30	47	62
2020-03-04 05:42:47	30	48	63
2020-03-04 05:42:45	30	49	63
2020-03-04 05:42:43	29	47	61
2020-03-04 05:42:41	29	47	59
2020-03-04 05:42:37	30	48	61
2020-03-04 05:42:33	30	51	65
2020-03-04 05:42:31	30	53	65
2020-03-04 05:42:29	31	53	65
2020-03-04 05:42:27	31	52	63
2020-03-04 05:42:25	30	49	62
2020-03-04 05:42:21	30	45	57
2020-03-04 05:42:19	32	46	58

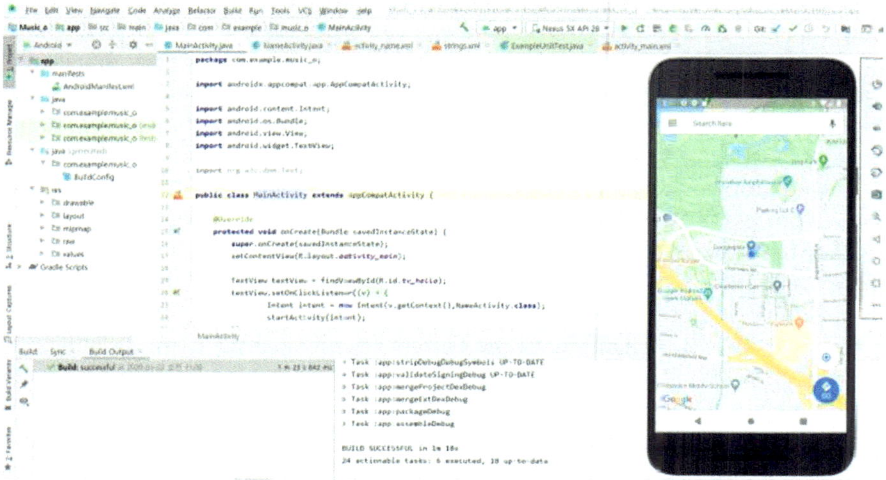

Fig. 7 Android studio

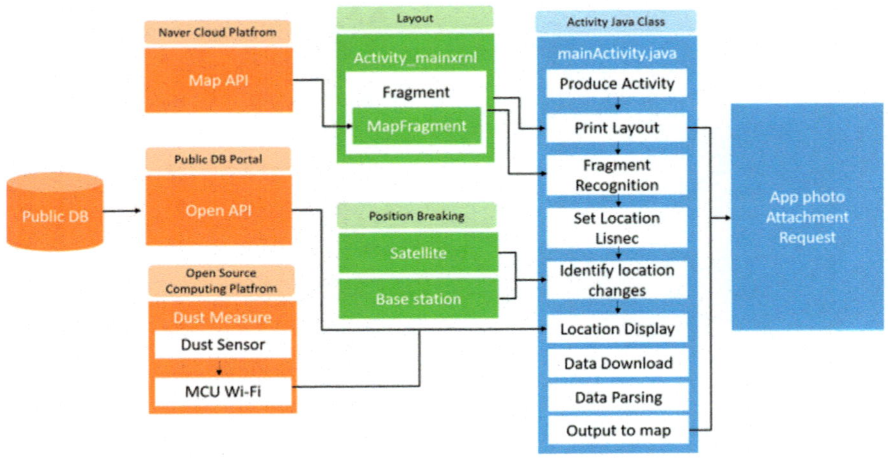

Fig. 8 Schematic diagram of smartphone apple

and API key settings are required, current location is displayed using GPS location tracking and location display, and address is printed.

Continuous monitoring of the current location identifies the changed location and, if the new location is changed, creates a new location and address mark. Public fine dust data and fine dust sensor values are printed on the position marker, and real-time change of location is called by location manager when the location changes, requiring GPS location and base station location change through GPS and NETWORK.

Figure 9 left shows the schematic of smart fine dust check. The main screen is configured through function (updateMap), and the main screen will be centered on the map, map size setting, address marking, and sensor data and public data output.

Figure 9 right is schematic diagram of this process representing the overall process flow of the application. This is the process in which the concentration of fine dust corresponding to the address mark and location is output to the sub-marker using caption after tracking the real-time location.

Figure 10 shows the flowchart of Naver map API and Naver map app screen output process. This process flowchart is printed by bringing in an open Naver map API. After project creation, maps can be printed out through the designation of Naver map storage path, the designation of Naver map SDK, and the setting of API key. GPS positioning is shown in Fig. 11.

This process flowchart shows the current location, displays the location, and then outputs the address. The changed position is identified through continuous monitoring of the current position. A new location also creates a new location and address mark. The flowchart of the fine dust API and measurement sensor value output is shown in Fig. 12.

This process flowchart outputs measured values through public fine dust APIs and direct fine dust sensors above the current location. Public fine dust DBs are

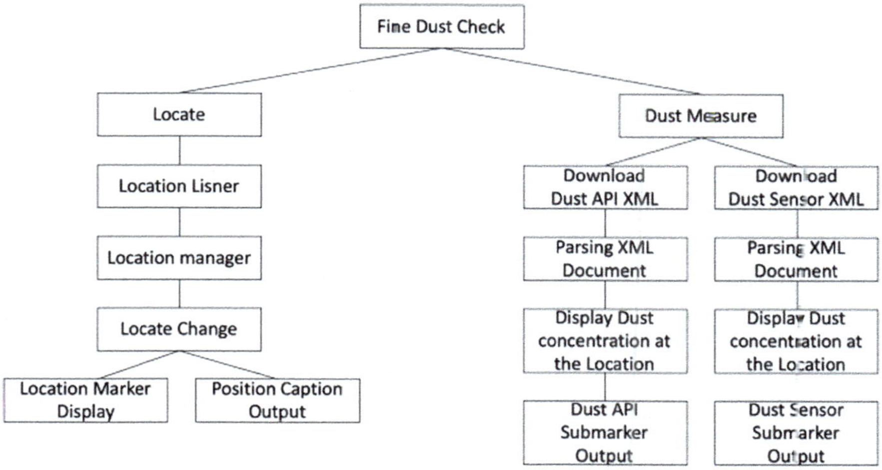

Fig. 9 Schematic diagram of smart fine dust check

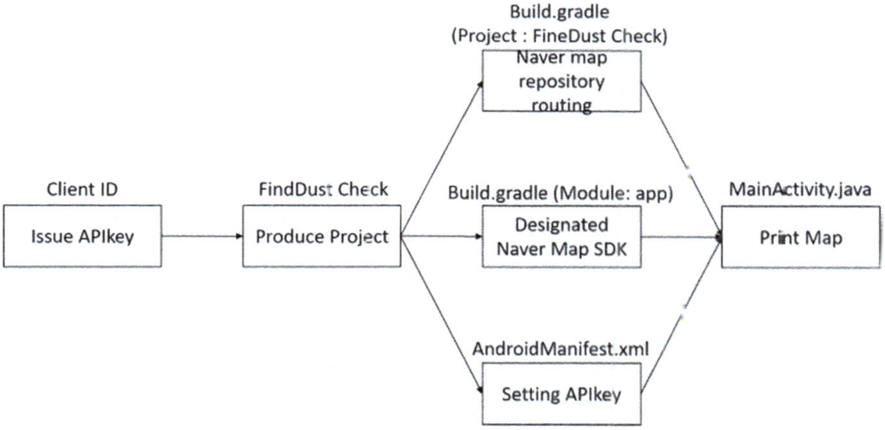

Fig. 10 Schematic diagram of smart fine dust check

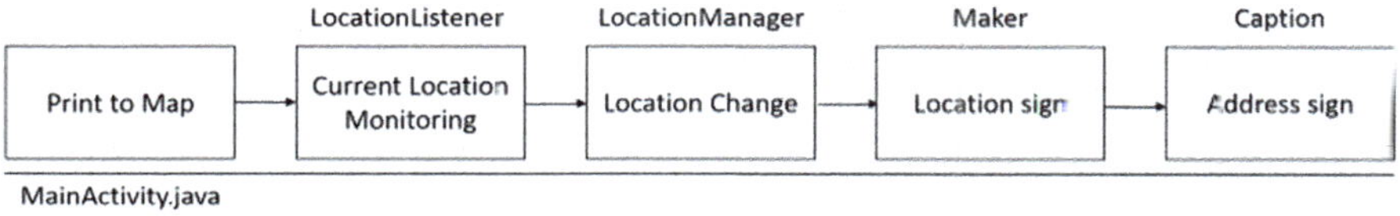

Fig. 11 Output the address

transmitted through fine dust APIs, and fine dust sensor values are transmitted through Wi-Fi servers built on orange boards. These two data are printed on the position marker via XML document parsing.

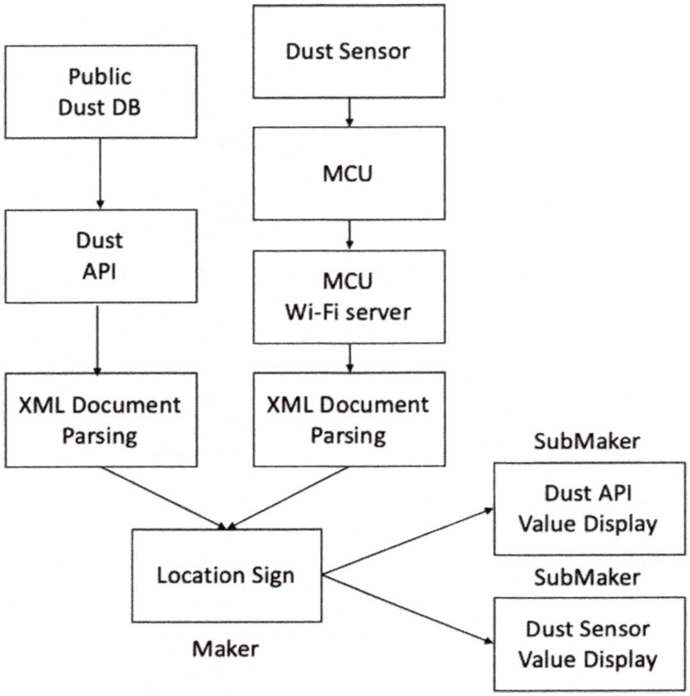

Fig. 12 Flowchart of the fine dust API and sensor value output

Figure 13 shows the map output and the current location verification process. It expresses functions (onCreate) that contain basic components that are needed when using a smartphone's app and objects that check the location in real time.

The real-time position change requires GPS location and base station location change via GPS_PROVIDER and NETWORK_PROVIDER when the location changes.

Figure 14 is a flowchart showing the process of the fine dust sensor measurement data download design. The app brings measured values and public data through sensors in URL and API formats, respectively. To determine the current location of the user, the GPS location is output if approved. Configure the main screen through function (updateMap). The main screen will have the current location centered on the map, the map size setting, address marking, and sensor data and public data. The app brings measured values and public data through sensors in URL and API formats, respectively. To determine the current location of the user, the GPS location is output if approved. The app brings measured values and public data through sensors in URL and API formats, respectively. To determine the current location of the user, the GPS location is output if approved. Download fine dust sensors and public data. The downloaded documents can bring values of fine dust sensors and public data through document parsing by storing buffers per text unit of

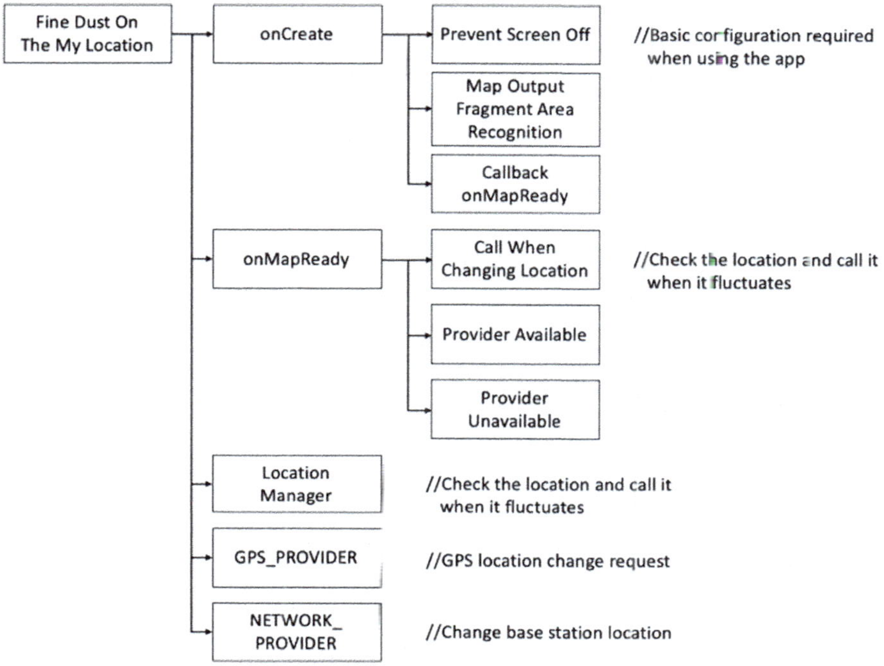

Fig. 13 Flowchart of the fine dust API and sensor value output

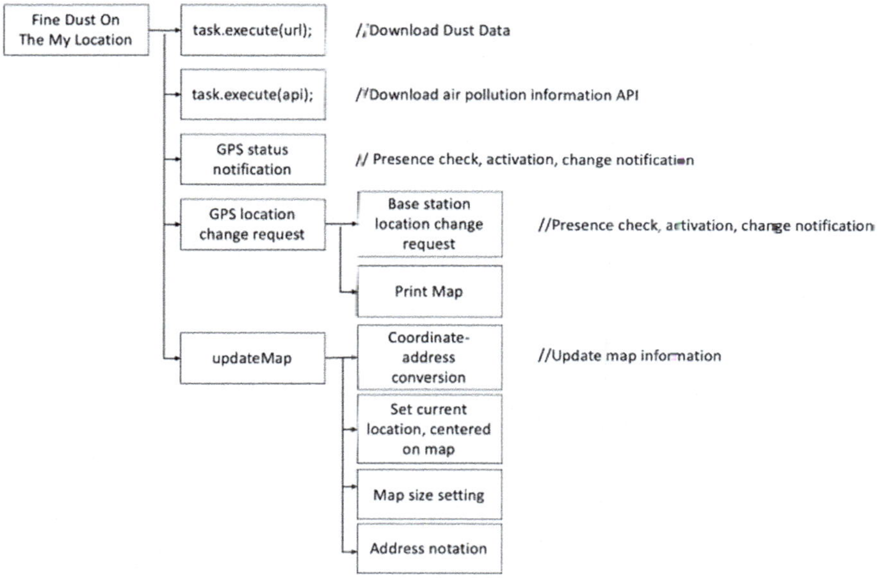

Fig. 14 Fine dust sensor measurement data download

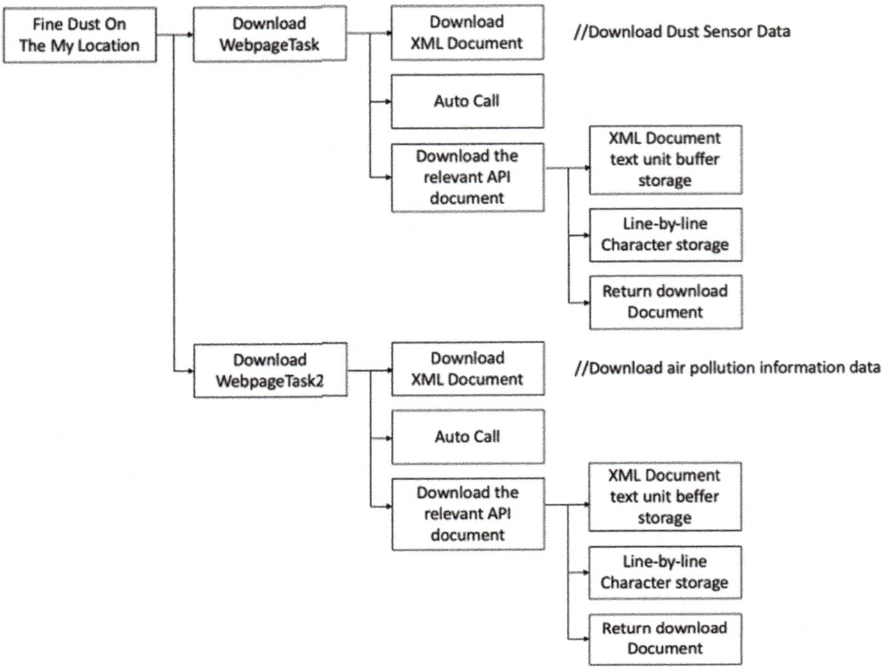

Fig. 15 Fine dust measurement public data download

XML documents, saving characters per line, and returning downloaded documents (Fig. 15).

Download fine dust sensors and public data. The downloaded data is stored in the XML document text unit buffer. The downloaded data can bring the value of fine dust sensors and public data through parsing.

4 Test and Result

Using fragment, the map was printed using the Naver API. Markers are output to the current location using GPS, and real-time fine dust sensor values and public DBs are output in the marker's information window (Fig. 16).

In this study, using fine dust sensors for accurate measurement and verification of fine dust, fine dust output and public database comparison application were developed at current location.

Through this study, sensor devices used in this study are small and inexpensive so many people can use them. It can check the fine dust concentration in real time through mobile phone application or Web page, which is thought to help the health of the people. Starting with this study, research on the reduction of fine dust will

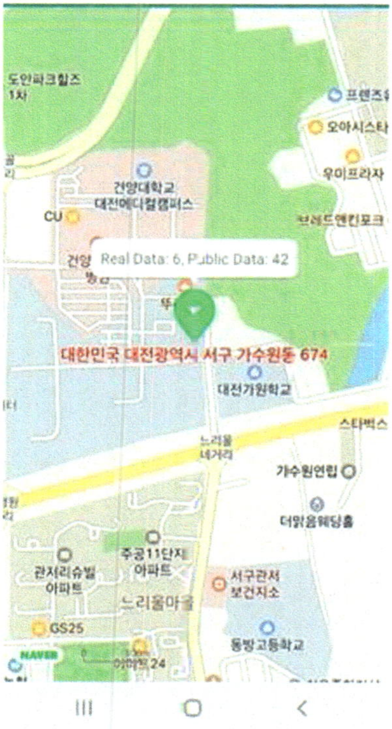

Fig. 16 Test result

have to be carried out continuously in the future. In this paper, the system proposed can quickly recognize the indoor environment, and it is judged that it will gradually help the health of individuals, and by checking data from outside, it can be able to cope with the current situation of fine dust concentration. Through this study, we can understand the basic system for receiving external signals in the Internet of Things environment and contribute to better environmental construction the use of various sensors as well as fine dust sensors in the future. It is also believed that it will be possible to use it in the field of education in the Internet of Things.

Dynamic Clustering Method for the Massive IoT System

Yunseok Chang

1 Introduction

Concepts such as smart buildings, smart factories, and smart campuses are gradually expanding in areas of multiple spaces, such as recently built buildings, factories, and universities. Various IoT devices are installed in spaces based on the state-of-the-art smart system according to the purpose or purpose of use, and these IoT devices are advanced IoT that collectively control IoT devices through mobile devices such as smartphones or smart pads. It is controlled by a centralized or distributed control method through a control system [1]. The collective control technology for a large number of IoT devices starts from a case where it is initially installed in a small space such as a home or office and has recently been applied in many fields using large-scale IoT sensors or devices. There are many application fields of collective IoT control technology. Recently, in the construction fields of smart factories, smart offices, and smart buildings, a smart city that applies environmental IoT sensors such as air pollution sensors and temperature and humidity sensors in large quantities has recently appeared. Meanwhile, the need to collectively operate numerous IoT devices and sensors is increasing [2].

In an environment in which a large number of IoT devices are collectively used, IoT devices perform IoT control in a segment or group unit grouped by space. This paper defines this as a static clustering technique. In a static clustering-based IoT system, the size of a segment or devices included in a group are determined and installed when the IoT system is initially designed so that the configuration of the cluster can be changed later, or a segment belonging to specific IoT devices. It is very difficult and expensive to change. In addition, since the IoT devices

Y. Chang (✉)
Department of Computer Engineering, Daejin University, Pocheon, Republic of Korea
e-mail: cosmos@daejin.ac.kr

K. Daimi et al. (eds.), *Advances in Security, Networks, and Internet of Things*, Transactions on Computational Science and Computational Intelligence,
https://doi.org/10.1007/978-3-030-71017-0_48

are connected to the control device and the physical network for each segment designed for the first time, the flexibility of the system structure decreases, and system expansion or modification can only be done within the designed segment. It is also very difficult to respond to changes in IoT system requirements following changes or changes in the purpose of use. In addition, the large-scale space such as a group of buildings or a smart city composed of groups beyond individual building units has the characteristics of changing in time and space as a living organism according to social changes such as population and role. IoT system technology is no longer a fundamental technology to efficiently and innovatively support ultra-scale IoT environments. In this paper, unlike the static clustering technique, we propose a dynamic clustering technique that logically configures IoT devices connected to the cluster. In the dynamic clustering technique, logical segments are composed, and logical segments regardless of the physical connection of IoT devices. B group control is performed on a group basis. The dynamic clustering technique has a flexible system reconfiguration ability compared to the existing static clustering technique and can efficiently perform group control logically without changing the physical network structure, so it is efficient and flexible against changes in the spatial structure to which the IoT is applied.

The composition of this paper is as follows. Section 2 presents the main features, configuration methods, and problems of the static clustering method based on IoT control system. Section 3 presents the concept and configuration methodology for the IoT control system based on dynamic clustering. Section 4 describes the evaluation elements and comparative evaluation of static clustering and dynamic clustering method and concludes in Sect. 5.

2 Static Clustering Method

2.1 Basic Concept

The static clustering method is a technique to be applied when constructing a typical large-scale IoT system. The entire system is composed of multiple IoT controllers and IoT devices physically connected to each controller. The IoT controller is usually a wired or wireless hub that functions as an AP, and the IoT device is connected to the controller through a wired or wireless network. IoT devices connected to one IoT controller are called segments, and IoT control (shown in Fig. 1) can be performed in units of segments or groups of randomly divided segments within a segment.

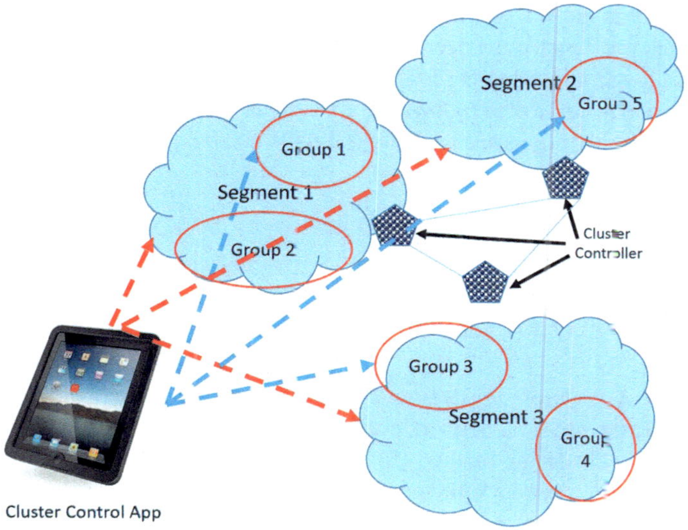

Fig. 1 The basic concept of the segment and group in the static clustering method

2.2 *Problems from the Static Clustering Method*

When the collective IoT control is operated in a group unit in a large-scale IoT system, the IoT controller may individually control IoT devices connected to it or perform group control in segments or groups [3]. However, to control a device connected to another segment, it may be impossible to use an additional device or have to go through a complicated control process. There are two reasons for this. First of all, existing large-scale IoT systems only use products of different manufacturers, and the connection between each manufacturer is not considered. Second, IoT systems independently install and operate IoT controllers according to their use, network addresses, and protocols for accessing devices connected to other controllers that are implemented exclusively. The collective IoT control method having these characteristics is called a static clustering method. In large buildings such as factories, schools, and buildings, the size and use of offices often change, so the internal structure of the building's compartments and walls also varies frequently.

When the IoT system is initially installed in a building, it is installed according to the office or compartment at the beginning of the design. Still, if the internal structure of the building changes or expands, the control range or the controller of the IoT system must be adjusted accordingly. However, the IoT system based on the static clustering method must be reconstructed to respond to the changes of the construction's usage. In other words, simple addition or reduction of devices requires repositioning of controllers or relocation of connections to reconfigure the previously physically installed IoT system. This method decreases the variability and flexibility of the IoT system, requires high reconfiguration cost and maintenance cost, and also suffers from a decrease in availability because it is difficult to use the system while reconfiguring the IoT system. After all, this problem appears because the IoT device is controlled by being connected to the controller subordinately, and a scheme capable of dynamic reconfiguration is needed to solve the problem [4].

3 Dynamic Clustering Method

3.1 Overview of the Dynamic Clustering Method

The concept of a network cluster that connects IoT terminals logically without being dependent on the physical network configuration is required to solve the problem of using the static clustering method [5]. It is called a logical IoT cluster (LIC) in this paper. In contrast, a network having a segment structure that is physically connected to a router or controller, as shown in Fig. 1, is called a physical IoT cluster (PIC).

In the logical IoT cluster, devices can be logically freely grouped, and it is possible to operate massive IoT control in the logical group units, as shown in Fig. 2. As the logical group is a concept of logically connected devices, a logical group can be reconfigured in software to create a new group as much as possible. Therefore, the problems of the IoT system based on static clustering footing a physical network can be easily solved. The IoT clustering method that enables the group control to be performed in units of logical groups is called a dynamic clustering method.

Fig. 2 Concept of the logical segment and logical group in the dynamic clustering method

3.2 Dynamic Clustering System Control Architecture

The dynamic clustering method does not collect IoT devices into PIC-based segments or groups. Still, it uses a logical-physical mapping (PLM) table that converts the address system of IoT devices used in PIC to the LIC address system. This table is composed of logical connection information of devices constituting a segment with data shared by dynamic cluster controllers in a cluster. The dynamic cluster controller maps the logical network connection to a physical single-multiple network connection, as shown in Fig. 3. If the user performs massive control for a specific logical group, the dynamic cluster controller executes a collective control command by connecting physical devices regarding the PLM table. When a new IoT device is added to or deleted from a group, the dynamic cluster controller, physically connected to the devices, registers or deletes the device in the physical-logical mapping table (PLM table). And it shares the modified PLM table with all dynamic cluster controllers in the cluster. If you change the logical group, you only need to change the logical-physical mapping table to reconfigure the logical connection. All dynamic cluster controllers need to share the new mapping table. A separate blockchain method can be applied to add/delete new IoT devices or change the mapping table to maintain the security of the shared PLM table [6].

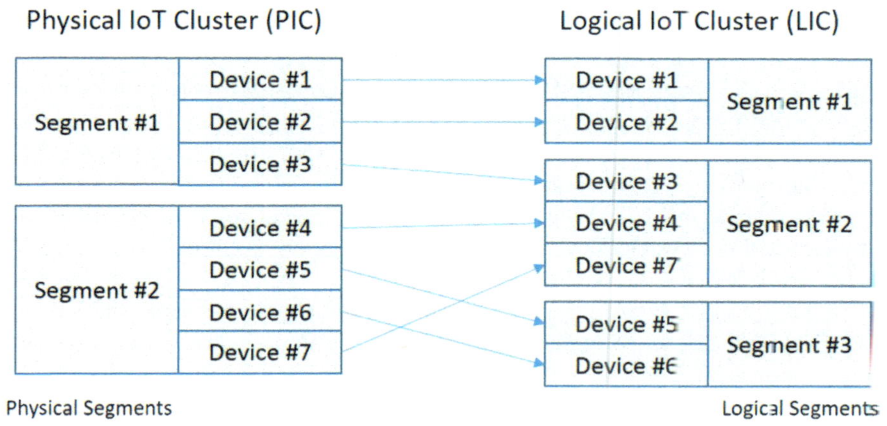

Fig. 3 Device mapping between the PIC and the LIC

4 Design of the Dynamic Clustering System

4.1 Cluster Control Structure

The core of the dynamic clustering architecture is to map physical network connections to logical network connections. In the static cluster structure, group control is performed on a group basis in a preset physical segment. However, the dynamic clustering method performs group control on logical group units regardless of the physical segment. To this end, the following functions must be implemented in the dynamic cluster controller used in the dynamic clustering method:

- Connection management between dynamic cluster controllers
- Control of physical connection with IoT devices
- Logical segmentation and logical grouping of IoT devices
- Logical-physical mapping of IoT devices included in a logical group
- Perform logical group unit control

These functions are mounted on the operating system of the dynamic cluster controller or in firmware. Sharing connection information between IoT devices and the dynamic cluster controller is needed for the implementation of dynamic clustering. The cluster controllers individually managed the network connection information between the IoT devices and the cluster controller when the IoT system was based on the existing static clustering method. In the dynamic clustering method, all dynamic cluster controllers included in the cluster share the PLM table. Each controller included in the cluster receives the collective control command at the same time, selects the nodes to be collectively controlled from the PLM table, and then passes the control command to the nodes by referring to the PLM information of the corresponding node. To perform this process, the dynamic cluster controller maps the IoT devices included in the logical segment or group to a physical IoT device regarding the PLM. Table 1 shows the structure of the PLM table with *k* devices.

In Table 1, PIC has *M* physical segments and *N* physical groups, and LIC has *R* logical segments and *S* logical groups. Valid indicates whether the PLM entry is

Table 1 PLM table entries

Device ID	PIC segment #	PIC group #	LIC segment #	LIC group #	Valid
#1	1	1	1	1	1
#2	1	2	1	1	1
#3	1	2	2	1	1
…..	….	….	….	….	
#k-2	M-1	N-1	R-1	S-1	0
#k-1	M	N-1	R	S-1	0
#k	M	N	R	S	1

valid. Among the entries in the table, the entries included in the deleted cluster are marked as invalid and can be used by nodes in other clusters.

4.2 Cluster Management

In the case of adding a cluster to a system, a new cluster is created by adding nodes having a unique cluster ID to the PLM table of the controller through the following procedure. When one controller is included in two or more clusters, entries for nodes having multiple cluster IDs exist in the controller, and massive control for each cluster is performed by the cluster ID:

1. Select nodes to be included in the cluster.
2. Discover controller information physically connected to the node.
3. Create a new PLM table with network connection information between node and controller.
4. Broadcast the PLM table to the controllers included in the cluster.
5. Add the delivered PLM table to the PLM table of each controller.

When deleting a cluster, the valid item of the entry having the cluster ID is marked as invalid so that the corresponding entries can be used when another cluster is added.

In the case of reconfiguring the cluster, all existing cluster information is deleted, and new clusters are created. First, the PLM tables of all controllers in the system are removed and initialized, and then the clusters are reconfigured through a cluster addition process. The controllers included in the reconstructed clusters share the newly constructed PLM table and operate the corresponding nodes along the network connection path to the nodes corresponding to the user's collective control command.

4.3 Comparative Evaluation of the Dynamic Clustering Method

As shown in Table 2, the dynamic clustering method has many differences compared to the static clustering method. The main advantage of the dynamic clustering method is that reconfiguring the cluster does not require the reinstallation of physical equipment. Therefore, once the IoT system is implemented with the dynamic clustering method, there is no additional cost for the change or reconfiguration of all subsequent clusters. This method can be beneficial when you build a large-scale IoT because you can both reduce the operating cost and increase the ease of maintenance from the viewpoint of operating a large-scale IoT system.

Table 2 Comparison between the static and dynamic clustering method

Characteristics	Static clustering	Dynamic clustering
Logical segmentation	X	O
Network routing information sharing	X	O
Software reconfiguration	X	O
Reconfiguration cost	High	Low
Control delay	1 hop	N hops
Controller rebuilding	No	Yes

The disadvantage of the dynamic clustering method is that the existing installed and used IoT system cannot be transferred to the dynamic clustering method. As mentioned in Sect. 3, to implement the dynamic clustering method, a dynamic cluster controller that performs a network connection control function through mapping and PLM table is required. Therefore, there is a disadvantage in that a transfer cost according to the replacement or upgrade of the cluster controller for moving the existing static clustering-based system to dynamic clustering occurs.

The main difference between the dynamic clustering method and the static clustering method is whether network connection information is shared between the devices constituting the cluster and the cluster controller. In the static clustering method, devices and the cluster controller are connected in an N:1 structure, and connection information has only that controller. However, in the dynamic clustering method, connection information for devices and the dynamic cluster controller is shared by the entire cluster. For example, if a building is composed of multiple physical segments, the physical network connection information between controllers and nodes in each segment exists only in the segment in the static clustering method. However, in the dynamic clustering method, the controllers across the building share the connection information between devices and logical groups in each physical segment. Therefore, when a logical group consisting of nodes in different physical segments is configured, devices included in other physical segments can be logically controlled within access rights in any group through physical-logical mapping.

5 Conclusion

The IoT system using the dynamic clustering method is physically implemented in the same configuration as the IoT system using the static clustering method. Still, it is not always the case that the IoT devices constituting the cluster must perform group control only in the cluster controller unit directly connected to the physical network. The dynamic clustering method proposed in this study can operate the group control by only a logical mapping in the same way as the static cluster system as the cluster shares the logical-physical network mapping table. Therefore, the IoT control system based on the dynamic clustering method does not need to be just a first designed cluster structure. Logical cluster control of IoT devices can be performed by dynamically reconfiguring the cluster according to user requests or changes in the environment or building structure.

Recently, as IoT devices' applications have expanded, the extensive use of multiple IoT devices, such as buildings, factories, and schools, has gradually increased. In the conventional static clustering method, once a cluster is designed and configured, it is complicated to change the cluster's size or configuration. However, since the dynamic clustering method can perform massive control by constructing a logical cluster regardless of the physical connection structure between the IoT devices and the controller, it can very flexibly cope with changes in the cluster control structure that changes from time to time. Therefore, the dynamic clustering method presented in this study is expected to be very useful in the future large-scale IoT collective control technology.

Acknowledgments This work was supported by the National Research Foundation of Korea (NRF) grant funded by the Korea government (MEST) (No. NRF-2020R1D1A1B03034804).

References

1. A survey on IoT security: Application areas, security threats, and solution architectures. IEEE Access **7**, 82721–82743 (2019)
2. IoT enabled adaptive clustering based energy efficient routing protocol for wireless sensor networks. Int. J. Ad Hoc Ubiquitous Comput. **32**(2), 133–145 (2019)
3. Evaluating performance of containerized IoT services for clustered devices at the network edge. IEEE Internet Things J. **4**(4), 1019–1030 (2017)
4. Dynamic IoT device clustering and energy management with hybrid NOMA systems. IEEE Trans. Industr. Inform. **14**(10), 4622–4630 (2018)
5. Dynamic clustering method based on power demand and information volume for intelligent and green IoT. Comput. Commun. **152**, 119–125 (2020)
6. Probability-based IoT management model using blockchain to expand multilayered networks. J. Korea Converg. Soc. **11**(4), 33–39 (2020)

A Network Traffic Reduction Method for a Smart Dust IoT System by Sensor Clustering

Joonsuu Park and KeeHyun Park

1 Introduction

The Smart Dust was defined as minute nodes that can do sensing and wireless communication. Thanks to the MEMS (microelectromechanical system) technologies, the Smart Dust has become inexpensive and easy to deploy [1–4]. However, once millions of the Smart Dust nodes are scattered over an area, particularly in areas that are difficult to access by human, controlling the Smart Dust nodes and gathering the sensed data are challenging tasks.

In this paper, we propose a network traffic reduction method for a dust IoT system by sensor clustering as an attempt to solve the control problem mentioned above. Having millions or billions of sensing nodes in a system, we should anticipate an enormous amount of sensed data coming from the nodes. For such a system, we have to come up with methods to lessen network traffic loads, to alleviate the bottleneck problem, and to process the data efficiently in the IoT server.

We had been working on these problems. In our earlier works [5, 6], we proposed a smart IoT system to process the sensed data efficiently in the IoT server. Based on the earlier works, we propose a method to lessen network traffic loads further and hence to alleviate the bottleneck problem by the clustering of smart nodes. Experiments show that the transmission data size of the proposed work is reduced by as much as 24% to 26% of that of our earlier work.

J. Park · K. Park (✉)
Department of Computer Engineering, Keimyung University, Daegu, Republic of Korea
e-mail: khp@kmu.ac.kr

K. Daimi et al. (eds.), *Advances in Security, Networks, and Internet of Things*,
Transactions on Computational Science and Computational Intelligence,
https://doi.org/10.1007/978-3-030-71017-0_49

2 Our Earlier Works

In our earlier studies, a three-layered remote monitoring system in a Smart Dust environment was proposed [6]. As shown in Fig. 1, the system consists of DD (Dust Sensor Device) layer, RDD (Relay Dust Sensor Device) layer, and the Smart Dust IoT Server layer. The DD is a Smart Dust Sensor Device scattered in wide area to sense climate data. The DD has low-computing/low-communicating power and hence can neither send sensed data in a long distance nor filter/compress data. The RDD, a special DD with a relatively higher-computing/higher-communicating power than normal DD, can process to transmit the data collected from normal DDs to the IoT server.

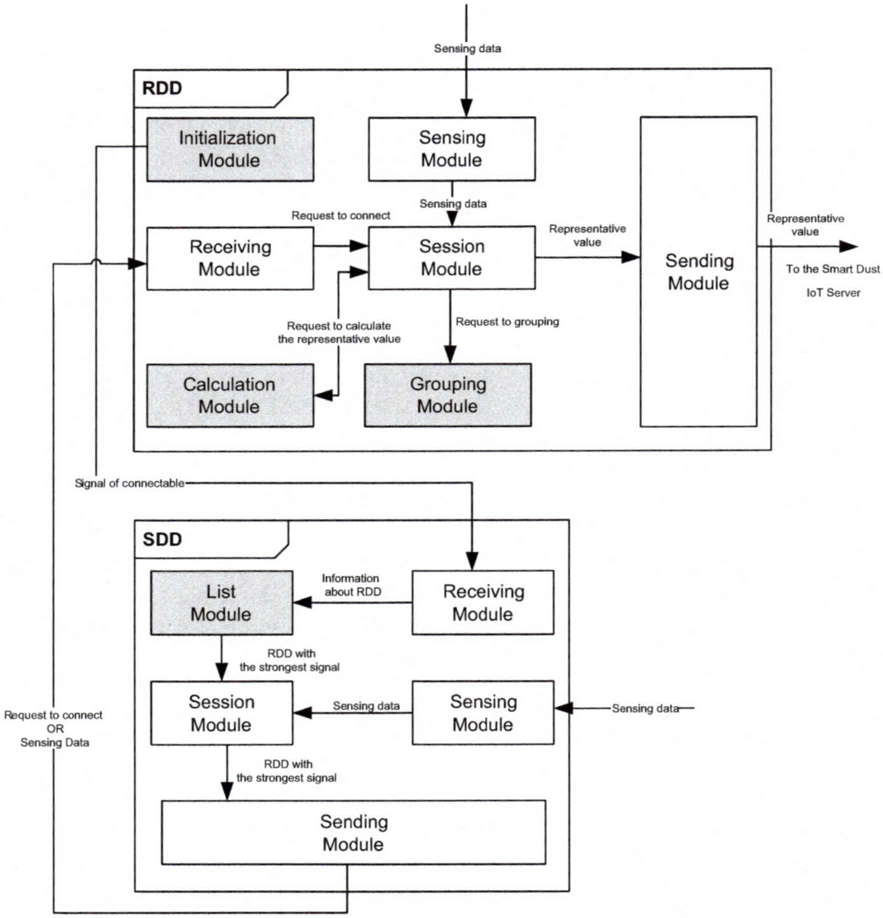

Fig. 1 Software module configuration diagram of an RDD and an SDD

In Fig. 1, the shaded boxes represent newly added or changed software modules for this study. The modules of an RDD are Initialization Module (to send connectable signal within its connection range), Calculation Module (to calculate the representative value for SDDs in its cluster), Grouping Module (to perform the SDD clustering), Session Module (to maintain a session with SDDs in its cluster), Sensing Module (to collect sensed data), and Receiving/Sending Module (to receive/send sensed data). The modules of an SDD are List Module (to manage a list of RDDs that have sent a connectable signal to the SDD) and Receiving/Sending Module (to receive/send sensed data).

3 Network Traffic Reduction Method

Figure 2 shows the process of the network traffic reduction method.

The method proposed in this paper reduces the size of transmission data from an RDD to the Smart Dust IoT Server by sending a single representative sensed data value, instead of sending the data from all the SDDs. The method is based on the assumption that the SDDs very close to each other may send the very similar sensed data values.

Once SDDs and RDDs hit the ground, every RDD sends a connectable signal to the neighboring SDDs. Once a SDD receives the signal, it creates a list of reachable RDDs sorted by signal strength. That is, the RDD having the strongest signal strength is located at the top of the reachable RDD list. Then, the SDD sends a connection request to the top RDD in the list in order to be included in the RDD's cluster. After the steps mentioned above, each SDD can transmit sensed data to the RDD. Then, the RDD receives the sensed data from the SDDs in its cluster and calculates the average value of the data for its cluster in order to transmit to the Smart Dust IoT Server.

Stages	Steps	Contents
Initialization stage	Step 1	Each A RDD sends a connectable signal to SDDs within its connection range.
	Step 2	Each SDD creates a list of the RDDs that sent a connectable signal to it.
	Step 3	Each SDD sorts the list created in Step 2 by signal strength.
Clustering stage	Step 4	Each SDD sends a connection request to the first RDD in the sorted list (the RDD with the strongest signal strength).
	Step 5	Each RDD performs the SDD clustering by accepting the connection requests from SDDs.
Data transmission stage	Step 6	Each RDD begins to receive the sensed data from the SDDs in its cluster.
	Step 7	Each RDD calculates a representative value (i.e., an average value) for the SDD cluster.
	Step 8	Each RDD sends the representative value calculated in Step 7 to the Smart Dust IoT server.

Fig. 2 The process of the network traffic reduction method

Table 1 Comparison of transmission data size between the earlier work and the proposed work

The number of DDs and RSDDs (unit)	100	500	1000
Transmission data size in the earlier work (bytes) - A	7038	35,190	70,380
Transmission data size in the proposed work (bytes) - B	5384	25,977	53,231
*(A-B) / A * 100 (%)*	23.49%	26.18%	24.37%

4 Experiment

We conducted some experiments to verify that the proposed method in this paper can effectively reduce the transmission data size to improve the stability of the system. In this experiment, RDDs and SDDs are generated as a process to perform the simulation; 90% (10%) of the generated device processes are assigned to SDDs (RDDs). In addition, it consists of four types of SDD sensors (i.e., temperature, humidity, illuminance, surface temperature sensors), and the position of each SDD and RDD is determined using the standard normal distribution.

It would be better to obtain enormous amounts of data collected over a long period of time with vast amounts of equipment. However, in consideration of the fact that such data collection is difficult in reality, this experiment verifies the effectiveness of the method using the official data of temperature, humidity, illuminance, and surface temperature collected in the real environment provided by the Korea Meteorological Administration [7] and the National Information Society Agency [8].

Table 1 shows the results of comparative experiments between our earlier works (i.e., without the network traffic data reduction method) and the proposed work in this paper. From the experiments, it can be confirmed that the transmission data size of the proposed work is reduced by as much as 24% to 26% of that of our earlier work.

5 Conclusion

In our earlier works, we proposed a smart IoT system to process the sensed data efficiently in the IoT server. Based on the earlier works, we propose a method to lessen network traffic loads further and hence to alleviate the bottleneck problem by the clustering of smart nodes. The method proposed in this paper reduces the size of transmission data from an RDD to the Smart Dust IoT Server by sending a single representative sensed data value, instead of sending the data from all the SDDs. The method is based on the assumption that the SDDs that are in very close proximity to each other may send the very similar sensed data values. Experiments show that the transmission data size of the proposed work is reduced by as much as 23% to 26% of that of our earlier work.

Acknowledgments This research was funded by the Basic Science Research Programs through the National Research Foundation of Korea (NRF), grant number funded by the Ministry of Education, Science, and Technology (No. NRF-2018R1D1A1B07043982).

References

1. J.M. Kahn, R.H. Katz, K.S. Pister, Next century challenges: Mobile networking for "Smart Dust", in *Proceedings of Proceedings of the 5th annual ACM/IEEE international conference on Mobile computing and networking*, pp. 271–278 (1999)
2. J.M. Kahn, R.H. Katz, K.S. Pister, Emerging challenges: Mobile networking for "smart dust". J. Commun. Netw. **2**, 188–196 (2000)
3. B. Warneke, M. Last, B. Liebowitz, K.S. Pister, Smart dust: Communicating with a cubic-millimeter computer. Computer **34**, 44–51 (2001)
4. L. Niccolai, M. Bassetto, A.A. Quarta, G. Mengal, A review of Smart Dust architecture, dynamics, and mission applications. Prog. Aerosp. Sci. **106**, 1–14 (2019)
5. J. Park, K. Park, A dynamic plane prediction method using the extended frame in smart dust IoT environments. Sensors **20**, 1364 (2020)
6. J. Park, K. Park, Construction of a remote climate monitoring system in a smart dust environment. J. Inf. Process. Syst. (2020), Accepted for publication
7. KMA, the Korea Meteorological Administration [online], https://data.kma.go.kr/, Accessed on: May, 14, 2020
8. NISA, National Information Society Agency [online], https://www.data.go.kr, Accessed on: May, 14, 2020

Part VII
Embedded Systems, Cyber-physical Systems, Related Tools, and Applications

On the Development of Low-Cost Autonomous UAVs for Generation of Topographic Maps

Michael Galloway, Elijah Sparks, and Mason Galloway

1 Introduction

Topographic maps can be defined as detailed and accurate two-dimensional representations of three-dimensional natural and human-made features on the Earth's surface [11]. Information detailed in these maps can be used by a large number of applications including but not limited to: defining small and large geographic watersheds [9], observation of sediment transport [5] and its effects, urban landscape development [14], navigation and vehicle route development [7], and hiking [10], hunting, and fishing [19].

Topographic maps are compiled into overlapping layers to give detailed geographical information over a surveyed area related to land cover, geographical structures and boundaries, hydrography, elevation by use of contour lines, optical imagery, and orthoimagery. The field surveying process is currently conducted manually with various equipment including a level tripod and rod, compass, tape measure, and notebook. Surveyors manually acquire topographic data in the field using these tools and record this data on a physical map drawing or in mapping software. The primary goal of our research is to replicate the surveying process using a UAV, making it fully autonomous.

In the past 20 years, a typical land survey team consisted of four people. Typically, two surveyors are active, and the other two are gaining field experience. One surveyor operates the Total Station, while the other holds and moves the Prism Pole in designated areas to measure distances and elevation.

M. Galloway (✉) · E. Sparks · M. Galloway
Western Kentucky University, Bowling Green, KY, USA
e-mail: jeffrey.galloway@wku.edu; elijah.sparks662@topper.wku.edu; mason.galloway118@topper.wku.edu

K. Daimi et al. (eds.), *Advances in Security, Networks, and Internet of Things*,
Transactions on Computational Science and Computational Intelligence,
https://doi.org/10.1007/978-3-030-71017-0_50

A topographic review of a surveyed area ranges from \$500 to \$1200 for areas less than 1/4 acre [6]. Our research proposes to create an alternative to the way topographical land surveying is performed by existing engineering and construction firms with the use of low-cost autonomous unmanned aerial vehicles (UAVs).

The land survey data acquired by the UAV will be transferred to a storage and processing server that can be accessed using any modern web browser. The development of software and mechanical tools for this research is separated into two categories: (1) UAV development and navigation and (2) environmental data acquisition, processing, and user interface. Topics of interest and requirements are as follows.

1.1 UAV Development and Navigation

- Development of the UAV. The UAV will have a sustained flight time of 10–15 min, based on environmental conditions. The UAV will be composed of the PixHawk [8] flight controller and two Raspberry pi 4 embedded computers (one for navigation and the other for environmental data acquisition).
- Autonomous flight control of the UAV. Users will set GPS way-points using a web-based graphical user interface. The UAV must have the capability of navigation to and between two-to-many GPS locations.
- UAV area coverage path planning. The UAV must have the ability to generate paths that ensure complete area coverage based on given GPS way-points as a perimeter.
- Development of a "Flight Plane." The UAV will be equipped with Global Navigation Satellite System (GNSS), GPS, and altimeter sensors for development and maintenance of spatial location. The flight plane relative altitude will be determined by the user when a job is submitted to the UAV. The UAV must stay within the vertical flight plane to generate elevational data of the coverage area.
- UAV obstacle avoidance. The UAV must avoid collision with all obstacles in the given environment. Obstacle avoidance is achieved with the use of ultrasonic range detection sensors.

1.2 Environmental Data Acquisition, Processing, and User Interface

- The primary data acquisition sensors on the UAV are Light Detection and Ranging (LiDAR) for generating elevational data and a CMOS image sensor camera for generating an image overlay layer for the area topographic map.
- The primary LiDAR and camera sensors will be mounted to the UAV frame using a two-axis real-time adjustment gimbal. This is needed for maintaining a precise

perpendicular angle from the LiDAR sensor to the ground, as the UAV moves through the coverage area.
- The UAV will be used for environmental data collection and temporary storage. Data processing for topographic maps will not occur on the UAV. Data from each job request will be transferred to a data processing, permanent storage, and web server.
- The dedicated processing, storage, and web server will be accessible using a web-based interface. Authenticated users will have the ability to submit requests to the UAV for environmental data collection. Data collected is viewable as a filtered layer topographic map.
- Elevational data is used to generate 2D or 3D models of the surveyed area using a point cloud [18] approach.
- Previously submitted jobs/criteria can be submitted again in the future. This allows for the analysis of specified coverage areas over time.

To validate the research and design concepts mentioned above, we developed a prototype UAV that allows users to gather elevational data of a coverage area specified by GPS way-points. During initial testing, we verified the functional requirements of the UAV in terms of flight time, navigation, and environmental data acquisition.

The next section covers the general overview and related works for creating topographic maps with and without the use of UAVs. Section 3 covers development of the UAV with considerations for being cost efficient. Section 4 describes autonomous flight of the UAV with obstacle avoidance and the approach to complete area coverage and path planning. In Sect. 5, we describe the need for the UAV flight plane. Section 6 describes LiDAR data and the development of two-axis real-time adjustment gimbal. Section 7 covers data transfer from the UAV, processing, storage, and user interactions. Finally, we present conclusions of this research and discuss future work in Sect. 8.

2 Related Works

Airborne and space LiDAR systems have been used to map terrain of the Earth and Moon since the first demonstration of the laser altimeter experiment conducted on the Apollo 15 Command and Service Module (CSM) in 1971 [12]. During this experiment, a laser was aimed from the Apollo CSM to the lunar surface. The time delta for the laser pulse to travel from the CSM to the lunar surface and back was then used to determine the spacecraft's altitude. As the Apollo spacecraft was in a stable orbit around the Moon, terrain elevations could be successfully calculated using this approach.

Since the Apollo mission, there has been increased research and development in this area [15], [21], and [3]. The authors of [15] used a fixed wing UAV for surveying which produced good results when the number of turns required is low.

Zainuddin et al. [21] propose a similar approach to topographic map generation using an autonomous quadcopter UAV but use an off-the-shelf quadcopter (DJI Phantom 2 Vision +) that costs twice as much as the one proposed in our research. The authors of [3] use a combination of aerial and terrestrial vehicles to develop topographic maps.

3 Development of a Cost-Efficient, Replicatable UAV

The major objective of this research is to develop an autonomous UAV that has the capability of generating data necessary to develop a topographic map of a specified environmental region. Table 1 gives the materials list to develop a single autonomous UAV needed for this research. Details for replication were recorded in the project documentation for future work efforts to apply swarm robotics approaches [2] for acquiring environmental data.

The system block diagram of the quadcopter UAV is shown in Fig. 1.

3.1 LiDAR Lite V3

Light Detection and Ranging (LiDAR) devices measure distance between the senor and an object by calculating ΔT of transmitting a near-infrared (0.75–1.4 μm) laser signal and its return from reflection of the target object. The LiDAR Lite V3 has the ability to generate distance acquisitions based on the following [4]:

Table 1 UAV list of materials

Item	Quantity	Cost
F450 UAV frame	1	$15
PixHawk 4 flight controller	1	$73
Raspberry Pi 4 w/ 64 GB onboard storage	1	$75
Brushless 1000 KV DC motors	4	$60
20 A electronic speed controllers	4	$40
10x4.5 CW/CCW quadcopter propellers	4	$3
3 Cell 11.1V 80c 2800 mAh Battery	1	$54
PRVDrone 8CH PPM encoder	1	$13
Zubax GNSS, GPs, GLONASS, Galileo receiver	1	$150
HC-SR04 ultrasonic sensor	6	$10
LiDAR Lite V3	1	$130
2-axis auto leveling Gimbal	1	$55
Miscellaneous connectors/cables	*n*	$50
	Total:	$728

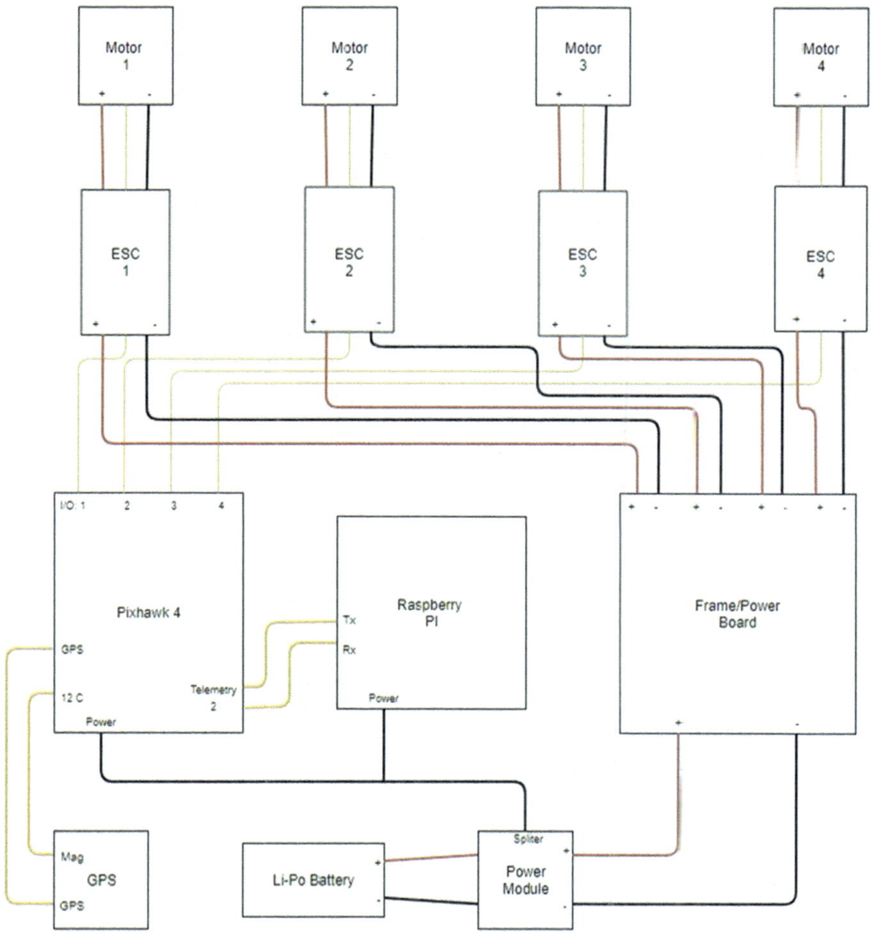

Fig. 1 UAV system block diagram

$$\text{Maximum Acquisition Count/s} = 1/n$$

$$\text{range} = n^{1/4}, \text{ where n is the number of acquisitions.}$$

Based on the above equations, the maximum distance acquisition count/sec is relative to the distance between the LiDAR Lite V3 sensor and target objects. Distance data acquisition frequency can be adjusted with this LiDAR sensor (with a trade-off of maximum range). The default distance data updates at a frequency of 270 Hz, which exceeds the goal of a class 1 [20] point density classification of 10–100 points/m^2 for this research. Point clouds are explained in detail in Sect. 6.

3.2 Raspberry Pi 4

The Raspberry Pi 4 Model B [16] is the current iteration of the popular general-purpose embedded computer system used in this research. This device is equipped with a Broadcom BCM2711 quad-core Cortex-A72 1.5 GHz CPU, 4 GB LPDDR4 RAM, IEEE 802.11ac and Gigabit Ethernet, and a Micro SD card slot for loading the operating system and data storage. The general-purpose input/output (GPIO) pins on this device accommodate many standard interface protocols, including Serial Peripheral Interface (SPI), Universal Asynchronous Receiver/Transmitter (UART), and Inter-integrated Circuit (I^2C).

All environmental sensors used for obstacle avoidance, elevational data acquisition, and optical image generation are connected to the Raspberry Pi's GPIO pins. Performance of the Raspberry Pi 4 GPIO sensor module read and write to disk rate [13] exceeds the need for this research.

3.3 Ultrasonic Range Sensors

Ultrasonic sensors operate by sending out a high-frequency ping whenever a high pulse is received on a trigger pin, then sending out a high pulse on an echo pin when the echo of the high frequency ping is received. The distance to the nearest obstacle in the direction the sensor was facing can be calculated by using the time difference between the high pulse being sent to the trigger pin and the high pulse being received on the echo pin, as well as the speed of sound (343 m/s at sea level). The following equation can be used by the ultrasonic sensors to determine distances from target objects:

$$\begin{aligned} \text{Speed(ofsound)} &= \text{Distance}/\text{Time} \\ 34,300\,\text{cm/s} &= \text{Distance}/(\text{Time}/2) \\ 17,150\,\text{cm/s} &= \text{Distance}/\text{Time} \\ 17,150\,\text{cm/s}^{*}\,\text{Time} &= \text{Distance} \end{aligned}$$

The HC-SR04 has a high probability distance range detection between 2 cm and 400 cm (1 inch to 13 feet). These sensors are not used for environmental data acquisition in this research. Instead, they are used for obstacle avoidance during the autonomous flight path movements through the specified coverage area.

Obstacle Avoidance Using SIGMOID Function Reactive obstacle avoidance was implemented using the distance to nearby objects and the estimated velocity in each direction. Pressures were generated in each of the four directions by plugging in the distance to a sigmoid function to estimate how strong of a "push" is needed away from that direction. A sigmoid function was chosen because of its "S"-shaped curve, which provided minor increases with increasing input values, then rapid growth around a certain parameter (the midpoint of the curve), and followed by minor increases again. The sigmoid function is expressed as

$$S(x) = \frac{a}{1 + e^{-b(x-c)}}$$

We used a variation of the logistic function that had three parameters, denoted as a, b, and c. Adjusting a altered the max value of the curve, b controlled its steepness, and c determined its midpoint. This bounded exponential growth created a smooth way to adjust the pressures with parameters that could be tuned empirically. The logistic function used is given below in Fig. 2.

Algorithm 1 explains the approach for obstacle avoidance using distance data generated by the HC-SR04 ultrasonic sensors.

3.4 DC Motors and Propellers

As a general rule, the DC motors selected for the quadcopter UAV should be able to generate twice as much thrust as the total weight of the UAV. As for our quadcopter UAV, DC motor and propeller selection was determined by the need for stability at a given maximum flight height of 35 m (115 feet) from the ground. This maximum height flight plane is based on the maximum distance measurement of the LiDAR Lite V3.

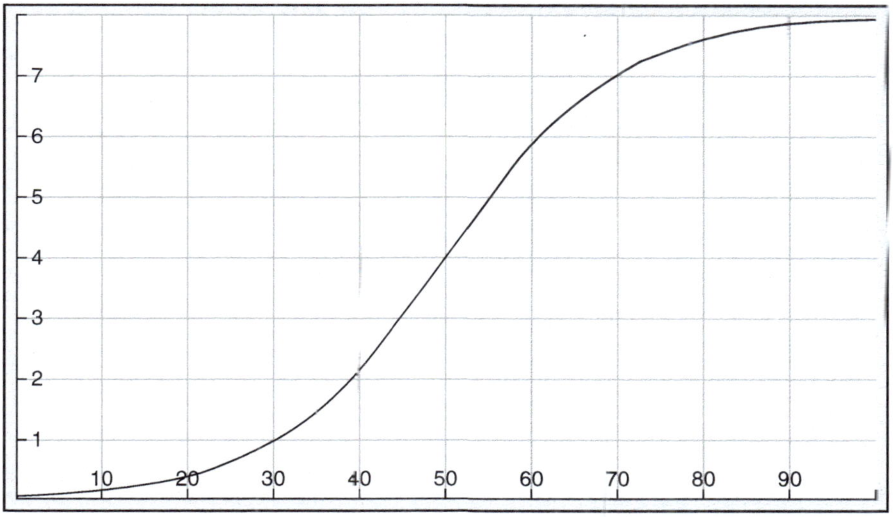

Fig. 2 Sigmoid function $S(x) = \frac{8}{1+e^{-0.1(x-50)}}$ over (0, 100) used by the UAV to determine "push" movement for obstacle avoidance

Algorithm 1 Obstacle avoidance using SIGMOID function

STATE airborne:
if $left\_distance < 100$: **then**
  $left\_push = SIGMOID(100 - left\_distance)$
else
  $left\_push = 0$
end if
if $right\_distance < 100$: **then**
  $right\_push = -1 * SIGMOID(100 - right\_distance)$
else
  $right\_push = 0$
end if
if $front\_distance < 100$: **then**
  $front\_push = SIGMOID(100 - front\_distance)$
else
  $front\_push = 0$
end if
if $back\_distance < 100$: **then**
  $back\_push = -1 * SIGMOID(100 - back\_distance)$
else
  $back\_push = 0$
end if
$left\_velocity = left\_distance - old\_left\_distance$
$right\_velocity = right\_distance - old\_right\_distance$
$front\_velocity = front\_distance - old\_front\_distance$
$back\_velocity = back\_distance - old\_back\_distance$

$aileron\_velocity\_push = (left\_velocity - right\_velocity) * CONSTANT$
$elevator\_velocity\_push = (front\_velocity - back\_velocity) * CONSTANT$

$aileron\_output = left\_push + right\_push + aileron\_velocity\_push$
$elevator\_output = front\_push + back\_push + elevator\_velocity\_push$

DC Motors and Propeller Flight Test The UAV DC motors were tested for efficiency and performance. Based on the 1000 KV rating for each motor, at full throttle, each propeller has a theoretical maximum angular speed of 1162.4 radians/s. Approximate Revolutions Per Minute (RPM) for the propeller can be calculated based on the following equation:

$$1 \text{ rad/s} = 30/\pi \text{ rpm}$$

$$1162.4 \text{ rad/s} = 1162.4 * 30/\pi \text{ rpm}$$

$$\approx 11{,}100 \text{ rpm (theoretical zero-load maximum)}$$

At full throttle, each motor has a maximum current draw of 10 A. Based on the 2800 mAh rating of our 11.1 V battery, the motors can sustain full throttle for approximately 4 min. Full throttle is not necessary for the UAV to sustain flight in the flight plane based on its weight and thrust produced from the DC motors. We have approximated 10–12 min total flight time based on an average of 40–50%

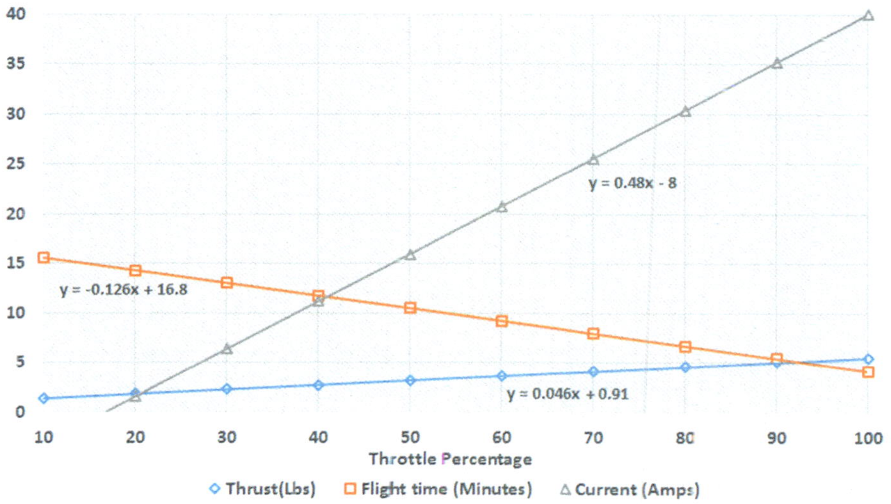

Fig. 3 UAV DC motor efficiency test

throttle needed to sustain the UAV in flight. Thrust, flight time, and current draw, as related to throttle percentage, are shown in Fig. 3.

4 UAV Area Coverage Approach

The UAV takes a complete target area coverage approach for scanning the entire area. It does this without the availability of *priori* information. The onboard environmental sensors (GPS, GNSS, ultrasonic distance sensors) are used to direct the UAV coverage operations.

Our approach for complete area coverage is to use an *exact cellular decomposition* of the total area. *Cellular decomposition* divides the total coverage area into cells, where path planning for total coverage of a specific cell is “simple” back and forth motions [1]. Specific locations within the cell are designated as GPS coordinates and are stored in an adjacency matrix. Complete coverage is ensured when all locations of the adjacency matrix have been visited (Fig. 4).

4.1 Obstacle Avoidance and Boustrophedon Cellular Decomposition

The only *priori* knowledge known is the GPS way-points that make up the perimeter angle locations of the coverage map. Since this is the case, we take a

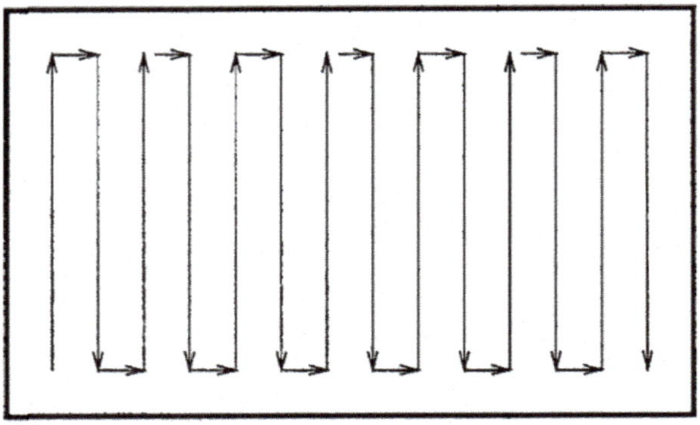

Fig. 4 UAV path taken within each cell to ensure complete cell coverage [1]

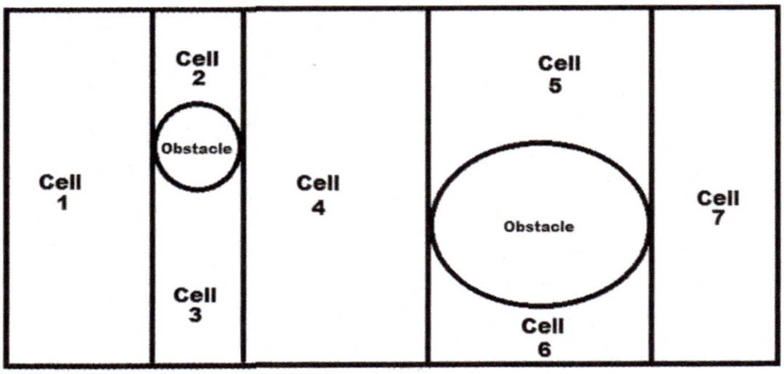

Fig. 5 Coverage area cellular decomposition using Boustrophedon approach

Boustrophedon decomposition approach to achieve complete area coverage while avoiding obstacles. This approach relies on the line segments generated by the sweeping motion of the UAV through the current cell. Once an object is encountered, the current line segment is broken and a new cell is created.

An example of this approach can be seen in Fig. 5. In this example, the UAV sweeping motions of the survey area are top to bottom and left to right. The UAV begins in the bottom-left position of Cell 1. At this time, the complete coverage area is one single cell. As the UAV progresses to the right with top-to-bottom, left-to-right sweeping motions, it will encounter the first obstacle, breaking the current line segment. Once this occurs, a new cell is created, Cell 2, where the UAV proceeds to achieve complete area coverage of this cell using the same approach as Cell 1. Once the UAV exceeds the barrier of the object and proceeds to reach the predetermined survey area perimeter, a new Cell is created. This approach continues until the UAV has successfully traversed the entire survey area.

5 UAV Navigation Flight Plane

Acquiring the main topological data, terrain elevation, requires the quadcopter UAV to fly at a constant altitude, the UAV flight plane as shown in Fig. 6, relative to a specific position, mean sea level. This is the average level of one or more bodies of water from which atmospheric pressure is calibrated, and altitude measurements can be calculated.

In this research, we calculate the altitude for the UAV flight plane using the barometric pressure sensor on the Zubax Robotics GNSS 2 Receiver, which also includes GPS, GLONASS, and Galileo satellite navigation systems.

5.1 Calculating Altitude Using the Barometric Sensor

Altitude data can be obtained by using a barometer to compare the current atmospheric pressure with a known pressure at mean sea level. Atmospheric pressure decreases as the UAV's altitude increases. It is also worth noting that current weather conditions will also affect air pressure.

The following equation can be used for calculating altitude using a barometric sensor, where A_c is the calculated altitude, c is a constant that depends on the acceleration of gravity and the molar mass of the air, T_a is the absolute temperature, P_o is the atmospheric pressure at mean sea level, and P is the pressure read by the Zubax barometer sensor:

$$A_c = cT_a \log(P_o/P) \text{ [17]}$$

Fig. 6 UAV flight plane

6 Elevational Data Acquisition

Accurate elevational LiDAR data depends on two variables: (1) the stability of the UAV's ability to maintain height within the flight plane and (2) the ability to always direct the LiDAR sensor perpendicular to the ground while flying through the survey area. An illustration of these problems can be seen in Fig. 7.

6.1 LiDAR Gimbal Development

This research identified two solutions for the second problem mentioned above. The first solution includes mounting the LiDAR sensor statically to the UAV. Since the UAV will tilt in both x and y directions due to acceleration, wind, etc., adjustments to the LiDAR data can be calculated in software using simple trigonometry when the x and y tilt angles are known. This approach was abandoned due to the direct manipulation of the LiDAR elevational data. The second solution that develops a dynamically updated 2-axis gimbal using two independent servo motors was used. By attaching one servo motor directly to the shaft of a second servo motor, we have the ability to adjust the position of the attached LiDAR in real time using the pitch and roll angles of the UAV. Figure 8 gives an illustration on the development of the 2-axis gimbal.

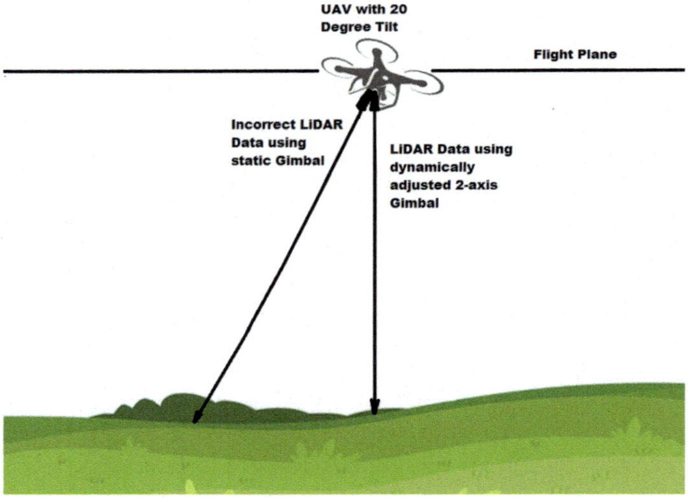

Fig. 7 UAV LiDAR data correction using 2-axis dynamically updated Gimbal

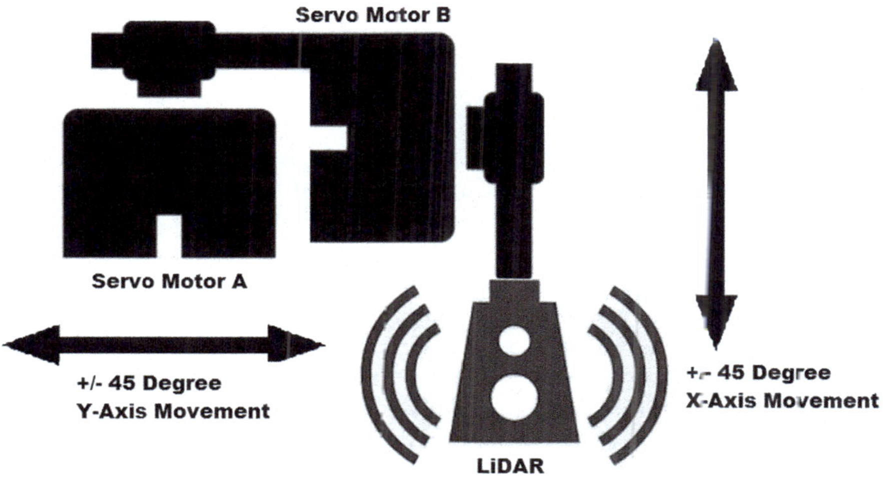

Fig. 8 Illustration of 2-axis gimbal using two servo motors

Table 2 LiDAR Aerial Survey (LAS) file format

Field	Description
Header w/metadata	Elevation range used for color gradient
X, Y coordinates	Latitude and longitude position
Elevation	Distance from target object to LiDAR sensor
Classification	Ground, water, vegetation, building
Returned pulse intensity	Grayscale image map overlay
RGB value	Interlaced with camera data
Return number	Total number of returns per transmitted pulse
Scan angle	+/− degree angle from nadir
Overlap points	Where two flight lines covered the same area

6.2 *Point Cloud Data*

This research stores elevational data as point clouds using the industry standard LiDAR Aerial Survey (LAS) format. The LAS file contains the following fields for each data point acquired by the UAV's LiDAR sensor (Table 2):

One concern with this research is the area that can be surveyed by the UAV based on the UAV's flight time. Since the LiDAR sensor is always facing perpendicular to the ground, the point density of the coverage area is proportional to the distance between straight line paths taken by the UAV discussed in Sect. 4. Decreasing the distance between paths allows for a higher point density, and therefore higher resolution, of the coverage area. The trade-off to this approach is reduced total coverage area. Increasing the distance between the paths will increase the total coverage area surveyed by the UAV, but at lower resolution.

7 Data Transfer, Storage, and Usage

Area survey jobs are generated in the web-based interface and transferred to the UAV using an ad hoc wireless network. This gives the user the ability to be close to the survey area with the UAV and wireless device, allowing the UAV maximum flight time. Once the job has been sent to the UAV, it begins flight to the nearest perimeter GPS way-point, noting the GPS position of the user. The UAV also navigates to the precalculated flight plane and begins sweeping the coverage area.

Once the UAV has finished sweeping the coverage area, or the battery state is too low to continue, the UAV will return to the user's GPS coordinates. The LAS file that has been generated is transmitted to the user's mobile device using the wireless ad hoc network. The mobile device places all LAS files into a Dropbox folder, which is synchronized with the storage and web server. The survey data is saved on the storage server and is used to generate the topographic map displayed in the web browser.

8 Conclusions and Future Work

We present in this research the development of an autonomous quadcopter UAV that is used for collecting environmental data needed to create topographic maps. Current trends for UAV development were followed, allowing the production cost to remain less than $800. Flight time is a major factor for this research and is the main limiting factor for total survey coverage area that can be mapped with a single UAV. We have used progressive research for developing autonomous flight with obstacle avoidance, area coverage path planning, and development of a flight plane.

The primary data acquisition sensors on the UAV for this research are LiDAR and CMOS image sensors. The LiDAR was mounted on a 2-axis dynamically updated gimbal, which allows the sensor to acquire data that is perpendicular to the UAV flight plane.

Data acquired during the UAV flight is stored temporarily on the UAV as a LAS file. Once the job is complete, the UAV returns to the user's coordinates close to the survey area and transmits the LAS file to a mobile device using an ad hoc wireless network. Finally, the LAS data is synchronized with a storage/web server that is used to hold LAS data and generate topographic maps using a web interface.

Future work for this research includes updating the LiDAR angle from the static nadir positing +/−20–30 degrees to allow for higher density and overlapping points within the LAS point cloud data file. Another future work effort will be to include autonomous UAV Swarm algorithms to allow higher density data for a surveyed area, or increased total surveyable area.

References

1. H. Choset, Coverage for robotics – a survey of recent results. Ann. Math. Artif. Intell. **31**(1), 113–126 (2001). ISSN: 1573-7470. https://doi.org/10.1023/A:1016639210559
2. S. Chung, A.A. Paranjape, P. Dames, S. Shen, V. Kumar, A survey on aerial swarm robotics. IEEE Trans. Robot. **34**(4), 837–855 (2018). ISSN: 1941-0468. https://doi.org/10.1109/TRO.2018.2857475
3. L. Garberoglio, P. Moreno, I. Mas, J.I. Giribet, Autonomous vehicles for outdoor multidomain mapping, in *2018 IEEE Biennial Congress of Argentina (ARGENCON)*, June 2018, pp. 1–8. https://doi.org/10.1109/ARGENCON.2018.8646054
4. Garmin International Inc., LiDAR Lite v3 operation manual and technical specifications. Technical report, 2017. https://static.garmin.com/pumac/LIDAR_Lite_v3_Operation_Manual_and_Technical_Specifications.pdf
5. A. Gyr, K. Hoyer, *Sediment Transport: A Geophysical Phenomenon*. Fluid Mechanics and Its Applications (Springer Netherlands, Dordrecht, 2006). ISBN: 140205016X, 9781402050169. https://books.google.com/books?id=veWBel6nrnkC
6. HomeAdvisor. Hire a land surveyor (2019). https://www.homeadvisor.com/cost/architects-and-engineers/hire-a-land-surveyor
7. J.G. Linders, The use of structured digital road network data bases for dispatching and routing of emergency service, in *Conference Record of papers presented at the First Vehicle Navigation and Information Systems Conference (VNIS '89)*, September 1989, pp. A54–A59. https://doi.org/10.1109/VNIS.1989.98824
8. L. Meier, P. Tanskanen, L. Heng, G.H. Lee, F. Fraundorfer, M. Pollefeys, Pixhawk: a micro aerial vehicle design for autonomous flight using onboard computer vision. Auton. Robots **33**(1–2), 21–39 (2012). ISSN: 0929-5593. https://doi.org/10.1007/s10514-012-9281-4
9. National Oceanic and Atmospheric Administration, What is a watershead? (2018). https://oceanservice.noaa.gov/facts/watershed.html. Accessed 25 Oct 2019
10. J. Neumann-Williams, A. Neumann, Interactive hiking map of Yosemite National Park, in *Proceedings SVGopen 2005*, 2005. http://www.svgopen.org/2005/papers/abstract_williams_yosemite_national_park/index.html
11. M. Pidwirny, *Topographic Maps. Fundamentals of Physical Geography*, 2nd edn. (2006). http://www.physicalgeography.net/fundamentals/2d.html Accessed 25 Oct 2019
12. F.I. Roberson, W.M. Kaula, Laser altimeter, NASA SP-289 Apollo 15 preliminary science report. Technical report, NASA-GSFC, 1972
13. Z. Scholl, Benchmarking the GPIO pins on a raspberry Pi (2015). https://rpiai.com/raspberrypi/gpio/
14. R.V. Sliuzas, M.J.G. Brussel, Usability of large scale topographic data for urban planning and engineering applications: examples of housing studies and DEM generation in Tanzania, in *ISPRS 2000 Congress: Geoinformation for All* Amsterdam, 16–23 July, 2000 (International Society for Photogrammetry and Remote Sensing (ISPRS), Vienna, 2000), pp. 1003–1010
15. A. Tariq, S.M. Osama, A. Gillani, Development of a low cost and light weight UAV for photogrammetry and precision land mapping using aerial imagery, in *2016 International Conference on Frontiers of Information Technology (FIT)*, December 2016, pp. 360–364. https://doi.org/10.1109/FIT.2016.072
16. The Raspberry Pi Foundation. Raspberry Pi 4 Model b. https://www.raspberrypi.org/products/raspberry-pi-4-model-b/
17. UAV Navigation: Cutting edge Autopilots. UAV navigation in depth: altimeters. https://www.uavnavigation.com/company/blog/uav-navigation-depth-altimeters
18. L. Wang, B. Yuan, Curvature and density based feature point detection for point cloud data. in *IET 3rd International Conference on Wireless, Mobile and Multimedia Networks (ICWMNN 2010)*, September 2010, pp. 377–380. https://doi.org/10.1049/cp.2010.0694

19. B.G. Whiteside, R.M. McNatt, Fish species diversity in relation to stream order and physicochemical conditions in the plum creek drainage basin. Am. Midland Nat. **88**(1), 90–101 (1972). ISSN: 00030031, 19384238. http://www.jstor.org/stable/2424490
20. H.-J. Yoo, F. Goulette, J. Senpauroca, G. Lepere, Simulation-based comparative analysis for the design of laser terrestrial mobile mapping systems. Boletim de Ciências Geodésicas **15**, 01 (2010)
21. K. Zainuddin, N. Ghazali, Z.M. Arof, The feasibility of using low-cost commercial unmanned aerial vehicle for small area topographic mapping, in *2015 IEEE International Conference on Aerospace Electronics and Remote Sensing Technology (ICARES)*, December 2015, pp. 1–7. https://doi.org/10.1109/ICARES.2015.7429825

Wireless Blind Spot Detection and Embedded Microcontroller

Bassam Shaer, Danita L. Marcum, Curtis Becker, Gabriella Gressett, and Meridith Schmieder

1 Introduction

Large blind spots present safety hazards when driving around large semitrucks as noted in Fig. 1. Many accidents occur every year when the driver of the semitruck is turning or changing lanes while vehicles are in these blind spots. Our goal is to implement a wireless blind spot detection system for semitrucks with a user-friendly interface. The sensors must be easily attached and removed without any permanent changes to the truck exterior. Blind spot detectors have been implemented in many passenger cars and SUVs these days by big name brands such as Hyundai, Chevrolet, and Mercedes. There is a similar product in the market made by company called Goshers that performs a similar function [7]. However, it requires a professional installation by a local installer, and it requires drilling into the truck's exterior.

This design will not require professional installation. The driver will simply turn on the sensor, link it with the receiver, and attach it to the vehicle; no additional help required. As a wireless system, there will be no drilling into the truck (tractor) and trailer. Many drivers do not own the trailers they are towing; they drop off one trailer, pick up another one, and proceed on their way. The sensor design of this project will be easily detachable when needed so that the system can also be used on other trucks rather than being permanently fixed to one truck. The small receiver, about the size of a GPS unit, will mount on the dash, between the steering wheel and the driver's window. This allows the driver to scan the display while looking to their mirrors.

B. Shaer (✉) · D. L. Marcum · C. Becker · G. Gressett · M. Schmieder
Emerald Coast, ECE-University of West Florida, FWB, Pensacola, FL, USA
e-mail: bshaer@uwf.edu; dmarcum@umf.edu

K. Daimi et al. (eds.), *Advances in Security, Networks, and Internet of Things*,
Transactions on Computational Science and Computational Intelligence,
https://doi.org/10.1007/978-3-030-71017-0_51

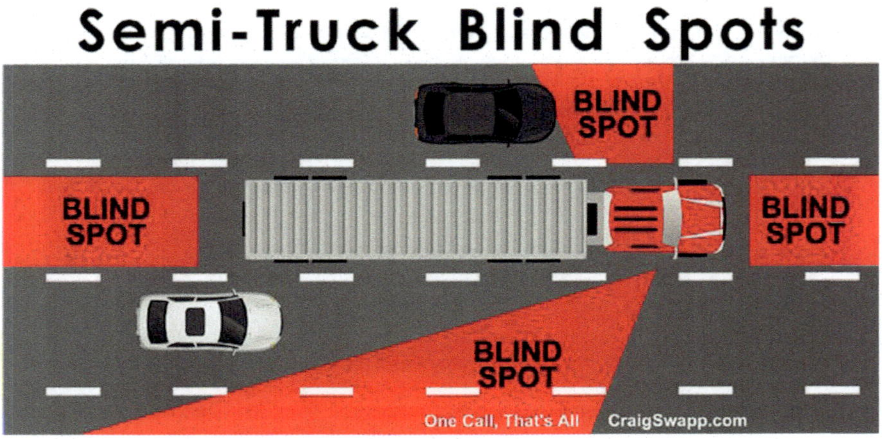

Fig. 1 Blind spot areas of a semitruck

A portable blind spot detection system for a semitruck that is easy to set up and operate is the goal. The entire system is wireless and rechargeable and can be set up in 5 minutes or less. The sensors are placed around the outside of the vehicle in multiple blind spot areas. LED indicators will appear on the receiver that is located inside the cab of the truck to alert the driver when an object has been detected in a blind spot area.

2 Technical Backgrounds

The main design objectives include the sensors, the receiver, and the wireless charging tray. The sensors will be wireless and located inside of a detachable unit that is rechargeable and will signal transmission when an object is in their blind spot. The receiver will contain compact LED indicators and will also be rechargeable. The wireless charging tray will charge all of the wireless parts of the system simultaneously just by placing each sensor and the receiver on the charging pad using the 12 volt system of the vehicle.

The receiver will be a small GPS-sized unit located inside of the vehicle with LED indications from the sensors whenever an object is detected inside the blind spot perimeter that the sensor is monitoring. The receiver in the cab will be located between the steering wheel and the window on the dash and will be attached by a suction cup. It will indicate when something is located within the range of that sensor. The receiver will display an outline of the tractor and trailer with LED indicators that light up when something is in that blind spot. This easy-to-read display will give the driver the information in a quick glance to limit the process time for the driver. The indicators will distinguish which sensor the data is coming from and if multiple sensors detect an object at the same time. The receiver can also be connected directly via USB for power during driving.

The sensors will be aerodynamic, rechargeable, and easily detachable from the outside of the vehicle. They also need to be able to withstand varying weather conditions without detaching or deteriorating. Charging all seven sensors and the receiver will be accomplished on a single charging tray that provides inductive charging for all eight items simultaneously. The sensors must be able to operate a minimum of 12 hours before recharging.

Seven sensors will be placed at key points on the outside of the vehicle to create a blanket around the vehicle. Each sensor will be made up of an Arduino Pro Mini, a JSN-SR04T Waterproof Ultrasonic Rangefinder, an XBee DigiMesh 2.4 RF Transmitter/Receiver module (XBee), a power charging circuit, and a 3.7 V lithium-ion (Li-ion) battery.

The receiver will be made up of an Arduino Uno Rev3, a 2.3" TFT LCD, an XBee, a power charging circuit, and a 3.7 V Li-ion battery.

An induction charging system for wirelessly charging all the devices simultaneously while the driver is stopped will be included. This system consists of a charging tray with sockets for seven sensors and the receiver, as alignment of the coils from tray to unit is necessary to consistent charging. For ease of use, the tray will operate using the 12 VDC from the vehicle plug-in cigarette lighter port.

3 The Proposed Approach

To better simplify and analyze the desired inputs/outputs and functionality of the system, a model to visualize the system's desired processes can be shown in a block diagram as shown in Fig. 2.

This diagram shows three major systems of the wireless blind spot detection system: the receiver, sensors, and wireless charging tray. Designed to be wireless, there is no physical connection between the receiver and each of the sensors. But the connection in the diagram between the receiver and sensor through the XBee components is utilized to express where these two major systems communicating back and forth with each other as objects are detected in the blind spot areas.

The connection between the induction input boards between the receiver and sensor to and charging tray also demonstrates where those components are interacting even though there is no physical connection in the design such as a wire. The receiver and sensors will be placed on top of the charging tray over each induction coil input board location. Although there are seven sensors to cover all of the blind spot areas around the semitruck, this diagram shows the connection between a single sensor and the receiver to prevent overcrowding the functional block diagram.

The wireless charging tray includes eight induction input boards, one for the receiver and seven for the sensors, connecting the 12 V car adapter to the cigarette lighter inside the cab of the truck for charging.

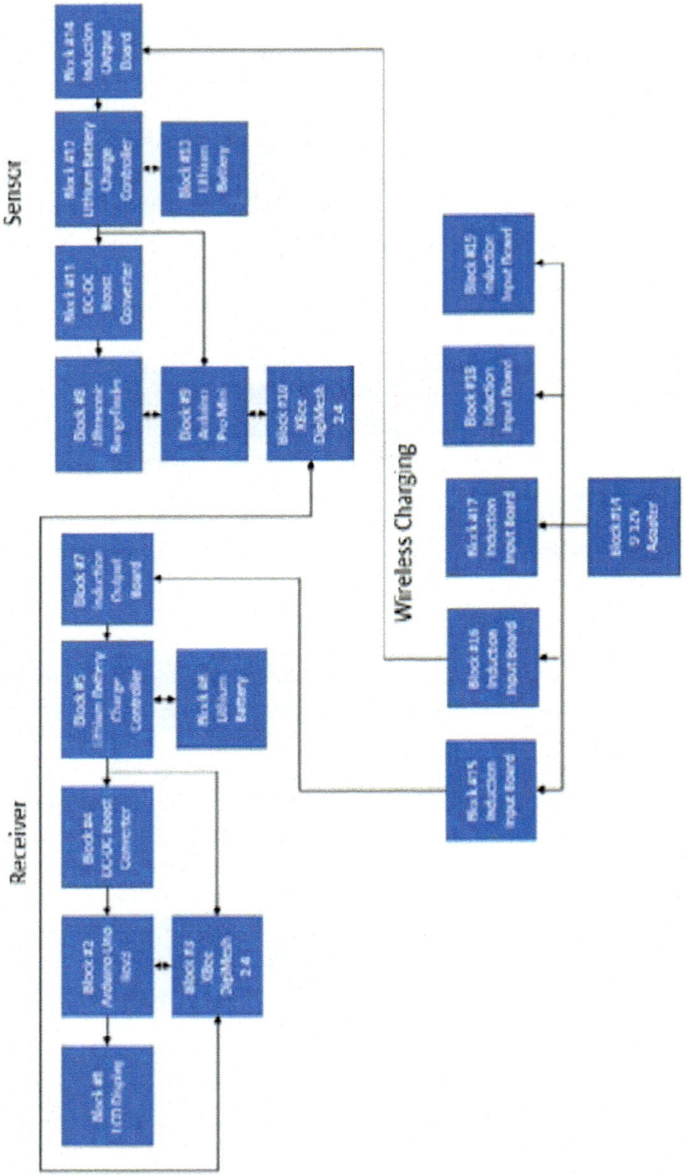

Fig. 2 Functional block diagram of system

The LCD display serves as the main display for the receiver and will require 3–5 V from the battery. The display will possibly be a touch screen. The operator will have to look at the display while driving to see where there is an object in the blind spot.

An Arduino Uno, the receiver system microcontroller, will require at least 5 V from the USB. It works with the display and the transmitter. Although Arduino IDE is used to program the receiver, no operator interaction with the program is necessary.

The XBee will require 3.3 V to operate. The XBee transmits and receives information between Arduinos in the receiver and the sensors. The operator will not have direct interaction with the transmitter. While some knowledge of transmitter protocols is expected, additional research was complete to ensure the correct transmitter/receiver was chosen [6].

The DC-DC Boost Converter provides a + 5 V DC step-up voltage with an input of 3–12 V.

The Li-ion battery charge controller is used for the battery in the receiver, controlling the charge received based on current.

The Li-ion battery is a rechargeable battery for the power system in the receiver and sensors with a maximum input charging current of 500 mA.

The induction output circuit card will be providing the voltage used by the Li-ion charger to charge the built-in battery.

The Arduino Pro Mini (Pro Mini) will be attached to the JSN-SR04T and is powered by a Li-ion battery. The Pro Mini will communicate with the receiver to indicate when the Ultrasonic Rangefinder is detecting a vehicle.

The JSN-SR04T is the waterproof sensor module that will be implemented to detect vehicle presence in a blind spot. Connected and powered by the Arduino Pro Mini, the JSN-SR04T will detect a vehicle and the vehicle's distance from the truck and send this information to the Pro Mini.

The wireless charging tray is the main charging platform for the system. Connected to a 12 V power source, the charging tray will charge the sensors and the receiver using an inductive charging circuit that will send a charge to the inductive circuit in the sensors and receiver. These internal circuits will charge the Li-ion battery inside the sensors and receiver.

4 Experimental Design

4.1 Ultrasonic Rangefinder: JSN-SR04T

A JSN-SR04T uses ultrasonic sound to detect distance from an object [1, 2]. This is done by transmitting/receiving ultrasonic signals through an ultrasonic transducer. Once the sound is reflected off an object back to the transducer, the sensor converts it into an electrical signal [2]. From the pulse width (pw) of this signal, the distance can be calculated as

$$\text{distance} = \text{speed} * \text{time} \quad [1]-[2]$$

$$\text{Speed of sound } (c) = 331.4 + (0.606 * T) + (0.0124 * 50) \tag{1}$$

At room temperature (23 °C) and average humidity (50%),

$$c = 331.4 + (0.606 * 23) + (0.0124 * 50) = 331.4 + 13.938 + 0.62 = 345.958\ m/s \tag{2}$$

If a 25 ms pulse is detected,

$$d = \frac{c * pw}{2} = \frac{345.958\frac{m}{s} * 0.025\ s}{2} = 4.36\ m \tag{3}$$

The distance travelled was divided by two because the sound is being reflected back, thus travelling twice the distance from the object.

For the JSN-SR04T sensor, the trigger pin and echo pins are connected to two separate digital pins on the Arduino Pro Mini [2], shown in Fig. 3. The trigger pin is then set as an output and made high for 10 ms. The sensor will then begin to transmit the eight 40 kHz ultrasonic pulses. The echo pin is set as an input, and the distance is calculated from the pulse width of the electrical signal [2]. Other types of sensors such as mmWave sensors [5] were considered but found to be cost prohibitive.

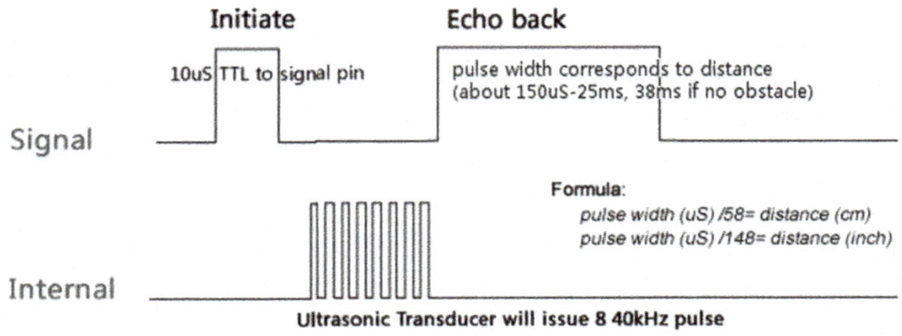

Fig. 3 SEN0208 signal properties (illustration from dfrobot.com) [2]

4.2 *LCD Display ILI9341*

To connect the LCD to the Arduino Uno, the LCD must be put into SPI Mode. Soldering the IM1, IM2, and IM3 pins on the back of the LCD breakout board closed enables SPI Mode while the CS connection is used for the SD card slot [3].

Serial Peripheral Interface (SPI) is a common interface used to communicate between devices [4]. In SPI connections, slave devices are connected to a master device and provided a serial clock (SC) [4]. A Master-Out Slave-In (MOSI) pin and a Master-In Slave-Out (MISO) are used to communicate, and a slave select (SS) pin is used to select which device to communicate with [4].

For the ILI9341 in SPI Mode, the devices are connected as follows:

- 3–5 V to Arduino Uno Rev3 5 V pin
- GND to Arduino Uno Rev3 GND pin
- CLK to Arduino Uno Rev3 SPI clock, digital pin 13 (D13)
- MISO to Arduino Uno Rev3 SPI MISO, digital pin 12 (D12)
- MOSI to Arduino Uno Rev3 SPI MOSI, digital pin 11(D11)
- CS to Arduino Uno Rev3 SPI CS, digital pin 10 (10)
- D/C to Arduino Uno Rev3 SPI data/command select (SS), digital pin 9 (D9)

4.3 *Receiver Flowchart*

The flowchart in Fig. 4 represents how the program works in the receiver. At startup, the program initializes and then displays the vehicle diagram on the LCD display screen.

The flowchart in Fig. 5 represents how the program works in the sensor. At startup, the program initializes the sensor, and then the Arduino Pro Mini sends a pulse to the ultrasonic transducer and receives the duration of the echo.

5 Results and Discussion

5.1 *Receiver*

Figure 6 shows the schematic drawing for the receiver board. The LCD is connected to the Arduino UNO in an SPI configuration. The Arduino Uno is then connected to the XBee through serial transmit and receive pins. The XBee is then connected to the Li-ion battery charge controller as a power source. Additionally, the Arduino Uno is being powered by a 3.7 V Li-ion battery with a 3.7 V DC to 5 V DC boost converter. The Li-ion battery charge controller is used to regulate the 5 V output of the induction board and to charge the Li-ion battery.

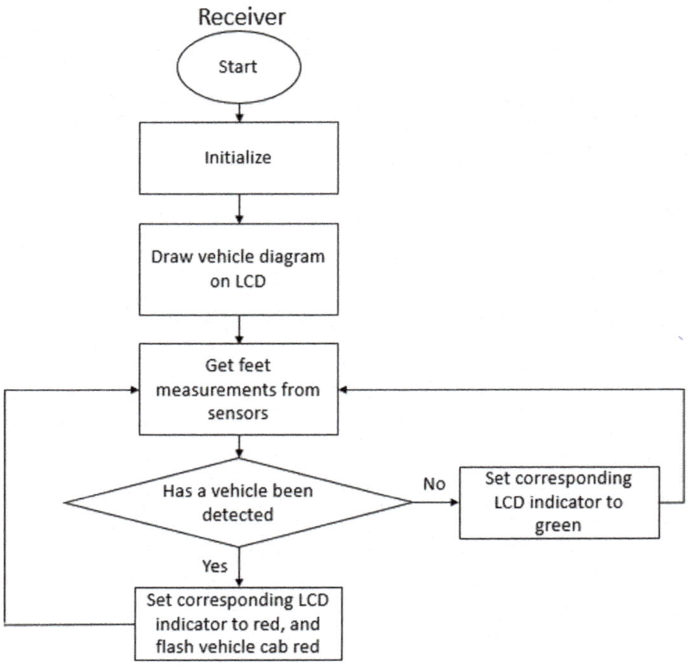

Fig. 4 Flowchart for the receiver board

5.2 *Sensor*

Figure 7 is the schematic drawing for the sensor. The power section contains the 3.7 V Li-ion battery, transfer coil and induction power circuit card, and the battery charging circuit card. The charging circuit limits the current to charge the battery without overcharging.

The power system provides the power for the Arduino Pro Mini and the Ultrasonic Rangefinder. The Ultrasonic Rangefinder trigger signal and echo use two digital pins. The Arduino Pro Mini is connected to the XBee, providing 3.3 V for power, and transmit and receive serial pins. The XBee provides communication with the receiver.

5.3 *Wireless Charging Tray*

The schematic of the wireless charging tray show, in Fig. 8, a copper coil inside of the tray for each of the seven sensors and the receiver resulting in eight total coils. Each coil is attached to a pulse width modulator to convert the 12 V DC to an AC signal, enabling inductive charging.

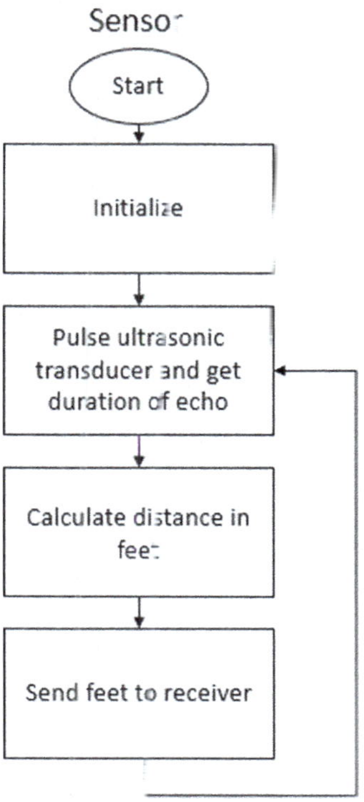

Fig. 5 Flowchart for the sensor board

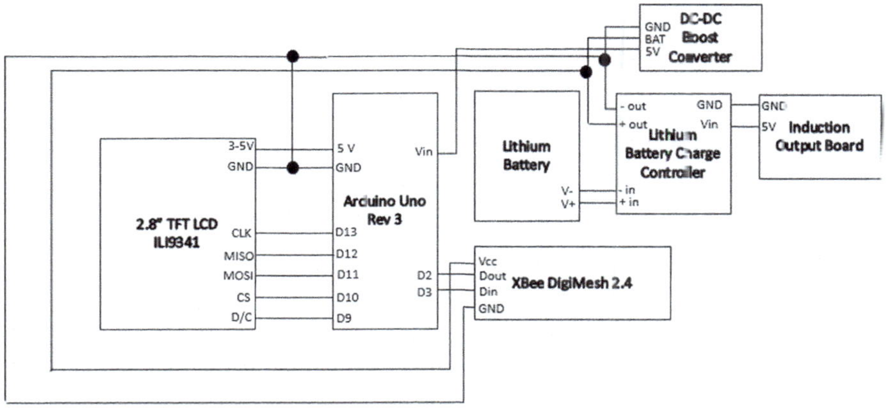

Fig. 6 Schematic for the receiver board

Each component that required charging is placed in the slot on the exterior surface of the charging tray in one of the coil locations (Fig. 9).

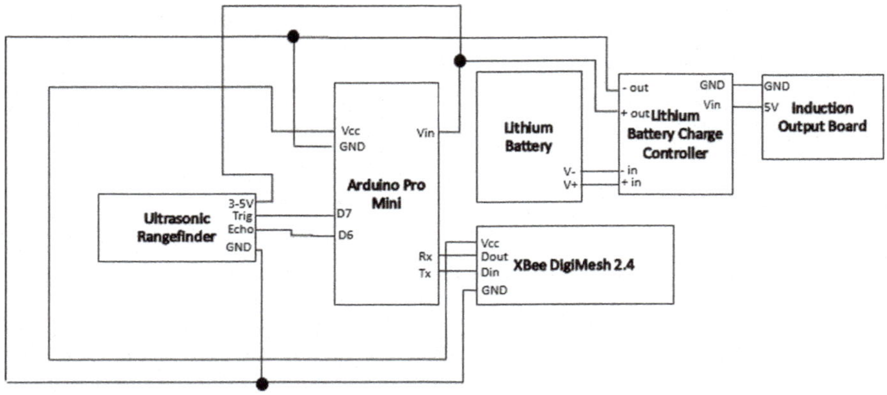

Fig. 7 Schematic for the sensor board

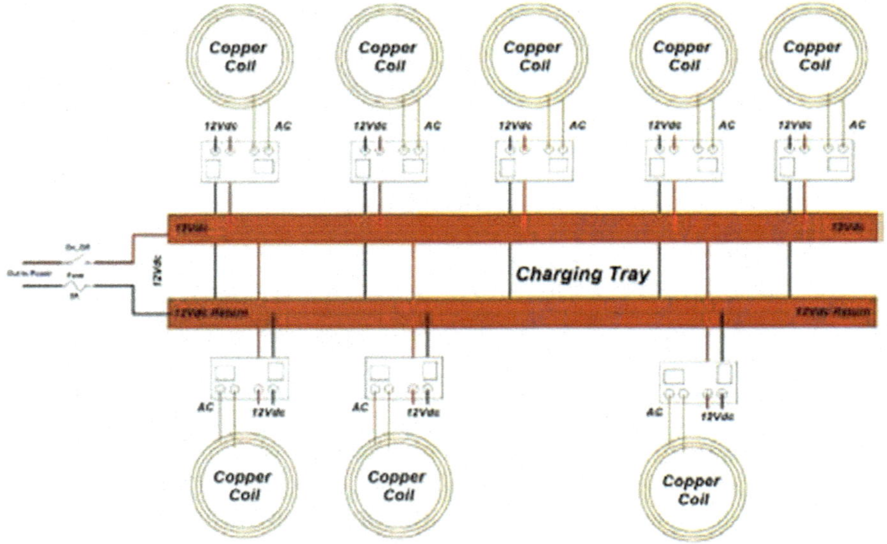

Fig. 8 Schematic for the charging tray

6 Operational Results and Testing

Figure 10 demonstrates the early testing for the receiver to process a signal from the sensor and display on the LCD screen. In the top left picture is the initial setup of the sensor system. The top right shows the initial program using four sensor points. Note that nothing is in the line of sight for the ultrasonic sensor while no indication is given of something. In the left bottom, the sensor now has something in its line of sight. In the bottom right, note the indication on the LCD screen in red. A physical

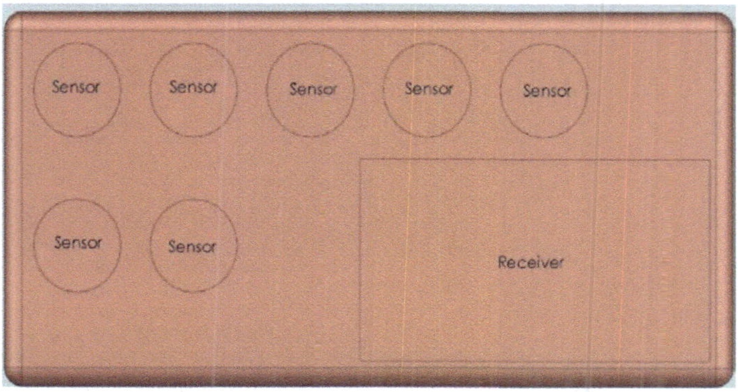

Fig. 9 Charging tray prototype

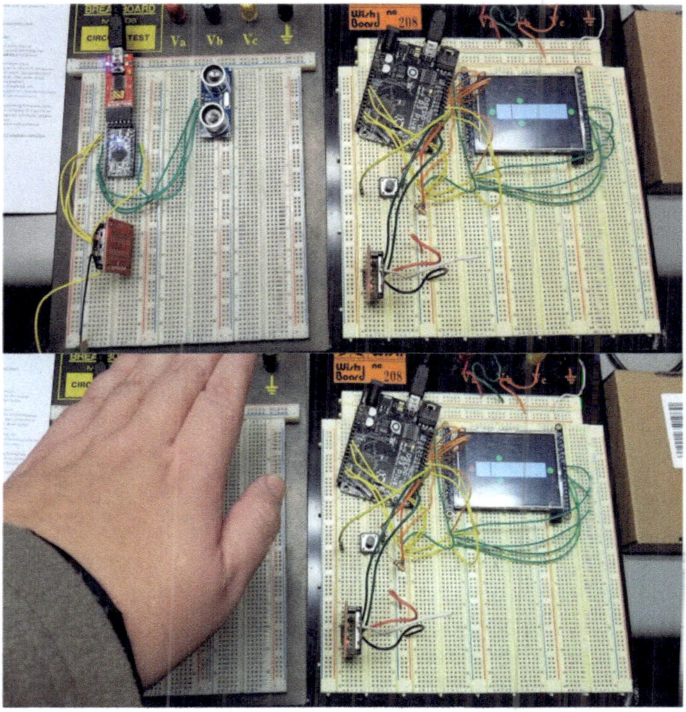

Fig. 10 Wireless blind spot detection system demonstration

demonstration of the working programming as the front sensor transmits a signal to the receiver board. As an object enters the path of the Ultrasonic Rangefinder, the indicator for the sensor in the front will change from green to red.

The final project design achieved is in correspondence with the original design specifications with only minor additions. The DC boost converter was removed from

Fig. 11 Bottom view of sensor

Fig. 12 Outside top view of sensor

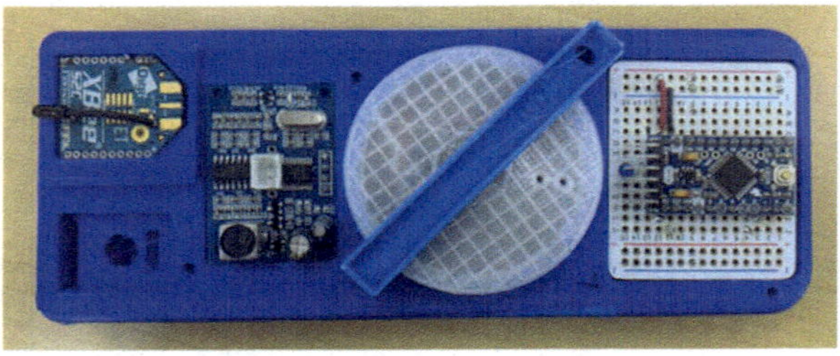

Fig. 13 Inside sensor

the sensor design because it was not needed as previously thought in the preliminary design.

It was also determined that using two strong earth magnets, as shown in Fig. 11, would be sufficient for each sensor to attach them to the truck and trailer. The spacers and metal rods were used inside of the housing to connect the housing and the bottom plate and to hold the inner components in place so that they could not move around inside of the housing. Figure 12 is the top outside view of the sensor, with the JSN-SR04T waterproof sensor visible. Figure 13 is the inside of the sensor before the top cover is installed. From left to right, XBee, JSN-SR04T Circuit Card, Power Module, and Pro Mini on Proto Board.

The receiver, as shown in Figs. 14 and 15, included an on/off switch for power as shown in which was added into the receiver design later. The display outlines a semitruck with trailer along with the sensor locations (green dots). When the dot changes from green to red, something is in the blind spot.

The housing for the system was 3D printed by members of the team.

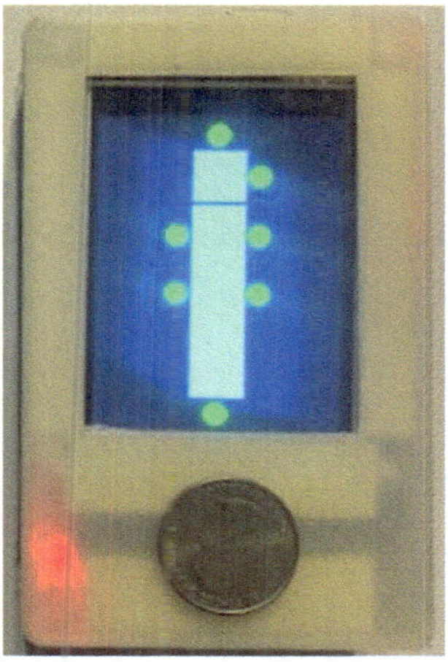

Fig. 14 Top of receiver with LCD display

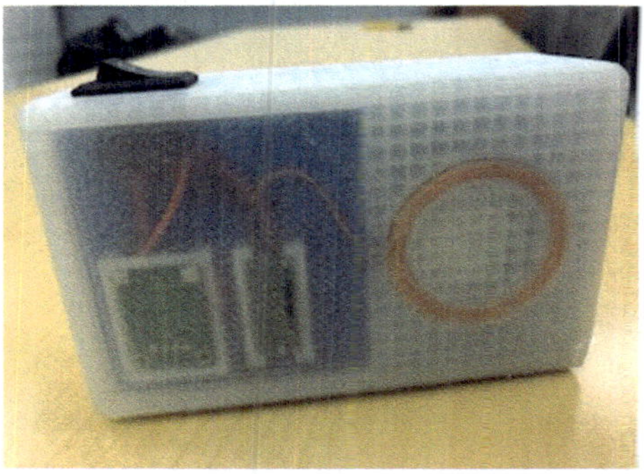

Fig. 15 Bottom view of receiver

7 Conclusion

This paper presented the design and implementations of a portable blind spot detection system for semitrucks. Overall, all seven sensors, the receiver, and charging tray are properly working. The sensors are able to detect objects around the vehicle in various blind spot areas simultaneously, and the receiver accurately displays these results without any delay. The charging tray is able to be powered in the truck through the cigarette lighter adapter and charges all seven removable sensor modules and the receiver at the same time. The most difficult issue with this project was with the 3D print quality, which required many parts to be reprinted multiple times or to find other resources for printing in order to have accurate tight fits. Future improvements could be made by using better 3D printers, materials, and overall outer sensor design to prevent the sensor from losing connection.

Future work would include creating custom circuits for a smaller footprint, especially the sensor. By reducing the size of the sensor, the size of the charging tray could also be reduced. A smaller receiver could be made thinner while increasing the display size.

References

1. How to Set up an Ultrasonic Range Finder on an Arduino, circuitbasics.com, [Online]. Available: http://www.circuitbasics.com/how-to-set-up-an-ultrasonic-range-finder-on-an-arduino/. [Accessed Nov. 22, 2018]
2. Weather - proof Ultrasonic Sensor with Separate Probe SKU: SEN0208, dfrobot.com, [Online]. Available: https://www.dfrobot.com/wiki/index.php/Weather_-_proof_Ultrasonic_Sensor_with_Separate_Probe_SKU_:_SEN0208. [Accessed Nov. 22, 2018]
3. L. Ada, Adafruit 2.8 and 2.2 Color TFT Touchscreen Breakout v2, adafruit.com, [Online]. Available: https://learn.adafruit.com/adafruit-2-dot-8-color-tft-touchscreen-breakout-v2. [Accessed Nov. 21, 2018]
4. SPI Tutorial, corelis.com, [Online]. Available: https://www.corelis.com/education/tutorials/spi-tutorial/. [Accessed Nov. 21, 2018]
5. Automotive mmWave sensors, ti.com, [Online]. Available: http://www.ti.com/sensors/mmwave/awr/overview.html. [Accessed Oct. 01, 2018]
6. Automotive mmWave sensors, electronicdesing.com, [Online]. Available: https://www.electronicdesign.com/what-s-difference-between/what-s-difference-between-ieee-802154-and-zigbee-wireless. [Accessed Nov. 21, 2018]
7. Goshers. [Online]. Available: http://www.goshers.com/blind-spot-detection-system-bsds-003306-for-small-bus-box-truck-and-commercial-fleets/

BumpChat: A Secure Mobile Communication System

Brian Kammourieh, Nahid Ebrahimi Majd, and Ahmad Hadaegh

1 Introduction

We live in a world where we are more connected than ever. Individuals are able to communicate with each other over a vast variety of services. Each service, however, comes with trade-offs, usually between ease of use and privacy. There is a subset of the global population whose privacy is constantly under threat. There are journalists and activists who can be persecuted for the news they share or publish. Some states have been known to access the private messages of their citizens to single out individuals to stop the spread of information. Whistleblowers also need a secure means of communication with journalists. A prime example is Edward Snowden, who implemented extreme measures to ensure that the information he possessed was able to get into the hands of journalists. The encrypted email service LavaBit was forced to shut down or else they would have been forced to turn over their private key that encrypted all of their user's private communications.

The goal of BumpChat is to provide a secure communication infrastructure that is both open source and privately hosted. An individual or group of individuals can spool up their own secure server on either their own hardware or through a cloud server. BumpChat is composed of a mobile messaging app for the Android platform and a backend API developed in PHP. Security is provided through the use of end-to-end encryption, at-rest encryption, RSA authentication, and AES message encryption. The addition of NFC to the pairing process removes the need to trust internet infrastructure with key data that underpins the security of communications.

B. Kammourieh · N. E. Majd (✉) · A. Hadaegh
Department of Computer Science and Information Systems, California State University, San Marcos, CA, USA
e-mail: kammo002@cougars.csusm.edu; nmajd@csusm.edu; ahadaegh@csusm.edu

K. Daimi et al. (eds.), *Advances in Security, Networks, and Internet of Things*, Transactions on Computational Science and Computational Intelligence,
https://doi.org/10.1007/978-3-030-71017-0_52

One of the downsides of the highly regarded Signal messaging app [1] is the need for a public identifier that is held on their servers. This allows a third-party entity to verify that a user is part of the network by entering either their phone number or their username into their contacts. Signal will verify that a user has an account with Signal without notifying the end user. With the design of BumpChat, it is impossible to determine or verify that two parties are communicating with each other without having both devices unlocked and in your hands.

The main benefits of using BumpChat over similar commercial applications (Signal, WhatsApp, Facebook Messenger) are as follows:

1. The key exchange is done person to person without a central authority involved, which puts the trust between users.
2. A major benefit to the key exchange process is the inability to identify if a user is using the system. With all other applications, there is some identifier (email or phone number) that another user or attacker can use to find someone. With BumpChat, the identifiers are cryptographically generated and only used for communication between two parties.
3. If the server database is ever compromised, the amount of meta-data leaked is minimal by design. It cannot be determined how many inboxes are owned by a single device or even from whom the messages are sent. The goal is to prevent meta-analysis from leaking data about the users.
4. By focusing on the worst-case scenario of infrastructure compromise, the system is designed to keep as much data confidential as possible. BumpChat focuses on anonymity and security.

Section 2 describes the architecture of client and server. Section 3 describes the structure of the designed protocol. Section 4 demonstrates the implementation of the protocol. Section 5 concludes the chapter.

2 Architecture

The system is split into two distinct applications. A native Android application developed in Java and a server-side API developed in PHP. Both the client and server have a local database that is encrypted at rest. The clients and server communicate over HTTPS to ensure that data is encrypted in transit as well. During the pairing process that links two clients together, they each register an inbox on the server using their individual public RSA keys. The hashed RSA public key is used as an inbox identifier that functions similar to an email address except that the client has no control over the choice of identifier. The RSA key also doubles as an authentication key when a user wants to register, retrieve, or send messages.

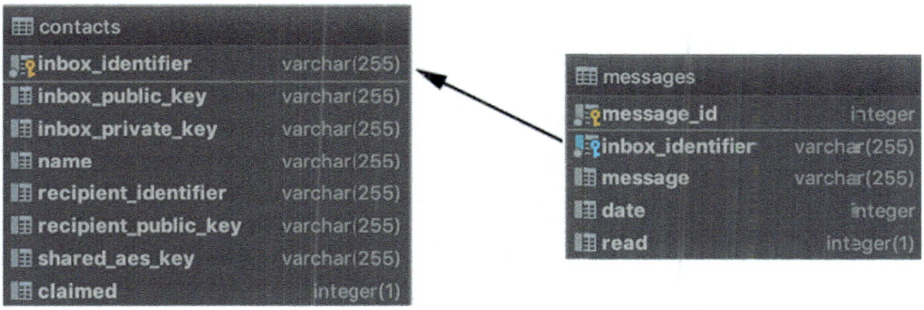

Fig. 1 Client database schema—SQLite

2.1 *Client Database*

The database on the Android client is built on SQLite running the SQLCipher extension. The SQLCipher extension allows the database to be encrypted at rest with AES-256. The database is unlocked at each application and opened with a user supplied password. The schema of the client database is shown in Fig. 1.

A contact represents a pairing between two devices. The inbox_identifier is an SHA-256 hash of the public RSA key. This allows sending and retrieving messages using a smaller identifier than the entire public RSA key. The inbox_public_key and inbox_private_key fields are the Base64 encoded strings of the respective RSA keys. The Base64 encoding allows the keys to be easily stored as string and converted back to a key when needed. The name field is name that the user chose for their recipient and is shown on the front end of the Android application. As the users must meet face to face to add each other as contacts, the name field can either be the other party's real name or an alias so the user knows who they are communicating with when selecting a chat. It is stored only in the encrypted database on the client and not sent to the server. The recipient_identifier and recipient_public_key are the counterparts for the recipient inbox. The shared_aes_key is derived from a Diffie–Hellman exchange during pairing that produces an AES-256 shared key that is used to symmetrically encrypt messages. The claimed field denotes that the inbox has been created on the server and verified with an initial challenge response. The server will not allow any actions to happen until the inbox has been claimed.

2.2 *Server Database*

The database on the server is built on MariaDB with Data-at-Rest encryption enabled. The server stages messages, which are deleted upon retrieval. The schema of the server database is shown in Fig. 2.

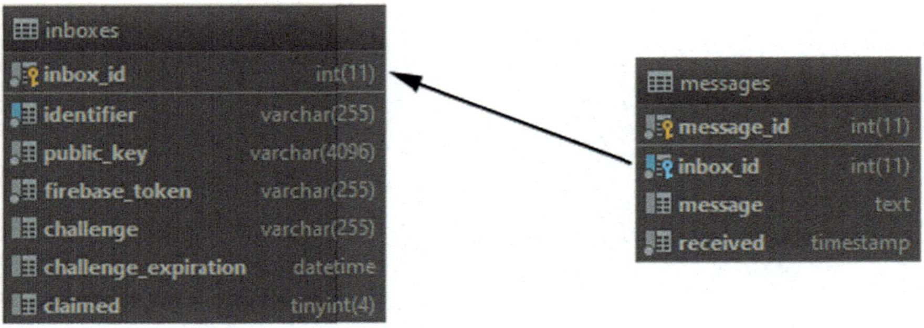

Fig. 2 Server database schema—MariaDB

There are two tables on the server, one for inboxes and the other for messages. The inbox_id is an auto-generated primary key and is not exposed to the client. The identifier is the SHA-256 hash of the RSA-2048 public_key field.

The firebase_token was initially designed to store a Google Firebase token that is used to send push notifications to a device. It was deemed to be a security risk, as the token is always the same for a device and would allow an attacker with access to the database to determine which inboxes are linked to a single device. The challenge and challenge_expiration fields are used for authentication. A random string is generated and encrypted using the public key for the inbox. The device must respond with the decrypted challenge within 5 min of generation. Once the challenge has been answered, the field is cleared to defend from replay attacks. The claimed field denotes that the inbox has been created on the server and verified with an initial challenge response. The server will not allow any actions to happen until the inbox has been claimed. The messages table is very simple with a foreign key inbox_id that references the inbox the messages were sent to. The message_id is an auto-generated primary key that is not exposed to the client. The message field holds the body of the message with a received timestamp that is sent to the client. Messages are cleared once they are successfully downloaded through an API call.

3 Communication Structures

3.1 Direct Client–Client Communication Over NFC

A key feature in the program is the client key sharing process that is achieved through NFC communication. NFC allows the passing of data wirelessly between two devices in extremely close proximity, usually within 4 cm with devices facing back to back. The key sharing process is explained in detail in Sect. 4.7. The goal of the key sharing process is for each user to share their RSA public key with their partner and generate a shared AES key to encrypt messages sent between parties.

The RSA public key is shared so the receiving party can hash it and derive the inbox_identifier of the sending party. An overview of the key sharing process is shown below.

1. Initial Key Generation—Each device generates an RSA-2048 key, a 2048-bit Diffie–Hellman key, and an HMAC-SHA256 salt.
    (a) The RSA key is used for ownership verification of an inbox with the server. It is also used to generate the inbox_identifier by hashing the public key as described in Sect. 2.2.
    (b) The Diffie–Hellman (DH) key is used in part to generate the AES shared key. The DH key allows both parties to securely generate the AES key even if the NFC communication is sniffed during the sharing process.
    (c) The HMAC-SHA256 salt is a random 32-byte array that is passed into the Key Derivation Function along with the DH key to generate the AES shared key.
2. Key Sharing—With NFC enabled on the devices, bringing them in proximity to one another will put the devices into Beam mode. This allows a party to touch their screen and send data one way to the recipient. The sharing process takes 3 Beams in order to complete. Each device tracks their progress in the sharing process using a bit-masked byte called TransferState. The TransferState holds three pieces of information: user sent keys, user received keys, and partner verified key receipt.
    (a) User sending to Partner
        i. User shares their TransferState, RSA public key, DH public key, and HMAC-SHA256 salt with the partner.
        ii. The first party to receive uses the HMAC-SHA256 salt of the other party. In this case, the partner overwrites their local salt with that of the user.
        iii. The partner cannot immediately reply back successful receipt of keys but is able to update their TransferState to denote that they successfully received keys. When the partner makes their transfer, the user will be notified that the keys were received.
    (b) Partner sending to User
        i. Partner shares their TransferState, RSA public key, DH public key, and HMAC-SHA256 salt with the user.
        ii. During this share, the user made aware through the TransferState that the partner successfully received their keys.
    (c) User sending to Partner (again)
        i. User shares their current TransferState to the partner.
        ii. This final share is needed to inform the partner that the user has successfully received the partner's keys.

3. AES-256 Key Generation—Now that both devices possess their partner's public DH key and a shared salt, the AES shared key can be generated. Each user follows a Diffie–Hellman key agreement using DH private key with partner's DH public key. This allows them both to derive the same 256-byte DH shared secret. The new DH shared secret along with the salt is passed into HMAC-SHA256 Key Derivation Function (as specified in RFC 5869). The AES shared key is solely used to encrypt messages sent between parties.

3.2 Client–Server Communication

Communication between the client and server is used to register inboxes, send messages, retrieve messages, and clear inboxes on the server. All communications between the client and server are over HTTPS. Below is the high-level overview of the sequence of events for primary actions and where they happen.

1. Registering an inbox—Inboxes are registered automatically during the key transfer process described in Sect. 4.7.
   (a) Sender device generates a public key pair for a new inbox locally.
   (b) Sender device sends the inbox public_key to the server to create a new inbox.
   (c) Server responds with a challenge encrypted with the public key.
   (d) Sender device sends the decrypted challenge to server to claim the inbox.
   (e) Server acknowledges that the inbox was successfully registered at the server.
2. Sending a message.
   (a) Sender device requests a challenge from the server by sending their inbox_identifier.
   (b) Sender device sends decrypted challenge, recipient_identifier, and the message.
   (c) Server acknowledges that the message successfully stored at the server's message table.
3. Retrieving messages—Messages are retrieved automatically. New messages are retrieved from the server and added to the local database.
   (a) Retrieving device requests a challenge from the server by sending their inbox_identifier.
   (b) Server returns a list of messages for the inbox.
4. Clearing messages—Messages are requested to be cleared by the client software automatically after the previous step succeeds. The message_id of the last message is used to clear all messages up to that one.
   (a) Retrieving device requests a challenge from the server by sending their inbox_identifier.
   (b) Server replies a challenge.

(c) Retrieving device sends decrypted challenge, their inbox_identifier, and the max id of the message to delete.
(d) Server acknowledges that the message was successfully deleted from the server messages table.

4 Implementation

4.1 Server Overview

The server software is written in PHP and is accessed over HTTPS. Currently it resides in an Ubuntu VPS that resides on my home server. HTTPS is achieved through a free 2048-bit LetsEncrypt SSL certificate. The server is accessed through a web API that supports 6 endpoints, which allow access to inboxes and messages. The database is accessed through the MySQLi database driver using parameter binding to prevent from SQL injection attacks.

4.2 Server API Endpoints

Each endpoint requires input data passed over HTTP POST. This ensures that any secret data does not get accidentally stored in web server logs as would be the case with HTTP GET requests. All endpoints return data in a JSON-encoded format. Some endpoints are protected, which require a challenge to be generated before the endpoint is accessed. When the user requests a challenge generation, the server will send a random challenge string encrypted with the inbox's public RSA key and the client must respond with the decrypted challenge before the server allows the action to take place. In this way, an attacker cannot request or forge messages on a user's behalf.

- Challenge/generate.php—This generates a new challenge for an inbox and saves it to the database. It is required to generate a challenge before any protected endpoint is accessed. Required POST data: inbox_identifier
- Inbox/register.php—This creates a new inbox using the RSA public key in PEM-encoded format. Returns the initial encrypted challenge that is used to verify the inbox. Required POST data: public_key_pem
- Inbox/register_verify.php—This finalizes the creation of an inbox. Required to be done before any protected endpoints are accessed. Required POST data: identifier, challenge_response
- Message/clear_all.php—Clears all messages in an inbox up to an upper message id. The upper message id is needed to protect the client from attempting to clear messages it has not retrieved yet. This is a protected endpoint. Required POST data: identifier, challenge_response, upper_message_id

- Message/retrieve_all.php—Returns all incoming messages currently staged on the server. Client is responsible for clearing messages after confirming successful delivery. This is a protected endpoint. Required POST data: identifier, challenge_response
- Message/send.php—Sends a message to a specified inbox. This is a protected endpoint. Required POST data: identifier, challenge_response, recipient_identifier, message

4.3 Android Client Overview

The client Android application is programmed in Java and targets Android 9 and above. The application is composed of a series of Android activities that are screens in the application's user interface. NFC hardware is required in the device for the application to function. Storage is done locally with an SQLite database file located within the apps storage area and is accessed through the Android Room Storage Persistence Library. The library acts as an abstraction layer over direct database access to the SQLite database. The library integrates with the SQLCipher encryption extension that is used to encrypt the database at rest. The app consists of three primary sections: database unlock, message inbox, and message list. Each section has activities that branch off to access specific functionality. The rest of this section details each activity individually. Currently the application is only available as an unsigned .apk file, but part of the future work entails releasing through the Google Play store.

4.4 Master Password Setup (Activity)

This screen comes up on the first time the application is opened. The user is prompted to create a 12-character minimum password. The input fields give visual feedback by turning green when the minimum character length has been reached and when the password confirmation matches. The master password is used to encrypt the SQLite database and is required to be entered on every application to open and decrypt the database. The user is sent to their inbox after successful decryption.

4.5 Database Unlock/Login (Activity)

The login screen displays when the application is opened after the master password has been set. Once the minimum character length of 12 is reached, the input field turns green and the UNLOCK DATABASE button is enabled. The user is sent to their inbox after successful decryption.

4.6 Message Inbox (Activity)

The message inbox is where all the contacts can be accessed. Contacts are ordered by the last message received and show a truncated last message. New contacts can be added in this inbox. Long pressing a contact will enable contact deletion mode.

4.7 Contact Creation

The contact creation process is the novel portion of our proposed protocol in comparison to the existing secure messaging protocols [2–6]. The devices pair by sending keys while in contact with one another over NFC. The steps of the pairing process are described below. The pairing screen is shown with the contact name filled in by the user in Fig. 3a. The contact name and checkboxes must all be filled checked for the SAVE CONTACT button to be enabled.

Step 1: Key Generation—The process starts with each device generating an RSA-2048 key, a 2048-bit Diffie–Hellman key, and an HMAC-SHA256 salt. Once keys are successfully generated, the first checkbox is marked as shown in Fig. 3a. The key generation happens automatically once the activity is started.

Step 2: Each device registers an inbox on the server using the Inbox/register.php endpoint. The RSA public key is sent to the server, and an encrypted challenge is sent back. The client sends the decrypted challenge back to the Inbox/register_verify.php endpoint. The server response is displayed with a small Toast message. On a successful response, the second checkbox is marked as shown in Fig. 3a.

Step 3: Key Sharing—When the phones are placed back to back in close proximity (<4 cm), they will enter an NFC beam mode that allows the one-way data transfer from the initiating device. A transfer is initiated by the user touching the screen as shown in Fig. 3b.

A NdefMessage object is sent from the initiating device to the partner. The message object contains an array of NdefRecord objects that are each a byte[] array. The payload consists of 4 NdefRecords. First is a single byte that holds the transfer state of the sending party that is bitmasked to show: user sent keys, user received keys, and user verified partner received keys. This first record is the one used to update the checkbox state on the UI. The next two records are the sending parties' public RSA key and DH key, respectively. Both keys are encoded as a byte array, which are decoded on the receiving partner's end. The last record is an HMAC-SHA256 salt that each user sends, but only one is used. The first person to receive keys uses the partner's salt instead of their own. This allows both parties to generate the same AES-256 shared key.

The AES-256 shared key is generated during at the end of sharing of the public keys using a Diffie–Hellman key agreement followed by an HMAC-based KDF. This is the key that actually encrypts the messages sent between devices, as it is much less computationally intensive to encrypt using AES than RSA.

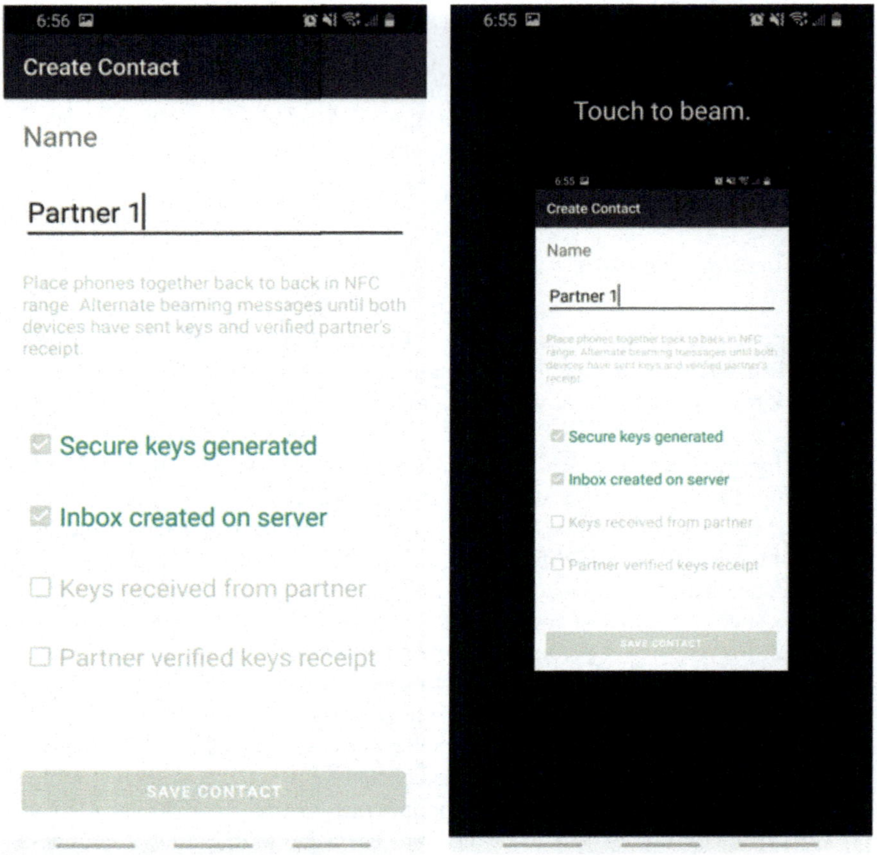

Fig. 3 Initialization. (**a**) Create contact screen with contact name entered and successful inbox creation. (**b**) NFC beam view from Partner 1 before any keys have been exchanged

The key sharing process requires 3 transfers of data in the following order: Partner 1 → Partner 2, Partner 2 → Partner 1, and Partner 1 → Partner 2. The first two transfers are for key and salt exchange. The last transfer is for Partner 2 to verify that Partner 1 received keys. Figure 4a shows Partner 2's UI state after receipt of data from Partner 1. Figure 4b shows Partner 1's UI state after receiving data from Partner 2. Partner 1 at this stage is now aware that Partner 2 has successfully received keys and salt.

After the second data transfer from Partner 1, which makes Partner 2 aware of successful key receipt from Partner 1, the last checkbox in Fig. 4a will be marked. At this stage, all keys and the salt have been exchanged and verified. Both devices' UIs will unlock their SAVE CONTACT buttons, which will create a new Contact object and persist it to the local database as described in Sect. 2.2.

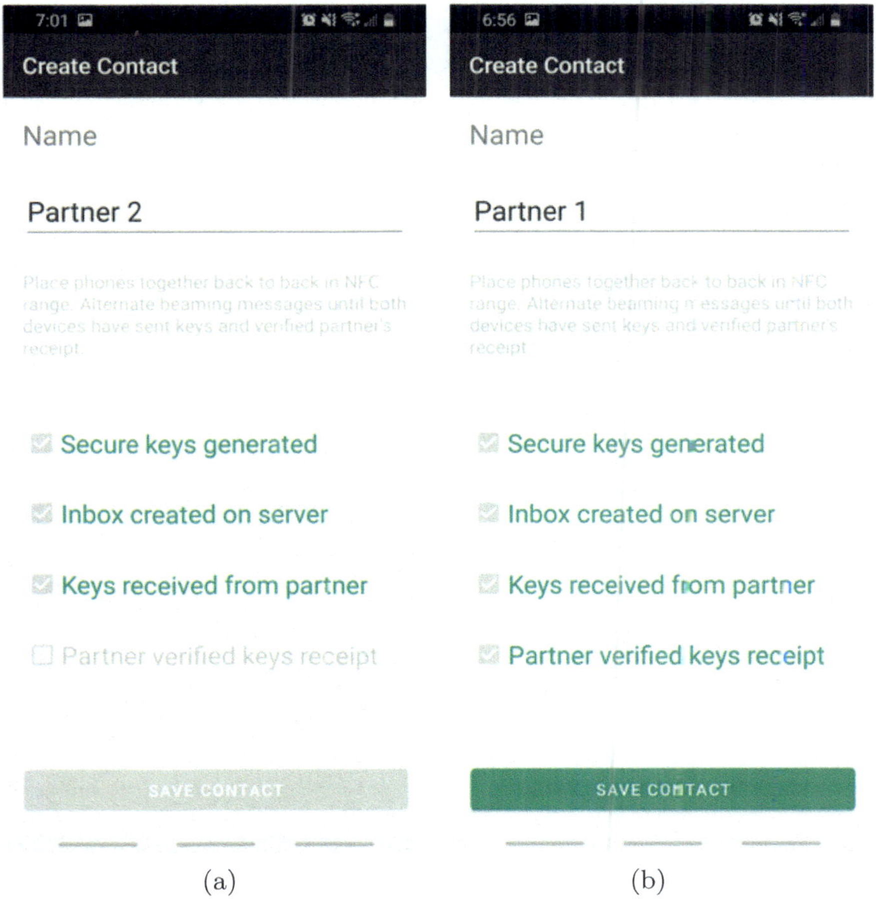

Fig. 4 Key exchange and pairing. (**a**) Pairing view from Partner 2 after keys have been exchanged. (**b**) Partner 1 after receipt of keys from Partner 2

4.8 Message List (Activity)

The message list screen shows all messages with a single contact. Messages are retrieved when the screen is first loaded and are retrieved in an interval as long as the app is in the foreground. Figure 5a shows the message history between two partners. Each message is tagged with a time difference in a readable format that changes from minutes to hours to the date. The sending of a message is shown in Fig. 5b. Pressing the paper airplane icon will encrypt the message with the shared AES-256 key. The message is addressed to the inbox identifier of the partner. The client generates a new challenge and sends it in the Message/send.php with the inbox_identifier, encrypted message, and decrypted challenge response. The triple

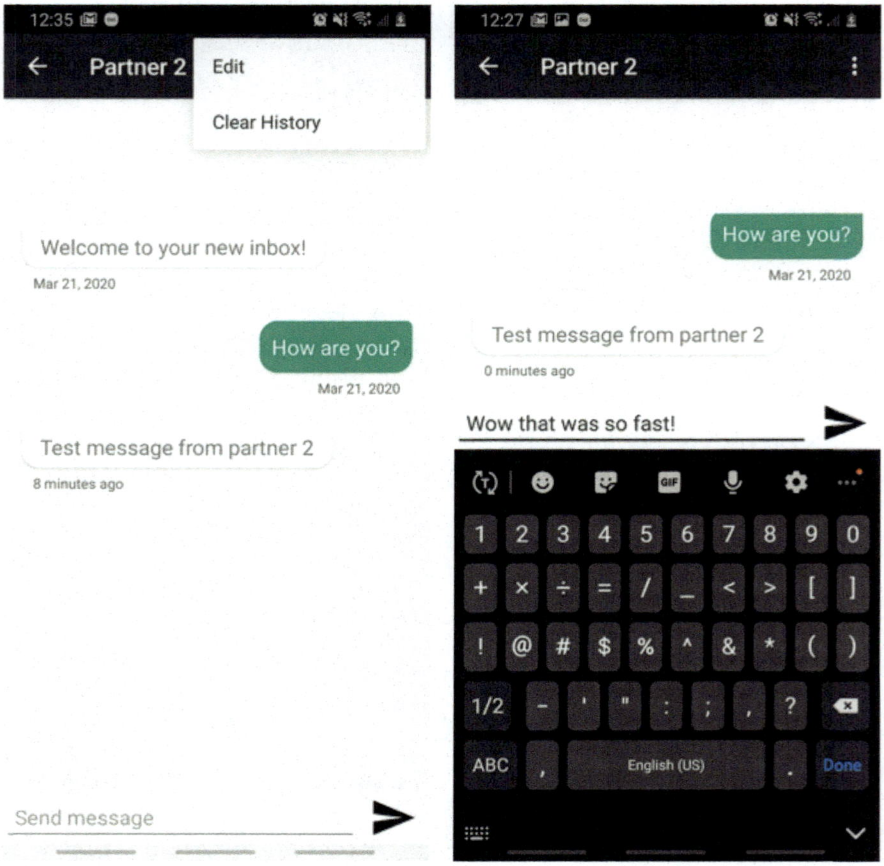

(a) (b)

Fig. 5 Secure messaging and options. (**a**) Message list option's menu. (**b**) User sending a message

dot icon in the top corner triggers the options menu when pressed as shown in Fig. 5a. The Edit option takes the user to the contact edit page where they can change the name of the contact. The Clear History option clears all messages and leaves the chat history with a single "Messages Cleared" message.

4.9 Contact Edit (Activity)

The contact edit page can be reached through the options menu on any chat history page. This page is solely responsible for updating the contact name that shows up in the message inbox and chat history page.

5 Conclusion

The primary goal of this project was to build a novel key sharing mechanism over NFC. This moves the chain of trust from digital space to physical space, as pairing requires physical contact between devices. This project was based on an idea to use an NFC implant to unlock the app and do the key sharing over the internet. However, it felt like the best use of NFC technology was for the key sharing and pairing process. The implementation was challenging. One of the main challenges was getting encryption consistent across Java and PHP. Formatting and encoding were critical in getting both systems to produce the same results with the same input data. The hashing of the RSA public keys is done independently on both the client and server to prevent from a chosen inbox identifier attack. The ability to generate an RSA key that hashes to a chosen inbox is outside the scope of current technology and should be for the foreseeable future. Generating 2048-bit RSA keys on Android devices is on the order of tenths of a second, which makes brute forcing not a feasible option.

Unlocking the app using either an NFC tag or an NFC tag as a secondary authentication method would extend the usage of NFC inside the client while adding extra security. Adding group communication will be an interesting challenge with this new model. The pairing process will need to be revamped to handle sharing keys between devices.

References

1. K. Cohn-Gordon, C. Cremers, B. Dowling, L. Garratt, D. Stebila, A formal security analysis of the signal messaging protocol, in *2017 IEEE European Symposium on Security and Privacy (EuroS&P)*, Paris, 2017, pp. 451–466. https://doi.org/10.1109/EuroSP.2017.27
2. N. Kobeissi, K. Bhargavan, B. Blanchet, Automated verification for secure messaging protocols and their implementations: a symbolic and computational approach, in *2017 IEEE European Symposium on Security and Privacy (EuroS&P)*, Paris, 2017, pp. 435–450. https://doi.org/10.1109/EuroSP.2017.38
3. S. Nayak et al., An application for end to end secure messaging service on Android supported device, in *2017 8th IEEE Annual Information Technology, Electronics and Mobile Communication Conference (IEMCON)*, Vancouver, BC (2017), pp. 290–294. https://doi.org/10.1109/IEMCON.2017.8117222
4. H. Chen, H. Wijayanto, C. Chang, F. Leu, K. Yim, Secure mobile instant messaging key exchanging protocol with one-time-pad substitution transposition cryptosystem, in *2016 IEEE Conference on Computer Communications Workshops (INFOCOM WKSHPS)*, San Francisco, CA (2016), pp. 980–984. https://doi.org/10.1109/INFCOMW.2016.7562224
5. Z. Wang, Z. Ma, S. Luo, H. Gao, Enhanced instant message security and privacy protection scheme for mobile social network systems. IEEE Access **6**, 13706–13715 (2018). https://doi.org/10.1109/ACCESS.2018.2813432
6. N. Unger et al., SoK: Secure messaging, in *2015 IEEE Symposium on Security and Privacy*, San Jose, CA (2015), pp. 232–249. https://doi.org/10.1109/SP.2015.22

Data Collection and Generation for Radio Frequency Signal Security

Tarek A. Youssef, Guillermo A. Francia, III, and Hakki Erhan Sevil

1 Introduction

Unmanned aerial systems (UASs) are rapidly evolving as alternative transport systems for a wide range of applications ranging from commercial to defence purposes. The proliferation of these systems ushers the critical call for their protection. With an ever-increasing number of UASs being interconnected and publicly exposed on the Internet, the need to address the security issues presented by these modern transport infrastructure vehicles has never been so great. There have been several reported adversarial instances of security-related incidents involving UAS. Prominent among these are the following:

- the disruption of the FBI's Hostage Rescue Team by a swarm of UASs in late 2017 [1];
- the trespassing incident in a nuclear facility by Greenpeace activists utilizing a UAS [2];
- the potential threats posed by UAS being used as reconnaissance instruments by criminals to watch Border Control officers [3];

T. A. Youssef
Department of Electrical and Computer Engineering, University of West Florida, Pensacola, FL, USA
e-mail: tyoussef@uwf.edu

G. A. Francia, III (✉)
Center for Cybersecurity, University of West Florida, Pensacola, FL, USA
e-mail: gfranciaiii@uwf.edu

H. E. Sevil
Department of Intelligent Systems and Robotics, University of West Florida, Pensacola, FL, USA
e-mail: hsevil@uwf.edu

K. Daimi et al. (eds.), *Advances in Security, Networks, and Internet of Things*, Transactions on Computational Science and Computational Intelligence,
https://doi.org/10.1007/978-3-030-71017-0_53

- the delivery of illegal items with UAS on prison yards in England in 2017 [4];
- during the battle at Mosul in Iraq, the enemies flew over 300 UAS missions with about a third being armed strike missions [5].

One approach to address the security issues of UAS is to compare their expected behaviour with their actual behaviour. Any discrepancy could be flagged as anomalous. This approach is widely presented in the literature. Most of the techniques presented rely on observer-based methods that are based on threshold logic and hypothesis testing [6]. Some of those methods are worth mentioning here, such as extended Kalman filter (EKF) [7], unscented Kalman filter (UKF) [8], statistical-based methods [9], and state space models [10]. Application areas and implementation platforms for these methods also vary, e.g., multi-rotor platform [11–13], fixed-wing aircraft [14–16]. In terms of multi-agent teams of UAS, distributed systems are widely adopted in recent years in UAS application fields [17], and thus their secure operation becomes crucial. Most of the presented studies are based on maximizing the global gain while having relatively simple individual members in the team [17], such as swarm of UASs [18], multi-agent UAS inspired by insect colony [19], and heterogeneous swarms [20]. However, little or no attention is provided to their secure operations.

A swarm of UASs can exceed over 100 vehicles with a common objective. This notion of group dynamics has been extensively studied (e.g., see [21]). As reported in experimental swarms, a single operator is able to control multiple UASs that are acting as a single entity [22]. Understanding the behaviour of the swarm is the key to effectively countering the entire group. For example, a swarm can have one leader that is directed to search the area and acts as a coordinator to accomplish the mission [23]. Furthermore, communications among swarm members can be either implicit or explicit [24]. While explicit communication may involve short messages, implicit messaging may use sensor signals, data from positioning systems, and/or visual perceptions. More advanced UAS may have the capability to dynamically reconfigure into a different swarm in cases of attrition or situational necessity.

1.1 Chapter Organization

The rest of the chapter is organized as follows. The following section provides a brief introduction to radio frequency (RF) communication followed by a literature review of RF security primarily those on UAS. The chapter then presents a section on RF signal dataset collection and generation with descriptions on the equipment and the software used to process the datasets. Following this segment is a presentation of the preliminary research results of our work on applied machine learning for pattern recognition of RF signals. We use these results to validate our proposed RF security research utilizing the RF signal datasets that we are currently collating. Finally, the chapter ends with some conclusions and directions for future research.

2 RF Signal Communication

Radio frequency (RF), in general, covers a wide range of signal with different applications, such as AM radio and wireless networks [25]. Basic RF communication includes simplex, half-duplex, and full-duplex systems [26]; and mainly, they differ from one another in terms of communication ways.s In simplex systems, the communication allows one-way communications, and contrarily, in half (not simultaneous)- and full (simultaneous)-duplex systems, each end can transmit and receive data [26]. Communication with RF for UAS application consists of transmitter, which is the controller, and the receiver, which is the UAS [25]. Receivers and transmitters share some common goals, conserve energy, and reject outside interference [27]. RF communications used in UAS operations are mostly in frequencies of 900 MHz, 1.3 GHz, 2.4 GHz, and 5.8 GHz. Whereas most of the commercial UASs use 2.4 GHz and 5.8 GHz for control or data streaming [28]. The distance for signal reception depends on various factors, e.g., power, interference, and it can range from 2 km to 75 km [28].

3 Security

3.1 UAS Security

One of the most important subjects in UAS security is the use of RF signals to detect UAS, which can lead to possibility to capture movements of the UAS [29]. The major hardware component needed for this approach is software-defined radios (SDRs) [29]. The changes in received signal strength indicator (RSSI) could result in inferring the UAS body movements [29]. Furthermore, one solution can be to collect unique RF patterns from widely used UAS, train a classifier, and then use that classifier real time to identify UAS of interest [30].

Using RF signals for the detection of UAS has its own challenges [31]. The existing UASs have some specific frequency bands; however, many other products nowadays use the same frequency bands [31]. One can develop a detection system based on monitoring these specific frequency bands, which will lead a high percentage of false alarms [31]. Another feasible technique can be identifying media access control (MAC) address of a UAS. The main challenge with this technique is that it only works if the UAS has an open MAC address. Additionally, a MAC address can be spoofed easily to avoid detection [31].

Geo-fencing is another method to provide RF-based security against UAS [32]. Considering that will create a virtual wall, it will be a challenge to differentiate between different UAS platforms [32]. More importantly, this approach will require the UAS having the necessary software installed by the manufacturer [32]. Most of the modern UASs use Wi-Fi protocol for media streaming [33]. With the assumption of UAS to be detected uses a Wi-Fi protocol, Wi-Fi fingerprinting techniques can be used for detection [33]. Basically, statistical parameters of Wi-Fi communication features, such as average packet time/duration, packet length, are used in machine learning (ML) algorithms to perform classification [33]. The major drawback of this method is the privacy [33]. There is one more aspect of this problem; the appearance and average flying properties of the UAS can be used in detection [34]. Although this approach cannot be successful if used alone, it can improve RF detection methods' accuracy if used in conjunction [34]. These properties include low altitude flying, small size, and high manoeuvrability [34]. For instance, a radar-based method, then, can distinguish between a UAS and a bird [34]. Internet protocol (IP) address-based methods are also presented in the literature. The main idea is to detect IP address that UAS uses, but again, the assumption here is that UAS is using an IP address to connect a router [35]. By itself, this method does not give sufficient security merits, and however, combining with another RF-based approach may lead to more robust detection [35]. Another idea is to use communication between UAV and radio controller, if the UAV is controlled by a radio [36]. There are signature voltage levels of the generated pulses from the controller which can be used for detection [36].

Due to their dynamic topology, UAS networks also raise unexplored security challenges. One approach to increase security in UAS networks and to detect a possibly compromised UAS swarm is to use peer-to-peer information inspired by blockchain [37]. Another security challenge in UAS development is the secure communication between ground station and the UAS itself [38]. In order to address this concern, Multiple Independent Levels of Security (MILS) software architecture is introduced in the literature to mitigate attacks such as Confidentiality, Integrity, Availability [39], Man in the Middle [40], Buffer Overflow, Denial of Service (DoS), Address Resolution Protocol (ARP), and Cache Poison [41]. In the literature, there are also detection systems introduced against various attacks, such as false information dissemination detection, GPS spoofing detection, and jamming detection [42].

3.2 RF Signal Attacks

Attack classes on RF signal-controlled devices have been documented by Andersson et al. [43]. The attack classes include the following:

- *Replay Attack.* RF packets are recorded and replayed to control the device in question.
- *Command Injection.* The attacker injects malicious packets using the required communication protocol format to control the device.
- *GPS Spoofing Attack.* The attacker maliciously sends false GPS information to disable or confuse the device.
- *Malicious Re-pairing.* The attacker clones a remote controller or its functionality to take control of the device.

– *Malicious Reprogramming.* The attacker maliciously alters the controller's firmware with the goal of enabling full remote control.

3.3 Security Enhancements Through Machine Learning

The main theme of this chapter is on the collection and generation of RF signal datasets. The quality of these datasets is the basis for meaningful studies in applied ML for RF signal security. In [44], deep neural networks are used for the detection and identification of drones using an RF signal database. In a related work Zhang et al. [45] used neural network-based detection algorithm for an unmanned aerial vehicle (UAV). In that study, the slope, kurtosis, and skewness of the RF signal are used. In [46], acoustic drone detection and identification using support vector machine (SVM) is presented. Convolutional neural network (CNN) is used for surveillance by detecting the presence of drones from closed-circuit television (CCTV) videos [47].

4 Data Collection and Generation

As previously mentioned, in order to perform a meaningful study of applied ML for RF signal security, particularly that in classification and identification, a reliable dataset is needed. This chapter is dedicated to that end. Similar work can be found in [44] and [48]. The URL of the NIST dataset is https://data.nist.gov/od/id/mds2-2116.

4.1 Methodology

Most of the commercial drones use 2.4 GHz for control communication between remote controllers and drone in addition to the streaming of live video. Some drones use the standard IEEE 802.11 protocol for communication, while others use proprietary protocols augmented with Direct-Sequence Spread Spectrum (DSSS) and/or Frequency-Hopping Spread Spectrum (FHSS) to reduce the impact of interference and ensure reliable communication. Capturing raw RF signals for transmitter utilizing FHSS or DSSS requires a receiver with sufficiently wide bandwidth to cover all possible communication channels. For example, FlySky FS-6IX remote controller works in the frequency range of 2408–2475 MHz While this band is divided into 135 channels, each remote controller uses only 16 channels with channel separation $\geq$1 MHz. The remote controller uses a hopping sequence to avoid jamming. In order to capture raw RF data for this remote controller, the receiver must have, at least, a 67 MHz bandwidth. Other remote controllers use standard Wi-Fi or Bluetooth protocol. The 2.4 GHz Wi-Fi uses a frequency range of 2412–2462 MHz in North America and 2412–2484 MHz in Japan, having a total bandwidth of 72 MHz. Bluetooth-controlled drones work at a frequency range of 2402–2480 MHz, having a total bandwidth of 78 MHz. A radio receiver with 78 MHz bandwidth will be able to capture the raw RF data for most of the commercial drones working at the 2.4 GHz band. RF receivers with such bandwidth are expensive. Thus, in order to reduce the cost, an open-source software-defined radio (SDR) will be used to collect the data. Preliminary experiments indicate the judicious use of multiple SDRs to cover the entire 72 MHz bandwidth could present extra benefits.

4.2 Equipment Setup

To capture the raw RF data, LimeSDR [49] has been selected as a radio receiver. LimeSDR is an open-source software-defined radio with a continuous frequency range from 100 kHz to 3.8 GHz and a maximum bandwidth of 61.44 MHz as shown in Table 1. Since the maximum bandwidth of the SDR receiver is less than 72 MHz, two SDR receivers are used to cover the full bandwidth. The two SDRs are connected to a Windows computer using a USB port. Raw RF data can be collected and stored using open-source SDR software framework such as GNU Radio or Pothosware [50]. Pothos flow graph has been created to capture the information from the two SDR receivers and store it in two separate files in binary format. A binary format is chosen for efficient size and writing speed. However, the binary files can be converted easily into a comma-separated values (CSV) format.

Different types of drones have been selected to generate the RF dataset. The selected drones and controllers' models have been chosen to cover proprietary and standard protocols. The selected drones include DJI Phantom 3 (Wi-Fi), T-65-x-wing (Bluetooth), and FlySky FS-6IX (proprietary protocol with hopping sequence).

4.3 Dataset Attributes and Descriptions

The captured raw RF data are saved in binary files with the following naming convention: *XYZIBN.bin*, where:

Table 1 LimeSDR specification

Frequency range	RF bandwidth	Sample depth	Sample rate	Interface
100 kHz–3.8 GHz	61.44 MHz	12 bits	61.44 MS/s	USB 3.0

- *X* indicates the protocol type: W for Wi-Fi, B for Bluetooth, and P for proprietary protocol;
- *Y* indicates the drone size: M for medium size, S for small size, and U for micro size;
- *Z* indicates the distance from the receiver when RF data are acquired: *L* for long-distance (>1 mile), *M* for medium distance (>0.5 mile and <1 mile), and *S* for short distance (<0.5 mile);
- *I* indicates the Drone ID field—a 3-digit number assigned to each drone model;
- *B* indicates the bandwidth: L for 2402–2441 MHz and *H* for 2441–2480 MHz; and
- *N* indicates a 3-digit assigned file number.

4.4 Dataset Snapshots

The following figures (Figs. 1 and 2) depict the captured RF signal on the 2402–2441 MHz spectrum.

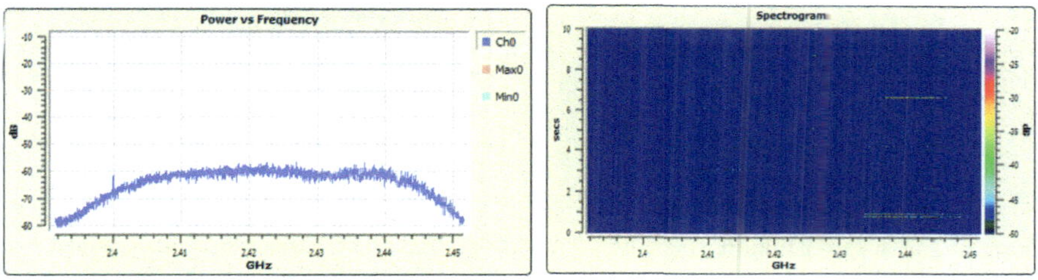

(a) (b)

Fig. 1 (**a**) Background noise spectrum (2402–2441 MHz). (**b**) 1-sec Waterfall background noise (2402–2441 MHz)

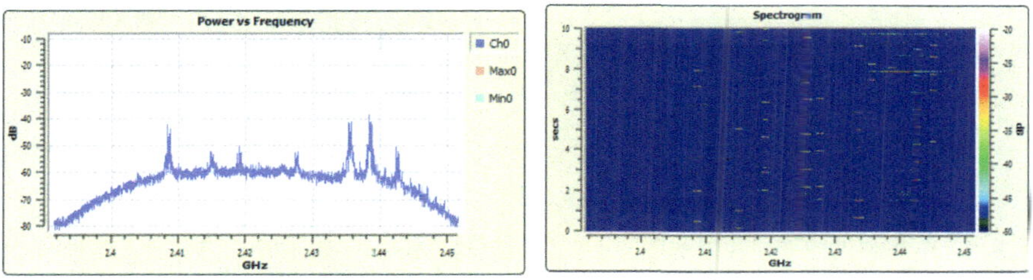

(a) (b)

Fig. 2 (**a**) RF spectrum FS-61X controller (2402–2441 MHz). (**b**) 1-sec Waterfall FS-61X Controller (2402–2441 MHz)

Figure 1a and b shows the spectrum and waterfall for the background noise in the 2402–2441 MHz band immediately prior to turning on the FS-I6X remote controller. Note that the FS-16X utilizes a proprietary protocol. The waterfall shows the spectrum history for a 1 s period. Figure 2a and b shows the spectrum for the FS-I6X remote controller in the 2402–2441 MHz band.

We next shifted our experiments to the high frequency band: 2441–2480 MHz. We collected the data while repeating the same steps. Figure 3a and b depicts the background noise spectrum and the 1-sec Waterfall background noise, respectively, just before turning on the controller. Figure 4a and b depicts the background noise spectrum and the 1-sec Waterfall background as soon as the FS-16X controller is turned on in the 2441–2480 MHz band.

We repeated the same experiments using a Bluetooth-controlled drone: T-65 micro drone. Similarly, the Bluetooth controller signal is captured at both the low spectrum band (2402–2441 MHz) and the high spectrum band (2441–2480 MHz). The results are depicted in Figs. 5 and 6.

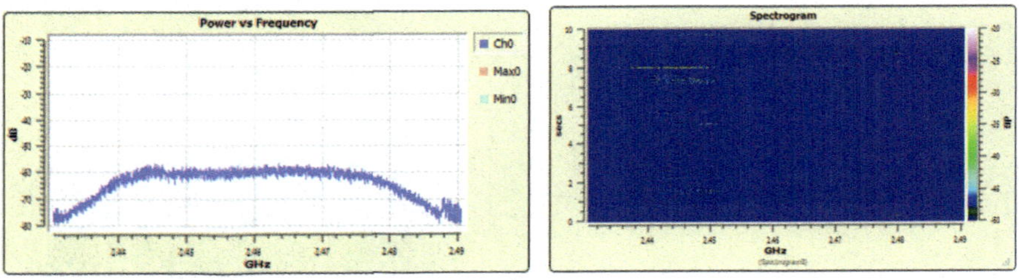

(a) (b)

Fig. 3 (**a**) Background noise spectrum (2441–2480 MHz). (**b**) 1-sec Waterfall background noise (2441–2480 MHz)

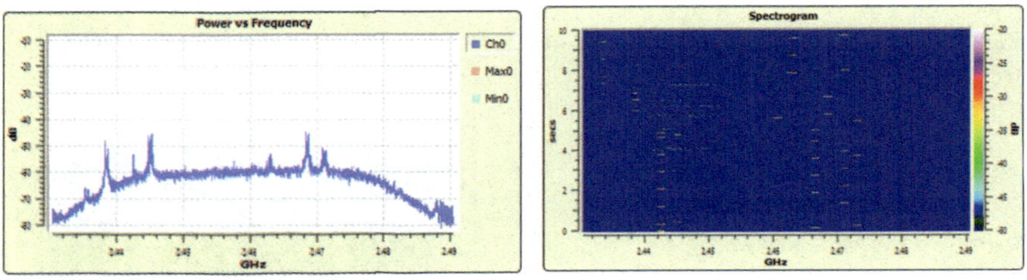

(a) (b)

Fig. 4 (**a**) RF spectrum FS-61X controller (2441–2480 MHz). (**b**) 1-sec Waterfall FS-61X Controller (2441–2480 MHz)

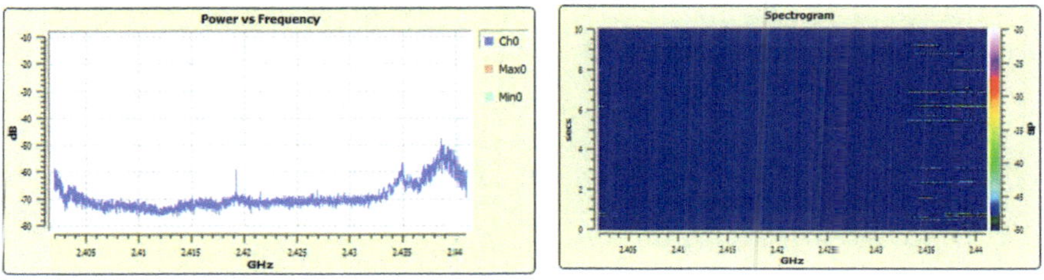

(a) (b)

Fig. 5 (**a**) RF Spectrum T-65 Bluetooth (2402–2441 MHz). (**b**) 1-sec Waterfall T-65 Bluetooth (2402–2441 MHz)

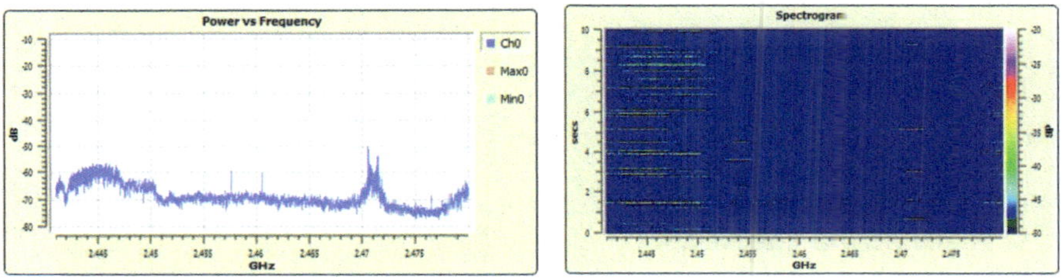

(a) (b)

Fig. 6 (**a**) RF Spectrum T-65 Bluetooth (2441–2480 MHz). (**b**) 1-sec Waterfall T-65 Bluetooth (2441–2480 MHz)

5 Machine Learning

While anticipating the completion of our RF dataset collection and generation activities, we embarked on some preliminary research work on applying machine learning techniques for the pattern recognition of RF signals. We considered two RF datasets that are available on the Internet: the drone dataset by Al-Sa'd et al. [44] and the synthetically generated RF dataset by Hall et al. [48].

The reference datasets on RF signal detection and classification that was created at the National Institute of Standards and Technology (NIST) focus on signals, schemes, systems, and environments found in spectrum sharing systems [48]. Most of the dataset is synthetically generated. Thus, we opted not to use this for our proof-of-concept work.

The RF dataset that was collected by Al-Sa'd et al. [44] is a collation of raw RF signals of drones by various manufacturers under different flight modes. During the collection process, two RF receivers were enabled to collect a maximum of 40 MHz bandwidth signals each with one set to capture the lower end and the other to capture the upper end of the spectrum. This dataset is akin to the dataset that we are currently compiling. Thus, we elected to use this dataset in our preliminary work.

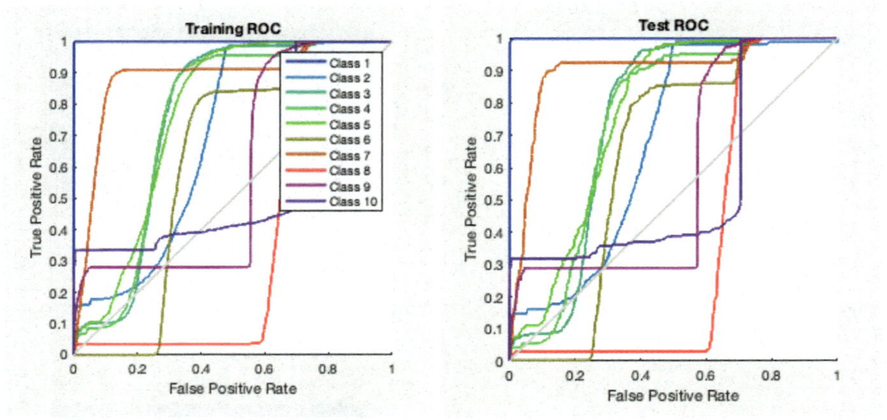

(a) (b)

Fig. 7 (**a**) Training phase ROC. (**b**) Testing Phase ROC

5.1 Preliminary Results

We used the scaled conjugate gradient (SCG) learning algorithm with the cross-entropy loss function on the dataset that was compiled by Al-Sa'd et al. [44]. The results of applying this ML algorithm are a mere 36% classification accuracy. This result suggests two things: the utilization of a deep neural network and the cleansing out of the dataset for outliers or anomalies.

Two of the Matlab-generated receiver operating characteristic (ROC) curves are depicted in Fig. 7. The ROC is a plot of the True Positive Rate (TPR) vs. the False Positive Rate (FPR). Notice that there are 3 classes that tend to pull a bias towards the FPR. Also, it should be noted that points above the diagonal line represent good classification and those points below represent bad classification.

We believe that this preliminary work is justified by the results that we obtained and can be used as a guide for future data collection and generation efforts.

6 Conclusion and Future Research Directions

This chapter presents a review of RF signal communication security approaches and our on-going research effort to produce RF signal datasets to be used in ML systems. As a proof of concept, we attempted to create a reliable dataset for RF signal security research, primarily on the application of ML. We selected three UAS platforms, DJI Phantom 3, T-65-x-wing, and FlySky FS-6IX, and successfully collected RF datasets for our future work on RF signal pattern recognition. Power, frequency, and spectrogram plots were presented for each UAS platform. Furthermore, as a proof of concept, we adopted two RF datasets that are available on the Internet. We used one of the datasets to test a Scaled Conjugate Gradient machine learning algorithm. The results yield a classification accuracy of a mere 36%. This accuracy level is particularly significant to point out the need for additional optimization techniques for classification, such as deep neural network. Additionally, it indicates the need for filtering of datasets for outliers or anomalies.

With the preceding discussions in mind, we offer the following future research directions:

- an expanded RF dataset collection and generation using a variety of UAS platforms in order to develop a more accurate, comprehensive, and robust classifier;
- Pre-processing the dataset for filtering out the signal anomalies;
- validation of ML techniques on both our compiled RF datasets and the RF datasets obtained from the Internet;
- the use of feature extraction and principal component analysis (PCA) to reduce the data attributes for more efficient cluster analysis; and
- the utilization of other ML techniques such as deep neural networks to accurately identify a UAS using its RF signal fingerprint.

Acknowledgments This work is partially supported by the Florida Center for Cybersecurity, under grant number 3901-1009-00-A (2019 Collaborative SEED Program) and the National Security Agency under grant number H98230-19-1-0333. The United States Government is authorized to reproduce and distribute reprints notwithstanding any copyright notation herein.

References

1. P. Tucker, A criminal gang used a drone swarm to obstruct an FBI hostage raid. Defense One (2018). Retrieved from: https://www.defenseone.com/technology/2018/05/criminal-gang-used-drone-swarm-obstruct-fbi-raid/147956/
2. M. Greenwood, *Greenpeace Activists Fly 'Superman' Drone into French Nuclear Site* (The Hill, Washington, 2018)
3. J. Cansler, N. Ruff, M. Schreiber, Drone use and defense by enterprise security management: UAV applications, concerns, and countermeasures, in *ASIS International 2017 Annual Seminar*, Dallas, TX (2017)
4. BBC News: Ten sentenced for smuggling drugs into prisons by drones. BBC (2017)
5. M. Pomerleau, How $650 drones are creating problems in Iraq and Syria. C4ISRNET-Media for the Intelligence Age Military (2018)
6. P.M. Frank, Advanced fault diagnosis techniques in aerospace systems, in *Proceedings. VLSI and Computer Peripherals. COMPEURO 89* (IEEE, New York, 1989), pp. 3–136
7. A. Alessandri, M. Caccia, G. Veruggio, Fault detection of actuator faults in unmanned underwater vehicles. Control Eng. Pract. **7**(3), 357–368 (1999)
8. J. Qi, Z. Jiang, X. Zhao, J. Han, UKF-based rotorcraft UAV Fault adaptive control for actuator failure, in *2007 IEEE International Conference on Robotics and Biomimetics (ROBIO)* (IEEE, New York, 2007), pp. 1545–1550

9. P.A. Samara, G.N. Fouskitakis, J.S. Sakellariou, S.D. Fassois, A statistical method for the detection of sensor abrupt faults in aircraft control systems. IEEE Trans. Control Syst. Technol. **16**(4), 789–798 (2008)
10. N. Léchevin, C.A. Rabbath, Decentralized detection of a class of non-abrupt faults with application to formations of unmanned airships. IEEE Trans. Control Syst. Technol. **17**(2), 484–493 (2008)
11. G. Heredia, A. Ollero, R. Mahtani, M. Béjar, V. Remuß, M. Musial, Detection of sensor faults in autonomous helicopters, in *Proceedings of the 2005 IEEE International Conference on Robotics and Automation* (IEEE, New York, 2005), pp. 2229–2234
12. A. Mancini, F. Caponetti, A. Monteriu, E. Frontoni, P. Zingaretti, S. Longhi, Safe flying for an UAV helicopter, in *2007 Mediterranean Conference on Control & Automation* (IEEE, New York, 2007), pp. 1–6
13. C. Rago, R. Prasanth, R.K. Mehra, R. Fortenbaugh, Failure detection and identification and fault tolerant control using the IMM-KF with applications to the Eagle-Eye UAV, in *Proceedings of the 37th IEEE Conference on Decision and Control (Cat. No. 98CH36171)*, vol. 4 (IEEE, New York, 1998), pp. 4208–4213
14. H.E. Sevil, A. Dogan, False fault detection in airdata sensor due to nonuniform wind in aerial refueling, in *AIAA Atmospheric Flight Mechanics Conference* (2011), p. 6446
15. H.E. Sevil, Airdata Sensor Based Position Estimation and Fault Diagnosis in Aerial Refueling. PhD Dissertation (2014)
16. H.E. Sevil, A. Dogan, Fault diagnosis in air data sensors for receiver aircraft in aerial refueling. J. Guidance Control Dyn. **38**(10), 1959–1975 (2015)
17. M.G. Hinchey, R. Sterritt, C. Rouff, Swarms and swarm intelligence. Computer **40**(4), 111–113 (2007)
18. A. Das, P. Kol, C. Lundberg, K. Doelling, H.E. Sevil, F. Lewis, A rapid situational awareness development framework for heterogeneous manned-unmanned teams, in *NAECON 2018-IEEE National Aerospace and Electronics Conference* (IEEE, New York, 2018), pp. 417–424
19. P. Dasgupta, A multiagent swarming system for distributed automatic target recognition using unmanned aerial vehicles. IEEE Trans. Syst. Man Cybernet. Part A Syst. Hum. **38** (3), 549–563 (2008)
20. D. C. MacKenzie, Collaborative tasking of tightly constrained multi-robot missions, in *Multi-Robot Systems: From Swarms to Intelligent Automata: Proceedings of the 2003 International Workshop on Multi-Robot Systems*, vol. 2 (2003), pp. 39–50
21. A. Huth, C. Wissel, The simulation of the movement of fish schools. J. Theor. Biol. **156**(3), 365–385 (1992)
22. D.S. Brown, M.A. Goodrich, S.-Y. Jung, S. Kerman, Two invariants of human-swarm interaction. J. Hum.-Rob. Interact. **5**(1), 1–31 (2016)
23. R. Tiwari, P. Jain, S. Butail, S.P. Baliyarasimhuni, M.A. Goodrich, Effect of leader placement on robotic swarm control, in *AAMAS* (2017), pp. 1387–1394
24. M. Haque, C. Ren, E. Baker, D. Kirkpatrick, J.A. Adams, Analysis of swarm communication models, in *Proceedings of the Twenty-second European Conference on Artificial Intelligence* (2016), pp. 1716–1717
25. National Instrument: Introduction to RF & Wireless Communications Systems [Online]. Available: http://www.ni.com/tutorial/3541/en/. Accessed 14 May 2020
26. D. Grini: RF Basics, RF for Non-RF Engineers [Online]. Available: http://www.ti.com/lit/ml/slap127/slap127.pdf?&ts=1589436629635/. Accessed 14 May 2020
27. K.S. Kundert, Introduction to RF simulation and its application. IEEE J. Solid-State Circ. **34**(9), 1298–1319 (1999)
28. NCC Group, Drones: detect, identify, intercept, and hijack, [Online]. Available: https://www.nccgroup.trust/uk/about-us/newsroom-and-events/blogs/2015/december/drones-detect-identify-intercept-and-hijack/. Accessed 14 May 2020

29. P. Nguyen, H. Truong, M. Ravindranathan, A. Nguyen, R. Han, T. Vu, Matthan: drone presence detection by identifying physical signatures in the drone's RF communication, in *Proceedings of the 15th Annual International Conference on Mobile Systems, Applications, and Services* (2017), pp. 211–224
30. I. Guvenc, F. Koohifar, S. Singh, M.L. Sichitiu, D. Matolak, Detection, tracking, and interdiction for amateur drones. IEEE Commun. Mag. **56**(4), 75–81 (2018)
31. X. Shi, C. Yang, W. Xie, C. Liang, Z. Shi, J. Chen, Anti-drone system with multiple surveillance technologies: architecture, implementation, and challenges. IEEE Commun. Mag. **56**(4), 68–74 (2018)
32. P. Nguyen, H. Truong, M. Ravindranathan, A. Nguyen, R. Han, T. Vu, Cost-effective and passive RF-based drone presence detection and characterization. GetMobile: Mob. Comput. Commun. **21**(4), 30–34 (2018)
33. M. Ezuma, F. Erden, C.K. Anjinappa, O. Ozdemir, I. Guvenc, Detection and classification of UAVs using RF fingerprints in the presence of Wi-Fi and Bluetooth interference. IEEE Open J. Commun. Soc. **1**, 60–76 (2019)
34. G. Ding, Q. Wu, L. Zhang, Y. Lin, T.A. Tsiftsis, Y.-D. Yao, An amateur drone surveillance system based on the cognitive Internet of Things. IEEE Commun. Mag. **56**(1), 29–35 (2018)
35. V. Sharma, M. Kumari, Drone detection mechanism using radiocommunication technology and Internet protocol address, in *2019 International Conference on Smart Systems and Inventive Technology (ICSSIT)* (IEEE, New York, 2019), pp. 449–453
36. A. Shoufan, H.M. Al-Angari, M.F.A. Sheikh, E. Damiani, Drone pilot identification by classifying radio-control signals. IEEE Trans. Inf. Forensic Secur. **13**(10), 2439–2447 (2018)
37. I. García-Magariño, R. Lacuesta, M. Rajarajan, J. Lloret, Security in networks of unmanned aerial vehicles for surveillance with an agent-based approach inspired by the principles of blockchain. Ad Hoc Netw. **86**, 72–82 (2019)
38. C. Constantinides, P. Parkinson, Security challenges in UAV development, in *2008 IEEE/AIAA 27th Digital Avionics Systems Conference* (IEEE, New York, 2008), pp. 1–C
39. A.Y. Javaid, W. Sun, V.K. Devabhaktuni, M. Alam, Cyber security threat analysis and modeling of an unmanned aerial vehicle system, in *2012 IEEE Conference on Technologies for Homeland Security (HST)* (IEEE, New York, 2012), pp. 585–590
40. N.M. Rodday, R. de O. Schmidt, A. Pras, Exploring security vulnerabilities of unmanned aerial vehicles, in *NOMS 2016–2016 IEEE/IFIP Network Operations and Management Symposium* (IEEE, New York, 2016), pp. 993–994
41. M. Hooper, Y. Tian, R. Zhou, B. Cao, A.P. Lauf, L. Watkins, W.H. Robinson, W. Alexis, Securing commercial WiFi-based UAVs from common security attacks, in *MILCOM 2016-2016 IEEE Military Communications Conference* (IEEE, New York, 2016), pp. 1213–1218
42. M. Sliti, W. Abdallah, N. Boudriga, Jamming attack detection in optical UAV networks, in *2018 20th International Conference on Transparent Optical Networks (ICTON)* (IEEE, New York, 2018), pp. 1–5
43. J. Andersson, M. Balduzzi, S. Hilt, P. Lin, F. Maggi, A. Urano, R. Vosseler, A security analysis of radio remote controllers for industrial applications. Technical report, Trend Micro, Inc., January 2019 [Online] (2019). Available: https://documents.trendmicro.com/assets/white_papers/wp-a-security-analysis-of-radio-remote-controllers.pdf
44. M.F. Al-Sa'd, A. Al-Ali, A. Mohamed, T. Khattab, A. Erbad, RF-based drone detection and identification using deep learning approaches: an initiative towards a large open source drone database. Fut. Gener. Comput. Syst. **100**, 86–97 (2019)
45. H. Zhang, C. Cao, L. Xu, T. Aaron Gulliver, A UAV detection algorithm based on an artificial neural network. IEEE Access 6, 24720–24728 (2018)
46. A. Bernardini, F. Mangiatordi, E. Pallotti, L. Capodiferro, Drone detection by acoustic signature identification. Electron. Imag. **2017**(10), 60–64 (2017)
47. G.J. Mendis, T. Randeny, J. Wei, A. Madanayake, Deep learning based Doppler radar for micro UAS detection and classification, in *MILCOM 2016-2016 IEEE Military Communications Conference* (IEEE, New York, 2016), pp. 924–929

48. T.A. Hall, R. Caromi, M. Souryal, A. Wunderlich, Reference datasets for training and evaluating RF signal detection and classification models, in *2019 IEEE Globecom Workshops (GC Wkshps)* (IEEE, New York, 2019), pp. 1–5
49. Lime Microsystems: LimeSDR, 2020 [Online]. Available: https://limemicro.com/products/boards/limesdr/. Accessed 16 May 2020
50. J. Blum, Phothosware, 2020. [Online]. Available: http://www.pothosware.com. Accessed 16 May 2020

Real-Time Operating Systems: Course Development

Michael Rivnak and Leonidas Deligiannidis

1 Introduction

Real-time operating systems (RTOS) are typically used in scenarios where low latency is the focus, generally in embedded systems. Traditional operating systems are not designed for time-sensitive operations. Instead, they focus on multiprocessing where multiple applications can be run simultaneously to ensure a full desktop experience for the user. Time sensitivity, in this case, means that the operations a process is attempting to execute must be completed by a certain deadline. This deadline, and the consequences associated with missing it, can be used to define real-time systems as soft or hard.

It is common for real-time systems to be used in embedded systems. In this case, there is limited user interaction with the system directly, and any operations that are to be executed are generally predetermined and static. The most basic embedded systems perform a single predetermined task and have no method of user interaction, while the most complex can approach the complexity of a desktop computer system [1]. The basic functions of an embedded system are preprogrammed into the firmware of the microcontroller, either a bespoke chip or a field-programmable gate array (FPGA), the latter which can be reprogrammed to simulate different microcontrollers without having to design a new electronics system.

Recently, there have been releases of the Linux kernel that include patches to allow the user to preempt the kernel scheduling model. The objective of these Linux patches is to provide a Linux-based alternative to traditional embedded real-time operating systems. It provides the ability for the task scheduler to act like a real-time operating system where execution order can be changed.

M. Rivnak · L. Deligiannidis (✉)
Wentworth Institute of Technology, Boston, MA, USA
e-mail: rivnakm@wit.edu; deligiannidisl@wit.edu

K. Daimi et al. (eds.), *Advances in Security, Networks, and Internet of Things*, Transactions on Computational Science and Computational Intelligence
https://doi.org/10.1007/978-3-030-71017-0_54

2 Operating System Fundamentals

2.1 What Makes an Operating System Real-Time?

A real-time system is designed to focus on timing correctness, being able to complete a task by a certain deadline or with certain regularity, in addition to computational correctness. This is defined as being able to execute tasks by a certain time. This could be a set deadline, such as a fixed point it time or a period used in repeating tasks. Periodic scheduling will be discussed later under rate monotonic scheduling. Typically, real-time systems are made up of many microprocessors arranged in clusters. This large-scale parallelism allows for individual tasks to be executed with higher timing precision. This is simply due to load balancing across the separate microprocessors so that any since microprocessor is not tasked with a load too large to execute accurately.

2.2 Soft Versus Hard Real-Time Operating Systems

Precision in RTOSs is defined by deadlines. Depending on the consequences of missing the deadlines, RTOSs can be classified as soft, firm, or hard [2]. This classification is an important component of real-time task scheduling. Different classifications can come with trade-offs. For example, a hard real-time system needs to be limited in terms of the amount of tasks it can run so that it is afforded more flexibility in task scheduling and thus the ability to meet deadlines. This flexibility results in lower system response time since there are fewer tasks running that could cause operation delays. Additionally, in the case of peak-load scenarios, a hard real-time system must still be able to meet deadlines, whereas a soft real-time system may allow some performance degradation in these situations [2].

2.3 Real-Time-Specific Functions

Fail-safe Versus Fail-operational In certain embedded systems, there must be defined error handling. Two methods are fail-safe and fail-operational. In a fail-safe application, there is a defined safe state to enter when there is a detected error in the system. This safe state generally implies a state which does not pose harm to operators or users, such as stopping an industrial machine. In some situations, there is not a defined safe state, and thus these applications must implement fail-operational error handling where if an error occurs, the system must continue working and optionally try to correct the error.

Task Triggers In embedded systems, there are two main types of triggers that cause a task to run, event-driven and time-driven. With event-triggered control, a task is run when the prerequisite event takes place, whether it be another task completing or an interaction from an operator, whereas with time-triggered control, the only interrupt in the system is the tick from a real-time clock. Ideally, this real-time clock is available and synchronized to all individual nodes in a real-time system.

2.4 *Threads and Tasks*

The two real-time operating systems mentioned in the paper use different methods of defining parallel operations. FreeRTOS, as an embedded solution, refers to its operations as tasks, where a task is a set of actions that need to be taken independent of other tasks within the system [3]. Threads are similar, but due to the nature of POSIX systems and the C programming language, they are not self-contained and may interact with other parts of the system in software.

3 FreeRTOS

FreeRTOS is a bare metal, real-time operating system designed for use in embedded systems. It uses a task system to schedule operations in a time-sensitive manner. Upon startup, tasks are initialized and then passed to the task scheduler which then runs them as needed. Because tasks run independently in FreeRTOS, they are given their own process stack which is suspended and resumed upon switching the task out of and into the execution state. This context switching paradigm allows all task data (registers, stack contents) to be protected when the task is switched out.

3.1 *Co-routines*

FreeRTOS also supports co-routines in addition to tasks. This, however, is not recommended for most applications and is not being further developed. Co-routines are meant for very small devices where memory is limited. This is accomplished by having all co-routines share a single-process stack. Of course, this limits the capabilities of each process, and each co-routine can only communicate among themselves.

3.2 *Arduino*

The Arduino port of FreeRTOS is the most accessible version for the average user looking to try out a real-time system. Out of the box, Arduino uses a single loop to process I/O data. This has the limitation of not allowing parallel processing and thus may hinder the temporal accuracy of operations. FreeRTOS allows different actions to be split into different tasks which can be run simultaneously.

3.3 Raspberry Pi

The Raspberry Pi is not officially supported by FreeRTOS, but there are a few community ports available. These ports, however, come with limited to no documentation associated with them and can often be missing features altogether. One such issue that we discovered is incorrect register accessing which compromises the use of the included general-purpose I/O pins on the board. Overall, there is not much utility to using a Raspberry Pi for FreeRTOS compared to the similarly popular Arduino. The port is not of the same quality, and the development process is slightly more complicated due to not having a dedicated integrated development environment for the platform.

4 RT Linux

The Linux foundation provides a set of community-supported patches that can be applied to the publicly available Linux kernel. This is (relatively) the most accessible real-time system for the average user as it does not require dedicated hardware and can be installed on traditional x86/x86_64 hardware as well as the ever-popular Raspberry Pi with minimal effort.

4.1 Multi-Environment Real Time

Multi-Environment Real Time (MERT) was a primitive dual real-time and UNIX operating system developed by Bell Labs in the 1970s for use in embedded systems. This differs from RT Linux with its use of virtualization rather than explicit modifications to either system. In this case, the RTOS is used as the host operating system with the UNIX system virtualized as the guest system [4, 5]. The main disadvantage of this implementation is that you incur all the overhead associated with virtualization. Essentially, this means that you have the UNIX system running on the real-time system as if it was a program itself. The result of this is that the system must deal with the workload of two systems rather than one, as well as running an entire operating system in software is considerably more inefficient than running it on hardware directly. Most modern virtualization programs do have some optimizations included, such as running as a hypervisor rather than a full emulator, but they are still not very efficient. Hypervisors can use some of the hardware directly instead of simulating the hardware with software, providing considerable efficiency gains but limiting the virtual machine to only the same architecture as the host machine.

4.2 Building a Kernel

With officially provided kernel patches on kernel.org, it is relatively easy to add real-time functionality to a traditional desktop system. For x86 systems, the patching process is fairly simple, only requiring a few commands that are all well documented. With the Raspberry Pi, specifically, there are wiki pages detailing how to build kernels. These resources were used in the testing below. Building a real-time kernel in this case only requires cloning a different branch from the Raspberry Pi GitHub repository that contains the kernel's sources. Many Linux distributions provide utilities that assist in building kernels from source. This generally includes configuration programs that will provide a default kernel configuration that should be stable enough to run without any modifications. With RT Linux, it is recommended to verify that these tools enable the `FULLY_PREEMPTIBLE_KERNEL` option in the configuration file. This can quickly be accomplished using `menuconfig` which is included in all the distributions mentioned as well as many others.

As far as compiling is concerned, a Linux kernel is built using make, similar to most Linux applications as shown below:

```
make -j(nproc)
sudo make install
```

These are the only commands required to compile a kernel, in this case using all of the available processing cores. On a small, low-power system like the Raspberry Pi, this will likely take a couple hours, but it is not anything prohibitively difficult.

4.3 Pre-patched Kernels

Many Linux distributions even provide official channels to download and install real-time kernels. The Raspberry Pi Foundation has a GitHub repository with all their officially supported kernels that are ready to be downloaded and installed. Some distributions offer officially or community supported packages for running a real-time kernel. Arch Linux, for example, can utilize the `linux-rt` kernel package on the Arch User Repository (AUR) that makes running a real-time kernel as simple as installing any other program [6]. This is facilitated by the fact that Arch is a rolling release distribution, so its kernel is a modular package that can be installed through the package manager. Manjaro, which is a derivative of Arch, allows users to choose their kernel during installation with their 'Manjaro Architect' installer, including versions with the real-time patch installed. Additionally, they provide a set of user tools that allow the user to easily modify integral parts of the system. A part of that is a kernel switching GUI, which also includes those real-time variants [7].

4.4 Consistency

The main advantage of using a real-time variant of Linux as your RTOS is the intercompatibility between systems. Any system that can run Linux can compile a custom kernel and use the same software across systems since the system calls are the same. During testing compute times, which will be discussed later, the same code was shared between an armhf and x86_64 system without any issues.

4.5 Dual Function

Using RT Linux as your RTOS provides the added benefit of being able to run real-time and non-real-time applications concurrently. This widely supported software base provides tools for developers who want to add real-time components to their preexisting application for functions that require lower latency in their operation.

4.6 Installing RT Linux

Raspbian The Raspberry Pi foundation provides detailed instructions on the fairly simple process of compiling and installing a new kernel. This process is exactly the same for a real-time kernel as it is for a traditional kernel. You clone the git repository with the Raspberry Pi kernel source code, then build with the `make` command, and finally move the boot files into the `/boot/` directory and reboot [8].

Arch Linux Using the GRUB bootloader switching kernels is quite an easy process. It does also require the user of an AUR helper such as yay to install the `linux-rt` kernel which is not included in the default Arch repository. This process only requires three commands in the terminal [6]:

```
pacman -R linux
yay -S linux-rt
grub-mkconfig -o /boot/grub/grub.cfg
```

This set of commands uninstalls the default kernel packages, installs a real-time kernel package, and reconfigures the bootloader to load the real-time kernel. This assumes that the Grand Unified Bootloader (GRUB) is used on this system.

Manjaro Manjaro provides the simplest method of installation for real-time kernels among Linux distributions. There is a utility included in Manjaro called Manjaro Hardware Detection (MHWD) that can be used for simple modification of firmware and drivers. Another function of this is the ability to quickly switch kernels. This can be done at the command line or with the Manjaro Setting Manager GUI. A single operation is all that is needed in order to switch to a real-time kernel with Manjaro [7].

Additionally, with the Manjaro Architect installer, the kernel version can be selected during the installation process. Real-time kernels are available during this step, making real-time kernels require no extra effort to install compared to a traditional kernel. One quirk to this installation is that the bootloader must be reconfigured to be able to boot to the real-time kernel. Both kernels will be installed; however, the bootloader is hidden by default so it must be shown on boot to be able to switch between kernels.

5 RT Linux Scheduling

The PREEMPT_RT patch set extends the default scheduling capabilities of Linux by adding three new scheduling paradigms. Each of these produces different results in terms of performance and timings and thus must be used in appropriate situations depending on the functional requirements of the system.

5.1 SCHED_OTHER

This is not a real-time scheduling policy, but it is important when discussing other policies. This is essentially the default scheduling policy; every thread has the same priority, and thus every thread gets its fair share of CPU time.

5.2 SCHED_FIFO

The first-in, first-out (FIFO) scheduling policy that is a part of the PREEMPT_RT patch is one of two priority-based policies. Priority-based mean that threads with a higher priority will preempt lower-priority threads. The FIFO aspect of the policy relates to how threads of the same priority are handled; in this case, same-priority threads are executed in the order that they are created in. This means that a thread gets run to completion before the next thread in the queue is started [9]

5.3 SCHED_RR

The round-robin (RR) policy is an extension of the FIFO policy mentioned previously. The key different between them, as noted earlier, is in the handling of same-priority threads. In this instance, similar threads are given split priority of the execution state. For example, if four threads of the same priority are all set to run, they will each be given 25% of the CPU time, assuming everything else about the threads is identical. This is similar to how the default scheduler works but with those four threads running in place of everything else on the system [9].

5.4 SCHED_DEADLINE

The SCHED_DEADLINE policy in RT Linux implements the earliest deadline first algorithm for scheduling tasks [10]. This algorithm will be discussed later in this paper. It does, however, use a dynamic priority system where the priority of each task is determined in real time, no pun intended. This means that it can function more efficiently [11]. But more importantly, it means that an RT Linux system can adapt to new processes being created and run. This is not always a consideration in embedded systems where the set of tasks is rigidly defined, but in a desktop system, it is incredibly important.

This policy is implemented fundamentally differently than the other two policies though. Because POSIX threads do not support the process attributes used by this policy, the attributes must be applied to the current process. In practical terms, this means you cannot create deadline threads explicitly but must instead create a standard thread and modify it. This is done using the `__NR_sched_setattr` system call and a struct that contains all the requisite parameters for deadline scheduling, among others.

6 Scheduling Algorithms

6.1 Priority-Based Scheduling

Fixed-priority scheduling assigns tasks with a priority that defines when they will be executed, with higher-priority threads being executed first. The Figs. 1 and 2 show two threads that each have worst-case execution times of 25 ms. In Fig. 1, t_1 can be executed before t_2 due to its higher priority, and thus this allows it to complete before its first deadline. Note that 25 ms is the *worst-case* execution time, and in the examples above, the second iteration of the task t_2 does not require the full 25 ms. This is important in Fig. 2 in which t_1 is not able to run first and thus misses its first deadline [12]. This is important because there is a missed deadline even though

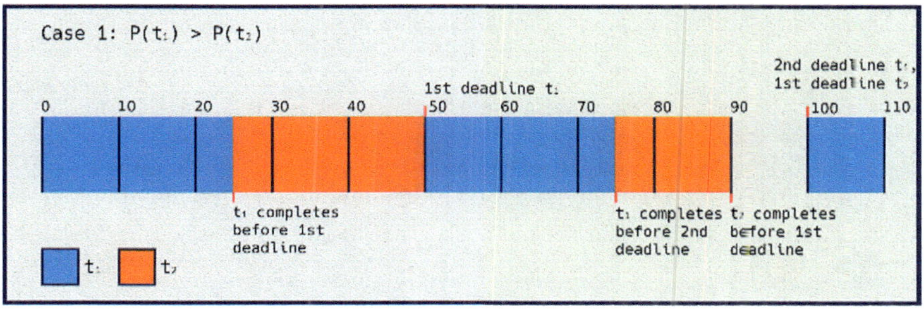

Fig. 1 Case showing task t_1 with high priority, able to meet all deadlines

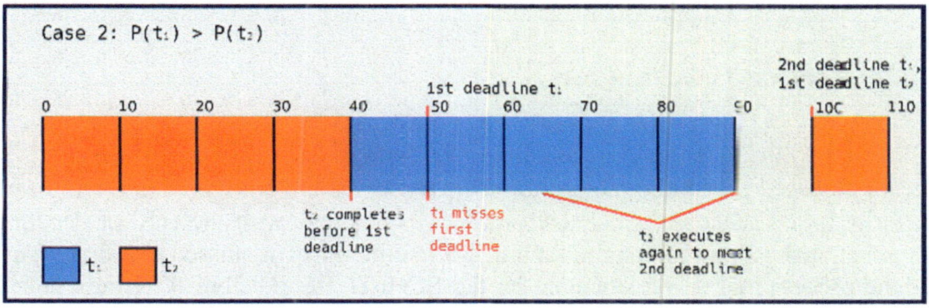

Fig. 2 Case showing task t_2 with high priority. Deadlines are missed

t_2 is not executing in worst-case time. Part of the reason that this is an issue in Fig. 2 is because the lower-priority thread has the earlier deadline. Because of this, unless the high-priority thread is sufficiently fast, it is unlikely or impossible for the lower-priority thread to meet its deadline.

6.2 *Rate Monotonic Scheduling*

Rate monotonic scheduling is a deadline-based scheduling algorithm that uses the runtime, or period, of a task in order to determine the priority of that task. It works out so that shorter tasks are given higher priority so that longer tasks do not clog up the scheduler while other tasks have yet to be run. This addresses the problem of long tasks preempting shorter ones and preventing the shorter ones from meeting their deadlines.

Figure 3 shows the worst-case scenario of a set of tasks that are not able to meet all deadlines. Because this is scheduled as rate monotonic, the longer task, t_2, is not able to finish before its deadline since it is interrupted at time point 50 by the other task. Even with the system only operating at 80% utilization, it is still not able to meet the deadlines. This is inherently a flaw with fixed-priority systems [12].

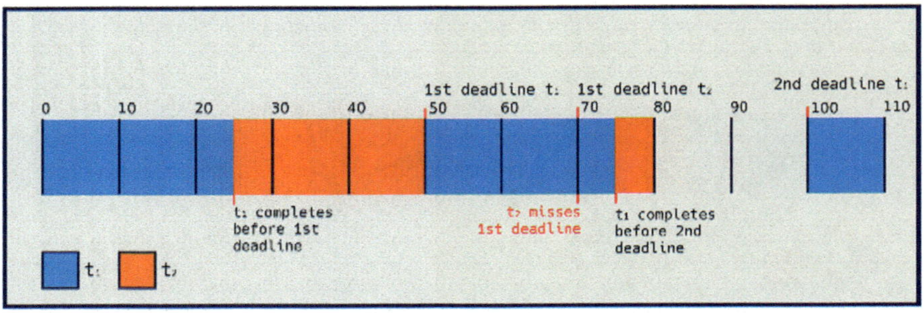

Fig. 3 Worst-case scenario for tasks t_1 and t_2. Deadlines are missed despite being below processing capacity

6.3 Earliest Deadline First

Similar to rate monotonic scheduling, the earliest deadline first algorithm is a deadline-based scheduling algorithm. This addresses the other problem with basic priority-based scheduling as shown in Fig. 2. The task with the earliest deadline is scheduled to run first, minimizing the likelihood of a missed deadline. This is the process that is implemented by the SCHED_DEADLINE RT Linux policy [10]. An issue with earliest deadline first scheduling is its propensity to allow error accumulation. When the system is overloaded, possibly due to environment errors causing too many tasks to be scheduled at once, this algorithm causes missed deadlines to compound since one missed deadline will cause subsequent tasks to miss theirs.

6.4 Overview

Despite these algorithms being based on the deadline and period of a task, it is not necessary to implement them using the SCHED_DEADLINE policy. They are able to be translated into priority-based algorithms based on their deadlines and be implemented using SCHED_FIFO [10].

7 Benchmarking and Results

7.1 Primitive Stress Testing

The first method of testing was a stress test to determine how real-time threads behave. This consisted of a loop to stress the integer and floating-point units of the

CPU as much as possible. Each thread was given this task as a way of providing some separation between processes. The justification for this was that there would not be much differentiation between different thread priorities, so adding a more computationally intensive task would address this.

```
RTThread thread0("OTHER");
RTThread thread1(1, "FIFO");
RTThread thread2(99, "FIFO");
```

Here is where each thread is created. Note the order of the threads and their scheduling policy and priority. The following is the output. Again note the order of completion.

```
Thread: 2
       Sched Policy: SCHED_FIFO
       Priority: 99
Thread: 1
       Sched Policy: SCHED_FIFO
       Priority: 1
Thread: 0
       Sched Policy: SCHED_OTHER
```

7.2 *Adding More Threads*

An issue with the previous method of testing was that it was inconsistent. Most of the time, it finished in the expected order, that is, with the highest priority threads first, but not every time. The most likely cause for this was that it was running multiple threads concurrently since it was running on a multicore system. To address this, more threads were added. The new tests consisted of 30 threads instead of 3, with 10 of each priority of thread like the previous. The result of this was a much more consistent output. Apart from one or two 1-priority threads, all the 99-priority threads finished first, then the 1-priority threads, then finally all the default threads. The one low-priority thread that finished ahead of the higher-priority threads most likely finished its work before it was able to be preempted. This fits observations from earlier testing where lighter workloads were not enough for real-time threads to distinguish themselves from non-real-time threads. These errant low-priority threads did influence the data, but knowing the cause of this allows the data to still be understood.

This issue manifests as two outliers in thread timing, mostly when using the SCHED_RR scheduling policy as ordinarily this would evenly divide CPU time for equal priority threads. Two outliers show up at the beginning of the test with a thread that ran very quickly since it was able to preempt all system processes but was not preempted by higher-priority threads and then another outlier after all the higher-priority threads have completed, which is a lower-priority thread with a significantly higher runtime since it was preempted by higher-priority threads after it had already started but had not yet completed. This appears to be consistent in many of the tests. These two outliers also appear to be mutually exclusive, and when there is a quick 1-priority thread before the 99-priority threads, there is not a slow one after them and vice versa. This can be explained by how the threads are created and started. The 1-priority thread is always created before the 99-priority thread. So when the 99-priority thread preempts the 1-priority thread, the 1-priority will have either completed, leading to the errant quick thread, or have started, leading to the long thread after the 99-priority threads have completed.

7.3 Restricting to One Core

Initial testing of core restriction was done with taskset in order to set the CPU core affinity for the process. This makes the process only run on the specified core(s). The issue with this method is that it adversely affects the real-time scheduling. The effects on real-time scheduling cause lower-priority threads to not be preempted, causing the real-time threads to still be run before the non-real-time threads, but the priority then has no effect. There are a couple ways to artificially limit the process to certain cores. It is possible to set the CPU affinity directly in the source code. The method that was used in this testing was using a virtual machine, in this case Oracle VirtualBox, to run tests. This allows the "hardware" to be modified to test different conditions, as well as possibly providing a more consistent environment since there will not be thermal factors due to the host machine having plenty of performance overhead.

An interesting result of limiting this test to a single core is that it looks like an inefficiency with the real-time scheduler getting exposed. In Fig. 4, the multicore tests seem to be around roughly the same value where the single core values are significantly higher. This extra overhead is from the real-time scheduler taking time away from the processes in order to manipulate the processing order. This looks to

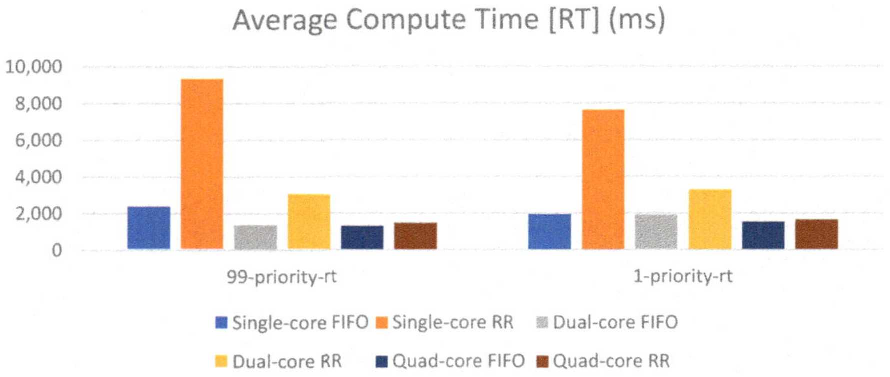

Fig. 4 Average compute times of a $O(10^9)$ stress test run with first-in first-out and round-robin policies on multiple core counts

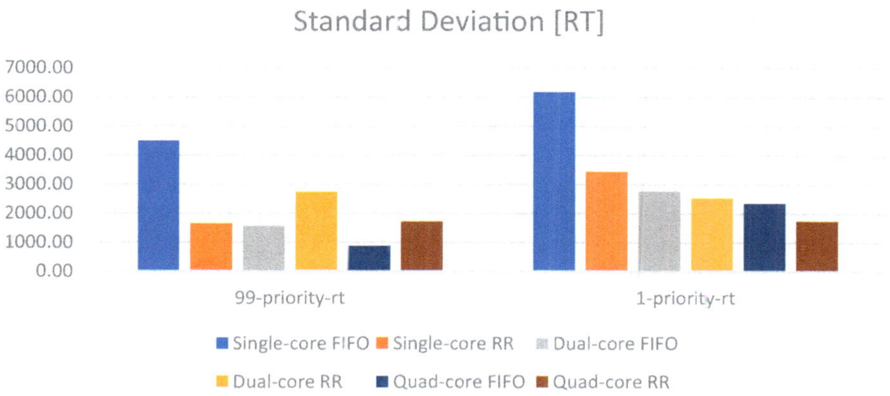

Fig. 5 Standard deviation of average compute times of a O(10^9) stress test run with first-in first-out and round-robin policies on multiple core counts

also affect the consistency of execution time with single core processes as shown in Fig. 5, although this does not appear to have the same effect on round-robin scheduled threads.

7.4 First-in First-out Scheduling

This is the simplest form of real-time scheduling included in RT Linux. Each thread is given a priority, and when a thread is run, it will preempt all lower-priority threads. What makes this a first-in, first-out algorithm is that the task that has preempted another will run to completion rather than being constantly switched out to allow other threads to run [9].

7.5 Round-Robin Scheduling

An interesting aspect of the round-robin scheduling that is a part of the RT Linux patch set is that these threads are not run concurrently in the same way that the default scheduler would. Without knowing that the threads are running in this mode, it would look identical to the FIFO policy if you were looking at a task manager like top or htop; only one thread at a time appears in the task manager. Comparatively, the default scheduler would display all the processes that are running. The cause of this is that the real-time scheduler is designed to run threads based on priority, and thus having a perfectly balanced workload is not prioritized. The result of this is that the execution time for the real-time threads is not as consistent as the traditional threads [9].

7.6 Default Scheduling

There are many different methods of process scheduling in a traditional operating system. They can, however, mostly be summarized with a single description. Schedulers on desktop systems attempt to simulate running every process concurrently by switching between them quickly. To minimize latency and processes taking up unnecessary CPU time, any processes that are waiting on resources or input/output are sent to a holding state until they are able to be run again. Processes that are waiting to be run are, generally, in a first-in first-out queue.

7.7 Comparison

Looking at the difference in average compute time and standard deviation in Fig. 6, you can see a clear advantage to using real-time threads. They execute significantly faster than traditional threads even when using round robin-scheduling which hinders the execution time. This is a result of the threads preempting every other thread on the systems and having exclusive access to the CPU. The consequence of this is that anything else that is trying to run at this time is halted. This includes system processes such as networking. Preempting network threads also broke any active secure shell connections during testing. It would be reasonable to assume this could affect TCP connections as well. This will affect systems that require networking to complete functions. This is the main drawback of using RT Linux for any nontrivial process. The effects of this were exaggerated for testing purposes, so it is possible that in a more realistic situation, this would not be an issue, and thus balancing the load over multiple systems in a cluster might be preferable here.

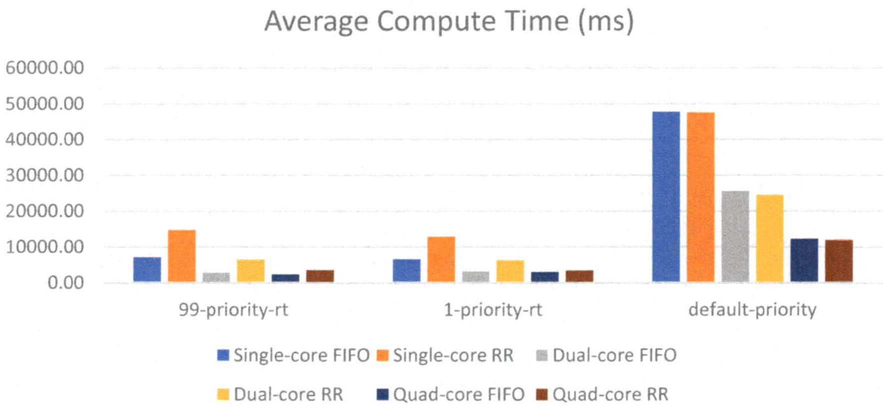

Fig. 6 Average compute times of a O(10^9) stress test run with first-in first-out and round-robin policies on multiple core counts as well as the same test run on generic POSIX threads

7.8 *Deadline Testing*

In testing deadline-based profiles, it is important to ensure that the deadlines are being met properly. This task posed the most difficulty of all the testing. This is due to the way that deadline tasks are created in RT Linux.

Having the data from the previous tests was helpful in creating these tests. It provided information on how long the task will generally run and the worst-case execution time.

In practice, this does not work very well as intended, and documentation is very light, so this method will likely not be included in the course. In testing the deadline, threads are being blocked by some other process, and they will only intermittently run. What is curious is that occasionally, the deadline threads will run without a problem, but most of the time, they will not run at all. There is also a sched_yield() function that is intended to tell the scheduler that a process has finished its work. It was expected that using this function may have alleviated deadline threads blocking themselves and possibly alleviating this issue; instead, it had the opposite effect, causing only the first thread to run before blocking the others.

8 Demonstration

8.1 *Course Background*

COMP3400 Operating Systems is focused on the fundamentals of operating systems. This includes process scheduling, context switching, system calls, and memory management. This paper details a real-time operating system module that is designed to fit into the curriculum for that class.

8.2 *Interactive Examples*

To demonstrate the functionality of real-time threads, it would be useful to have each student run a virtual machine with a real-time operating system. The simplest way to accomplish this while focusing on the functionality of the RTOS rather than the installation process would be to use Manjaro. It is a standard x86_64 operating system so it can run in any standard hypervisor program such as Oracle VirtualBox. This class already requires the use of a virtual machine as it stands using it for this application would not be a departure from this. Additionally, Manjaro has the simplest installation process for a real-time kernel, so this would be ideal for students who are not as familiar with the Linux operating system and how to switch kernels on different versions.

8.3 Installation Documentation

In a case where many students are all independently installing Linux and changing kernels, it is imperative to create as consistent an installation process as possible. It cannot be assumed that some, if any, of the students have experience configuring Linux kernels and bootloaders. The current process that will be used for this course utilized the default bootloader included with Manjaro. Unfortunately, the default bootloader must be configured to be able to switch kernels since out of the box it is hidden. This, though, is still a simple enough process for students to figure out. The current process is as follows:

```
$ sudo mhwd-kernel -li
```

This will list the current available kernels:

```
$ sudo mhwd-kernel -i linux54-rt
```

Install the Linux 5.4.X kernel with real-time patch

```
$ sudo nano /etc/default/grub
```

Change "`GRUB_TIMEOUT_STYLE=hidden`" to "`#GRUB_TIMEOUT_STYLE=hidden`"
Change "`GRUB_HIDDEN_TIMEOUT=10`" to "`#GRUB_HIDDEN_TIMEOUT=10`"
Open the GRUB bootloader configuration file and comment out the timeout style and timeout lines. This will unhide the bootloader menu on boot, allowing students to pick which kernel to boot from.

```
$ sudo update-grub
```

This will sync the above changes to the default internal GRUB config file.
Reboot.
On boot, select "Advanced options for Manjaro Linux."
Select one of the options with "rt."

References

1. T.J. Dingwall, N. Kumar, *United States Patent No. US5903752A* (1999, May 11)
2. Embedded Staff. *Introduction to Rate Monotonic Scheduling*. (Embedded) (2002, February 28). Retrieved March 25, 2020, from https://www.embedded.com/introduction-to-rate-monotonic-scheduling/
3. *FreeRTOS - Market leading RTOS (Real Time Operating System) for embedded systems with Internet of Things extensions* (n.d.). Retrieved March 19, 2020, from https://www.freertos.org/index.html
4. H. Kopetz, *Real-Time Systems: Design Principles for Distributed Embedded Applications*, 2nd edn. (Springer eBooks, Boston, 2011)
5. C.M. Krishna, K.G. Shin, *Real-Time Systems* (McGraw-Hill, 1997)
6. H. Lycklama, D.L. Bayer, UNIX time-sharing system: The mert operating system. Bell Syst. Tech. J. **57**(6), 2049–2086 (1978). https://doi.org/10.1002/j.1538-7305.1978.tb02142.x

7. *Manjaro Kernels - Manjaro Linux* (n.d.). Retrieved February 24, 2020, from https://wiki.manjaro.org/index.php?title=Manjaro_Kernels
8. *Realtime Kernel Patchset - Arch Wiki* (n.d.). Retrieved February 24, 2020, from https://wiki.archlinux.org/index.php/Realtime_kernel_patchset
9. S. Rostedt, *Using SCHED_DEADLINE: Controlling CPU Bandwidth* (n.d.). (The Embedded Linux Wiki) Retrieved April 4, 2020, from https://elinux.org/images/f/fe/Using_SCHED_DEADLINE.pdf
10. *The Linux Foundation Wiki - Real-Time Linux* (2020, January 2). (The Linux Foundation) Retrieved February 24, 2020, from https://wiki.linuxfoundation.org/realtime/start
11. The Raspberry Pi Foundation, *Kernel Building - Raspberry Pi Documentation* (n.d.). Retrieved March 19, 2020, from https://www.raspberrypi.org/documentation/linux/kernel/building.md
12. *What is an Embedded System? Definition and FAQs | OmniSci* (n.d.). Retrieved March 25, 2020, from OmniSci: https://www.omnisci.com/technical-glossary/embedded-systems

Piano Player with Embedded Microcontrollers

Bassam Shaer, Garrick Gressett, Phillip Mitchell, Joshua Meeks, William Barnes, and Stone Hewitt

1 Introduction

Pianos have been around for centuries and have always held the highest regard within the music industry. When one walks into the Peabody hotel in Memphis, Tennessee, one of the first things seen is a grand piano with a human performer—the signature of American class and prestige. This is common in many prestigious hotels and common areas. The Arduino Piano player was inspired by this, and its main goal was to produce a low-cost way for businesses and organizations to utilize a piano without the requirement of a pianist.

The Arduino Piano player is intended to be relatively low cost compared to other self-playing pianos in the industry, while also allowing for more choice in the type of piano it plays. Since it can play any midi file, the piano player has an unlimited amount of play time and music to operate on. Unlike humans, the Arduino Piano player offers better efficiency, timing, and more choice in when it can play giving businesses more freedom and customers more enjoyment. Currently, the Arduino Piano player can only play 800 notes, but further work on the device would produce longer note playtime.

Competitors in self-playing piano devices include the PianoDisc company which fail in comparison to the Arduino project due to its difficult installation. Additionally, the consumer must have a pre-owned piano that is prepared for installation. The setup and installation of this device are difficult. The consumer must ship their pianos to the company for installation [3]. The average cost for shipping pianos is $5.56 per mile for less than 100 miles and $0.87 per mile for

B. Shaer (✉) · G. Gressett · P. Mitchell · J. Meeks · W. Barnes · S. Hewitt
Electrical and Computer Engineering Department, University of West Florida, Fort Walton Beach, FL, USA
e-mail: bshaer@uwf.edu

K. Daimi et al. (eds.), *Advances in Security, Networks, and Internet of Things*, Transactions on Computational Science and Computational Intelligence,
https://doi.org/10.1007/978-3-030-71017-0_55

more than 100 miles [4]. The price further increases with larger pianos. A consumer could spend over $1000 dollars for shipping a piano for professional installation, and that's excluding the actual installation costs, which could add hundreds of dollars extra. Additionally, using this service, a person must pay for the songs that want to use, which can cost anywhere from $29.99 to $61.99 for songs that they might already own.

Another possible competitor to our device is the QRS Music Technologies PNOmation Playback piano device. This is another automatic piano playing system. It is installed directly into a chosen system by the consumer. The installation kit sells for $2499.99 dollars [3]. The setup will prove to be complex for most people, and most might opt for the professional installation offered. This kit does come with a few songs, but if the user wants to add some songs, they will have to pay for them. Compared to this device, the Arduino Piano player would be much cheaper with easy installation and be easily implemented with the users that already own a library of songs for their piano.

Overall, the Arduino Piano player has the potential to outcompete similar self-playing piano devices while imitating the same prestige of piano music that has been around for centuries. The rest of this paper is organized as follows. Section 2 discusses system specification overview. Section 3 presents the technical approach. Section 4 presents the testing approach. Section 5 presents the final product/project results. Section 6 presents results and discussion with conclusions and final remarks.

2 System Specifications Overview

The Computer GUI is a C# program that can translate downloaded midi files in order to properly play them on the piano. The program will then send the translated data to the Arduino. There will also be a MATLAB program that reads basic sheet music and produces midi files [5–9]. The case will be placed on top of the piano and, when lined up properly, will be high enough for the solenoids to fire but low enough in order for all the solenoids to make proper contact with the piano. N-type MOSFESTs will be used in order to activate each individual solenoid. Once the Arduino sends the digital signal, the MOSFET will switch on, allowing the power supply to power the solenoid.

The Arduino used to control the process is a 2560 Mega. The reason for choosing the 2560 Mega is the number of utility ports that allowed control over the 32 individual components. Two Arduino will be used in order to accommodate for the total number of keys. The solenoids are Gangbei 12 V solenoids. These solenoids met the design requirements in that the piston travels 10 mm while providing 5 N of force when activated. The 5 N is above the 2 N required to depress the piano keys with a margin for error.

3 Technical Approach

Figure 1 presents the overall functional block diagram of the new device. Each block represents a functional segment of the system. Each colored segment of blocks is a fundamental basis of the overall device.

The gray blocks represent the parts of the program that are run using a computer. The computer program GUI is where the operator has the highest level of interaction with the system. The operator can choose what midi file to play, which leads to the image recognition or midi file conversion system. When selecting the sheet music recognition option, the program will read the jpeg picture and produce a midi file based on the notes put into the song. The notes are then sent to the Arduino to be played. Otherwise, the midi file can be received from the Internet and then the program reads the corresponding notes and sends them to the Arduino.

The blue blocks contain the master and slave Arduinos. Once the song is sent to the master Arduino from the PC program, the master will then send the same notes over to the slave Arduino. This is performed since a single Arduino does not have enough digital outputs in order to play the entire piano. The slave will then send the master a ready signal, and the song will start to play. The Arduinos will then start to power the MOSFET switches.

The green blocks contain the power supply, MOSFETs, and resistors. The power supply is rated for 200 watt and powers the slave Arduino and the solenoids when the switches are on. The N-type MOSFETs are used in order to supply power to the solenoids when the Arduino sends the digital signal. The schematic of the wiring can be seen in Fig. 6. The heat dissipating resistors are there in order to reduce the heat produced by the solenoids with continuous use.

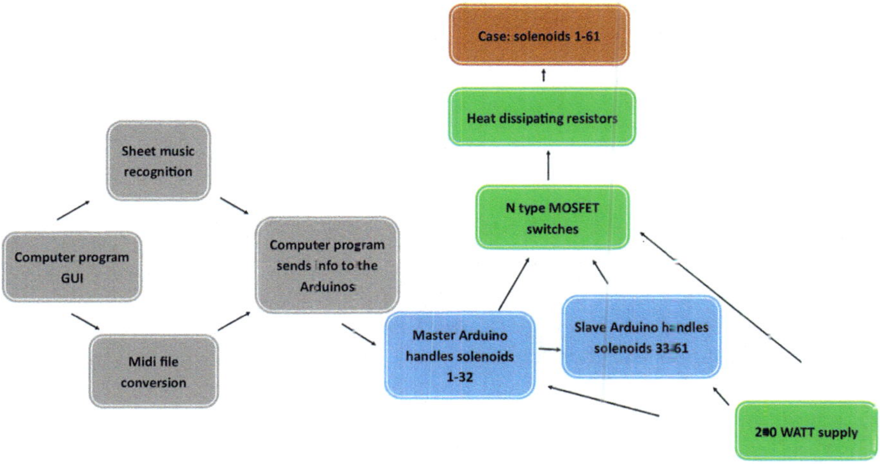

Fig. 1 Function block diagram

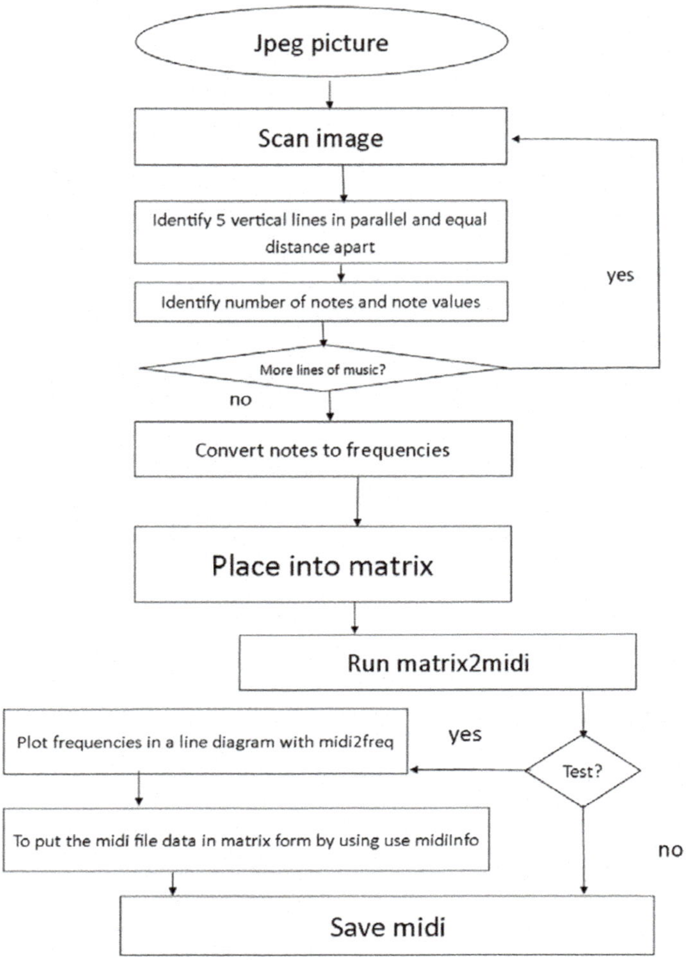

Fig. 2 Image recognition diagram

The brown block represents the casing, where the solenoids, Arduinos, resistors, and MOSFETs are all held.

Figure 2 presents a flowchart illustration for the image recognition program. For the first step, the user inserts a music sheet [9] jpeg file within a 1600 x 960 pixel image. The program will then identify where the music begins by finding the first set of five vertical parallel bars and use these bars as a point of reference. The program will then search for the first circle in the sheet. The circles found in the sheet are interpreted as notes, and its position is measured in relation to the bottom horizontal bar with the *x* value representing time and the *y* value representing the value of the note itself whether it is A sharp or a B minor. Once the program reaches the end of the line, it will continue down the image to identify any more lines of music and repeat the process. When the program reaches the end of the jpeg image, it will return the image to the user and show where it had identified the notes present in the jpeg image. Once the notes are listed, the notes are then converted to their appropriate frequencies [1–8] and placed into the matrix in column 3 in the matrix2midi program. The matrix2midi program will then convert the matrix into a midi format. At this point, the file can be tested using the midi2freq to ensure that all the notes were properly analyzed, though it should be noted that the sounds generated from the midi2freq will not accurately represent the notes played on the piano player. The file can also be analyzed using midi Info, where the time of the midi file, the size of the file, as well as the frequencies for individual notes of the midi file are displayed. The midi file can now be saved and used as a song input for the PC program GUI.

Fig. 3 C# computer code

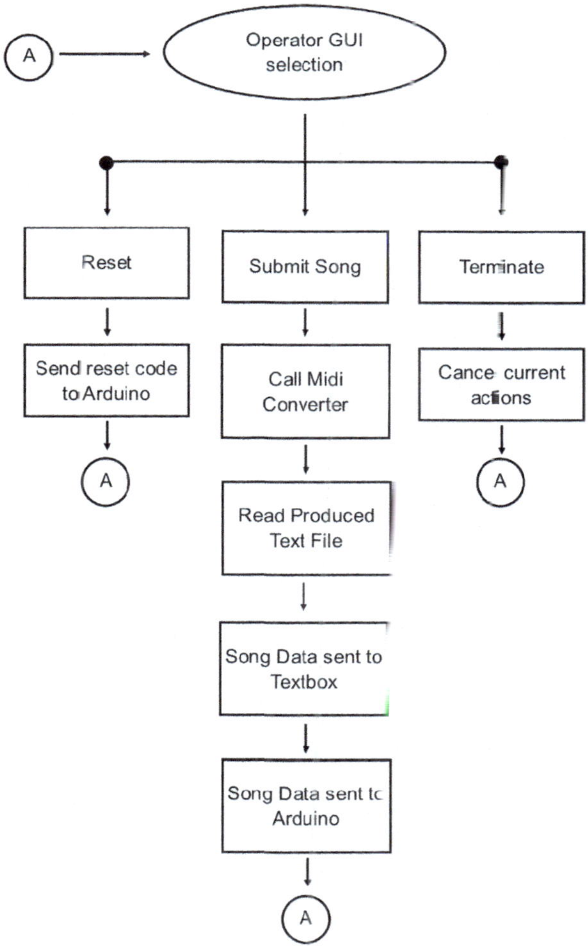

Figure 3 presents an illustrated flowchart of our C# program that runs on the computer connected to the master Arduino. As described in Fig. 3, the operator has three choices for operation. They have the choice to reset the program, which will send a reset code to the Arduino and reset all variables in order to prepare to play another song. Another option for the operator is to select a song to play. For this option, the user selects a midi file to play. The selected file is then sent through the midi file converter, with the text file output then being interpreted by our program with the data then being relayed to the operator, displayed on textbox. The data is then sent to the Arduino. The last option is to terminate and cancel the playing song.

Figures 4 and 5 are very similar with Fig. 4 being for the master Arduino and Fig. 5 for the slave. The master receives commands from the connected computer and then relays those commands to the slave. If a one is received, all the variables are reset in order to prepare to receive a new song. When a two is received, the total number of notes is set in order for the program to recognize when the song is done sending. When a number greater than 10 is sent, the program interprets it as a note and will then store it in the song array. For the master, once this array reaches the upper limit, the song is then sent to the slave and the master waits for the return signal. Once the slave reaches the upper limit, it sends the start signal to the master, and the song will start to play. If it is time for a note to be updated, the song will engage or disengage the corresponding note. The timer is then incremented and sent back to the top of the loop.

The circuit diagram [2] can be seen in Fig. 6. It uses an N-type MOSFET with the source connected to the ground, the drain connected to our resistor and solenoids along with the 24 V power supply, and the gate connected to our 5 V Arduino output. When the 5 volts is received, the MOSFET creates a short between the drain and the source causing current to flow through and activate the solenoid. The 10-ohm resistor is used in order to reduce the heat produced by the solenoid.

4 Testing Approach

A 32-key keyboard was used as a proof of concept. Using SolidWorks, the first iteration of the chassis was designed and modeled. This is shown below in Fig. 7. The first chassis was the initial design for how the solenoids were going to be laid out and the layout of the wiring.

Each solenoid was held in a 3D-printed housing that was designed to suspend the solenoid over the key and able to slide up and down and left and right throughout the case to be adjustable. Iteration 1 for the solenoid holder for the Mk. 2 and Mk. 3 chassis is shown in Fig. 8.

The chassis for the third iteration was 100% constructed from 3D-printed high-temperature PLA filament. Each portion of the chassis had the ability to be disassembled and packed in a lightweight and small footprint. The peg and socket connections that were modeled in SolidWorks allowed for the pieces to be broken

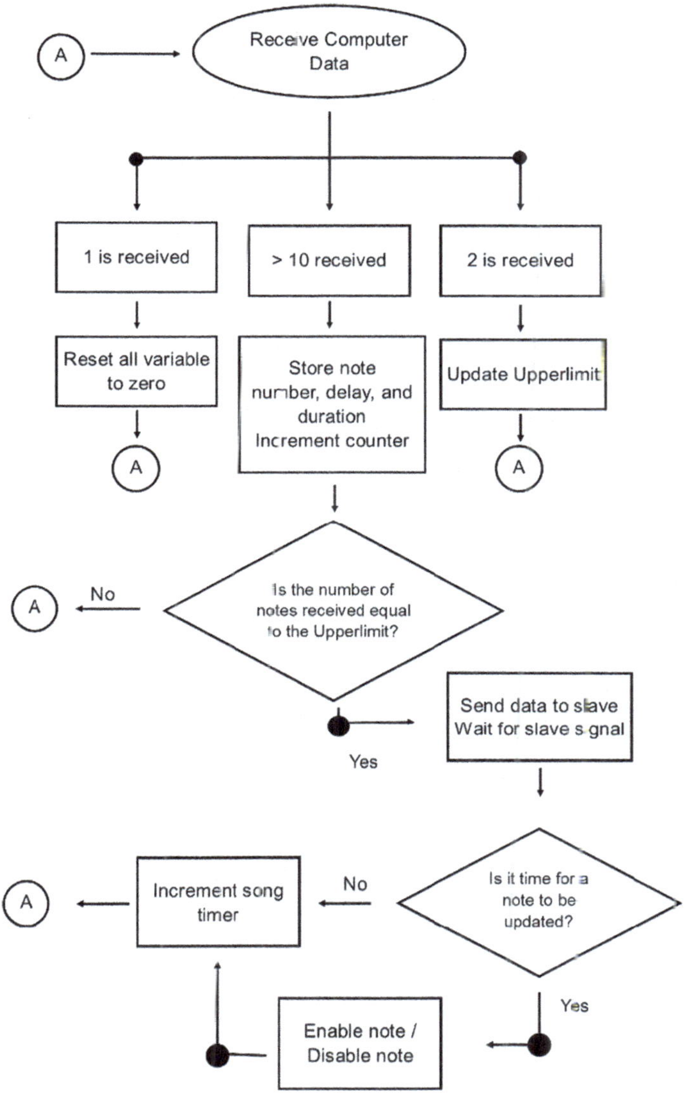

Fig. 4 Master Arduino code

down. The chassis took around 50 hours total to print. The Chassis Mk 3 is shown below in Fig. 9.

The final iteration of the chassis was a modified version of the first chassis built. Version 4 had a stepped layout to accommodate the height differences between the black and white keys. This final chassis is shown below in Fig. 10.

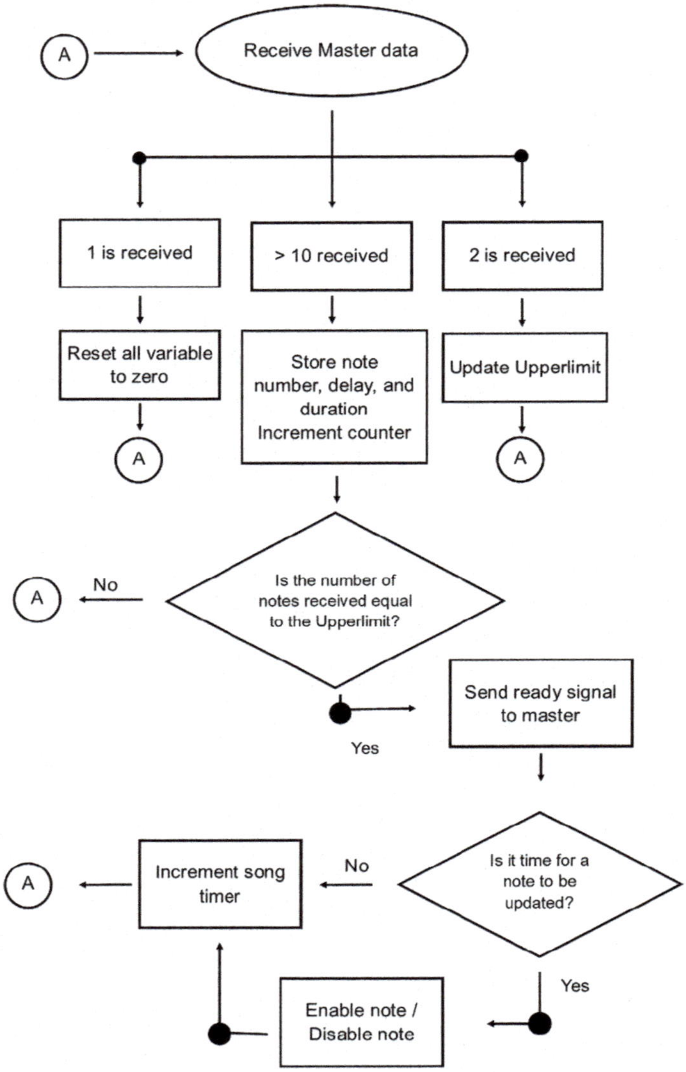

Fig. 5 Slave Arduino code

For initial testing, the two Arduinos were sent a variety of songs and data in order to illustrate that they have received the same data points and will be able to play the notes in the song synchronously. Once this test was completed, the second test was performed by ensuring that every solenoid had successfully made contact with its corresponding key. The device was then tested on a 61-key piano using a test program in order to demonstrate each key pressed from high to low across the keyboard for one second each. This test program was used to establish the baseline

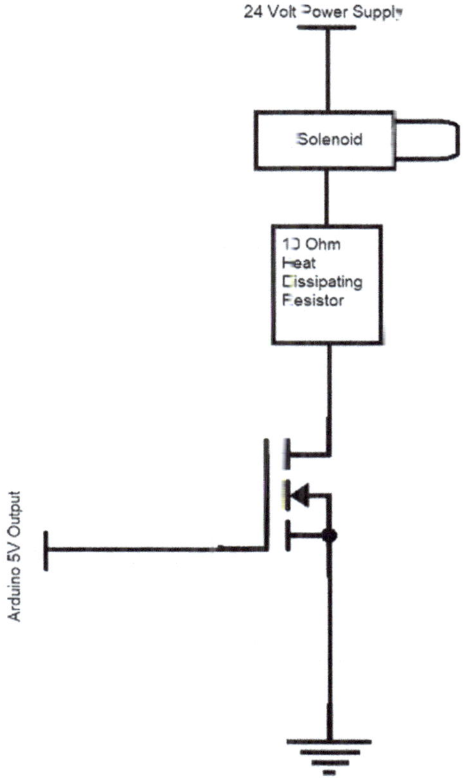

Fig. 6 Solenoid wiring diagram

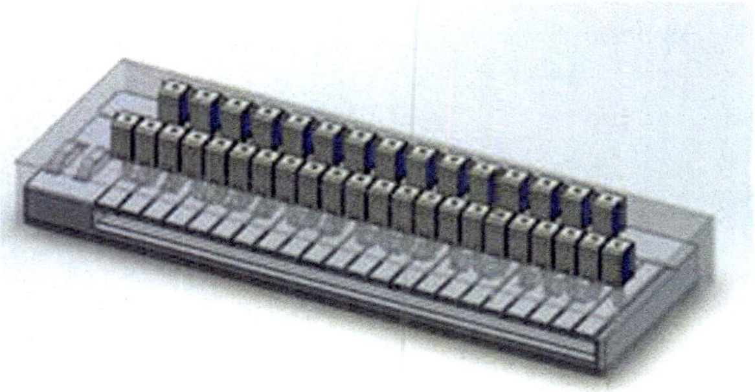

Fig. 7 32-key keyboard chassis design isometric view

from which more advanced songs could be played from. This program allowed us to identify if the system contained any wiring issues or bridges in soldering that needed to be fixed before continuing. The next step was to test playing a full completed song on the piano.

The device was able to play certain songs wonderfully. *Mary Had a Little Lamb* plays exactly how one would expect it to. It was a simple song and perfectly demonstrated the project's ability to recognize midi files and play the song from

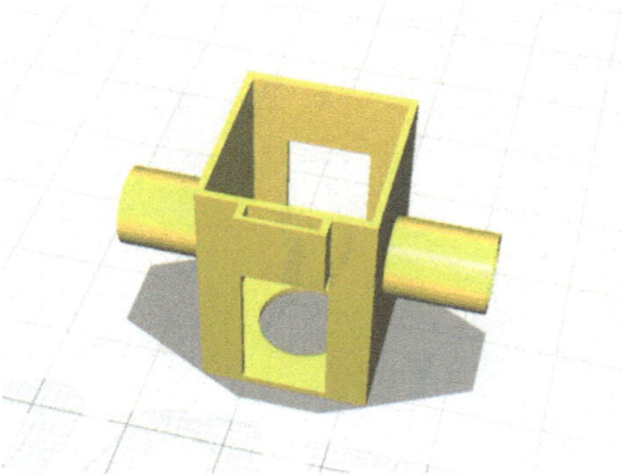

Fig. 8 Solenoid holder for the Mk. 2 and Mk. 3 chassis, iteration 1

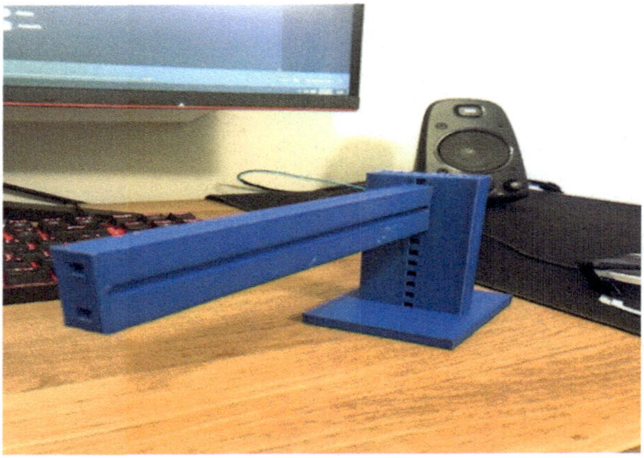

Fig. 9 Disassembled portion of the Mk. 3 chassis support crossbar

Fig. 10 Final chassis design with partial circuitry installed

start to finish. Once *Mary Had a Little Lamb* worked, more complex songs were tested. *The Entertainer* is a good example of a complex song. It has many different notes being played simultaneously and quickly switching from note to note. For this song, the timing was a little off, but it hit all the correct notes in the correct order,

Fig. 11 Final project view

Fig. 12 Project back view

and the song was still highly recognizable. Several other songs were also able to play to great success, including *If You're Happy and You Know It, Für Elise,* and *Green Sleeve*.

5 Final Product/Project Results

The final product shown in Figs. 11 and 12 is a working 61-key piano player that is positioned over a 61-key piano and plays a complete song from start to finish using solenoids in order to engage each individual key. The system is controlled by a GUI program that is connected to the system via a computer connected to the master Arduino. The workload of the song is split in half, and the lower notes are controlled by the master Arduino while the slave Arduino plays the higher notes. The project accepts midi files for input. These midi files can be retrieved from online websites or can be created by the image recognition program. The image recognition program is a MATLAB program [7] that converts sheet music into a playable midi file.

The piano player has successfully played *Mary Had a Little Lamb*, *The Entertainer*, as well as several songs that demonstrate the project capabilities. With the case being wooden, the project is not as adaptable as initially expected. The current design works with 61-key pianos; however, the solenoids are not movable, which limits the adaptability even more.

6 Conclusions and Final Remarks

We have successfully created an Arduino Piano player that has a wooden shell with its own power supply and an image recognition system. In order to control the piano player, we developed a hardware software integration between two individual microcontrollers and a C# program. The piano player works by receiving midi files that are either generated in two methods: the image recognition system, which scans sheet music from an image file, then translates the information into a midi file, and saves it, or a MIDI file from the Internet downloaded by the user. After receiving or downloading the midi file, the C# program sends the song information to the master Arduino which then sends the same information to the slave Arduino. After the slave Arduino has received all the song data, a handshake is achieved between the two Arduinos confirming the reception of all data, and then the song begins. The 61-key prototype that has been developed and constructed by the students was used for testing.

Future improvement for this project includes making the device modular in order to be able to be played on different pianos and keyboards. The proof-of-concept keyboard that was used in the beginning of the project to simulate future builds was not representative of many of the keyboards more commonly used; the flat deck layout of the keys allowed all the solenoids to be in line and staggered with each other. The black keys and the white keys of the midi controller were of the same height and do not protrude higher than one another. This in turn allowed all the solenoids to be easily installed on the same plane or piece of wood.

The later iterations of the keyboard chassis proved to be much more difficult to design and create as the black and white keys on the 61-key keyboard were spaced much further apart and had a significant difference in key height and stroke. Due to the 1 cm travel of the solenoid's piston, the white and black solenoids needed to be offset on different planes to account for the varying travel distances.

Another improvement is the image recognition to be more accurate and use either C programming or Python to run the software. Having the devices use a camera and an already loaded library that can be sent to the Arduino separate from a computer can be a future goal for this project.

References

1. 2019 Piano Tuning Costs. Available: https://homeguide.com/costs/piano-tuning-cost
2. Electrical Safety NFPA 70E. Available: https://www.une.edu/sites/default/files/electrical_safety.pdf
3. What is PianoDisc. Available: https://prodigy.pianodisc.com/what-is-pianodisc/
4. Easily Ship Your Piano. Available: https://www.uship.com/piano-movers/
5. QRS Music Technologies, Inc. Available: https://www.qrsmusic.com

6. 23.2.7.4 Analysis of Music, Musical Notation, Music Scores. Available: https://www.britannica.com/art/musical-notation
7. MATLAB and MIDI. Available: https://kenschutte.com/midi#Writing%20MIDI
8. Frequencies of Musical Notes. Available: https://pages.mtu.edu/~suits/notefreq446.html
9. Sheet Music Reader. (Feb 12, 2018). Available: https://github.com/doctor-entropy/Sheet-Music-Reader

Software-Defined Global Navigation Satellite Systems and Resilient Navigation for Embedded Automation

Jeffrey Wallace, Angelica Valdivia, Srdjan Kovacevic, Douglas Kirkpatrick, and Dubravko Babic

1 Virtualization and Navigation Systems

Virtualization appears in many disciplines due to improvements in processing and networking technology, and this effort establishes the first Software-Defined Global Navigation Satellite Systems (SDGNSS and Software-Defined Navigation (SDNav) products in the market. The motivation is similar: more user functionality, lower-per-cost user, improved size/weight and power (SWaP) profile, and software/firmware updates vice new hardware that makes incremental technology upgrades possible. In addition to the technology, how to package resilient navigation systems remains a practical challenge and to do so for drones further compounds the SWaP constraints.

The Dual-Frequency Resilient Position, Navigation, and Timing (DF RPNT) system combines a best of breed set of algorithms proven in multiple European

J. Wallace (✉) · A. Valdivia
Rocket Technology Systems, LLC, Washington, DC, USA
e-mail: jeff@roc-tec.systems

S. Kovacevic
Orqa d.o.o, Osijek, Croatia

D. Kirkpatrick
Eridan Communications, Inc., Mountain View, CA, USA

D. Babic
Eridan Communications d.o.o, Zagreb, Croatia

K. Daimi et al. (eds.), *Advances in Security, Networks, and Internet of Things*, Transactions on Computational Science and Computational Intelligence,
https://doi.org/10.1007/978-3-030-71017-0_56

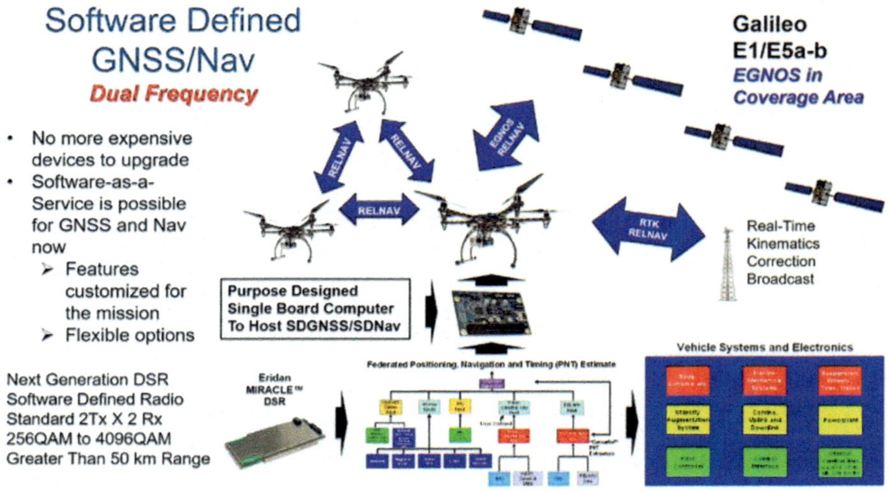

Fig. 1 Dual-Frequency Resilient Positioning, Navigation, and Timing (DF RPNT) block diagram

technology competitions[1,2] to implement the first truly resilient PNT software and hardware framework for drones that utilizes Galileo E1/E5a/b shown in Figs. 1 and 2. The open application programming interface (API) software core of the RPNT framework represents the next generation of PNT packaging and new product offering created by this project. This allows the integration of an Inertial Measurement Unit (IMU) appropriate for the mission with Galileo (or any GNSS) input, implemented as a software-defined radio (SDR) application, alongside a wide variety of alternative navigation data sources customizable based on the drone's specific systems.

This represents a departure from today's Global Navigation Satellite System (GNSS)/Inertial Navigation System (INS) in more than one way. First, decoupling the GNSS from the INS means the GNSS signal can drop out, and the overall system will still produce a navigation solution whose quality metrics are exposed, suitable for human and machine interpretation. Second, this approach avoids the need for a complex faceplate/plug system on some new navigation device and devising some means of programmability through it to integrate organic sources of PNT data present on almost every useful drone system. Minimally, a speed indicator, input from visualization sensors (Vision Nav), or exploiting the communications system for relative navigation (RELNAV) inputs represent a key set of alternative navigation (Alt-Nav/AltNav) data sources. Whether one is in a remote canyon, natural or man-

[1]Galileo Services Administration, "Safewayz, your safety app. Winner of Hackathon 2017", https://galileognss.eu/safewayz-your-safety-app-winner-of-hackathon-2017/.

[2]Galileo Services Administration, "Platform for GNSS/PNT/Intelligent Vehicle Systems", https://galileo-masters.eu/winner/platform-for-gnsspntintelligent-vehicle-systems/.

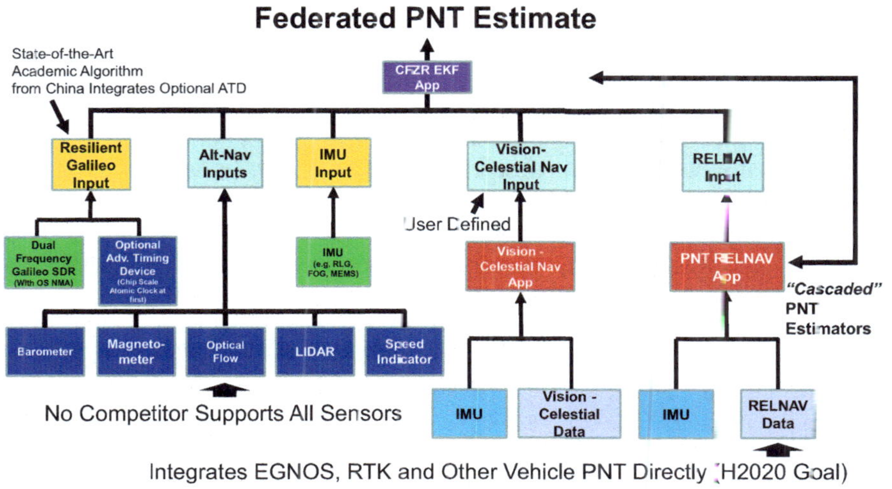

Fig. 2 Dual-Frequency Resilient PNT operational architecture

made, far out at sea, or even underground, the DF RPNT product will reliably operate with industry leading performance.

2 Software-Defined Dual-Frequency GNSS

This section describes software-defined GNSS implementation that can be ported to any capable SDR, such as a LimeSDR,[3] but in this case aimed at the Eridan MIRACLE Digital Sampling Radio (DSR) described later based on the Eridan MIRACLE transceiver.[4] The existing DF RPNT prototype must be ported to different hardware and operating system combinations since it is not a generalist SDR package like GNU Radio and the associated GNSS SDR [1]. The code was built to be optimized on a given set of hardware and support scientific data collection and dissemination, such as accurate magnetic field strength data.

The basic functionality of the Galileo receiver [2] can be broken into a few different steps: signal acquisition, signal tracking, data demodulation, and calculation of the navigation solution. In order to implement a dual-frequency Galileo receiver, this must be done for the E1 and E5a/E5b signal for each satellite. The hardware portion of the receiver includes an antenna configured to receive Galileo signals and will be connected to the inputs on the Eridan MIRACLE DSR [3] which will simultaneously capture at least the two channels necessary to demonstrate signal

[3]Lime Microsystems, https://limemicro.com/products/boards/limesdr/.
[4]Eridan MIRACLE Transceiver, http://eridan.io/product/.

acquisition at E1 (1575.42 MHz) and E5a (1176.45 MHz) or E5a/b (1207.14 MHz) Galileo carrier frequencies. The instantaneous bandwidth of the receiver is sufficient to capture the E1 and E5a/b bands (greater than 25 MHz), which are direct-sequence spread-spectrum coded modulated (CDMA).

The first processing step in a Galileo receiver consists of decoding the CDMA in the signal-processing portion of the receiver where the receiver measures the time delay and frequency shift of incoming signals from four distinct and visible satellites. In a single-frequency receiver, only one signal from each satellite is acquired, and a dual-frequency receiver acquires two signals from each satellite [4].

Once the receiver has measured the time delay and frequency shift associated with each carrier frequency from each satellite, the Galileo receiver uses the information to begin tracking each of the incoming signals. For each incoming signal, the receiver generates a model of the expected signal based on the measured time delay and frequency shifts. The expected signal—as well as a slightly time advanced and a slightly time delayed version of the expected signal—is then compared with the incoming signal. The time delay and frequency shift measurements are continuously updated so that the expected signal best correlates with the incoming signal. Once the receiver is appropriately tracking all signals, it must recover the data sent from each satellite. Since the data bits sent on the E1 and E5a/E5b bands from each satellite are the same, only the data from one frequency must be demodulated. Thus, data demodulation is identical in both single-frequency and dual-frequency Galileo (or any GNSS) receivers.

Finally, a Galileo receiver generates a location estimation using the time delay and frequency shift measurements from the tracking phase. The advantage of using a dual-frequency receiver comes from being able to extract the signal while eliminating the random delay acquired by satellite signal traversing the atmosphere and encounter zones of air that contains high concentrations of free electrons, causing the signals to be deflected and delayed, namely, the effects of ionospheric scintillation. Since the scintillation is inconsistent for each signal, it can cause errors in the Galileo receiver's original location estimate, which is based off the time delays and frequency shifts of the incoming satellite signals. Eliminating it is essential to realizing resilient navigation.

By measuring two frequencies from each satellite, the Galileo receiver can compare the time delay and frequency shift measurements for each satellite and correct for the error caused by ionospheric scintillation. Thus, a dual-frequency receiver can generate more accurate location estimations than a single-frequency receiver. Using the more accurate location estimation and the data from the satellite signal, the dual-frequency receiver generates its final position, velocity, and time more rapidly than a single-frequency receiver. The dual-frequency correction algorithm is implemented in the same function that calculates the receiver's location estimation. The receiver uses the pseudoranges from the tracking of each frequency on a given satellite to generate a more accurate approximation of the distance between the satellite and the receiver using the following equation:

$R(k) = \{(f_1{}^2/f_2{}^2)R_1(k) - R_2(k)\}/\{(f_1{}^2/\mathrm{f}_2{}^2) - 1\}$where f_1 and f_2 are the two carrier frequencies received (in this case 1575.42 MHz and 1176.45 MHz or 1207.14 MHz),

$R_1(k)$ and $R_2(k)$ are the pseudoranges generated from each signal from the kth satellite, and $R(k)$ is the more accurate pseudorange from the kth satellite. From the corrected pseudoranges, the receiver can calculate a more accurate location in the same manner as a single-frequency receiver. However, the dual-frequency receiver will converge more quickly than a single-frequency receiver.

Additionally, the software-defined property of the Galileo receiver means that the receiver can change the way it operates depending on the environment. Sometimes, not all satellites emit both frequencies, limiting the satellites that a dual-frequency receiver communicates with and possibly causing the precision of the receiver to be reduced. To minimize the area of uncertainty in the measurement, the satellites should be as diverse as possible. The on-off capability is also critical if the receiver cannot find four or more satellites that emit both frequencies. Turning off the dual-frequency capability allows the receiver to operate as a single-frequency receiver to find its location. Since the receiver is software defined, it will be able to operate even in environments where there are not enough visible satellites that emit both E1 and E5a/E5b signals.

3 Inertial Navigation Systems and Inertial Measurement Unit Processing

There is a considerable body of work on orientation estimation, or attitude reconstruction, for robotics and control applications. The standard approach is to use extended Kalman filter estimation techniques, but a lesser known alternative is to use deterministic complementary filter and nonlinear observer design techniques [5]. Recent work has focused on some of the issues encountered for low-cost IMU systems as well as observer design for partial attitude estimation. It is also worth mentioning the related problem of fusing IMU and vision data that is receiving recent attention in addition to the problem of fusing IMU and GNSS data. Parallel to the work in robotics and control, there is a significant literature on attitude heading reference systems (AHRS) for aerospace applications.

The interest in small low-cost robotic vehicles has led to a renewed interest in lightweight embedded IMU systems. For the low-cost lightweight systems considered, linear filtering techniques have proved extremely difficult to apply robustly and linear single-input single-output complementary filters are often used in practice. A key issue is online identification of gyro bias terms. An important development that came from early work on estimation and control of satellites was the use of the quaternion representation for the attitude kinematics. The nonlinear observer designs that are based on this work have strong robustness properties and deal well with the bias estimation problem.

However, there appears to be almost no work that considers the formulation of nonlinear attitude observers directly on the matrix Lie-group representation of SO(3) and employs nonlinear attitude observers on SO(3) in a general setting.

The observers are termed complementary filters because of the similarity of the architecture to that of linear complementary filters although for the nonlinear case, there is no frequency domain interpretation. A general formulation of the error criterion and observer structure is based on the Lie-group structure of SO(3). This formulation leads to two nonlinear observers on SO(3), termed the direct complementary filter and passive complementary filter. We do not know of a prior reference for the passive complementary filter.

The passive complementary filter has several practical advantages associated with implementation and low sensitivity to noise. In particular, the filter can be reformulated in terms of vectoral direction measurements such as those obtained directly from an IMU system, a formulation that is termed the explicit complementary filter. The explicit complementary filter does not require online algebraic reconstruction of attitude, an implicit weakness in prior work on nonlinear attitude observers due to the computational overhead of the calculation and poor error characterization of the constructed attitude. As a result, the observer is ideally suited for implementation on embedded hardware platforms.

Furthermore, the relative contribution of different data can be preferentially weighted in the observer response, a property that allows the designer to adjust for application-specific noise characteristics. Finally, the explicit complementary filter remains well defined even if the data provided is insufficient to algebraically reconstruct the attitude. This is the case, for example, for an IMU with only accelerometer and rate gyro sensors. The direct complementary filter is closely related to recent work on invariant observers and corresponds to nonlinear observers using the quaternion representation.

4 Software-Defined Navigation: The Multi-PNT Integration and Estimation Framework

The cascaded, federated zero-reset (CFZR) as defined in Groves [6] is a multi-PNT estimation software framework that uses an Extended Kalman Filter (EKF) algorithm to process sensor measurements from which PNT data can be derived and provide an estimate of the following states:

- Quaternion defining the rotation from North, East, Down local earth frame to X, Y, Z body frame
- Velocity at the IMU—North, East, Down (m/s)
- Position at the IMU—North, East, Down (m)
- IMU delta angle bias estimates—X, Y, Z (rad)
- IMU delta velocity bias estimates—X, Y, Z(m/s)
- Earth magnetic field components—North, East, Down (gauss)
- Vehicle body frame magnetic field bias—X, Y, Z (gauss)
- Wind velocity—North, East (m/s)

The EKF runs on a delayed "fusion time horizon" to allow for different time delays on each measurement relative to the IMU. Data for each sensor is FIFO buffered and retrieved from the buffer by the EKF to be used at the correct time. The delay compensation for each sensor is controlled by two parameters.

Figure 3 shows a traditional control system representation of the framework. A complementary filter is used to propagate the states forward from the "fusion time horizon" to current time using the buffered IMU data. The time constant for this filter is controlled by the EKF2_TAU_VEL and EKF2_TAU_POS parameters. The "fusion time horizon" delay and length of the buffers are determined by the largest of the EKF2_*_DELAY parameters.

If a sensor is not being used, it is recommended to set its time delay to zero. Reducing the "fusion time horizon" delay reduces errors in the complementary filter used to propagate states forward to current time. The position and velocity states are adjusted to account for the offset between the IMU and the body frame before they are output to the control loops. The position of the IMU relative to the body frame is set by the EKF2_IMU_POS_X,Y,Z parameters (Fig. 3).

The EKF uses the IMU data for state prediction only. IMU data is not used as an observation in the EKF derivation. The EKF has different modes of operation that allow for different combinations of sensor measurements. On start-up, the filter checks for a minimum viable combination of sensors and, after initial tilt, yaw, and height alignment is completed, enters a mode that provides rotation, vertical velocity, vertical position, IMU delta angle bias, and IMU delta velocity bias estimates. This mode requires IMU data, a source of yaw (magnetometer or external vision), and a source of height data. This minimum data set is required

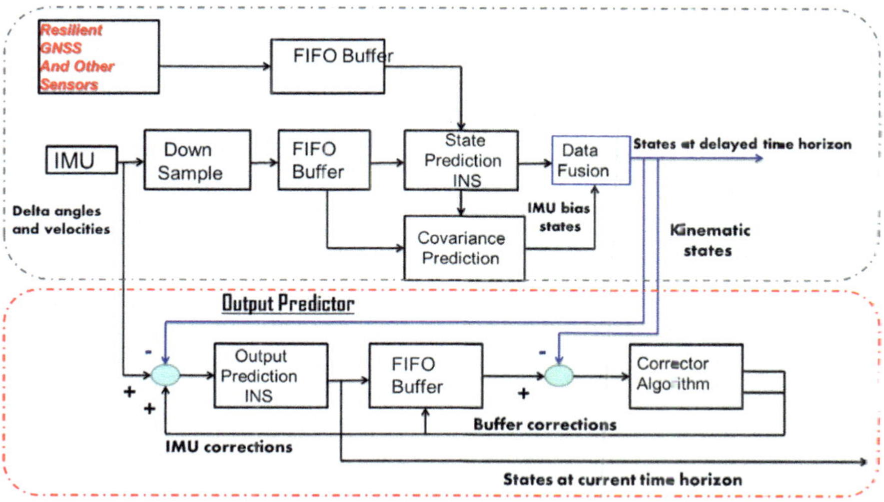

Fig. 3 Multi-PNT estimation framework

for all EKF modes of operation. Other sensor data can then be used to estimate additional states.

5 Relative Navigation

In the RELNAV concept [7], shown below in Fig. 4, a user passively ranges sequentially to several other systems and determines their position from the measured ranges using a form of multilateration [6]. Thus, if the user were in perfect synchronization with their sources, three suitable range measurements could determine three-dimensional position. However, if the user has a synchronization error or time bias (as is normally the case), a sequence of such passive range measurements on signals from higher time-quality sources will provide a continuously updated measure of both position and time bias. (This time bias determination is also called passive synchronization.) In this sense, the RELNAV process is a form of pseudoranging since the range measurements are made with respect to the user's own clock rather than on an absolute or round-trip basis, and the user's clock bias is inherently determined (Fig. 4).

All active units transmit periodically (e.g., once every cycle) a position and status message (P-message, or PPLI), which contains the source's position, speed, course, and altitude, as well as its position quality, tune quality, and relative grid azimuth quality. (The latter is an estimate of own heading error with respect to the grid north direction.) Using these data and appropriate source selection logic,

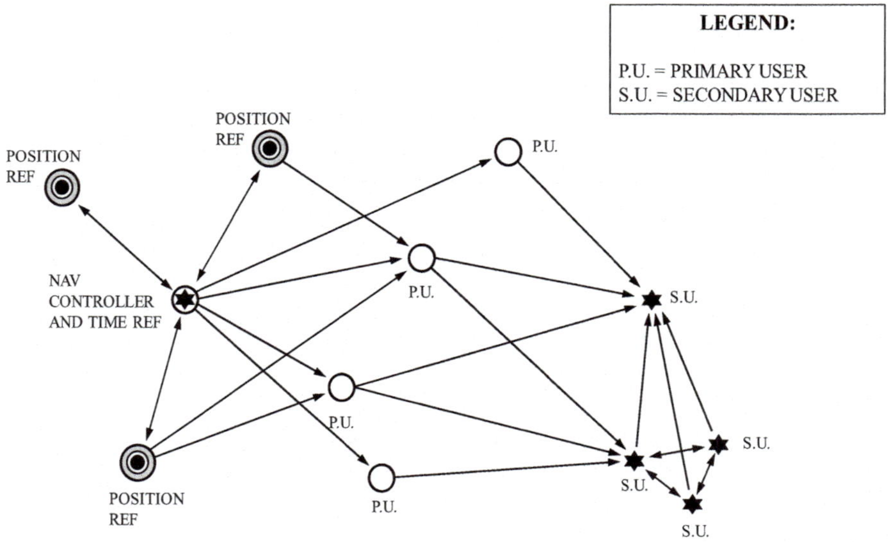

Fig. 4 PNT network RELNAV concept of operation

based on the quality levels of the source and the user terminal, the user terminal selects the desired sources, calculates the predicted range, and compares it to the measured range as obtained from the TOA. By means of a recursive (e.g., Kalman) filter mechanization in the user's processor, the unit then continuously updates its position, velocity, and time bias (with respect to system time) on the basis of these sequential TOA measurements. In most applications, dead reckoning data from such sensors as inertial platforms, Doppler radars, air data systems, electromagnetic logs, and heading references are used to extrapolate the TOA-derived data between filter updates and to optimally mix the data from the two types of sensors.

In order to maintain proper system operation over an extended period of time, some users are also permitted to perform periodic, infrequent, active round-trip timing (RTT) to the time reference or highest time-quality source since this process inherently provides an excellent measure of the user's own time bias. In addition, some users are permitted to RTT relatively frequently. In this manner, the time quality of several units in the net is always maintained at a very high level so that these units can act as navigating sources to other units. This hierarchical organization is discussed more fully in the next section.

RELNAV operation may be either in relative grid coordinates, wherein all units determine their position in a coordinate system established by one member, called the navigation controller, or in an absolute, geographic, Earth-fixed coordinate system (latitude and longitude). In either case, one member of the net is designated as time reference, establishes system time, and is assigned the highest time quality. For the relative grid operation, the navigation controller, which may be an aircraft or a ship, establishes the origin and north direction of a tangent plane grid, whose origin is at sea level and assumed stationary. (Actually, the grid origin and north may be slowly moving as a result of the dead reckoning errors of the navigation controller.)

All net members determine their position with respect to these grid coordinates. The navigation controller is assigned the highest relative position quality. For operation in geographic coordinates, terminals possessing high-accuracy absolute position information are designated as position references and are assigned the highest absolute position quality. In the figures below the references and the navigation controller, there are two classes of users: primary and secondary users.

The primary users (PUs) are permitted to use RTTs for clock synchronization at relatively frequent intervals, i.e., whenever their time quality falls below a certain level (as estimated by the terminal's filter). These (relatively few) units, having excellent time quality, are used as primary navigation references. Secondary users (SUs) do not perform RTTs frequently and, in fact, must be capable of performing clock synchronization and relative navigation completely passively, i.e., without recourse to RTT transmissions. Within the two classes of users, quality levels are established based on accuracy estimated by the terminal's filter. Sources election algorithms can be established, which determine the rules of data exchange, for example, SUs use other SUs as sources only if the latter have higher position and time quality, while PUs may follow somewhat different rules.

Such rules are designed to assure that users' position and time errors do not diverge because of using sources of inferior quality. Through this architecture, users who are not in line of sight of the time reference, the position references, or the navigation controller are able to navigate by passively ranging to other sources, for example, primary users, who are in line of sight of either the references or other primary users. The navigation coverage of the RELNAV net is extended well beyond the horizon (light of sight), despite the radio frequencies used.

The one-sided arrows between the PUs and SUs in the figure denote the directions in which PNT messages (called PPLI) might be used for passive ranging as a result of the assumed higher-quality levels of the PUs. In addition, the SU on the right side is assumed to operate in a fully passive (radio-silent) mode. If users are in line of sight of position references, they can determine their position in geographic coordinates.

The RELNAV function is normally a software module which resides within the navigation program. It is coupled to the TOA measurement function in the signal processor (SP) and the communication processing function within the CP. The basic RELNAV software subfunctions are (1) initialization, (2) source selection, (3) dead reckoning data processing, (4) Kalman filtering, (5) data extrapolation, (6) quality level conversion, and (7) coordinate conversion.

It is also possible to implement RELNAV with TOA-derived data only, i.e., without any dead reckoning data. Figure 5 below shows a typical RELNAV software functional flow diagram. In operation, source data in the incoming P-messages and the TOAs for these are received from the SP and processed first in the communication processing modules of the navigation software. The source quality levels are compared to the terminal's own quality levels in the source selection

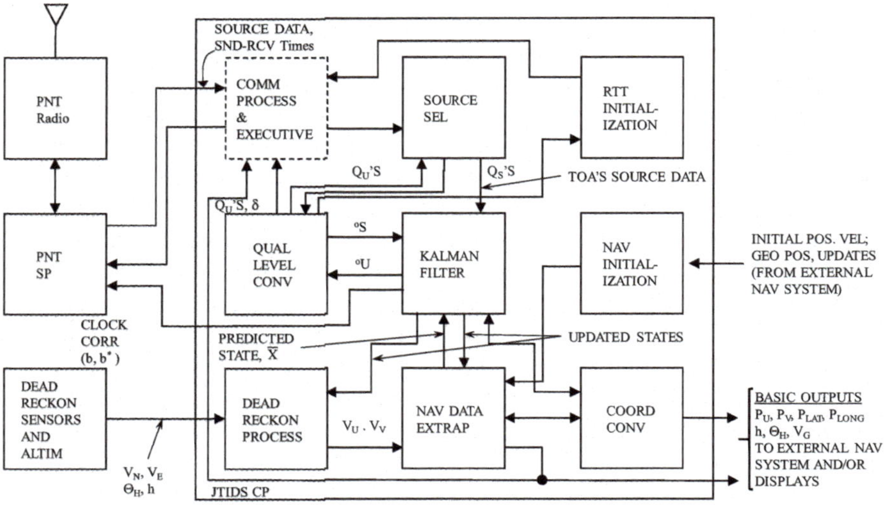

Fig. 5 RELNAV processing architecture

module to determine whether the source should be used. The position and TOA data on the selected sources are then provided to the Kalman filter for processing. Source qualities are also converted into variances for use by the Kalman filter, as described in more detail later.

The filter uses the TOA for the observed range, which is differenced with the predicted value, modified by the appropriate Kalman gains, and added to the predicted state vector, thereby generating updated values for the various states. The covariance matrices are also updated in every filter cycle, in a conventional Kalman filter manner. The updated states, many of which may be error states (e.g., airspeed scale factor, heading error, altitude error), are fed to the dead reckoning processing and navigation data extrapolation modules, as appropriate. Ideally, a new filter update of the states is obtained every time a new TOA measurement is made (on selected-source P-messages (PPLI)), and this is typically once every few seconds.

In the interval, the required navigation data is extrapolated and read out at a much higher data rate, typically several times per second, in order to provide these data to the user at the required rate. Since it may be more convenient to mechanize the Kalman filter in Cartesian coordinates and in view of the dual-grid operation of RELNAV, several coordinate conversion functions may be required before the data is fed to the user, for example, U, V, W position to latitude, longitude, and altitude above the Earth. The dead reckoning (DR) data is received from the DR sensors on the vehicle and processed using the latest estimates of the correction states from the filter and extrapolated to generate the required outputs. The extrapolated values are also used to generate the predicted states.

6 Results and Summary

To fully exercise the overall framework, an eight-blade drone with a 1-meter diameter was constructed as the test platform and flown at the Floyd Bennett Memorial Airport (GFL). In addition to the dual-frequency SDGNSS and IMU inputs, the environment alt-nav inputs—barometer, magnetometer, and optical flow and speed indicator—are included as the baseline.

The RELNAV system used four ground transmitters. A geo-registered vision navigation application was integrated into the vision navigation input that used a FLIR™ brand camera that produced the same CCD data output as the F-35 for realism. Geo-registered VisionNav has a set of landmarks or features in the mission area that can be recognized by a computer vision system, and the coordinates have been measured or otherwise derived and known. As the mission proceeds, the geo-registered landmark/feature is overflown and recognized, and the location is fed into the navigation filter with an effectiveness whose influence can be seen in the results in Fig. 6.

The GNSS dropout occurs at time equal to 60 seconds, and drift occurs until a high-accuracy geo-registered set of features are overflown. The positional error

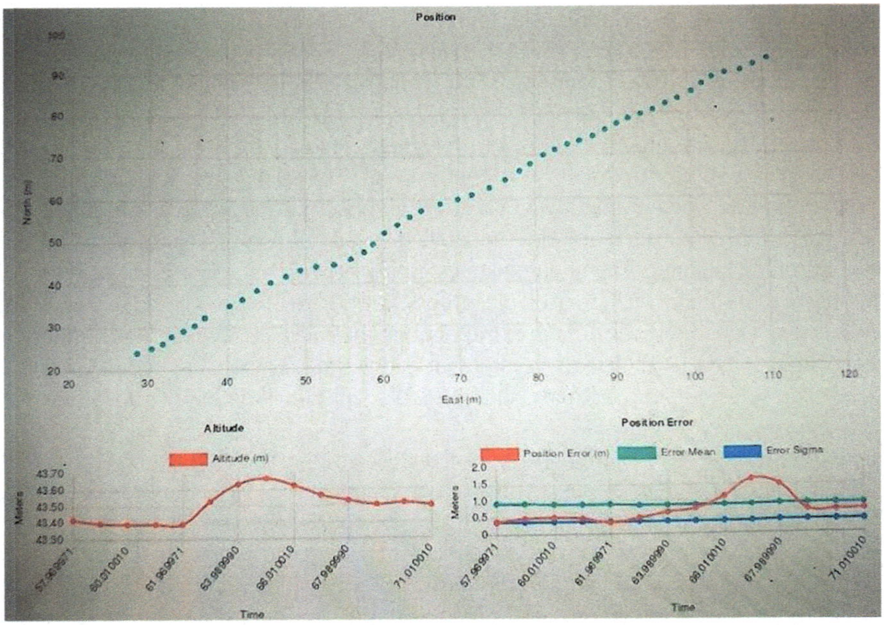

Fig. 6 Test results

is slightly without GNSS but still quite reasonable for an initial prototype. The number of PPLI messages was low, and more of these would improve performance for a minimal amount of bandwidth. In the initial experiments, a very conservative number of alt-nav sensors measurements were made to minimize processor load trading-off accuracy.

This extended the flying time and demonstrates another dimension along which mission needs can influence tuning for specific requirements, further illustrating the value of software-defined navigation components and systems. Using more powerful processors like the newest low-power/high-performance offerings from Xilinx and NSX will allow higher update rates on the alt-nav sensors sampling and processing, enabling higher accuracy and minimizing power consumption to not adversely affect mission capability.

References

1. C. Fernández-Prades, Documentation. (2017). Retrieved November 13, 2017, from http://gnsssdr.org/docs/
2. K. Borre, *A Software-Defined GPS and Galileo Receiver: A Single-Frequency Approach* (Birkhauser, Boston, 2007) Retrieved from http://link.springer.com/10.1007/978-0-8176-4540-3

3. E. McCune, *Dynamic Power Supply Transmitters: Envelope Tracking, Direct Polar, and Hybrid Combinations*, 1st edn. (Cambridge University Press, 2015)
4. Z. Yao, M. Lu, Dual-frequency constant envelope multiplex with non-equal power allocation for GNSS. IET Electron. Lett. **48**(25), 1624–1625 (2012)
5. R. Mahony, T. Hamel, J.-M. Pflimlin, Nonlinear complementary filters on the special orthogonal group. IEEE Trans. Automat. Contr. **53**(5), 1203–1218 (2008)
6. P. Groves, *Principles of GNSS, Inertial, and Multisensor Integrated Navigation Systems, Second Edition* (Artech House, 2013) ISBN: 9781608070053
7. J.F.O. Ranger, Principles of JTIDS relative navigation, J. Navig. (Cambridge University Press, 2009), pp. 22–35

Smart Automation of an Integrated Water System

F. Zohra and B. Asiabanpour

1 Introduction

With the growing population globally, the developing countries are facing insecurity with growing food and supplying fresh water to meet the demand [1]. Almost 1.2 billion people in the world do not even have access to safe water [1, 2]. Figure 1 shows how much the water withdrawal percentage has increased around the world in color-coding from 1995 and a prediction for 2025 [3]. Notice the span of orange- and yellow-colored regions increased significantly over time.

Figure 2 shows that agriculture is considered the highest consumer of the world's fresh water, about 70% [4, 5]. Forty percent of the world's population faces water scarcity at least for a month each year [6]. The planet has to prepare to feed around nine billion people by 2050, which requires an anticipated 50% increase in cultivated production along with a 15% increase in freshwater withdrawals [4].

With modern vertical farming systems such as hydroponics, higher yield is possible with much less water and fertilizer than conventional farming [7, 8] and is considered as a viable approach. In such soilless farming, plants are grown inside a controlled environment, and plant roots are submerged into nutrition solutions and placed in vertical racks to utilize space (Fig. 3).

The benefits of this farming method are almost no weeds, no pests, double growth rate, nutrition efficiency, and only 20% of water usage compared to conventional farming. However, this system needs a consistent water supply to produce yields year-round and perform better with the use of pure water as they provide control over the pH level, which can facilitate maximum nutrition absorption by the plants and results in good yield [11].

F. Zohra · B. Asiabanpour (✉)
Ingram School of Engineering, Texas State University, San Marcos, TX, USA
e-mail: ba13@txstate.edu

K. Daimi et al. (eds.), *Advances in Security, Networks, and Internet of Things*,
Transactions on Computational Science and Computational Intelligence,
https://doi.org/10.1007/978-3-030-71017-0_57

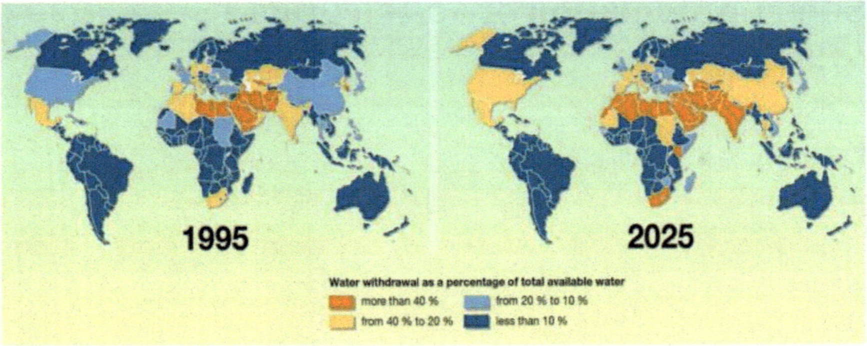

Fig. 1 Water withdrawal percentage of total available water [3]

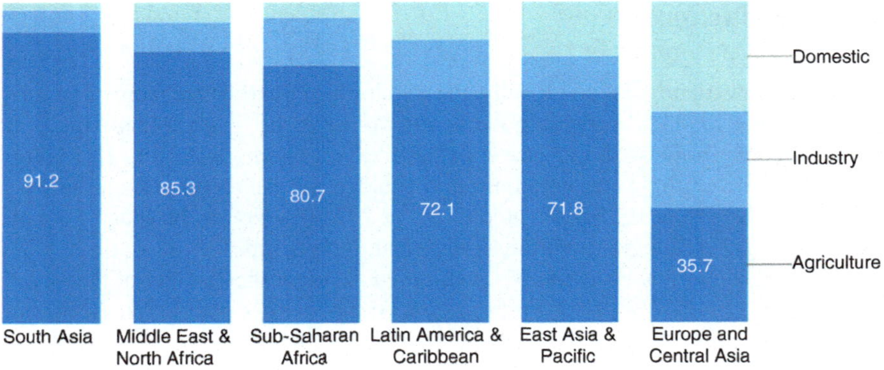

Fig. 2 Consumption of fresh water by % around the world [4]

Fig. 3 (**a**) Hydroponic farming [9], (**b**) vertical farming [10]

Rainwater harvesting has been used for many years for irrigation and other non-potable use [12]. Highly populated states in the United States were taken into consideration to assess the demand for outdoor water usage vs. supply of water that can be collected from rooftops. Except for arid regions, it was found that it is possible to meet 100% of this *non-potable* demand (even for Arizona, Texas, etc.) only by collecting rainwater from the roof-space [13].

As for the regions where rain is insufficient to meet either or both potable and non-potable demands, such as Texas (insufficient to meet both kinds of demands) [14], alternative, independent sources such as Atmospheric Water Generator (AWG) system can be introduced as a complementary source. At present, AWG products are available in the market that look like typical water dispensing machines, and its working principle is similar to a conventional refrigeration system. These devices cool down the air passing through its air filter and condensate the moisture present in the air and collect it. The collected water is run through a set of filters to ensure that the water is disinfected and safe to drink. However, as the production largely depends on the atmospheric condition, a prediction curve (Fig. 4) has been developed to estimate the amount of water production at different combinations of temperature and relative humidity [15].

Smart automation of a water system indicates that the system will perform automatically and will take the decision to turn on or off its subsystems, taking input form real-time feedback or analyzing previous data. So far, many technologies have been developed for automation of different water facilities, water treatment plants, and even for the storage tank and water pumps. Raspberry Pi [16], Arduino [17], a simple algorithm for water pump [18], smart metering [19], supervisory control and data acquisition (SCADA) system are some recent technologies that are being implemented worldwide. Apart from these, many commercially available sensors such as DUV 2/0,005-5 (Kazakhstan), INNOLevelECHOIL-EC-A (Russia), and Siemens Sitrans Probe LU (Germany) are used for recording the water level by measuring the distance between the sensor and surface water surface [20]. With the aid of Raspberry Pi or a typical computer and Python software, Arduino

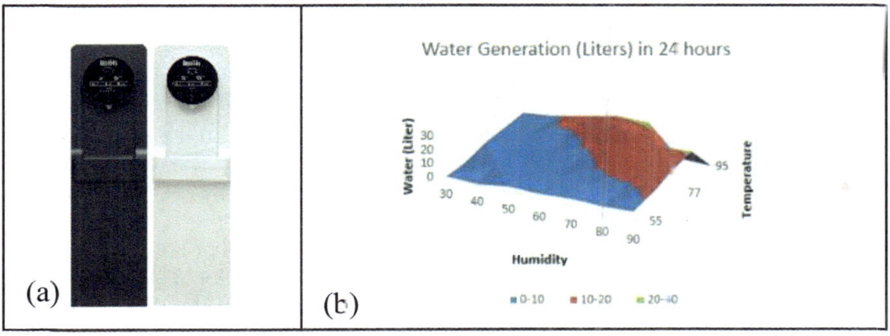

Fig. 4 (**a**) AquaBoy Pro-II [15]; (**b**) estimated water production [14]

microcontrollers can be reprogrammed and modified for a program-based control system which is effective and cost-efficient.

Preliminary studies based on local historical rainwater and atmospheric water generation showed that a smart integrated RHS-AWG would provide stable and sustainable water [21]. This article explains the mechanical, electrical, network, and data logic design and implementation of this integrated system. The rest of this paper is organized as the following. Section 2, system development methodology including mechanical and electrical subsystems and data logic of integrating the components for both water generation/collection and delivery/consumption aspects of the system are explained. Results, including design details, fabrication, and system operation, are illustrated in Sect. 3. Discussion and conclusions are presented in Sect. 4.

2 System Development Methodology

RHS and AWG are evaluated independently to identify and study their essential subsystems such as collection and generation, storage, replenishment, filtration, preservation, and delivery. The water systems contain few similar components such as filter, storage tank, treatment system, pump, power supply, water flow line, etc. These standard components have been combined with being unique instead of having them in two separate systems. The whole system is divided into two sections for a better explanation: (1) collection/generation side and (2) delivery/consumption side. Design, coding for Arduino Uno, electrical connection modification, and testing are performed separately for both sides of the system.

2.1 Collection/Generation Side

2.1.1 Mechanical System

The RHS is installed in the Freeman Center (San Marcos, TX) with a 3000-gallon capacity reservoir, a filtration unit consisting of the pressure tank, carbon filter, reverse osmosis, and ultraviolet unit and a gutter system along the roof that collects the rainwater directly from the catchment area and delivers it to the main reservoir. The following steps were taken to establish the mechanical systems suitable for automation.

(a) A float sensor is installed inside the main reservoir so that when the water level goes below 1000 gallons, it turns on a 110 V power outlet. This power outlet is used to power the electrical system for the generation side. So the electrical system in this part stays off if when there is no immediate demand for water, i.e., the water level is above 1000 gallons.

(b) Two commercial Atmospheric Water Generators "AquaBoy Pro-II" made by Atmospheric Water Solutions, Inc. has been chosen to integrate with the RHS. Each can produce 2–5 gallons of water per day depending on the atmospheric conditions and has a warranty of 2 years (Fig. 4).
(c) The AWG is designed so that the generated water keeps circulating inside its closed system, and the water can be retrieved from a tap by pushing a touch pad. Hence, the modification of this closed system is done to let the generated water come out automatically to deliver it to the main reservoir. The waterline from the bottom tank of AWG is disconnected from its system and redirected to an outer secondary storage tank. All the level sensors are disabled after performing the first. This modification eliminates unnecessary energy consumption by its water pump operation, filtration, and recirculation.
(d) A wooden infrastructure has been fabricated to place the AWGs on the top shelf and the temporary collection reservoir in the bottom to aid the flow of the generated water by gravity.
(e) The integrated system is designed and drawn in a 3D CAD model using SolidWorks software in a personal computer for a complete and better visualization.

2.1.2 Electrical System

To introduce automation to the generation side, the information from the estimated water production plot shown in Fig. 4(b) is used to decide on the range of temperature (55–120 °F) and relative humidity (30–100%) combination. The setup of all the electrical components, test, and debugging of this electrical system is carried out and installed on-site.

(a) A schematic diagram is drawn to illustrate all the connections (power supply, digital signal, and water flow) in the collection/generation side.
(b) Arduino coding was developed according to the selected range of atmospheric conditions to operate AWG based on the real-time feedback system. If the atmospheric condition does not match the defined range, it stays off, and the logic repeats itself after 15 minutes to check.
(c) A float and a submersible water pump are installed in the secondary tank that collects the AWG generated water. Upon receiving the level up the signal from the float of this tank, the Arduino turns on the pump through a relay switch and runs the pump for 20 minutes to deliver the collected water to the main reservoir. Otherwise, the logic delays for 15 minutes before repeating itself.
(d) The logic flowchart is developed according to the complete coding.
(e) The detailed wiring schematic is drawn using diagrams.net (Google Drive-integrated diagramming application).
(f) The float structure is designed and drawn in SolidWorks software and fabricated using the PVC pipe and fittings available on-site.

2.2 *Delivery/Consumption Side*

2.2.1 Mechanical System

(a) To facilitate water to all eight supply tanks, a water flow distributor is added with 12 outlet lines with manual shutoff valves. A solenoid is installed on each line for respective tanks.
(b) A similar float structure with a PVC pipe (Fig. 5) is installed in the supply tanks with a float height of 13 inches from the bottom of the tanks.

2.2.2 Electrical System

(a) A schematic diagram is drawn to illustrate all the connections (power supply, digital signal, and water flow) in the delivery/consumption side using diagrams.net.
(b) Arduino programming is developed, uploaded, and tested in an experimental setup to ensure that it works as expected. Since 16 pins are required for a total of eight input pins from floats and eight output for the tanks and the Arduino having the limitation of 12 digital input/output pins, two Arduino Uno boards are considered for connecting four floats and four solenoids to each.

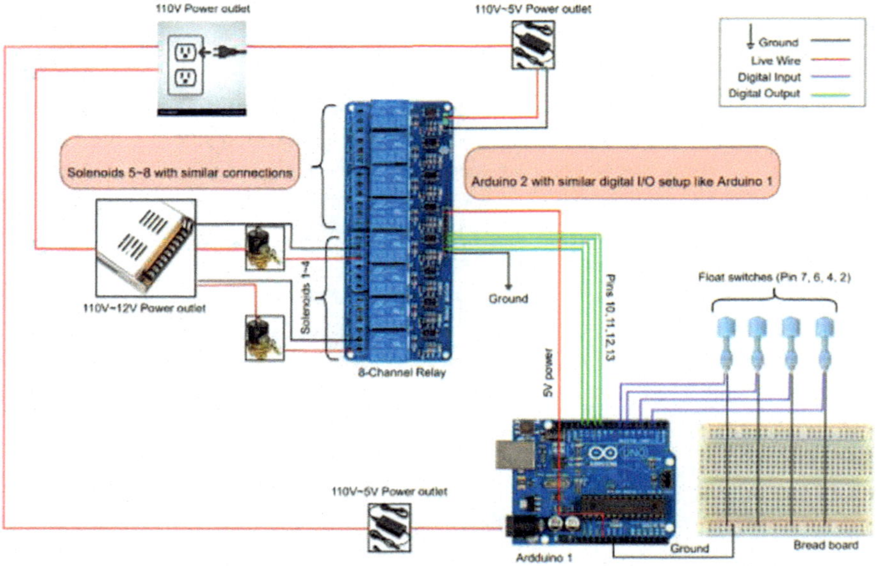

Fig. 5 Circuit diagram for wiring of all the components on the consumption side

(c) The logic flowchart for one tank with a float-solenoid system is developed based on the coding in diagrams.net.
(d) A schematic for the circuit diagram showing all the wiring of all the components is drawn in diagrams.net. Test and debugging are carried out in an experimental setup, and then the system was introduced on-site.
(e) Similar float and fittings (Fig. 5) are installed in each tank to send the Arduino that controls and controls solenoids to refill water.

3 Results

3.1 Collection/Generation Side

3.1.1 Mechanical System

(a) Figure 6 shows the rainwater harvesting system and the vertical farming unit in the Freeman Center, San Marcos, TX.

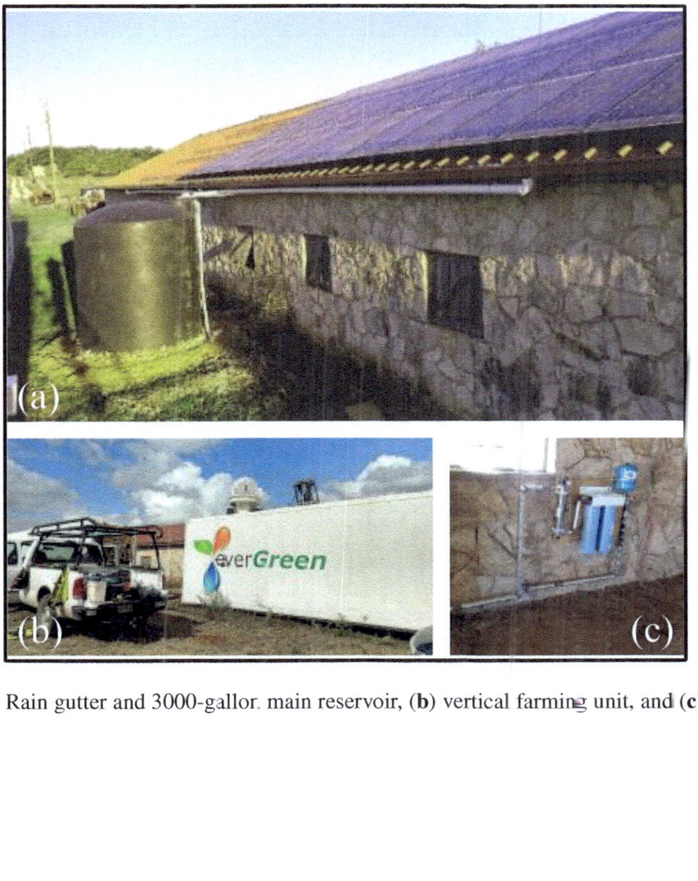

Fig. 6 (**a**) Rain gutter and 3000-gallon main reservoir, (**b**) vertical farming unit, and (**c**) filtration unit

(b) Figure 4 shows the selected AWGs and the plot of estimated water production based on temperature and humidity. Similar operation parameters are also provided in the manufacturer's manual.
(c) The 3D CAD models in the following figures show different components of the smart integrated water system (Figs. 7 and 8).

3.1.2 Electrical System

(a) The schematic drawing of the integrated system for collection/storage side is shown in Fig. 9.
(b) Arduino program for operating the AWG based on the real-time feedback logic system first assesses if the environmental condition is in favor or not to run the AWG and decides upon turning on/off the AWG and thus avoids unnecessary consumption of energy. Therefore, it becomes a smart AWG in terms of operation.
(c) The Arduino program (Fig. 10) using feedback from the float switch to turn on the water pump is shown below. Once the "Water level is up" signal is received, the pump is turned on for 20 minutes to transfer the water from this secondary reservoir to the main reservoir.
(d) The logic flowchart for the AWG generation and for the float system is shown in Fig. 11.
(e) Details of established connections among the electrical components (Fig. 12) are shown below. In Arduino, pin 2 is the input from the atmospheric sensor, and pin 7 is from the float sensor. Pin 13 and pin 8 are the outputs to operate AWG and the pump, respectively.

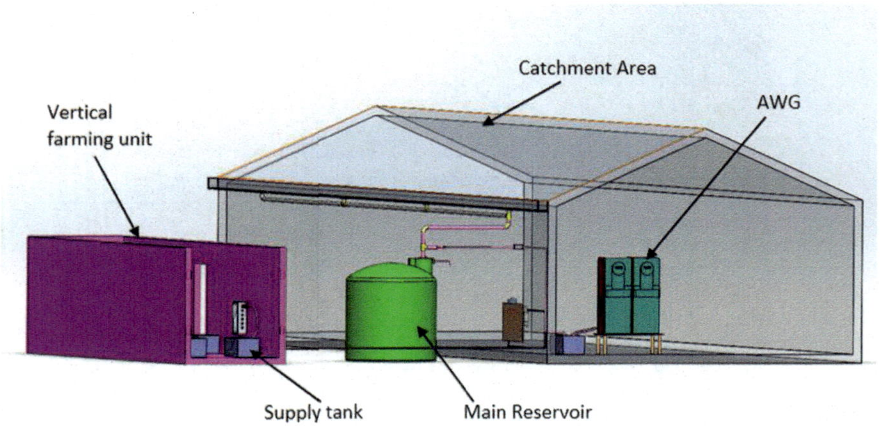

Fig. 7 The integrated water system for the vertical farming unit

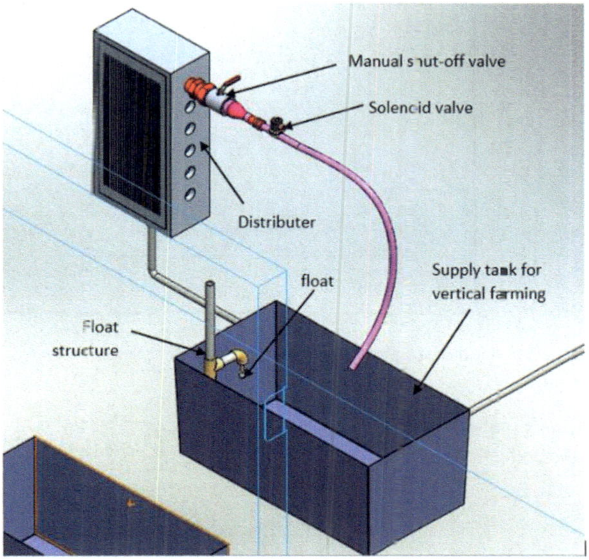

Fig. 8 Supply tank and the distribution with manual valve, solenoid, and float

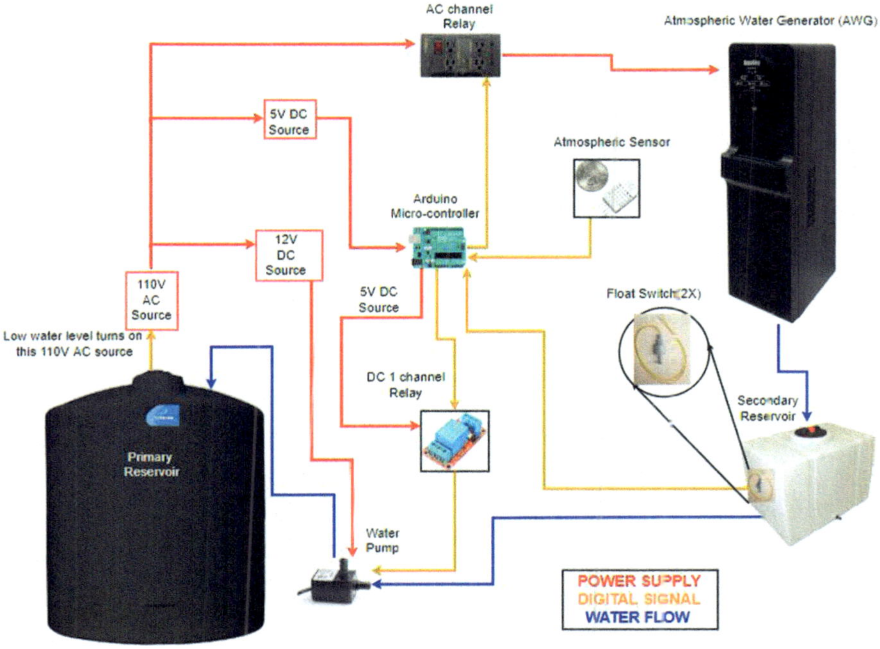

Fig. 9 Schematic drawing of the generation/collection side with automation

```
//controlling the pump by the float
   if(digitalRead(FLOAT_SENSOR) == LOW)//water level is up
  {
       Serial.print("Water level is up");
    digitalWrite(pump_1, HIGH);// start the pump
      }

  else
   delay(20*60*1000UL); // pump runs for 20 min, then off
 digitalWrite(pump_1, LOW);
}
```

Fig. 10 Coding to pump the generated water from AWG into the main reservoir by a submersible pump

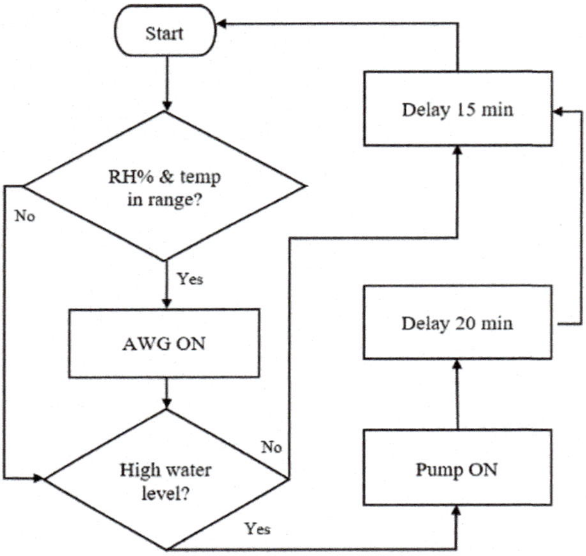

Fig. 11 Logic flowchart for the collection/storage side

(f) The support structure for the float is shown in Fig. 5. The fabrication process of this structure is easy to achieve and has the following advantages:

- Convenient to assemble or disassemble into the tank.
- No intensive machining is required.
- Sturdy and easily replicable.
- Protection for float wiring from water damage (Fig. 13).

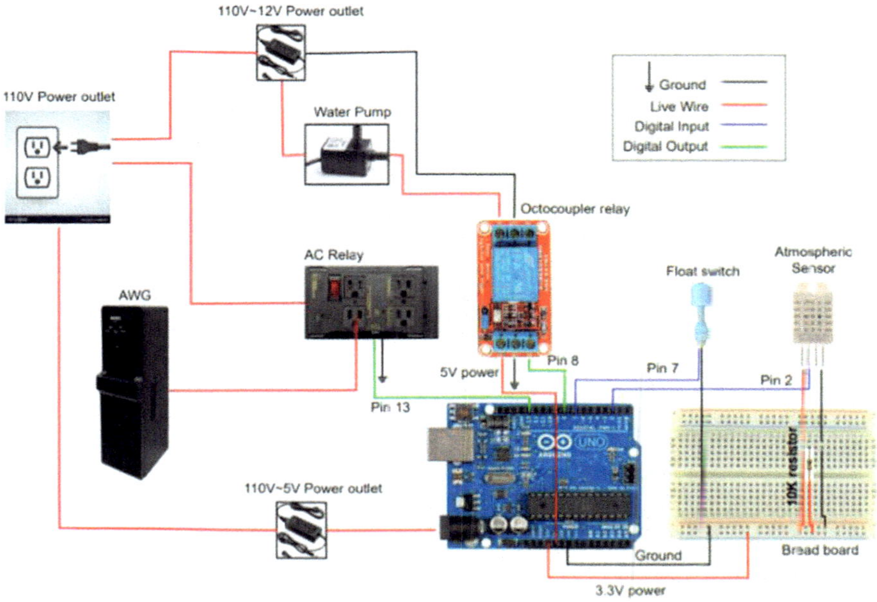

Fig. 12 Circuit diagram for generation side developed in diagrams.net

Fig. 13 3D CAD model for the float with tank

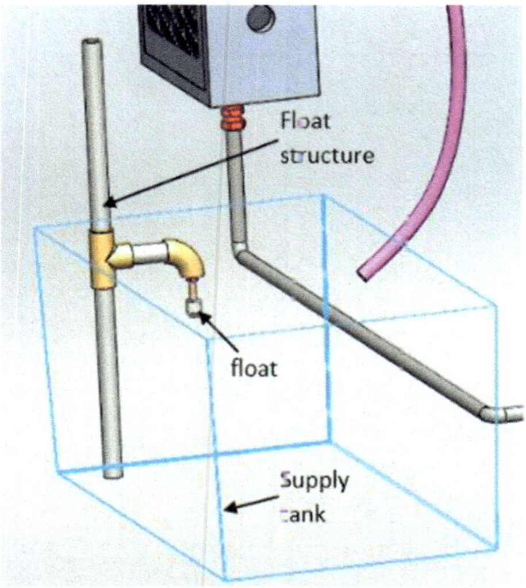

3.2 Delivery/Consumption Side

3.2.1 Mechanical System

(a) The delivery system consists of a waterline that goes through the pressure tank and filtration set and finally to the vertical farming unit. The line is then divided into a total of 12 outlet lines with manual shutoff valves that can deliver water directly to the supply tanks or can be used for cleaning or other purposes. The eight outlets connected to the eight supply tanks also have electromagnetic solenoid shutoff valve to control the water refill process to the respective tanks automatically. Figure 8 illustrates how the solenoid valve is installed between the waterline to control the water refilling process.

3.2.2 Electrical System

(a) The schematic drawing of the integrated system for the delivery/consumption side is shown in Fig. 14. Automation in the consumption side is introduced to

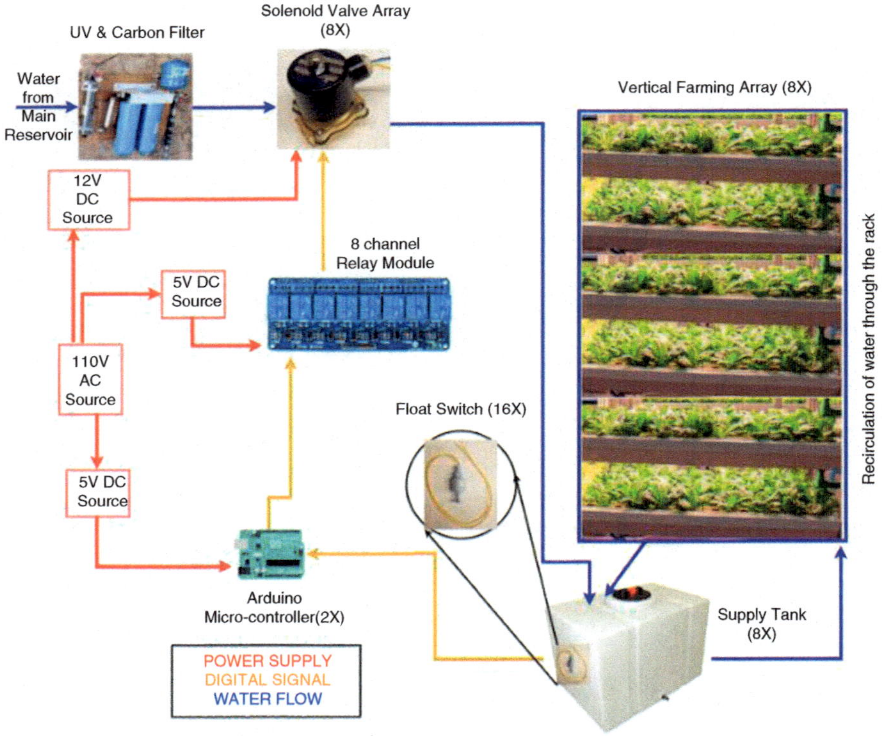

Fig. 14 Schematic diagram for the delivery side using diagrams.net

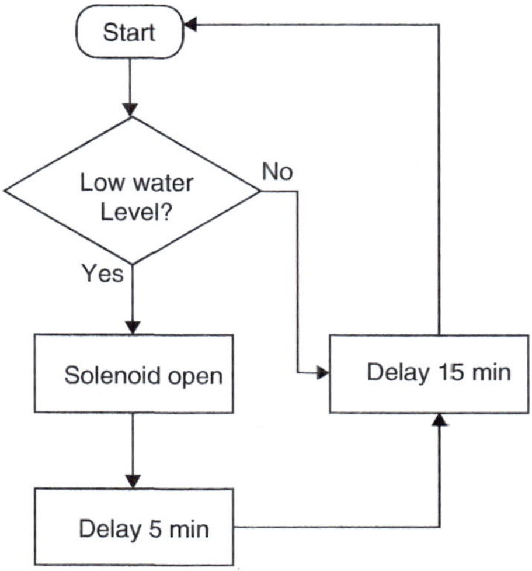

Fig. 15 The logic flow diagram for a single float-solenoid-tank

maintain a constant level of water in the supply tank. Since the consumption of water in eight racks of vertical farming might vary due to different watering schedules, different species, the number of plants for each rack, etc., each tank might need refilling in a different frequency.

(b) Two Arduino is necessary to control eight solenoids, and each Arduino controls four solenoids taking input for respective float valves for respective supply tanks. Similar program is loaded to each Arduino to control the second set of four solenoids.

(c) The logic flow diagram for the consumption side for one supply tank is shown in Fig. 15. This logic is similar for all the supply tanks respective to their floats and solenoids.

(d) Details of established connections among the electrical components are shown in Fig. 5. Pins 2, 4, 6, and 7 are the input pins, and pins 10, 11, 12, and 13 are the output for the floats from tank 1–4, respectively, which switches on/off the eight-channel relay to control the respective solenoids. Connection for only one Arduino and individual solenoids is shown in that figure to avoid the cluster of similar components.

4 Discussion and Conclusion

This smart integrated water system is developed in a way that it can be operated autonomously. The experimental setup is done first to ensure all the components with logic control is working as expected and then installed on-site at the Freeman Center, San Marcos, TX. Observing the operation of this system clarified that such an integrated system is very location dependent. Available area/rooftop for rainwater catchment, available reservoir capacity, energy price, and degree of risk that consumers can take about water shortage are among factors that may affect the scale of the system and the ratio between RHS and AWG.

Acknowledgments This work has been completed with funding from the US Department of Agriculture (grant number 2016-38422-25540). The authors would like to thank the USDA, Freeman Center, and Ingram School of Engineering at Texas State University for providing funding and access to infrastructure and laboratories. Sponsors are not responsible for the content and accuracy of this article.

References

1. M. Hameed, P. Abbaszadeh, A. Ahmadalipour, A. Alipour, H. Moftakhari, H. Moradkhani, A review of the 21st century challenges in the food-energy-water security in the Middle East. Water **4**, 11 (2019)
2. D.J. Garcia, F. You, The water-energy-food nexus and process systems engineering: A new focus. Comput. Chem. Eng. **91**, 49–67 (2016)
3. L. Swatuk, M. McMorris, C. Leung, Y. Zu, Seeing 'invisible water': Challenging conceptions of water for agriculture, food and human security. Can. J. Dev. Stud. **36**(1), 24–37 (2015)
4. Chart: Globally, 70% of freshwater is used for agriculture. [Online]. Available: http://blogs.worldbank.org/opendata/chart-globally-70-freshwater-used-agriculture. [Accessed: 01-Jan-2020]
5. P. Pingmuanglek, N. Jakrawatana, S.H. Gheewala, Freshwater use analysis of cassava for food feed fuel in the Mun River basin, Thailand. Int. J. Life Cycle Assess. **22**(11), 1705–1717 (2017)
6. R. Kaewmai, T. Grant, S. Eady, J. Mungkalasiri, C. Musikavong, Improving regional water scarcity footprint characterization factors of an available water remaining (AWARE) method. Sci. Total Environ. **681**, 444–455 (2019)
7. J. He, Farming of vegetables in space-limited environments. Cosmos **11**(01), 21–36 (2015)
8. H.S. Grewal, B. Maheshwari, S.E. Parks, Water and nutrient use efficiency of a low-cost hydroponic greenhouse for a cucumber crop: An Australian case study. Agric. Water Manag. **98**(5), 841–846 (2011)
9. Is Hydroponic Plantation Beneficial? | Everchem Fertilizer Company. [Online]. Available: https://everchem.com.my/is-hydroponic-plantation-beneficial/. [Accessed: 29-May-2020]
10. J. Birkby, Vertical farming. Natl. Cent. Appropr. Technol. **88**(12), 1173–1176 (2016)
11. J.J.A. Dominguez, C. Inoue, M.-F. Chien, Hydroponic approach to assess rhizodegradation by sudangrass (Sorghum x drummondii) reveals pH- and plant age-dependent variability in bacterial degradation of polycyclic aromatic hydrocarbons (PAHs). J. Hazard. Mater. **387**, 121695 (2019)
12. M.P. Jones, W.F. Hunt, Performance of rainwater harvesting systems in the southeastern United States. Resour. Conserv. Recycl. **54**(10), 623–629 (2010)
13. K. Tamaddun, A. Kalra, S. Ahmad, Potential of rooftop rainwater harvesting to meet outdoor water demand in arid regions. J. Arid Land **10**(1), 68–83 (2018)
14. B. Asiabanpour, N. Ownby, M. Summers, F. Moghimi, Atmospheric water generation and energy consumption: An empirical analysis, in *2019 IEEE Texas Power and Energy Conference, TPEC 2019*, 2019
15. AWS - Home. [Online]. Available: https://www.atmosphericwatersolutions.com/. [Accessed: 29-Dec-2019]
16. B.I. Farhan, Design and construct intelligent tank 'water level sensor'. J. AL-Qadisiyah Comput. Sci. Math. **10**(3), 1–8 (2018)

17. H. Ahmad, U. Atikol, A simple algorithm for reducing the operation frequency of residential water pumps during peak hours of power consumption. Energy. Sci. Eng. **6**(4), 253–271 (2018)
18. Z. Yuan et al., Sweating the assets - The role of instrumentation, control and automation in urban water systems. Water Res. **155**, 381–402 (2019)
19. T. Imanaliyev tonimontana, O. Karlykhanov, M. Li, N. Bakbergenov, A. Zhakashov, D. Ponkratyev, Automation of water facilities in Kazakhstan and its solutions. Contemp. Dilemmas Mag. Educ. Polit. Values **7**(1), 1–30 (2019)
20. M. Li, O. Karlykhanov, D. Ponkratyev, T. Imanaliyev, T. Tazhiyeva, Automatic water meter for gauging stations of irrigation canals. SPACES Mag **34**, 39 (2018)
21. F. Zohra, B. Asiabanour, F. Moghimi, *Design and Development of an Integrated Water System Combining Rainwater Harvesting System and Atmospheric Water Generation* (2020 IISE Annual Conference, LA, 2020)

Quadratic Integer Programming Approach for Reliability Optimization of Cyber-Physical Systems Under Uncertainty Theory

Amrita Chatterjee and Hassan Reza

1 Introduction

Cyber-physical systems are composed of both hardware and software components. Apart from integration of these two components, they comprise elements like actuators, sensors, and processors [9, 11, 23]. A cloud-edge computing framework including the design of CPS has been put forward in [29]. Data processing and its various challenges faced have been discussed in [28] while exploring a systematic big data-as-service framework for CPS.

Among the challenges present in CPS design, the most important one is to make the system reliable. In our day-to-day life, there are a lot of uncertainties related to environment, weather, and other disturbances. Finding a software/hardware partition of a CPS is an ongoing challenge. Automation of this problem has been already researched, and exact partitioning [4, 20, 22] or heuristic partitioning models have been deployed [5, 6, 30]. However, few algorithms work under the real-world constraint that it is impossible to estimate the time and cost of the components of the system accurately. Also, analyzing the attribute of reliability is intricately connected to the hardware/software partition during the design of the CPS. Unless these challenges are addressed, existing partitioning algorithms are limited to its applications in the real-world systems. Therefore, various methods and proposals are on the rise to bridge this gap. To achieve this, an uncertain programming model has been developed recently in [14], which characterizes reliability and includes uncertainty compliance.

A. Chatterjee (✉) · H. Reza
School of Electrical Engineering and Computer Science, University of North Dakota, Grand Forks, ND, USA
e-mail: amrita.chatterjee@und.edu; hassan.reza@und.edu

K. Daimi et al. (eds.), *Advances in Security, Networks, and Internet of Things*, Transactions on Computational Science and Computational Intelligence,
https://doi.org/10.1007/978-3-030-71017-0_58

In [14], the partitioning problem has been depicted as a mathematical optimization equation. The cost-related objectives and delay-related constraints are stated in terms of uncertain variables. The characteristics of reliability are depicted as task graph of the system. This design has higher degree of assurance in terms of safety and security under the circumstances of uncertainties. The authors of [14] attempt to solve the formulated optimization problem with a simulated annealing heuristic implemented with genetic algorithm.

As a novel contribution of this work, we explore alternative methods to solve the optimization problem: we pose it as a 0–1 integer quadratic programming problem, which is known to be NP-hard [3], and explore heuristic methods to solve this problem. We settle on a heuristic method to solve this problem. Preliminary results indicate that solutions with improved values of the reliability metric compared to genetic algorithm solution are feasible.

2 Related Work

The existing models and algorithms for partitioning can be broadly differentiated as exact partitioning and heuristic partitioning models. The exact algorithms comprised branch and bound [4], integer linear programming [22], and dynamic programming [15, 20], whereas the heuristic algorithm consists of simulated annealing [8, 21, 26, 27] and genetic algorithm.

These algorithms work at a smoother pace when put in their co-design environments. The parameters of all the components are deterministic, meaning for some specific inputs, it will give definite outcomes that are already predefined. However, in the design phase, the cost and time of the software components remain hard to estimate. Some works assume these factors as subjective probabilities and make use of this assumption to perform system-level partitioning [1]. However, some of these assumptions do not hold when we consider the system as a whole from the project management viewpoint.

Reliability has been taken to be the prioritized attribute in terms of partitioning according to [12, 13, 25]. The imprecise quantities presented in [17] are devoid of fuzziness or randomness. Therefore, based on some preliminary idea in [10], system partitioning is conducted, taking into consideration the reliability factor. Belief degree of uncertain events is measured relying on the uncertainty theory [16–18].

3 Uncertainty Model

A formal exhibition of the software/hardware partitioning issue encountered in cyber-physical system is provided in this section. This formulation including the various metrics and optimization problem definition is obtained from the extensive body of work spanning many years from Jiang et al. [10, 14]. This formulation targets to reduce system cost and execution time while enhancing reliability under uncertainty theory.

3.1 Problem Setup

The cyber-physical system under design is modeled as a directed acyclic graph $G(V, E)$, where V is the set of nodes $\{v_1, v_2, \ldots, v_n\}$ and E is the set of edges $\{e_{ij} | 1 \leq i < j \leq n\}$.

1. γ_i^h denotes the additional cost of the hardware implementation of node i in the system, over a software implementation.
2. δ_{ci}^h denotes the linear uncertainty distribution of γ_i^h, denoted by $\lambda(p_{ci}^h, q_{ci}^h)$, where p_{ci}^h and q_{ci}^h are non-negative real numbers.
3. t_i^h denotes the execution time of node i if implemented in hardware, while t_i^s denotes the execution time of an equivalent software implementation.
4. δ_{ti}^h and δ_{ti}^s are uncertainty distributions of uncertain variables t_i^h and t_i^s, respectively, denoted by $\lambda(p_{ti}^h, q_{ti}^h)$ and $\lambda(p_{ti}^s, q_{ti}^s)$, where p_{ti}^h, q_{ti}^h, p_{ti}^s, and q_{ti}^s are non-negative real numbers.
5. c_{ij} is the communication time between nodes i and j, and it is given a non-zero value only if nodes i and j differ in implementation choice.
6. r_i^h denotes the reliability of node i if implemented in hardware, while r_i^s denotes the reliability of an equivalent software implementation.

The hardware–software partitioning problem is defined as finding a bipartition P of this graph, where $P = (V_h, V_s)$ such that $V_h \cup V_s = V$ and $V_h \cap V_s = \Phi$. In addition, T_0 is the given execution time upper bound, while R_0 is the given reliability lower bound. We can represent such a partition in terms of a n-wide vector of binary index variables $x = \{x_1, x_2, \ldots, x_n\}$. If node i is implemented in software, then it is assigned a value of 1, else it is assigned a value of 0. The objective is to find a partition P such that $T(x) \leq T_0$ and $R(x) \geq R_0$, while the total cost $H(x)$ is minimized.

3.2 Optimization Problem Metrics

Evaluation of a partition is based on three primary metrics, reliability, execution time, and cost discussed in [14]. The total cost includes both software and hardware components of the CPS. Similarly, the execution time is the sum of the execution time of each node and the communication time between nodes. The reliability is assumed to be the probability that the system will perform its intended function accurately, which requires all nodes to function appropriately.

Cost Metric The total cost of the system can be represented by the sum of the additional cost of the hardware implemented nodes (with an inherent assumption that hardware implementation is more costly than software implementation) for the purposes of optimization. Hence, the cost $H(x)$ of the partition x can be formulated as

$$H(x) = \sum_{i=1}^{n} \gamma_i^h (1 - x_i). \tag{1}$$

Time Metric Time metric is addition of two parts: execution time of each node and communication time between nodes. Again, it is a rational assumption that two nodes that do not alter in implementation choice will have zero communication cost. The total time is the sum of total execution time and total communication time, which is expressed as follows:

$$T(x) = \left[\sum_{i=1}^{n} t_i^s x_i + t_i^h (1 - x_i)\right] + \left[\sum_{i=1}^{n-1} \sum_{j=i+1}^{n} c_{ij}[(x_i - x_j)^2]\right]. \tag{2}$$

Reliability Metric Nodes that are devoid of any outgoing edges are the destination/system output nodes, while nodes that have zero incoming edges are regarded as start/system input nodes. The adjacency matrix $Adj[n][n]$ represents the dependency edges of parent to child nodes in the directed acyclic graph. Based on the theory of fault tree and reliability block diagram, the reliability of the task node x_i is the product of its parent nodes and reliability of itself. Hence, the system reliability $R(x)$ is the summation of the reliability of all output nodes. The reliability of each output node is obtained in a recursive manner as shown in Algorithm 1.

Algorithm 1 System reliability $R(x)$

Input: $Adj[][]$ Adjacency matrix of the DAG $G(V, E)$
    binary vector x
Output: $R(x)$
Init: $R(x) \leftarrow 0$

Function *Recursive(m)* **is**
    for $i \leftarrow 1$ **to** n **do**
        if *Adj[i][m] == 1* **then**
            $R_m \leftarrow R_m.(r_m^s x_m + r_m^h (1 - x_m)).Recursive(i)$
        end
    end
    return R_m
end
foreach *output node m in the DAG G(V,E)* **do**
    $R(x) \leftarrow R(x) + Recursive(m)$
end
return $R(x)$

3.3 Optimization Problem Formulation

Based on the metric definitions in the previous section, the given constraint M on total execution time, and a lower bound R_0 on the reliability, the hardware/software partitioning problem is modeled as given in P_0.

$$P_0 : \begin{cases} \text{minimize } H(\mathrm{x}) \\ \text{subjectto } \; T(\mathrm{x}) \leq M \\ \qquad\qquad\quad R(\mathrm{x}) \geq R_0 \\ \qquad\qquad\quad \mathrm{x} \in \{0, 1\}^n. \end{cases} \tag{3}$$

Since minimizing the value of $H(x)$ is equivalent to maximizing the value of $\sum_{i=1}^{n} \gamma_i^h x_i$, problem P_0 can be reduced to problem P_1.

$$P_1 : \begin{cases} \text{maximize } \sum_{i=1}^{n} \gamma_i^h x_i \\ \text{subjectto } \; \sum_{i=1}^{n-1} \sum_{j=i+1}^{n} c_{ij} \left[\left((x_i - x_j)^2 \right) \right] + \\ \qquad \sum_{i=1}^{n} \left(t_i^s - t_i^h \right) x_i \leq M - \sum_{i=1}^{n} t^h \\ \qquad 1 - R(x) \leq 1 - R_0 \\ \qquad \mathbf{x} \in \{0, 1\}^n. \end{cases} \tag{4}$$

The uncertain objective function is targeted first. It is known that if $\gamma_1, \gamma_2 \cdots \gamma_n$ are uncertain variables with uncertain distributions $\delta_1, \delta_2 \cdots \delta_n$ and the function $f(\mathbf{x}, \gamma_1, \gamma_2 \cdots \gamma_n)$ is strictly increasing with respect to $(\gamma_1, \gamma_2 \cdots \gamma_m)$ and strictly decreasing with respect to $(\gamma_{m+1}, \gamma_{m+2} \cdots \gamma_n)$, then the converted expected objective function can be calculated as

$$\begin{aligned} E\left[f(\mathrm{x}, \gamma_1, \gamma_2 \cdots \gamma_n)\right] = \int_0^1 f\Big(&\mathrm{x}, \delta_1^{-1}(\alpha), \delta_2^{-1}(\alpha) \\ &\cdots \delta_m^{-1}(\alpha), \delta_{m+1}^{-1}(1-\alpha) \cdots \delta_n^{-1}(1-\alpha)\Big)\, d\alpha, \end{aligned} \tag{5}$$

where $\delta_i^{-1}(\alpha)$ is $(1-\alpha)\left(p_{ci}^h - q_{ci}^s\right) + \alpha\left(q_{ci}^h - p_{ci}^s\right)$. The mathematical exposition in [14] describes how this problem can eventually be converted to the following reliability maximization problem

$$P_{final} : \begin{cases} \text{maximize } R(x) \\ \text{subjectto } \; \sum_{i=1}^{n} \left(q_{ti}^s - p_{ti}^h \right) x_i + \\ \qquad \sum_{i=1}^{n-1} \sum_{j=i+1}^{n} c_{ij} \left(x_i - x_j \right)^2 \leq M - \sum_{i=1}^{n} p_{ti}^h \\ \qquad \sum_{i=1}^{n} \left[\left(\int_0^1 \delta_i^{-1}(a) d\alpha \right) (1 - x_i) \right] \leq N \\ \qquad x_i \in \{0, 1\};\, i = 1, 2 \cdots n. \end{cases} \tag{6}$$

4 Research Contribution: Solution with 0–1 Quadratic Integer Programming Heuristic

The final reliability maximization problem is an instance of a multi-objective 0–1 integer quadratic programming problem, which is known to be an NP-hard [3], and hence only heuristic-based algorithms are applicable. Several general-purpose heuristic algorithms discussed in Sect. 2 can be implemented to solve the resultant optimization problem. The authors of [14] chose a genetic algorithm with simulated annealing heuristics. However, we realized that since this is a special case of 0–1 integer quadratic programming, more specialized heuristics such as [19] exist and can be applied to this problem.

Quadratic 0–1 integer programming is an actively explored research area. While the general problem is NP-hard, certain classes of this problem are even amenable to exact solutions [2]. However, vast majority of problem instances are only solvable with heuristic methods. We explored a number of heuristic solution methods for this problem including but not limited to genetic (GA) [19], alternating direction method of multipliers (ADMM) [24], and convex hull-based [7] heuristics.

The structure of the particular problem instance at hand lends itself to be suitable to the convex hull heuristic (CHH)-based approach [7], since it is a nonlinear optimization problem with linear constraints. This heuristic is based on simplicial decomposition (SD), which is applied repeatedly to obtain feasible solutions for a nonlinear integer programming problem with linear constraints. For this problem, SD is started each time from a different feasible point. This decomposition method allows us to use the linear programming tools such as CPLEX on each iterative step of the process.

5 Preliminary Results

Although we are still exploring the space of specialized heuristics for this problem, we implemented a heuristic algorithm similar to as described in [7], and the very preliminary experiments on random graphs show some improvement over the genetic algorithm solution. Table 1 shows the improvement over the genetic algorithm solution on the reliability metric.

Table 1 Preliminary results on randomly generated DAGs

Name	Node	Edge	Gain over GA
random1	500	1,000	1.21
random2	100	1,200	1.37
random3	1,500	3,000	−0.32
random4	2,000	4,000	2.14
random5	2,500	5,000	1.79
random6	3,000	6,000	1.48

6 Conclusion

This work leverages the uncertainty theory-based formulation of the CPS hardware–software partition problem as developed in [14]. The resultant optimization problem is treated as a 0–1 quadratic integer programming problem, and a heuristic-based solution method is applied. Preliminary results promise some improvement over genetic algorithm-derived solution.

References

1. J. Albuquerque, C. Coelho, C.F. Cavalcanti, D. Cecilio da Silva, A.O. Fernandes, System-level partitioning with uncertainty, in *Proceedings of the Seventh International Workshop on Hardware/Software Codesign*, pp. 198–202 (1999)
2. S.T. Chakradhar, M.L. Bushnell, A solvable class of quadratic 0–1 programming. Discrete Appl. Math. **36**(3), 233–251 (1992)
3. W.A. Chaovalitwongse, I.P. Androulakis, P.M. Pardalos, *Quadratic Integer Programming: Complexity and Equivalent Forms* (Springer US, 2009), pp. 3153–3159
4. K.S. Chatha, R. Vemuri, Hardware-software partitioning and pipelined scheduling of transformative applications. IEEE Trans. Very Large Scale Integr. (VLSI) Syst. **10**(3), 193–208 (2002)
5. R.P. Dick, N.K. Jha, Mogac: a multiobjective genetic algorithm for hardware-software cosynthesis of distributed embedded systems. IEEE Trans. Comput. Aided Des. Integr. Circuits Syst. **17**(10), 920–935 (1998)
6. P. Eles, Z. Peng, K. Kuchcinski, A. Doboli, System level hardware/software partitioning based on simulated annealing and tabu search. Des. Autom. Embeded Syst. **2**, 5–32 (1997)
7. M. Guignard, A. Ahlatcioglu, The convex hull heuristic for nonlinear integer programming problems with linear constraints and application to quadratic 0–1 problems. J. Heuristics (2020)
8. R.K. Gupta, G. De Micheli, Hardware-software cosynthesis for digital systems. IEEE Design Test Comput. **10**(3), 29–41 (1993)
9. S. Jeschke, C. Brecher, H. Song, D.B. Rawat, *Industrial Internet of Things* (Springer International Publishing, 2017)
10. Y. Jiang, H. Zhang, X. Jiao, X. Song, W.N.N. Hung, M. Gu, J.G. Sun, Uncertain model and algorithm for hardware/software partitioning, in *2012 IEEE Computer Society Annual Symposium on VLSI*, pp. 243–248 (2012)
11. Y. Jiang, H. Song, R. Wang, M. Gu, J. Sun, L. Sha, Data-centered runtime verification of wireless medical cyber-physical system. IEEE Trans. Ind. Inf. **13**(4), 1900–1909 (2017)
12. Y. Jiang, M. Wang, H. Liu, M. Hosseini, J. Sun, Dependable integrated clinical system architecture with runtime verification, in *Proceedings of the 36th International Conference on Computer-Aided Design*, pp. 951–956 (2017)
13. Y. Jiang, H. Song, Y. Yang, H. Liu, M. Gu, Y. Guan, J. Sun, L. Sha, Dependable model-driven development of CPS: From stateflow simulation to verified implementation. ACM Trans. Cyber Phys. Syst. **3**(1) (2018)
14. Y. Jiang, M. Wang, X. Jiao, H. Song, H. Kong, R. Wang, Y. Liu, J. Wang, J. Sun, Uncertainty theory based reliability-centric cyber-physical system design, in *2019 International Conference on Internet of Things (iThings) and IEEE Green Computing and Communications (GreenCom) and IEEE Cyber, Physical and Social Computing (CPSCom) and IEEE Smart Data (SmartData)*, pp. 208–215 (2019)
15. P. Knudsen, J. Madsen, Pace: A dynamic programming algorithm for hardware/software partitioning, in *CODES* (IEEE Computer Society, 1996), pp. 85–93

16. B. Liu, Fuzzy random dependent-chance programming. IEEE Trans. Fuzzy Syst. **9**(5), 721–726 (2001)
17. B. Liu, Uncertainty theory, in *Uncertainty Theory* (Springer, Berlin, Heidelberg, 2010), pp. 1–79
18. B. Liu, Y.K. Liu, Expected value of fuzzy variable and fuzzy expected value models. IEEE Trans. Fuzzy Syst. **10**(4), 445–450 (2002)
19. A. Lodi, K. Allemand, T.M. Liebling, An evolutionary heuristic for quadratic 0-1 programming. Eur. J. Oper. Res. **119**(3), 662–670 (1999)
20. J. Madsen, J. Grode, P.V. Knudsen, M.E. Petersen, A. Haxthausen, Lycos: The Lyngby co-synthesis system. Des. Autom. Embedded Syst. **2**(2), 195–235 (1997)
21. R. Niemann, P. Marwedel, Hardware/software partitioning using integer programming, in *Proceedings of the 1996 European Conference on Design and Test*, p. 473 (1996)
22. R. Niemann, P. Marwedel, An algorithm for hardware/software partitioning using mixed integer linear programming. Des. Autom. Embedded Syst. **2**(2), 165–193 (1997)
23. H. Song, D.B. Rawat, S. Jeschke, C. Brecher, *Cyber-Physical Systems: Foundations, Principles and Applications* (1st edn.) (Academic Press, 2016)
24. R. Takapoui, N. Moehle, S. Boyd, A. Bemporad, A simple effective heuristic for embedded mixed-integer quadratic programming (2015)
25. S. Tosun, N. Mansouri, E. Arvas, M. Kandemir, Y. Xie, W. Hung, Reliability-centric hardware/software co-design, in *Sixth International Symposium on Quality Electronic Design (ISQED'05)*, pp. 375–380 (2005)
26. F. Vahid, D.D. Gajski, Clustering for improved system-level functional partitioning, in *Proceedings of the 8th International Symposium on System Synthesis* (Association for Computing Machinery, New York, NY, USA, 1995), pp. 28–35
27. F. Vahid, D. Gajski, J. Gong, A binary-constraint search algorithm for minimizing hardware during hardware/software partitioning, in *Proceedings of the Conference on European Design Automation*, pp. 214–219 (1994)
28. X. Wang, L.T. Yang, L. Huazhong, M.J. Deen, A big data-as-a-service framework: State-of-the-art and perspectives. IEEE Trans. Big Data, 1–1 (2017)
29. X. Wang, L.T. Yang, X. Xie, J. Jin, M.J. Deen, A cloud-edge computing framework for cyber-physical-social services. IEEE Commun. Mag. **55**(11), 80–85 (2017)
30. X. Zhao, H. Zhang, Y. Jiang, S. Song, X. Jiao, M. Gu, An effective heuristic-based approach for partitioning. J. Appl. Math. (2013)

Brief Review of Low-Power GPU Techniques

Pragati Sharma and Hussain Al-Asaad

1 Introduction

Graphics Processing Units, or GPUs, have become a major force in today's modern computing and cloud revolution. High-performance computing (HPC) and deep learning applications in computer vision, health care, genomics, and Natural Language Processing are gaining traction. GPUs are naturally well-fitted to run these applications with their multiple computing cores and ability to run kernels in parallel, thereby maximizing performance. However, as technologies improve and we strive to enhance the computing horsepower in these GPUs, there has been a significant increase in the power consumption which impacts their economic feasibility, performance per watt scaling, and reliability [1], thus making it imperative to come up with novel techniques to reduce power.

This recent interest can also be attributed to the consumer, who has become more aware of the power and performance aspect of an electronic gadget. Users want higher battery life along with greater performance, the specifications that unanimously makes a product best in the market. There is a renewed focus on various power management techniques available for GPUs and has therefore spiked interest in the research space to come up with novel solutions.

The most important metric when trying to optimize power is the energy efficiency (or performance per watt) which accounts for both power and performance at the same time. Majority of the techniques that are most effective in improving energy efficiency of a GPU are implemented at an architectural or application level, as the scope becomes too tedious to control at circuit level, and in most cases, it may be too

P. Sharma · H. Al-Asaad (✉)
Department of Electrical and Computer Engineering, University of California, Davis, San Jose, CA, USA
e-mail: psharm@ucdavis.edu; hsalasaad@ucdavis.edu

K. Daimi et al. (eds.), *Advances in Security, Networks, and Internet of Things*, Transactions on Computational Science and Computational Intelligence,
https://doi.org/10.1007/978-3-030-71017-0_59

late for the product life cycle. In any system, the peak performance is nearly most of the times capped by power along with the limitations caused by memory bottleneck. Therefore, for all practical purposes, performance per watt directly translates into peak performance of any system. In this paper, we focus on techniques that directly help reduce GPU power.

This paper begins by reviewing basic GPU architecture for the uninitiated. Subsequent sections discuss power basics, further followed by detailed analysis of common power-saving techniques. The final sections venture into advanced low-power techniques for GPU and concludes with recommendation for future work.

2 Understanding the GPU Architecture

To analyze and study various novel techniques to reduce power in a GPU, it is imperative that we clearly understand the GPU architecture and its corresponding pipeline. A classic GPU pipeline is composed of the following stages. First is the *vertex shader*, which processes and generates vertices of the objects in a scene. This is followed by triangle formation, which essentially uses these vertices to create complex shapes and objects using polygon primitives like triangles or lines. Further down the pipeline, the *pixel shader* uses 3D scene to 2D rasterization mapping information for texturing, shading, pixel-level detailing, and coloring. Then comes the penultimate step of blending all these fragments or shapes together to create a scene as close to the reality as possible. Finally, this information is sent to the frame buffers (or onboard/chip memory) and then further to display on screen [2].

With the advent of DX10, the GPU pipeline has moved from a classic implementation of separate vertex and shader processing units to a unified shader architecture [2]. This unified shader architecture allows the overall pipeline for more scope of optimization and efficiency for different types of workloads. The shader processors are essentially just a series of floating-point ALUs with input, temporary, and output register banks. As the workload varies from vertex-heavy to pixel-heavy and vice versa, accordingly more or less shader units can be made active, hence making the overall architecture more efficient [3].

In a typical GPU pipeline to render a 3D scene, we have vertex coordinates transformation, lighting computations, and assembling of vertices to triangle primitives. These triangle primitives are then used to form complex structures and textures. Each pixel is further applied color and then sent out to frame buffer to render the scene on a screen [4]. Those computations are all so well-fitted for the GPUs by the virtue of them being high latency and high throughput. Those highly parallel systems work by extracting data parallelism in SIMD (Single Instruction and Multiple Data) instructions. The kind of data it can handle and process at a given time is huge, and that is how it hides the processing latency – by operating on huge data sets in parallel for a given instruction with really high throughput [2]. That is how the GPUs are able to render frames after frames with such ease in a graphical setting.

GPUs are also used for non-graphical applications such as high-precision fast computations; these are mainly used in servers and clouds. The graphical pipeline is programmable and therefore allows the compute workloads to run on them [5]. Taking a precise look at the hardware of vertex, texture, and blending engines, we are essentially looking at a bare minimum of huge floating-point processing horsepower, which can be used to perform basic multiply and accumulate (MAC) instructions in a compute-intensive environment [5]. In the above case, the data being processed will not be pixels but rather just numbers in a matrix multiplication for instance – which is a very popular application in HPC. This helps us appreciate the versatility of GPUs and why they are applied in such various fields.

The memory hierarchy in a GPU (shown in Fig. 1) is also important to understand as it will lead us to a complete and clear picture of the GPU architecture. GPUs have off-chip memory in the form of DRAMs. Commonly used types are GDDR5 and GDDR6. In recent state-of-the-art chips, companies have used HBM and HBM2 memories, which are on-chip, stacked memory banks sitting very close to the GPU chip such that they are both manufactured together. Since these memories sit close to the processor, we can have high number of pins or connections to the chip, and even though the pins can operate at lower bandwidth, a good net data transfer rate is observed. Typically, GPUs have three cache levels, L1, L2, and L3, and a set of internal registers [6]. The computing cores of the GPU are formed by the streaming multiprocessors (SM). These SMs are where the computation actually take place. These SMs have multiple processing blocks. Inside each processing block, there is a register set and L1 instruction cache which is private to an SM [6]. Among all the SMs, there is an L2 instruction and data cache shared by all SMs in a GPU. Then there may be a shared unified L3 (data and instruction) cache before the final level of memory hierarchy, which typically is the off-chip DRAM [6].

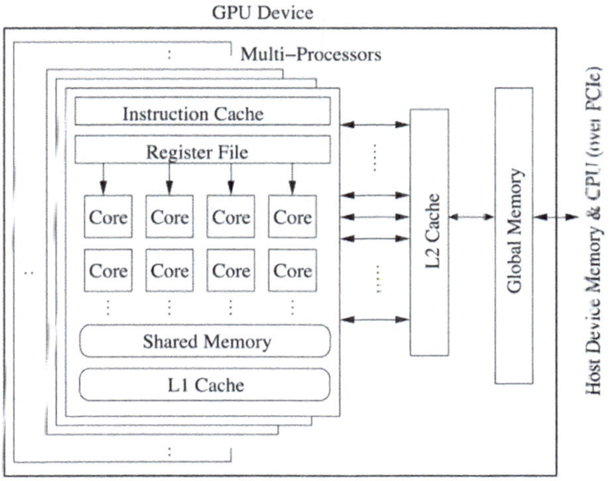

Fig. 1 Typical block diagram of a GPU

3 Background

3.1 Power Consumption – An Overview

To begin our review of low-power design techniques, it is imperative we understand all the components of CMOS power – broadly classified as dynamic and leakage power. Dynamic power consumption occurs during the time of switching, i.e., when the input to the transistor switches, and therefore, the output capacitor is continuously charged and discharged amounting to power consumption (see Fig. 2). Dynamic power is dependent on the switching activity profile, frequency, and square of operational voltage.

On the other hand, leakage power is that component of power which burns when there is no activity going on in the circuit. In other words, the transistors are ON with no activity (or quiescent state), and they still leak current and burn power. Major components of leakage current include subthreshold leakage, drain-induced barrier leakage, junction reverse bias leakage, and gate-induced leakage currents (see Fig. 3). These effects become more pronounced as the channel length decreases or when we go to a lower process node. Leakage power was not a main concern back in the old days, but now as we see the technology shrinking smaller and smaller, as low as 5 nm in production in near future, these currents have now become substantial and call for our attention to mitigate them.

Leakage power varies with process, voltage, and temperature or the PVT corners in an ASIC chip. Usually, it is seen that any low-power architecture pertains to a very specific block of hardware implemented for a given application. This directed approach helps designers to optimize in terms of power, area, and timing for an ASIC chip. So as we now understand the GPU architecture, the next logical step is to understand the major sources of power dissipation in a GPU. It will help one follow a directed approach and focus on relevant information to come up with novel techniques for reducing GPU power.

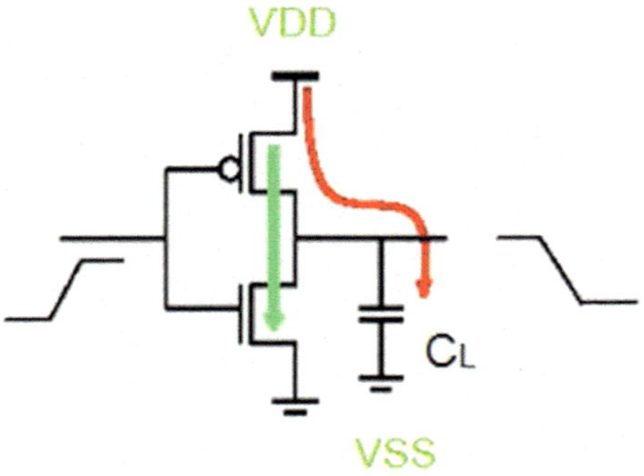

Fig. 2 Dynamic power consumption in a transistor

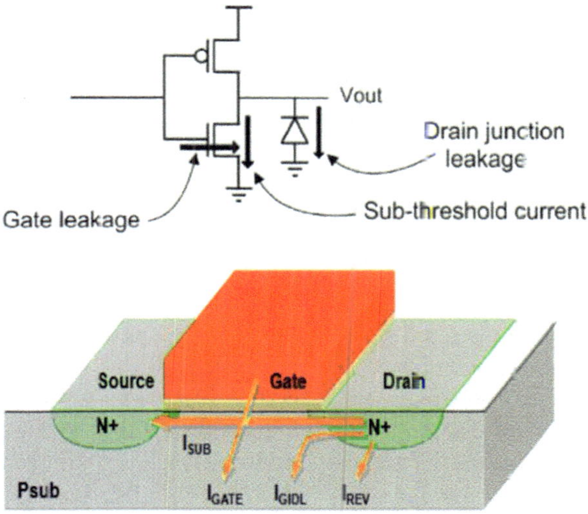

Fig. 3 Leakage power consumption in a transistor

3.2 *GPU Power Analysis and Breakdown*

A GPU consists of a large number of streaming multiprocessor (SM) units, which essentially are its main computing cores. These basic computation core units consist of ALUs, dynamic schedulers, load/store units, register files, caches, shared memory, and so on [1]. GPUs are based on the concept of data parallelism, where a large data set is divided into multiple small chunks of data and given to these computing units to independently operate on.

A very good example to understand this process is through a matrix multiplication application; the computations here very closely replicate the basic multiply and accumulate (MAC) instructions. Each element in the resultant output matrix depends only on certain specific elements in the multiplicand matrices for computation. Thus, the complete matrix multiplication operation can be broken down into smaller independent operations consisting of partial multiplicand data and eventually having all of them to come together for computation of the whole matrix as compared to sequentially multiplying them in a traditional way for each output element, thereby making the whole process a lot faster. Furthermore, these computing units have little to no data reuse as the computations are parallel for most part; hence, the requirement of smaller cache suffices along with its own memory for storing data and instructions [1]. While executing these kinds of matrix multiplication applications, the input matrix A and B are generated and stored in the main memory. To compute C = A * B, the GPU fetches partial data sets of multiplicand from the main memory and subsequently stores the fetched data in caches and vice versa while writing back the results. The ALUs inside the SMs operate on them, and the result is stored back into the memory.

So it is very clear that the majority of dynamic power and energy consumption in a GPU comes from these computing cores and the memory subsystem consisting of caches and accesses made to GPU memory. Therefore, it only makes sense to come up with techniques focusing particularly on these aspects to reduce net power consumption – which will be discussed in subsequent sections.

4 Common Low-Power Techniques

One of the firsthand issues with higher power consumption in GPUs is obviously high temperatures which call for sophisticated and costly cooling mechanisms like liquid- or nitrogen-based cooling equipment. Additionally, there is cost overhead in terms of designing the required Power Delivery Network to sustain high-current, direct by-product of high power. Hence, it is imperative that efficient techniques should be employed to reduce overall GPU chip power consumption. Architects have been using different power-saving techniques to optimize system efficiency. Some of the more popular ones are discussed in this section.

4.1 DVFS

One of the most popular techniques is DVFS (Dynamic Voltage Frequency Scaling) – this technique dynamically adjusts frequency and thus the voltage of a system depending on type of workload running on the chip. If it is an idle workload, the system drops the frequency and voltage values, thereby reducing dynamic power consumption. This mechanism happens on the fly as the GPU chip operates, thus proving to be a "dynamic" mechanism. This is a simple yet powerful technique since dynamic power is directly proportional to frequency and square of voltage – any reduction there, especially voltage, directly sees reduction in dynamic power.

In an exhaustive quantitative study published by Mei et al. in [7], it was observed that by scaling down frequency and voltage, they achieved on an average of about 19% energy reduction while having 4% of performance impact. The study was conducted with 37 benchmarking application on a GPU platform. Mei et al. were experimentally able to establish that GPU DVFS is indeed an effective approach to saving system energy, thus power. Scaling down core voltage when working at an appropriate frequency is a very effective way to save run-time power [7]. As dynamic power is proportional to square of voltage, so even a slight drop in voltage does the trick. While memory frequency and voltage scaling also contribute to power savings, it is highly application-dependent. Interestingly, the GPUs at any given point may be performance limited due to the following factors – computation horsepower (core capacity), memory bandwidth bottleneck, and power capping [8].

Abe et al. in their study quantified that for a given application, an optimum scaling of core and memory frequency will help reduce GPU power with minimal performance impact [9]. They found that voltage and frequency scaling offer significant reduction in power, and the results obtained demonstrated about 28% reduction in system power while retaining performance dip within 1% [9]. Abe et al. also suggested developing an efficient DVFS algorithm which controls core clock for a memory-intensive workload and memory clock for a core-intensive workload for optimum results.

Lee et al. came up with a technique to increase GPU throughput by incorporating both frequency/voltages scaling and dynamically scaling the number of cores, which resulted in an increase in throughput of about 20% on average [10]. Depending on the application characteristics, for a given power budget, if only 75% of the cores are operating, then we can optimize the system by running all cores at a lower frequency, thereby improving net performance. So, in this case, the throughput is increased compared to the baseline case for same power constraint. Therefore, the energy efficiency is improved as power consumed per operation is reduced drastically. This can also be looked at in terms of resource allocation management directly impacting performance at a given power budget and improving energy efficiency.

Nam et al. in [11] suggested implementing DVFS in a module-wise manner. The proposal is to implement triple-domain DVFS solution in each of the three modules – objects processing, vertex processing, and pixel processing in the 3D graphics pipeline—and these are managed independently depending on the workload dynamically. The study focused on handheld GPUs for wireless applications and found that there was about 50% reduction in power numbers with additional $\approx$2.5x performance improvement [11]. Deriving a cue from this, one can experiment trying this technique at multiple blocks in a mainstream GPU.

4.2 *Clock and Power Gating*

Another popular mainstream method to lower power consumption in a chip is to use gating mechanisms. Clock and power gating are among the most popular methods employed along with a bunch of other novel techniques recently proposed in this domain (discussed in subsequent parts).

Clock gating is a power-saving technique where the clock is temporarily disabled at times it is not required to be present. For instance, if the input data pin of the flip-flops in a design does not change, then it means that there is no new data to be captured, and the clock can be disabled during that time. It reduces dynamic power as disabling the clock for that moment reduces the activity or the net switching in a chip. The downside to this power-saving feature is that it needs an additional circuitry to determine when to disable/enable the clocks and also an extra circuit to prevent any glitches while transitioning from clocks OFF to ON – thus having a trade-off between low power and high area. Figure 4 shows different ways in which clock gating can be employed. Another interesting method is to allow clock toggles but adding a knob to control the data flow. If disabled, the data does not unnecessarily toggle and dissipate dynamic power.

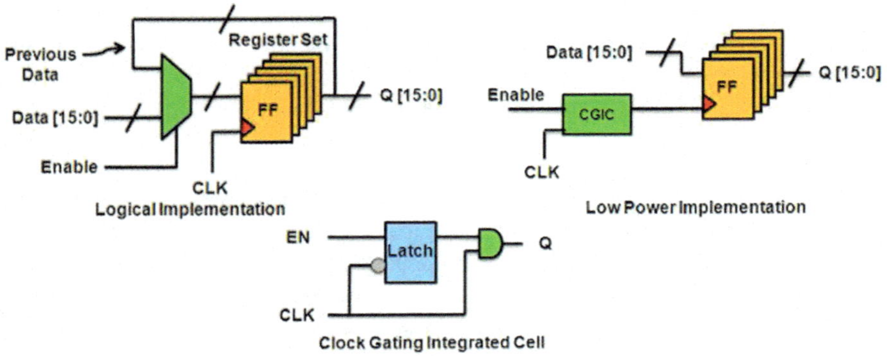

Fig. 4 Clock gating techniques

Power gating on the other hand is a technique which cuts off the power supply to the operating transistors – thereby shutting down the power-consuming transistors and saving power. Power gating saves leakage power too unlike clock gating which just saves dynamic power (clock temporarily stops toggling but the transistors are still ON and burn leakage power), thereby making it a very effective power-saving technique. Most of the times, the GPU computing cores are power gated when they are not in use along with other additional circuitry to reduce net power consumption.

The power gating feature could be either state-retentive or not. For a state-retentive gating, the current state of the chip needs to be preserved before power gating is enabled, and after coming out, this last known state needs to be restored for smooth operation. So there is a latency cost of this whole operation – going into the power-down mode and coming out of it; the delay of which might not be practical in some cases and thus resulting in a trade-off. In a non-state-retentive power gating, the block is just powered down plain and simple, there is no state retention, and it needs to be powered up fresh.

Figure 5 below depicts the implementation of power gating by either having the control switch at header or footer. These gate controls break the current flow path, thereby totally shutting off the logical block. There is some dynamic switching overhead involved when we enable or disable power gating which is usually overcome by letting the block be gated for some minimum time called *break-even* period [12].

Clock and power gating can be implemented at different levels in a chip – like a single unit, a block, or at an engine level resulting in varying degrees of power saving and latency overhead, which are needed to be factored in while designing the GPU product.

Another commonly used method for power reduction is to use multiple clock domains (MCD) in a chip. The idea is to have different clock speeds for different functional units [13]. For instance, the following are the different clock domains in a GPU – Core, DRAM, PCIe, and IO. All of them have separate range of operating frequencies if adhered to, in turn reducing dynamic power. Although there are power

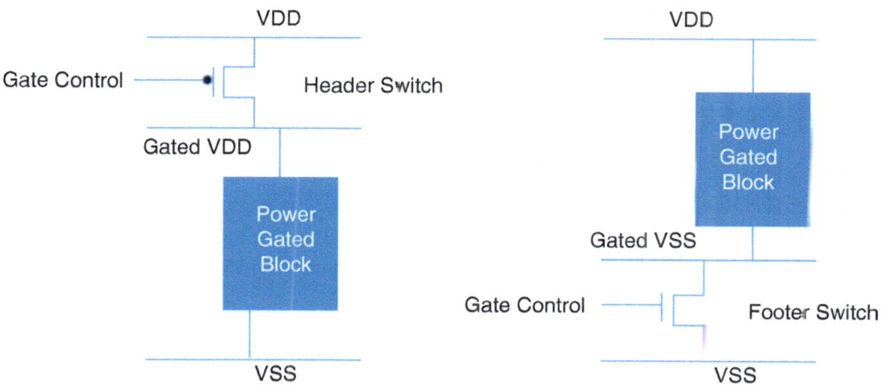

Fig. 5 Implementation of power gating

benefits using MCD, it requires designers to have chips with independent voltage islands using level shifters (increased area) and independent clock domains [14], which typically increases verification and validation costs. Therefore, an optimized lower frequency for a functional block is operated at a lower voltage thereby reducing both dynamic and leakage power dissipation.

For low-power GPUs in mobile applications, Sohn et al. in [15] suggested the use of fixed-point graphics processing instead of floating-point processing. The small screens of mobile terminals have an unnoticeable effect due to loss of accuracy and precision. Ergo, techniques discussed in this section summarizes the commonly implemented power-saving techniques for the reader to comprehend.

5 Advancements in Low-Power Architecture Design

Researchers have always tried to come up with exciting solutions for low-power GPUs. Some of these techniques at times use a combination of power or clock gating along with other ingenious ways for maximum power-saving benefits.

Jiao et al. in [16] studied three different computationally diverse GPU applications – compute-intensive (matrix multiplication), memory-intensive (matrix transpose), and hybrid (FFT). They observed that the power consumption for any given type of workload is primarily dependent on the rate of instructions issued and the ratio of global memory transactions to compute instructions. We can identify the type of applications based on these two metrics, and then the CPU core and memory frequency can be adjusted to an optimized point to effectively lower power consumption. The idea is that a memory-intensive workload can be operated at a minimum core frequency but optimal memory frequency to prevent any bottlenecks and preserve performance, vice versa for the implementation of core-intensive applications.

The method by Lin et al. in [17] focuses on the use of software pre-fetching to enhance performance together with the use of DVFS to increase energy efficiency. Software pre-fetching in GPUs involve adding pre-fetch instructions into the program. So the processor fetches data required by the program into registers or caches ahead of time, and this increases power overhead. Therefore, it is important to have optimal pre-fetching every time considering the memory and loop execution latency [17]. Timing is very important here. Data need not be fetched too early such that it gets evicted before use nor too late such that the pre-fetch operation is nulled. This increase in power due to pre-fetching is then evened out by iteratively reducing frequency and voltage such that desired performance is met while saving maximum power.

Another sophisticated technique proposed by Wang et al. in [4] is based on the fact that different scenes in a graphical application have different complexities like number of objects, types, etc. Thus, the computing resources vary frame by frame. They suggest a method of history-based predictor to predict the estimated GPU shader processor resources for future frames and shut down the rest using power gating. The idea is to optimize the number of shader and non-shader resources and turn off others while idle and return back to processing within the timing constraint of the arrival of next batch of frames. They observed about 46% leakage power savings with negligible performance hit. Again, this whole method may become more effective if used alongside DVFS.

Some techniques discussed below make architecture-level changes in GPUs and leverage-specific GPU usage patterns to make optimizations.

Wang et al. in [12] propose a run-time power gating of caches in GPU that can potentially save leakage power. They suggest switching L1 cache to a certain sleep mode whenever the pipeline is stalled. It saves leakage power as well as contents are retained when the stall ends. The power gate controller knows the number of stall cycles, and it can decide if power gating can be enabled or not (such that the timing is met). The L1 cache is private to each SM, so when an SM has completed execution of its own share of workload and waits for other SMs to finish, its L1 cache can be switched to off mode. The L2 cache is not accessed as frequently as L1 cache; thus, there is a scope of more savings there [12]. So whenever the L2 is sitting idle and serving no memory accesses, it can be switched to sleep mode. This mode switch latency is hidden in the following manner: whenever there is a request to or from L2, the request first goes a queue, and hence, L2 array need not be active during fetch, and therefore, the wake-up latency is effectively hidden.

A unique approach to determining the memory access granularity and adjust it to exploit optimum spatial locality is proposed in [18]. Rhu et al. observe that not all applications have high spatial locality and thus do not require the standard fetching of all four 32-byte sectors of the 128-byte cache block. For an application with irregular memory access pattern, it is a waste of energy to fetch all the data sectors at once. They add control mechanism at sector level in a cache line to enable the flexibility of being either coarse-grained or fine-grained memory access at a given time and propose a locality-aware memory hierarchy (LAMAR). Each cache has a granularity decision unit (GDU) which determines the access granularity for each cache miss to determine the granularity best suited for an application [18]. Based on this, a simple low-cost hardware predictor adaptively adjusts this access granularity at run-time without any intervention from a programmer [1]. This method resulted in an average of 17% increase in energy efficiency [18].

GPUs hide memory latency by using many threads (or parallelized piece of code). Usually, these threads are stored on a large on-chip storage, and they also require a complex thread scheduler, which has high latency and is an energy-consuming process. A new twofold technique proposed by Gebhart et al. [19] focuses on reducing energy in the core data path of the GPU [1]. Firstly, they propose having a register file caching in place which can replace access to the main register file instead access to a smaller storage containing only the registers needed for immediate active threads. Secondly, they suggest having a two-level thread scheduler logically dividing active threads (short latency operations and currently issuing) and pending threads (waiting on long-memory latencies) It helps reduce thread scheduling space to just the active threads. Both these techniques reduce the energy consumed by register file by 36% (translates to about ≈4 watts of total GPU power) [19].

Another interesting energy-saving technique is proposed by Lashgar et al. in [20]. In GPUs, it is seen that the instruction temporal locality is high, which is because the GPU operates on multiple threads of the same instruction context on different cores. So essentially the same instruction is being fetched repeatedly (SIMD). Therefore, during short execution cycles, a few numbers of instructions result in significant fetching and decoding, thereby increasing the likelihood that the same instruction will be fetched again in near future [20]. They propose a simple direct-mapped filter cache hardware as part of the current memory subsystem to filter out these recurring instructions and cache them. This directly reduces the number of accesses to the instruction caches and results in increased energy efficiency. Their study found that this method can reduce about 30% to 100% of I-cache accesses and save up to 19% of the energy [20].

The above section sheds some light on some of the interesting ideas waiting to be tapped and made into more commercially viable techniques.

6 Conclusions and Future Work

The idea of various ingenious ways to save power is indeed very enticing to implement in a next-generation GPU architecture. As a good designer, one should never ignore the implementation overhead in terms of timing (latency) and area. Also, it is equally important to have a deep understanding of the micro-architecture and the end product application which can help make good decisions in terms of employing low-power implementation methodologies. It is usually seen that a lot of power-saving mechanisms do indeed make their way into silicon and are thoroughly tested out during the post-silicon validation phase to evaluate their practicality in terms of performance and latency impact. Once there is a thorough understanding, in a most likely case, many of these features actually do not make it to the released product. Only a few handfuls of them do. Thus, there is an onus on the GPU community and companies to invest more into the R&D of their own respective GPU implementations to further make an impact and have a product exceeding expectations on the energy efficiency front.

The next logical step in this study is to be able to work on actual GPU products and measure power numbers for real-world graphics, compute applications, and known benchmarks. A practical study of that kind will give more confidence in the feasibility of suggested power-saving techniques. Empirical study and quantification of performance and power numbers shall give us a good basis to suggest novel techniques that are indeed effective in a real-world scenario. Furthermore, it will be a great way to analyze the practical feasibility of any power-saving technique by having related area and latency studies as well. Finally, collaboration of researchers in this field with the major GPU chip-producing companies would seem an ideal way to step closer to this growing demand for low-power and high-performance parallel systems.

References

1. S. Mittal, J.S. Vetter, A survey of methods for analyzing and improving GPU energy efficiency. ACM Comput. Surv. (CSUR) **47**(2), 19 (2015)
2. A. Rege, An introduction to modern gpu architecture. En ligne (2008)
3. V. Moya et al., Shader performance analysis on a modern GPU architecture, in *Proceedings of the 38th Annual IEEE/ACM International Symposium on Microarchitecture*, (IEEE Computer Society, 2005)
4. P.-H. Wang et al., Power gating strategies on GPUs. ACM Trans. Archit. Code Optim. (TACO) **8**(3), 13 (2011)
5. E. Kilgariff, R. Fernando, The GeForce 6 series GPU architecture, in *ACM SIGGRAPH 2005 Courses*, (ACM, 2005)
6. Z. Jia et al., Dissecting the nvidia volta gpu architecture via microbenchmarking. *arXiv preprint arXiv:1804.06826* (2018)
7. X. Mei et al., A measurement study of GPU DVFS on energy conservation, in *Proceedings of the Workshop on Power-Aware Computing and Systems*, (ACM, 2013)
8. R. Ge et al., Effects of dynamic voltage and frequency scaling on a k20 gpu, in *2013 42nd International Conference on Parallel Processing*, (IEEE, 2013)
9. Y. Abe et al., Power and performance analysis of GPU-accelerated systems, in *Presented as part of the 2012 Workshop on Power-Aware Computing and Systems,* (2012)
10. J. Lee et al., Improving throughput of power-constrained GPUs using dynamic voltage/frequency and core scaling, in *2011 International Conference on Parallel Architectures and Compilation Techniques*, (IEEE, 2011)
11. B-G. Nam et al., A low-power handheld GPU using logarithmic arithmetic and triple DVFS power domains, in *SIGGRAPH/EUROGRAPHICS Conference On Graphics Hardware: Proceedings of the 22 nd ACM SIGGRAPH/EUROGRAPHICS Symposium on Graphics Hardware.* vol. 4(05) (2007)
12. Y. Wang, S. Roy, N. Ranganathan, Run-time power-gating in caches of GPUs for leakage energy savings, in *2012 Design, Automation & Test in Europe Conference & Exhibition (DATE)*, (IEEE, 2012)

13. J.W. Sheaffer, K. Skadron, D.P. Luebke, Studying thermal management for graphics-processor architectures, in *IEEE International Symposium on Performance Analysis of Systems and Software, 2005. ISPASS 2005*, (IEEE, 2005)
14. J.W. Sheaffer, D. Luebke, K. Skadron, A flexible simulation framework for graphics architectures, in *Proceedings of the ACM SIGGRAPH/EUROGRAPHICS Conference on Graphics Hardware*, (ACM, 2004)
15. J.-H. Sohn et al., A 155-mw 50-m vertices/s graphics processor with fixed-point programmable vertex shader for mobile applications. IEEE J. Solid State Circuits **41**(5), 1081–1091 (2006)
16. Y. Jiao et al., Power and performance characterization of computational kernels on the gpu, in *Proceedings of the 2010 IEEE/ACM Int'l Conference on Green Computing and Communications & Int'l Conference on Cyber, Physical and Social Computing*, (IEEE Computer Society, 2010)
17. Y. Lin, T. Tang, G. Wang, Power optimization for GPU programs based on software prefetching, in *2011 IEEE 10th International Conference on Trust, Security and Privacy in Computing and Communications*, (IEEE, 2011)
18. M. Rhu et al., A locality-aware memory hierarchy for energy-efficient GPU architectures, in *2013 46th Annual IEEE/ACM International Symposium on Microarchitecture (MICRO)*, (IEEE, 2013)
19. M. Gebhart et al., Energy-efficient mechanisms for managing thread context in throughput processors, in *2011 38th Annual International Symposium on Computer Architecture (ISCA)*, (IEEE, 2011)
20. A. Lashgar, A. Baniasadi, A. Khonsari, Inter-warp instruction temporal locality in deep-multithreaded GPUs, in *International Conference on Architecture of Computing Systems*, (Springer, Berlin, Heidelberg, 2013)

Ethical Issues of the Use of AI in Healthcare

Suhair Amer

1 The Use of Artificial Intelligence in Healthcare

Artificial intelligence (AI) can be described as "*the science and technology that seeks to create intelligent computational systems*" [1]. The goal of artificial intelligence, or AI, is to build computer systems that can represent or replicate the human thought process. One industry that has taken an interest in the use of AI in daily practice is the healthcare industry. The goal of AI in healthcare is to assist doctors with their jobs rather than replace them. For the past several years, research has been aimed toward demonstrating the ways that AI can help on an entirely new level [2].

AI has the potential to benefit humanity especially the healthcare system. It is true that human doctors are usually overworked, under slept, etc., but an artificially intelligent doctor can interact with the physical world and know every single sequence of your genome. For example, IBM Watson Oncology can decide on treatment drugs for cancer patients with equal or better efficiency than human experts. Yielding such results already although this technology is in its beginnings is phenomenal progress. Stanford's radiology algorithm, also, identified pneumonia better than human radiologists. While in diabetic retinopathy challenge, the computer was like expert ophthalmologists in making a referral decision. This indicates that an AI is a great candidate for diagnosis. Some researchers say that once the initial costs come down for AI services and advanced, it is difficult for humans to compete, nor should they [3]. Sani writes that "*Now, AI is being employed in diagnosing cancer, tuberculosis, skin, eyes, stroke and other conditions, and it is more precise, accurate, faster and cheaper*" [4]. The variety of diagnosis that an AI can successfully carry

S. Amer (✉)
Department of Computer Science, Southeast Missouri State University, Cape Girardeau, MO, USA
e-mail: samer@semo.edu

K. Daimi et al. (eds.), *Advances in Security, Networks, and Internet of Things*, Transactions on Computational Science and Computational Intelligence, https://doi.org/10.1007/978-3-030-71017-0_60

can leverage AI to unveil its potential, dispel the prejudices around it and build awareness among professionals. To achieve this, the industry will have to set ethical standards for organizations and metrics to measure AI system performance" [22]. This also ties into supporting regulatory bodies that have a strong code of ethics. Addressing an issue like this from a top-down perspective is likely the best way to combat ethical problems when they arise. Relying on a company to care about ethics when they are more concerned about profit will always lead to ethical dilemmas. However, having a body enforcing and ensuring that ethics is considered when developing software will assist those companies in producing a better product.

Some researches propose more extreme solutions. Some state that if explicit consent and notice have not been given, then all de-identified data should come into public domain and be published to keep a check on illegal proprietary exploitation of the data. In addition, human physicians must be informed of all reasons and decisions taken by the machine, and a human operator must possess a veto power or a manual override. Even such a solution has some concern especially if the machine refuses to respond to the human's decision or override [3].

Artificial intelligence has the potential to enhance medical treatments provided by hospitals to patients. It is important to ensure that if AI is self-learning that their needs to be ways to ensure that it does not start acting in an unethical and immoral way. There are many different ethical guidelines that artificial intelligence systems need to follow. For example, a human expert would be monitoring the system as it works. At first, the machines could be restricted to their virtual world until they develop enough tendencies to be trusted and used in the real world on real-world systems. Then embed them into real system to develop and become as efficient as it can be to provide the best care possible to the patient. With an expert monitoring, the system ensures it is running as it is supposed to. This can decrease the likelihood of a mishap and can help ensure that the artificial intelligence is always going to be acting in an ethical and safe way for everyone that is involved [31].

Some countries already started creating effective legislation, for example, the European Union's Directive for General Data Protection Regulation, also known as GDPR. This directive regulates the disclosure of decisions made by artificial intelligence systems. This grants a right to explain to citizens of the countries of the EU which necessitates that data controllers of these artificial systems disclose information on decisions made by these systems. This includes disclosing the logic and importance of the processing involved in such decisions and addresses the black box issue of artificial intelligences. This legislation would necessitate that software engineers develop artificial intelligence in a way that eliminates or significantly reduces the black box aspect of AI. This allows patients to request and understand the decisions made by AI that affect everyone [13].

Other legislation passed by the European Union that affects AI would be the Cybersecurity Directive, the Medical Devices Regulation, and the In Vitro Diagnostic Medical Device Regulation. This legislation includes AI. This legislation extends the scope of regulatory frameworks, extends legal reliability regarding defective products, and the increasing of the traceability of medical devices, among other things. Comparatively, the United States is far behind the European Union on out is rapidly growing and becoming more accurate which is a great reason to keep expanding the scope of what the AI doctor can do and challenge its limits. Recent research is also investigating "*3D bioprinting technique that prints human organs for replacing damaged organs. This means humans can now live longer*" [5].

In several studies, AI has been used to assist with detecting tuberculosis on chest radiographs. Out of 150 studies that were conducted, AI was found to "*accurately classify TB at chest radiography with an AUC of 0.99*" [6]. This can help reduce waiting room times for patients because the quicker patients are diagnosed, the quicker they can receive treatment [2].

AI can take a large influx of information and summarize it very quickly. For example, rather than doctors spending long time reading charts, notes, and conducting routine tests such as blood work, AI can assist. With some adjustments made along the way, AI can put all the information into easy-to-read formats for doctors to evaluate quickly [2].

Across the globe, several AI robots are used in surgical procedures by assisting doctors. They provide capabilities, such as precision and sight, which doctors are incapable of having with the naked eye. AI can keep track of potential drug interactions as there are many facts and important details humans cannot keep track of [7].

AI is also able to analyze diagnostic imagery such as MRIs, X-rays, and PET with better resolution, better accuracy, and faster than human doctors. This allows earlier detection of disease, meaning that doctors can treat patients before their condition gets too serious and would negate so of the issues that can arise from doctor error. This may make the patient be less stressed about whether the diagnosis is correct or not [8, 9].

There has been some usage of AI for certain neurological procedures, like brain tumor ablation. AI analyses the patient's brain and figures out the exact shape and size of the affected part of the brain which aids in removing the tumor with a laser. For surgeries, the mechanical hands of a robot or an AI could be much more precise and in some cases capable than surgeons, which can reduce patient deaths. The only issue with AI-powered robots currently is their inability of tying knots when it is time to sew the patient back up. AI can also help with early detection of strokes. The US Food and Drug Administration (FDA) approved an mHealth app that uses AI software to analyze CT scans for signs of a stroke and sends a text message to the neurologist patients to get help sooner and allow their neurologists to get the needed resources quicker [10].

AI can also help the elder population of a community that are 60 and above. Elder population represent 10% of the world, and it is projected to double by the year 2050 [11]. A study involving the elderly, AI was compared to four radiologists trying to detect pneumonia in patients. The study showed that AI outperformed the medical professionals [12].

One prominent uses of AI in healthcare is as a medical decision support system or MDSS [13]. The system aids physicians in making decisions regarding diagnosing and treating medical conditions. These systems are usually constructed as neural networks that can quickly analyze more information. The use of MDSSs also reduces healthcare costs because of increased efficiency. They can also be used and serve rural communities because they can be used in conjunction with a general practitioner in areas where specialized physicians are not available. This allows rural residents get access to the healthcare they need [14].

There is also a great potential for AI to improve healthcare in the field of mental health. Sophisticated AI algorithms can be used to substitute therapists in situations where access to mental healthcare professionals is not available. AI can be trained to respond to patients like therapists by guiding the patient. In addition, chatbots allow individuals to talk to these bots to alleviate depression and anxiety resulting from loneliness. Social robots can be used in a similar way; however, they have a physical presence. Social robots like the Paro and eBear have already been tested. Their primary purpose is to provide companionship to elderly patients who have limited social interactions due to their condition [15].

Surgical robots aid surgical procedures and are divided to two categories. One category are robots that merely aid doctors in surgical procedures, and the second category are robots that operate independently. The MAKO robot and the Da Vinci robot are examples of surgical robots that aid doctors [16].

The AI algorithm PREDICT is a simultaneously decision support software and an ongoing, open cohort, prospectively designed study. PREDICT integrates with the practice management system to retrieve patient data along including that is entered interactively to provide a personalized estimate of the probability of a cardiovascular disease event in the next 5 years [17].

Another example are the continuous glucose monitoring systems that generate giant amounts of blood glucose measurements. Newer systems can display, in real time, the glucose levels on a smartphone or other reader. The AI can interpret larger amounts of data so that clinicians can better understand the complexity in simpler terms. Chatbots are also being used for glycemia management. A chatbot is a software program that mimics a human voice through audio and text. Scientists are still researching the uses of chatbots to understand and motivate people with diabetes [18].

2 Ethical Issues Related to the Use of AI in Healthcare

Along with the positive aspects of using AI in the healthcare system, ethical issues come with it. Some are legal issues because ethics and legality, in some circumstance, go hand in hand. One example is making certain that the machine being used does not hurt the patient in any way. Another issue is ensuring that the AI-operated device or software operates safely as it approaches humans. It is also important to recognize that it is debatable whether the artificial intelligence has any moral values like a regular human would. Without moral values, an AI may deem a person unworthy to live and ultimately terminate their life because their conclusion indicated that this was their best choice when in reality, it may have not have been the best choice morally [19].

Another ethical issue is making healthcare provider's jobs obsolete and needing to deal with conflicting patient wishes and interests. AI could end up taking several healthcare provider's jobs, which becomes prevalent across several different less skilled fields such as manifesting. The development of robots powered by AI will end up leaving several surgeons out of a job because an AI will be able to do the jobs better and for cheaper. AI-powered robots are currently performing delicate procedures such as removing brain tumors. Also, many wonders whether such systems will consider patient wishes even if it conflicts with what is right or wrong and what is morale or not. For example, an elderly patient may wish not to be treated even though a cure is available. Or vice versa, an elderly would request trying untraditional treatments although success rate is very low. Other patients may refuse treatment because of religious reasons. Doctors work on convincing the patient and follow local laws, whereas it is not clear how will an AI doctor react which may lead to legal ramifications [16, 20].

AI systems can diagnose skin cancer more accurately than a board-certified dermatologist and are able to do it faster and more efficiently, only requiring a training data set as opposed to years of expensive and labor-intensive medical education. Looking closer at this technology and the role it plays raises ethical concerns that is related to patient privacy and confidentiality, informed consent, and patient autonomy. It is crucial to stress the importance of proper informed consent and responsible use of AI, stating the potential harms related to AI, and this must be transparent to all individuals involved [21].

Another ethical issue is the role that AI can play in medical education which involves preparing future physicians for a career integrating AI. Also, ethical issues arise from directly using AI in the education of medical students. With the rise of AI, medical education should be refocused, with focus shifting from knowledge recall to training students to interact with and manage AI machines. This would require more diligent attention to the ethical and clinical complexities that might arise among patients, caregivers, and the machines [21].

The use of AI in healthcare is a black box aspect especially when using machine learning algorithms. A popular method of creating AI is through machine learning algorithms that train a neural network with vast amounts of training data. The resulting technology is essentially a black box, where information is fed and a relevant output is generated, but the process in between is not well-known. Some argue that it is difficult to trust the decisions of these algorithms especially if it is hard to understand how they were reached, and it is important to understand the process by which these algorithms make decisions [13].

Another important ethical issue is how to protect patients' privacy and security. AI systems can be used to analyze existing information in databases to draw new connections that were not immediately obvious. Hospitals may take data from multiple databases, which could generate a concern over the centralization of data into one area. This is a concern because it has the potential to violate patients' privacy by using their information in ways that was not explicitly authorized by patients. Also, storing large amounts of medical information in a centralized location increases security concerns because unauthorized access to a single database could compromise all data [13].

AI systems are built by software engineers and computer scientist that are not experts in principles relating to autonomy, non-maleficence, and beneficence They also required to build the system with fairness in mind. However, fairness is already an ill-defined concept and is something that a society in general does not always agree on. So one way for software engineers to implement the principle of fairness would be to ensure that the training data used to generate AIs is not biased. Biased training data has negatively influenced decision-making of AI in the past. For example, a program that aided judges in predicting recidivism of an offender was discriminatory as a direct result of biased training data [9]. It is important to ask, “What if the algorithms see a pattern where none exist or acquire a bias because of partial or skewed data?” [22]. Craft in his four pillars of AI governance proposal, the second is “clinical and scientific verification and valuation to confirm that the AI algorithm has been tested on a valid data set” [23]. These ideas of verifying the inputs to AI healthcare systems are already being explored. Craft’s paper addresses “ethical evaluation and usage guidelines to determine whether or to what extent patients are informed about the role AI is playing in their diagnosis and treatment” [23]. This information must be provided to a patient who should be able to increase or decrease the role of that AI in their treatment. In addition, AI technologies lack the ability to recognize and ignore human biases. When these AI machines are built and if there is only one group doing the programming, then ethnical biases could be built in. The types of big data these machines handle need to be factual in every way. For example, “an algorithm designed to predict outcomes from genetic findings may be biased if there are no genetic studies in certain populations” [9].

Another indirect effect of the implementation of AI in healthcare would be the replacement of trained physicians and psychiatrists in the field of healthcare. AI has the capacity to eliminate many jobs in healthcare in the field of mental health. Arguments are made that access to virtual psychotherapists is enough even though virtual psychotherapists are not meant to replace actual therapists. This could contribute to social inequality as it relates to access to effective mental health treatments [15].

Currently, nobody owns DNA; however, DNA testing companies have claimed the rights to the genetic information given to them. In a 2018 Gizmodo article written by Kristen Brown, “Testing companies can claim ownership of your DNA, allow third parties to access it, and simply by virtue of possessing it make your DNA vulnerable to hackers” [24]. Some claim that an AI doctor can know a patient’s genome and transfer the ownership of that DNA to their parent company. In addition, the legal system has and always will move slower than the AI industries.

The ethical issue of profit also comes into consideration. Historically, many algorithms have been designed to maximize profit above potential negative results. If AI is set up with the goal of achieving profit rather than maximizing health, the healthcare industry could suffer. Hypothetically speaking, AI systems could be designed to recommend treatments that would generate the healthcare system more profits but not necessarily reflect better healthcare [9].

3 Legal Issues Related to the Use of AI in Healthcare

Healthcare is a profession governed with law and is becoming more dependent on technology every day. Technology is easily outpacing the human ability to create laws which is expected to become an issue when implementing AI within the field. This will become more evident when recursive self-improving AI is created. Price notes that "new and emerging medical technologies and devices are typically regulated for safety and efficacy by the Food and Drug Administration" [5]. Regulation is a double-edged sword as it is important to know that the technologies used in the medical field are safe but also want them rapidly advance. It is impossible to run over every line of source code to verify that the medical technology is theoretically safe and then put it in testing. Price realizes the shortcomings of rapid development because "if there are flaws built into the algorithms themselves, or if regulation fails to ensure that algorithms are high quality, then the developers of algorithms (or technologies that rely on them) might become liable under tort law" [5].

Intellectual property law is another area in which the legality of implementing AI in healthcare will be challenged. Intellectual property does not work with black box medicine. Patents are used to protect technological innovation but do not provide strong incentives for black box medicine. Some believe that this issue needs to be addressed as developers of these technologies should be encouraged considering the overall benefit. One thing that is obvious is that conversations about this topic should be happening sooner rather than later to provide a safe service to patients and provide profit motivation to the business sector. Vast amount of data is available but with privacy laws such as HIPAA makes this data unavailable to private firms. One way would be to have these data sets anonymized so that they could be used to benefit AI. Price says that "de-identified information is not governed by the Privacy Rule (though it raises its own concerns about data aggregation and the possibility of re-identification)" [5].

It is important to resolve legal definitions. For effective legislation to be passed, a clear understanding of basic definitions is necessary when creating these laws. Right now, the definitions relating to artificial intelligence are fuzzy and ill-defined. AI's legal law is an example of a fuzzy and ill-defined. The lack of a concrete definition of intelligence hinders people from creating effective policy [9].

Another legal issue that needs to be resolved is whether artificial intelligence software should be classified as a medical device. Medical devices are already regulated, and the inclusion of artificial intelligence software into this category could potentially aid in the regulation of such software. In the European Union, artificial intelligence software falls under the definition of medical devices; however, in the United States, it is not so clear cut. This hinders the development of such systems in the United States because it is harder to get regulatory approval from the FDA if the software is autonomous [9].

An example of a real-world legal issue regarding artificial intelligence in healthcare would be when London hospitals violated patients' privacy by sharing their records with Google. These hospitals violated civil law when they shared records with Google to aid in the development of their Google-developed artificial intelligence subsidiary DeepMind. Google wanted to use this data to test their artificial intelligence system in diagnosing kidney injuries. However, the way the data was transferred to Google's DeepMind was not done according to regulations. Such illegal breaches of patient data could become more common in the future as companies become more eager to use such data to train and test their AI systems [13].

Sometimes, legality and ethics go together especially in how the patient is treated. One issue is related to being difficult to hold accountable a machine when it hurts a patient. If a human hurt a patient intentionally rather than helping, they would lose their license and face time in prison, but this is impossible for a machine. A machine can be taken out of production and no longer used, but this does not help the individual who was affected. The company that created the AI or used it can be held financially liable for the damages of patients and their family. In addition, it is also possible for the AI to cause harm to the individual who is it [19]. It is important to note that as of right now when an error is made by AI, the healthcare provider relied on such information, and a patient gets hurt because a physician prescribed the wrong dosage of medicine, the physician is held accountable [5].

The rapid growth and speed of innovation in the field of AI diagnostics raises the legal concerns of whether patent law can keep up. In the United States, the US Supreme Court has invalidated patents related to diagnostics and computer-implemented inventions in the following cases: Mayo Collaborative Servs. v. Prometheus Labs Inc. in 2012 and Alice Corp. Pty Ltd. v. CLS Bank Int'l in 2014. In Mayo, the plaintiff, Prometheus Laboratories, patented a suite of tests for assessing the proper dosages of certain Crohn's disease medications. When Mayo Clinic developed similar tests and started using them, Prometheus sued Mayo for infringement. The Supreme Court invalidated Prometheus's method patent and laid out framework for determining patent eligibility in two steps. The Supreme Court then went on to clarify Mayo in Alice, which resulted in the Alice/Mayo test. As part of the two-step test, it is first determined whether the claims are a natural phenomenon, a law of nature, or an abstract idea. Since then, the United States Patent and Trademark Office has developed several iterations of guidelines for evaluating eligibility. Because different examiners within technology centers applied inconsistent standards due to the lack of clarity of how the judicial exception to the eligibility should be applied raised many concerns. Demand for clarity and consistency led to the release of the 2019 guidance [25].

According to the AMA Journal of Ethics, liability for medical errors falls under tort law which allows patients to receive financial compensation after a malpractice due to physicians, healthcare organizations, or other similar groups liable for their physical well-being. Applying the current liability doctrines to AI can be difficult because of the unclear nature and unforeseeable results. For example, if the designer of the AI could not have predicted the actions it would take, how can the individual be held liable? Keeping this in mind, if the law continues to make exceptions for liability resulting from AI, then patients will be left with fewer opportunities for keeping responsible parties accountable for their actions [26].

Solutions to Ethical Dilemmas Resulting from Using AI in Healthcare

Ethical solutions to problems resulting from the use of artificial intelligence in healthcare will not be quick to implement; however, some steps can be taken now to begin the process. For example, physicians and healthcare providers should be involved in the decisions AI makes, keeping in mind that human interaction cannot be replaced in the field of healthcare [21]. It is important to involve and consider all ethnic groups in the decision-making process. Thorough tests should be conducted to ensure all genetic backgrounds are considered before implementing AI. In addition, AI should be considered as a tool, not a replacement for critical thinking [9].

It is also important that physicians who practice medicine using AI understand how the decision-making algorithms work and that conclusions drawn by an AI should be supported by an individual with medical training. Physicians should not only rely on AI decisions and question a judgment that the machine makes [27]. "Advanced technologies used to delivery healthcare should be designed and operated to be compatible with ideals of human dignity, rights, freedoms, and cultural diversity" [28].

New European data protection guidelines have provided those affected by automated decision-making systems a right to obtain meaningful information about the logic involved with the decision-making. New Zealand's Privacy Commissioner and Chief Government Data Steward have issued a set of principles on the use of data and analytics. These principles state that explanations of decisions and the analytical activities behind them should be clear to understand. In addition, General Data Protection Regulation provides those affected by the decision-making systems a right to not be subject to a decision just based on the automated processing. New Zealand's principles for safe and effective use of data and analytics do specify that analytical processes are a tool to inform human decision-making and should not be used as a replacement for human oversight [29].

Another legal issue is the problem of who owns the data and the AI software. The manufacturing companies designing the AI technology does not own the data; therefore, it is critical the data held within stays anonymous. In addition, just as patents, copyrights, and trade secrets protect software programs, there is still debate on algorithms installed within AI. There are a lot of potential legal cases that arise regarding the ownership and redistribution of AI technologies. HIPAA laws should not be violated under any circumstance. All patients should be aware and able to give or deny consent when it comes to their personal data being used for big data analysis. They should keep the right to remain anonymous if they choose to [27, 30].

Many researchers believe that AI governance must be implemented as a formal set of guidelines with enterprise-level authority. By creating firm rules within this space will reduce ethical issues where they may arise. Shukla agrees with a more balanced approach saying, "As an early technology adopter, the healthcare industry ters relating to the regulation of AI in healthcare. US lawmakers could benefit greatly from studying such EU legislation and its effects [9].

References

1. H.T. Tavani, *Ethics and Technology: Controversies, Questions, and Strategies for Ethical Computing* (John Wiley & Sons, Inc., Hoboken, 2013), pp. 355–363
2. V.H. Buch, I. Ahmed, M. Maruthappu, Artificial intelligence in medicine: Current trends and future possibilities. Br. J. Gen. Pract. **68**(668), 143–144 (2018)
3. J. Bali, R. Garg, R. Bali, Artificial intelligence (AI) in healthcare and biomedical research: Why a strong computational/AI bioethics framework is required? Indian J. Ophthalmol. **67**(1), 3 (2019) http://library.semo.edu:2275/10.4103/ijo.IJO_1292_18
4. T. Sani, Artificial intelligence: AI in healthcare: Will AI replace doctors? Electronics for you, (2019, Apr 01). Retrieved from https://library.semo.edu:2443/login?url=https://library.semo.edu:4836/docview/2212787461?accountid=38003
5. W.N. Price, Artificial intelligence in health care: Applications and legal issues. Scitech Lawyer **14**(1), 10–13 (2017) Retrieved from https://library.semo.edu:2443/login?url=https://library.semo.edu:4836/docview/2043221836?accountid=38003
6. P. Lakhani, B. Sundaram, Deep learning at chest radiography: Automated classification of pulmonary tuberculosis by using convolutional neural networks. Radiology (Thomas Jefferson University Hospital) **284**(2), 574–582 (2017)
7. T. Davenport, J. Glaser, Just-in-Time Delivery Comes to Knowledge Management, https://hbr.org/2002/07/just. 7 Jan 2020, (Harvard Business Review, Boston, 2002)
8. Icahn school of medicine at mount sinai to establish world class center for artificial intelligence - hamilton and amabel james center for artificial intelligence and human health, Targeted News Service (2019, June 11). Retrieved from https://library.semo.edu:2443/login?url=https://library.semo.edu:4836/docview/2240407128?accountid=38003
9. F. Pesapane, C. Volonté, M. Codari, F. Sardanelli, Artificial intelligence as a medical device in radiology: Ethical and regulatory issues in Europe and the United States. Insights Imaging **9**(5), 745–753 (2018). https://doi.org/10.1007/s13244-018-0645-y
10. K. Ganapathy, S. Abdul, A. Nursetyo, Artificial intelligence in neurosciences: A clinician's perspective. Neurol. India **66**(4), 934–939 (2018) http://library.semo.edu:2275/10.4103/0028-3886.236971in-time-delivery-comes-to-knowledge-management
11. M.E. Pollack, Intelligent technology for an aging population: The use of AI to assist elders with cognitive impairment. AI Mag. **26**(2), 9–9 (2005)
12. E.J. Topol, High-performance medicine: The convergence of human and artificial intelligence. Nat. Med. **25**(1), 44 (2019)
13. A. Vellido, Societal issues concerning the application of artificial intelligence in medicine. Kidney Dis. **5**(1), 11–17 (2018). https://doi.org/10.1159/000492428
14. R. Shah, A. Chircu, IOT and AI in healthcare: A systematic literature review. Issues Inf. Syst. **19**, 33–41 (2018) Retrieved from http://www.iacis.org/iis/2018/3_iis_2018_33-41.pdf
15. A. Fiske, P. Henningsen, A. Buyx, Your robot therapist will see you now: Ethical implications of embodied artificial intelligence in psychiatry, psychology, and psychotherapy. J. Med. Internet Res. **21**(5), e13216 (2019). https://doi.org/10.2196/13216
16. M. Shi, Z. Zhao, The impact of intelligent medicine on health care against the backdrop of big data, in *Proceedings of the 2018 Joint International Advanced Engineering and Technology Research Conference* (JIAET, 2018). https://doi.org/10.2991/jiaet-18.2018.84
17. J. Warren, How will AI change health care delivery? N. Z. Med. Student J. **28**, 10–12 (2019) Retrieved from http://search.ebscohost.com/login.aspx?direct=true&db=a9h&AN=138364986&site=ehost-live

18. A. Unnikrishnan, Artificial intelligence in health care: Focus on diabetes management. Indian J. Endocrinol. Metab. **23**(5), 503–506 (2019) https://doi.org/10.4103/ijem.IJEM_549_19
19. N. Bostrom, E. Yudkowsky, The ethics of artificial intelligence, in *The Cambridge handbook of artificial intelligence*, vol. 316, (2014). p. 334
20. R. Chakraborty, M. Sabharwal, Human intelligence better than artificial intelligence: New technologies will displace jobs but also create jobs. Business Today, (2019, Apr 07). Retrieved from https://library.semo.edu:2443/login?url=https://library.semo.edu:4836/docview/2195292184?accountid=38003
21. M. Rigby, Ethical dimensions of using artificial intelligence in health care. AMA J Ethics **21**(2), 121–124 (2019)
22. A. Shukla, The age of AI in healthcare: Disrupting efficiency & impacting ethics. Express Computer, (2017). Retrieved from https://library.semo.edu:2443/login?url=https://library.semo.edu:4836/docview/1927197401?accountid=38003
23. L. Craft, The need for AI governance in healthcare. Express Computer, (2019). Retrieved from https://library.semo.edu:2443/login?url=https://library.semo.edu:4836/docview/2269403092?accountid=38003
24. K.V. Brown, Why a DNA Test Is Actually a Really Bad Gift, (2018, December 19). Retrieved December 24, 2019, from https://gizmodo.com/why-a-dna-test-is-actually-a-really-bad-gift1820934113
25. R. White, Medical Ai: Can patent law keep up with the trajectory of innovation? Australas. Biotechnol. **29**(1), 22–24 (2019) Retrieved from http://search.ebscohost.com/login.aspx?direct=true&db=a9h&AN=136376341&site=ehost-live
26. H.R. Sullivan, S.J. Schweikart, Are current tort liability doctrines adequate for addressing injury caused by AI? AMA J. Ethics **21**(2), 160–166 (2019)
27. P. Hannon, Researchers Say Use of Artificial Intelligence in Medicine Raises Ethical Questions, https://med.stanford.edu/news/all-news/2018/03/researchers-say-use-of-ai-in-medicine-raises-ethical-questions.html, 7 Jan 2020, (Stanford School of Medicine, Stanford, 2018)
28. E.T. Stefanie Valentic, Organization develops safety guidelines for artificial intelligence in healthcare. EHS Today, (2019). Retrieved from https://library.semo.edu:2443/login?url=https://library.semo.edu:4836/docview/2255274744?accountid=38003
29. T. Dare, Ethics of artificial intelligence and health care. N. Z. Med. Student J. **28**, 5–7 (2019) Retrieved from http://search.ebscohost.com/login.aspx?direct=true&db=a9h&AN=138364984&site=ehost-live
30. D. Ross, V. Surgernor et al., Artificial Intelligence and Healthcare – FAQ's, https://businesslawtoday.org/2019/02/artificial-intelligence-healthcare-faqs/, 7 Jan 2020, (American Bar Association: Business Law Section, 2019)
31. R. Yampolskiy, J. Fox, Safety engineering for artificial general intelligence. Topoi **32**(2), 217–226 (2013)

Index

K. Daimi et al. (eds.), *Advances in Security, Networks, and Internet of Things*, Transactions on Computational Science and Computational Intelligence,
https://doi.org/10.1007/978-3-030-71017-0

B

M

Printed by Printforce, the Netherlands